ORDINATION AND CLASSIFICATION OF COMMUNITIES

HANDBOOK

OF

VEGETATION SCIENCE

Editor in Chief

Reinhold Tüxen

1973

Springer-Science+Business Media, B.V.

PART V

ORDINATION AND CLASSIFICATION OF COMMUNITIES

edited by

Robert H. Whittaker

1973

Springer-Science+Business Media, B.V.

ISBN 978-94-010-2703-8 ISBN 978-94-010-2701-4 (eBook)
DOI 10.1007/978-94-010-2701-4

Originally published byOriginally published by Dr. W. Junk b.v.Publishers,
Robert H. Whittaker
Softcover reprint of the hardcover 1st edition 1973

Cover design: Charlotte van Zadelhoff

FOREWORD

During the International Botanical Congress in Edinburgh, 1964, Mrs. I. M. Weisbach-Junk of The Hague discussed a plan for preparation by her publishing company (Dr. W. Junk B.V.) of an international *Handbook of Vegetation Science.* She proposed a series that should give a comprehensive survey of the varied directions within this science, and their achievements to date as well as their objectives for the future. The challenge of such an enterprise, and its evident value for the further development of vegetation research, induced the undersigned after some consideration to accept the offer of the honorable but also burdensome task of General Editor.

The decision was encouraged by a well formulated and detailed outline for the Handbook worked out by the Dutch phytosociologists J. J. Barkman and V. Westhoff. A circle of scholars from numerous countries was invited by the Dr. Junk Publishing Company to The Hague in January 1966 to draw up a list of editors and contributors for the parts of the *Handbook.* The outline and list have served since for the organization of the *Handbook*, with no need for major change.

The different burdens of editors and authors have compelled quite different timings for completion of the individual sections. It was consequently decided by Dr. W. Junk B.V. that the parts of the *Handbook* would be published separately as they were ready. Thanks to the tireless and determined service of Prof. Robert H. Whittaker as editor and to the work of his collaborators as well, the firm can now publish the first part of the *Handbook* to appear, Part V on ordination and classification as means of synthesizing information from a number of communities toward the understanding of vegetation.

The *Handbook* has as its purpose the presentation, through contributions of numerous collaborators from varied fields, of a comprehensive picture of modern vegetation science, including its development, its methods, its discoveries, and its goals, in approaches ranging from plant sociology as a descriptive science based on sharp observation of communities as concrete objects of study in the field (Braun-Blanquet), through the inquiry into casually significant relationships that for many represents its principal goal, to abstract vegetation science that seeks a detached and generalized understanding based on quantitative relationships that may be represented in diagrams, equations, and models.

In this the *Handbook* should reveal the fundamental value of vegetation science for the other disciplines with which it is linked in many ways – not only such other areas of botany as plant geography and systematics, paleobotany and palynology, genetics and evolution, and others, but also soil science and the applied areas of forest, grassland, and wildlife management, water and fishery management and coastal protection, and interpretation of the land's potential for agricultural and other use by man. Not last in this is the decisive importance of vegetation science for nature and environmental conservation.

All these applied fields use vegetation, whether as an object of either harvest or preservation (or both), or as the essential context of the other phenomena of their direct concern, or as an indicator of relationships with which they deal. They all are in some way directed toward the care and maintenance of the green cover of our earth that they seek to use, wisely. For their purposes understanding of plant communities as functional systems and expressions of life-phenomena and the laws that govern them is valuable, or even quite indispensable. The plant cover of the earth is and must remain the basis for all other life.

So we hope to show that for men of many aptitudes and interests concern with the plant cover and its social units is rewarding – not merely in the sense of precedence and preference among individuals, but rather on the basis of the mutual concern of different fields and research purposes that complement and enrich one another. May our *Handbook* be a portrait – a composite portrait as painted by many scientists but harmonious as a whole – or one of evolution's achievements, the plant community as itself a harmoniously functioning, living system.

All our thanks go to the initiators, the editors, and the contributors, as well as Dr. W. Junk, B.V., Publishers, of the Hague. They together are making a lasting contribution to our science; may it also contribute in some measure to solving the grave problems for life on our earth.

Todenmann über Rinteln R. Tüxen

ORDINATION AND CLASSIFICATION OF COMMUNITIES

CONTENTS

Authors and addresses

VERA D. ALEKSANDROVA
Komarov Botanical Institute, Popova Str. 2, Leningrad 22, U.S.S.R.

JAN BARKMAN
Laboratorium voor Plantensystematiek en -Geographie van de Landbouwhogeschool, Biologisch Station, Kampsweg 27, Wijster (Dr.), The Netherlands

JOHN S. BEARD
Director, Royal Botanic Gardens, Sydney, N.S.W., Australia 2000

GRANT COTTAM
Department of Botany, University of Wisconsin, Madison, Wisc. 53706, U.S.A.

PIERRE DAGNELIE
Faculté des Sciences Agronomiques, Statistique, Avenue de la Faculté 59, Gembloux, Belgique

TOOMAS E. A. FREY
Institute of Zoology and Botany, 21 Vanemuise St., 202400 Tartu, Estonian S.S.R., U.S.S.R.

HUGH G. GAUCH, JR.
Ecology and Systematics, Cornell University, Ithaca, N.Y. 14850, U.S.A.

F. GLENN GOFF
Ecology Section, Oak Ridge National Laboratory, Oak Ridge, Tenn. 37830, U.S.A.

DAVID W. GOODALL
Ecology Center, Utah State University, Logan, Utah 84321, U.S.A.

EDDY VAN DER MAAREL
Botanical Laboratory, Geo-Botany Division, University of Nijmegen, Toernooiveld, Nijmegen, The Netherlands

NILS MALMER
Department of Plant Ecology, Lunds University, Östra Vallgatan 14, 223 61 Lund, Sweden

ROBERT P. MCINTOSH
Department of Biology, University of Notre Dame, Notre Dame, Ind. 46556, U.S.A.

LASZLO ORLOCI
Department of Plant Sciences, University of Western Ontario, London 72, Ont., Canada

L. N. SOBOLEV
Institute of Geography, USSR Academy of Sciences, Staromonetny 29, Moscow, U.S.S.R.

HANS TRASS
Department of Plant Taxonomy and Ecology, Tartu State University, 40 Michurin St., Tartu, Estonian S.S.R., U.S.S.R.

V. D. UTEKHIN
Institute of Geography, USSR Academy of Sciences, Staromonetny 29, Moscow, U.S.S.R.

VICTOR WESTHOFF
Botanical Laboratory, Geo-Botany Division, University of Nijmegen, Toernooiveld, Nijmegen, The Netherlands

ROBERT H. WHITTAKER
Ecology and Systematics, Cornell University, Ithaca, N.Y. 14850, U.S.A.

GEORGE M. WOODWELL
Biology Department, Brookhaven National Laboratory, Upton N.Y. 11973, U.S.A.

1 INTRODUCTION

ROBERT H. WHITTAKER

1 INTRODUCTION

Over the land surface of the world there extends a mantle of vegetation, a living fabric of plant communities that is diverse and subtle in its response to environment, varied in its structure and composition, expressive of the biological productivity of the land, and interesting and esthetically appealing to man. The mantle of vegetation is part of man's life in evolution and in the present; we see the mantle now in many areas subject to accelerating destruction and cannot yet know the implications of that destruction for man's environment and the psychology of a species that evolved in natural landscapes. The destruction may increase the importance of one of science's endeavors – knowledge and understanding of plant communities. It is always of interest and often of value to know the plant communities of an area, a country, or a continent. We wish more than the description of those communities; we seek understanding – of how the different communities relate to one another in the mantle of vegetation and express their environments, of how plant species evolve and distribute themselves in relation to one another in communities, and of how communities develop and are organized and function as living systems.

Especially, for the concerns of this book, we desire to understand the relations of plant communities to one another and the particular environments (biotopes) those communities occupy. Understanding is based on abstraction. From phenomena that are often intricate and obscure, significant relationships are to be detected, embodied in concepts, related to one another, and tested and revised in the systems of abstraction that are science. As much as any other phenomena, plant communities offer complexities that challenge ourefforts at abstraction and understanding. In dealing with these complexities students of vegetation seek ways of relating to one another information on three levels – environmental factors, species populations, and community characteristics. We may distinguish two broadly conceived research methods as ways of abstracting and of seeking relations among these (WHITTAKER 1962, 1967).

One, the more traditional and familiar, is classification. When plant communities are classified, a number of samples representing communities are grouped together on the basis of shared characteristics into an abstract unit or *class* of plant communities. Throughout this discussion such units of classification, on any level, by any definition of shared characteristics, may be referred to as *community-types*. When vegetation samples have been classified into a community-type one may then determine for this type the ranges of environmental factors, species compositions, and community characteristics that its samples represent. Through classification we relate kinds of communities to kinds of environments as an approach to understanding the landscape and its vegetation.

The alternative method deals not primarily with classes (which may be treated as discontinuous with one another) but with continuities and gradient relationships. Study of communities in terms of gradients of phenomena on the three levels is termed *gradient analysis* (WHITTAKER 1951, 1967). Samples from plant communities may be arranged in sequence by their positions along a gradient of environment (or of community characteristics). In this sequence of samples (a *transect*) changes in species populations and community characteristics are related to changes in environment. In many cases the samples are arranged in relation to a coordinate system of two or more environmental or community gradients. The arrangement permits us to interrelate samples (and species populations) as parts of an abstract pattern of variation in species composition and other characteristics of communities, and to seek understanding of vegetation in terms of the major directions of variation in that pattern. If the samples are arranged in terms of one or more environmental gradients accepted as given, as a basis of the study, the study is *direct* gradient analysis. Alternatively, measurements of sample similarity or species correlation may be used to derive abstract axes or directions of community variation. If the samples are arranged in terms of such abstract axes (which may or may not correspond to environmental gradients), the study is one of *indirect* gradient analysis. Direct and indirect gradient analysis are not sharply separated from one another; various intermediates and combinations are possible. They represent different perspectives on the solution of vegetation problems, however, and will be separately treated here.

The process of arranging samples (or species) in relation to one or more environmental gradients or axes is termed *ordination* (GOODALL 1954). Ordination is an essential technique of gradient analysis but is not (or should not be) its purpose. Classification likewise should not be an objective in itself but a means of ordering information, describing, interrelating, and generalizing about phenomena. In dealing with individual organisms the more technical approach to classification is called *taxonomy*; whereas the broader concern with relationships of organisms, particularly on the basis of evolution, is called *systematics*. In dealing with plant communities broader approaches to description and understanding that are based on classification are often termed *phytosociology* (BRAUN-BLANQUET 1932, 1964). Use of this term has varied between English-language and Continental countries (EGLER 1942, WESTHOFF 1970); and for some Continental students phytosociology represents the whole of vegetation science. Phytosociology may be distinguished, however, from plant geography with its concern with floristics and relations of vegetation structure to environment on a continental or world-wide scale. Within the study of natural communities (biocoenology, symbiology, synecology in its Anglo-American use) we may well regard phytosociology as the approach to vegetation that emphasizes the classificatory perspective and the species composition of communities, the approach we think most characteristic of phytosociology in western Europe. We may then set in approximate parallel three pairs of terms, each including an essential procedure and the approach to understanding of which that procedure is part: taxonomy and systematics, community classification ('syntaxonomy') and phytosociology, ordination and gradient analysis.

The present articles are concerned more with procedures than with application to particular communities; this is an account of ordination and classification, rather than of gradient analysis and phytosociology. Some of the results from gradient analysis must, however, be considered as a basis for understanding both ordination and classification of communities. The articles that follow will proceed from direct gradient analysis and some of its results, through indirect ordination (from the less to the more mathematical forms of this), to the basis of community classification, to reviews of various major approaches to classification. Articles have been arranged by their primary emphasis, but the reader will find that a number of them concern both ordination and classification. Ordination and classification are in no wise antagonistic. Many studies in gradient analysis use a classification to present some of their results, and studies based on classification may approach the interrelations of community-types and species through gradient analysis. These two major approaches are thus complementary and can be combined in various ways for the most effective study and interpretation of plant communities and their relations to environment.

SUMMARY

Two major ways of relating the plant communities of a landscape to one another may be distinguished. Classification groups communities into classes (community-types) that may be characterized and treated as discontinuous with one another. Gradient analysis studies vegetation in terms of gradients of environment, species populations, and community characteristics in relation to one another. The arrangement of samples in relation to environmental gradients as a basis for gradient analysis is termed *ordination.* Ordination may be either direct (arranging samples by positions along one or more environmental gradients accepted as given), or indirect (from comparisons of samples are derived compositional axes, directions of change in community composition, along which the samples are arranged.) The articles of this book are arranged in sequence from direct gradient analysis, to indirect ordination, to major approaches to the classification of communities.

ZUSAMMENFASSUNG

Beim Studium der Beziehungen zwischen den Pflanzengesellschaften einer Landschaft können zwei Hauptmethoden unterschieden werden. Die *Klassifikation* gruppiert die Gesellschaften in Klassen (Gesellschaftstypen), die charakterisiert und unabhängig voneinander betrachtet werden können. Die *Gradienten-Analyse* betrachtet die Vegetation auf der Basis der Gradienten des Standorts, der Art-Populationen und der Gesellschaftsmerkmale in ihren gegenseitigen Beziehungen. Die Anordnung der Aufnahmen auf Grund der Standortsgradienten als Grundlage für die Gradientennalyse wird *Ordination* genannt. Die Ordination kann entweder direkt erfolgen (Anordnung der Aufnahmen längs eines oder mehrerer Gradienten, die als gegeben vorausgesetzt werden), oder indirekt vorgenommen werden (aus dem Vergleich der Aufnahmen untereinander werden "Kompositions-Achsen" abgeleitet, Variationsrichtungen der Gesellschaftsstruktur, entlang derer die Aufnahmen angeordnet werden.) Die Beiträge in diesem Band sind in der Reihenfolge von direkter Gradienten-Analyse über indirekte Ordination zu der Hauptabhandlung über Gesellschafts-Klassifikation angeordnet.

(The Editor is indebted to E. VAN DER MAAREL for his translations of the summaries).

References for articles 1 and 2 are given with those for article 3.

2 DIRECT GRADIENT ANALYSIS: TECHNIQUES

ROBERT H. WHITTAKER

Contents page

2 DIRECT GRADIENT ANALYSIS: TECHNIQUES

2.1 Samples

2.1.1 Sample Characteristics

Community samples ('Aufnahmen,' relevés) are the essential working material for gradient analysis. The procedures to be described below can (with some limitations) be applied to any kind of community sample. Thus one can, in principle, use for gradient analysis samples of the vascular plants of forests, lichens on the barks of trees, singing bird pairs of grasslands, marine zooplankton, or soil microarthropods. We shall concern ourselves primarily, however, with samples that include the vascular plants (with or without accompanying data on thallophytes) from land communities. A sample will normally include (i) a list of plant species present in a given study area, the plot or quadrat, (ii) some indications of relative importance of these species (and usually of their growth-forms, heights, and stratal relationships), and (iii) supporting information on environment, soil, community structure, and evidences of disturbance and community change.

There are a wide range of kinds of samples that can be applied to plant communities and used for gradient analysis. Some considerations affecting choice and kind of sample are: (i) The sample should be large enough in area, or should include counts of a sufficient number of plants, to represent effectively the composition of the plant community. If the samples are too large, however, difficulties are encountered in meeting the two following considerations. (ii) The sample should be homogeneous – there should be no trend of change in community composition or structure from one edge of the sample to another. (iii) The sample should be efficient. Since considerable numbers of samples may be needed, the samples should be designed to obtain and record rapidly the kinds of information regarded as most important. (iv) The sample should be appropriate. Among the many kinds of information that might be gathered on a plant community, some are more interesting or significant, appropriate to the character of the community, and informative in relation to time spent and the purposes of gradient analysis than others. These considerations of minimum area, homogeneity, efficiency, and appropriateness of samples are to be discussed further elsewhere in the *Handbook* (Part IV). Certain types of samples have been most widely used for gradient analysis of plant communities. The most widely used, and from many standpoints the most efficient, of all types of samples for classification is the relevé of the School of BRAUN-BLANQUET (1932, 1964, ELLENBERG 1956, and article 20 in this volume). There is no reason the BRAUN-BLANQUET relevé cannot be used also for gradient analysis (MAAREL 1969, MOORE et al. 1970). In American work in gradient analysis more strongly quantitative samples, with additional time spent to obtain more effective measurements on plant populations, have been preferred. The measurements that express relative abundance, or conspicuousness, or massiveness, or vigor of species in natural communities are termed 'importance values.' Among the various possible importance values the following are of most interest to us: (i) *Density* is the number of individuals of a species per unit ground surface area (or other spatial measurement). (ii) *Coverage* is the percentage of ground-surface (or other substrate) area above which foliage of a given species occurs. (iii) *Basal area* is the area occupied by cross-sections of stems at 1.3 m above the ground (or other plant basal measurements) per unit ground surface area. (iv) *Frequency* is the percentage of small subquadrats within a larger sample plot, in which a given species is present. (v) Species *biomass* is the total mass (usually dry weight of organic matter) present at a given time per unit ground surface area. (vi) Species net *production*, for plants, is total dry weight growth (or its energy equivalent), or organic matter production by photosynthesis minus the plants' own respiration.

These and other measurements are often divided by the total of the same kind of importance values for all species in the sample; the resulting percentages are *relative* importance values. When two or more importance values are combined, the result is a *synthetic* importance value; for example the Wisconsin synthetic importance value (CURTIS & MCINTOSH 1951, GOFF & COTTAM 1967) sums for each species its relative density, relative frequency, and relative basal area. Two types of samples used in the United States for efficient measurement of importance values are:

2.1.2 QUADRAT SAMPLES

For the tenth-hectare strip quadrat (WHITTAKER 1960, WHITTAKER & NIERING 1965) a 50-meter tape is laid out on the ground, and a rectangular plot extending 10 meters on each side of the tape is marked out. All trees rooted within the plot are tallied by species and diameters at breast height (1.3 m). Shrub individuals (and tree seedlings in the shrub stratum) are tallied for the same area or, if very dense, in a narrower inner strip along the tape. Herbs (and tree and shrub seedlings in the herb stratum) are sampled in 25 subquadrats. The subquadrats may be either (i) bounded by eye and a meter stick, on alternate sides of the tape, every other meter, or (ii) marked out with ring-stakes tied into cord (with the cord measured to produce 1×1 or 0.5×2.0 m^2 areas) at random numbers of meters distant from the central tape, on alternate sides of that tape at 2 meter intervals. Occurrence and coverage (visually estimated by dm^2 units) of plant species in these subquadrats are recorded; in some studies densities in the subquadrats are also recorded. Tree and shrub coverages are measured either for the 100 points at corners of the subquadrats, or as intercept coverage along the 50-meter tape, or both. (In open vegetation intercept coverage should be taken along the long sides as well as the center of the quadrat, hence for 150 m in all). Lichen and moss occurrence in the plot should be recorded. The sample may be completed with information on elevation, topographic position, parent material and soil, location, disturbance and developmental trends, and voucher specimens of plant species as needed.

Such samples are effective for obtaining measurements of density, frequency, and coverage of plant species of different sizes in a wide range of vegetation types (WHITTAKER & NIERING 1965). The 0.1 hectare (20×50 m) size has been found suitable for forest, woodland, grassland, and desert samples; but the quadrats have been expanded to 0.2 or 0.5 ha in forests of large trees, and contracted to 0.05 or 0.01 ha in some dense shrub and pygmy forest communities. When procedures are established and plant species are known, samples from most communities (that are not too rich floristically) can be taken in 45 minutes to 1.5 hour by an investigator with an assistant. It is thus possible to take several samples a day for the systematic sampling of a vegetation pattern.

2.1.3 PLOTLESS OR POINT-CENTERED SAMPLES

The Wisconsin School (COTTAM & CURTIS 1949, 1956, CURTIS 1959, and article 8) have developed a quite different set of sample procedures. In a tract of forest judged homogeneous a number of points (often 50 or 100) are located by random or systematic procedures. Each of these points (or the tree nearest the point) is used as sample-point, and distances from these sample-points to trees nearest them, and species and diameters of these trees, are recorded. Undergrowth subquadrats may be located at the same sample-points. Details of the various techniques cannot be described here, but are discussed by GREIG-SMITH (1964). The distance measurements permit calculation of tree densities; from densities together with diameters may be calculated basal areas; and from occurrence at sample-points as subquadrats may be calculated frequencies.

The point-centered systems are usually (though not necessarily for all applications) more time-demanding than the tenth-hectare quadrats. Point-centered samples, with systematic or random location of points through the sample area and undergrowth information and environmental data, generally take several hours. The point-centered samples are more effective in representing the composition of a large forest tract (of the order of 10 ha or more); the tenth-hectare samples are more effective in representing in greater detail community composition within a limited area. Plotless or point-centered samples were used as standard for sampling forest tracts on the more level topography of Wisconsin; the tenth-hectare samples have been used in mountains and other areas where steep environmental gradients make point-centered samples inappropriate because of the large areas needed for such samples. The two sampling approaches have different advantages for different applications; but the tenth-hectare samples are more widely applicable and for many purposes more efficient.

Point-centered samples can be applied to grassland tufts and desert shrubs, but it is not clear that such application is advantageous. Applied to a grassland without trees or shrubs, the tenth-hectare plot becomes a sample of herbs in 25, 1-m^2 subquadrats. A number of gradient analyses of grasslands and desert have been based on rectangular, square, or circular subquadrats of varying numbers and sizes (CURTIS 1955, DIX 1959, PERRING 1959, ITOW 1963, GITTINS 1965, DIX & SMIENS 1967, BEALS 1969b). Further discussions of quantitative sample procedures are given in the books by BROWN (1954), CAIN & OLIVEIRA (1959), PHILLIPS (1959), GREIG-SMITH (1964), and GOUNOT (1969), and in the *Handbook* Part IV.

2.1.4 SAMPLE SETS

Whatever the type of sample used, a study in gradient analysis is based on a sample *set* – a number of samples taken with similar procedures to represent a range of variation in environment and community composition. It is natural to think that samples of the set might best be taken from random locations in a landscape. Formal randomization of sample location has been abandoned by most who have tried it (cf. MOORE et al. 1970), because location of the sample points in the field with surveying techniques is time consuming, because a high proportion of the samples obtained may be heterogeneous or disturbed, and because randomization may be inefficient to represent the range of variation of communities in the landscape. Procedures often used are: (i) Along a well-defined gradient of vegetation and environment, samples are taken at fixed intervals (for example, of elevation up a mountain slope, of compass degrees around a hill, of distance down a desert bajada or from a water body outward through its bordering vegetation). (ii) In a substantially disturbed landscape, samples may be taken from all, or many, of the tracts of vegetation that are sufficiently large, undisturbed, and homogeneous to give useful samples. Such has been a procedure of the Wisconsin school. (iii) In an area of complex environmental variation, such as mountain topography, samples are taken at frequent but unspecified intervals as the investigator encounters new combinations of community composition and environment. The investigator may walk through the mountains taking samples with each change of 50 m elevation, or change of compass direction of exposure, as he goes, excluding samples which are heterogeneous (or, if succession is not being studied, disturbed). Such sampling may be continued until the samples are judged to represent adequately the range of communities and environments that has been observed and is to be studied. If some topographic positions of limited areal extent (e.g. ravines) are poorly represented, samples of these may be specially sought (WHITTAKER 1956, 1960, WHITTAKER & NIERING 1965).

Numbers of samples desired for sets may vary widely. In procedure (i) relatively few samples, of the order of 5 to 20, may serve the purposes of some studies. In procedure (ii) in the Wisconsin school, sample sets have been of the order of 50 to 100 samples. For procedure (iii) the author has used as a standard 50 to 60 samples from each 300-meter or 1000-foot elevation belt on a given parent material in mountains, hence several hundred samples to represent the full range of variation of the mountain vegetation. For phytosociological studies samples are often chosen to represent community types already recognized by the investigator. For gradient analysis it is better to avoid such bias, and to base sample choice on environmental variations and not primarily on community composition according to recognized types. (If samples have been taken to be representative of community-types, they cannot then be used to test the continuity or discontinuity of those types.) All three of the procedures suggested can provide for sample choice that is unbiased in this respect, though not formally random.

2.2 Transects of Single Complex-Gradients

2.2.1 Simple and Composite Transects

Our concern in direct gradient analysis is with the way in which species are distributed and vegetation changes along environmental gradients. These relationships are best observed when a set of vegetation samples are arranged in sequence along an environmental gradient; such a sequence of samples forms a *transect*. In some cases the simplest approach to an environmental gradient – a single series of samples taken at intervals along it – suffices to reveal what the investigator wants to know. Such simple or field transects have been used for study of elevation gradients (Mark 1963, Mark & Sanderson 1962, Scott et al. 1964, Wells & Mark 1966, Beals 1969a), topographic moisture gradients (Horikawa & Okutomi 1955, Okutomi 1958, Mowbray 1966, Mowbray & Oosting 1968), and the vegetation surrounding a swamp or water body (Beschel & Webber 1962).

In many cases, however, such a simple transect provides data that are too limited or too irregular and subject to sample error. It is often desirable to group a larger number of samples, taken from similar environmental conditions at different locations within the study area, into a *composite transect*. Samples may be grouped by sets of five, say, for 30-meter elevation intervals, or for positions along a topographic moisture gradient. Such grouping tends to average out the irregularities affecting individual samples, and to produce a clearer, more easily interpreted picture of species distributions and community change along the gradient. Table I is such a composite transect. Each column in the table is an average of tree-species densities in five, tenth-hectare samples that are similar in composition and represent similar positions along the complex-gradient. The manner in which species populations rise and fall along the gradient in Table I is representative of the way species populations are distributed in a large number of transects, simple and composite, direct and indirect, that have been studied (Whittaker 1967).

2.2.2 Complex-Gradients and Environmental Ordination

Environmental gradients are complex phenomena. Along a given spatial gradient, for example of elevation up a mountain slope,

Table I. A composite transect of the topographic moisture, 1830-2140 m elevation, in the Santa Catalina Mountains, southern Arizona (Whittaker & Niering 1965). Fifty samples are ordinated by weighted average indexes (first column) and grouped in subsets of five per transect step. Topographic positions range from mesic ravines (step 1) to xeric open southwest-facing slopes (step 10). Data are numbers of tree stems over 1 cm dbh in 0.5 ha. The communities are pine-evergreen oak woodlands. See also Fig. 1.

	Species Weight	Transect steps 1	2	3	4	5	6	7	8	9	10	Mean Density /ha	Mean Coverage %	Transect Constancy %
Acer glabrum var. neomexicanum	1	6										1.2	0.1	8
Alnus oblongifolia	1	53										10.6	3.0	10
Salix lasiolepis	1	55										11.0	0.6	10
Abies concolor	1	57	1									11.6	0.4	8
Quercus gambelii	1	32	10									8.4	0.05	6
Juglans major	1	7	6	1								2.8	0.08	16
Pinus strobiformis	1	39	12	2								10.6	0.36	18
Pseudotsuga menziesii	2	44	65	27	1							27.4	1.79	32
Robinia neomexicana	2	8	5	2	1							3.2	0.06	16
Rhamnus californica ssp. ursina	2	3	13	4	1							4.2	0.31	28
Prunus virens	2	6	38	21	6	1						14.4	0.85	28
Pinus ponderosa	2	18	322	236	106	24	3	1				142.0	10.1	52
Quercus rugosa	2	126	93	58	36	22	12	16	16	11	5	79.0	2.5	40
Quercus hypoleucoides	3	2	223	335	526	536	362	124	53.	62	11	446.8	19.1	90
Arbutus arizonica	3		2	8	27	16	8					12.2	0.27	32
Pinus chihuahuana	3		1	4	10	32	36	7				18.0	0.85	24
Juniperus deppeana	4		1	4	4	13	46	51	39	29	25	42.4	1.95	72
Quercus arizonica	4		9	18	47	72	98	100	82	25	19	94.0	3.03	62
Pinus cembroides	4			1	3	13	29	93	161	114	85	99.8	6.0	70
Garrya wrightii	4			1	4	9	5	13	68	57	10	33.4	17.3	66
Arctostaphylos pringlei	4				6	21	134	54	154	154	34	111.4	9.8	50
Arctostaphylos pungens	4				1	10	17	14	5	12	7	13.2	6.9	50
Quercus emoryi	5						5	8	18	14	76	24.2	0.77	22
Total stems/0.5 ha		456	801	722	779	769	755	481	596	478	272			
Mean weighted average		1.46	2.27	2.54	2.89	3.12	3.43	3.67	3.89	3.85	4.20			
Percentage similarity to step 1.		100	31.8	20.2	10.5	7.02	2.81	4.06	3.42	2.78	1.92			
" " " " 10.		1.92	4.85	8.04	10.1	19.4	27.3	54.2	48.8	56.0	100			
Mean coverage %, trees		59.	96.	80.	76.	60.	44.	41.	26.	25.	25.			
shrubs		0.5	1.6	1.5	3.2	22.	34.	42.	33.	39.	33.			
herbs		6.4	3.9	1.3	0.3	1.6	1.5	0.5	0.8	1.2	2.5			
Mean vascular plant species per sample		28.5	24.8	17.8	16.4	21.0	16.6	13.6	17.4	15.2	17.2			

many environmental factors of precipitation and humidity, temperature and wind velocity, and soil characteristics change together. An environmental gradient may be conceived as a *complex-gradient* comprising gradients of many particular environmental factors, or characteristics of environmental complexes, that change together in space (WHITTAKER 1956, 1967). Gradient analysis may be based on arrangement or *ordination* of samples either (i) in relation to complex-gradients that are accepted as given, in direct gradient analysis, or (ii) along axes of variation in composition of samples, axes derived from study of the samples themselves, in indirect gradient analysis. The latter axes may or may not correspond to complex-gradients. (In either case some gradients or axes may represent disturbance or successional time, rather than spatial variation.) In direct gradient analysis ordination may be based either on environmental measurements, or on indices expressing relative position in a gradient of community composition as this changes in response to an environmental complex-gradient.

For some complex-gradients a convenient environmental measurement is available by which samples may be arranged. Elevation provides an index of sample position along a complex-gradient, even though elevation itself has no bearing on plant physiology. Soil characteristics can also be used as indices of ordination for some studies; for example moisture conditions, acidity, and other qualities of agricultural soils (ELLENBERG 1950, 1952), nutrient levels and acidity along a forest gradient of increasingly intense leaching (WESTMAN (1971), salinity or mechanical analysis of particle sizes along a desert slope or bajada (YANG & LOWE 1956). For the topographic moisture gradient samples may be arranged by topographic positions–from deep ravines or valley bottoms, through smaller canyons and lower sheltered slopes, to open slopes of varying exposures (in the northern hemisphere, from northeast and north, through northwest, east, and west, to southeast, south, and southwest—note the horizontal axis of Fig. 5). Topographic positions are crude indices of position along the complex-gradient; other approaches to topographic moisture gradients are discussed by LOUCKS (1962).

2.2.3 COENOCLINES AND COMPOSITIONAL ORDINATION

There are other ways of arranging samples along complex-gradients when no single environmental measurement provides a satisfactory index. Underlying such possibilities is a principle: Community composition expresses the characteristics of the community environment or biotope. (Community composition may express disturbance and time-gradients of succession as well as stable environmental factors.) Corresponding to, and expressing, a complex-gradient is a compositional gradient of species importance values in communities (see Table I). The community gradient may also be termed a *coenocline* (WHITTAKER 1960, 1967, MAAREL 1960, MAAREL & LEERTOUWER 1967). The complex-gradient and coenocline together form a gradient of communities-and-environments, or of ecosystems, an *ecocline* (Clements 1936, WHITTAKER 1960, 1967, 1970a). Since the community gradient expresses the environmental gradient, we may use relative positions of samples in the community-gradient as indices of their relative positions along the complex-gradient. There are two major approaches to ordinating samples by their composition, when an environmental complex-gradient is accepted as given:

2.2.3.1 *Weighted averages*

Fig. 1 gives smoothed curves for the species populations along an environmental gradient (data of Table I). Note that, as indicated by the numbers above the centers or modes for these species, the species can be grouped by relative positions of their centers along the gradient. For the transect of Fig. 1 and Table I species were grouped and numbered by relative position along the topographic

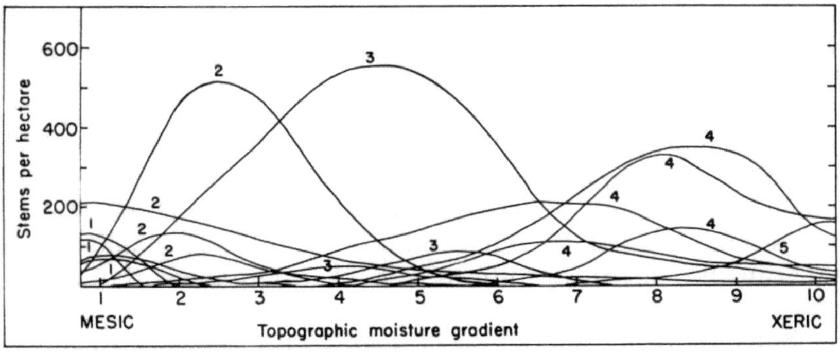

Fig. 1. Species populations and ecological groups in a coenocline. Population distributions are shown for major tree species in the 1830–2140 m elevation belt in the Santa Catalina Mountains, southern Arizona from mesic ravines (left) to xeric southwest slopes (right) (Table I and WHITTAKER & NIERING 1965). Ecological groups for these species in relation to the topographic moisture gradient are indicated and used for the weighted averages of Table I; the groups are: mesic (1), submesic (2), subxeric (3), xeric (4), and more strongly xeric (5).

moisture gradient as: moist or mesic (1), submesic (2), subxeric (3), dry or xeric (4), and one more strongly xeric species (5) (WHITTAKER & NIERING 1965). Such species groupings, based on distributional similarity, are termed *ecological groups* (DUVIGNEAUD 1946, ELLENBERG 1950, 1952, TANGHE 1964, WHITTAKER 1967). The technique of arranging samples by weighted averages was independently discovered by several authors and has been applied to both environmental gradients (ELLENBERG 1948, 1950, 1952, WHITTAKER 1951, 1954a, 1956, 1960, WHITTAKER & NIERING 1965, 1968a, 1968b, CURTIS & MC INTOSH 1951, CURTIS 1955, 1959, ROWE 1956, KNIGHT 1965, WARING & MAJOR 1964, BRAY 1956, 1960, MAYCOCK & CURTIS 1960, DIX & SMIENS 1967, BEALS 1969b, see also MCINTOSH 1967, GOFF & COTTAM 1967) and disturbance gradients (DYKSTERHUIS 1948, 1958, DIX 1959, ITOW 1963).

The ecological groups are derived from field data – either extensive observation of species behavior (ELLENBERG 1950, 1952, CURTIS 1955), or field transects (WHITTAKER 1956), or preliminary composite transects (WHITTAKER 1960, WHITTAKER & NIERING 1965, cf. BROWN & CURTIS 1952, DIX & SMIENS 1967). For analysis of the topographic moisture gradient, for example, a preliminary composite transect may use topographic positions to arrange the samples. Distributions of species in this preliminary transect are used to classify them into ecological groups on which a more effective, weighted-average ordination may be based (WHITTAKER 1960, WHITTAKER & NIERING 1965). If species distributions are used to arrange samples for the study of species distributions, the method involves a clear circularity (WHITTAKER 1956), and this circularity has been part of criticisms of gradient analysis (DAUBENMIRE 1966, LANGFORD & BUELL 1969, GOUNOT 1969). The circularity does not apply to transects of elevation gradients. For topographic gradients and other transects using compositional ordination, the circularity does not invalidate the method when ordination is based on correctly observed gradient relations of species and is used to refine sample arrangements and results that can be obtained also by use of environmental measurements.

Since representation of the ecological groups changes continuously along the coenocline, weighted averages expressing representation of the ecological groups can be used as indices of relative position along the coenocline (and, hence, the complex-gradient). For example, in the first column of Table I, stems of species designated as mesic (249 stems in all) are multiplied by 1 as a weight, submesic stems (205) by 2, subxeric stems (2) by 3, and xeric stems (0) by 4. The sum, $249 \times 1 + 205 \times 2 + 2 \times 3 + 0 \times 4 = 665$, is divided by the unweighted sum (456) to give a quotient, the weighted average 1.46. The 50 individual samples were arranged in sequence by weighted averages, and then grouped in sets of five to produce the composite transect summarized in Table I. The progressive change in weighted averages along the coenocline is indicated in a bottom row of Table I. It is sometimes desirable to apply independent weightings to different strata of a set of samples, and to use the two weighted averages as axes of a chart (Fig. 2). The two weightings provide a check on each other; and samples which are deviant for the set (affected by disturbance or environmental factors other than the complex-gradient under study) may appear above or below the main axis of the chart (WHITTAKER 1960, WHITTAKER & NIERING 1965).

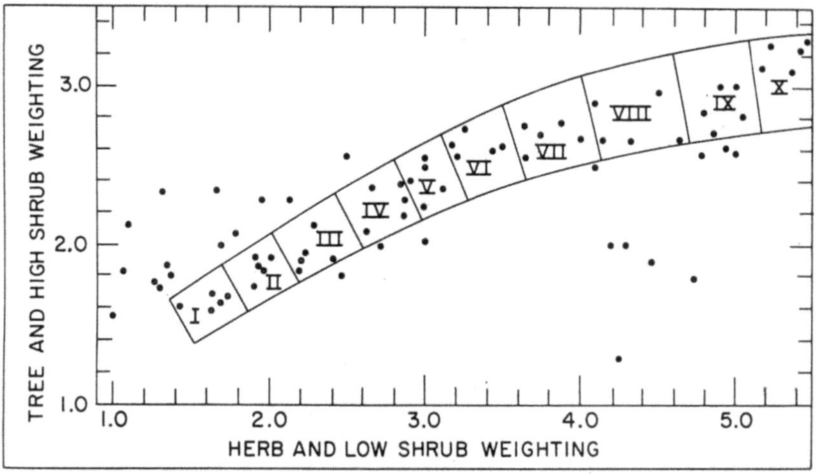

Fig. 2. A double-weighted-average ordination of Sonoran desert samples from upper (low indices) to lower (high indices) valley plain or bajada below the Santa Catalina Mountains, Arizona (WHITTAKER & NIERING 1965). The scale on the ordinate applies weights to occurrences of tree and arborescent shrub species in 0.1 ha quadrats, that on the abscissa to frequencies of herbs and shrubs in meter-square subquadrats. Points representing samples are grouped by fives into the ten steps of a composite transect. Samples which are deviant in relation to this coenocline from upper to lower bajada lie to one side of the axis. The samples above the axis on the left are from less xeric bajada disturbed by grazing, those below the axis on the right are from desert washes.

2.2.3.2 *Sample comparisons*

If we start with samples representing one end of a coenocline, other samples will be increasingly unlike these end-point samples with increasing distance along the gradient. We can thus use quantitative comparison of samples to express distance of samples from one another along a coenocline (compositional distance or ecological distance, WHITTAKER 1952, 1967, article 4). There are a number of ways of measuring sample similarity, discussed in article 6; one of the simplest is percentage similarity in the form, $PS = 100 - 0.5\,\Sigma\,|\,a - b\,| = \Sigma \min\,(a, b)$, in which a and b are, for a given species, the relative importance values for that species in two samples. End-point samples (or averaged groups of samples) representing the opposite extremes of an environmental gradient may be chosen. Using percentage similarities or related measurements we can arrange samples from those most like the first end-point, through various intermediate positions, to those most like the opposite end-point. The manner in which such values change along a gradient is indicated by the percentage similarities for the transect steps, compared with steps 1 and 10, at the bottom of Table I. (The arrangement may be based on the same calculations

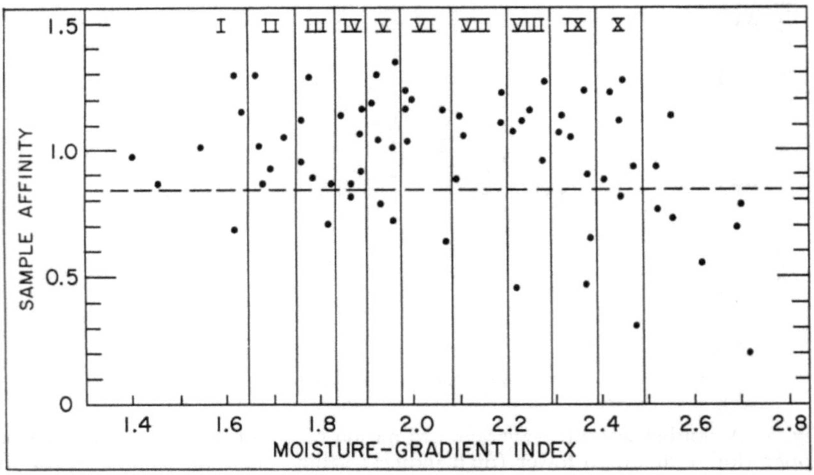

Fig. 3. A sample-similarity ordination technique, applied to low-elevation (610–910 m) forests in the Siskiyou Mountains, southwestern Oregon (WHITTAKER 1960). Three reference points were used, each an average of 5, 0.1–ha samples: most mesic ravines (1), intermediate, open east-facing slopes (2), and most xeric south to southwest-facing slopes (3). These numbers were used as weights. For each sample its similarity values with the three reference points were first summed to give the 'sample affinity' index on the ordinate. The similarity values were then multiplied by the weights, and the sum of weighted values was divided by the unweighted 'sample affinity' to give the moisture–gradient index on the abscissa, by which the samples are arranged in sequence from those most like (1) to those most like (3). Samples with low sample affinity indexes were deviant in some respect (of parent material or disturbance) from the main body of samples for the coenocline. Samples with affinity values above 0.85 were grouped by fives into ten steps for a composite transect of the topographic moisture gradient.

as those used in Wisconsin comparative and other indirect ordinations (Bray & Curtis 1957, Beals 1960, Loucks 1962, Whittaker 1967, articles 8 and 11). The procedure then differs from indirect ordination only in that the end-point samples have been chosen to represent a complex-gradient accepted as given.) Fig. 3 represents a variant of this ordination technique (Whittaker 1960). Again, samples have been ordinated and grouped in sets of five, to form a composite transect.

2.2.4 Effectiveness of Ordination

Ways in which the relative effectiveness of different ordinations of the same set of samples can be compared are discussed by Whittaker (1960, 1967), Bray (1961), Loucks (1962), and article 11. Three approaches to ordination effectiveness are:

(i) It is assumed that increased effectiveness of ordination will be expressed in reduced dispersions of species distributions through the intervals of the transects. The dispersions can be measured as standard deviations if the data are adequate; but it is simpler, and may suffice, to compare the numbers of transect intervals through which species distributions extend. These numbers are averaged for two transects being compared, using the same lists of species that are represented by sufficient measurements to be useful and that occur in several but not all steps of the transect (Whittaker 1960, Bray 1961).

(ii) It is assumed that more effective ordination will be expressed in closer approach to binomial patterns of species importance values along the gradient, and hence greater consistency in decline of importance values away from the peak or modal values. Differences in species importance values in successive transect steps are averaged to give an index that decreases with increasing effectiveness of ordination (Whittaker 1960). For a measure that is probably somewhat more sensitive, the importance value differences that decline in sequence away from the peak value are summed, and from this sum are subtracted the amounts by which values that increase away from the peak would have to be reduced to make them consistent with the trends (Bray 1961).

(iii) It is assumed that more effective ordination will result in closer mean similarities of adjacent samples in the transect. Loucks (1962) used a statistical test of sample similarity; but percentage similarity or other sample similarity measurements (articles 6, 11) can be averaged for the sets of samples combined into transect steps, or for adjacent transect steps, to express the ordination's success in placing similar samples next to one another.

Ordinations by composition (weighted averages or sample comparisons) are in many cases more effective than ordination by any single environmental measurement. Such is particularly true of topographic moisture gradients, for which samples are best ordinated by composition; elevation in contrast, may provide as effective an index for ordination as any index based on composition. Between the two major means of compositional ordination – weighted averages and sample comparisons – the advantage is less clear. Weighted averages are probably more effective when the range of community change is large along a complex-gradient that can be accepted as given, hence for many direct gradient analyses. Ordination by sample comparison may be equally effective for shorter ranges of community variation, and may be preferred when the major environmental variables are less easily recognized in the field. Sample comparison may also be preferred when, because the number of samples is limited or because their composition is affected by more than one complex-gradient, it is difficult to assign species to ecological groups. Though it is possible to derive weighted averages from indirect techniques (Bray 1956, 1960, Dix & Butler 1960, Beals 1969b), for most indirect ordinations other means of arranging samples (or species) in relation to axes of community composition have been preferred, as described in articles that follow.

2.3 Analysis of Intercommunity Patterns

Most areas have a complex mantle of vegetation affected by more than one complex-gradient. Means are needed by which effects of these different gradients on community composition can be determined and related to one another. A number of techniques will be described as approaches to intercommunity patterns (that is, landscape patterns involving different kinds of communities and their relations to one another, as distinguished from intracommunity patterns involving contagion and species association within a given community).

2.3.1 Crossed Transects

In some cases two complex-gradients are predominant in their effects on communities of a landscape. Such may be the case with the elevation and topographic moisture gradients in moutains, and it may then be reasonable to analyze mountain vegetation in terms of these two gradients (if disturbance and parent material differences do not introduce major complications). Vegetation of the Great Smoky Mountains was analyzed (WHITTAKER 1956) through four transects of the topographic moisture gradient, holding the elevation gradient constant in each (within 1000-foot elevation belts), and four transects of the elevation gradient, holding 'constant' topographic moisture conditions. (Weighted averages were used to classify the samples into mesic, submesic, subxeric, and xeric groups, and each of these four groups was used for an elevation transect.) The pattern is thus approached through a grid of crossed transects representing effects of the major gradients separately and in different combinations. Where elevation difference is not a concern, direction of exposure and steepness of slope may be used as axes (PERRING 1959, 1961, AYYAD & DIX 1964). It is possible also to use ordinations by composition to treat vegetation in relation to two or more axes. In some cases it is feasible to classify species in the same vegetation into different sets of ecological groups by their responses to different environmental gradients. The same species, as differently classified in relation to different gradients, can then be used for weighted-average ordination of samples in relation to each of these gradients (ELLENBERG 1950, 1952, WARING & MAJOR 1964). Ordination by sample comparisons should be effective for sample sets that are not extensive enough to permit weighted-average treatment or to justify a transect grid. End-point samples may be chosen for two or more complex-gradients, and the samples may be ordinated into composite transects in relation to each of these. The samples may also be arranged in a coordinate system defined by these axes with the techniques of BRAY & CURTIS (1957, article 8) used for, in this case, 'direct' gradient analysis.

2.3.2 COENOCLINE COMPARISONS

A related approach uses community-gradients as units of comparison between different landscapes (or elevation belts, or parent materials, or disturbance conditions). Coenoclines can be compared with one another in a variety of ways: (i) Relative importances and distributions along the gradients of species populations. Fig. 4, for example, illustrates a few species populations in their distributions along topographic moisture gradients on three parent materials, under similar climates at low elevations in the Siskiyou Mountains (WHITTAKER 1954b, 1960, cf. WHITTAKER & NIERING 1968a). Striking differences appear in the relative importances and distributions of species on the three parent materials.

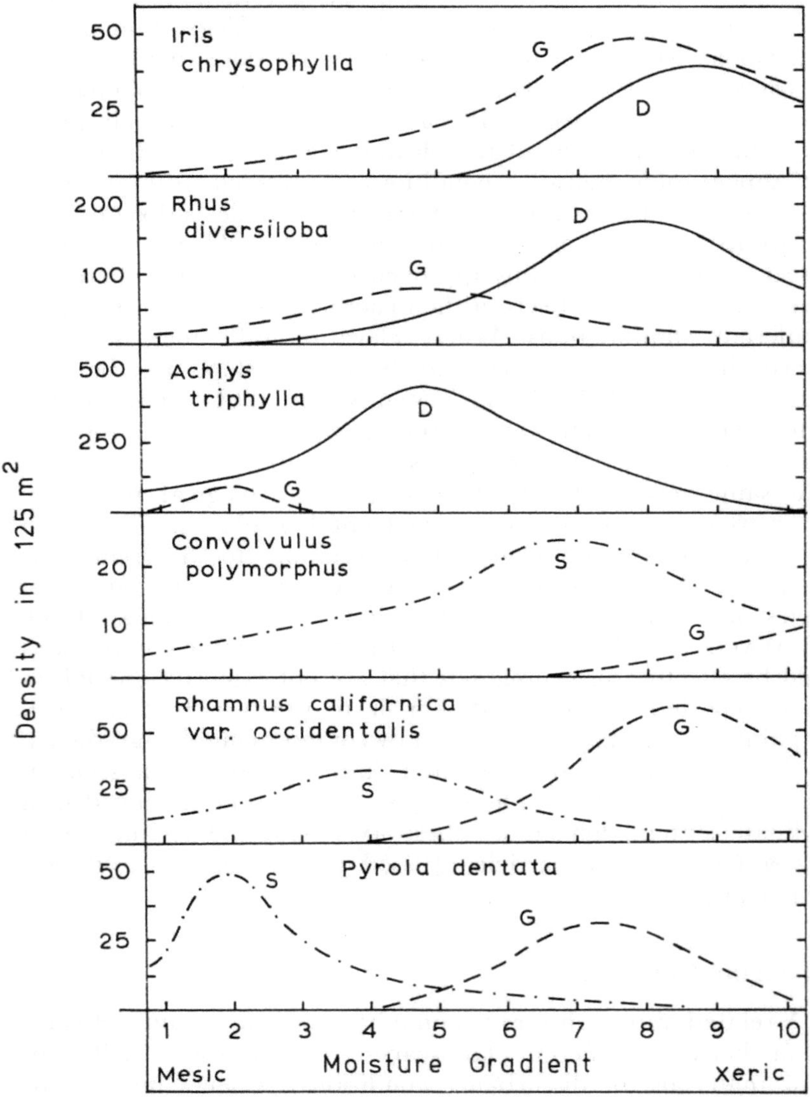

Fig. 4. Population shifts toward the mesic in response to parent materials, at low elevations in the Siskiyou Mountains, southwestern Oregon (WHITTAKER 1960). In each case the population distribution is shifted to the left (more mesic in terms of topographic positions) from acid parent material (D = quartz diorite) to basic parent material (G = olivine gabbro), or from basic parent material to ultrabasic (S = serpentinite and peridotite). The three study areas are near one another in similar climates, but comparable shifts are observed on a given parent material following species from a more to a less humid climate.

(ii) Character and composition of stands from the mid-points of the coenoclines. Mid-point samples, if undisturbed, may be used to represent 'intermediate climax composition' (WHITTAKER 1960); along topographic gradients such mid-point samples may be from open east and west-northwest slopes. (iii) Representation of community-types. As species change in relative importance and distribution from one coenocline to another, so must the community-types they characterize. On a 10-point topographic transect scale a given kind of community may occupy positions 3 to 9 in a more humid climate and appear reduced in distribution and displaced toward the mesic end, occupying positions 1 to 2 or 2 to 4, in a more arid climate. (iv) Floristic composition of the coenoclines as wholes, combining all samples for each of these. To the coenocline floras may be applied such characterizations as RAUNKIAER life-form spectra, representation of areal-types, and over-all species diversity ('gamma diversity') (WHITTAKER 1960, WHITTAKER & NIERING 1965, 1968b). (v) Degree of floristic differentiation of communities along the gradient. Coenoclines differ in the average width of species distributions along the gradient, and consequently in the degree of turnover or replacement of species populations from one end-point sample to the other. Degrees of compositional change along coenoclines have been measured as 'beta diversity' (WHITTAKER 1960, 1965, 1967, WHITTAKER & NIERING 1965, 1968b, article 4).

2.3.3 MOSAIC CHARTS

Two complex-gradients may be used as axes of a chart, as in Fig. 5. The samples from the landscape are classified into community-types, or transitions between these; and each sample is plotted on the mosaic chart according to its elevation and topographic position. Boundaries are drawn around community-types, thus outlining the ranges of environments these occupy, and showing how they relate to one another and the landscape pattern as a whole. (Because of the crudity of topographic position as an expression of moisture conditions, the samples for different types may overlap broadly, and mean positions must be sought for the boundaries.) Fig. 5 gives two such mosaic charts, by which major characteristics of vegetation patterns may be compared for acidic (diorite) and ultrabasic (serpentinite and peridotite) soils within the same mountain range in southern Oregon (WHITTAKER 1960, cf. WHITTAKER 1951, 1956, 1967, WHITTAKER & NIERING 1965, 1968b, WARING & MAJOR 1964).

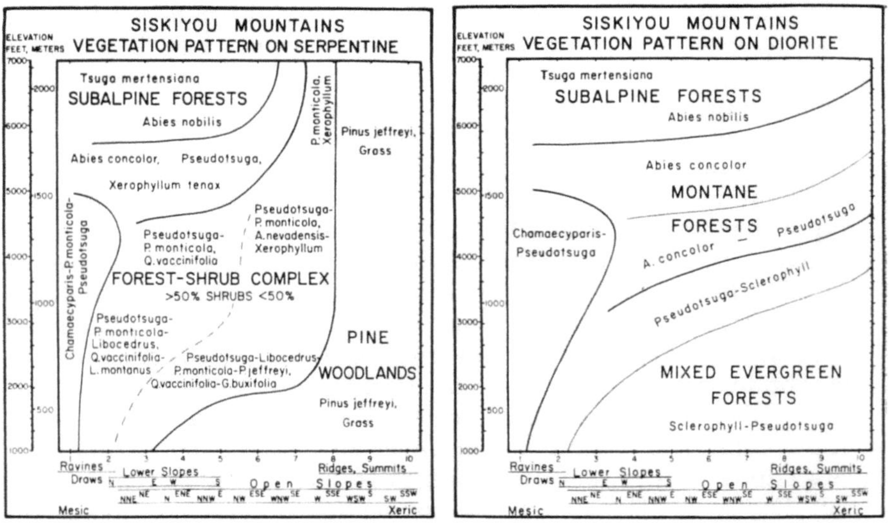

Fig. 5. Mosaic diagrams for vegetation on quartz diorite (right) and serpentines (left) in the Siskiyou Mountains, southwestern Oregon (Whittaker 1960, cf. Whittaker 1956, Whittaker & Niering 1965, 1968b). Vegetation samples were classified (into dominance-types) and plotted by elevation and topographic position on the chart. Boundaries for community-types were drawn at average positions of transitions between types. Vegetation on each parent material is considered a complex pattern of continuously intergrading communities (Whittaker 1960).

2.3.4 Nomograms

Further data can be plotted at the points on the mosaic chart that represent samples. (It is usually advantageous to group samples in sets of 5 or 10, average the composition for a group, and plot the data at the mean positions on the chart for the groups.) When importance values for a species have been plotted at points on the mosaic chart, isodems may be drawn to show the distribution of the population. Such population charts typically appear as elliptical hills (Fig. 6, upper left); but genetically complex species may show more complex patterns, as in the other three charts of Fig. 6. Community-level characteristics may be plotted and outlined in the same manner, as illustrated in Fig. 7 (Whittaker 1952, 1956, Whittaker & Niering 1965).

2.4 Conclusion

The diagrams represent ways in which techniques serve the purposes of gradient analysis. A first purpose is ordering of infor-

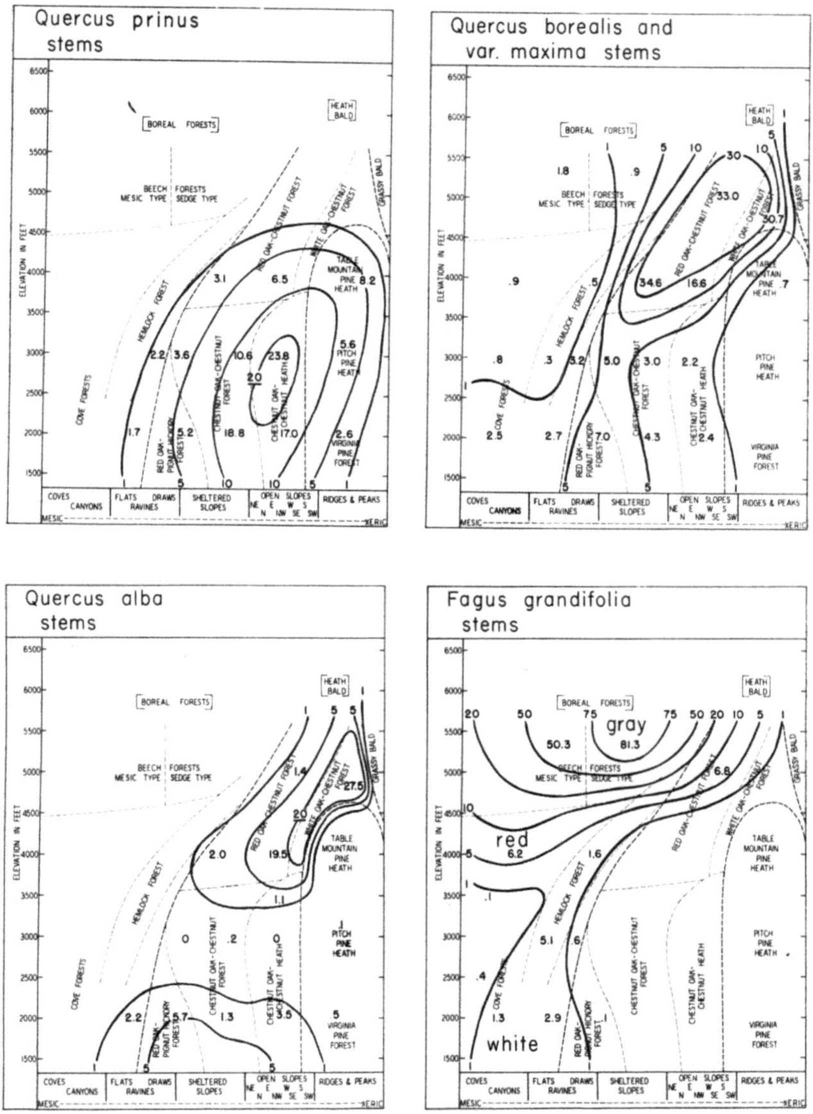

Fig. 6. Population nomograms for four tree species in the Great Smoky Mountains, Tennessee (WHITTAKER 1956). Per cents of tree stems in composite samples of about 1000 stems were plotted on a mosaic chart, and the values bounded by population lines (isodems). *Quercus prinus* shows a typical hill-shaped population, with a 'peak' in the chestnut oak heath, and densities decreasing in all directions away from this. The other three species show more complex distributions reflecting their genetic complexities.

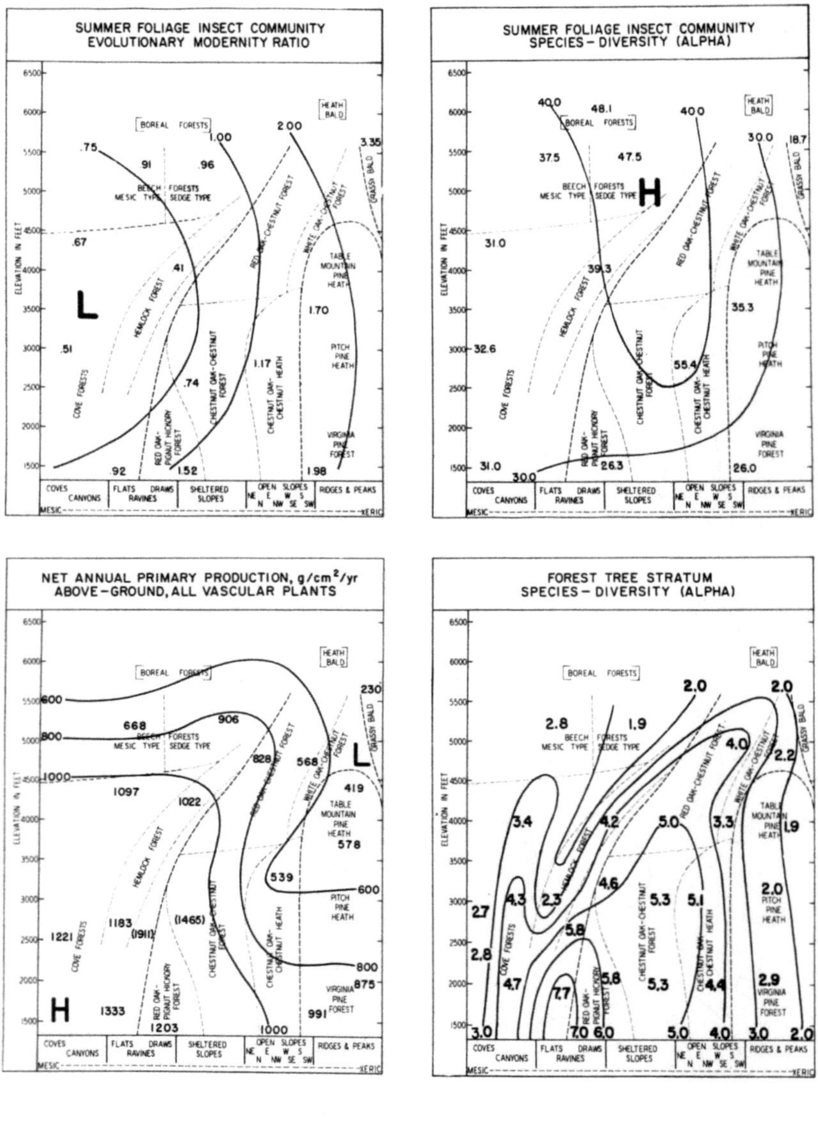

Fig. 7. Nomograms for community characteristics in the Great Smoky Mountains Tennessee. Upper left, the ratio of more to less modern insect groups in sweep samples, showing concentration of more primitive forms in the more humid environments; upper right, species diversities of sweep samples of insects (WHITTAKER 1952). Lower left, primary production of forest types (WHITTAKER 1966); lower right, species diversity of composite samples of the forest tree stratum (WHITTAKER 1956).

mation, a purpose shared with classification, though the ordering is based on gradients and continuities rather than classes. A transect table is a most effective means of organizing information about communities in relation to an environmental gradient and one another. A second purpose is revealing relationships to one another of gradients on three levels of ecological phenomena: environment, species populations, and community characteristics. The various types of figures are means of displaying some of these relationships in one direction of environmental variation, or two directions, among them: (i) species populations in relation to environment and one another, Figs. 1, 4, and 6; (ii) community characteristics in relation to environment, Fig. 9 in the following article and Fig. 7; (iii) community-types in relation to environment, Fig. 5; and (iv) species populations and community characteristics in relation to community-types, Figs. 6 and 7. Many of the figures in following articles represent analogous ways from different techniques, for illustrating these same relationships. Third, these techniques, in revealing relationships that are otherwise obscure in the complexity of vegetation, suggest directions for more intensive research into causal or functional meaning of the relationships. Finally, gradient analysis makes possible understanding of vegetation as a pattern of populations and communities in relation to environments, as developed further in the following article.

2.5.1 Summary

Direct gradient analysis is based on samples of vegetation, taken to represent the composition of plant communities and the characteristics of their environments at different points in the landscape. Samples should be taken from communities that are homogeneous, should be large enough to represent composition of communities effectively, and should be efficient in obtaining information. Three kinds of samples are in wide use: (i) the relevé of the school of Braun-Blanquet (see article 20), (ii) quadrat samples based on a tenth hectare or other fixed area, used for many ordination studies, and (iii) the plotless or point-centered samples of the school of Wisconsin.

For direct gradient analysis samples are ordinated either on the basis of measurements of environment, or by means of other indices that use community composition to express relative position along environmental gradients. Two most useful means of ordination by composition are: (i) use of weighted averages based on 'ecological groups'–sets of species whose distributions are centered in the same parts of an environmental gradient, and (ii) comparison of samples with end-point samples representing the extremes of the environmental gradient. Relative effectiveness of different ordinations can be measured. By ordination samples may be arranged into a transect, and their data then compiled into a transect table representing change in species populations and community characteristics along the gradient.

By other techniques more complex vegetation relationships to more than one environmental gradient may be analyzed: (i) Crossed transects may be used to study the relation of species and communities to two environmental gradients affecting the vegetation of an area. (ii) Coenoclines or community-gradients, particularly those along topographic moisture gradients, may be compared from one elevation belt, parent material, or climate to another. (iii) Mosaic charts outline the distribution of community-types in relation to two environmental gradients as axes. (iv) Nomograms show the relation of species populations and community characteristics to the range of environment and communities represented in the mosaic chart. By these means it is possible to understand the vegetation of a landscape as a pattern of species populations and communities.

2.5.2 Zusammenfassung

Die direkte Gradienten-Analyse geht von Vegetationsaufnahmen aus, welche die Zusammensetzung der Pflanzengesellschaften und die Standortseigenschaften an verschiedenen Punkten des Geländes wiedergeben. Die Aufnahmen sollen aus homogenen Beständen stammen, ihre Fläche soll groß genug sein und die gesamte Arten-Verbindung enthalten und sie sollen ergiebig anInformation sein. Drei Analysen-Methoden sind üblich:

1. die Aufnahme-Methode der Braun-Blanquet-Schule (s. Artikel 20). 2. Quadratanalyse, ausgehend von fixierten Oberflächen, z.B. 1/10 ha oder anderen Gröen, und 3. die Probenflächenlose Punkt-Proben-Methoden der Wisconsin-Schule.

Für die direkte Gradienten-Analyse werden die Aufnahmen entweder nach den Messungen von Standortsfaktoren oder mittels anderer Indices geordnet, welche die Gesellschaftszusammensetzung betreffen, um die Lage auf den Standorts-Gradienten zu bestimmen. Die brauchbarsten Ordinations-Methoden nach der Zusammensetzung sind: 1. Die Anwendung von gewogenen Mittelwerten von ökologischen Gruppen, d.h. Artengruppen gleicher Hauptverbreitung längs der Standorts-Gradienten, und

2. Vergleich der Aufnahmen mit 'Vergleichs-Aufnahmen', welche die Extrempunkte des Standorts-Gradienten darstellen. Der jeweilige Aussagewert der verschiedenen Ordinationen kann bestimmt werden. Mittels der Ordination können die Aufnahmen in einem Transekt angeordnet und ihre Daten wiederum in einer Transekt-Tabelle zusammengestellt werden, welche die Verschiebungen innerhalb der Arten-Verbindungen und der Gesellschaftseigenschaften längs der Gradienten darstellen. Durch andere Techniken können komplexere Beziehungen zwischen der Vegetation und mehreren Standorts-Gradienten analysiert werden:

1. Mit gekreuzten Transekten kann die Beziehung von Arten und Gesellschaften zu zwei die Vegetation eines Gebietes beeinflußenden Standorts-Gradienten studiert werden.

2. Coenoclines oder Gesellschafts-Gradienten, vor allem solche, die längs topographischen Feuchtigkeits-Gradienten verlaufen, können nach Höhenstufen, Muttergestein oder Klima miteinander vergleichen werden.

3. Mosaik-Karten bezeichnen die Verbreitung von Gesellschaftstypen in Beziehung au zwei Standorts-Gradienten, die als Achsen dienen.

4. Nomogramme zeigen die Beziehung von Art-Populationen und Gesellschaftsmerkmalen zur Amplitude der Standorte und Gesellschaften in der Mosaik-Karte. Auf diese Weise kann die Vegetation einer Landschaft als Strukturmuster von Art-Populationen und Gesellschaften verstanden werden.

3 DIRECT GRADIENT ANALYSIS: RESULTS

ROBERT H. WHITTAKER

Contents

3 DIRECT GRADIENT ANALYSIS: RESULTS

3.1 Introduction

There may be reason to judge that research in gradient analysis has led to significant advances in the understanding of natural communities. Results from gradient analysis bear in several ways on theory of communities, for they offer insight into the manner in which species evolve in relation to one another in communities, the character of the organization of species into communities, and the way communities as assemblages of species respond to environmental gradients. The ways species are distributed, and communities change, along environmental gradients have in turn wide implications for the practical problems of both classification and ordination. Results from gradient analysis have been discussed in some detail in the reviews of WHITTAKER (1967, 1970b) and MC INTOSH (1967). The figures in the preceding article on technique illustrate some of the essential findings, and some implications of these will be summarized here.

3.2 Forms of Species Distributions

The general form for distribution of a species population along an environmental complex-gradient is the bell-shaped, apparently Gaussian or binomial curve illustrated in Table I and Fig. 1 (WHITTAKER 1951, 1956, 1967, BROWN & CURTIS 1952). In response to two gradients, most populations form Gaussian surfaces, as illustrated in Fig. 6, upper left, with contours outlining a population "hill" any cut through which forms a Gaussian curve (WHITTAKER 1956). The form of distribution may be generalized to *n* dimensions as an "atmospheric distribution" (BRAY & CURTIS 1957) with a center of maximum density in a hyperspace of *n* complex-gradients as axes, away from which center density decreases in Gaussian curves in all directions. The center or mode of a species population along a complex gradient is not its physiological optimum but is a center of maximum population success in competition with other species populations. The population center may be in a less "favorable" part of the gradient than the physiological optimum defined by maximum growth-rate.

Some species depart from the simple Gaussian hill toward more complex forms (Fig. 6). The different centers of these bimodal or polymodal distributions represent different ecotypes, different adaptive peaks of genetic response to environment. As shown in Fig. 6, these peaks are in some cases connected and in some cases discontinuous with one another. In most species morphological differentiation can be detected when plants from different peaks are compared in detail with one another. When the peaks are connected by a "saddle" or col of intervening populations (Fig. 6, *Q. borealis*), the intermediate populations may show a continuous cline in morphological characteristics (WHITTAKER 1956). Width and form of species population distributions express some genetic characteristics of species, and results from gradient analysis are thus linked with those of genecology (WHITTAKER 1956).

The oblique relation of the population hills of Fig. 6 to the environmental complex-gradients may be noted. Proceeding in mountains from higher elevations toward lower elevations (usually with drier climates), the distribution of a species population is normally displaced toward the left (the more mesic) along the topographic moisture gradient. Similar displacements toward the mesic along topographic gradients affect species populations as one moves at low elevations from more humid toward drier climates (BOYKO 1947, WALTER & WALTER 1953, PERRING 1960). Such displacements appear also (Fig. 4) comparing areas of more and less favorable soil parent materials under the same climate–diorite versus serpentine, or diorite versus limestone (WHITTAKER 1960, WHITTAKER & NIERING 1968a, 1968b). More complex displacements of their different ecotypes appear in species with bimodal or polymodal distributions (WHITTAKER 1960).

3.3 Species Individuality

The centers of species populations are scattered along a complex-gradient in an apparently random manner (Fig. 1). When species are arranged in relation to one another and more than one environmental gradient or axis, in indirect gradient analysis, their positions are scattered rather than clustered (WHITTAKER 1967, articles 7 and 9). Species do not form well-defined groups of associates with similar distributions, clearly separate from other such groups, but are distributed according to the principle of species individuality: Each species is distributed in its own manner, according to its own genetic, physiological, and population response to environmental factors that affect it (including effects of other species) (RA-

MENSKY 1924, cf. GLEASON 1926, 1939, LENOBLE 1926, 1928). Species can be assigned to ecological groups and character-species groups by similarity of distribution, but the limits of such groupings are essentially arbitrary.

3.4 Community Continuity

Along a complex-gradient species populations, with their scattered centers and tapering, broadly overlapping distributions, form a population continuum or compositional gradient (WHITTAKER 1951, 1956, 1967, CURTIS & McINTOSH 1951, McINTOSH 1967). Results from gradient analysis thus support a second principle stated by RAMENSKY (1924) and GLEASON (1926) (see also article 5), that of the continuity of plant communities with one another (when that continuity is not interrupted by abrupt environmental change or disturbance). In relation to more than one complex-gradient, vegetation may be conceived as a *complex population continuum*, which to varying degrees in different landscapes is interrupted by discontinuities of environment and disturbance (WHITTAKER 1951, 1956, 1967). Physiognomic change in communities, as well as compositional change, is in general continuous along environmental gradients (BEARD 1955, WHITTAKER & NIERING 1965, 1968b). The response of communities to an environmental gradient may thus be approached either as a coenocline or community-gradient, or as a chain or sequence of (arbitrarily bounded) community-types that make up the gradient. A sequence of community-types along an environmental complex-gradient is an *ecological series* (CAJANDER 1903, KELLER 1925-6, SUKACHEV 1928, 1932, articles 5, 15, 17). A sequence of formations (as physiognomically defined community-types along a complex-gradient) is a *formation-series* (BEARD 1944, 1955, and article 13, WHITTAKER & NIERING 1968b).

Certain situations occur in which, in the absence of environmental discontinuity or disturbance, the principle of community continuity must be modified (Fig. 8). The distribution illustrated has been termed a "plateau distribution" (WHITTAKER 1956, 1967) and interpreted as a vertical truncation of the Gaussian distribution of a strongly dominant species. Such a distribution implies relative discontinuity of community composition (as regards the dominant species, at least) at the edge of the plateau distribution. Other observations on plateau distributions are: (i) Observed cases are very few in number, compared with the thousands of bell-shaped distributions observed in research in gradient analysis. (ii) These

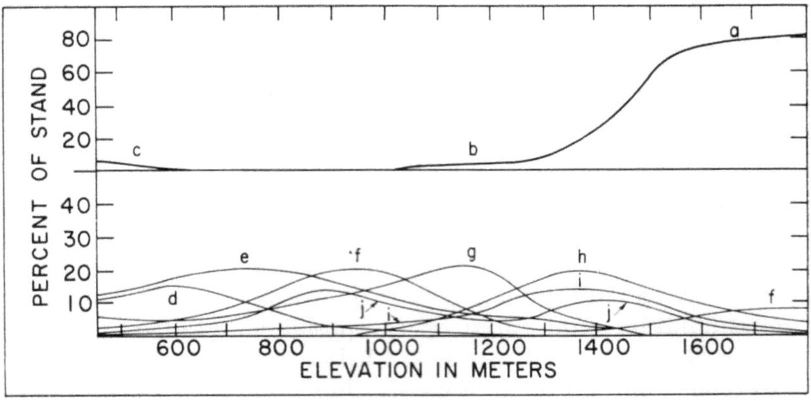

Fig. 8. Tree species distributions in an elevation transect of mesic forests, Great Smoky Mountains, Tennessee (WHITTAKER 1956). Beech (*Fagus grandifolia*) shows a 'plateau' distribution, which may be contrasted with the more typical bell-shaped curves for other tree species in the lower panel. The three ecotypes of *Fagus* (cf. Fig. 6) are indicated – gray (a), red (b) and white (c). Other species – *Acer rubrum* (d), *Tsuga canadensis* (e), *Halesia monticola* (f, bimodal), *Tilia heterophylla* (g), *Acer spicatum* (h), *Aesculus octandra* (i) and *Betula allegheniensis* (j, bimodal). Data are percentages of tree stems over 1 cm dbh in composite samples at 100-meter elevation intervals. *Fagus* shows a relatively steep change in population density between 1350 and 1500m, in south-facing canyons, but the distributions of many other species extend continuously through these elevations.

cases have been limited to more extreme environments, occupied by communities of strong single-species dominance. BEALS (1969a) observed a greater tendency toward such distributions along a steep elevation gradient in mountains than along a more gentle one. (iii) A community with a relatively discontinuous boundary along one complex-gradient may intergrade continuously with other communities along other gradients (WHITTAKER 1956). (iv) In some cases, such as the high-elevation beech forests (WHITTAKER 1956, 1967, cf. MARK 1963, SCOTT et al. 1964), the single dominant shows a relatively discontinuous, plateau distribution, but other species are continuously distributed through the edge of its distribution. The situation may then be one of apparent discontinuity, affecting a dominant population, superimposed on an underlying largely or fully continuous gradient of over-all floristic composition. Plateau distributions appear to represent local and partial qualifications of the principle of community continuity.

3.5 Community Trends

Many community-level characteristics show continuous trends along complex-gradients. Fig. 9 shows relations of RAUNKIAER life-

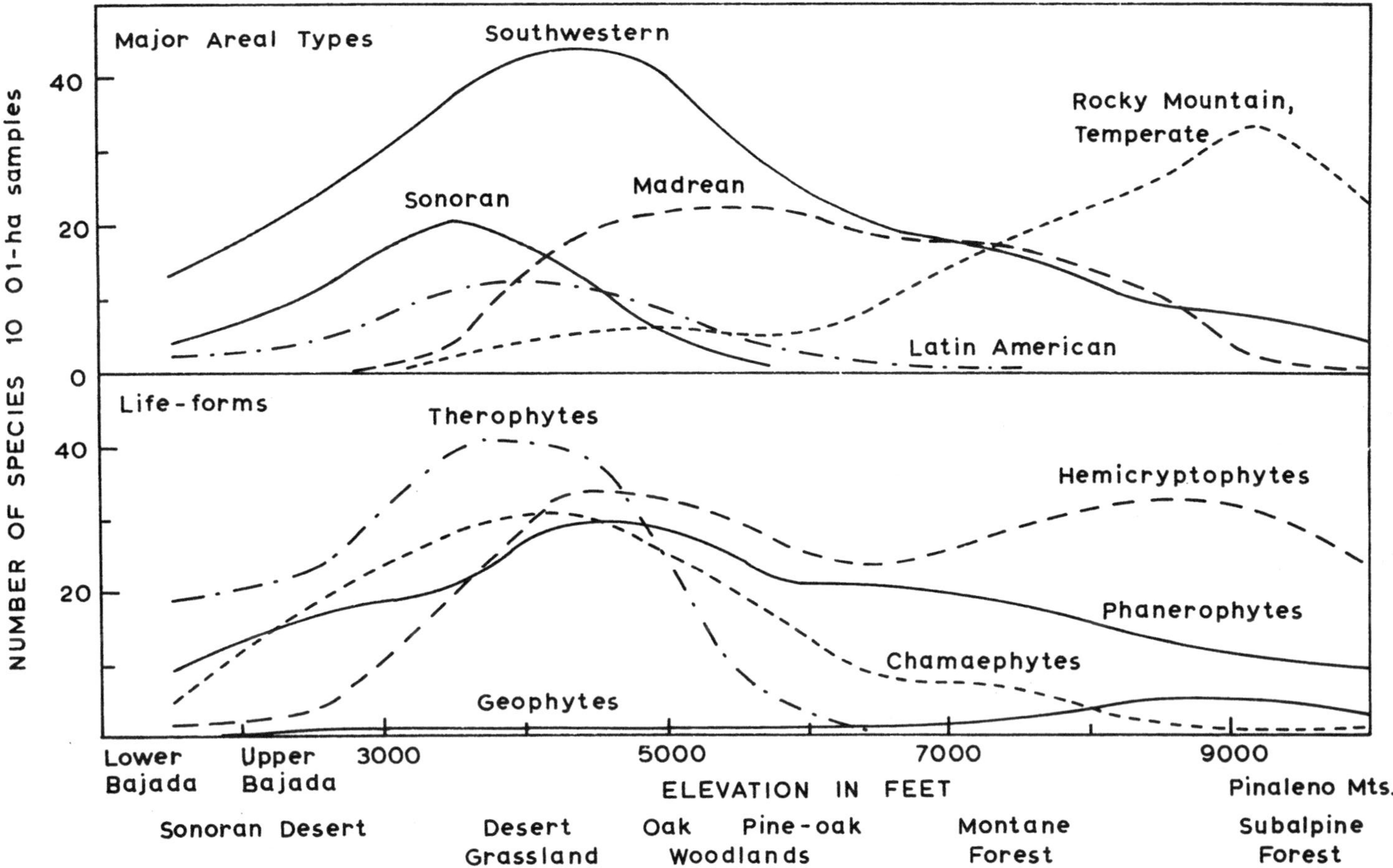

Fig. 9. Trends in community composition in relation to elevation, Santa Catalina Mountains, southeastern Arizona (WHITTAKER & NIERING 1964). Data are numbers of species by areal types and life-forms, in composite samples (of 10, 0.1 ha samples) from open intermediate and southerly slopes on the south side of the range.

forms and areal types (geographic distributions) of species to the elevation gradient in the Santa Catalina Mountains, Arizona. A life-form composition that is characteristic for a cool-temperate forest, with hemicryptophytes predominant, changes toward low elevations to the chamaemphyte- (and therophyte-) dominated life-form composition of the Sonoran desert. In high-elevation forests species with northern and Rocky-Mountain distributions predominate; at middle elevations in woodlands Madrean species (with the main areas of their distributions in Mexican mountains) are important; at low elevations southwestern and Sonoran species (with their major areas in the deserts of Sonora, Arizona, and California) prevail. Coverages of plant strata show trends along coenoclines (Table I, WHITTAKER 1956, 1960, WHITTAKER & NIERING 1965). Some strata show more complex changes along a gradient than a simple rise or fall; and in some cases the coverages of the tree vs. the shrub or the shrub vs. the herb stratum are inversely related to one another (WHITTAKER 1956). Canopy height, community biomass, and primary productivity show trends along the elevation and topographic moisture gradients in mountains (WHITTAKER 1966).

3.6 Species Diversity

Diversity relations to gradients have special interest. *Alpha* or within-habitat diversity refers to richness in species of particular communities, or individual samples from them. The major geographic relationship of alpha diversity for many groups of organisms is its increase toward more favorable, and particularly more stable environments, especially those of tropical rainforests (DOBZHANSKY 1950, ASHTON 1969, MACARTHUR 1969, WHITTAKER 1969). Along local environmental gradients, however, diversity change is often not a simple decrease toward less favorable environments. Maximum diversity of vascular plant species is in the middle, not at the mesic extreme, of some topographic moisture gradients (WHITTAKER 1956, 1969); in some cases maximum diversity is at intermediate pH values (MAAREL & LEERTOUWER 1967). Diversity generally increases toward less acid and more fertile soils (DAHL 1957, MONK 1965, FRYDMAN & WHITTAKER 1968, WHITTAKER 1969, PRUSINKIEWICZ 1970), but in the Siskiyou Mountains diversity is higher on the chemically extreme serpentine soils than on the more normal diorite soils (WHITTAKER 1954b, 1960). Diversities of different life-forms and strata of plant communities are relatively independent of one another. In the Siskiyou Mountains maximum tree species diversities are at low, but maximum herb species diversities at middle, elevations (WHITTAKER 1960, cf. DAUBENMIRE & DAUBENMIRE 1968). Herb diversities are higher in broadleaved deciduous than in needleleaved evergreen forest in similar habitats, and they tend to be higher in open woodlands with small trees and an open canopy admitting light to the herb stratum than in closed forests (WHITTAKER 1965, WHITTAKER & NIERING 1968b). Over-all diversities of vascular plants are in consequence generally higher in temperate woodlands than in temperate forests. Diversity shows no clear relation to productivity (WHITTAKER 1965, 1969).

Measurements can be applied to determine beta diversity, or degree of floristic differentiation, of coenoclines (WHITTAKER 1960, WHITTAKER & NIERING 1965). *Beta* or between-habitat diversity refers to the degree of change in species composition in the samples taken along a coenocline, or degree of contrast in species composition among samples of a set taken from a landscape. Beta, like alpha, diversity can be interpreted as a consequence of the accumulation through evolutionary time of species that have evolved toward reduction of competition between them by difference in the way they relate to environment and other species. Alpha diversity results from evolution into a community of species that differ in niche, that is in function, position, or relation to other species within that community. Beta diversity in a coenocline results from accumulation of species differing in habitat, that is in distributional relation to the environmental gradient, and from the narrowing of species amplitudes along the gradient by increasing competition from increasing numbers of species (WHITTAKER 1969, 1970a). Since both alpha and beta diversity express increase in species numbers through evolutionary time, they might be expected to vary in parallel along major environmental gradients. Both may be expected to decrease toward more rigorous, and especially more unstable, environments.

Both alpha and beta diversity appear to increase (but with various complications affecting different communities and strata) from arctic and alpine to tropical lowland communities. Both alpha and beta diversity increase from maritime climates inland to continental climates in the Temperate Zone (WHITTAKER 1960, 1969, WHITTAKER & NIERING 1965). Responses of beta diversity (as coenocline differentiation along the topographic moisture gradient) to parent material contrasts are less consistent. On serpentine beta diversity is as high as, or higher than, on diorite in the Siskiyou Mountains (WHITTAKER 1960); but beta diversity is markedly lower on limestone than on diorite in the Santa Catalina Mountains (WHITTAKER & NIERING 1968b).

3.7 Patterns of Community Characteristics

Measurements of community characteristics can be plotted on a mosaic chart, as in Fig. 7, to observe the manner in which these change through the vegetation pattern of a landscape (Whittaker 1952, 1956, Whittaker & Niering 1965). Of these patterns of community characteristics one may observe: (i) They form (as suggested by their behavior along particular complex-gradients) surfaces of complex, but generally continuous slope. In some cases the surfaces are bimodal or polymodal. (ii) The community characteristics consequently form in many cases continuous gradients through the communities of the landscape, but a community characteristic may not show a simple and linear relation to any particular complex-gradient as normally recognized. (iii) Like species populations, growth-forms and life-forms, the community characteristics are "individualistic"; no two of these show the same pattern of variation in relation to communities and environments.

3.8 Criticisms of Gradient Analysis

Such patterns for species and community characteristics emphasize the coherence of the vegetation of a landscape, the interrelation and intergradation of its communities in the pattern as a whole. This conception of community intergradation, which is shared by some phytosociologists, has been rejected by others whose emphasis is on community-types as distinct assemblages of species separated by relatively sharp boundaries from other such assemblages. History of the argument has been reviewed elsewhere (Whittaker 1962, article 18); recent criticisms of gradient analysis are Daubenmire (1966), Langford & Buell (1969), and Gounot (1969), see also McIntosh (1967-8). Some of these criticisms are, first, in the area of method, with the arguments: (i) That the apparent continuity of communities is an artifact resulting from the circularity of method that is inherent in certain approaches to ordination (see section 2.2.3.1), but not in others; (ii) That the community samples used for gradient analysis are not sufficiently homogeneous, and that the apparent continuity of vegetation in a transect results from blurring of community boundaries by heterogeneous samples; (iii) That successional relationships are inadequately considered, and inclusion in ordinations of unstable stands obscures boundaries that would be evident in stable communities; and (iv) More broadly, that studies in gradient analysis have imposed on their research an assumption of vegetational continuity that has prevented recognition of discontinuity. Second, in the area of observation and theory it is asserted: (v) That the occurrence in the field of combinations of species that repeat themselves from place to place, and in some cases seem to be sharply bounded, is in conflict with the results of gradient analysis, and (vi) That the evolution of species toward adaptation to one another, and especially of subordinate species toward adaptation to the community dominants, must imply the formation of relatively discrete groups of co-adapted species that would appear in the field in this self-repeating manner.

The number of studies seems now adequate to establish that, in relation to points (i) to (iii), the most essential results of gradient analysis – the tapered and scattered curves of species populations, individuality of species distributions, and continuity of undisturbed communities along continuous environmental gradients – are not artifacts of any particular kind of sample or technique of ordination. Undisturbed communities are in general more clearly continuous than disturbed ones. It is, furthermore, possible to illustrate these essential results with tables compiled by phytosociologists employing classification when the tables arrange community-types along environmental gradients (WHITTAKER 1962). The last statement is true despite the fact that samples taken on the assumption of discontinuity should not be used to test for continuity. In contrast and in relation to point (iv), samples taken without bias and ordinated should, if discontinuities are present, reveal those discontinuities. Freedom from bias for or against discontinuity has been an objective of sample procedure in most research in gradient analysis, and some studies have had as an objective the unbiased testing of continuity vs. discontinuity of dominance-types or associations (WHITTAKER 1956, 1960, BEALS 1969a, BESCHEL & WEBBER 1962) or of formations (WHITTAKER & NIERING 1965). Such work supports the conclusion that the response of plant communities to environmental gradients is predominantly continuous even though, because of the effects of topographic, edaphic, and disturbance discontinuities in the field, vegetation in the field is often a complex mixture of continuity and discontinuity (WHITTAKER 1956, cf. DAHL 1957).

Repetition of similar species combinations in similar environments (v) is equally to be expected whether species are individualistic or form co-adapted groups, whether communities are continuous or discontinuous. The more significant question is that of the direction of species evolution. There is, at first glance, a paradox in the evolution of species both toward co-adaption within the community, and toward distributional individuality (WHITTAKER 1962). The paradox is resolved by recognition that species evolve toward the reduction of competition by both niche difference (difference in position or role in the community as a basis of co-adaption) and habitat difference (difference in distribution, or location of population centers, along environmental gradients). The fact species differ in niche implies that they will not in general exclude one another's populations along sharp boundaries, but may form the gradually tapered and broadly overlapping distributions observed in gradient analysis. Species do not in general evolve toward formation of discrete and clearly bounded groups as suggested by point (vi), but toward distributional difference and toward accomodations to one another that imply, for most species, continuously overlapping distributions (WHITTAKER 1962, 1967, 1970b). There is no conflict between such reasoning on theory of species evolution, and the results of gradient analysis.

3.9 Conclusion

From these results may be drawn a conception of the spatial structure of vegetation (WHITTAKER 1956, 1965, 1967, 1970b). Numerous plant species occur in the landscape and have evolved toward difference in their environmental relations. This evolution implies (i) difference in species' intracommunity positions, or niches, (ii) scattering of their population centers to different locations along environmental (habitat) gradients, and (iii) broadly overlapping distributions of these populations, with overlap in distribution made possible by niche difference. The species that make up the vegetation pattern freely and variously combine with one another in communities that predominantly intergrade with one another, forming a complex and potentially continuous but locally interrupted population pattern. It is a purpose of gradient analysis to understand this pattern in terms of gradient relations of species populations, environmental factors, and community characteristics, and as a design of interrelated parts.

The complexities of topography and other environmental factors in space make the pattern complicated and difficult to recognize in the field. A vegetation pattern is to some degree implicated—folded in upon itself— by the alternation and repetition of its parts from one valley or slope exposure to another. Gradient analysis is an approach to explication of the pattern, in the sense of unfolding its spatially complex relationships. Gradient analysis seeks to abstract from the landscape a coherent and comprehensible pattern of community composition in relation to environmental gradients.

The axes of this abstract pattern are ecoclines–directions of variation in the landscape pattern, correlations of complex-gradients and coenoclines that repeat themselves in the landscape. The abstract pattern may be conceived also as a coordinate system, a field defined by two, or a hyperspace defined by *n* ecoclines as axes. Community samples, and centers of species populations, may be located in relation to one another in this field or hyperspace. The research approach of indirect gradient analysis, which derives a coordinate system from the compositional relations of samples and then seeks to relate the compositional axes to complex-gradients, will be discussed in articles that follow. The results of indirect gradient analysis in general converge with those of direct gradient analysis summarized here, and support the conception of vegetation suggested (McIntosh 1967, Whittaker 1967, 1970b).

3.10.1 Summary

Significant results from direct gradient analysis include: (i) Distributions of species populations along environmental gradients typically appear as broadly overlapping, bell-shaped curves. (ii) Along a given gradient the centers or modes of species populations are scattered, and no two species have the same distribution. (iii) Most natural communities (if they have not been disturbed) intergrade continuously along uninterrupted environmental gradients. (iv) Species diversity and other community characteristics change along environmental gradients through different kinds of communities as continuous trends, or more complex curves. (v) These characteristics form varied response surfaces on a mosaic chart representing a pattern of environment and vegetation, and in general no two community characteristics show the same pattern in relation to environment in such charts. (vi) Results from gradient analysis lead to a conception of the vegetation of a landscape as a complex continuum, an abstract pattern of interrelated gradients of environment, species populations, and community characteristics. The direction of species evolution is toward the scattered or "individualistic" and broadly overlapping distributions that are responsible for the complexity and continuity of vegetation patterns.

The complex continuum may be approached through a coordinate system using as axes the major directions of community variation in the landscape. The axes then define an abstract vegetational "space" in which samples, and species centers, may be located. Indirect ordination seeks to derive such a space from the characteristics of community samples as a basis for the arrangement of samples, and species, in relation to one another.

3.10.2 Zusammenfassung

Die direkte Gradienten-Analyse liefert als wichtige Ergebnisse:

1. Die Verteilung von Art-Populationen längs eines Standorts-Gradienten erscheint als einander breit überlappende glockenförmige Kurven.

2. Längs eines bestimmten Gradienten zeigen die Zentren ("modes") der Art-Populationen eine Streuung; nicht zwei Arten haben diselbe Verbreitung.

3. Die natürlichen Gesellschaften verlaufen, falls sie nicht gestört sind, kontinuierlich längs ununterbrochenen Standorts-Gradienten.

4. Die Art-Diversität und andere Gesellschaftsmerkmale ändern sich längs Standorts-Gradienten in den verschiedenen Gesellschaften in kontinuierlichem Verlauf oder in komplexeren Kurven.

5. Diese Eigenschaften bilden verschiedene "Responds-Flächen" in Mosaik-Karten, die ein Muster von Standort und Vegetation darstellen. Im allgemeinen zeigen in solchen Karten nicht zwei Gesellschaftsmerkmale die gleichen Verbreitungsbilder in Beziehung zum Standort.

6. Die Ergebnisse der Gradienten-Analyse führen zur Auffassung der Vegetation oder der Landschaft als ein komplexes Kontinuum, ein abstraktes Muster von untereinander zusammenhängenden Standsortgradienten, Art-Populationen und Gesellschaftsmerkmalen. Die Evolution der Arten strebt einer zerstreuten oder "individualistischen" sich breit überlappenden Verteilung zu, die für die Komplexität und Kontinuität der Vegetationsstruktur verantwortlich ist.

Das komplexe Kontinuum kann durch ein Koordinatensystem dargestellt werden, bei dem die Achsen in den Hauptrichtungen der Gesellschaftsvariation verlaufen. Diese Achsen begrenzen einen abstrakten Vegetations-"Raum", in den Aufnahmen und Verbreitungszentren der Arten eingeordnet werden können. Mit Hilfe der indirekten Ordination versucht man einen solchen Raum aus den Gesellschaftsmerkmalen abzuleiten, in dem Aufnahmen und Arten in ihren Beziehungen zueinander dargestellt werden können.

References

Ashton, P. S., – 1969 – Speciation among tropical forest trees: some deductions in the light of recent evidence. *Biol. J. linn. Soc. Lond.* 1: 155–196.

Ayyad, M. A. G. & R. L. Dix, – 1964 – An analysis of a vegetation-micro-environmental complex on prairie slopes in Saskatchewan. *Ecol. Monogr.* 34: 421–442.

Beals, E. W., – 1960 – Forest bird communities in the Apostle Islands of Wisconsin. *Wilson Bull.* 72: 156–181.

Beals, E. W., – 1969a – Vegetational change along altitudinal gradients. *Science, N.Y.* 165: 981–985.

Beals, E. W., – 1969b – Vegetation of the Erer-Gota Plain, Ethiopia. *J. Ecol.* 57: 655–667.

Beard, J. S., – 1944 – Climax vegetation in tropical America. *Ecology* 25: 127–158.

Beard, J. S., – 1955 – The classification of tropical American vegetation-types. *Ecology* 36: 89–100.

Beschel, R. E. & P. J. Webber, – 1962 – Gradient analysis in swamp forests. *Nature, Lond.* 194: 207–209.

Boyko, H., – 1947 – On the role of plants as quantitative climate indicators and the geo-ecological law of distribution. *J. Ecol.* 35: 138–157.

Braun-Blanquet, J., – 1932 – Plant Sociology, the Study of Plant Communities. McGraw-Hill, New York. 439 pp.

Braun-Blanquet, J., – 1951 – Pflanzensoziologie, Grundzüge der Vegetationskunde. 2nd ed. Springer, Wien. 631 pp. 3rd ed., 1964, 865 pp.

Bray, J. R., – 1956 – A study of mutual occurrence of plant species. *Ecology* 37: 21–28.

Bray, J. R., – 1960 – The composition of savanna vegetation in Wisconsin. *Ecology* 41: 721–732.

Bray, J. R., – 1961 – A test for estimating the relative informativeness of vegetation gradients. *J. Ecol.* 49: 631–642.

Bray, J. R. & J. T. Curtis, – 1957 – An ordination of the upland forest communities of southern Wisconsin. *Ecol. Monogr.* 27: 325–349

Brown, Dorothy., – 1954 – Methods of surveying and measuring vegetation. *Bull. Commonw. Bur. Past. Fld. Crops*, (Hurley, Berks, England) 42: 1–223.

Brown, R. T. & J. T. Curtis, – 1952 – The upland conifer-hardwood forests of northern Wisconsin. *Ecol. Monogr.* 22: 217–234.

Cain, S. A. & G. M. de Oliveira Castro, – 1959 – Manual of Vegetation Analysis. Harper, New York. 325 pp.

Cajander, A. K., – 1903 (1906) – Beiträge zur Kenntniss der Vegetation der Alluvionen des nördlichen Eurasiens. I. Die Alluvionen des unteren Lena-Thales. *Acta Soc. Sci. fenn.* 32(1): 1–182.

Clements, F. E., – 1936 – Nature and structure of the climax. *J. Ecol.* 24: 252–284.

Cottam, G. & J. T. Curtis, – 1949 – A method for making rapid surveys of woodlands by means of pairs of randomly selected trees. *Ecology* 30: 101–104.

Cottam, G. & J. T. Curtis, – 1956 – The use of distance measures in phytosociological sampling. *Ecology* 37: 451–460.

Curtis, J. T., – 1955 – A prairie continuum in Wisconsin. *Ecology* 36: 558–566.

Curtis, J. T., – 1959 – The Vegetation of Wisconsin: An Ordination of Plant Communities. Univ. Wisconsin, Madison. 657 pp.

Curtis, J. T. & R. P. McIntosh, – 1951 – An upland forest continuum in the prairie-forest border region of Wisconsin. *Ecology* 32: 476–496.

Dahl, E., – 1957 – Rondane: mountain vegetation in South Norway and its relation to the environment. *Skr. norske Vidensk-Akad.*, Mat.-naturv. Kl. 1956(3): 1–374.

Daubenmire, R., – 1966 – Vegetation: identification of typal communities. *Science, N.Y.* 151: 291–298.

Daubenmire, R. & Jean B. Daubenmire, – 1968 – Forest vegetation of eastern Washington and northern Idaho. *Tech. Bull. agric. Exp. Stn. Washington St.* 60: 1–104.

Dix, R. L., – 1959 – The influence of grazing on the thin–soil prairies of Wisconsin. *Ecology* 40: 36–49.
Dix, R. L. & J. E. Butler, – 1960 – A phytosociological study of a small prairie in Wisconsin. *Ecology* 41: 316–327.
Dix, R. L. & F. E. Smiens, – 1967 – The prairie, meadow, and marsh vegetation of Nelson County, North Dakota. *Can. J. Bot.* 45: 21–58.
Dobzhansky, T., – 1950 – Evolution in the tropics. *Am. Scient.* 38: 209–221.
Duvigneaud, P., – 1946 – La variabilité des associations végétales. *Bull. Soc. r. Bot. Belg.* 78: 107–134.
Dyksterhuis, E. J., – 1949 – Condition and management of range land based on quantitative ecology. *J. Range Mgmt.* 2: 104–115.
Dyksterhuis, E. J., – 1958 – Ecological principles of range evaluation. *Bot. Rev.* 24: 253–272.
Egler, R. E., – 1942 – Vegetation as an object of study. *Philosophy Sci.* 9: 245–260.
Ellenberg, H., – 1948 – Unkrautgesellschaften als Mass für den Säuregrad, die Verdichtung und andere Eigenschaften des Ackerbodens. *Ber. Landtech.* 4: 130–146.
Ellenberg, H., – 1950 – Landwirtschaftliche Pflanzensoziologie. I. Unkrautgemeinschaften als Zeiger für Klima und Boden. Ulmer, Stuttgart. 141 pp.
Ellenberg, H., – 1952 – Landwirtschaftliche Pflanzensoziologie. II. Wiesen und Weiden und ihre standörtliche Bewertung. Ulmer, Stuttgart. 143 pp.
Ellenberg, H., – 1956 – Aufgaben und Methoden der Vegetationskunde. In 'Einführung in die Phytologie,' ed. H. Walter, vol. IV, pt. 1. Ulmer, Stuttgart. 136 pp.
Frydman, I. & R. H. Whittaker, – 1968 – Forest associations of southeast Lublin Province, Poland (Germ. summ.) *Ecology* 49: 896–908.
Gittins, R., – 1965 – Multivariate approaches to a limestone grassland community. I. A stand ordination. *J. Ecol.* 53: 385–401.
Gleason, H. A., – 1926 – The individualistic concept of the plant association. *Bull. Torrey bot. Club* 53: 7–26.
Gleason, H. A., – 1939 – The individualistic concept of the plant association. *Am. Midl. Nat.* 21: 92–110.
Goff, F. G. & G. Cottam, – 1967 – Gradient analysis: the use of species and synthetic indices. *Ecology* 48: 793–806.
Goodall, D. W., – 1954 – Vegetational classification and vegetational continua. (Germ. summ.) *Angew. PflSoziol.* Wien, Festschr. Aichinger 1: 168–182.
Gounot, M., – 1969 – Méthodes d'étude quantitative de la végétation. Masson, Paris. 314 pp.
Greig-Smith, P., – 1964 – Quantitative Plant Ecology. 2nd ed. Butterworths, London. 256 pp.
Horikawa, Y. & K. Okutomi, – 1955 – The continuum of the vegetation on the slopes of Mt. Shiroyama, Iwakuni City, Prov. Suwo (Japan. with Engl. summ.) *Seibutsugakkaishi, Hiroshima* 6: 8–17.
Itow, S., – 1963 – Grassland vegetation in uplands of western Honshu, Japan. II. Succession and grazing indicators. *Jap. J. Bot.* 18: 133–167.
Keller, B. A., – 1925–6 – Die Vegetation auf den Salzböden der russischen Halbwüsten und Wüsten. *Z. Bot.* 18: 113–137.
Knight, D. H., – 1965 – A gradient analysis of Wisconsin prairie vegetation on the basis of plant structure and function. *Ecology* 46: 744–747.
Langford, A. N. & M. F. Buell, – 1969 – Integration, identity and stability in the plant association. *Adv. ecol. Res.* 6: 83–135.
Lenoble, F., – 1926 – À propos des associations végétales. *Bull. Soc. bot. Fr.* 73: 873–893.

LENOBLE, F., – 1928 – Associations végétales et espèces. *Archs Bot.*, Caen, *Bull. mens.* 2(1): 1–14.
LOUCKS, O. L., – 1962 – Ordinating forest communities by means of environmental scalars and phytosociological indices. *Ecol. Monogr.* 32: 137–166.
MAAREL, E. VAN DER, – 1960 – Rapport inzake de vegetatie van het duingebied van de Stichting 'Het Zuid-Hollands Landschap' bij Oostvoorne. Report Z.H.L., Delft.
MAAREL, E. VAN DER, – 1969 – On the use of ordination models in phytosociology. (Germ. summ.) *Vegetatio* 19: 21–46.
MAAREL, E. VAN DER & J. LEERTOUWER, – 1967 – Variation in vegetation and species diversity along a local environmental gradient. *Acta bot. neerl.* 16: 211–221.
MACARTHUR, R. H., – 1969 – Patterns of communities in the tropics. *Biol. J. linn. Soc. Lond.* 1: 19–30.
MARK, A. F., – 1963 – Vegetation studies on Secretary Island, Fiordland. 3: The altitudinal gradient in forest composition, structure and regeneration. *N. Z. Jl. Bot.* 1: 188–202.
MARK, A. F. & F. R. SANDERSON, – 1962 – The altitudinal gradient in forest composition, structure and regeneration in the Hollyford Valley, Fiordland. *Proc. N.Z. ecol. Soc.* 9: 17–26.
MAYCOCK, P. F. & J. T. CURTIS, – 1960 – The phytosociology of boreal conifer–hardwood forests of the Great Lakes region. *Ecol. Monogr.* 30: 1–35.
MCINTOSH, R. P., – 1967 – The continuum concept of vegetation. *Bot. Rev.* 33: 130–187. Responses, *ibid.* 34: 253–332, 1968.
MONK, C. D., – 1965 – Southern mixed hardwood forest of northcentral Florida. *Ecol. Monogr.* 35: 335–354.
MOORE, J. J., P. FITZSIMMONS, E. LAMBE & J. WHITE, – 1970 – A comparison and evaluation of some phytosociological techniques. (Germ. summ.) *Vegetatio* 20: 1–20.
MOWBRAY, T. B., – 1966 – Vegetational gradients in the Bearwallow Gorge of the Blue Ridge Escarpment. *J. Elisha Mitchell scient. Soc.* 82: 138–149.
MOWBRAY, T. B. & H. J. OOSTING, – 1968 – Vegetation gradients in relation to environment and phenology in a southern Blue Ridge gorge. *Ecol. Monogr.* 38: 309–344.
OKUTOMI, K., – 1958 – A forest continuum in Is. Ujina, Hiroshima. (Japan. with Engl. summ.) *Bull. Fukuoka Gakugei Univ.* 8: 75–83.
PERRING, F., – 1959 – Topographical gradients of chalk grassland. *J. Ecol.* 47: 447–481.
PERRING, F., – 1960 – Climatic gradients of chalk grassland. *J. Ecol.* 48: 415–442.
PHILLIPS, E. A., – 1959 – Methods of Vegetation Study. Holt, New York. 107 pp.
PRUSINKIEWICZ, Z., – 1970 – Die Zahl der auf dem Minimi-Areal vorkommenden Gefässpflanzenarten als Mass für die Fruchtbarkeit der Waldböden. (Engl. summ.) In 'Gesellschaftsmorphologie (Stukturforschung)' ed. R. Tüxen. *Ber. Symp. int. Ver. Vegetationskunde*, Rinteln, 1966: 282–296.
RAMENSKY, L. G., – 1924 – Die Grundgesetzmässigkeiten im Aufbau der Vegetationsdecke. (In Russian) *Vêst. opŷt. Dêla sred.-chernoz. Obl.*, Voronezh (Abstr. *Bot. Zlb.* N.F. 7: 453–455, 1926).
ROWE, J. S., – 1956 – Uses of undergrowth plant species in forestry. *Ecology* 37: 461–473.
SCOTT, G. A. M., A. F. MARK & F. R. SANDERSON, – 1964 – Altitudinal variation in forest composition near Lake Hankinson, Fiordland. *N. Z. Jl. Bot.* 2: 310–323.
SUKACHEV, V. N., – 1928 – Principles of classification of the spruce communities of European Russia. *J. Ecol.* 16: 1–18.

Sukatschew, W. N., – 1932 – Die Untersuchung der Waldtypen des osteuropäischen Flachlandes. *Handb. biol. ArbMeth.* 11, 6: 191–250.

Tanghe, M., – 1964 – Contribution à l'étude de la végétation forestière de la Haute Belgique. 2. Utilisation d'un transect topographique des environs de Muno, pour la délimitation des groupes écologiques forestiers de la Gaume et de l'Ardenne. *Bull. Soc. r. Bot. Belg.* 98: 5–22.

Walter, H. & E. Walter, – 1953 – Einige allgemeine Ergebnisse unserer Forschungsreise nach Südwestafrika 1952/53: Das Gesetz der relativen Standortskonstanz; das Wesen der Pflanzengemeinschaften. *Ber. dt. bot. Ges.* 66: 228–236.

Waring, R. H. & J. Major, – 1964 – Some vegetation of the California coastal redwood region in relation to gradients of moisture, nutrients, light, and temperature. *Ecol. Monogr.* 34: 167–215.

Wells, Judith A. & A. F. Mark, – 1966 – The altitudinal sequence of climax vegetation on Mt Anglem, Stewart Island. I. The principal strata. *N.Z. Jl Bot.* 4: 267–282.

Westhoff, V., – 1970 – Vegetation study as a branch of biological science. (Germ. summ.) In 'Vegetatiekunde als synthetische wetenschap,'ed. H. J. Venema, H. Doing, & I. S. Zonneveld. *Misc. Pap. Landbouwhogeschool Wageningen,* 5: 11–30.

Westman, W. E., – 1971 – Production, nutrient circulation, and vegetation-soil relations of the pygmy forest region of northern California. *Thesis, Cornell Univ., Ithaca.*

Whittaker, R. H., – 1951 – A criticism of the plant association and climatic climax concepts. *NW. Sci.* 25: 17–31.

Whittaker, R. H., – 1952 – A study of summer foliage insect communities in the Great Smoky Mountains. *Ecol. Monogr.* 22: 1–44.

Whittaker, R. H., – 1954a – Plant populations and the basis of plant indication. (Germ. summ.) *Angew. Pflsoziol.,* Wien, Festschr. Aichinger 1: 183–206.

Whittaker, R. H., – 1954b – The ecology of serpentine soils. IV. The vegetational response to serpentine soils. *Ecology* 35: 275–288.

Whittaker, R. H., – 1956 – Vegetation of the Great Smoky Mountains. *Ecol. Monogr.* 26: 1–80.

Whittaker, R. H., – 1960 – Vegetation of the Siskiyou Mountains, Oregon and California. *Ecol. Monogr.* 30: 279–338.

Whittaker, R. H., – 1962 – Classification of natural communities. *Bot. Rev.* 28: 1–239.

Whittaker, R. H., – 1965 – Dominance and diversity in land plant communities. *Science, N.Y.* 147: 250–260.

Whittaker, R. H., – 1966 – Forest dimensions and production in the Great Smoky Mountains. *Ecology* 47: 103–121.

Whittaker, R. H., 1967 – Gradient analysis of vegetation. *Biol. Rev.* 42: 207–264.

Whittaker, R. H., – 1969 – Evolution of diversity in plant communities. *Brookhaven Symp. Biol.* 22: 178–196.

Whittaker, R. H., – 1970a – Communities and Ecosystems. Macmillan, New York. 162 pp.

Whittaker, R. H., – 1970b – The population structure of vegetation (Germ. summ.) In 'Gesellschaftsmorphologie (Strukturforschung)' ed. R. Tüxen. *Ber. Symp. int. Ver. Vegetationskunde,* Rinteln 1966: 39–62.

Whittaker, R. H. & W. A. Niering, – 1964 – Vegetation of the Santa Catalina Mountains, Arizona. I. Ecological classification and distribution of species. *J. Ariz. Acad. Sci.* 3: 9–34.

Whittaker, R. H. & W. A. Niering, – 1965 – Vegetation of the Santa Catalina Mountains, Arizona. (II.) A gradient analysis of the south slope. *Ecology* 46: 429–452.
Whittaker, R. H. & W. A. Niering, – 1968a – Vegetation of the Santa Catalina Mountains, Arizona. III. Species distribution and floristic relations on the north slope. *J. Ariz. Acad. Sci.* 5: 3–21.
Whittaker, R. H. & W. A. Niering, – 1968b – Vegetation of the Santa Catalina Mountains, Arizona. IV. Limestone and acid soils. *J. Ecol.* 56: 523–544.
Yang, T. W. & C. H. Lowe, Jr., – 1956 – Correlation of major vegetation climaxes with soil characteristics in the Sonoran Desert. *Science, N.Y.* 123: 542.

4 RETROGRESSION AND COENOCLINE DISTANCE

Robert H. Whittaker and George M. Woodwell

Contents page

4 RETROGRESSION AND COENOCLINE DISTANCE

4.1 Introduction

In succession community and environment are coupled together as parts of a functional system in which the interplay among species populations and the interactions between community and environment determine the course of development. Development continues until a stable community or climax, adapted to steady-state function and self-maintenance of its species in its biotope as affected by itself, results. It seems reasonable to consider the community changes during succession a community-gradient or coenocline (even though some replacements of species populations may be relatively abrupt). The progressive changes in environment during succession may be considered a seral complex-gradient (though successions differ widely in the degree of environmental modification they entail). A succession is thus an ecocline in time, comprising a parallel and functionally related coenocline and complex-gradient.

The perspectives of gradient analysis are appropriate to research on succession. There has been little application of them to such problems, except for the appearance of successional or disturbance factors among the axes of indirect ordinations. In one area of research in ecology, however, this direction can be pursued. This is the study of retrogression: effects of chronic stress in reducing community structure and changing species composition in ways suggestive of succession in reverse. Technological civilization is producing widespread retrogressive changes in communities—toward simplified structure, loss of specialist species and reduction of species diversity, loosening of nutrient cycles, and decrease in stability (Woodwell 1970). We shall consider the character of retrogressions, measurements of degree of retrogression, and the nature of ecological distance along retrogressive and other coenoclines.

4.2 Observations on Retrogression

4.2.1 Grazing

From overgrazing of grasslands a familiar sequence of changes may result: (i) First, the more sensitive grass and herb species are decreased in cover or eliminated, with some compensatory increase in less sensitive species. (ii) Second, the more sensitive species are largely eliminated, while less sensitive ones assume dominance and weedy or woody invaders appear in a community now of reduced total coverage. (iii) Third, continued grazing reduces these less sensitive dominants, until the invaders assume dominance of a community of gravely reduced coverage and productivity, with much of its soil surface exposed to erosion. (iv) Fourth, combined effects of grazing, trampling, and erosion reduce the cover of even the invaders that remain until (v) Finally, on vulnerable slopes, there are few plants left on an eroded rocky slope or mudfield. One notes, as expressions of 'retrogression' in this process, the reduction of community coverage, mass and productivity, and ultimately of soil quality. Of the extensive American literature on the evaluation of range condition and trend (ELLISON 1949, 1954, STODDART & SMITH 1955), most to the point of our purpose is the approach of DYKSTERHUIS (1949). Three groups of plants are recognized: (i) decreasers: the the more sensitive climax species that are reduced in coverage by even moderate grazing pressure, (ii) increasers: species present in the climax grassland that increase in coverage with moderate grazing, and (iii) invaders: species that are not present (or are restricted to local disturbed areas) in the climax grassland and that enter the community in response to grazing. It is reasonable to regard the sequence of changes in response to grazing as a community-gradient or coenocline, and to treat these three groups as ecological groups in the sense of DUVIGNEAUD (1946), ELLENBERG, (1950, 1952) and others. Weighted averages using these three groups can be used as indices of position along the gradient of community disturbance. DIX (1959) and ITOW (1963) have used weighted averages for indication of grazing effects. ITOW's study shows also the decrease in species diversity with increasing pressure, although light or moderate grazing in some cases increases diversity (HARPER 1969). DYKSTERHUIS (1949), used a form of percentage similarity as a measurement of retrogression under grazing. Given composition of the climax as a standard of comparison, he summed for a given sample the percentage of its cover made up of decreasers, plus the percentage of increasers up to but not beyond their coverage in the climax, as an index of degree of departure from climax composition.

4.2.2 POLLUTION

A second group of observations involve responses of terrestrial communities to air pollution. It has long been known that epiphytic

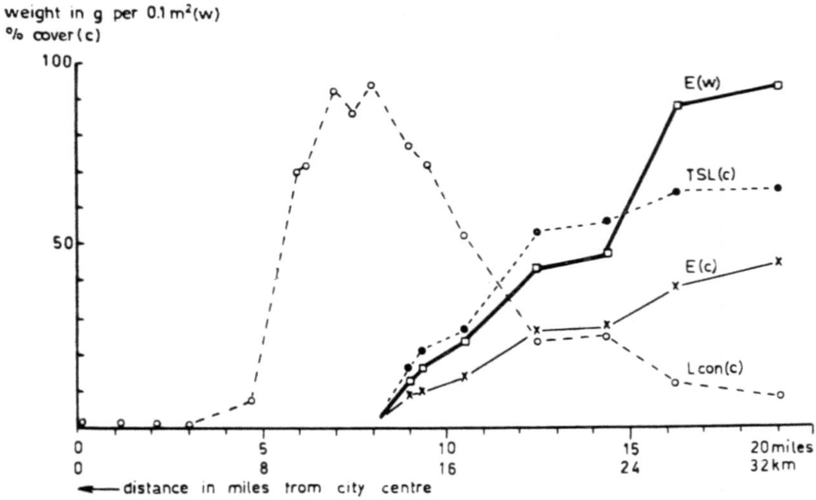

Fig. 1. Reduction in numbers of lichen and bryophyte species along a gradient of increasing sulfur dioxide pollution in England (GILBERT 1969). The lichen species *Lecanora conizaeoides* increases as other species are reduced at intermediate pollution intensities.

lichens and mosses are affected by the air pollution of cities; these effects, familiar in Western Europe (HILITZER 1925, BARKMAN 1958, 1969, SKYE 1958, 1968, GILBERT 1965, 1968), are known also from Montreal (LEBLANC 1969, 1971) and other American cities. Indicator species – decreasers and increasers – can be recognized among lichens and mosses. Particularly to the point are such observations as those of GILBERT (1969) in Fig. 1, showing both decrease in numbers of species and the behavior of an increaser, *Lecanora conizaeoides*, in response to an air pollution gradient. This, and other data for species populations along successional gradients, give bell-shaped curves suggestive of those for stable populations along environmental gradients (preceding article). The forms and dynamics of the successional curves, the extent to which they are skewed on a time-axis, and the possible occurrence of plateau distributions among them, have apparently not been studied. GORDON and GORHAM (1963) describe effects of an iron sintering plant releasing sulfur dioxide in Ontario. Along a gradient of increasing damage to forest communities, the dominant trees (*Picea glauca* and *P. mariana*) were eliminated about 25 km from the plant; and between 25 and 7 km the plant species diversity of the remaining vegetation was progressively reduced, until from 2 to 7 km from the plant only two resistant species (*Polygonum cilinode* and *Sambucus pubens*) remained in most samples. Fig. 2 represents

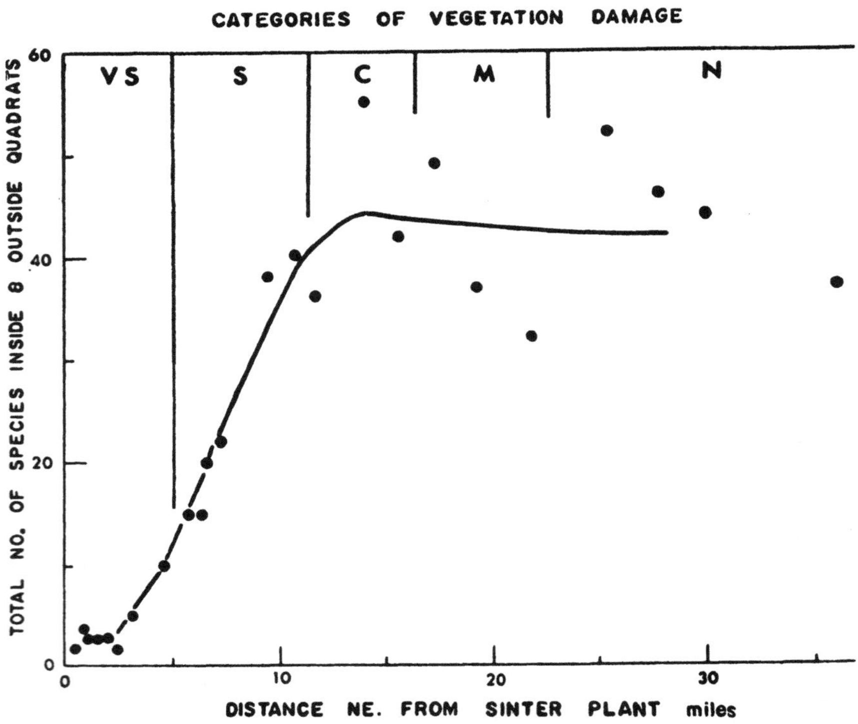

Fig. 2. Reduction in species diversity of vascular plants along a gradient of increased sulfur dioxide pollution in Ontario (GORDON & GORHAM 1963).

the effects on species-diversity with increasing proximity to the sulfur dioxide source. RAO and LEBLANC (1967) studied effects of the same pollution gradient on epiphytic lichens and mosses and observed increasing sensitivity with plant height, from crustose to foliose to fruticose forms among lichens. HADJÚK (1970) and HADJÚK & RUŽIČKA (1959) describe reductions in species diversity by air pollution. Retrogression of terrestrial communities under pollution is thus expressed in reduction of stature and coverage, living mass and productivity, and species diversity.

A third group of observations concern the effects of pollution on aquatic communities. The most extensive literature deals with effects of various kinds of pollution on animal groups, particularly benthic forms (PATRICK 1951, HYNES 1963, INGRAM et al. 1966, WILHM & DORRIS 1968). GORHAM and GORDON (1963) have shown the decrease in species diversity of aquatic vegetation with sulphate pollution. Effects on phytoplankton communities have been studied by PATRICK (1963, PATRICK et al. 1954) and others (STAUB et al. 1970). PATRICK has shown that plankton diatom communities may

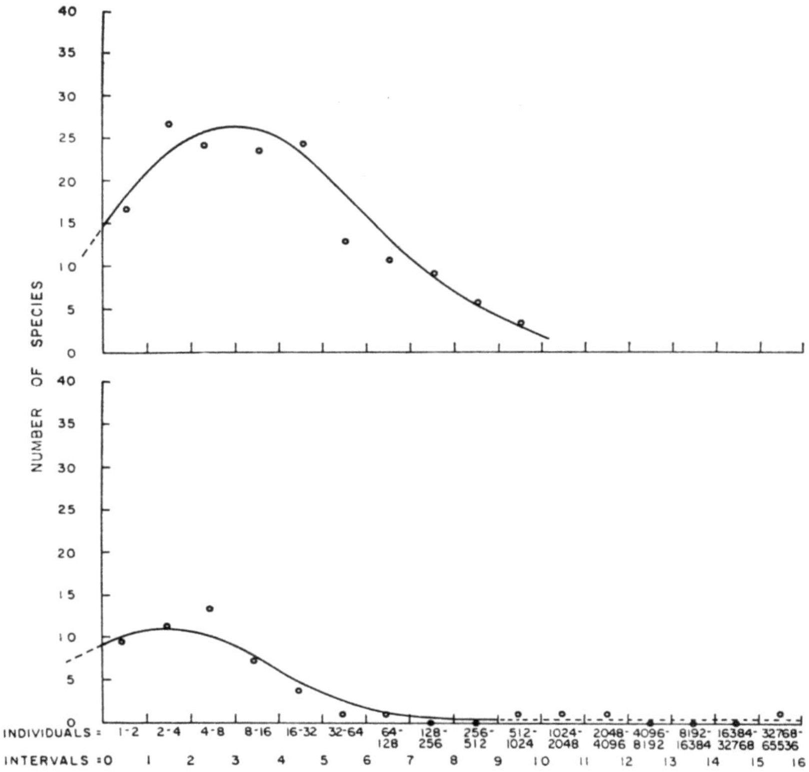

Fig. 3. Lognormal distributions of diatom species on glass slides suspended in streams, an unpolluted stream (Ridley Creek, Pennsylvania) above, and a polluted stream (Back River, Maryland) below (PATRICK 1963).

be described by the lognormal distribution of PRESTON (1948). When species in a large community sample are classed into octaves or doubling ranges of importance values, they form a frequency distribution with many species of intermediate abundance, few very common ones, and few very rare ones. The curves of Fig. 3 are truncated or broken by the limits of sample adequacy on the left, but one may estimate the total numbers of species in communities by extending the curves to the left. The upper curve is for a diatom community established on glass slides immersed in an unpolluted stream; with it may be compared the curve for a polluted stream below. The polluted stream has a lower number of species per octave, a wider dispersion of the species through octaves, and a much lower total richness in species (PATRICK 1963).

4.2.3 Irradiation

An oak-pine forest at Brookhaven National Laboratory was subjected to chronic gamma irradiation (Woodwell 1967, 1970, Woodwell & Whittaker 1968). With increasing intensity of irradiation the forest was progressively reduced. First the pine (*Pinus rigida*) was eliminated, then the canopy oaks (*Quercus coccinea, Q. alba*) to leave a community dominated by vacciniaceous shrubs (*Gaylussacia baccata, Vaccinium vacillans*). With more intensive irradiation the shrubs were eliminated and a sedge (*Carex pensylvanica*) normally present in the forest at low coverage acted as an increaser, forming a sedge-dominated zone. At still more intense irradiation the sedge was eliminated, and of the plant community close to the gamma source only lichens on the bark of dead trees remained. Fig. 4 represents the effect of irradiation on vascular plant species diversity of the forest. Fig. 5 shows by coefficients of community the increasing degree of change in forest composition with increasing irradiation.

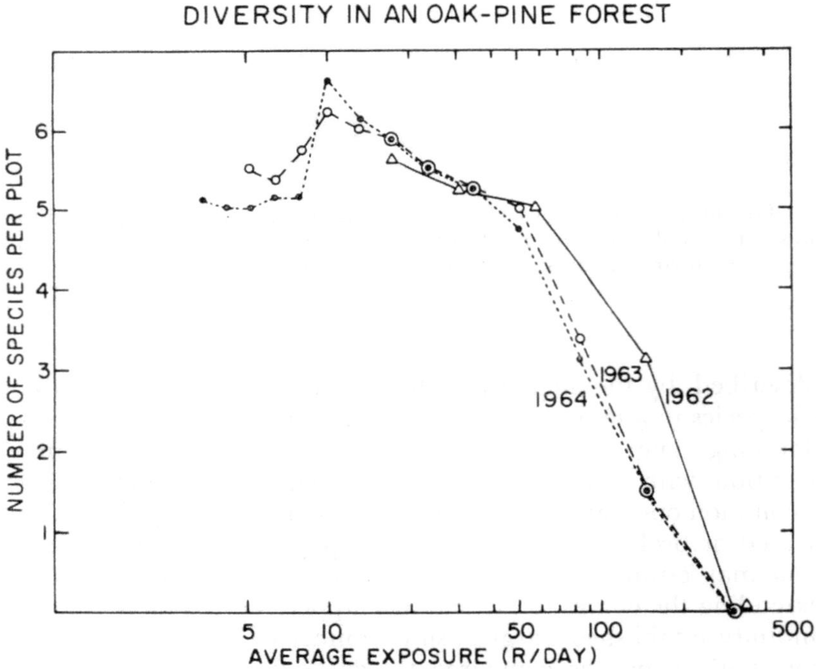

Fig. 4. Reduction of species diversities of vascular plants along a gradient of increasing gamma irradiation in the Brookhaven forest after one, two, and three years' exposure (Woodwell & Rebuck 1967).

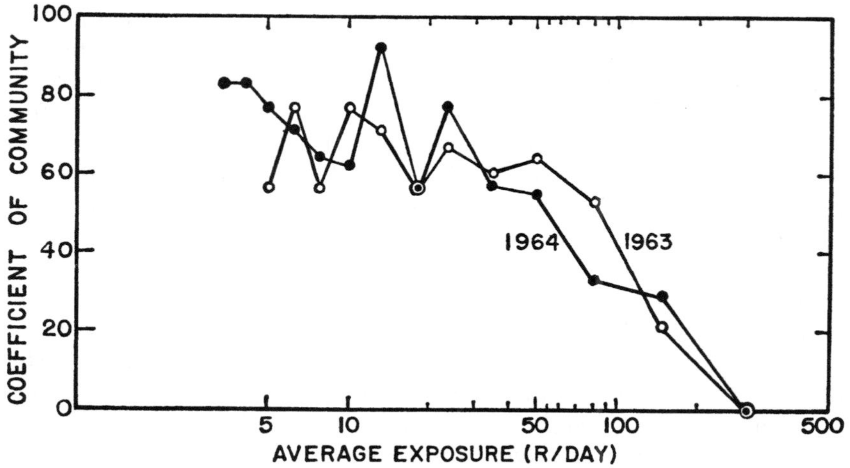

Fig. 5. Decrease in sample similarity (coefficient of community for vascular plants) along a gradient of increasing gamma irradiation in the Brookhaven forest (WOODWELL & REBUCK 1967).

The Brookhaven studies included observations on three different plant communities in response to irradiation: the forest vascular plants, the forest lichens, and the herbs of old-field successions (WOODWELL & OOSTING 1965, WOODWELL & GANNUTZ 1967, WOODWELL & REBUCK 1967, WOODWELL & WHITTAKER 1968). In the responses of these communities some parallelisms and contrasts of interest were observed. (i) The coenocline of forest response to irradiation showed marked similarity to the retrogressive effects of repeated fires, to the successional sequence for forest recovery from fire, and to the climatic gradient from Brookhaven north to the Arctic (WOODWELL 1967, 1970). Implications of plant stature (and chromosome volumes) for sensitivity to stress may underlie these parallels. (ii) In all three of the communities, there is correlation of plant height with sensitivity, and consequent replacement of taller forms – trees, erect herbs, fruticose and foliose lichens – with lower or prostrate forms with increasing intensity of irradiation. (iii) Species diversity and coefficient of community approach a linear relation to the logarithm of radiation intensity. This relation implies that the community response is proportional to the relative increase in radiation ($\Delta R/R$) rather than to the absolute value of the increase (ΔR) in Roentgen units. The community relation to a logarithmic, rather than a linear environmental gradient has been observed for salinity in inland water bodies (WHITTAKER & FAIRBANKS 1958) and may apply to chemical pollutants (GORHAM & GORDON 1963, Fig. 2, HADJÚK 1970, Fig. 1).

(iv) The three kinds of communities differ markedly in sensitivity. As a criterion of sensitivity we used the exposure to irradiation that produced a fifty per cent reduction in either species diversity, or coefficient of community, from that of the unirradiated community (Woodwell & Whittaker 1968). The approach parallels on the community-level the 50 % lethal dose (LD_{50}) for expressing relative toxicity of different substances for organisms of a given species. The irradiation intensities for fifty-per cent reduction were: for species diversity 150 R/day for the forest vascular plants, 1000 R/day for the first-year old field herbs, and 2700 R/day for the lichens; for coefficient of community 120 R/day for the forest vascular plants, 200 R/day for the herbs, and 1500 R/day for the lichens. These three groups as wholes thus show increasing sensitivity with increasing height above the substrate. Lichens, with their low-growing forms, relatively simple tissue organization, and small chromosomes are most resistant to irradiation among these groups. They are, in contrast, because of their direct subjection to the chemistry of precipitation, most sensitive to air pollution.

4.3 Measurement of Retrogression

4.3.1 Successional Trends

Successions normally involve trends of increase in a number of community dimensions (Whittaker 1953, Margalef 1963, Odum 1969). During succession on land we expect: (i) increasing productivity, supporting increasing biomass, (ii) a consequent increase in the ratio of biomass to productivity, and increase in the ratio of total community respiration to gross productivity toward 1.0, (iii) increasing stock of nutrients held in the organisms and organic matter, and increasing ratio of the community's nutrient stock to that in environment, (iv) increasing soil organic content and horizon development, increasing microclimatic modification by the increasingly massive plant community, (v) increasing community height and stratal differentiation, increasing species diversity, (vi) increasing longevity of dominant populations and increasing stability of the community. Most of these reverse themselves during some successions, and it may often be the case that productivity and species diversity are higher in a subclimax stage than in the climax. Probably biomass and the biomass accumulation ratio (of biomass to net annual primary production) are least subject to reversal, most reliably maximal in the climax.

In principle a decrease in any of these trends, particularly decrease in biomass, might serve as an index of retrogression. In practice most of these may be set aside for difficulty of measurement or unreliability as indicators of retrogression. In terrestrial communities biomass may be an insensitive measurement of retrogression because killing of more sensitive species may occur before the mass of tissue has been significantly reduced. In aquatic communities both biomass and production may be ineffective measurements of disturbance or retrogression, for both may increase in response to some, but decrease in response to other kinds of pollution. We desire measurements that can express any pollution disturbance including unnatural eutrophication that increases productivity, whether or not we interpret eutrophication as retrogression.

4.3.2 Taxon Sensitivities

Such generalized approaches to the measurement of retrogression must probably rely on species composition. Many pollution indices depend on calibration by species sensitivities, on specification of particular species or broader taxa that may be used to express effects for a particular stress and kind of community. Two purposes of retrogression measurement are in this connection to be distinguished. The first is detection of early retrogressive effects; such detection must generally depend on knowledge of sensitive species. The second is measurement of degrees of retrogression, which may be well advanced. The measurement should be independent of calibrations by the sensitivity of particular taxa, and should permit comparisons of degrees of retrogression for different kinds of communities and stresses. Three approaches that may serve in this way follow.

4.3.3 Weighted Averages

The standard ordination technique of weighted average calculation (Ellenberg 1950, Whittaker 1954, 1967) can be used as an effective index. The index values are dependent on definition of the ecological groups for a particular retrogression gradient. Weighted average indices consequently may not fit the stipulation of independence of the species of particular communities, except as the definitions of decreasers, increasers, and invaders provide a kind of standard that is consistent from one retrogressive coenocline to another.

4.3.4 SPECIES DIVERSITY

Reduction of richness in species is a general and effectively measurable expression of retrogression. Numbers of species in standard quadrats or sample counts may provide more sensitive measurement than the Shannon-Wiener information index and other indirect expressions. Reduction of species diversity to one half (or some other fraction) of that of the undisturbed community or control sample provides a standard by which one can compare relative sensitivities – of different communities to the same stress, or of the same community to different stresses. It should be recognized as a limitation of diversity as an index, that there are cases in which light or moderate disturbance increases species diversity before heavier disturbance decreases it.

4.3.5 SAMPLE SIMILARITIES

Degree of departure from composition of undisturbed communities can be measured by various similarity indices. The work at Brookhaven (WOODWELL 1967) suggests that coefficient of community, based on species presence and absence, may be preferable to percentage similarity and Euclidean distance, which are based on species importance values and are much affected by irregularity of unstable populations. (These measurements are defined and discussed in articles 6 and 11.) Reduction of coefficient of community to 50 % of that for comparisons of controls provides a standard by which sensitivities may be compared. Fifty per cent of control does not imply a coefficient of community of 50 %. When replicate samples from undisturbed communities are compared, their coefficients of community are not 100 %, but are lower values, in this case about 60 %. The standard of comparison, CC_{50}, should thus be a reduction of coefficient of community of samples along the retrogression coenocline to one half the value for undisturbed or control samples compared with one another. A standard of comparison other than 50 % of control similarity might, of course, be used for other studies.

4.4 Coenocline Distance

4.4.1 MEASUREMENT

The measurement of retrogression by sample similarity relates to a more general problem of ordination: the nature of 'distance' in coenoclines and community patterns. WHITTAKER (1952) stated the concept of ecological distance as degree of community separation, expressed in change of community composition in response to environmental change. We shall term the idea "coenocline distance" here to distinguish this, as a measurement applied to communities, from separation along an environmental complex-gradient as expressed by an environmental measurement.

Two approaches to coenocline distance suggest themselves. It may, first, be expressed as a multiple or fraction of mean species amplitude in the coenocline. The mean spread of species populations through the ten steps of Table I, in article 2, is 5.3 steps (including the incomplete distributions of species whose full distributions extend beyond the ends of the transect). Using this amplitude as a unit, the coenocline distance from one extreme of this transect to the other would be 10/5.3 steps or 1.9. The amplitudes of species populations in transects are, however, affected by sample size; as sample size increases the apparent amplitudes increase. It may be better, when feasible, to use as a unit mean standard deviation for these bell-shaped curves (Fig. 1, article 2). Distances in the simulated coenoclines to be discussed below (article 11) are thus defined .It is difficult, however, to compute standard deviations for the species in transects from field data, for the distributions of many species are cut off by the ends of the transects.

The alternative is use of sample similarity measurements, as applied above to retrogression. The simplest use of such measurements— the similarity value for samples from the extremes of a coenocline— is inadequate as an expression of coenocline distance. It is not the case that a similarity measurement of 25 % represents twice the coenocline distance of one of 50 %, because: (i) It is the complement of the similarity measurement that expresses distance hence (if 100 % is used to calculate the complements) the similarities just given would become percentage differences of 75 % and 50 %. (ii) But 100 % is not the appropriate basis of calculation. As indicated above, similarity values for replicate samples are less than 100 %; such similarity or "internal association" values are often in the range of 60 to 90 % (WHITTAKER 1952, 1960, BRAY & CURTIS 1957). In general replicate similarities are lower, the smaller the sample and the higher the species diversity (WHITTAKER 1952). If 85 % is the replicate similarity, sample similarities of 25 % and 50 % become percentage differences of 60 and 35 %. (iii) Even if the appropriate similarity is used, percentage difference of samples bears a curvilinear relation to environmental separation (Fig. 6). Use of sample similarities for coenocline distance has a further limitation (iv) the fact that similarity mea-

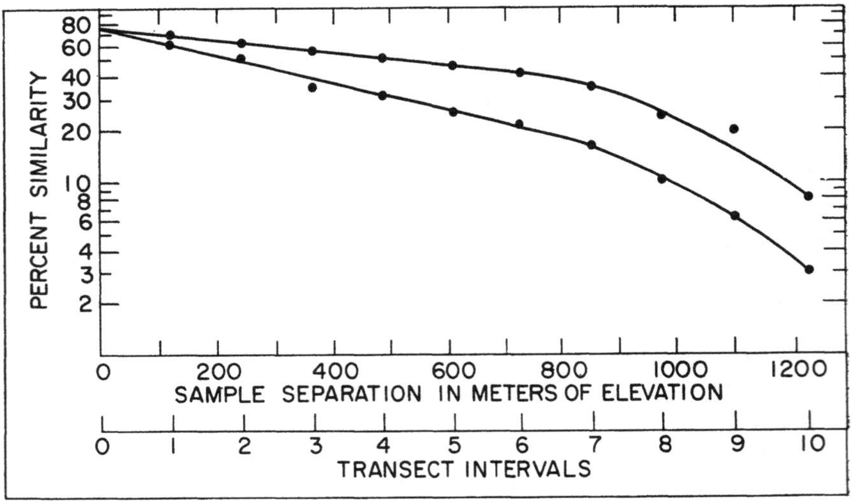

Fig. 6. Decrease in mean sample similarity measurements with increasing sample separation along the elevation gradient in the Great Smoky Mountains (WHITTAKER 1960, Fig. 18, 1967, Fig. 9). The upper curve is for coefficient of community, the lower for percentage similarity, applied to samples of forest trees, on the same logarithmic scale on the ordinate. The abscissa indicates the separation of the samples being compared in terms of elevation in meters and transect intervals of 122 m. The curves are smoothed by averaging a number of sample pairs for a given number of transect intervals by which the samples are separated.

surements become increasingly insensitive expressions of coenocline distance as the values become small until, when the values are zero, no measurement of coenocline distance is possible.

Fig. 6 illustrates the form of sample similarity decrease with separation along an environmental gradient. The first parts of many such curves approach straight lines when the logarithm of sample similarity is plotted against environmental separation. All such curves, however, slope off with increasing steepness for coenocline distances that exceed mean species amplitudes. The forms of the curves were used by WHITTAKER (1960, WHITTAKER & NIERING 1965) to obtain "half-changes" as units of coenocline distance. A straight line is fitted to the initial portion of the curve. The y intercept of this line, for zero separation along the environmental gradient, is used as the replicate similarity value. The line is extrapolated also to the full length of the coenocline for which a distance value is desired; the upper curve of Fig. 6 thus extrapolated for 10 transect intervals gives a similarity value of 30 % (in contrast with the actual similarity measurement for extreme samples of 8 %). Half-changes (β), as a measurement of beta diversity, may then be computed as, $z=a\times 2^{-\beta}$, $\beta=(\log a-\log z)\,/\log 2$, in which *a* is the replicate similarity (75 %) and *z* the estimate for extreme similarity (30 %). On this basis half-changes for the elevation transect illustrated are 1.32 for the upper curve (and 3.06 for the lower curve), inplying rates of community change in half-change units per 1000 m of 1.08 when measured through coefficient of community, 2.5 when measured through percentage similarity. The latter rate is steeper because it reflects change in quantitative representation of species, whereas coefficient of community expresses only species presence and absence. The technique has been used to compare beta diversities for topographic moisture gradients (WHITTAKER 1960, WHITTAKER & NIERING 1965, 1968) and, as indicated, degrees of retrogression.

4.4.2 THE FORM OF SAMPLE SIMILARITY DECREASE

More recently the form of sample similarity decrease in relation to environmental separation has been investigated (GAUCH 1972). Computations are based on the assumption that the coenocline includes numerous species with Gaussian distributions, their modes in scattered positions along the gradient. The curve formed by similarity values is then a complement of the non-standardized error function,

$$PS_{(s)} = 100 \left[1 - ERF \frac{s}{2\sqrt{2}}\right] \text{ where } ERF_{(x)} = \frac{2}{\sqrt{\pi}} \int_0^x e^{-u^2}\, du,$$

and *s* is separation in units of average standard deviation of species along the coenocline. The curves are of the same form for the different sample similarity measurements investigated (coefficient of community, percentage similarity, and the complement of Euclidean distance). An ideal curve, for coenocline data unaffected by sample error, is indicated in Fig. 7, *a*; simulated coenocline data have been used to check the fit of the relationship given.

A curve for simulated coenocline data affected by sample error or population irregularity is given as Fig. 7, *b*. It is a fair approximation of the forms observed for field data. One notes that the 50 % point is not the same for the ideal curve and that affected by sample error. The half-change technique uses a quasi-linear initial slope that is a consequence of the effect of sample error on field data. In principle the error function should be used to correct sample similarity measurements to give coenocline distances (article 11); but the form of field data is not that of the ideal error function curve.

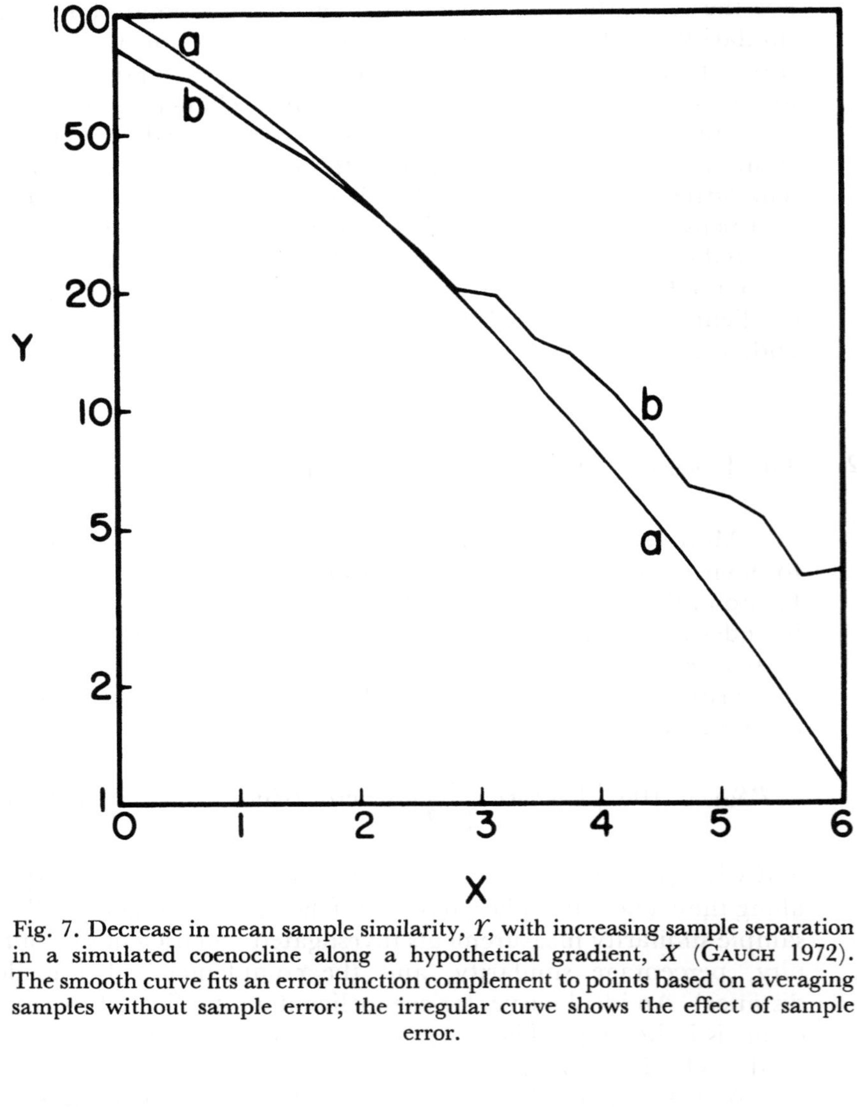

Fig. 7. Decrease in mean sample similarity, Y, with increasing sample separation in a simulated coenocline along a hypothetical gradient, X (GAUCH 1972). The smooth curve fits an error function complement to points based on averaging samples without sample error; the irregular curve shows the effect of sample error.

4.5 Conclusion

Limitations inherent in the character of community data affect all these measurements of retrogression and coenocline distance. The use of such measurements for ordination is to be further discussed below. For retrogression there seems to be no single, fully satisfactory measurement. We find instead a group of measurements that are reasonably suitable, affected by different merits, limitations, and appropriateness to different research problems. If such measurements are not ideal, they may still be sufficiently effective to express the degrees of the disturbance and retrogression that man is increasingly imposing upon many of the natural and seminatural communities of the world.

4.6.1 Summary

Chronic disturbances or stresses applied to natural communities by man can produce changes that are in some respects the reverse of succession, changes termed "retrogression". Among these are the reduction of grasslands by overgrazing, of terrestrial and aquatic communities by air and water pollution, and of forests by ionizing irradiation.

Reduction in any community characteristic that increases through succession might be used as a measure of retrogression, but many such measurements are ineffective. Reduction of particular sensitive species or other taxa may indicate beginning retrogression. For more general measurements three approaches seem most useful: (i) weighted averages, which may use "decreasers, increasers, and invaders" as ecological groups, (ii) reduction of species-diversity from that of a control or undisturbed sample, and (iii) coefficients of community for disturbed compared with undisturbed samples. The second and third of these make possible definition of a 50 % change in the community with retrogression, by which may be compared the responses of a given community to different stresses, or of different communities to a given stress.

Measurement of retrogression relates to the problem of coenocline distance— degree of difference in community composition in response to degree of separation along an environmental (or successional) gradient. Coenocline distance can be approached as extent of change in community composition, as expressed in measurements of sample similarity. These similarity measurements bear a complex and curvilinear relation to sample separation; the relation can be fitted by the complement of a non-standardized error function. Mean amplitudes of species populations and half-changes in community composition provide units by which coenocline distance can be expressed.

4.6.2 Zusammenfassung

Anhaltende anthropogene Störungen oder Beeinflussungen von natürlichen Pflanzengesellschaften können darin Veränderungen erzeugen, die in gewissem Sinne die Sukzession umkehren. Solche Veränderungen werden "Retrogression" genannt. Beispiele dafür sind der Abbau von Grünlandgesellschaften durch Überbeweidung, von terrestrichen und Wasser-Gesellschaften durch Luft- und Wasserverschmutzung und von Waldgesellschaften durch ionisierende Strahlung.

Der Rückgang eines mit der Sukzession zunehmenden Gesellschaftsmerkmals kann als Maß für die Retrogression verwendet werden; jedoch sind viele solcher Messungen unzureichend. Der Rückgang bestimmter empfindlicher Arten oder anderer Taxa kann eine beginnende Retrogression anzeigen. Für allgemeinere Messungen scheinen drei Methoden am geeignetsten: 1. Gewogene Mittelwerte der abnehmenden, der zunehmenden und der neu auftretenden Arten als ökologische Gruppen. 2. Rückgang der Art-Diversität in Vergleich zu derjenigen eines Kontroll- oder ungestörten Bestandes und 3. Vergleich der Gemeinschafts-Koeffizienten zwischen gestörten und ungestörten Beständen. Die zweite und dritte Methode ermöglichen die Feststellung einer 50 %-igen Änderung der Gesellschaft als Folge einer Retrogression. Damit kann entweder die Reaktion einer bestimmten Gesellschaft auf verschiedene störende Einflüsse, oder von verschiedenen Gesellschaften auf einen bestimmten Einfluß erfaßt werden.

Die Bestimmung der Retrogression ist mit dem Problem des Coenocline-Abstandes verknüpft. Darunter wird der Unterschied in der Gesellschaftszusammensetzung im Verhältnis zum Trennungs-Grad (Abstand) längs eines Standorts- oder Sukzessions-Gradienten verstanden. Der Coenocline-Abstand kann als Maß der Veränderung in der Gesellschaftsstruktur gedeutet werden, das durch den Ähnlichkeitswert von Aufnahmen ausgedrückt wird. Diese Ähnlichkeitswerte haben eine komplexe und kurvenlineare Beziehung zur Entfernung der Bestände voneinander. Diese Beziehung kann mathematisch beschrieben werden durch das Komplement einer "nicht standardisierten Fehler-Funktion." Mit Hilfe von Durchschnittsamplituden der Art-Populationen und 50 %-iger Änderung der Arten-Verbindung kann der Coenocline-Abstand ausgedrückt werden.

Acknowledgment

Research carried out at Brookhaven National Laboratory un-the auspices of the U.S. Atomic Energy Commission.

Barkman, J. J., – 1958 – Phytosociology and Ecology of Cryptogamic Epiphytes. Van Gorcum, Assen. 628 pp.
Barkman, J. J., – 1969 – The influence of air pollution on bryophytes and lichens. (French & Germ. summs.) In 'Air Pollution,' *Proc. First European Congr. on the Influence of Air Pollution on Plants and Animals*, Wageningen 1968, pp. 197–209. Centre for Agricultural Publishing and Documentation, Wageningen.
Bray, J. R. & J. T. Curtis, – 1957 – An ordination of the upland forest communities of southern Wisconsin. *Ecol. Monogr.* 27: 325–349.
Dix, R. L., – 1959 – The influence of grazing on the thin-soil prairies of Wisconsin. *Ecology* 40: 36–49.
Duvigneaud, P., – 1946 – La variabilité des associations végétales. *Bull. Soc. r. Bot. Belg.* 78: 107–134.
Dyksterhuis, E. J., – 1949 – Condition and management of range land based on quantitative ecology. *J. Range Mgmt* 2: 104–115.
Ellenberg, H., – 1950 – Landwirtschaftliche Pflanzensoziologie. I. Unkrautgemeinschaften als Zeiger für Klima und Boden. Ulmer, Stuttgart. 141 pp.
Ellenberg, H., – 1952 – Landwirtschaftliche Pflanzensoziologie. II. Wiesen und Weiden und ihre standörtliche Bewertung. Ulmer, Stuttgart. 143 pp.
Ellison, L., – 1949 – The ecological basis for judging condition and trend on mountain range land. *J. For.* 47: 786–795.
Ellison, L., – 1954 – Subalpine vegetation of the Wasatch Plateau, Utah. *Ecol. Monogr.* 24: 89–184.
Gauch, H., – 1972 – The relationship between sample similarity and ecological distance. *Ecology*. (in press.).
Gilbert, O. L., – 1965 – Lichens as indicators of air pollution in the Tyne Valley. In 'Ecology and the Industrial Society,' ed. G. T. Goodman, R. W. Edwards, & J. M. Lambert. *Brit. Ecol. Soc. Symp.* 5: 35–47.
Gilbert, O. L., – 1968 – Bryophytes as indicators of air pollution in the Tyne Valley. *New Phytol.* 67: 15–30.
Gilbert, O. L., – 1969 – The effect of SO_2 on lichens and bryophytes around Newcastle upon Tyne. (French & Germ. summs.) In 'Air Pollution' *Proc. First European Congr. on the Influence of Air Pollution on Plants and Animals*, Wageningen 1968, pp. 223–235. Centre for Agricultural Publishing and Documentation, Wageningen.
Gordon, A. G. & E. Gorham, – 1963 – Ecological aspects of air pollution from an iron-sintering plant at Wawa, Ontario. *Can. J. Bot.* 41: 1063–1078.
Gorham, E. & A. G. Gordon, 1963 – Some effects of smelter pollution upon aquatic vegetation near Sudbury, Ontario. *Can. J. Bot.* 41: 371–378.
Hadjúk, J., – 1970 Einwirkung von Industrie-Exhalationen auf die Struktur der Phytocoenosen. (Engl. summ.) In 'Gesellschaftsmorphologie (Strukturforschung),' ed. R. Tüxen. *Ber. Symp. int. Ver. Vegetskunde*, Rinteln 1966: 340–349.
Hajdúk, J. & M. Ružička, – 1969 – Das Studium der Schäden an Wildpflanzen und Pflanzengesellschaften verursacht durch Luftverunreinigung. (French & Engl. summs.) In 'Air Pollution,' *Proc. First European Congr. on the Influence of Air Pollution on Plants and Animals*, Wageningen 1968, pp. 183–192. Centre for Agricultural Publishing and Documentation, Wageningen.
Harper, J. L., – 1969 – The role of predation in vegetational diversity. *Brookhaven Symp. Biol.* 22: 48–62.
Hilitzer, A., – 1925 – Étude sur la végétation épiphyte de la Bohême. *Spisy vydáv. přír. Fak. Karl. Univ.* (*Publ. Fac. Sci. Univ. Charles*), Praha 41; 1–202.
Hynes, H. B. N., – 1963 – The Biology of Polluted Waters. Liverpool Univ. Press, Liverpool. 202 pp.

Ingram, W. M., K. M. Mackenthun, & A. F. Bartsch, – 1966 – Biological field investigative data for water pollution surveys. U.S. Dept. Interior, Washington. 139 pp.

Itow, S., – 1963 – Grassland vegetation in uplands of western Honshu, Japan. II. Succession and grazing indicators. *Jap. J. Bot.* 18: 133–167.

LeBlanc, F., – 1969 – Epiphytes and air pollution. (French & Germ. summs.) In 'Air Pollution.' *Proc. First European Congr. on the Influence of Air Pollution on Plants and Animals*, Wageningen 1968, pp. 211–221. Centre for Agricultural Publishing and Documentation, Wageningen.

LeBlanc, F., – 1971 – Possibilities and methods for mapping air pollution on the basis of lichen sensitivity. In 'Methoden zur Erkennung und Beurteilung forstschädlicher Luftverunreinigungen' XV. IUFRO-Kongress, 1971. *Mitt. Forstl. BundVersAnst. Wien* 92: 103–126.

Margalef, R., – 1963 – On certain unifying principles in ecology. *Am. Nat.* 97: 357–374.

Odum, E. P., – 1969 – The strategy of ecosystem development. *Science, N.Y.* 164: 262–270.

Patrick, Ruth, – 1951 – A proposed biological measure of stream conditions. *Verhandl. int. Ver. theor. angew. Limnol.* 11: 299–307.

Patrick, Ruth, – 1963 – The structure of diatom communities under varying ecological conditions. *Ann. N. Y. Acad. Sci.* 108: 359–365.

Patrick, Ruth, M. H. Hohn & J. H. Wallace, – 1954 – A new method of determining the pattern of the diatom flora. *Notul. Nat.*, Acad. Nat. Sci. Philad. 259: 1–12.

Preston, F. W., – 1948 – The commonness, and rarity, of species. *Ecology* 29: 254–283.

Rao, D. N. & F. LeBlanc, – 1967 – Influence of an iron-sintering plant on corticolous epiphytes in Wawa, Ontario. *Bryologist* 70: 141–157.

Skye, E., – 1958 – Luftföroreningars inverkan på busk- och bladlavfloran kring skifferoljeverket i Närkes Kvarntorp. (Engl. summ.) *Svensk bot. Tidskr.* 52: 133–190.

Skye, E., – 1968 – Lichens and air pollution: a study of cryptogamic epiphytes and environment in the Stockholm region. *Acta phytogeogr. suec.* 52: 1–123.

Staub, R., J. W. Appling, A. M. Hofstetter, & I. J. Haas, – 1970 – The effects of industrial wastes of Memphis and Shelby County on primary planktonic producers. *BioScience* 20: 905–912.

Stoddart, L. A. & A. D. Smith, – 1955 – Range Management. 2nd ed. McGraw-Hill, New York. 433 pp.

Whittaker, R. H., – 1952 – A study of summer foliage insect communities in the Great Smoky Mountains. *Ecol. Monogr.* 22: 1–44.

Whittaker, R. H., – 1953 – A consideration of climax theory: the climax as a population and pattern. *Ecol. Monogr.* 23: 41–78.

Whittaker, R. H., – 1954 – Plant populations and the basis of plant indication. (Germ. summ.) *Angew. PflSoziol.*, Wien, Festschr. Aichinger 1: 183–206.

Whittaker, R. H., – 1960 – Vegetation of the Siskiyou Mountains, Oregon and California. *Ecol. Monogr.* 30: 279–338.

Whittaker, R. H., – 1967 – Gradient analysis of vegetation. *Biol. Rev.* 42: 207–264.

Whittaker, R. H. & C. W. Fairbanks, – 1958 – A study of plankton copepod communities in the Columbia Basin, southeastern Washington. *Ecology* 39: 46–65.

Whittaker, R. H. & W. A. Niering, – 1965 – Vegetation of the Santa Catalina Mountains, Arizona. (II.) A gradient analysis of the south slope. *Ecology* 46: 429–452.

WHITTAKER, R. H. & W. A. NIERING, – 1968 – Vegetation of the Santa Catalina Mountains, Arizona. IV. Limestone and acid soils. *J. Ecol.* 56: 523–544.

WILHM, J. L. & T. C. DORRIS, – 1968 – Biological parameters for water quality criteria. *BioScience* 18: 477–481.

WOODWELL, G. M., – 1967 – Radiation and the patterns of nature. *Science N.Y.* 156: 461–470.

WOODWELL, G. M., – 1970 – Effects of pollution on the structure and physiology of ecosystems. *Science, N.Y.* 168: 429–433.

WOODWELL, G. M. & J. K. OOSTING, – 1965 – Effects of chronic gamma irradiation on the development of old field plant communities. *Radiat. Bot.* 5: 205–222.

WOODWELL, G. M. & T. P. GANNUTZ, – 1967 – Effect of chronic gamma irradiation on lichen communities of a forest. *Am. J. Bot.* 54: 1210–1215.

WOODWELL, G. M. & A. L. REBUCK, – 1967 – Effects of chronic gamma radiation on the structure and diversity of an oak-pine forest. *Ecol. Monogr.* 37: 53–69.

WOODWELL, G. M. & R. H. WHITTAKER, – 1968 – Effects of chronic gamma irradiation on plant communities. *Q. Rev. Biol.* 43: 42–55.

5 RUSSIAN (RAMENSKY) APPROACHES TO COMMUNITY SYSTEMATIZATION

L. N. SOBOLEV AND V. D. UTEKHIN

Contents

5 RUSSIAN (RAMENSKY) APPROACHES TO COMMUNITY SYSTEMATIZATION

5.1 Introduction

In this paper we summarize views of L. G. RAMENSKY, one of the greatest Russian geobotanists, on the systematic analysis of vegetation. RAMENSKY's views were underlain by the concept of continuity of the vegetation cover. Since such ideas were expressed by him as far back as the beginning of the twentieth century (RAMENSKY 1910), he has clear priority over American authors (WHITTAKER 1951, 1956, CURTIS & MCINTOSH 1951) for the concept of the vegetation continuum (ALEKSANDROVA 1969). RAMENSKY considered arrangement of plant communities into gradient series to be the best method for systematizing data on continuous vegetation. So far as the second way of systematizing objects – namely classification – is concerned, RAMENSKY did not rule it out; but he criticized contemporary methods of classification of plant communities and asserted that it must not be separated from classification of other components of the biocoenose-ecosystem.

It is noteworthy that RAMENSKY considered vegetation a part of the natural complex and opposed separation of vegetation from the whole for fully independent study. He possessed the knowledge of both a geobotanist and a soil-scientist, and applied all his ideas equally to vegetation and to soils. It is not by chance that his main work (1938) is entitled "*Introduction to the Complex Soil-Geobotanic Investigation of Lands.*" These ideas are very close to what WHITTAKER (1962) characterized as the "tradition of multifactoral or landscape units." RAMENSKY connected his teaching with practical evaluation of productivity of lands for agricultural purposes, and in this respect he was close to ELLENBERG's (1952) "*Landwirtschaftliche Pflanzensoziologie.*" He aspired to create a "production typology of lands" for the foundation of which he developed the "natural types of land-units," the latter representing an idea very close to the ecosystem of TANSLEY or biogeocoenose of SUKACHEV, although the latter concept was strongly criticized by RAMENSKY in many respects. RAMENSKY objected particularly to SUKACHEV's idea of the biogeocoenose as a sum of three independent parts: phytocoenose, zoocoenose and biotope; he pointed out that there existed interrelations between every element of these parts.

Despite the range and importance of RAMENSKY's work, much of that work has not been made available in Western languages (see also, however, RAMENSKY 1926, 1930, 1932). We shall here treat his concepts and techniques in application to geobotanic material but it should be borne in mind that they apply to the broader problems of natural complexes.

5.2 Ideas on the Plant Community

5.2.1 Minimal Area

RAMENSKY was one of the first investigators to pay attention to the statistical character of species abundance. Even in a homogeneous habitat the abundances of species can vary substantially due to chances and conditions that influence the result of competition (RAMENSKY 1924). As size of the sample area is enlarged, effect of such variation on the sample decreases, and in the end it becomes negligible. The minimum area for which variation of abundance of a species in samples does not exceed some value was called by RAMENSKY a minimal area of abundance for that species. Generalizing this notion he spoke of minimal areas for quantitative proportions of species, for crop production measurements, etc. (RAMENSKY 1924, 1937a). This idea of minimal area may be parental to those in use among Western authors (DU RIETZ 1921, GREIG-SMITH 1964), article 18).

5.2.2 Dependence of Vegetation upon Biotope

When investigating vegetation with samples of at least minimal area, one would expect vegetation to show strict dependence upon conditions of the biotope or habitat. A number of circumstances should, however, be taken into account here. First, vegetation response to the biotope is complicated by the influence of two factors – historical causes and availability of disseminules. The presence of traces of a former state of the vegetation, or the absence of some normal components of a community because disseminules fail to reach it, can weaken tightness of the bond of the vegetation composition with conditions of the environment. From normal historical causes (involving past community development) may be distinguished violation or destruction of vegetation by irregular influences of hazardous forces of nature or man. Regular influences, for example annual spring flood or haymaking, may be regarded as factors of the biotope.

Second, different characteristics of external environment have different implications for vegetation. RAMENSKY warned against confusing 'ecology with topology' and insisted upon separating 'conditions of the biotope' (habitat or ecotope) from 'conditions of the locality' (entopion). Briefly speaking, the distinction between these lies in the difference between directly acting factors and the factors that act indirectly. Heat, light, water, salt and other such qualities of the location that act directly on plants are conditions of the biotope. Geomorphological conditions, geological, hydrological, macroclimatic and other conditions, for example, relief, mechanical composition of the soil, parent material, the depth of ground waters below plant roots, and so on, are conditions of the locality. The importance of distinguishing between factors of the biotope and factors of the locality is conditioned by the fact that there is no simple dependence between the first factors and the second ones: various combinations of factors of the locality can stipulate the same biotope conditions, and dissimilar biotope conditions can occur in similar localities.

Thus, vegetation samples should display close relation with conditions of the environment, but only when, (i) the environment is characterized by effective factors of the biotope, (ii) the vegetation sample is adequate as regards minimal area, and (iii) the character of the phytocoenose is not unduly distorted by the influence of historical causes, or as a result of inaccessibility to disseminules or violation by disturbance. However, even when these conditions are observed the connection between vegetation and conditions of the environment is not always obvious. This can be explained by the complexity of the connection itself, when the vegetation shows multidimensional changeability in relation to a large number of environmental factors (RAMENSKY 1924, 1926).

5.2.3 ECOLOGICAL INDIVIDUALITY OF SPECIES

Biotopes, in the main, change continuously; moreover, they change differently in respect to different factors. This creates a continuous diversity of biotopes, full use of which by plants is possible only because of the diversity of environmental requirements of species. This last diversity should be such that any combination of conditions of the biotope will be favorable for the development of some species, from the many present in the landscape. Species differ in their requirements and consequent distributions in relation to biotopes; species show ecological peculiarity or ecological individuality. Ramensky illustrated the ecological individuality of species with numerous examples, particularly for meadow vegetation (RAMENSKY 1924, 1938, RAMENSKY et al. 1956).

5.2.4 CONTINUITY OF THE VEGETATION COVER

The idea of continuity of the vegetation cover is a logical outcome from the ecological individuality of species. Vegetation must continuously change with continuous change of the environment. Hence it follows that vegetation can be the best integrator and indicator of conditions of the habitat (RAMENSKY 1924). Indicator properties of vegetation were a major interest of RAMENSKY 's (1938, RAMENSKY et al. 1956) research.

We must ask how absolute was his view of the continuity of the vegetation cover. Two kinds or levels of continuity are to be distinguished: the spatial (between communities actually in contact along a habitat gradient in the field) and typological (between samples, that may have been taken from different areas not in contact, and are compared in the laboratory). RAMENSKY considered continuity to be a natural feature of the vegetation cover. He did not, however, deny the existence of gaps in continuity where sharp changes in the environment occur. At the same time he pointed out that even in such cases interpenetration of species from neighboring communities produces a distinctive transitional zone ("marginal deviations"). Marginal communities can be well developed and of extent surpassing the minimal area. The idea of marginal deviations seems to be very close to the ideas of "ecotone" and "edge effect" among American authors (ODUM 1953, WHITTAKER 1956).

So far as the typological continuum is concerned, RAMENSKY speculated that some breaks might occur in this continuum in ways suggestive of, as the author puts it, "inversion points of physicochemical systems (water-ice-vapour, two hydrates of Glauber salt in one solution)" (RAMENSKY 1924). Later on Ramensky suggested a simple method of detecting whether breaks in continuity of gradient series of communities actually occur (see 5.4.4.).

5.3 Ordination as a Method

5.3.1 SAMPLE CHARACTERISTICS

The vegetation is to be analyzed through a set of samples (lists, relevés) that represent its variation. The selection of samples from various types of localities in an area should be as uniform as possible. At the same time RAMENSKY considers sample sets which are not too heterogeneous to be the most suitable for analysis. Separate investigations of rather small territories with relatively minor variations of the environment are most effective.

The size of sample quadrats (plots) must correspond as far as possible to the minimal area for a sufficient number of species. The species number and minimal area are established empirically and can be different in different landscapes. For meadows RAMENSKY recommended quadrats of 100 m^2, for desert communities he quotes the figure of 1000 m^2, and for forest 2500 m^2 (RAMENSKY 1937a). The shape of the test area can be different, and instead of one large plot a series of smaller ones can be used. Intentional selection of 'typical' or 'average' conditions is accepted. The number of sites, equal in area to the minimal area, for each type of locality should be no less than 8 to 10.

RAMENSKY demanded that determination of plants in the samples described be as full and exact as possible. To make possible determination of plants at life stages other than the flowering, he compiled a guide-determiner by vegetative parts (FLOROVA & RAMENSKY 1932, 1938).

5.3.2 COVERAGE MEASUREMENT

Out of several measures of the role of plants in a community (importance values – density, frequency, biomass, coverage, etc.) RAMENSKY preferred to single out coverage or "projective abundance," i. e. proportion of the earth's surface covered by plant parts, as vertically projected onto the ground (RAMENSKY 1929b, 1932), as a characteristic with a number of advantages. First, coverage has biological significance in relation to area of light-use by plants (RAMENSKY 1937b); second, it is easier to determine than biomass or density and can be better determined than these; third, in case of necessity it is possible to estimate biomass and density from coverage since the three indices are interrelated. The RAMENSKY school applies visual evaluation of coverage on a wide scale. A number of devices were developed to make visual determinations more objective: scales in the form of a net through which the vegetation cover is viewed, forks placed amid the standing grass, and artificial standards of coverage degrees based on depiction of plants' leaves, and so on (RAMENSKY 1929a,b, 1937b, 1938). Applications of such devices can provide quite reliable coverage determinations. For further treatment of samples, however, actual coverage percentages are replaced by coverage classes. The scale of coverage degrees (RAMENSKY et al. 1956) is the following: *s* is less than 0,1 %; *p* is equal to 0,1–0,2 %; *n* is 0,3–2,5 %; *c* is 2,5–8 %; and *m* is more than 8 %. Ranges of coverage values increase approximately in a geometrical progression, as the "octaves of numbers" of PRESTON (1948) might suggest is appropriate.

5.3.3 ORDINATION

The thesis that the vegetation cover is predominantly continuous suggests the need for ways of systematizing plant communities consistent with this thesis. This method, the basic principles of which were formulated by RAMENSKY in the 1920's, was named by the author arrangement into rows or row coordination (RAMENSKY 1924, 1929a 1930). Later in the West it was called *ordination* (Latin 'ordo' – row, order; 'ordination' hence denotes 'arrangement into rows'). Of late, the term ordination has gained wide use in Russia owing to its brevity; and it will be used here. The simplest methods of ordination (ecological series) were used in geobotany long before RAMENSKY, but the history of systematic ordination begins with this investigator.

The essence of ordination as a method lies in locating plant communities along abstract gradients of environmental factors or in a theoretical space. This location can in principle be found with any precision desired. It is in this that ordination has advantage over classification; the precision with which plant communities may be located in a classification is limited by the fact that each unit of the classification comprises a range of conditions. Classification and ordination are by no means mutually exclusive; and RAMENSKY did not seek to refute classification, while decisively preferring ordination. It is worthwhile to note that, even though vegetational continuity implies the appropriateness of ordination, RAMENSKY (1924) did not consider continuity a necessary condition for ordination to be effective.

5.3.4 THE IDEA OF STANDARD LANDMARKS

Since ordination is location of samples (or species) in relation to coordinates, a question arises: what shall the axes for such a system be? It is natural that indices of biotope conditions should be involved in the coordinates, but these indices are not always to be identified with elementary physical characteristics (temperature, moisture content of soil, light intensity in luxes, pH and so on). Some of these characteristics (e. g. year-around moisture regime) are very difficult to measure in a large number of samples to be ordinated. There are both theoretical and practical difficulties that prevent all-sided investigation of the environments of many phytocoenoses as a basis of ordination.

Proceeding from this difficulty, RAMENSKY (1938) suggested an approach to ordination based on *standard landmarks*. The essential idea is that there are to be found for a given area of vegetation a certain number of standard, normal, or typical community samples, to which correspond certain intensities of environmental factors. These standards may be either actual samples, or provisional approximations to them for positions along environmental gradients. Given these standards, then the samples of a set may be arranged according to the degree of their proximity to them, and departure from them along axes of environmental and community variation. Such arrangement is done separately for every one of the major factors of the environment. As a result it becomes possible to show the position of each community sample in the coordinate axes as described by a first, second, third factor, etc., and, in the end, in a multidimensional coordinate system. This approach both has some things in common with, and in other respects is quite different from, that of the American school of CURTIS (BRAY & CURTIS 1957, CURTIS 1959, see also GREIG-SMITH 1964, WHITTAKER 1967).

5.3.5 SUMMARIZED ECOLOGICAL SERIES

Our description of the method may be clearer if we begin, not with the multidimensional relationships, but with a particular environmental gradient and ordination axis. We shall come back (5.3.9) to the origin of the standard landmarks. A sequence of standard landmarks, that is of samples lying directly along a given axis of the ordination system, is called a *summarized monofactor ecological series* (RAMENSKY et al. 1956). Making such series is, according to RAMENSKY, the principal basis of ordination as a method. To explain the meaning of a summarized ecological series it is useful to compare it with the ideas of "concrete" and "assembled" ecological series (RAMENSKY 1938, cf. the "field transect" and "composite transect" of WHITTAKER 1956, 1967, article 2). The term *ecological series* was introduced into Russian scientific literature by KELLER (1909), who so designated a sequence of communities following one another along a gradient of the environment (for instance, a series from desert through littoral to aquatic communities on a particular shore of a desert lake). Such a series can be called a concrete ecological series. It is possible to make an abstract series by grouping into each step of the series several similar but non-neighboring communities; the result is an assembled ecological series. A generalized ecological series can, finally, describe a community gradation by mean characteristics of communities at a number of points along the gradient. Concrete ecological series very often imply changes not of one but of several interrelated factors. RAMENSKY reasoned that it would be most effective to build the ordination system so that the axes of the coordinates corresponded as much as possible to changes of each factor of the biotope separately, with relative constancy of other factors.

Ordination is possible not only by gradients of factors of biotopes but also by the position of communities in successional series (RAMENSKY et al. 1956). In fact, successions are regarded as a variety of ecological series ('temporal ecological series'), for in the time-gradient of succession changes in vegetation occur in parallel with changes in environment (cf. preceding article).

5.3.6 THE IDEA OF THE FUNCTIONAL MEAN

Perhaps the most important theoretical idea of RAMENSKY is that of the *functional mean* (RAMENSKY 1929a, 1930). The concept of the functional mean is rather simple in principle: if one index depends

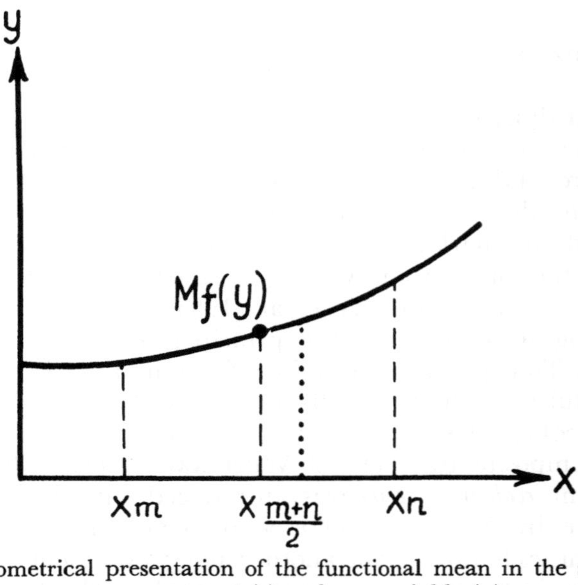

Fig. 1. Geometrical presentation of the functional mean in the simplest case: one argument (x) and one variable (y).

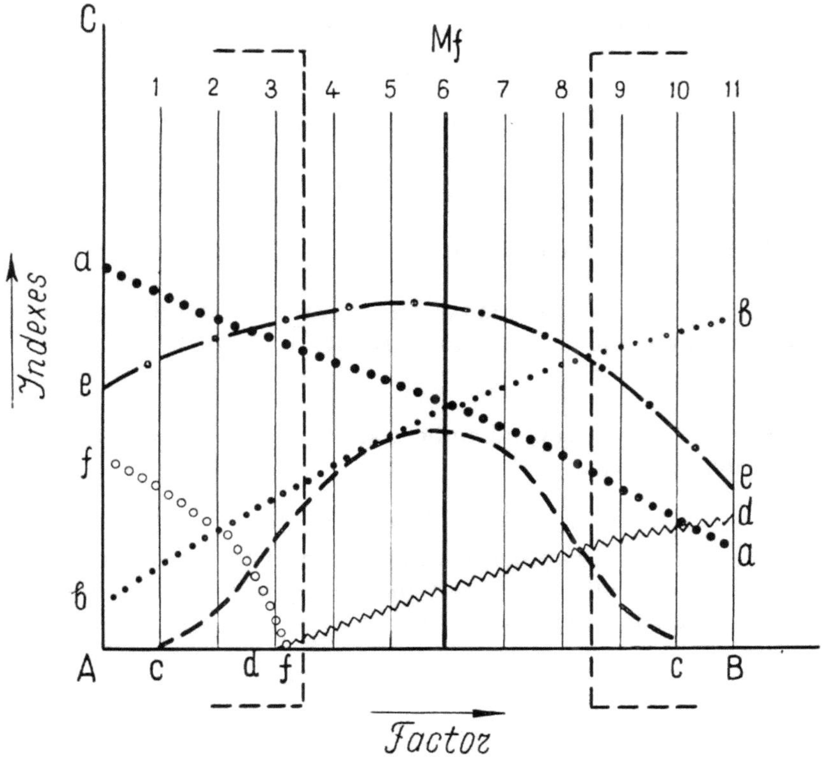

Fig. 2. The position of the functional mean M_f in the series of eleven plant communities varying in response to change in an environmental factor. Species *a* and *f* decrease in abundance; species *b* and *d* increase, whereas species *e* and *c* find their optima near the middle of the range of communities and the functional mean that represents that median position.

on a second index, then the value of the first index which corresponds to the mean value of the second index is its functional mean (Fig. 1). In other words, the functional mean is the value of a function or dependent variable, corresponding to the mean value of the argument or independent variable, for a set or subset of samples for which the latter is averaged.

The functional mean for a series of vegetation variants is depicted by a simple sketch (Fig. 2). On the vertical axis are indicated the abundance indices for species in a series of community variants located along the horizontal axis of the factor AB. The median position among these variants is occupied by sample 6, which represents the functional mean.

For the further account we introduce a mathematical symbol

that is not RAMENSKY's. What is said above can be written:

$$y = f(x)$$

If x is to be found in the interval between "m" and "n" then

$$y = f\left(\frac{m+n}{2}\right) = M_f,$$

where M_f is the functional mean.

The concept of the functional mean can be extended to the dependence of one index on several variables, i. e., when

$$y = f(x_1, x_2, \ldots x_n),$$

and to the case when several indices are dependent upon several variables, namely when

$$(y_1, y_2, \ldots y_n) = f(x_1, x_2, \ldots x_n).$$

The last case with the number of arguments over three is impossible to imagine graphically. In Fig. 3 there is shown the position of functional mean for the dependence

$$(y_1, y_2, y_3, y_4) = f(x_1, x_2),$$

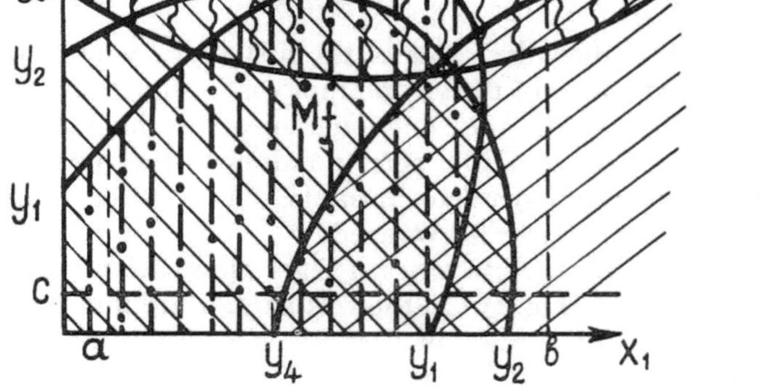

Fig. 3. The position of the functional mean for the function $\{y_1, y_2, y_3, y_4\} = f\{x_1, x_2\}$ in the interval of x_1 from a to b, x_2 from c to d. Areas are hatched for each dependent variable where it is not equal to zero.

where x_1 assumes any value from a to b, x_2 from c to d, and y_1, y_2, y_3, y_4 can assume only one out of two possible values (equal to zero, non-equal to zero).

The concept of the functional mean was used to prepare a monofactor summarized ecological series in the following way. As was noted above, a monofactor series is a series of communities corresponding to the gradation of a certain factor of the habitat, with other factors being constant. This condition will be met when a series of functional mean values are derived for successive sections of the gradient of a given environmental factor, while other factors remain invariable along the whole gradient. Change in communities along a monofactor ecological series is represented in a series of functional mean values for sections or steps of the gradient. RAMENSKY suggested an "elective" method for finding a functional mean when the values of independent variables are not known.

5.3.7 THE IDEA OF THE ELECTIVE MEAN

The election method is based upon several considerations (RAMENSKY 1929a, 1930). The functional mean should, for one thing, occur in the geometrical centre of the (abstract, compositional) space for a set or subset of samples (Fig. 3). Thus, samples of vegetation which correspond to the functional mean must have the largest sum of similarity indices with all other samples, of those under consideration. Consequently, the sample with the highest similarity sum with other samples must approximate the functional mean and is termed the "elective mean." Without a computer calculation of community indices for all possible pairs of samples is laborious. RAMENSKY offered also a simpler, semi-empirical method by successive removal of "peripheral" samples. Besides election RAMENSKY suggested that for a homogeneous set of samples the medians of abundance values for species in the set may be taken as representing the functional mean.

RAMENSKY (1929a, 1930) did not provide a detailed mathematical substantiation of the election methods, and two questions were left open for investigation. The first concerns the dependence of the position of the elective mean on the form of the function $y=f(x)$. Must the sample corresponding to the functional mean in all cases have the highest sum of affinity indices with the other samples, that have other values of the function? The second question is connected with the dependence of the elective mean on the character of a sample set for which it is being calculated. RAMENSKY (1929a, 1938) proceeded from the assumption that all values of the function for which an elective mean is determined are equally probable. But some variants of environmental conditions occur more often in the field than others, and with random or systematic selection of samples some combinations of species will occur in sample sets more often than others. RAMENSKY was conscious of the possibility of biasing the election results in this way, and he therefore recommended selecting samples so that the range of environmental conditions is filled uniformly.

5.3.8 ELECTION BY REJECTING DEVIANTS

As was mentioned, the essence of the method is in successive removal from a set of samples of those which are in some way deviant from the rest. The technique under description includes also some simplifications (inter- and extrapolation) at the initial stage of forming a generalized ecological series.

We shall now regard modes of work on a simple specific example. Let us imagine that we have a group of only 20 samples collected in relatively similar biotope conditions. Among them we must find the most "characteristic" or "central" samples, i. e. samples corresponding to the average conditions of the habitat for the group. The following operations are performed with these samples:

i) Projective coverage of every species in a group should be transformed into classes of coverage as mentioned above.

ii) All samples of the set should be compiled into a table (TableI).

iii) All the abundance indices of the species should be divided into three groups. The first group comprises indices which occur often, in no less than in 75 % of the samples as does, for example the index 0 for the species *Festuca supina.* (The percentage boundary to be used is arbitrary and may depend on homogeneity of the material; a value other than 75 % may be preferred.) Occurrence in a sample of one of these indices is regarded as a positive indication implying that species in that abundance is characteristic for the sample set. A second group of indices occur rarely, in no more than in 25 % of lists, say, for example the indices *n* and *p* for the species *Andromeda polifolia*. Occurrence in a sample of one of these indices is considered a negative indication, implying that species in that abundance is not characteristic for the sample set. The remaining indices form a third group considered neutral as indicators.

iv) Positive data are underlined in one way (solid line), negative data in another way (dotted line), or the use of colored pencils is helpful.

v) After this it is possible to reject samples with larger numbers

Table I. Determination of the elective mean with the help of the rejection metho

Species	First evaluation and rejection															
	Number of lists															
	1	(2)	3	(4)	(5)	6	(7)	8	(9)	(10)	11	12	(13)	14	15	16
Andromeda polifolia	s	0	n	0	0	s	s	0	n	0	s	0	p	s	s	s
Arctostaphylos uva ursi	0	n	s	c	n	0	0	0	s	n	0	0	0	0	0	0
Carex limosa	c	0	p	0	0	c	m	c	n	0	n	0	n	m	c	m
Cetraria islandica	n	m	m	m	m	n	p	p	p	m	p	m	p	p	p	n
Cladonia rangiferina	0	n	p	m	m	n	p	0	0	m	p	c	p	n	m	s
Eriophorum angustifolium	n	0	0	0	0	p	c	c	s	0	n	0	s	n	p	n
Festuca supina	0	n	0	n	c	0	0	0	0	m	0	0	0	0	0	0
Ledum palustre	n	s	n	0	p	n	p	n	c	s	n	n	p	p	p	s
Pleurozium schreberi	c	n	c	c	c	n	0	n	c	n	n	c	m	n	n	n
Polygonum viviparum	0	p	0	s	0	0	0	0	0	s	0	0	0	0	0	0
Polytrichum commune	n	n	n	n	p	0	0	m	m	c	n	m	m	n	n	n
Rubus chamaemorus	c	s	c	p	c	c	n	c	c	s	n	n	s	c	n	n
Sphagnum angustifolium	m	0	m	0	0	m	m	m	m	0	m	0	c	m	m	m
Sphagnum fuscum	c	0	c	0	0	m	m	m	m	0	m	0	0	m	c	m
Vaccinium uliginosum	n	0	c	0	0	c	n	p	c	s	c	c	s	c	c	c
Vaccinum vitis idaea	s	p	p	n	n	0	s	0	s	n	0	c	p	s	0	s
The number of positive indications	2	0	2	0	1	2	2	2	2	0	2	2	2	2	2	2
The number of negative indications	4	6	4	7	5	3	5	3	6	7	2	2	7	3	3	4
The difference	-2	-6	-2	-7	-4	-1	-3	-1	-4	-7	0	0	-5	-1	-1	-2

							Second evaluation and rejection															
	Occurrence with the given coverage						Number of lists										Occurrence with the given coverage					
) 20	m	c	n	p	s	0	(1)	(3)	6	(8)	11	(12)	14	15	16	(18)	m	c	n	p	s	0
s	0	0	10	10	45	35	s	n	s	0	s	0	s	s	s	p	0	0	10	10	60	20
0	0	5	15	0	15	65	0	s	0	0	0	0	0	0	0	0	0	0	0	0	10	90
m	20	25	15	10	0	30	c	p	c	c	n	0	m	c	m	c	20	50	10	10	0	10
n	30	5	25	40	0	0	n	n	n	p	p	m	p	p	n	p	10	0	40	50	0	0
p	25	10	15	30	5	15	0	p	n	0	n	c	n	n	s	p	0	10	30	30	10	20
c	0	20	20	10	15	35	n	o	p	c	n	0	n	p	n	c	0	20	40	20	0	20
0	5	10	10	0	0	75	0	0	0	0	0	0	0	0	0	0	0	0	0	0	0	100
n	0	10	40	30	15	5	n	n	n	n	n	n	p	p	s	c	0	10	60	20	10	0
0	10	35	40	5	0	10	c	c	n	n	n	c	n	n	n	c	0	40	60	0	0	0
0	0	0	0	5	20	75	0	0	0	0	0	0	0	0	0	0	0	0	0	0	0	100
m	25	15	40	5	0	15	n	n	0	m	n	m	n	n	n	0	20	0	60	0	0	20
m	5	30	30	5	25	5	c	c	c	c	n	n	c	n	n	n	0	50	50	0	0	0
n	55	5	5	5	0	30	m	m	m	m	m	0	m	m	m	m	90	0	0	0	0	10
n	35	15	5	0	0	45	c	c	m	m	m	0	m	c	m	0	50	30	0	0	0	20
n	5	45	15	5	10	20	n	c	c	p	c	c	c	c	c	c	0	80	10	10	0	0
s	0	15	15	15	30	25	s	p	0	0	0	c	s	0	s	0	0	10	0	10	30	50
2							4	4	5	4	5	4	5	5	5	5						
7							3	6	2	5	1	9	2	3	3	5						
-5							+1	-2	+3	-1	+4	-5	+3	+2	+2	0						

ird evaluation and jection				The ranged series for deduction of the median					The median
mber of lists									
11	14	15	16						
s	s	s	s	s	s	s	s	s	s
0	0	0	0	0	0	0	0	0	
n	m	c	m	m	m	c	c	n	c
p	p	p	n	n	n	p	p	p	p
p	n	n	s	n	n	n	p	s	n
n	n	p	n	n	n	n	p	p	n
0	0	0	0	0	0	0	0	0	
n	p	p	s	n	n	p	p	s	p
n	n	n	n	n	n	n	n	n	n
0	0	0	0	0	0	0	0	0	
n	n	n	n	n	n	n	n	0	n
n	c	n	n	c	c	n	n	n	n
m	m	m	m	m	m	m	m	m	m
m	m	c	m	m	m	m	m	c	m
c	c	c	c	c	c	c	c	c	c
0	s	0	s	s	s	0	0	0	

of negative indications and smaller numbers of positive ones. For this purpose, the number of negative indications is subtracted from the number of positive ones, and the difference is entered as a positive or negative value below each sample in the table.

vi) Samples are now rejected, beginning with those most deviant as indicated by the largest negative values at the bottom of the table. The samples deleted may be no more than 30 % of the total if the material is fairly homogeneous, but up to 50 %, if the set is less homogeneous, as in our example. The numbers of the samples rejected are in parentheses.

vii) The remaining samples are compiled into a new table. It will be seen that the abundance indices of the species are more consistent in the samples, and that this new sample set is more homogeneous or homotonous.

viii) The same operations are performed with the new table, after which we have left 5 samples which are quite consistent in composition. We can now derive a typical index of coverage for each species in these samples using the arithmetical mean, even better, the median as such an index.

Groups of lists considerably larger than the example can be subjected to similar treatment.

5.3.9 Ordination on the Basis of Election

We now return to the problem of ordering a larger set of samples, representing a broader vegetation pattern and a number of directions of community variation in response to environment. For each environmental factor we shall partition the samples into steps, by ranges of magnitude of that biotope factor (irrespective of the value of other factors). The sequence of steps represents the gradation of the factor, and if we derive the elective mean for each step we then obtain a monofactor summarized ecological series. These elective means serve as standard landmarks by which the whole set of samples may now be ordinated by relative positions along this series as an ordination axis. The problem lies in how to partition the samples into steps when effective measurements of the environmental variable are not available. However, when investigating vegetation it seems always possible to single out several or, at least two, rough gradations of a major biotope factor from the description of the locality (conditions of the relief, proximity of ground waters, etc.) or other ecological data (Ramensky 1929a, 1938). Given two (or more) groups of samples representing difference in community response to the environmental gradation, refinement of the community gradation (ecological series) can be accomplished by election.

In a large number of samples collected in a region, at least two groups A and B are singled out as of contrasting composition in response to a given environmental factor. Elective means with their standards are derived for these groups. We can now characterize an artificial intermediate sample C between them. For instance, if the coverage standard for a given species is m in group A and n in group B, then the index should be c for the intermediate C between A and B. We may choose further to characterize more extreme samples by extrapolation. The same species should have index values that extrapolate down to p for a sample E, more extreme than B; in the opposite direction the index, that is m or maximal in sample A, should be m also in a more extreme sample D.

As a result we obtain 5 standard landmark samples (D, A, C, B, E). Three samples, obtained artificially, are only provisional; but the sequence characterizes an ecological series in relation to the environmental factor with which wé are concerned. All the samples available are compared with these five and classified into five groups according to which one they most nearly correspond to. The elective mean is again determined for each group. (If groups are homogeneous the abundance medians or means for species may be determined without going through the rejection process.) We have now an ecological series of five steps of which none are artificial. Between the 5 elective means as actual, not provisional, standards we can establish new provisional intermediates and new actual intermediate elective means. The procedure can be repeated to divide the ecological series as finely as we desire.

5.3.10 Multidimensional Ordination

After such ordination by one factor we may ordinate by another factor, and still a third and so on with the use of the same technique. With this there will be obtained series of values of elective means for a number of biotope factors, by which samples may be located in the multidimensional ordination system. Not only intensity of each factor taken separately, but also the effects of their combinations are often of importance for vegetation. In this connection Ramensky suggested a further method of multidimensional ordination: building up of monofactor series for successive factors separately, for various intensities of the preceding factors (Ramensky 1938, cf. the transect grid of Whittaker 1956, article 2). Fig. 4 represents the approach to two factors, soil humidity and fertility (Ramensky 1938). On the basis of such an approach for two factors

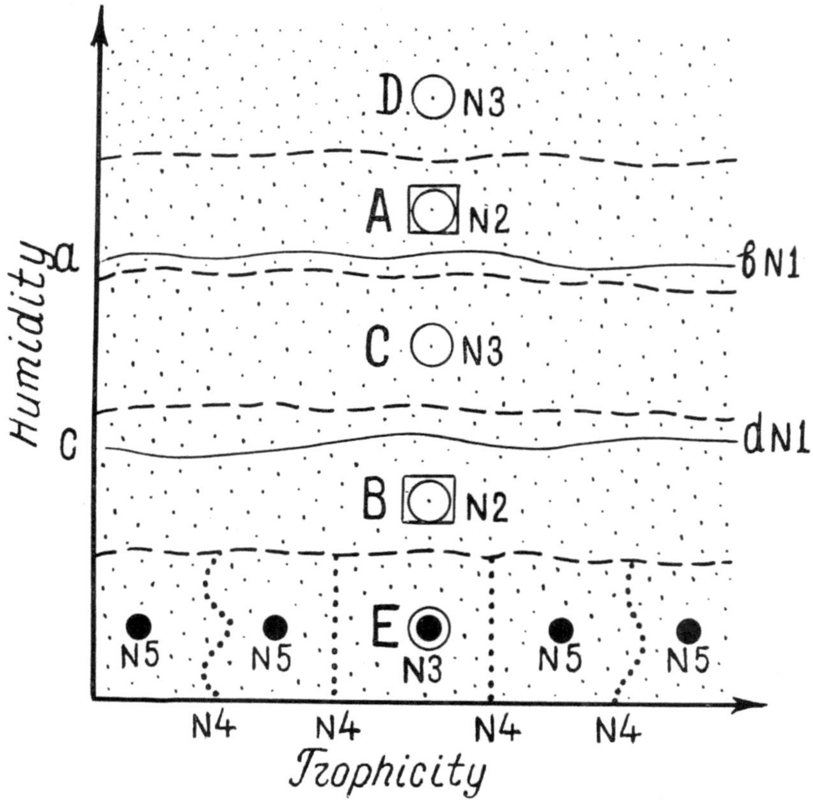

Fig. 4. Scheme for the successive deduction of ecological series for two factors. The positions of separate samples of plant communities in the 'space' of two factors (of humidity and trophicity of soils) are marked with points. Numbers show the succession of actions. Number I (N1) shows singling out of two groups by contrast of soil moisture (their boundaries, which are very approximate, are are suggested by lines ab and cd). Number 2 is the determination of the functional mean by the election method in each of these groups (the mean value is marked with a square). Number 3 is the division of all lists into 5 groups with the help of interpolation (*C*) and extrapolation (*D* and *E*). With further determination of the elective means or medians (marked with circles) a monofactor ecological series for the humidity gradient is obtained. Following this a similar division of the samples by the second factor (trophicity, fertility) is possible. It can be done irrespective of the division by the first factor or, as shown for group *E* (Number 4), for each of the gradations of soil moisture separately. Determination of the functional means of this division is shown, finally, by black circles (Number 5) which form a monofactor ecological series for the fertility gradient.

one can prepare charts on which samples can be located and species distributions outlined by isopleths (isodems, lines bounding equal importance values for a species population). Relations of three species populations to a coordinate system are illustrated in Fig. 5.

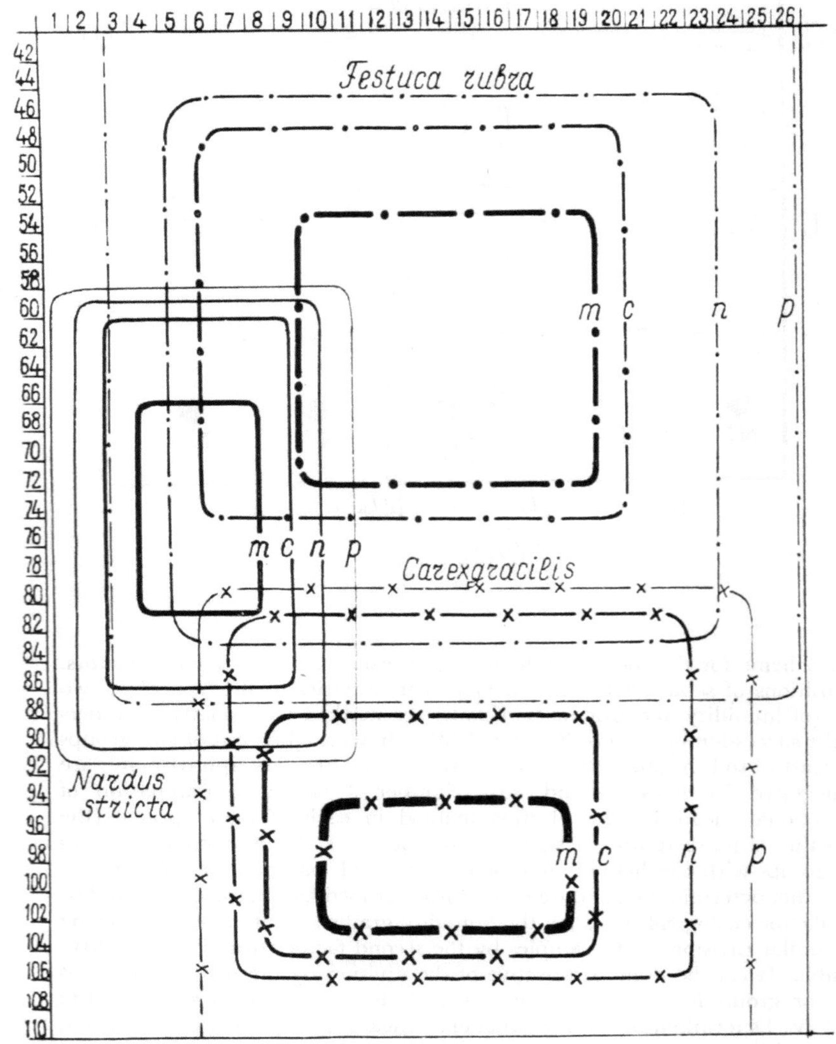

Fig. 5. Isopleths (isodems) of three species depicting their distribution in ecological series of humidity and richness of the soil or trophicity (*m* - mass distribution, *c* - abundant, *n* - moderate, *p* - small).

5.3.11 Location of Samples in an Ordination

In a monofactor ordination series each species, and each level of abundance for a species, has a definite band distribution. If there is a sample in which species *a*, *b*, *c* and *d* are represented (for the sake of simplicity we shall think here that only presence or absence of a species is taken into account), and these species have the distribution amplitudes equal to AB, CD, EF and GH, then the position of this sample should be delimited by the range of overlap of the distributional amplitudes of all the species. What is said above can be written down with the help of symbols of mathematical logic:

$$R_{\text{phe}} = R\ (a,\ b,\ c,\ d) = R(a) \cap R(b) \cap R(c) \cap R(d),$$

where R is a segment of an ecological series in which the phytocoenose under investigation can be located. This approach permits one to locate in the ordination system samples taken after this system has been developed. For the system of multifactor ordination the position of a sample will be determined by the overlap in two or more dimensions of the ecological areas of species (see Fig. 5).

Table 11 Determination of the stage of an ordination series for the cenosis by the scale of richness of the soil with the help of the notch method.

Species	coverage	Notches on stages of richness of the soil: 1 2 3 4 5 6 7 8 9 10 . . . .
Vaccinium vitis idaea	m	
Vaccinium uliginosum	m	
Ledum palustre	n	
Carex limosa	n	
Rubus chamaemorus	n	
Sphagnum fuseum	c	
Deduction of the stage for the cenosis		2.5

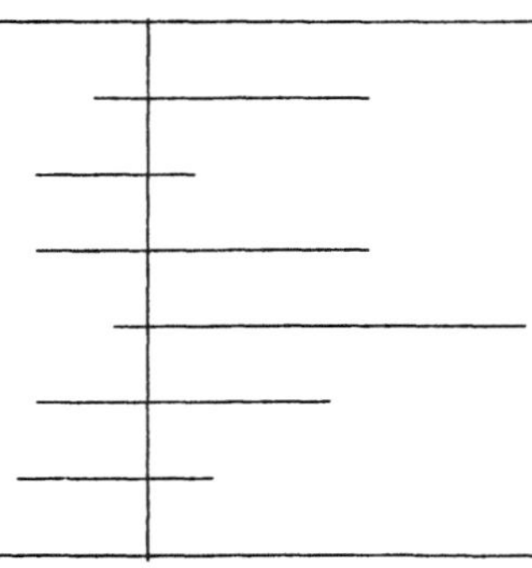

Two techniques for locating an individual sample along an ecological series may be briefly described.

Table II gives a list of 6 species and, on the right side the stages of an ecological series (in this case richness of soil). If for the first species (*Vaccinium vitis idaea*) its coverage (*m*) is characteristic of stages 2-6 we mark this range with a line or "notch." We further mark with such lines ranges for all species of the sample. Evidently, the sample fits into the ecological series where there is the largest number of overlapping lines–hence at 2.5.

Alternatively, coverage values for the species are converted into index categories *m*, *c*, *n* (species with abundance *p*, *s* can be best left unused). Ranges of an ecological scale for which these indices are characteristic are written as two vertical columns (Table III). These numbers are now rearranged in descending and ascending orders, and the first pairs are averaged to derive the sample position as shown in Table III.

Table III Determination of the stage of an ordination series by the limitation method.

Species	Coverage	Limitating stages		Limitating stages in the order of subsiding and increasing		Deduction of the mean from the closest stages
Vaccinium vitis idaea		2	6	2	3	(2+3) : 2=2.5
Vaccinium uliginosum		1	3	2	3	(2+3) : 2=2.5
Ledum palustre		1	6	1	5	
Carex limosa		2	9	1	6	
Rubus chamaemorus		1	5	1	6	
Sphagnum fuseum		1	3	1	9	

Here the account of principal foundations of the method of Ramensky comes to an end. In the next sections we give accounts of some practical research methods worked out by Ramensky and his scholars.

Table IV Fragment of a standard ecological scale for humidity and richness of soil. The values show the ranges of scales for environmental factors, in which the species may occur in the coverage index at the head of column.

Species	Factors	Gradations of projective coverage and stages of factors corresponding to them				
		m (more than 8%)	c (2.5-8%)	n (0.3-2.5%)	p (0.1-0.2%)	s equal to 1 (single)
Vaccinium uliginosum	Humidity	80-89	74-89	72-90	67-96	65-10
	Trophicity	1-3	1-4	1-5	1-6	?
Vaccinium vitis idaea	Humidity	61-83	59-87	58-89	55-90	? -92
	Trophicity	2-6	1-7	1-9	1-10	1-10

5.4 Other Applications

5.4.1 Ecological Scales

The monofactor generalized ecological series obtained is a series of standards showing the influence of different gradations of an environmental factor upon the composition of communities. Such data compiled for different species (Table IV) have been termed *standard ecological scales* (Ramensky 1938, Ramensky et al. 1956). For making series of standard scales for different biotope factors Ramensky used very extensive material collected from all zones of the plains of the Soviet Union, except northern taiga, tundra, and forest tundra, in all about 20,000 samples. Results of this work by a group of collaborators between 1925 and 1950 were published in 1956, after death of the main author (Ramensky et al. 1956). This book also contains an important theoretical introduction. The book gives scales of humidity, richness of the soil (trophicity and salinity, the latter regarded as a surplus trophicity), changeability of humidity, alluviality (intensiveness of silt deposition), and pasture retrogression. V. A. Sorokina (1953, 1955) worked out the scale of altitude (regarding altitude as a complex-gradient of environment, not as a purely topographical characteristic) for pastures of the Northern Caucasus. Of late standard ecological scales for various factors of the biotope, including altitude, have been published by TSATSENKIN (1967) for mountains of Middle Asia, the Urals, and the Altai Mountains, as well as for the Caucasus (1968). Such scales have value for purposes other than the ordination of particular samples. They permit one to locate in the system of ecological coordinates any unit of geobotanical classification that is of interest, and they permit separate evalution of ecological relations of strata or other structural parts of communities.

5.4.2 APPLICATIONS TO SOILS

RAMENSKY justly considered his technique of ordination applicable not only to plant communities; he regarded it as a general method of systematization of "objects defined by several factors which act differently" (RAMENSKY 1929a, 1930). Of other categories of objects RAMENSKY was especially concerned with soils, for which a series of experiments in ordination and making of ecological scales were undertaken by him and his scholars (RAMENSKY 1938). RAMENSKY also used mean values of soil characteristics for the groups of samples on which were based the standards in ecological series. Thus correspondence of vegetation gradation and soil factors was established (RAMENSKY 1937a, 1938, SOBOLEV 1969).

5.4.3 INVESTIGATION OF VEGETATION DYNAMICS

The method which enables us to determine the position of any geobotanic object in a system of ecological coordinates allows investigation of vegetation changes through time. Evaluation by ecological scales of different time-stages of the same site provides us with an indication of direction and degree of change in the community of that site. Analyses of meadow vegetation in the Voronezh Reservation on this basis were carried out by NIKOLAYEVSKAYA (1949). The author annually evaluated samples from the same sites by ecological scales; changes in the (humidity factor) positions of sites in the ordination system were shown and clearly expressed climatic fluctuation. Later UTEKHIN (1969) studied vegetation changes during longer-term development, using manuscript materials of NIKOLAYEVSKAYA. Samples taken in 1936-7 and 1965-6, were compared, both treated by ecological scales. Significant changes in positions of communities in relation to the humi-

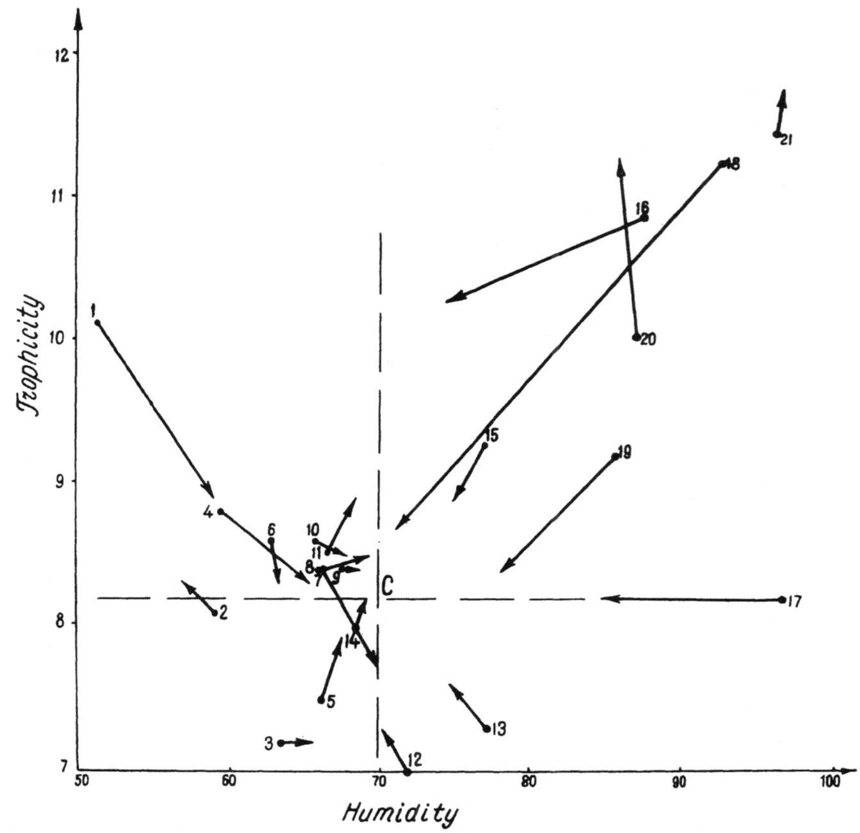

Fig. 6. Dynamics of forest communities of the Voronezh Reservation, for 30 years, shown in the space formed by the ordination axes for factors of humidity and trophicity (Utekhin 1969). Numbers beside the tails of arrows denote different types of communities and are put in the general order of communities of decreasing moisture availability.

dity and fertility scales had occurred in the 30-year period, and would have been hard to discover by other means. Utekhin's results are shown in Fig, 6 and indicate, in particular, rapprochement of different communities in the ordination system with time when man does not interfere.

5.4.4 An Experiment in Application to Classification

While not simply opposed to classification of plant communities Ramensky preferred that it be based on ordination. Continuity of the vegetation cover is not monotonous. In some cases vegetation may change more rapidly in certain sections of ecological series than in others. If such sections of more rapid change occur, they may naturally be used as boundaries between types of plant communities. RAMENSKY (1938, cf. WHITTAKER 1960) suggested a simple method for locating such boundaries: summing the differences in projective coverage (by the RAMENSKY scale) of species in the standard samples for successive pairs of stages of an ecological series. This method of establishing types of communities was applied by UTEKHIN (1967) to vegetation of the Central-Chernozem Reservation near Kursk.

5.4.5 METHODS OF RAMENSKY IN PHYTOGEOGRAPHY

Application of RAMENSKY's methods in phytogeography is being worked out by L. N. SOBOLEV (KUBANSKAYA & SOBOLEV 1969). SOBELEV's approach involves ordering the full ranges of community-types peculiar to different regions in a universal system of ecological coordinates, one embracing all habitats that are possible in regions compared. The system of coordinates in this case is made rough and generalized, and it is divided into discrete sections forming a grid in the cells of which plant communities are placed. Such a method, though strongly schematic, has advantage in allowing interrelation of geographically diverse vegetation cover in a single coordinate system, facilitating comparison of vegetation units, and clarifying the relations of zonal and localized community-types.

The extent of this paper does not permit us to develop further the application of RAMENSKY's concepts and methods in diverse areas of phytocoenology and related fields. But we wish recognition that RAMENSKY's ideas not only antedate gradient analysis in the West, but are part of a continuing tradition being actively advanced by the investigations of his disciples.

5.5.1 SUMMARY

The approach to vegetation of L. G. RAMENSKY was based upon the ecological individuality of species, the continuity of communities, and the systematization of communities by their arrangement into ecological series. The essential concepts and method are those of gradient analysis, as it is called in the West, but the Russian development of this approach preceded the American by WHITTAKER, CURTIS, et al.

Vegetation samples are arranged (ordinated) along an environmental gradient; communities thus arranged in sequence form an ecological series. Positions along an environmental gradient may be described by the compositions of actual or ideal samples, used as standard landmarks. A sequence of such landmarks along a given ordination axis (representing an environmental gradient) describes major features of an ecological series. When a set of samples have been ordinated by comparing them with standard landmarks, each section of the environmental gradient will contain a number of samples. For each section of the gradient a mean or most representative sample is to be chosen. The concept underlying the choice is that of the functional mean: One seeks the value of a dependent (vegetational) variable corresponding to the mean value of an independent (environmental) variable. For a section of the gradient the sample(s) most nearly representing the functional mean is chosen by a process of election, by which deviant samples are progressively removed until only the most central or characteristic samples remain. Such a sample is an elective mean, and its composition (serving as a standard landmark) describes a position along the environmental gradient. A sequence of elective means along a given environmental gradient form a monofactor summarized ecological series.

For analysis of a large number of samples from a region, these may first be divided into two or more groups by vegetational responses to a given environmental factor. If the set is divided into two groups, elective means may be determined for each of these; and the means may then be used to specify characteristics of five standard landmarks (of which one is intermediate and two more extreme than the two means). These five standards (or these and additional intermediates) may be used to ordinate the sample set in relation to the first factor, and for each section of the gradient elective means may be derived to replace the first sequence of standards and form a monofactor summarized ecological series. The sample set may now be treated in the same way for a second and additional environmental factors to produce a multidimensional ordination, in which individual samples may be located by their species compositions in relation to each environmental factor as an ordination axis.

Other applications include: (i) Standard ecological scales that summarize species' responses to intensities of environmental factors, (ii) Correlation of soils characteristics with ecological series, (iii) Treatment of successional sequences as ecological series, (iv) Expression of relative rates of change in community composition along the environmental gradient in different sections of an ecological series, and (v) Broad systematization of community-types in relation to gradients over wide geographic areas.

5.5.2 Zusammenfassung

Der Vegetationsauffassung von L. G. Ramensky lagen die ökologische Individualutät der Arten, die Kontinuität der Gesellschaften und ihre Systematisierung durch Anordnung in ökologischen Reihen zugrunde. Die wesentlichen Begriffe und Methoden sind diejenigen der Gradienten-Analyse, wie sie im Westen genannt wird; jedoch ging die russische Entwicklung dieser Betrachtungsweise der amerikanischen Arbeiten von Whittaker, Curtis et al. voraus.

Vegetationsaufnahmen werden längs eines Standorts-Gradienten angeordnet (ordiniert); die so geordneten Gesellschaften bilden eine ökologische Reihe. Die Lagen auf dem Standorts-Gradienten können durch die Zusammensetzung aktueller oder idealer Aufnahmen beschrieben werden, die als Richtbaken gelten. Eine Reihenfolge solcher Richtbaken längs einer bestimmten Ordinations-Achse (die einen Standorts-Gradienten darstellt) beschreibt die wichtigsten Merkmale einer ökologischen Reihe. Wenn eine Anzahl von Aufnahmen auf diese Weise geordnet worden ist, enthält jede Sektion des Standorts-Gradienten eine Anzahl von Aufnahmen. Für jede Sektion des Gradienten muß eine Durchschnitts- oder repräsentative Aufnahme gewählt werden. Die Grundlage dieser Auslese ist der Begriff des funktionellen Durchschnitts: Man sucht den Wert einer abhängigen (Vegetations-) Variabeln, der mit dem Wert einer unabhängigen (Standorts-) Variabeln übereinstimmt. Die Aufnahmen, die für eine Sektion des Gradienten dem funktionellen Mittel am nächsten kommen, werden ausgewählt, indem nach und nach die abweichenden Aufnahmen ausgeschlossen werden, bis die mittleren und am meisten charakteristischen Aufnahmen übrigbleiben. Eine solche Aufnahme ist ein ausgewähltes Mittel und ihre Arten-Verbindung beschreibt ihre Lage auf einem Standorts-Gradienten. Eine Folge von ausgewählten Mitteln längs eines Standorts-Gradienten bildet eine als Monofaktor zusammengefaste ökologische Reihe.

Für die Analyse einer gröseren Zahl von Aufnahmen aus einem Gebiet zuerst auf Grund ihres Verhaltens gegen einen bestimmten ökologischen Faktor in zwei oder mehrere Gruppen geteilt werden. Wenn zwei Gruppen unterschieden worden sind, werden die ausgewählten Mittel für jede dieser Gruppen bestimmt: mit diesen Mittelwerten können dann die spezifischen Eigenschaften von fünf Richtbaken gefunden werden (von denen eine in der Mitte und zwei mehr außerhalb als die andern liegen). Mit diesen fünf Richtbaken (oder diese und weitere dazwischen liegenden Punkte) ist es möglich alle Aufnahmen in Beziehung zum ersten Faktor zu ordnen, und für jede Gradienten-Sektion können ausgewählte Mittel abgeleitet werden, welche die erste Stufenfolge ersetzen und zusammenfassend einen Monofaktor ökologischen Reihen bilden. Die Aufnahmen-Reihe kann dann in der gleichen Weise für einen zweiten und weitere Standortsfaktoren beschrieben werden um zu einer multidimensionalen Ordnung zu gelangen, bei der Einzelaufnahmen auf Grund ihrer Arten-Verbindung in Beziehung zu jedem Standortsfaktor, der als Achse dient, angeordnet werden können.

Andere Anwendungen sind: 1) Standardisierte ökologische Skalen, in denen das Verhalten der Arten gegenüber den Intensitäten von ökologischen Faktoren zusammengefast wird. 2) Aufstellung von Beziehungen zwischen Bodeneigenschaften und ökologischen Reihen. 3) Behandlung von Sukzessions-Folgen als ökologischen Reihen. 4) Ausdruck der relativen Änderungsgeschwindigkeit in der Gesellschaft-Zusammensetzung längs des Standorts-Gradienten in verschiedenen Sektionen einer ökologischen Reihe. 5) Allgemeine Systematisierung von Gesellschaftstypen in Beziehung zu Gradienten über weite geographische Gebiete.

REFERENCES

ALEKSANDROVA, VERA D., – 1969 – Classification of Vegetation: Principles of Classification and Classification Systems of Various Phytocoenological Schools (In Russian). Publ. House Nauka, Leningrad. 275 pp.

BRAY, J. R. & J. T. CURTIS, – 1957 – An ordination of the upland forest communities of southern Wisconsin. *Ecol. Monogr.* 27: 325–349.

CURTIS, J. T., – 1959 – The Vegetation of Wisconsin: An Ordination of Plant Communities. Univ. Wisconsin, Madison. 657 pp.

CURTIS, J. T. & R. P. MCINTOSH, – 1951 – An upland forest continuum in the prairie–forest border region of Wisconsin. *Ecology* 32: 476–496.

DU RIETZ, G. E., – 1921 – Zur methodologischen Grundlage der modernen Pflanzensoziologie. Holzhausen, Wien. 267 pp.

ELLENBERG, H., – 1952 – Landwirtschaftliche Pflanzensoziologie. II. Wiesen und Weiden und ihre standörtliche Bewertung. Ulmer, Stuttgart. 143 pp.

FLOROVA, V. M. & L. G. RAMENSKI, – 1932 – Identification Guide to Plants in a Non-Flowering State for the Middle Part of the USSR (In Russian). Moscow-Leningrad. 256 pp. 2nd ed. 1938.

GREIG-SMITH, P., – 1964 – Quantitative Plant Ecology. 2nd ed. Butterworths, London. 256 pp.

KELLER, B. A., – 1909 – Grassland, semidesert, and desert formations, pp. 7–17 in: Preliminary Programs of Botanical-Geographical Investigations (In Russian). Pechatnyi Trud, Petrograd.

KUBANSKAYA, Z. V. & L. N. SOBOLEV, – 1966 – The system of plants and soils of the Betpak-Dala desert (In Russian). *Trudy̆ Inst. Bot. Akad. Nauk Kazakh. SSSR*, Alma-Ata 23: 54–80.

NIKOLAYEVSKAYA, M. V., – 1949 – Dynamics of the vegetation cover of the Voronezh Reservation (In Russian). *Nauchno-Metod. Zapiski Glavn. Uprav. po zapovednikam* 12: 126–135.

Odum, E. P., – 1953 – Fundamentals of Ecology. Saunders, Philadelphia. 384 pp. 2nd ed., 1959, 546 pp. 3rd ed., 1971, 574 pp.

Preston, F. W., – 1948 – The commonness, and rarity, of species. *Ecology* 29: 254–283.

Ramenski, L. G., – 1910 – On the comparative method for ecological study of plant communities (In Russian). *Dnevnik S'ězda russk. Estestvoisp. Vrach.* 2(9): 389–390.

Ramenski, L. G., – 1924 – The main regularities of the vegetation cover (In Russian). *Věstnik opȳtnogo děla Sredne-Chernoz. Obl.*, Voronezh pp. 37–73.

Ramensky, L. G., – 1926 – Die Grundgesetzmässigkeiten im Aufbau der Vegetationsdecke. *Bot. Zbl.*, N.F. 7: 453–455.

Ramenski, L. G., – 1929a – On the method of comparative treatment and systematization of lists of plants and other objects determined by several factors with unlike actions (In Russian). *Trudȳ Sovesch. geobot.-lugov., sozvan. (Gos.) lugovoi Inst.* 15–20, 1928, pp. 11–36. (Transl. in *Beitr. Biol. Pfl.* 18: 269–304, 1930).

Ramenski, L. G. – 1929b – Projective counts and description of vegetation (In Russian) Moscow. (Transl. in *Handb. biol. ArbMeth.* 11, 6: 137–190, 1932).

Ramensky, L. G., – 1930 – Zur Methodik der vergleichenden Bearbeitung und Ordnung von Pflanzenlisten und anderen Objekten, die durch mehrere, verschiedenartig wirkende Faktoren bestimmt werden. *Beitr. Biol. Pfl.* 18: 269–304.

Ramensky, L. G., – 1932 – Die Projektionsaufnahme und Beschreibung der Pflanzendecke. *Handb. Biol. ArbMeth.* 11, 6: 137–190.

Ramenski, L. G., – 1937a – Inventarization of natural hay-lands and pastures and methodological foundations of nature-productivity typology of lands (In Russian). *Trudȳ Vses. Akad. Sel'. Khoz. Nauk*, Moscow 31(2): 11–36.

Ramenski, L. G., – 1937b – Counts and Description of Vegetation (on the Basis of the Projective Method) (In Russian). Selkhozgiz, Moscow. 100 pp.

Ramenski, L. G., – 1938 – Introduction to the Complex Soil-Geobotanical Investigation of Lands (In Russian). Selkhozgiz, Moscow. 620 pp.

Ramenski, L. G., – 1953 – On ecological study and systematization of plant communities (In Russian). *Byull. mosk. Obshch. Ispȳt. Prir.*, otd. biol. 58(1): 35–54.

Ramenski, L. G., I. A. Tsatsenkin, O. N. Chizhikov & N. A. Antipin, – 1956 – Ecological Evaluation of Pasture and Fodder Lands by the Vegetation Cover (In Russian). Selkhozgiz, Moscow. 472 pp.

Sobolev, L. N., – 1969 – Dependence of productivity of plants, communities, and their geographical systems on factors of the environment (three levels of investigation of productivity) (In Russian). In, 'Obshchije teoreticheskije Problemy biologicheskoj Produktivnosti', pp. 70–74. Nauka, Leningrad.

Sorokina, V. A., – 1953 – The experience in application of the methods of L. G. Ramensky (In Russian). *Bot. Zh. SSSR* 38 (5) 718-728.

Sorokina, V. A., – 1955 – Some results of applying the methods of L. G. Ramensky *Herbage Abstr.* 25(4): 210–218.

Tsatsenkin, I. A., – 1967 – Ecological Scales for Plant Species of Pastures and Haylands of Mountainous and Plains Regions of the Middle Asia, Altai, and Urals (In Russian). 'Donish,' Dushanbe. 226 pp.

Tsatsenkin, I. A., ed., – 1968 – Ecological Evaluation of Pasture and Range and Fodder Lands of the Caucasus by the Vegetation Cover (In Russian). Moscow.

Utekhin, V. D., – 1967 – The vegetation of the Central-Chernozem Reservation and its productivity (In Russian). *Mater. Moskov. Fil. Geogr. Obshch. SSSR*, Biogeography, phenology, vyp 1: 18–21. Moscow.

Utekhin, V. D., – 1969 – Vegetation changes in the Voronezh Reservation during 1936–1966 (In Russian). *Mater. Moskov. Fil. Geogr. Obshch. SSSR*, Biogeography, vyp 3: 15–17.

Whittaker, R. H., – 1951 – A criticism of the plant association and climatic climax concepts. *NW. Sci.* 25: 17–31.

Whittaker, R. H., – 1956 – Vegetation of the Great Smoky Mountains. *Ecol. Monogr.* 26: 1–80.

Whittaker, R. H., – 1960 – Vegetation of the Siskiyou Mountains, Oregon and California. *Ecol. Monogr.* 30: 279–338.

Whittaker, R. H., – 1962 – Classification of natural communities. *Bot. Rev.* 28: 1–239.

Whittaker, R. H., – 1967 – Gradient analysis of vegetation. *Biol. Rev.* 42: 207--264.

6 SAMPLE SIMILARITY AND SPECIES CORRELATION

David W. Goodall

Contents

6 SAMPLE SIMILARITY AND SPECIES CORRELATION

6.1 Introduction

Correlations between species most often arise from biotope differences affecting their distributions. They are of interest for the light they throw on species biology, as a tool in the classification or ordination of plant communities, and as a means of testing the homogeneity of vegetation (GOODALL 1953, DAGNELIE 1962a, b). Similarity between vegetation samples may also be used as a tool in ordination or classification. Similarity between samples and correlation between species are two sides of the same coin. It is by their occurrence in a group of similar communities that a group of correlated species may be recognized, and it is through the fact that stands include a number of correlated species that they are recognized as similar. General discussions of these questions will be found in a number of books, such as those of GREIG-SMITH (1964), PIELOU (1969b), GOUNOT (1969) and VASILEVICH (1969).

As suggested in the previous paragraph, measures of correlation and similarity find their main use in application to the ordination and classification of vegetation samples. Other articles are devoted to these topics. Here we shall deal with measures of similarity and correlation in their own right, irrespective of the particular uses to which they may be put.

6.2 Distributional Similarity of Species

6.2.1. RELATIONS BETWEEN PAIRS OF SPECIES

6.2.1.1 *Interpretation of Interspecific Relationships*

Two species may be positively associated because one is dependent on the other (as parasite and host, tree and epiphyte) or because both are similarly affected by some third factor in the biotic or inanimate environment. They may be negatively associated through mere spatial exclusion, through competition or allelopathic effects, or through differing environmental requirements.

Whatever the nature and causes of the relationship between them, it is essentially dependent on the area of study. It is virtually meaningless to say simply that two species are correlated without indicating the area covered by this generalization. Taking the earth's surface as a whole, all European plants, however diverse in their habitat requirements, will be positively correlated in the sense (for instance) that the distance from a plant of one of these species to the nearest plant of another will be, on average, much shorter than an arc joining two random points. On the other hand, if one is studying a particular rock surface by a grid of small samples, two lichen species may be negatively correlated through spatial exclusion, even though they are both confined to limestone and consequently normally found side by side. Thus the degree of correlation between two species, and even its sign, may depend on the range of biotopes occurring within the area studied.

Moreover, if association is being studied by a sampling method, the size of sample will also affect the results (BOURDEAU 1961, BYER 1970). If samples in a bog are small compared with the size of the hollows, two *Sphagnum* species may be negatively associated, whereas quadrats comparable in size with the pattern of hummocks and hollows might show them as positively associated.

In homogeneous vegetation, one may expect interspecific correlation to change with increases in sample size up to a certain point, beyond which it would remain constant. This sample size, reflecting the scale of local heterogeneity in the vegetation, might be a suitable choice as a minimal area for sampling (GOUNOT 1961). It does not follow, though, as GOUNOT suggests, that correlation studies on a smaller scale are not useful. They can in fact be valuable in illuminating the pattern of the vegetation. This is illustrated by KERSHAW's (1961) application of analysis of covariance to the study of joint distribution of pairs of species in nested samples of different sizes. In consequence of these effects of sample size, a statement about the correlation of two species can have precision only if it incorporates a statement of the universe sampled and of the sampling method – such as, "When Tregaron Bog was sampled by random metre-square quadrats, a negative correlation was shown between *Calluna vulgaris* and *Erica tetralix*."

Correlations among species are most commonly studied by recording their presence or absence in a series of samples, or by estimating some quantitative measure of their abundance in each sample. A third approach, less commonly used, depends on measurement of distance between individuals of different species. The relative importance of presence on the one hand and quantitative measures (density, cover, basal area, etc.) on the other has been somewhat debated. KASSAS & EL-ABYAD (1962/3) point to the inadequacy of presence records to distinguish types of desert vegetation, and many workers have expressed the view that quantitative measures add appreciably to the useful information about vegetation samples contained in species lists. DAGNELIE (1962b) has stated that correlation results based on presence and on quantitative measures are usually concordant. On the other hand, KOLBE (1956), for instance, regards correlation studies on the quantities of two species as appropriate only if they have first been shown to be correlated in respect of presence.

6.2.1.2 *Presence and Absence Data*

We first consider measures of species association (SA) based on records of presence or absence only. For two species, these may be expressed in the form

		Species A		
		Present	Absent	Total
Species B	Present	a	b	$a+b$
	Absent	c	d	$c+d$
	Total	$a+c$	$b+d$	M

where the total number of samples M, is partitioned into those with both species (a), those without both (d), and those with one or the other only (b and c).

Numerous ways of expressing the degree of association between the two species have been proposed and used. The data are in fact the usual 2×2 contingency table, and the measures of association based on such tables for use in completely different contexts are also available here. Most of these measures range from 1 (in the case of perfect positive association) to 0 or -1 (in the case of perfect negative association).

Measures of association and tests of their statistical significance are often confused. It may be important to determine whether two species are distributed independently of one another, or whether on the other hand they are positively or negatively associated. In this case, a significance test is required. More often, the hypothesis of independence is unrealistic, and a test of it has little interest; one is concerned rather with expressing the *degree* of association than with questioning whether it exists. The two questions are discussed separately in the following sub-sections.

6.2.1.2.1 Significance tests

A significance test for interspecific association is based on the null hypothesis that the probability of encountering each species is uniform throughout the set of samples, irrespective of whether the other species is present or not. It goes without saying that significance tests are relevant only if the samples have been taken randomly from the universe they represent.

The null hypothesis has normally been tested by calculating χ^2 with YATES's (1934) correction for continuity:

$$\chi^2 = \frac{M(|ad-bc|-M/2)^2}{(a+b)(c+d)(a+c)(b+d)} \tag{1}$$

or from the rather more accurate alternative expression (GILBERT & WELLS 1966)

$$\chi^2 = \ln \frac{M^3}{2\pi\,(a+b)(c+d)(a+c)(b+d)} + 2\ln \frac{M!\,a!\,b!\,c!\,d!}{(a+b)!\,(a+c)!\,(b+d)!\,(c+d)!} \tag{2}$$

and comparing with the theoretical distribution of χ^2 with one degree of freedom. This is an approximation to the hypergeometric distribution, which may also be calculated exactly for the same purpose. The probability of the particular set of figures observed, given the marginal totals, is

$$\frac{(a+b)!\,(c+d)!\,(a+c)!\,(b+d)!}{a!\,b!\,c!\,d!\,M!} \tag{3}$$

and the probabilities for the observed set, and other less probable sets consistent with the marginal totals, are summed to provide a significance test. The use of this exact distribution has been recommended when the minimum product of the marginal totals is less than $5M$.

It has been pointed out by PIELOU (1969a,b) that the assumptions on which these tests are based are, to say the least, questionable. They assume that the marginal totals are fixed — in other words they give the probability that, in a set of M samples of which $(a+c)$ contained Species A and $(a+b)$ contained Species B, as many (or as few) as a would contain both if their occurrence were independent. This is not really the question the investigator usually wants answered. He wants to know whether, if Species A and Species B were independently distributed with their actual population frequencies and patterns, they would give by chance a set of samples with this apparent degree of association. In other words, the marginal totals are themselves random variables — sample estimates of population frequencies — and not fixed as assumed in the usual tests. The probability of the contingency table as calculated from (1), (2), or (3) above is greater than the true probability taking variation in marginal frequencies into account, and the test as normally performed has consequently a conservative bias. PIELOU describes a modified form of the χ^2 test taking variation in marginal frequencies into account. In most cases, the adjustment in the probability estimate will be relatively small.

PIELOU (1969a,b) has pointed out that, if the mean distance between samples is less than the scale of pattern in the vegetation, significance tests of association may be vitiated by lack of independence among the samples. If, for instance, two species occur in patches large compared with the mean distance between samples, the fact that one sample contains both species enhances the probability that neighbouring samples will also contain both species, and the association between them may appear significant even where the patches are in fact random and independently distributed.

6.2.1.2.2 Measures of interspecific association

A significance test such as is described in the preceding section indicates only *whether* the two species are associated, and provides no measure of the *degree* to which they are associated.

If M is large, a relatively small degree of association may be demonstrated at a high significance level. If M is small, even a large degree of association may not be demonstrable. On the other hand, most measures of association bear little or no relation to the significance of the relationship in the sample observations, and it is perhaps debatable whether this should be regarded as a drawback. It is perhaps best to regard the significance test and the measure of association as being separate, and in the latter to seek rather the property of being independent of the intensity of sampling – unbiased, with sampling variance progressively reduced as the size of sample increase. The various indices[1] used to express species correlation have been reviewed by GOODALL (1952), DAGNELIE (1960a, 1965), GREIG-SMITH (1964), and WHITTAKER (1967), and statistical aspects are considered by GOODMAN & KRUSKAL (1954, 1959).

We will first consider the non-centred indices, as DAGNELIE (1960a) calls them — those taking only non-negative values. Among

[1]) Most of these indices have from time to time occurred in two forms – as a proportion, and as a percentage. The expressions given will all be in the proportion form.

them are the association coefficient of AGRELL (1945) and IVERSON (1954), also called 'percentage co-occurrence' by WHITTAKER & FAIRBANKS (1958) and 'spezifische Gemeinschaftskoeffizient' by ELLENBERG (1956):

$$a/(a+b+c) \tag{4}$$

and the coincidence index introduced by DICE (1945) — though he subsequently (1952) discarded it — and also used by BRAY (1956):

$$2a/(2a+b+c) \tag{5}$$

both of which take into account the samples in which either species is present, and ignore those in which both are absent. MORISITA (1959) has shown that both these indices are markedly affected by sample size.

Another worker who considered that samples from which both species were absent should be ignored was FAGER (1957, 1963), who proposed the index:

$$\frac{a}{\sqrt{(a+b)\,(a+c)}} - \frac{1}{2\sqrt{\min\,\{(a+b),\,(a+c)\}}} \tag{6}$$

as a means of expressing the extent to which each of the species was normally a part of the other's environment. GILBERT & WELLS (1966) also argued that samples from which both species were absent were irrelevant, and fitted a model (GILBERT, 1963) which we may express:

$$\log a/M = \log q_i + \log q_j + p_i p_j \tag{7}$$

a and M having the meanings with which we are familiar, while q_i and q_j are the population frequencies of the two species, and p_i, p_j are parameters expressing their degree of association.

FORBES (1907), MOZLEY (1936) and BARNARD (1964) used

$$Ma/\{(a+b)\,(a+c)\} \tag{8}$$

which is the ratio of the observed to the expected number of joint occurrences. TARWID (1960) put forward a similar index in which the divisor was the integral part of the expected number of joint occurrences.

Closely related is the expression

$$C = (a+b)\,(a+c)/a \tag{9}$$

used by MARGALEF (1958) as an inverse measure of association. Since only positive association is of interest, the fact that the index could take values of infinity was not regarded as important. The centred form

$$\log\ (M/C) \tag{10}$$

was also proposed.

HACKER's (1921) "association value"

$$a/(a+b), \text{ and } a/(a+c) \tag{11}$$

also used by DICE (1945, 1952), GARDNER (1951) and CASSAGNAU (1961), takes different values according to which species is being considered as the independent variable, and expresses the probability that one will occur where the other is present, rather than their association as usually understood.

The indices mentioned so far use joint presence (a) but not joint absence (d) as evidence of association between the species. Other indices treat joint presence and joint absence equally. The matching coefficient of SOKAL & MICHENER (1958)

$$(a+d)/M \tag{12}$$

widely used in numerical taxonomy, does not yet seem to have been used as a measure of interspecific association. BEALS (1965b) used

$$5-\sqrt{\chi^2} \tag{13}$$

where χ^2 is derived from expression (1), as an inverse measure of association ranging from a low value where the species are very closely associated (the actual minimum depends on M and on the marginal totals) to a maximum close to 5 if their relationship is negative.

Other indices taking joint presence and absence into account are centred — taking negative or positive values according to the sign of the association between the species. One of the first was that proposed by MICHAEL (1920):

$$\frac{4\{Ma-(a+b)\,(a+c)\}}{(a+d)^2+(b+c+2d)^2} \tag{14}$$

but it has been little used since. Much more common is the point correlation coefficient

$$\phi = \frac{ad-bc}{\sqrt{(a+b)\,(a+c)\,(b+d)\,(c+d)}} \tag{15}$$

which has been used in ecology by DAGNELIE (1962a) and FREY (1963), among others.

Since where b and c are unequal ϕ can never attain unity, COLE (1949, 1957) proposed that the point correlation coefficient should be modified by using the ratio

$$\phi/\phi_{max} \tag{16}$$

where ϕ_{max} is the largest value of ϕ, with the same sign, consistent with the marginal totals. Similar considerations led FORBES (1925) to propose the expression

$$\frac{aM - (a+b)(a+c)}{M \min\{(a+b), (a+c)\} - (a+b)(a+c)} \tag{17}$$

and KOLBE (1956) to suggest the ratio of the deviation of a from expectation to the maximum possible deviation in the same direction consistent with the marginal totals.

HURLBERT (1969) tested COLE's index (16) in the form

$$\sqrt{\text{obs } \chi^2/\text{max } \chi^2} \tag{18}$$

with the sign of $(ad-bc)$, where obs χ^2 is calculated in the usual way for 2×2 contingency tables and max χ^2 is the largest value the marginal totals permit. He found this expression to be markedly influenced by the species frequencies, and suggested it be improved by taking into account the fact that, in general, the lower limit for χ^2 is not zero but a somewhat higher figure (min χ^2), owing to the data being integral. His expression accordingly became

$$\sqrt{\frac{\text{obs } \chi^2 - \text{min } \chi^2}{\text{max } \chi^2 - \text{min } \chi^2}} \tag{19}$$

again with the sign of $(ad-bc)$, and this he found a much more sastisfactory expression of association than COLE's index.

Information theory has contributed ideas in this field, as to other aspects of statistical ecology, and GODRON (1966) has suggested the information statistic

$$\log_2 \frac{M}{a!\,b!\,c!\,d!} \tag{20}$$

as a measure of association between the two species.

To a large extent, the choice of index is a matter of taste and of one's particular purpose, for each index has its own properties, and they may give very different results (see, for instance, MORISITA 1959, MIYATA 1966, CANCELA DA FONSECA 1966). A point for debate is, however, whether joint absence should be taken into account — whether the degree of association should be assessed solely on those quadrats where at least one of the species is present, (4)—(11), or whether the whole set of samples should be taken into account, including perhaps some that are completely bare of vegetation, (12)—(20). As GILBERT & WELLS (1966) pointed out, association measures including d may be greatly affected by its magnitude, which seems to introduce an arbitrary element into their determination. But the arbitrary element intrudes only in the delimitation of the universe to which the conclusions apply; once this is decided, measures including d will be no more arbitrary than those excluding it.

There are advantages in selecting an index whose possible range is independent of whether the frequencies of the two species are equal or not. Cole's index (16) and that of Kolbe, for instance (which take joint absence into account) range from −1 through 0 (where the two species are unrelated) to +1. Of the indices proposed which ignore joint absence, none has this property except perhaps that of Gilbert (7). One could suggest, for instance,

$$\frac{a}{a + \min(b, c)} \tag{21}$$

Pielou (1969b) distinguishes between "complete" and "absolute" association, the latter meaning that the two species occur only in one another's presence, while in the former case they may differ in frequency, so that Species A occurs only in the presence of Species B but not *vice versa*. Some indices, like that of Cole (16), can attain their maximum value if association is complete, while most of those used in ecology require it to be absolute. Pielou emphasises the importance of making this distinction. While not quarreling with this point of view, one might perhaps suggest that it would be useful to supplement the symmetrical indices which have this property with asymmetrical indices more suitable for handling relations between common and infrequent species. An example is the "association value" (11) of Hacker (1921). It may be of much greater interest to know whether, and how much, the chance of encountering the infrequent species differs according to whether the common one is present or not, than to have a value expressing their mutual association. The distinction is parallel with that between correlation and regression, the latter expressing the quantity of Species A in terms of Species B (or *vice versa*), the former the closeness of relationship between the two.

Little information is to be found in the literature on the sampling properties of these indices. Accordingly, an empirical study of the distribution of some of them has been carried out, by computing them for repeated samples from a population with known and constant cell frequencies a, b, c, and d. For 20 such sets of cell frequencies, 100 samples of ten quadrats were taken, 100 of 20 quadrats, 100 of 50, and 100 of 100 each. The results are given in Table I. It will be seen that the Iverson index (4) and the point correlation

Table I. Empirical estimates of mean and standard deviation of different association indices based on sets of 100 samples from known populations.

Population Proportions		a	.01	.02	.02	.05	.10	.10	.10	.20	.20	.30
		b	.29	.03	.03	.20	.02	.05	.10	.05	.10	.10
		c	.30	.45	.30	.35	.18	.25	.10	.25	.30	.10
		d	.40	.50	.60	.40	.70	.60	.70	.50	.40	.50
Simple Matching Coefficient (12)	N=10	Mean	.372	.223	.394	.371	.551	.610	.674	.684	.572	.804
		SD	.118	.269	.302	.200	.367	.310	.326	.222	.201	.119
	N=20	Mean	.414	.339	.575	.431	.711	.693	.749	.706	.585	.812
		SD	.102	.255	.213	.122	.252	.161	.183	.097	.113	.083
	N=50	Mean	.404	.472	.614	.442	.797	.693	.799	.707	.593	.789
		SD	.062	.148	.067	.065	.062	.066	.060	.069	.082	.065
	N=100	Mean	.405	.516	.626	.452	.797	.703	.798	.705	.698	.806
		SD	.048	.050	.044	.049	.037	.047	.043	.045	.053	.038
	Population Value		.410	.520	.620	.450	.800	.700	.800	.700	.600	.800
Iverson Coefficient (4)	N=10	Mean	.025	.050	.039	.065	.292	.295	.385	.430	.346	.622
		SD	.066	.103	.109	.108	.273	.252	.323	.253	.210	.212
	N=20	Mean	.016	.042	.057	.076	.300	.276	.327	.400	.321	.627
		SD	.035	.063	.084	.077	.202	.190	.202	.160	.134	.164
	N=50	Mean	.014	.041	.058	.080	.329	.246	.323	.413	.335	.581
		SD	.023	.034	.048	.045	.118	.093	.126	.109	.090	.106
	N=100	Mean	.013	.043	.050	.087	.329	.254	.344	.410	.328	.611
		SD	.013	.026	.036	.038	.092	.068	.084	.073	.064	.067
	Population Value		.013	.040	.050	.083	.333	.250	.333	.400	.333	.600
Dice Coefficient (5)	N=10	Mean	.041	.080	.058	.106	.387	.401	.478	.557	.478	.744
		SD	.108	.159	.154	.167	.319	.293	.347	.261	.236	.183
	N=20	Mean	.028	.073	.097	.132	.427	.399	.458	.553	.470	.757
		SD	.063	.108	.140	.127	.235	.227	.236	.168	.158	.137
	N=50	Mean	.026	.076	.106	.144	.484	.386	.475	.576	.495	.729
		SD	.043	.061	.084	.077	.134	.120	.143	.110	.102	.088
	N=100	Mean	.026	.081	.094	.157	.488	.400	.507	.578	.491	.757
		SD	.026	.048	.065	.064	.104	.085	.095	.075	.072	.052
	Population Value		.033	.077	.095	.154	.500	.400	.500	.571	.500	.750
Point Correlation Coefficient (15)	N=10	Mean	-.354	-.013	-.096	-.269	.344	.320	.411	.436	.202	.601
		SD	.237	.215	.212	.288	.323	.332	.376	.301	.351	.259
	N=20	Mean	-.382	-.034	-.051	-.258	.402	.295	.355	.415	.183	.611
		SD	.127	.178	.197	.216	.242	.261	.274	.204	.231	.183
	N=50	Mean	-.400	-.041	-.068	-.250	.450	.268	.361	.421	.211	.559
		SD	.085	.123	.129	.105	.138	.142	.173	.129	.158	.135
	N=100	Mean	-.400	-.031	-.082	-.228	.447	.282	.384	.413	.216	.597
		SD	.061	.097	.095	.087	.103	.105	.113	.091	.104	.079
	Population Value		-.392	-.032	-.086	-.236	.455	.279	.375	.406	.218	.583
Cole Coefficient (16)	N=10	Mean	-.806	-.058	-.454	-.579	.524	.465	.477	.676	.324	.755
		SD	.409	.623	.620	.544	.589	.665	.658	.508	.612	.379
	N=20	Mean	-.907	-.140	-.414	-.552	.713	.443	.434	.673	.287	.725
		SD	.213	.688	.705	.430	.441	.587	.531	.352	.402	.214
	N=50	Mean	-.913	-.157	-.341	-.532	.788	.478	.426	.659	.330	.625
		SD	.146	.589	.503	.233	.209	.256	.237	.190	.254	.161
	N=100	Mean	-.913	-.134	-.403	-.489	.768	.493	.430	.637	.328	.641
		SD	.087	.459	.390	.192	.155	.187	.125	.135	.162	.095
	Population Value		-.893	-.149	-.375	-.500	.768	.487	.375	.636	.333	.583

coefficient (15) are in general unbiased, even with samples of only ten quadrats; the simple matching coefficient (12) and the Dice index (5) estimated from small samples tend to underestimate the population value; in the Cole (16) index, significant discrepancies between sample and population values occur even up to sample sizes of 50, but the sign and extent of the discrepancy vary and appear to depend on the degree of association of the species, and the inequality of their total frequencies.

6.2.1.3 *Quantitative Data*

Where information about the samples includes some quantitative measure for the various species, and is not limited to their presence and absence, a variety of other measures of association become available. Such measures are referred to elsewhere in this volume as species correlation (SC).

In the first place, one may sacrifice part of the information by converting the quantitative data into a 2×2 table, the range of values for each species being divided at a convenient point, and the allocation of the sample to a cell of the table being determined by whether the quantity of each of the two species exceeds this arbitrary point or not. As a suitable point of division KOLBE (1956) used the geometric mean ignoring zero values, the Dutch grassland workers (VRIES 1953, 1954, VRIES et al. 1954) used the median, while TARWID (1960) selected a density value exceeded by 30-40 % of the samples.

Where an arbitrary division of this sort is used to produce a 2×2 table for testing, it is conceivable that the degree or sign of association could be affected by the choice of a point at which this arbitrary division is made. Consider, for instance, the following hypothetical data for numbers of quadrats in different cover categories for two species:

Table II

% cover, Species A

		0	1—10	11—20	21—30	31—40	41—50	>50
	0	28	1	2	0	1	0	0
	1—10	0	2	1	0	7	5	11
% cover,	11—20	2	11	0	2	4	3	7
Species	21—30	1	1	0	3	5	4	4
B	31—40	0	2	4	1	0	2	3
	41—50	0	8	12	2	0	0	2
	>50	0	10	7	0	1	0	1

On the basis of presence and absence only, this gives

28	4
3	115

showing a marked positive association.

If the data are split at 40 % cover for each species, the 2×2 table is

68	39
40	3

showing a *negative* association, which is also marked if quadrats in which either species is absent are omitted:

40	35
37	3

On the other hand, if the whole data are split at 10 % cover, the 2×2 table becomes

31	27
25	67

which shows a *positive* association.

These different results are not mutually contradictory – they merely reflect a non-linear relation between the quantities of the two species, or indicate that their presence or absence may depend on factors differing from those which affect their cover when they are present.

While one recognizes that the association observed will depend on the arbitrary break-point selected for each species, its significance may be determined by any methods appropriate to presence and absence data. If a significant association is demonstrated by this rough-and-ready method, its reality need not be doubted, though its interpretation may call for more study.

Akin to the conversion of quantitative data into a 2×2 table is RAUNKIAER's (1928) "degree of codominance", which is the proportion of the area studied in which both species have frequencies in excess of 80 % Rather similar is IVERSON's (1936) "Affinitätskoeffizient", which averages the frequency of species B in areas where species A is dominant, and that of species A where species B is dominant.

Other measures of correlation between species make direct use of the quantities of each of them in replicate samples. ELLENBERG (1956) proposed as a measure of quantitative association between species the "spezifische Massen-Gemeinschaftskoeffizient".

$$\sum_{i \varepsilon T} (x_i+y_i) \;/\; [\sum_{i \varepsilon T}(x_i+y_i)+2(\sum_{i \varepsilon U} x_i+\sum_{i \varepsilon V} y_i)] \qquad (22)$$

where x_i and y_i are measures (biomasses) of species A and B in the *i*th sample, T is the subset of samples containing both species, and U and V are the subsets containing respectively species A or species B but not both.

WHITTAKER & FAIRBANKS (1958) used a related measure as "percentage similarity of distribution" (cf. (51)),

$$\sum_i \min (x_i, y_i) = 1 - 0.5 \sum_i |x_i - y_i| \qquad (23)$$

in which x_i and y_i are measures for species x and y in the i th sample, when those measures have been divivided by the row sums of a community table (i.e. for each species they are the decimal proportions of the sum of that species' measures in all samples of the set). MONK (1968) derives a measure of correlation between species, not from their joint occurrence in the same samples, but from the similarity of the environmental factors in the samples where they occur; in other words, it is an *ecological* correlation in the narrow sense. For each environmental factor, the mean is calculated for those subsets of samples in which each of the species occurs. These means are then used as the variables x_i, y_i in calculation of a similarity index (expression (50), section 6.3.3.2 below.)

A very commonly used measure of association making good use of measurement data, but depending on the assumption of linear relations between species, is the correlation coefficient:

$$r = \frac{\sum_i (x_i - \bar{x})(y_i - \bar{y})}{\sqrt{\sum_i (x_i - \bar{x})^2 \sum_i (y_i - \bar{y})^2}} \qquad (24)$$

where x_i y_i are the quantities of the two species, $\bar{x}$, $\bar{y}$ their means, and summation is over all samples. If presence of the species is scored as 1, and absence is 0, this reduces to the point correlation coefficient (13). Where there is no association between the two species, and the quantity measures of each are normally distributed, the expectation of r is zero, and its sampling distribution is well known. Ecological data are often very far from normally distributed — in fact, a distribution curve with a mode substantially below the mean is usual — and in this case, though the correlation coefficient may provide an acceptable measure of association, its distribution is unknown, and to rely on published tables may lead one far astray. Moreover, non-linear relations such as those in Table II are probably far from uncommon, and will not be revealed by a correlation coefficient. For this purpose, tests of quadratic or higher-order regression may be required.

To avoid the drawbacks of unknown but certainly non-normal distributions, the use of rank-order correlation has sometimes been suggested. The actual measures are replaced by the ranking of the sample in respect of each species, and then the SPEARMAN (or other) rank correlation coefficient is calculated without reference to the measurements (FAGER 1957, NUMATA & MITSUDERA 1960). Alternatively, the ranks may be used as the basis for a normalizing ("rankit") transformation, and a correlation coefficient calculated for the transformed data in the usual way — a coefficient which is now assured of having the properties and distribution expected for the correlation coefficient if the null hypothesis is true.

Information measures offer another way of avoiding distribution problems. MORISITA (1959) has proposed an index of interspecific overlapping calculated as follows: let the number of individuals of species x in the i th sample be n_{xi}, and the total individuals of this species in all samples be N_x. Then one defines

$$\delta_x = \frac{\sum_i n_{xi}(n_{xi}-1)}{N_x(N_x-1)} \tag{25}$$

and the index of interspecific overlapping between species x and y is

$$C_\delta = \frac{2\sum_i n_{xi}n_{yi}}{(\delta_x+\delta_y)\,N_x N_y} \tag{26}$$

taking values of approximately unity where the two species vary in proportion in the different samples, and zero where they do not occur together. This index ignores those samples where neither species is present, and MORISITA was able to prove by empirical trials that it is almost independent of N_x and N_y.

In order to get an equivalent index of interspecific correlation (i.e. taking quadrats from which both species were absent into account), MORISITA defined

$$R'_\delta = C_\delta - \frac{2}{(\delta_x+\delta_y)M} \tag{27}$$

where M is the number of samples. Then the correlation index is given by

$$\begin{aligned} R_\delta &= R'_\delta & R'_\delta > 0 \\ R_\delta &= \frac{M\sum_i n_{xi}n_{yi}}{N_x N_y} - 1, & R'_\delta < 0 \end{aligned} \tag{28}$$

which allows it to range from -1 to $+1$. Like C_δ, MORISITA found it almost independent of N_x and N_y; HURLBERT (1969), however, was able to show that for certain types of vegetation pattern it may give very different results with quadrats of different size, and that some of these results may be seriously at variance with commonsense views.

MORISITA extended this treatment from density to cover data, so that indices of interspecific overlapping and correlation for cover were

$$C_{\delta(p)} = \frac{M \sum_i p_{xi} p_{yi}}{(\delta_{x(p)} + \delta_{y(p)}) M^2 \bar{p}_x \bar{p}_y} \qquad (29)$$

and

$$R'_{\delta(p)} = C_{\delta(p)} - \frac{2}{(\delta_{x(p)} + \delta_{y(p)}) M} \qquad (30)$$

where p_{xi} is the cover of the x th species in the i th sample; $\bar{p}_x$ is the mean cover of the x th species; and

$$\delta_{x(p)} = \frac{\sum_i p^2_{xi}}{M^2 \bar{p}_x^2} \qquad (31)$$

ESTABROOK (1967) also suggested an information measure for associations where the values observed could be expressed as a frequency table — e.g. the numbers of quadrats for which the quantities of the two species in question fell into each of a limited number of categories. If the probability with which the quantity of species j in a quadrat falls into the i th category is P_{ij}, then the information content of the species j data is

$$H_j = -\sum_i P_{ij} \log P_{ij} \qquad (32)$$

By finding the information content for species j within each category of species k (or vice versa) one can then determine the information content of j which is not shared with k, $H(j/k)$. If then we define

$$D(j,k) = \frac{H(j/k) + H(k/j)}{H(j\,.\,k)} \qquad (33)$$

where $H(j.k)$ is the total information in the two species considered together, we have a measure of the independence of the two species. This may readily be converted into a measure of association:

$$S(j,k) = \sqrt{1 - D^2(j,k)} \qquad (34)$$

which will range between 0 and 1. As ORLOCI (1968) suggests, this measure could be applied directly to count or frequency data in ecology or (by arbitrary categorization) to continuous measures. It should be remembered, though, that any information measure of this sort takes no account of the fact that the categories are ordered.

6.2.1.4 *Species Correlation from Distance Measures*

Association between species can also be studied on the basis of distance between individuals. The first to do this was Dice (1952). He took the less abundant of two species, and measured the distance from each individual of this species to the nearest individual of the other species in each sextant around it. He showed that the square roots of these distances would be distributed approximately normally if the two species were uncorrelated, and that the skewness parameter g would be positive if they tended to occur together, negative if they tended to be separated.

Pielou (1961) recorded random individuals together with their nearest neighbours, and studied the specific composition of the pairs thus formed. For every two species a 2×2 table is set up, partitioning the pairs according to whether species A was or was not the random individual, and whether B was or was not the nearest neighbour. The significance of the relation between the species is tested by the usual methods for 2×2 tables, and the measure

$$S = 1 - \frac{(b+c)M}{(a+b)(b+d) + (a+c)(c+d)} \qquad (35)$$

is used as a coefficient of segregation, showing the extent to which individuals tend to have as nearest neighbours others of the same species rather than a different species.

Goodall (1965) proposed two methods using distance measurements, one based on the recognition of mutually nearest neighbours, the other on comparison of distances between individuals of the two species with distances from a random point. These techniques were primarily intended for significance tests; but the second could also, if desired, provide a measure of association – for instance, by the ratio of the random-point distance to the nearest-neighbour distance.

A rather similar method has been proposed by Yarranton (1966) with special reference to lichen communities, in which a random point defines a pair of species — that of the individual on which the point falls, and the nearest other species in contact with it. A species x species matrix showing the number of times each possible pair-wise record occurs can then form the basis for association measures, just like the presence of species in other types of sample. It should be noted, however, that the observational matrix will not be symmetrical.

6.2.2 Relations Among More than Two Species

Sometimes the relations studied are not between species in the narrow sense at all, but rather between groups of species with similar biological characteristics, which may behave in a largely vicarious way. It was shown, for instance (Goodall 1965), that there was a close correlation in the mulga vegetation of Australia between a group of species of taller shrubs on the one hand, and another group of much lower-growing shrubs, which seemed to need protection at some stages of growth. Similarly, McDonough (1963) showed a relationship between some cactus and shrub species in the Sonoran Desert. In neither of these cases, though, was there any specific relation between pairs of species to be found.

One may wish to know, not whether particular pairs of species are correlated, but whether a whole group of species tend to occur together. For instance, correlations among groups of species provide an objective basis for the recognition of "ecological groups" or "sociological groups". This concept, originated by Duvigneaud (1946, 1953) and developed further by Ellenberg (1950) and Gounot (1961) among others (article 2.2.3.1), has affinities with the "characteristic species" of associations recognized by the Zürich-Montpellier School (article 20). A group of species with similar environmental requirements tend to occur together, and a given vegetation type is identifiable from the presence of one or more such ecological groups. A numerical method for recognizing correlated groups of species may accordingly be useful in vegetation classification.

The recognition of groups of correlated species is the counterpart of the problem of classifying stands (article 19), and many of the same methods can be applied. For instance the technique known as "inverse association analysis" (Williams & Lambert 1961, Lambert & Dale 1964) separates groups of species acording to whether or not they occur in certain key stands, whereas "inverse information analysis" (Williams et al. 1966) separates them according to the independence of their information content.

Another suitable approach is that of multivariate factor analysis (Dagnelie 1960a, b, 1965, article 9). Principal component analysis, for instance, discussed in connexion with stand ordination in article 10, is a procedure for finding orthogonal linear functions of the different variables (quantities of different species) which account for the maximum possible proportion of their variation. The species with large coefficients (of the same sign) in one of these functions then constitute a correlated group, whose structure may often be clarified by a rotation of axes. With presence and absence data, the principal component method may also be used, if presence and absence are recorded as "1" and "0".

Pair-wise correlations among species may serve as a basis for recognizing larger groups. WELTEN & HEGG (1958) in fact regarded correlations between single pairs of species as likely to be fortuitous unless they were supported by mutual correlations with other pairs of species, which thus constitute an ecological group. BEALS (1965b) used measures of "distance" (i.e., inverse correlation) between species pairs to ordinate a set of species in two dimensions, and then divided them into seven ecological groups from inspection of this ordination. KLEINER & HARPER (1966) distinguished groups of species on the basis of the number of significant positive correlations between pairs within the group.

Various ways have been proposed to express the degree of correlation among a number of species forming a group. HOPKINS (1957) and MC INTOSH (1962), for instance, used the expression

$$2A/\{s(s-1)\} \tag{36}$$

where the group consists of s species, and A of the pair-wise correlations among them reach significance at a selected level. FAGER (1957) suggested the use of KENDALL's (1955) measure of "concordance" — analogous with rank-order correlation. MORISITA (1959) presented extensions of his C_δ, R_δ indices (expressions (26) and (28) above) to cases where several species are involved, each with the same total number of individuals. MARGALEF (1958) generalized the expression (9) he used for association between two species, in the form

$$a^{-1}(m_1 \,.\, m_2 \,.\, m_3 \ldots\ldots m_s)^{s-1} \tag{37}$$

as a measure of association between s species present respectively in $m_1, m_2, \ldots.. m_s$ samples, and jointly present in a samples.

Questions regarding the significance of association among a group of species may be framed in different ways. Considering first presence and absence data, in the $2 \times 2 \times 2$ table

		Species A present		Species A absent	
Species B:		+	−	+	−
Species C	+	a	b	a'	b'
	−	c	d	c'	d'

one may wish to know whether the number of samples, a, in which all three species occurred is greater than was to be expected given the frequency of the species singly and in pairs. Or one could ask whether the eight cell frequencies are consistent with an assumption of complete independence of the species. Or, as a third possibility, one could consider the three species as a group which may occur in any combinations, and ask whether the four cell values *a*, *b*, *c*, and *a'* collectively are larger than expected, given the frequencies of the three species separately. All three questions are legitimate, and the answers will differ.

Taking GREIG-SMITH's (1952, 1964) figures for *Amaioua corymbosa*, *Lacistema aggregatum* and *Alibertia acuminata* in Trinidad as A, B, and C above, one may tabulate them:

23	6	6	7
14	15	11	18

GREIG-SMITH chooses to answer the second question raised above, and obtains a value for χ^2 of 16.69, with four degrees of freedom. He points out that this result may be ascribed to the large excess over expectation of *a* — in other words, that the negative answer to the second question depends on a positive answer to the first, though this is not explicitly answered. If, on the other hand, we ask the third question, we find that the four cells with two or more of the species contain *fewer* quadrats than would have been expected from their individual frequencies (49, as against 52.06, giving a non-significant χ^2 value of 0.38 with one degree of freedom); the low figures for quadrats with two of the three species counterbalance the high figure for quadrats with all three.

PIELOU & PIELOU (1967) proposed a significance test for "coexistence" among a group of species based on the number of samples without any of the group. Combinatorial expressions are given for the probability of an observed figure, taking the frequency of species into account, but will not be reproduced here. The test does not distinguish between cases where all of the group tend to occur together, and those in which the positive coexistence is confined to a subset. A normal approximation is also given for cases where the number of species in the group is small, the number of samples large.

6.2.3 PARTIAL CORRELATIONS

Where more than two species are consistently associated, the relationships among them may often be elucidated by calculating partial correlation coefficients (e.g. DAWSON 1951, VASILEVICH

1970). This is a well-established statistic for describing interrelations among normally-distributed variables. It expresses and tests the linear interrelation between two variables, after eliminating that part of their correlation which is attributable to a mutual linear association with one or more other variables. In the simplest case, it is expressed as

$$r_{ab \cdot c} = \frac{r_{ab} - r_{ac}\, r_{bc}}{\sqrt{(1-r_{ac}^2)\,(1-r_{bc}^2)}} \tag{38}$$

where a and b are the variables under study and c is the variable the relations with which are to be eliminated, while r_{ab} is the product-moment correlation (expression (24)) between a and b. The significance of this partial correlation may be tested against tabulated values showing the distribution of sample values from an uncorrelated population — the same as that of the product-moment correlation, with one fewer degrees of freedom.

COLE (1957) proposed a similar test for partial association between species, where presence and absence only had been recorded. Apart from relations among three or more species, however, he pointed out that the third variable might well be a habitat factor, so that the test would then examine the relation between the two species when this environmental factor was held constant. In this connexion, tests of partial association or correlation seem to have more ecological relevance; to know whether two species are still correlated in the absence of differences in an environmental factor which may affect them is likely to be more valuable in interpretation than to find the effects of removing mutual correlation with a third species, relations with which are less likely to be directly causal.

6.2.4 NON-LINEAR RELATIONS AMONG SPECIES

Most of the methods of identifying and measuring the correlation between species which have been described above are based on a linear model, or assume a substantial linear component in the relationship. It is by no means clear that this assumption is generally justified. It has been shown (WHITTAKER 1967, article 3) that, where observations cover a fairly wide range of ecological conditions, different species may occur in maximum quantity at different environmental values, and decrease both above and below this optimum. Where the modes of two species do not coincide, this means that a linear model may fail to reveal the relationship between them. If a species with a narrow range occurs with a broad-ranging one, there may be a real relationship between their quantities which could be shown only by studying higher-order terms.

Consider, for instance, the hypothetical set of data:

Species A, cover %:	0	3	6	15	30	42	60
Species B, cover %:	0	9.5	18	37.5	50	42	0

which could arise, for instance, if the environmental range of A was much wider than that of B. There is an exact mathematical relation between the two series of figures ($18B=60A-A^2$), but usual measures of association will not show this. The correlation coefficient r (expression (24), for instance, is +.169. SPEARMAN's rank correlation coefficient is —.229. If these two are the only species present in the quadrats, WHITTAKER's index (23) will be .33. In such cases, a regression approach may be appropriate; a test of quadratic regression would demonstrate the relationship, and its closeness could, for instance, be expressed by R^2, the proportion of the variance of B accounted for by the regression on A.

Where the relation between two species is studied, not for its own sake, but as part of an ordination procedure, the use of a linear model for essentially non-linear relations may lead to considerable obscurity. If principal components analysis is applied, for instance, to data for species each showing a separate mode along a single axis, each of these modes will be represented by a distinct principal component, and the simple underlying model of a single axis with non-linear relations of each species to it will not be revealed (cf. article 11). Such models have been the subject of some study by pyschologists, and attempts are currently being made to apply non-linear models for principal components analysis to ecological data.

6.3 Similarity Between Vegetation Samples

6.3.1 INTRODUCTION

A problem converse to interspecific correlation is that of establishing and expressing the degree of similarity or dissimilarity between samples of vegetation — or, indeed, between entire stands. A matrix of similarity is, for instance, often required as a preliminary to the classification or ordination of a set of vegetation samples. Assessments of similarity can be based on any observable features of the vegetation — or even of the biotope — according to the basic assumptions and intentions underlying the investigations. Growth-forms or structural features might well be so used, for instance. In practice, however, almost all investigators have used floristic criteria, and it is with these that we shall concern ourselves.

A distinction needs to be made between measures of similarity which are context-dependent, and those which are, in a sense, absolute. "Similarity" is sometimes regarded as a concept with a fixed value and a fixed expression, so that the degree of similarity between two objects can be stated without reference to any others. Sometimes, on the other hand, the degree of similarity is considered in reference to a particular range of variation, either concretely represented in a set of objects, or present conceptually in the mind assessing it; thus a black mouse and a white mouse may be regarded as highly dissimilar within a population of mice, very similar in a mixed population of rodent species. Examples of both types of similarity concept will be found below.

Though formally the problem of inter-sample similarity seems to be on all fours with that of interspecific correlation already discussed, there are important differences between the two situations, as may be made evident by comparing the significance tests appropriate in each of the two cases.

6.3.2 Significance Tests of Similarity

For species association, the null hypothesis adopted was that each of the two species had a probability of being recorded which remained uniform throughout the set of samples, and was independent of whether the other species occurred in the sample or not. These probabilities are estimated from the marginal totals. In comparing the set of species occurring in two quadrats with a view to judging their similarity, the appropriate null hypothesis is clearly that the two quadrats were samples of the same vegetation type, and that consequently, for each species, the probability of occurrence was the same for both samples. But records in two quadrats only of the presence or absence of a species provide no basis for estimating its probability of occurrence — beyond the trivial conclusion that the probability is non-zero if the species is recorded at all. Only where the two are part of a larger collection in which, for the purposes of the null hypothesis, the probability for each species can be regarded as uniform does the basis exist for estimating the probability, and hence for a test of the null hypothesis.

An insensitive significance test may however be applied to the records in two samples only, without the need for frequency estimates based on a larger set of samples, by comparing the number of species present in one but not in the other (GOODALL 1969). If they are from the same community, the expectation is that these numbers will be equal for the two samples; and — on the same assumption as above, of independence — the distribution of unequal numbers will follow the binomial distribution. Particularly where the two samples come from vegetation types differing markedly in floristic diversity, significant dissimilarity may often be revealed by this test.

An assumption of independence between species is almost inevitable in testing a hypothesis regarding similarity of samples based on species presence and absence. Where the vegetation sampled is uniform, and the sample size is sufficient to cover the largest scale of pattern, this assumption seems reasonable (GOODALL 1953), apart from occasional instances of obligate relations between particular pairs of species, such as parasite and host. In the absence of such an assumption, there seems no basis for significance tests of similarity based on qualitative criteria.

The fact that a 2×2 table for similarity between quadrats or stands combines the records for a large number of species each with its own frequency in the population means that methods (such as the χ^2 test) suitable for testing association are inapplicable. The model underlying the χ^2 test is of a number of events all alike, and falling with constant frequency into the two classes of each of two binary classifications. Were the frequencies of all species the same, the same technique could be applied to tests of sample similarity. In practice, this requirement is never even approximately met, and, if a χ^2 statistic were calculated, its distribution would differ widely from tabulated values even if the samples compared were replicates from the same community. Thus, a frequency curve for χ^2 between samples differing from the standard distribution (JUHÁSZ NAGY 1964) does not necessarily imply heterogeneity among the samples.

Where quantitative variables are used (such as density, cover, basal area, etc. for each species), the distribution of any index of similarity between samples would depend on the multivariate distribution of the variables included in the index. In GOODALL's (1966) probabilistic index this is taken into account, and the distribution is rectangular. For other indices an exact solution would be extremely difficult, and it might be best to approach the problem empirically by studying the distribution of calculated similarities in the whole matrix of sample comparisons, and finding where the index in question fell in the tail of the observed distribution.

It has so far been assumed that a particular pair of samples is of special interest, and that the similarity between this pair only is to be tested. This is not the usual situation. Usually all pair-wise similarities within a set of samples are studied simultaneously, and the question is rather whether the maximum or minimum among them is significant, or whether the whole set differ from the expected distribution. The latter may generally be tested by calculating χ^2 after arbitrary division into classes, or (better) by the KOLMOGOROV-SMIRNOV test. A test of the maximum or minimum necessitates an adjustment of significance levels. If the probability with which the null hypothesis may be falsely rejected when true is α, and n independent values have been tested, the maximum value calling for rejection of the null hypothesis is such as would be exceeded by individual values with a probability

$$1-(1-\alpha)^{1/n}$$

The fact that a set of similarity values may not be fully independent will usually make this test over-conservative.

6.3.3 MEASURES OF SAMPLE SIMILARITY

6.3.3.1 *Presence and Absence Data*

Where only presence or absence of species is recorded, a series of coefficients corresponding with the association indices mentioned above (Section 6.2.1.2) can be derived from the resulting 2×2 tables:

		Sample X	
		+	–
Sample Y	+	a	b
	–	c	d

where a, b, c and d are the numbers of species present or absent in the samples in question. Many of these indices have also been used in numerical taxonomy, and are discussed in this connexion by SOKAL & SNEATH (1963).

The oldest similarity index – that used by JACCARD (1901) in comparisons of species lists from different geographical areas in the Alps — is

$$a/(a+b+c) \tag{39}$$

which is analogous with expression (4) above for interspecific association, while SØRENSEN (1948) proposed the index

$$2a/(2a+b+c) \qquad (40)$$

analogous with expression (5). SØRENSEN's index has been used very widely for expressing similarity, and nomograms and tables for simplifying its calculation have been published by FALIŃSKI (1958). Expressions (39) and (40) have been called "coefficients of community", and are referred to elsewhere in this volume as CC. The complement of either is a coefficient of community difference (CD).

KULCZYNSKI (1928) proposed two indices, which may be represented

$$a/(b+c) \qquad (41)$$

and

$$\frac{1}{2}\left(\frac{a}{a+b}\right)\left(\frac{a}{a+c}\right) \qquad (42)$$

while OCHIAI (1957) and BARKMAN (1958) used

$$\frac{a}{\sqrt{(a+b)(a+c)}} \qquad (43)$$

MAAREL (1966) suggested an index which is the square of (43), but in fact used that of SØRENSEN because of its greater simplicity.

WILLIAMS (1944, 1947a, b) proposed as an index of similarity the ratio of the number of species present in two sample areas to the number expected from a species-area curve if they had been samples of the same homogeneous area.

WILLIAMS et al. (1966) used an index based on information theory (see (64) below) which, in the case of data limited to presence and absence of species in the two samples under comparison, reduces to:

$$2\ (b+c)\ \log 2$$

As FIELD (1969) points out, this tends automatically to give species-poor samples greater similarity than species-rich ones, which does not always correspond with intuitive views of the concept.

Similarity indices giving equal weight to presence and absence, such as the matching coefficient (12), have been little used in plant sociology. One example is the work of CRAWFORD et al. (1970), who applied the index of RUSSELL & RAO (1940) to comparison of samples of scrub vegetation in Peru. In our terms, this index was a/S, where S is the total species number, $a+b+c+d$; d is thus included in the denominator, but not in the numerator.

The limited use in plant sociology of indices taking d into account is understandable because of their great dependence on the less common species absent from both samples. It might equally be argued that the presence in both samples of ubiquitous species is also uninformative. One way of surmounting this difficulty is by the probabilistic similarity index (GOODALL 1966), which gives equal weight to presence in both the quadrats of a rare species, and absence from both of a ubiquitous one. In the calculation of this index (which can use binary, ranked or quantitative data), all possible pairs of values for each variable are placed in order of similarity, the probability of the observed or any greater degree of similarity is estimated for each variable, and these estimates are then combined.

It is clear that even replicate samples from the same homogeneous area of vegetation are unlikely to have exactly the same species complements, and consequently the values of any of these similarity indices will only exceptionally reach their theoretical maxima. The expectation and distribution of values between replicate samples have been little studied. The probabilistic index (GOODALL 1966) has of course a retangular distribution with an expectation of 0.5; and it has been shown (GOODALL 1967) that the matching coefficient has an expectation of

$$\frac{1}{S}\sum_{i=1}^{S}(1-2f_i+2f_i^2) \qquad (45)$$

where f_i is the proportional frequency in the samples of the i th species (out of a total of S species), and that its distribution may be approximated by the binomial. Likewise, the expectation of the JACCARD and SØRENSEN coefficients may be expressed respectively as

$$\sum_{i=1}^{S} f_i^2 \sum_{i=1}^{S} (2f_i-f_i^2) \qquad (46)$$

and

$$\sum_{i=1}^{S} f_i^2 \sum_{i=1}^{S} f_i \qquad (47)$$

LOOMAN & CAMPBELL (1960) attempted to set significance limits for the SØRENSEN coefficient by deriving them from χ^2 tests; they recognized, however, that it is necessary to take into account d, which is excluded from the SØRENSEN coefficient. Its distribution was studied empirically by GOUNOT & CALLEJA (1962) in samples from a *Brachypodium ramosum* grassland. They showed that the mean value increased and the variance decreased as the sample size increased, but that even with a 32 m² sample, for which the mean was .827, the confidence limits were .750 and .904. They accordingly recommended the use of multiple samples from each stand. The mean similarity coefficient decreased as the distance between samples increased — at least up to a distance greater than 4m — which fits in with observations on variance of quantities of individual species (GOODALL 1961).

WILLIAMS (1949, 1964) showed that the expectation of the JACCARD coefficient (39) between samples of equal size from homogeneous vegetation was simply related to the index of diversity α, on the assumption that the species densities followed the logarithmic series. If the mean number of species per quadrat is S, the expected value is

$$\frac{S - \alpha \log_e 2}{S + \alpha \log_e 2} \tag{48}$$

This implies that the value of the index, within homogeneous vegetation, will increase as the sample size increases.

6.3.3.2 *Quantitative Data*

We turn now to similarity indices based on quantitative measures of the contribution of each species to the vegetation – cover, biomass, basal area, frequency, etc.

A similarity index proposed by GLEASON (1920), may be expressed as

$$\frac{\sum_{i \varepsilon T} (x_i + y_i)}{\sum_{i} (x_i + y_i)} \tag{49}$$

where x_i, y_i are the quantities of the i th species in the two samples, and T is the sub-set of species present in both. This index, closely analogous with ELLENBERG's "spezifische Massen-Gemeinschafts-koeffizient" for interspecific correlation (22 above), has the drawback that the index may be affected greatly by a minimal occurrence of a species. Consider the two sets of data in Table III:

Quadrat		Species	Quadrat	
X	*Y*		*X*	*Y*
99	1	1	99	0
10	0	2	10	0
0	5	3	0	5
80	80	4	80	80

The GLEASON index is 0.82 for the data on the left-hand side of the Table, 0.41 for those on the right; yet the only difference is the presence of 1 % of Species 1 in quadrat *Y* in the former case.

CZEKANOWSKI's (1909) 'percentage similarity'

$$PS = \frac{2 \sum_i \min (x_i, y_i)}{\sum_i (x_i + y_i)} \tag{50}$$

was first introduced into plant sociology by DAHL & HADAČ (1941)[1], and subsequently used extensively by the Wisconsin school (e.g. CULBERSON 1955). If x_i, y_i are expressed as proportions of the total quantities of all species in samples, *X* and *Y*, index (50) reduces to

$$PS' = \sum_i \min (x_i, y_i) = 1-0.5 \sum_i |x_i-y_i| \tag{51}$$

in which form it was used by RENKONEN (1938, 1944) and WHITTAKER (1952) for comparisons of insect samples. Among others who have made use of percentage similarity, with various measures of sample composition, are DAHL (1957), CLAUSEN (1957), BRAY & CURTIS (1957), WHITTAKER & FAIRBANKS (1958), WHITTAKER (1960) and RAMSAY (1964a, b). It is analogous with index (23) for interspecific correlation.

Complements of percentage similarity are "percentage difference", which ODUM (1950) used to compare bird communities:

$$PD = \sum|x_i-y_i|/\sum(x_i+y_i) \tag{52}$$

KULCZYNSKI's (1928) "floristic degree of resemblance" was

$$\tfrac{1}{2}\left(\frac{1}{\sum_i x_i} + \frac{1}{\sum_i y_i}\right) \sum_i \min (x_i, y_i) \tag{53}$$

[1]) Their description of the index is ambiguous, but examination of their tables shows that it was calculated by (50), x_i and y_i being the mean value in the DOMIN scale for the stands in each of the two association tables compared.

Ružička's (1958) index was

$$\frac{\sum_i \min(x_i, y_i)}{\sum_i x_i + \sum_i y_i - \sum_i \min(x_i, y_i)} \tag{54}$$

and Pandeya (1961) used one very similar to that of Gleason (49), viz.

$$\frac{\sum_{i \varepsilon T} (x_i + y_i)}{\sum_i (x_i + y_i) + \sum_{i \varepsilon T} |x_i - y_i|} \tag{55}$$

where T is the sub-set of species present in both samples.

Forms of the correlation coefficient (24) have also been used as indices of similarity. Some investigators (e.g. Motomura 1952, Fresco 1969) have calculated the coefficient from unstandardized values of the variable; apart from the inapplicability of ordinary significance tests, this method will have the effect of giving much greater weight to similar values for the abundant species than to those less abundant in the vegetation.

A rather similar method was adopted by Orloci (1966) as a preliminary to ordination. He took deviations from the species mean without standardizing and showed that the scalar product

$$\sum_i (x_i - \bar{z}_i)(y_i - \bar{z}_i) \tag{56}$$

where $\bar{z}_i$ is the mean over all quadrats for the i th species, is a suitable measure of the similarity between quadrats X and Y for use in principal components analysis.

In calculating any of these indices there is, of course, no obligation to use the same measure of quantity for each species. Indeed, Beals & Cottam (1960) included two measurements — density and dominance — for each tree species, and one — frequency — for each herb species. But the similarity values obtained will, of course, be markedly affected by such arbitrary weightings.

Normally, like Motomura's "correlation coefficient", all these indices give the various species weight in proportion to the quantities in which they occur. This may or may not be desirable in expressing similarity, but many, like Kontkanen (1950), have felt that some redressing of this inequality in weighting is desirable. One way of doing this is to rank the species in each sample, and calculate a rank correlation coefficient, as was done by Ghent (1963).

If the values for each variable are first standardized by converting to zero mean and unit variance, and these standardized values are used to calculate a "correlation coefficient" between stands as an index of similarity, values of the index will follow the tabulated distributions on the assumption of species independence and normal distribution. If the distributions deviate widely from normality, a rankit transformation may be preferable to standardization. Either of these procedures will give all species equal weight.

In the Wisconsin modification (BRAY & CURTIS, 1957) of expression (50), the observations for each species are expressed as a percentage of the maximum for that species over all samples; these figures for each sample are then adjusted to a total of 100 before calculation of the coefficient. This modification, like those of the previous paragraph, has the effect of giving equal weight to all species, and making the similarity depend on their relative rather than absolute contribution to the vegetation, so that two samples in which the totals differed by a large factor would have 100 % similarity if their quantitative floristic composition were identical. As WHITTAKER (1967) points out, this procedure has the further advantage that different types of measurement (e.g. basal area and density) may be selected for different species. BRAY & CURTIS claim that this coefficient differentiated pairs of samples in the middle range of similarity better than the correlation coefficient.

HALL (1970) considered that the contribution of different species to a similarity function should be weighted in accordance with their mean quantity or frequency, within the set in which the similarity is being assessed. He was concerned with the more general case of similarity among a number of samples (see below, Section 6.3.5), but if only two samples are compared his index becomes

$$\frac{m + m^2/n}{m + n} \tag{57}$$

where m and n are the smaller and larger of the two values for each species.

ELLENBERG (1956) used a "Massen-Gemeinschaftskoeffizient"– corresponding with the expression (22) already cited for interspecific correlation —

$$\frac{\sum_{i \varepsilon T} (x_i + y_i)}{2 \sum_{i \varepsilon U} x_i + 2 \sum_{i \varepsilon V} y_i + \sum_{i \varepsilon T} (x_i + y_i)} \tag{58}$$

where T is the subset of species occurring in both samples, U and V those occurring in one or the other only.

MORISITA (1959) proposed a measure of community similarity, C_λ, calculated in exactly the same way as his index of interspecific overlapping C_δ (expression (26) above), and having the same independence of the total numbers of individuals in the samples except when they are small. He also extended this expression to weight (or cover), on the assumptions that the weight per individual within a species can be treated as constant, and that the number of individuals is large. The appropriate expression is then

$$C_{\lambda(w)} = \frac{2 \sum_i w_{1j} w_{2j}}{(\lambda_{(w)1} + \lambda_{(w)2}) W_1 W_2} \tag{59}$$

where w_{ij} is the weight of the j th species in the i th quadrat, W_i the weight of all species, and

$$\lambda_{(w)i} = \sum_j w_{ij}^2 W_i^2 \tag{60}$$

Beals (1965a) introduced a new consideration into similarity by taking growth-form into account. His modification of expression (50) for lichen and moss communities was

$$\frac{3 \sum_i \min (x_i, y_i) + \sum_j \min (a_j, b_j)}{2 \sum_i (x_i + y_i)} \tag{61}$$

where x_i, y_i were the frequencies of the i th species and a_j, b_j were the sums of species frequencies for the j th growth-form in the two samples respectively.

As with the similarity indices based on presence and absence, it is clearly relevant to consider, for those based on quantitative data, their expectation and distribution when the two samples are replicates from the same homogeneous community. Whittaker (1952) found that, in an insect community, values of expression (51) ranged from .5 to .8 in comparisons between replicate samples, while Bray & Curtis (1957) found that the Wisconsin coefficient averaged .82 between replicates. It is not, however, to be expected that this average value would be independent of the floristic composition of the vegetation. In fact, it is clearly desirable that every use of similarity coefficients should be accompanied by a study of the distribution of the same coefficient in replicate samples, for the cases where this distribution is known *a priori* are few.

6.3.3.3 *Expressions of Sample Distance*

An inverse measure of similarity between samples is provided by the compositional distance between them (compare "taxonomic distance" as discussed by Sokal & Sneath, 1963.) If we consider a multidimensional system in which each axis represents the quantity of a species, and a point represents the floristic composition of each vegetation sample, then the distance between the points representing two samples is their compositional distance. This may be calculated by the usual Euclidean distance expression:

$$\mathrm{ED} = \sqrt{\sum_i (x_i - y_i)^2} \qquad (62)$$

where x_i, y_i are the quantities (using any appropriate measure) of the i th species in samples X and Y. This was used, for instance, by FIJALKOWSKI (1954) and VASILEVICH (1963), and by NEWBOULD (1960) under the designation Ω^2, without taking the square root.

ORLOCI (1967, 1968) suggested replacing the absolute Euclidean distance by a relative distance, each observation being standardized by dividing by the length of the stand vector. This means that the values of the distance measure will be between 0 and $\sqrt{2}$, and will be zero if all species occur in the same proportions in the two stands; thus, the effects of varying quantities of the vegetation as a whole — of varying productivity between stands — are eliminated from the measure. Another way to improve Euclidean distance as an index of inter-stand relationship would be to replace the original variables by an orthogonal system based on them, such as is provided by principal component analysis; one would thus avoid giving excessive weight to a highly correlated group of species just because they happen to be numerous. This is closely akin to the generalized measure of distance (MAHALANOBIS, 1936; see also RAO, 1952), which has been used in ecology by HUGHES (1954), HUGHES & LINDLEY (1955), and HARBERD (1962). It may be expressed as

$$D^2 = \sum_i \sum_j a_{ij}\, d_i d_j \qquad (63)$$

where d_i is the difference between the values for the i th species, $|x_i - y_i|$ in the notation of (62), and a_{ij} is an element of the dispersion matrix.

Distance measures may also be based on presence and absence of species, values of "0" or "1" replacing the usual quantitative measures (WILLIAMS et al. 1966). The Euclidean distance (62) is here simply the square root of the number of species not common to the two samples. Again, however, standardization or — better — replacement by an orthogonal uncorrelated system of variables is a desirable improvement.

6.3.4 SIMILARITY BETWEEN TWO SETS OF SAMPLES

To comparisons between two sets of samples – two association tables, for instance — much the same considerations apply as between a pair of samples. If the number of samples in the two sets differ, this may have marked effects where the comparisons are based on numbers of species, as in the JACCARD (39) and SØRENSEN (40) indices (DAHL 1957, MORISITA 1959). The effects of this have been reduced by limiting the comparison to species with > 10 % presence (BARKMAN 1958) or > 50 % constancy (NORDHAGEN 1943, DAHL 1957), or by using a fixed number of species (RAABE 1950). Quantitative measures are not subject to the same difficulties. Percentage similarity (50, 51) and related measures can be applied to presence or constancy values (DAHL 1957, FRYDMAN & WHITTAKER 1968, MAAREL 1969) or mean density values (WHITTAKER & FAIRBANKS 1958) of species in two sets of samples being compared. WILLIAMS et al. (1966) calculated the Euclidean distance between the centroids of the clusters of points representing the composition of the different samples in the set. BARKMAN (1958) used

$$A_2 = \frac{\sum_j \min(x_{1j}, x_{2j})}{\sqrt{\sum_{i \varepsilon T} x_{1i} \sum_{i \varepsilon U} x_{2i}}} \tag{64}$$

where x_{1j} is the mean % cover of the j th species in the i th set of samples, T is the set of species for which $x_{1j} > x_{2j}$, and U the set for which $x_{1j} < x_{2j}$.

ČEŠKA (1966) proposed an index based on SØRENSEN's presence similarity index (40), viz.

$$\frac{2 \sum_j C_{1j} \cdot C_{2j}}{\sum_j C_{1j} + \sum_j C_{2j}} \tag{65}$$

where C_{ij} is the proportion of samples within the i th set including species j.

In a subsequent paper (ČEŠKA 1968), he gave an alternative expression:

$$\frac{\sum_j C_{1j} \cdot C_{2j}}{\sum_j C_{1j} + \sum_j C_{2j} - \sum_j C_{1j} \cdot C_{2j}} \tag{66}$$

but gave no arguments in favour of it.

6.3.5 MULTI-SAMPLE SIMILARITY

In inter-sample comparisons, as in interspecific association, it is possible to generalize consideration to subsets larger than two.

But even less attention has been given to multi-sample comparisons than to multiple relations among species.

The problem here is of testing for clustering, or devising a cluster index — one which will express the similarity among a group of more than two vegetation samples. In numerical taxonomy (SOKAL & SNEATH 1963), considerable attention has been given to methods of recognizing clusters — groups of taxa which are more similar to one another than to others not within the cluster. Some of these techniques have also been applied in plant sociology; further information will be found in article 19. Most of the methods, however, are highly arbitrary and dependent upon the particular similarity index used. Significance of the clustering — whether it could be the result of random assortment of attributes — has rarely been considered.

ČEŠKA (1966, 1968) derived a measure of mean similarity within a set of samples from SØRENSEN's presence similarity index (40). It may be expressed as:

$$\frac{\sum a_i(a_i-1)}{(M-1)\sum a_i} \qquad (67)$$

where the i th species occurs in a_i out of a total of M samples. This expression excludes self-comparisons.

HALL (1967a, b) proposed as a measure of heterogeneity (i.e. an inverse measure of similarity) among a group of samples

$$H_t = \frac{1}{S}\sum_{j=1}^{S}\frac{a_{jr}}{M} \qquad (68)$$

where a_{jr} is the number of samples (out of a total of M) with the rarer state (presence or absence) for the jth out of S species. For quantitative data, this was replaced by

$$H_{qm} = \sum_{j=1}^{S}\frac{\sum_{i=1}^{M} x_{ij}}{\sum_{j=1}^{S}\sum_{i=1}^{M} x_{ij}}\left(1-\frac{\sigma_j}{\sigma_M}\right) \qquad (69)$$

where x_{ij} is the value for the j th species in the i th sample, σ_j is the standard deviation of these data, and σ_M is the hypothetical maximum standard deviation if a group of M observations for this species were maximally heterogeneous — were divided as equally as possible between zero and the largest possible value (clearly this implies a type of measurement which has a natural upper bound).

MORISITA (1959) showed that his C_λ index of community similarity could be extended to comparisons among several community samples, provided the numbers of individuals were the same in each. WILLIAMS et al. (1966) used another information statistic as a measure of the differences among a group of quadrats, viz.

$$I = SM \log M - \sum_{j=1}^{S} [a_j \log a_j + (M - a_j) \log (M - a_j)] \qquad (70)$$

where S species are recorded in M samples, and the j th species is present in a_j samples. As mentioned above, this reduces to (44) where only two samples are under comparison.

ORLOCI (1968) suggested that the overall disorder among c samples could be measured, for count or frequency data summarized in contingency tables, by the minimum discrimination information statistic:

$$I = \sum_i \sum_j x_{ij} \log x_{ij} + x_{..} \log x_{..} - \sum_i x_{i.} \log x_{i.} - \sum_j x_{.j} \log x_{.j} \qquad (71)$$

where the cells of the table are x_{ij}, the row totals $x_{i.}$, the column totals $x_{.j}$, and the overall total $x_{..}$. Yet another information statistic:

$$\frac{1}{S} \sum_{j=1}^{S} \log_2 {}^{M}C_{a_j} \qquad (72)$$

was proposed by GODRON (1966) as an inverse measure of similarity among samples. Where $M=2$, this is the complement of JACCARD's similarity coefficient (39).

6.4 Conclusion

6.4.1 GENERAL PERSPECTIVES

In the first place, I repeat that there is nothing absolute about correlation between species — it is entirely context-dependent. Statements that a particular pair of species are positively correlated, negatively correlated, or uncorrelated may all be correct in different contexts.

Interspecific correlation means little or nothing if divorced from considerations of scale. When measured in small quadrats within a reasonably homogeneous community, it depends upon – and is indeed a good method of studying — the pattern of this community. If large quadrats are used in such a community, most interspecific correlations will disappear. This fact can be used at the same time for testing homogeneity and for determining the minimal area of the community, in a very useful sense of the term.

If one compares different communities by this size of quadrat or larger (or compares complete stands of different communities), interspecific correlations are generated by a different set of factors operating on a larger scale. They depend, not on the pattern within the community, but on the biotope differences leading to the establishment of the different communities, or resulting from them. These are the correlations which permit the recognition of ecological or sociological groups of species, so important for ordination and classification of vegetation and for indicating environmental conditions or successional status.

If small quadrats only are used in heterogeneous vegetation one risks confounding correlations from these two sources. The data will reflect both the pattern within each community, and the differences in composition between communities. These elements may be separated if correlations are computed separately for sets of quadrats within limited portions of the area covered and for comparisons between such sets, from which the effects of correlations within the sets can then be eliminated. But it may be simpler to use different sizes of quadrats for these two purposes.

The same remarks apply, *mutatis mutandis*, to correlations based on distance measurements — though in general these will be concerned with small-scale correlations. If one were interested in the larger-scale relationships, nearer neighbours would have to be ignored in order to avoid confounding these relationships with the effects of local pattern.

The preceding paragraphs assume a vegetational structure in which reasonably discrete communities are recognizable and transition zones can be ignored. Where variation in the vegetation is continuous, no "minimal area" can be found (Goodall 1961), and satisfactory representation of interspecific relationships may then take the form of a continuous function expressing correlation in terms of quadrat size and interspace. For any given quadrat size, for instance, one might develop a transition probability function giving the probability of finding different quantities of the two species in one quadrat in terms of the known composition of another quadrat at a known distance from it. More simply, correlation in quadrats of a given size could be expressed as a continuous function of the interspace between them.

Apart from their use in ordination and classification, studies of interspecific correlation have often been concerned with significance tests. One wonders whether this emphasis may not be misplaced. Underlying any significance test is a null hypothesis -- which in this case is that the species in question are, at the scale tested, distributed independently of one another. But is this null hypothesis ever ecologically reasonable? One knows that the establishment and growth of every species are affected, to a greater or lesser extent, by every factor of the environment. Consequently, the null hypothesis is strictly incredible, and a test of it is pointless. Estimates of correlation should be the ecologist's aim rather than tests of their significance — though these estimates should be associated with intervals expressing their uncertainty due to sampling errors.

In the field, pair-wise correlations between species are usually artificial isolates from a more complex situation, better represented in terms of overlapping groups of ecologically related species ("ecological groups" or "sociological groups") or, better still, as a multidimensional continuum. If the main axes of variation within this continuum can be identified and the association of each species with each of the main axes (including possible non-linear relationships) can be defined, the correlations between pairs or groups of species follow directly. This type of analysis of vegetation is dealt with elsewhere (articles 2,7,9).

While correlation between species is a relationship between populations which can be sampled, and one can consider the correlation index obtained as an estimate of a true value obtaining between the populations, the situation is somewhat different for similarity indices between quadrats or stands. One has two actual, concrete quadrats existing at a fixed time and place, and the calculated similarity between them is not subject to sampling error. The quadrats, on the other hand, may themselves be regarded as samples of larger areas of vegetation, and in this case one may be interested in the way in which the error-free similarity indices between pairs may vary from pair to pair. This variation will in part be due to heterogeneity (at the scale of quadrat size used) within each of the vegetation areas sampled, in part to differences which would arise even if a completely homogeneous area were sampled. There have been some tests of the variation of similarity indices within actual areas of vegetation assumed homogeneous. But it might well be easier and more satisfactory to study this in artificial populations, the properties of which can be defined exactly, for in real-life vegetation complete homogeneity is probably never attained.

Similarity between stands need not be context-dependent as is interspecific correlation. It can be, and often is, defined in absolute terms. This is possible because the variables on which it is based – the species present – are a definite and enumerable set, at least when species absent from both stands are ignored. As pointed out above, though, to treat similarity as an absolute relationship between two stands does not square well with a common-sense approach, according to which two objects which may be regarded as similar within a very broad context are considered quite dissimilar when the focus narrows.

Unless similarity measurements have some ulterior purpose (such as ordination or classification), the weights given to the different species are purely arbitrary. As we have seen, many indices take account of species presence or absence only, which amounts to giving all species equal weight. This means that the chance presence of the rarer species in one or both samples will play a part in determining similarity which could be reasonably regarded as excessive. The same is true of indices using quantitative data standardized for each species.

More in accordance with ecological thinking are indices in which each species is weighted according to some quantitative measure of its contributions to the vegetation – cover, basal area, biomass, etc. Which of these quantitative measures is used will affect the value of the indices obtained, and is an arbitrary decision. Even more arbitrary is a decision, say, to give trees double weight compared with herbs. Weighting by some measure of species contribution which is of biological importance at least gives an index which can claim to be natural and meaningful. It could be argued, however, that this over-weights any group of highly-correlated species, the information from which includes a large share of redundancy. Such an argument could be answered by using as an inverse measure of similarity the Euclidean distances between points in an orthogonal system provided by principal components analysis – *ex hypothesi* uncorrelated.

Discussion of pair-wise similarity of stands leads naturally to the question of similarities between and within groups of stands. Some of the indices mentioned cannot be extended to cover similarities among a group of stands larger than two; for others, this extension would appear quite natural. For comparison *between* two groups of stands, though, any of the similarity indices available for pair-wise comparisons among stands can be used. If they are based on presence and absence of species, different numbers and sizes of samples in the two groups will affect the results, just as two stands of different size cannot very satisfactorily be compared by these indices. If, however, the index is based on quantitative measures of the species present, there is no reason why the mean of these values within each group should not be used for computation of inter-group similarity. These questions, however, are closely associated with the use of quantitative methods for classification of vegetation, and will be discussed further in article 19.

6.4.2 Choice of Indices

One of the purposes of this article is to display to the interested reader the wide range of indices available in the literature. There has at present, however, been far too little informed comparison of their merits, and consequently choice among them is still largely a matter of taste. The different indices, however, mean different things. It is quite possible to find populations where the true values of the matching coefficient (12), say, are identical, while those of the IVERSEN coefficient (4) differ, and *vice versa*.

In choosing among the range of indices that have been proposed and used, the following guidelines may be suggested:

1. One is not usually concerned with statistical tests of the significance of correlation between species, or differences between samples, so much as with quantitative expression of these relationships.
2. Other things being equal, a simpler index is to be preferred.
3. Appreciation of the meaning of an index is easier if it has known upper and lower limits (e.g. $+1$ and -1, or $+1$ and 0); for interspecific association and correlation, it is also helpful if it has a known central value (for instance zero, where the limits are ± 1) when the species are independently distributed.
4. Since the index will have to be estimated from samples its sampling properties are of some importance. It is desirable, in particular, that a sample estimate should be unbiased, and should be independent of sample size once a 'minimal area' is exceeded. Few sampling studies of association and correlation indices for use in ecology have been performed. The example in Table I shows that some indices – the matching coefficient (12), for instance – may be markedly biased in small samples, while others are not. Among the indices for which estimates are unbiased, choice would often depend on sensitivity in relation to sampling variance.

None of the indices of species association (those based on presence and absence) is, in general, independent of sample size. Consider two situations where samples of given size have expected proportional frequencies as follows:

		Species A	
		+	−
Species B	+	.2	.2
	−	.3	.3

		Species A	
		+	−
Species B	+	.7	.7
	−	.1	.1

If the sample size is doubled (the two halves of the sample being assumed independent of one another in respect of species occurrence) the frequencies become respectively

		Species A				Species A	
		+	—			+	—
Species B	+	.48	.16	Species B	+	.93	.03
	—	.27	.09		—	.03	.01

In the first case (where the species are unassociated) the expectation of the point correlation coefficient (15) is zero in each case, but the expectation of the IVERSEN index (4) is .29 and .53 respectively. In the second case, where there is a strong positive association, both indices differ with sample size, the expected values of the point correlation coefficient being .38 and .22 respectively, and those for the IVERSEN coefficient .78 and .94.

These considerations argue strongly for the use of quantitative data if interspecific indices independent of sample size are required. If, however, one is content to work within the framework of a single fixed sample size, the great simplicity of the indices based on presence and absence, and the ease of collecting the field data needed, make them attractive. This applies particularly to such an index as that of IVERSEN, but the fact that the point correlation coefficient has a well-defined expectation where the species are uncorrelated seems to make the slight extra effort in computing it worth while.

Where quantitative data are being used, expression (23) has the advantage of simplicity, but the product-moment correlation (24) may again be preferable because of its known zero expectation if the species are uncorrelated. (Further note: Product-moment correlation is most appropriate when positive and negative values express direction of slope of monotonic distributions of species in relation to the environmental gradients represented in a set of samples. The wider the range of environments and communities sampled, the more the curvilinear and non-monotonic features of species distributions may affect the measurement, and the less interpretable may be 'positive,' 'neutral,' and 'negative' correlations. The centred measures of species association and correlation should in general be most useful for narrow ranges of sample difference; for wide ranges of samples they may express relative distributional overlap of species less effectively than the non-centred measures — the Editor).

In selecting an index of similarity between stands similar considerations apply. Presence-and-absence indices (coefficients of community) like those of JACCARD (39) and SØRENSEN (40) are less affected by sample size than the interspecific indices mentioned above. The influence of sample size depends here on the form of the species-area curve in the two samples compared; if the proportion of common species is the same in different parts of the curve, then the value of the similarity index will be almost independent of sample size. Broadly, however, presence-and-absence indices should be chosen only if samples of the same size are to be compared. Since no special intuitive significance attaches to a null similarity (as it does to a null correlation), there seems little reason to prefer a centred index of similarity, and the SØRENSEN index has the merit of simplicity.

Where similarity is considered inadequately represented by species presence and absence, or where samples of different size are to be compared, an index based on measures of the quantities of the species present is to be preferred. Percentage similarity is an appropriate and simple index, either in its absolute form (50), or in the relative form (51) preferred by some authors – the choice depending on whether the total quantity of plant material present is regarded as a relevant consideration in assessing the similarity of two samples. Euclidean distance (62) is an alternative appropriate to the assumptions of some ordination techniques. Sample similarity measures are further discussed elsewhere in this volume – in their mathematical aspects in article 10, their effects on ordination performance in article 11.

If, for some special reason, significance tests are required, the choice of index will be different. For presence-and-absence data, interspecific association may be examined by the tests for 2×2 tables, either exact (3) or approximate (1); for quantitative data, the rank correlation coefficient seems best. For similarity, the probabilistic index (GOODALL 1966) may be used with either presence-and-absence or quantitative data.

It is admitted that indices not mentioned is this sub-section may have advantages, at least in some circumstances. No clear advantages have been demonstrated, though, and until this has happened the choice should go to the simplest index which appears to meet the particular need.

6.5.1 SUMMARY

Similarity of distribution between species, and similarity in composition between samples, have much in common; they are in a sense opposite faces of the same coin. Both are of use in the ordination and classification of vegetation, and also for testing homogeneity of samples or sets of samples. Statistical tests for significance of differences can be applied to species distributions and composition of samples, but such tests are not usually appropriate. The problem is usually one of measuring relative similarity, rather than testing a null hypothesis.

A range of expressions are reviewed which espress relative similarity of species distributions. Some of these measure species association in terms of presence or absence in samples, with or without taking into account samples from which both species are absent. Other expressions of species correlation are based on quantitative measures of the contributions of two species to the different samples in which they occur. Measures of both types may be either non-centred (ranging from 1.0 to 0), or centred (ranging from +1.0 through 0 to −1.0 for positive, neutral, and negative distributional relationships). The measures can also be applied to groups of more than two species, and they then link with the concept of 'ecological groups'.

Various expressions of relative similarity of vegetation samples are also discussed. Some of these are based on presence and absence of species in samples; others are based on quantitative measures of the contribution of each species to the samples being compared. In both types of index, a distinction is drawn between measures based wholly on characteristics of a given pair of samples being compared, and measures which relate a given pair to one another within the context of a broader set of samples. Complements of sample similarity express relative difference of samples from one another in respect of floristic composition. The concept of similarity can be applied to sets of samples as well as to individual samples, and can be extended to relations between more than two entities, whether these entities are individual samples or sets of samples. Here similarity measurement impinges directly on the classification of vegetation.

Various merits and limitations affect the different similarity measures. Choices of measures may differ with research purposes and kinds of samples, but the following are suggestions of indices of wide utility, with abbreviations used in other articles and the numbers of the formulas in this article:

i) Species distributional similarity
  a) Species association (SA), from presence and absence data
    Percentage co-occurrence (4), non-centred
    Coefficient of point correlation (15), centred
  b) Species correlation (SC), from quantitative data
    Percentage similarity of distribution (23), non-centred
    Coefficient of correlation (24) or rank correlation, centred

ii) Sample similarity of composition
  a) Presence and absence data
    Coefficient of community (CC) of Sørensen (40) and its complement, coefficient of difference (CD)

b) Quantitative data

Percentage similarity (PS), (50, 51) and its complement percentage difference (PD)

Euclidean distance (ED),(62).

6.5.2 Zusammenfassung

Die Ähnlichkeit der Verteilung zwischen Arten und die Ähnlichkeit zwischen Aufnahmen, haben vieles gemeinsam. Sie sind gewissermaßen die beiden Seiten derselben Medaille. Beide sind von Nutzen für die Ordination und für die Klassifikation der Vegetation, und auch für die Prüfung der Homogenität von Aufnahmen oder Gruppen von Aufnahmen. Statistische Prüfungen der Signifikanz von Unterschieden können für die Verteilung der Arten und die Zusammensetzung von Aufnahmen verwendet werden, aber solche Prüfungen sind gewöhnlich nicht geeignet. Gewöhnlich kommt es eher darauf an die relative Ähnlichkeit zu messen als eine Null-Hypothese zu prüfen.

Eine Reihe von Formeln wird besprochen welche die relative Ähnlichkeit der Arten-Verteilung wiedergeben. Einige dieser Formeln messen die Vergesellschaftung von Arten auf Grund ihrer Anwesenheit oder ihrer Fehlen in den Aufnahmen, mit oder ohne Berücksichtigung von Aufnahmen in denen beide Arten fehlen. Andere Formeln der Arten-Korrelation sind begründet auf quantitativen Bestimmungen von zwei Arten in verschiedenen Aufnahmen in denen sie vorkommen. Die Messungen von beiden Typen können entweder nicht-zentriert (laufend von 1.0 zu 0) oder zentriert (laufend von +1.0 über 0 zu −1.0 für positive, neutrale oder negative Beziehungen) sein. Die Messungen können auch auf Gruppen von mehr als zwei Arten angewandt und dann mit dem Begriff der 'ökologischen Gruppen' verknüpft werden.

Auch verschiedene Formeln der relativen Ähnlichkeit von Vegetationsaufnahmen werden besprochen. Einige davon beruhen auf Anwesenheit und Fehlen von Arten in Aufnahmen, andere auf quantitativen Messungen des Anteils jeder Art in den zu vergleichenden Aufnahmen. In beiden Index-Typen wird Unterschieden zwischen Messungen, die völlig auf den Merkmalen eines bestimmten Paares der vergleichnen Aufnahmen beruhen, und solchen welche ein gegebenes Paar zu einem anderen in Merkmals-Begriffen für eine größere Reihe von Aufnahmen in Beziehung setzen. Durch das Komplement der Ähnlichkeit kann die relative Abweichung von Aufnahmen als Ergebnis von Standortsunterschieden ausgedrückt werden. Der Begriff der Verwandtschaft kann sowohl auf

Gruppen von Aufnahmen als auch auf einzelne Aufnahmen angewandt werden. Es kann erweitert werden für Beziehungen zwischen mehr als zwei Einheiten, seien es einzelne oder Gruppen von Aufnahmen. Hier nähert sich die Ähnlichkeits-Bestimmung unmittelbar auf der Klassifikation der Vegetation.

Die einzelnen Ähnlichkeitsmaßstäbe haben verschiedene Werte und Grenzen. Die Wahl der Maßstäbe wird vom Untersuchungszweck und von der Art der Bestände bestimmt. Dort seien die folgenden Indices für weite Brauchbarkeit empfohlen; die Abkürzungen sind anderen Beiträgen, die Ziffer den Formeln in diesem Beitrag entnommen.

1. Ähnlichkeit in der Verteilung von Arten
   a. Verbindung von Arten (SA), nach Vorkommen und Abwesenheit
      Prozentuales gemeinsames Vorkommen (4) nicht-zentriert
      Korrelationskoeffizient (15), zentriert
   b. Korrelation zwischen Arten (SD), nach quantitativen Daten
      Prozentuale Ähnlichkeit der Verteilung (23) nicht-zentriert
      Korrelationskoeffizient (24) oder Rangkorrelation, zentriert

2. Ähnlichkeit der Zusammensetzung von Aufnahmen
   a. Vorkommen und Abwesenheit
      Gemeinschaftskoeffizient (CC) nach Sørensen (40) und sein Komplement, Differenzkoeffizient (CD)
   b. Quantitative Daten
      Prozentuale Ähnlichkeit (PS) (50, 51) und sein Komplement
      Prozentuale Differenz (PD)
      Euklidischer Abstand (ED) (62)

REFERENCES

Agrell, I., – 1945 – An objective method for characterization of animal and plant communities. *K. fisiogr. Sällsk. Lund Förh.* 15: 63—67.

Barkman, J. J., – 1958 – Phytosociology and Ecology of Cryptogamic Epiphytes, Including a Taxonomic Survey and Description of their Vegetation Units in Europe. Van Gorcum, Assen. 628 pp.

Barnard, J. L., – 1964 – Marine Amphipoda of Bahia de San Quintin, Baja California. *Pacif. Nat.* 4: 55—139.

Beals, E. W., – 1965a – Ordination of some corticolous cryptogamic communities in south-central Wisconsin. (Russ. summ.) *Oikos* 16: 1—8.

Beals, E. W., – 1965b – Species patterns in a Lebanese Poterietum. *Vegetatio* 13: 69—87.

Beals, E. W. & G. Cottam, – 1960 – The forest vegetation of the Apostle Islands, Wisconsin. *Ecology* 41: 785—790.

Bourdeau, P., – 1961 – L'outil statistique en écologie et sociologie végétales *Biométr.-Praxim.* 2: 193—216.

Bray, J. R., – 1956 – A study of mutual occurrence of plant species. *Ecology* 37: 21—28.

Bray, J. R. & J. T. Curtis, – 1957 – An ordination of the upland forest communities of southern Wisconsin. *Ecol. Monogr.* 27: 325—349.

Byer, M. D., – 1970 – Interspecific plant association as influenced by three variables. *Ecology* 51: 103—112.

Cancela da Fonseca, J. P., – 1966 – L'outil statistique en biologie du sol. III. Indices d'intérêt écologique. *Revue Écol. Biol. Sol* 3: 381—407.

Cassagnau, P., – 1961 – Écologie du sol dans les Pyrénées centrales: Les biocénoses de Collemboles. Hermann, Paris. 235 pp.

Češka, A., – 1966 – Estimation of the mean floristic similarity between and within sets of vegetation relevés. *Folia geobot. phytotax. bohemoslovaca* 1: 93—100.

Češka, A., – 1968 – Application of association coefficients for estimating the mean similarity between sets of vegetational relevés. *Folia geobot. phytotax. bohemoslovaca* 3: 57—64.

Clausen, J. J., – 1957 – A phytosociological ordination of the conifer swamps of Wisconsin. *Ecology* 38: 638—646.

Cole, L. C., – 1949 – The measurement of interspecific association. *Ecology.* 39: 411—424.

Cole, L. C., – 1957 – The measurement of partial interspecific association. *Ecology* 38: 226—233.

Crawford, R. M. M., D. Wishart, & R. M. Campbell, – 1970 – A numerical analysis of high altitude scrub vegetation in relation to soil erosion in the Eastern Cordillera of Peru. *J. Ecol.* 58: 173—191.

Culberson, W. L., – 1955 – The corticolous communities of lichens and bryophytes in the upland forests of northern Wisconsin. *Ecol. Monogr.* 25: 215—231.

Czekanowski, J., – 1909 – Zur differential Diagnose der Neandertalgruppe. *Korrespbl. dt. Ges. Anthrop.* 40: 44—47.

Dagnelie, P., – 1960a – Contribution à l'étude des communautés végétales par l'analyse factorielle. (Engl. summ.) *Bull. Serv. Carte phytogéogr.* Sér. B. 5: 7—71, 93—195.

Dagnelie, P., – 1960b – Quelques problèmes statistiques posés par l'utilisation de l'analyse factorielle en phytosociologie. *Gembloux Inst. Agron. Stat. Rech., Bull. Hors Sér.* 1: 430—437.

Dagnelie, P., – 1962a – Étude statistique d'une pelouse à *Brachypodium ramosum.* III. les liaisons interspécifiques. Première partie. *Bull. Serv. Carte phytogéogr.* Sér. B, 7: 87—97.

Dagnelie, P., – 1962b – Étude statistique d'une pelouse à Brachypodium ramosum. V. Les liaisons interspécifiques. Deuxième partie. *Bull. Serv. Carte phytogéogr.* Sér. B, 7: 149—160.

Dagnelie, P., – 1965 – L'étude des communautés végétales par l'analyse statistique des liaisons entre les espèces et les variables écologiques: principes fondamentaux. *Biometrics* 21: 345—361.

Dahl, E., – 1957 – Rondane. Mountain vegetation in South Norway and its relation to the environment. *Skr. norske Videns-Akad.*, Mat.-naturw. Kl. 1956 (3): 1—374.

Dahl, E. & E. Hadač – 1941 – Strandgesellschaften der Insel Ostøy im Oslofjord. *Nytt Mag. Naturv.* 82: 251—312.

Dawson, G. W. P., – 1951 – A method for investigating the relationship between the distribution of individuals of different species in a plant community. *Ecology* 32: 332—334.

Dice, L. R., – 1945 – Measures of the amount of ecologic association between species. *Ecology* 26: 297—302.

Dice, L. R., – 1952 – Measure of the spacing between individuals within a population. *Contrib. Lab. vertebr. Biol. Univ. Mich.* 55: 1—23.

DUVIGNEAUD, P., – 1946 – La variabilité des associations végétales. *Bull. Soc. Bot. Belg.* 78: 107—134.

DUVIGNEAUD, P., – 1953 – Les savanes du Bas-Congo. Essai de phytosociologie topographique. *Lejeunia* 10: 1—92.

ELLENBERG, H., – 1950 – Landwirtschaftliche Pflanzensoziologie. I. Unkrautgemeinschaften als Zeiger für Klima und Boden. Ulmer, Stuttgart. 141 pp.

ELLENBERG, H., – 1956 – In, 'Einführung in die Phytologie,' by H. WALTER, Vol. 4, Part I. Aufgaben und Methoden der Vegetationskunde. Ulmer, Stuttgart. 136 pp.

ESTABROOK, F. F., – 1967 – An information theory model for character analysis. *Taxon* 16: 86—97.

FAGER, E. W., – 1957 – Determination and analysis of recurrent groups. *Ecology* 38: 586—595.

FAGER, E. W., – 1963 – Communities of organisms. In, 'The Sea', ed. M. N. HILL. Vol. 2: 415—437. Interscience, New York & London.

FALINSKI, J. B., – 1958 – Nomogramy i tablice współczynników podobieństwa między zdjęciami fitosocjologicznymi według wzoru Jaccarda i Steinhaus (Engl. summ.: Nomograms and tables for similarity coefficients). *Acta Soc. bot. pol.* 27: 115—130.

FIELD, G., – 1969 – The use of the information statistic in the numerical classification of heterogeneous systems. *J. Ecol.* 57: 565—569.

FIJALKOWSKI, D., – 1954 – Szata roślinna wawozów okolie Lublina na tle niektorych warunków siedliskowych. (Russ. & Engl. summs.: Vegetation of loess ravines near Lublin on the background of some environmental conditions) *Annls Univ. Mariae Curie-Skłodowska* Sect. B, 9: 125—215.

FORBES, S. A., – 1907 – On the local distribution of certain Illinois fishes: an essay in statistical ecology. *Bull. Ill. St. Lab. nat. Hist.* 7: 273—303.

FORBES, S. A., – 1925 – Method of determining and measuring the associative relations of species. *Science, N. Y.* 61: 524.

FRESCO, L. F. M., – 1969 – Factor analysis as a method in synecological research. *Acta bot. neerl.* 18: 477—482.

FREY, T. E. A., – 1963 – On the application of the correlation coefficient of the species interrelations (In Russian). *Bot. Zhur. SSSR* 48: 235—239.

FRYDMAN, I. & R. H. WHITTAKER, – 1968 – Forest associations of southeast Lublin Province, Poland. (Germ. summ.) *Ecology* 49: 896—908.

GARDNER, J. L., – 1951 – Vegetation of the creosotebush area of the Rio Grande Valley in New Mexico. *Ecol. Monogr.* 21: 379—403.

GHENT, A. W., – 1963 – Kendall's 'Tau' coefficient as an index of similarity in comparisons of plant or animal communities. *Can. Ent.* 95: 568—574.

GILBERT, N., – 1963 – Non-additive combining abilities. *Genet. Res.* 4: 65—78.

GILBERT, N. & T. C. E. WELLS, – 1966 – Analysis of quadrat data. *J. Ecol.* 54: 675—685.

GLEASON, H. A., – 1920 – Some applications of the quadrat method. *Bull. Torrey bot. Club* 47: 21—33.

GODRON. M., – 1966 – Application de la théorie de l'information à l'étude de l'homogénéité et de la structure de la végétation. *Oecol. Pl.* 1: 187—197.

GOODALL, D. W., – 1952 – Quantitative aspects of plant distribution. *Biol. Rev.* 27: 194—245.

GOODALL, D. W., – 1953 – Objective methods for the classification of vegetation. I. The use of positive interspecific correlation. *Aust. J. Bot.* 1: 39—63.

GOODALL, D. W., – 1961 – Objective methods for the classification of vegetation IV. Pattern and minimal area. *Aust. J. Bot.* 9: 162—196.

GOODALL, D. W., – 1965 – Plot-less tests of interspecific association. *J. Ecol.* 53: 197—210.
GOODALL, D. W., – 1966 – A new similarity index based on probability. *Biometrics* 22: 883—907.
GOODALL, D. W., – 1967 – The distribution of the matching coefficient. (French summ.) *Biometrics* 23: 647—656.
GOODALL, D. W., – 1969 – Procedure for recognition of uncommon species combinations in sets of vegetation samples. *Vegetatio* 18: 19—35.
GOODMAN, L. A. & W. H. KRUSKAL, – 1954 – Measures of association for cross-classifications. *J. Am. statist. Ass.* 49: 732—764.
GOODMAN, L. A. & W. H. KRUSKAL, – 1959 – Measures of association for cross-classifications. II. Further discussion and references. *J. Am. statist. Ass.* 54: 123—163.
GOUNOT, M., – 1961 – Les méthodes d'inventaire de la végétation. *Bull. Serv. Carte phytogéogr.*, Sér. B, 6: 7—73.
GOUNOT, M., – 1969 – Méthodes d'étude quantitative de la végétation. Masson, Paris. 314 pp.
GOUNOT, M. & M. CALLEJA, – 1962 – Coefficient de communauté, homogénéité et aire minimale. *Bull. Serv. Carte phytogéogr.* Sér. B, 7: 181—200.
GREIG-SMITH, P., – 1952 – Ecological observations on degraded and secondary forest in Trinidad, British West Indies. II. Structure of the communities. *J. Ecol.* 40: 316—330.
GREIG-SMITH, P., – 1964 – Quantitative Plant Ecology. 2nd ed. Butterworths, London. 256 pp.
HACKER, H. P., – 1921 – On the distribution of anophelines in the Brickfields Road area of Kuala Lumpur, November, 1917. *Rep. Fed. Malay States Malaria Bur.* 2: 3—28.
HALL, A. V., – 1967a – Methods for demonstrating resemblance in taxonomy and ecology. *Nature, Lond.* 214: 830—831.
HALL, A. V., – 1967b – Studies in recently developed group-forming procedures in taxonomy and ecology. *Jl S. Afr. Bot.* 33: 185—196.
HALL, A. V., – 1970 – A computer-based method for showing continua and communities in ecology. *J. Ecol.* 58: 591—602.
HARBERD, D. J., – 1962 – Application of a multivariate technique to ecological survey. *J. Ecol.* 50: 1—17.
HOPKINS, B., – 1957 – Pattern in the plant community. *J. Ecol.* 45: 451—463.
HUGHES, R. E., – 1954 – The application of multivariate analysis to (a) problems of classification in ecology; (b) the study of the interrelationships of the plant community and environment. *8th Int. bot. Congr., Paris, Rapp. Commis.* Sect. 7 & 8: 16—18.
HUGHES, R. E. & D. V. LINDLEY, – 1955 – Application of biometric methods to problems of classification in ecology. *Nature, Lond.* 175: 806—807.
HURLBERT, S. H., – 1969 – A coefficient of interspecific association. *Ecology* 50: 1—9.
IVERSEN, J., – 1936 – Biologische Pflanzentypen als Hilfsmittel in der Vegetationforschung. Ein Beitrag zur ökologischen Charakterisierung und Anordnung der Pflanzengesellschaften. *Medd. Skalling-Lab.* 4: 1—224.
IVERSEN, J., – 1954 – Über die Korrelationen zwischen den Pflanzenarten in einem grönländischen Talgebiet. *Vegetatio* 5—6: 238—246.
JACCARD, P., – 1901 – Distribution de la flore alpine dans le Bassin des Dranses et dans quelques régions voisines. *Bull. Soc. vaud. Sci. nat.* 37: 241—272.
JUHÁSZ NAGY, P., – 1964 – Continuum studies on meadow vegetation. (Russ. summ.) *Acta bot. hung.* 10: 159—173.

KASSAS, M. & M. S. EL-ABYAD, – 1962/3 – On the phytosociology of the desert vegetation of Egypt. *Ann. arid Zone* 1: 54—83.
KENDALL, M. G., – 1955 – Rank Correlation Methods. 2nd ed. C. Griffin, London. 196 pp.
KERSHAW, K. A., – 1961 – Association and co-variance analysis of plant communities. *J. Ecol.* 49: 643—654.
KLEINER, E. F. & K. T. HARPER, – 1966 – An investigation of association patterns of prevalent grassland species in Red Butte Canyon, Salt Lake County, Utah. *Proc. Utah Acad. Sci.* 43: 29—36.
KOLBE, W., – 1956 – Korrelationsstatistische Probleme der Grünlandsoziologie. *Inaug-Diss, Landw. Fak., Rheinischen Friedrich-Wilhelms-Universität, Bonn.* 91 pp.
KONTKANEN, P., – 1950 – Quantitative and seasonal studies on the leafhopper fauna of the field stratum on open areas in North Karelia. (Finn. summ.) *Annls zool., Soc. Zool.-bot. fenn. Vanamo* (*Suomal. eläin-ja kasvit. Seur. van. eläin. Julk.*) 13 (8): 1—91.
KULCZYŃSKI, St., – 1928 – Die Pflanzenassoziationen der Pieninen. *Bull. int. Acad. pol. Sci. Lett.*, Cl. Sci. Math. Nat., sér. B. 1927 (Suppl. 2): 57—203.
LAMBERT, J. M. & M. B. DALE, – 1964 – The use of statistics in phytosociology. *Adv. ecol. Res.* 2: 59—99.
LOOMAN, J. & J. B. CAMPBELL, – 1960 – Adaptation of Sorensen's K (1948) for estimating unit affinities in prairie vegetation. *Ecology* 41: 409—416.
McDONOUGH, W. T., – 1963 – Interspecific association among desert plants. *Amer. Midl. Nat.* 70: 291—299.
McINTOSH, R. P., – 1962 – Pattern in a forest community. *Ecology* 43: 25—33.
MAAREL, E. VAN DER, – 1966 – Over vegetatiestructuren, -relaties en -systemen in het bijzonder in de duingraslanden van Voorne. *Thesis, Univ. Utrecht.* 170 pp.
MAAREL, E. VAN DER, – 1969 – On the use of ordination models in phytosociology. (Germ. summ.) *Vegetatio* 19: 21—46.
MAHALANOBIS, P. C., – 1936 – On the generalized distance in statistics. *Proc. natn. Inst. Sci. India* 2: 49—55.
MARGALEF, T., – 1958 – Distribución de los crustaceos en las aguas continentales españolas. Grado de associación entre las especies en relación con factores ecologicos e historicos. *Publnes Inst. Biol. apl., Barcelona* 27: 17—31.
MICHAEL, E. L., – 1920 – Marine ecology and the coefficient of association: a plea in behalf of quantitative biology. *Ecology* 8: 54—59.
MIYATA, I., – 1966 – Ecological studies on the vegetation of Akiyoshi-dai limestone plateau (2) Structure of the reserved forest developed on the upper slopes of Akiyoshi-do cave. *Jap. J. Ecol.* 16: 241—247.
MONK, C. D., – 1968 – Successional and environmental relationships of the forest vegetation of North Central Florida. *Amer. Midl. Nat.* 79: 441—457.
MORISITA, M., – 1959 – Measuring of interspecific association and similarity between communities. *Mem. Fac. Sci. Kyushu Univ.*, Ser. E (Biol) 3: 65—80.
MOTOMURA, I., – 1952 – Comparison of communities based on correlation coefficients (In Japanese). *Ecol. Rev.*, Sendai 13: 67—71.
MOZLEY, A., – 1936 – The statistical analysis of the distribution of pond molluscs in Western Canada. *Am. Nat.* 70: 237—244.
NEWBOULD, P. J., – 1960 – The ecology of Cranesmoor, a New Forest valley bog. I. The present vegetation. *J. Ecol.* 48: 361—383.
NORDHAGEN, R., – 1943 – Sikilsdalen og Norges fjellbeiter. En plantesosiologisk monografi. *Bergens Mus. Skr.* 22: 1—607.
NUMATA, M. & M. MITSUDERA, – 1960 – Discrimination of the rank of species in a plant community. *Biol. J. Nara Women's Univ.* 19: 57—62.

Ochiai, A., – 1957 – Zoogeographical studies on the soleoid fishes found in Japan and its neighbouring regions. *Bull. Jap. Soc. scient. Fish.* 22: 526—530.
Odum, E. P., – 1950 – Bird populations of the Highlands (North Carolina) Plateau in relation to plant succession and avian invasion. *Ecology* 31: 587—605.
Orloci, L., – 1966 – Geometric models in ecology. I. The theory and application of some ordination methods. *J. Ecol.* 54: 193—215.
Orloci, L., – 1967 – An agglomerative method for classification of plant communities. *J. Ecol.* 55: 193—206.
Orloci, L., – 1968 – Definitions of structure in multivariate phytosociological samples. *Vegetatio* 15: 281—291.
Pandeya, S. A., – 1961 – On some new concepts in phytosociological studies of grassland. II. Community coefficient (F×C) ICC. *J. Indian bot. Soc.* 40: 267—270.
Pielou, D. P. & E. C. Pielou, – 1967 – The detection of different degrees of coexistence. *J. theoret. Biol.* 16: 427—437.
Pielou, E. C., – 1961 – Segregation and symmetry in two-species populations as studied by nearest-neighbour relationships. *J. Ecol.* 49: 255—269.
Pielou, E. C., – 1969a – Association tests versus homogeneity tests: their use in subdividing quadrats into groups. *Vegetatio* 18: 4—18.
Pielou, E. C., – 1969b – An Introduction to Mathematical Ecology. Wiley-Interscience, New York. 286 pp.
Raabe, E. W., – 1950 – Über die 'Charakteristische Arten-Kombination' in der Pflanzensoziologie. *Schr. naturw. Ver. Schlesw.-Holst.* 24: 8—14.
Ramsay, D. M., – 1964a – An analysis of Nigerian savanna. II. An alternative method of analysis and its application to the Gombe sandstone vegetation. *J. Ecol.* 52: 457—466.
Ramsay, D. M., – 1964b – Some recent work on savanna vegetation-soil relationships. *Proc. 1st Nigerian For. Conf., Kaduna*, 156—159.
Rao, C. R., – 1952 – Advanced Statistical Methods in Biometric Research. John Wiley, New York; Chapman & Hall, London. 390 pp.
Raunkiaer, C., – 1928 – Dominansareal, Artstaethed og Formationsdominanter. *Biol. Meddr.*, K. dansk Vidensk. Selsk. 7 (1) :1—47.
Renkonen, O., – 1938 – Statistisch-ökologische Untersuchungen über die terrestrische Käferwelt der finnischen Bruchmoore. (Finn. summ.) *Annls. zool., Soc. Zool. -bot. Fenn. Vanamo (Suomal. eläin-ja kasvit. Seur. van. eläin. Julk.)* 6 (1): 1—231.
Renkonen, O., – 1944 – Die Carabiden- und Staphylinidenbestände eines Seeufers in SW Finnland. Ein Beitrag zur Theorie der statistischen Insektensynökologie. *Ann. ent. fenn. (Suom. hyont. Aikak.)* 10: 33—104.
Russell, P. F. & T. R. Rao, – 1940 – On habitat and association of species of anopheline larvae in south-eastern Madras. *J. Malar. Inst. India* 3: 153—178.
Ružička, M., – 1958 – Anwendung mathematisch-statistischer Methoden in der Geobotanik (Synthetische Bearbeitung von Aufnahmen). *Biológia, Bratisl.* 13: 647—661.
Sokal, R. R. & C. D. Michener, – 1958 – A statistical method for evaluating systematic relationships. *Kans. Univ. Sci. Bull.* 38: 1409—1437.
Sokal, R. R. & P. H. A. Sneath, – 1963 – Principles of Numerical Taxonomy. Freeman, San Francisco & London. 359 pp.
Sørensen, T., – 1948 – A method of establishing groups of equal amplitude in plant sociology based on similarity of species content. *Biol. Skr..*, K. danske Vidensk. Selsk. 5 (4): 1—34.
Tarwid, K., – 1960 – Szacowanie zbierznośći nisz ekologicznych gatunków droga oceny prawdobieństwa spotykania sie ich w połowach (Engl. summ.:

Assessment of the coincidence of ecological niches of species by estimating the probability of their concurrence in samples). *Ekol. pol.*, Ser. B, 6: 115—130.
VASILEVICH, V. I., – 1963 – A tentative morphological analysis of the meadow continuum (In Russian), *Bot. Zhur. SSSR.* 48: 1653—1659.
VASILEVICH, V. I., – 1969 – Statistical Methods in Geobotany (In Russian). Nauka, Leningrad. 231 pp.
VASILEVICH, V. I., – 1970 – Partial linkages among species in a meadow community (In Russian). *Dokl. Akad. Nauk SSSR* 190: 1246—1249. *Dokl. Bot. Sci.* 190—192: 5—7.
VRIES, D. M. DE, – 1953 – Objective combinations of species. *Acta bot. neerl.* 1: 497—499.
VRIES, D. M. DE, – 1954 – Ecological results obtained by the use of interspecific correlation. *European Grassl. Conf.*, Paris, paper 224/53, 11 pp.
VRIES, D. M. DE, J. P. BARETTA, & G. HAMMING, – 1954 – Constellation of frequent herbage plants based on their correlation in occurrence. *Vegetatio* 5: 105—111.
WELTEN, M. & O. HEGG, – 1958 – Vorschlag für die statistiche Erfassung pflanzensoziologischer Zusammenhange. *Verh. schweiz. naturf. Ges.* 1958: 127—129.
WHITTAKER, R. H., – 1952 – A study of summer foliage insect communities in the Great Smoky Mountains. *Ecol. Monogr.* 22: 1—44.
WHITTAKER, R. H., – 1960 – Vegetation of the Siskiyou Mountains, Oregon and California. *Ecol. Monogr.* 30: 279—338.
WHITTAKER, R. H., – 1967 – Gradient analysis of vegetation. *Biol. Rev.* 42: 207—264.
WHITTAKER, R. H. & C. W. FAIRBANKS, – 1958 – A study of plankton copepod communities in the Columbia Basin, southeastern Washington. *Ecology* 39: 46—65.
WILLIAMS, C. B., – 1944 – Some applications of the logarithmic series and the index of diversity to ecological problems. *J. Ecol.* 32: 1—44.
WILLIAMS, C. B., – 1947a – The logarithmic series and the comparison of island floras. *Proc. Linn. Soc. Lond.* 158: 104—108.
WILLIAMS, C. B., – 1947b – The logarithmic series and its application to biological problems. *J. Ecol.* 34: 253—272.
WILLIAMS, C. B., – 1949 – Jaccard's generic coefficient and coefficient of floral community in relation to the logarithmic series and the index of diversity. *Ann. Bot.*, N. S. 13: 53—58.
WILLIAMS, C. B., – 1964 – Patterns in the Balance of Nature and Related problems in Quantitative Ecology. Academic Press, London & New York. 324 pp.
WILLIAMS, W. T. & J. M. LAMBERT, – 1961 – Multivariate methods in plant ecology. III. Inverse association-analysis. *J. Ecol.* 49: 717—729.
WILLIAMS, W. T., J. M. LAMBERT, & G. N. LANCE, – 1966 – Multivariate methods in plant ecology. V. Similarity analyses and information-analysis. *J. Ecol.* 54, 427—445.
YARRANTON, G. A., – 1966 – A plotless method of sampling vegetation. *J. Ecol.* 54: 229—237.
YATES, F., – 1934 – Contingency tables involving small numbers and the χ^2 test. *Jl. R. statist. Soc., Suppl.* 1: 217—235.

7 MATRIX AND PLEXUS TECHNIQUES

ROBERT P. MC INTOSH

Contents

7 MATRIX AND PLEXUS TECHNIQUES

7.1 Introduction

For several decades ecologists have concerned themselves with devising intelligible and mutually compatible methods of describing, analyzing and synthesizing the often bewildering complexity of plant and animal communities. GAMS (1961) and WHITTAKER (1967) consider many of these. Similar techniques have been developed or adapted, independently of ecologists, by others concerned with the analysis of character variation in diverse objects of study. CROVELLO (1970) has provided an enlightening and useful compendium and codification of the many methods of character variation analysis which have been used in ecology and systematics. This paper is concerned with some of the relatively simple techniques of arrangement and graphic representation which may reveal relationships between community samples (and their species).

7.2 Primary and Secondary Matrices

7.2.1 COMMUNITY TABLES

Phytosociologists traditionally have worked with a two-dimensional table consisting of lists of species, of measured quantities of species, or of environmental measurements, for community samples. The table may include data collected in a series of sample quadrats in one stand of a community, or in different stands of the same or of different community-types. The table thus represents some range of variation in characteristics of the samples; the range may be narrow or wide. Each column (object) of the table is a sample plot list or stand record (the relevé of the BRAUN-BLANQUET school of phytosociology). The rows (characters) correspond either to species or to environmental characteristics. The phytosociological table forms a primary matrix from which the ecologist hopes to synthesize a description of the pattern of variation in nature and, subsequently, to explain it. Description of pattern may be approached through recognition of groups of similar samples (classification), or directions (gradients or axes) of variation interrelating the samples (ordination).

The traditional uses of such tables in various schools of ecology are reviewed by WHITTAKER (1962), while BECKING (1957) and ELLENBERG (1956) give accounts of the methods of the most widely known of these, the BRAUN-BLANQUET school (cf. BRAUN-BLANQUET 1951, POORE 1955, DAGNELIE 1961, and article 20 in this volume). The basic approach (used also by other disciplines with multiple lists to order) is to rearrange the sequence of the plot lists or relevés, both horizontally and vertically, to form blocks of characteristic or differentiating species (or combinations of species and site characteristics) which best distinguish vegetational units by their occurrence in many or all of the samples representing a unit. In the process, some plots may be excluded. The final arrangement commonly is subjective, based upon the experience of the phytosociologist. Two major difficulties of such approaches to interpreting the primary matrix are that the arrangements of the matrix may, and commonly do, differ from botanist to botanist and that, in some instances, the stands sampled or relevés included in the primary matrix are subjectively selected by some predetermined criterion. Visual rearrangements of a primary matrix of taxonomic characters have been used by systematists in similar ways (SOKAL & SNEATH 1963, CARMICHAEL & SNEATH 1969). Whether such tables are used casually, as many ecologists have done, or in the more formalized manner of the BRAUN-BLANQUET school, a major problem is the specification of the characters (characteristic, constant, diagnostic, or dominant species) or other attributes of the objects (plot lists or relevés) which are to be used to differentiate the community-types in the primary matrix (WHITTAKER 1962, POORE 1955). MOORE et al. (1967) developed a computer method for objectively arranging phytosociological tables of presence-absence data which was designed to minimize the presence-absence exchanges and to provide an objective sorting of the relevés.

7.2.2 SIMILARITY INDICES

The difficulty in comparing visually with one another numerous lists representing different sample plots or communities has led to use of numerical indices. There are several such numerical coefficients or indices which attempt to measure objectively the relative similarity (affinity, association, correlation, or conversely, dissimilarity, difference, distance) between pairs of samples or communities as columns of the primary matrix, or between species or environmental characteristics as rows of the primary matrix. These numerical coefficients are not unique to ecology, and similar coefficients based on relative overall similarity have been developed and used in systematics, psychology and other disciplines. SOKAL & SNEATH (1963) and BOYCE (1969) reviewed various coefficients used for assessing phenetic similarity in numerical taxonomy and those (some identical) used in ecology are discussed by COLE (1949, 1957), CLAUSEN (1957a), BECKING (1957), DAHL (1957), MORISITA (1959), DAGNELIE (1960), GREIG-SMITH (1964), WHITTAKER (1967), ČEŠKA (1968), HURLBERT (1969), and GHENT (1972).

ORLOCI (1972) compares distance, probability, and information coefficients and provides arithmetic examples of their calculation. Among the numerous coefficients which have been used are coefficients of community (CC), percentage similarity (PS), Euclidean distance (ED), species association using presence and absence data (SA), and species correlation using quantitative data (SC). GOODALL reviews these in the preceeding article of this volume so they are not discussed or evaluated here. It is important to observe, however, that different coefficients differently assess the relationships of samples or of species to one another, and that choice of the coefficient may influence the results of the study. There is often advantage in seeing how more than one coefficient expresses the relationships; in particular it may be desirable to compare results with coefficients using quantitative data or presence and absence only (PS or ED vs. CC, and SC vs. SA).

7.2.3 SECONDARY MATRICES

The similarity coefficient, in effect, reduces the comparison between two samples or species to a single numerical value or to a point in a multidimensional character space. Similarity coefficients (or their complements, distance measures) may be calculated between all pairs of objects (sample plots, relevés, community-types), or between all pairs of characters (species, environmental characteristics) in the primary data matrix. The similarity coefficients are arranged in new data arrays in tables we may term here secondary matrices. Two secondary matrices (a 'Q' matrix of similarities between samples as objects, and an 'R' matrix of similarities between species or environmental characters) may be derived from a given community table as primary matrix, and similar kinds of analyses can be applied to the two secondary matrices.

The matrix may be used in various ways to elucidate the relations between the objects (or characters), since it contains a similarity value for each object of the original matrix with every other object. The problem of the ecologist is to find order (structure, pattern) in the matrix representative of the order in the natural system from which it was abstracted. He seeks to identify orders or clusters of the samples, communities, species or environmental attributes which reveal a pattern of relationships among them. The secondary matrix of similarity coefficients is the basic substrate from which many different techniques attempt to extract relationships (Mc Intosh 1967b, Whittaker 1967, Crovello 1970). In this paper we shall consider further only techniques based on rearrangements of the secondary matrix, and 'plexus' methods using graphical representation of relationships in the secondary matrix by points and lines in a two- or three-dimensional space.

7.3 Matrix Arrangement

Unless the object or characters of the primary data matrix were ordered in some way, there will be no evident pattern in the secondary similarity matrix.

In some cases the matrix may be simply inspected and the incidence and degree of significant similarities, or their absence, noted (Greig-Smith 1952, Whittaker 1952, Shanks 1953, Jones 1955, Préscényi 1958, Kershaw 1959, 1960, Ono 1961, Dagnelie 1962, Cook & Hurst 1963, Mc Donough 1963, Beaman & Andresen 1966). In the effort to clarify relationships the ecologist may choose one of several additional ways of examining or reordering the secondary matrix.

7.3.1 Sample and Species Ordering

Sanders & Hessler (1969) ordered a matrix of faunal similarity coefficients (CC) on an external criterion, latitude, (Fig. 1) using a common method of differential shading of similarity classes to make for easy visualization of the pattern. They subjectively interpreted the diagonal pattern from upper left to lower right as an indication that high faunal indices are always with neighboring stations and that there is a gradual and continuous change along the transect.

A more familiar method is to rearrange the secondary matrix on internal criteria, by trial and error, to bring the highest similarity values nearest the diagonal. This method, essentially an ordering of the matrix, was apparently originated independently by Polish

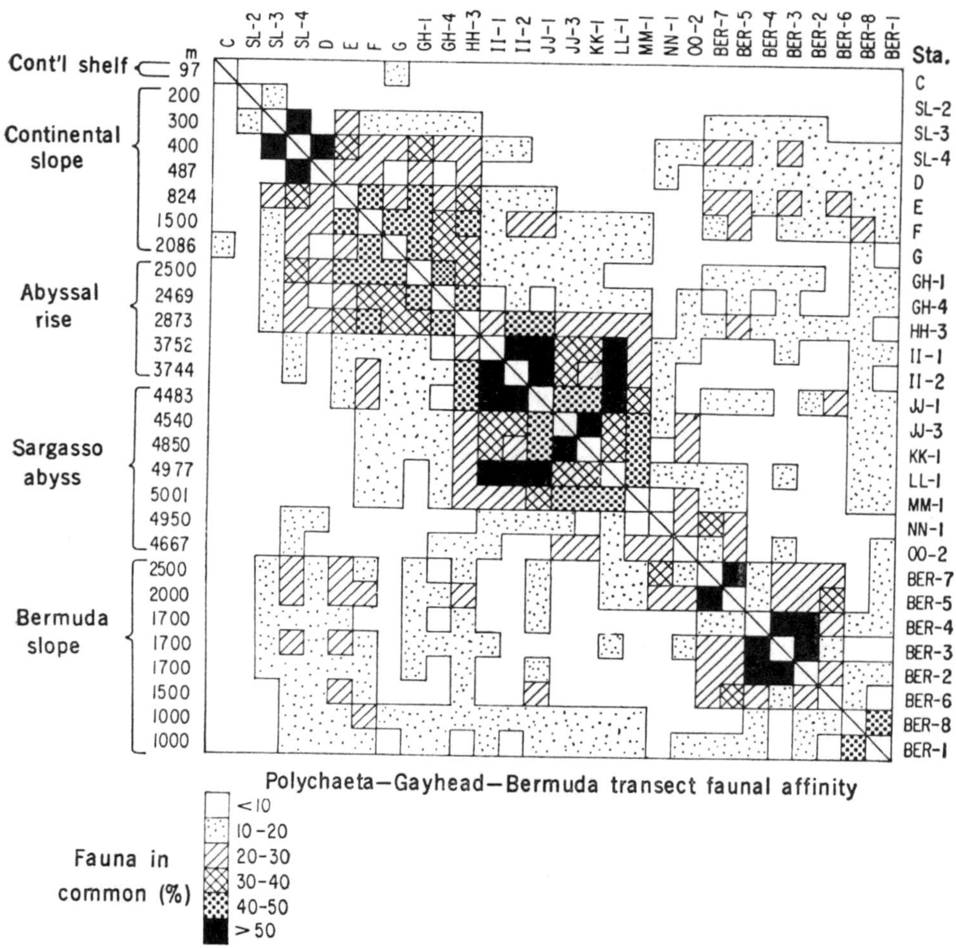

Fig. 1. A matrix diagram showing faunal similarity among sampling stations for marine bottom communities at different latitudes on a north-south axis (SANDERS & HESSLER 1969, Fig. 3).

and Danish ecologists and has been widely used by ecologists, systematists and psychologists. An extreme individual sample of a matrix is located on one end of the sequence and the sample having the highest similarity measure with it is located next to it, the third having maximum similarity to the second is then located next and the process continued up to the last sample which, presumably, has the minimum similarity value with the first. The numbers in each element of the resulting matrix may be replaced with a coded scale of shadings the highest generally being darkest (SOKAL & SNEATH

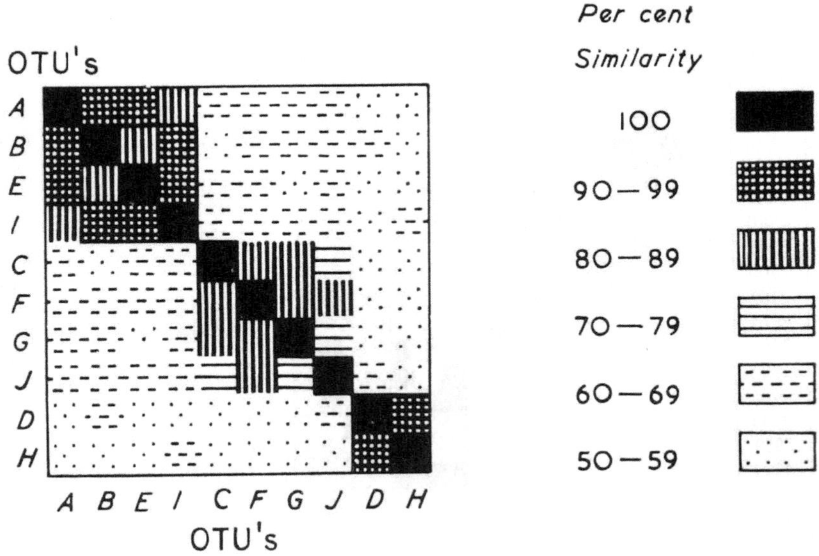

Fig. 2. A symmetric matrix diagram showing high similarity classes on the principal diagonal (SOKAL & SNEATH 1963, Fig. 7—4).

1963). By rearranging the numbers or shaded squares and trying different choices of the first sample, the matrix is rearranged until the highest values, or darkest shadings, are moved toward the diagonal placing the plots with high similarity values close to each other. Fig. 2 is an idealized symmetrical matrix produced by this method. Matrices of this type may be constructed ordering species as well as samples or stands (GREIG-SMITH 1952, GILBERT & CURTIS 1953, BRAY 1956, WHITTAKER & FAIRBANKS 1958). The ordered matrix or 'trellis diagram' has been widely used by plant ecologists; (KULCZYŃSKI 1928, MATUSZKIEWICZ 1948, SØRENSEN 1948, MOTYKA et al. 1950, GREIG-SMITH 1952, GILBERT & CURTIS 1953, GUINOCHET 1955, HALE 1955, BRAY 1956, CLAUSEN 1957a, DAHL 1957, GUINOCHET & CASAL 1957, DIX & BUTLER 1960, BRODO 1961, RUŽICKA 1961, STRUIK & CURTIS 1962, IZDEBSKI 1962, JUHÁSZ-NAGY 1963, YARRANTON 1967, YARRANTON & BEASLEIGH 1968, VASILEVICH 1969); animal ecologists, particularly entomologists, (RENKONEN 1938, 1949, MACFADYEN 1954, 1963, KONTKANEN 1957, FAGER 1957, WHITTAKER & FAIRBANKS 1958, BALOGH 1958, AGRELL 1963, HAGEMEIER 1966) and systematists (SOKAL & SNEATH 1963, MOSS 1967, TAYLOR & CAMPBELL 1969). Sometimes, especially in psychology, the matrix is arranged following a factor analysis by grouping the elements which are highly correlated with each of the principal factors in sequence. This places the high correlations along the diagonal (T. CROVELLO, pers. comm.)

Commonly, the matrix is rearranged by inspection to produce orders or clusters. Visual rearrangement of the matrix, however, is always subject to the subjective interpretation of the arranger and large matrices become very unwieldy, so a number of methods of evaluating similarity matrices in objective and repeatable fashion have been proposed (SØRENSEN 1948, RENKONEN 1949, CLAUSEN 1957a, FAGER 1957).

A simplified method, devised by J. R. BRAY, described by CLAUSEN (1957a, b) used a partial matrix of the similarity values of each of the samples (or species) of the complete matrix with the two extremes (the samples related by the lowest similarity value in the full matrix). The extreme or end-point samples are placed at the ends of a line, and each of the other elements is placed along this line by the formula

$$X = \frac{ICC + CRCC}{2}$$

Where: X= number of units along the line from the first extreme community.

ICC= the similarity coefficient value of the element with the second extreme community.

$CRCC$ = maximum similarity coefficient of the incomplete matrix minus RCC.

RCC = the similarity coefficient value of the element with the first extreme community.

The samples are arranged in sequence by values of X to form an ordered matrix in which, again, high similarity values should lie on or near the diagonal.

This method suffered from the difficulty of later methods of ordination using extreme end stands that, if these stands were very different, the remainder clustered toward the center. The positions of the communities were also weighted only by their similarity with the extremes and not with each other. Despite the difficulties, this technique (or that of placing most similar samples next to one another) will in some cases produce an effective ordination of the samples along a major environmental complex-gradient. Such is most likely to be the case if one environmental gradient overshadows all others in determining variation among the samples. Whether or not the arrangement is an effective ordination, it may be of interest to observe the groups of closely related samples that may be revealed in it.

7.3.2 Sample and Species Grouping

Sorensen (1948) arranged the elements of the secondary matrix into groups such that each member resembled all others of the group at an arbitrarily selected value of the similarity coefficient (CC). The smallest (most similar) groups had coefficient values > 50. Groups so defined could be assembled into larger, less restrictive, groups with lower values of the similarity coefficient, forming a hierarchy (Sørensen 1948, Hosakawa et al. 1957). Clausen (1957a) compared the results of Bray's and Sørensen's methods, preferring the latter, and Kontkanen (1957) compared Sørensen's method with Renkonen's (1938) and also prefered Sørensen's. Looman & Campbell (1960) criticized Sørenson's method because of his choice of an arbitrary value of the coefficient for establishing groups and developed a new coefficient by combining Sørenson's with Kendall's rank correlation coefficient (*Tau*). This is used to calculate an expected value of the similarity coefficient, based on the smallest number of species in common, which would be required if two sites are to be considered similar at the .01 level of significance. They use a grouping based on this coefficient to validate the results of a conventional association-table analysis. Ramsey (1964) and Greig-Smith (1964) questioned their detailed results because they referred to a chi-square table with two degrees of freedom rather than to the one appropriate to a 2×2 table, but the method has the merit of providing a statistical measure of significance.

Fager (1957) developed a method of producing 'recurrent groups' of species from a similarity matrix (SA). He defined a recurrent group as one in which:

1. The coefficient of similarity is significant at the .05 level for all pairs of species within the group.
2. The group includes the greatest possible number of species.
3. If several groups with the same number of members are possible, those are selected which will give the greatest number of groups without members in common.
4. If two or more groups with the same number of species and with members in common are possible, the one which occurs as a unit in the greatest number of samples is chosen.

Fager & Mc Gowan (1963) identified recurrent groups of zooplankton species; Davies (1971) identified groups of algae in beach sands using Fager's method.

Dahl (1957) used matrices and clusters based on the Sørensen coefficient (CC) to verify the results of a subjective analysis of association tables according to the methods of Continental ecological schools. Results may be subject to dispute as RENKONEN (1949) points out. Different methods may agree in broad features, which suggests some basic regularity. RENKONEN also warns against blind dependence on a statistical method and urges that the whole primary data base be available so that it can be used for comparisons as new and, hopefully, better methods of analysis become available (*cf.* CROVELLO 1970). Given the expense it may be impossible to publish entire primary data matrices, but these should be deposited in appropriate repositories for the use of later investigators. SOKAL & SNEATH (1963) discussed other methods of identifying clusters in a matrix which have not been used in ecology. MOSS (1967) used a shaded matrix (or 'trellis diagram') approach to compare results of matrices derived from correlation coefficients (r) and the Euclidean distance measure at three levels of similarity. There have been few comparative studies of the several methods described and these are only the simplest and most graphic ways in which patterns may be sought in a matrix of similarity coefficients. One difficulty is that these similarity or distance coefficients may be influenced by the diversity of the communities being compared. It is possible for two stands that share all or many of their species to have a greater interstand distance than two other stands having no species in common (MC INTOSH 1967a, NEWSOME & DIX 1968, BANNISTER 1968, GHENT 1972, ORLOCI 1972).

Objective comparisons of clusters or ordinations of ecological data produced by matrix methods have not usually been attempted. CROVELLO (1969) reviews several summary statistics for objective comparison of different analyses used in numerical taxonomy. The cophenetic correlation coefficient (SOKAL & SNEATH 1963) compares the similarity coefficients and the cophenetic values which are a matrix of coded similarity values extracted from the phenogram (diagram of the clusters) constructed from the similarity matrix. A perfect correlation shows that the original matrix is faithfully reproduced by the phenogram. The correlation between two or more similarity matrices may be calculated and the correlation between two or more cophenetic value matrices may also be useful, as CROVELLO suggests, in providing additional support for visual methods of comparing similarity matrices or patterns of relationship abstracted from them.

7.4 Plexus Techniques

It is natural to use a matrix of similarity coefficients to construct a graphic representation of the relations between the objects or characters. Samples or species then become points in a spatial pattern representing relative similarity by the arrangement of these points. The points may or may not be connected by lines. Quantitative relationships of the samples may be indicated by the spacing of the points, or by making the length or thickness of the lines proportional to the degree of similarity or dissimilarity (distance). Similarity coefficients are usually transformed because their values are inverse to measures of distance; these similarities are often subtracted from 100 %, or 85 %, or some other maximum value. The diagrams are variously known in ecology as dendrites (Faliński 1960, Izdebski 1963, Vasilevich 1969); pleiades or constellations (Vries 1953, Vries et al. 1954); correlata (Juhász-Nagy 1963); plexuses (Whittaker & Fairbanks 1958, Whittaker 1967) or, more generally, graphs.

7.4.1 Sample and Species Grouping

A plexus diagram is a logical extension of the traditional phytosociological concept of groups or 'associations' of species (or stands) with strong ecological ties with one another or to a common environment. As ecologically related species are expected to be spatially proximate on the ground, and related stands are expected to occupy similar environments, so they should be represented close together in abstract vegetation space. Plexus diagrams may be used to identify groups of associated or correlated species, or clusters of similar samples or stands, or to relate to one another groups of species and stands (Hopkins 1957, Whittaker & Fairbanks 1958). The latter approach to classification resembles Lambert & Williams, (1962) identification of noda from a matrix of chi-square values as species as species-in-habitat units.

The existence of natural groupings of species (recurrent arrays, imbricated ecological groups, iterative groups, associations) is a common premise which has permeated both plant and animal ecology. But there has been little agreement on the identity, or the methods of definition and delimitation, of these natural groups. A clear natural group would be one in which each member has greater affinity for all other members of the group than for any nonmember, producing a marked discontinuity between groups. Commonly, biological groups are not so clearly circumscribed, and manifestly, if the criteria for recognition of the group are broadened the scope and discreteness of the group will change (Sørensen 1948, Wirth et al. 1966, Gould & Garwood 1969). In some uses a simple linkage by any criterion is considered

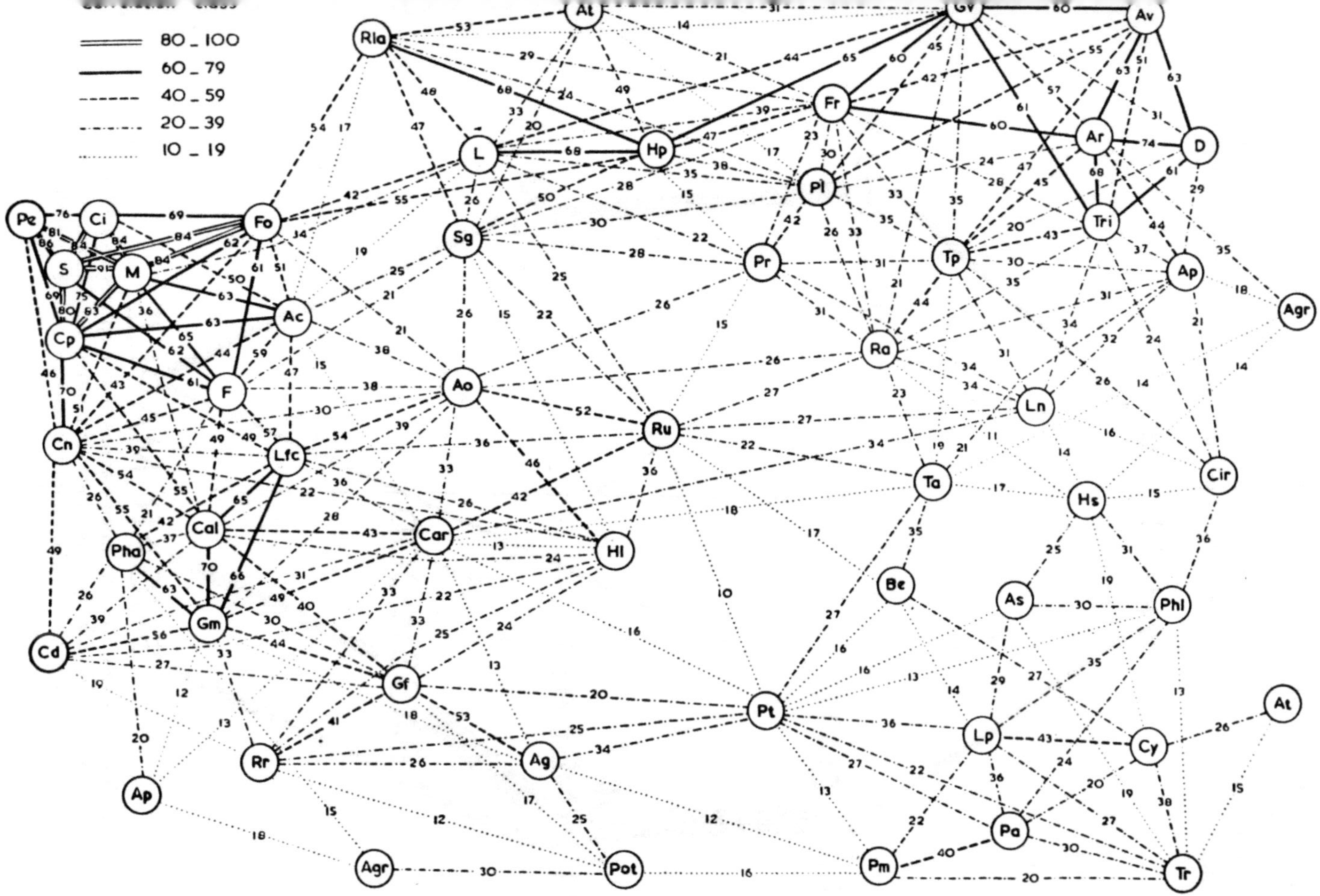

Fig. 3. A constellation or plexus diagram of species in Dutch grasslands based on correlation coefficients × 100 for species co-occurrence (Vries et al. 1954, Fig. 2). Drought-tolerant species appear in the upper part of the figure, with calciphile species to the right and acidiphile species to the left. Mesic grassland species are on the lower right, and wet-soil species on the left of the figure.

sufficient to include a species as a member of a group while a more restrictive usage (complete linkage) requires that it be linked with each member of the group (SOKAL & SNEATH 1963). In some single linkage groups the extreme members may be less similar to each

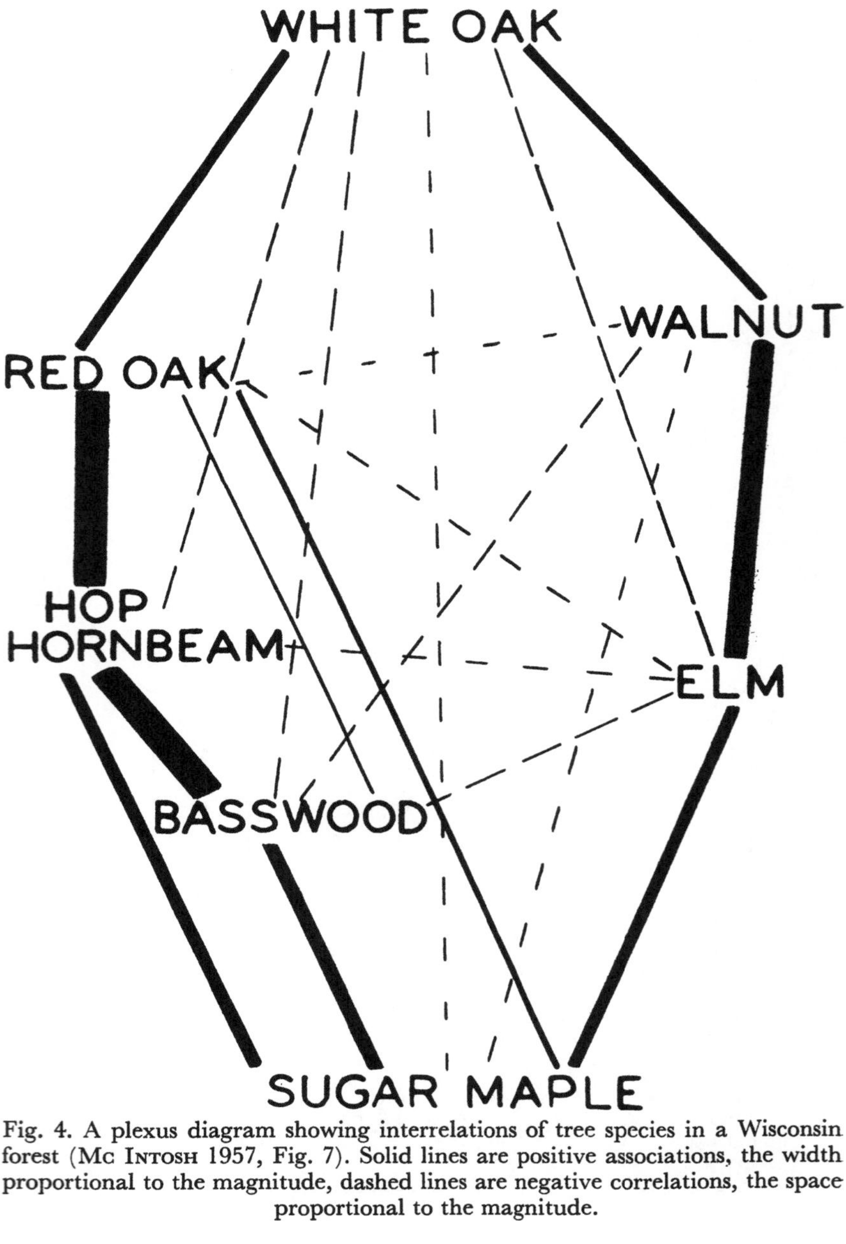

Fig. 4. A plexus diagram showing interrelations of tree species in a Wisconsin forest (MC INTOSH 1957, Fig. 7). Solid lines are positive associations, the width proportional to the magnitude, dashed lines are negative correlations, the space proportional to the magnitude.

other than either is to a nonmember of the group, whereas completely linked groups are more compact.

VRIES (1953) and VRIES et al. (1954) produced a plexus or constellation figure from a matrix of similarity coefficients (SA) by subjectively placing species with high positive correlation close together, and negatively correlated species far apart (Fig. 3). Because it was not possible to place the species correctly in two dimensions, classes of correlation were indicated by different types of lines. VRIES et al. recognized various areas of the diagram and clusters of species as being related to different site conditions. Similar methods relating species or plots have been used by (MACAN 1954, MC INTOSH 1957, HOSOKAWA et al. 1957, BARKMAN 1958, 1965, AGNEW 1961, QUARTERMAN & KEEVER 1962, JUHÁSZ-NAGY 1963, LOOMAN 1963, RAMSEY & DE LEEUW 1964, RAMSEY 1964, BEALS 1965, KLEINER & HARPER 1966, GODRON 1967, YARRANTON 1967, FRYDMAN & WHITTAKER 1968). Commonly only positive coefficients were used, but in some (Fig. 4) an attempt was made to illustrate negative values as well (MC INTOSH 1957).

The plexus diagrams most commonly used showed a pattern of species linked by positive similarity coefficients (WELCH 1960, AGNEW 1961, JUHÁSZ-NAGY 1963). The diagrams were usually interpreted as suggesting groups or clusters of related species as represented by positions in a single plane (Fig. 5). LOOMAN (1963) arranged species (by SA) in three dimensions and photographed the plexus to show relationships that could not be represented on a plane (Fig. 6). HEGG (1965) arranged species (by SA) in a plexus on which linkage-groups of species (ecological groups) on two levels are encircled (Fig. 7). The linkage-groups of the outer ring characterize lower-level community-types, those of the inner ring higher-level community-types, and each of the inner groups is linked with a number of the outer groups. Many species did not fit into the linkage-groups but could be given positions in the pattern of relationships the plexus represents. HEGG's treatment made possible both effective relation of species association to community classification, and indication of a number of directions of community relationships in two dimensions by the radial arrangement. Patterns of relationships of the plexus diagram are commonly interpreted, and 'groups' recognized subjectively, by occurrence of more links or higher coefficients among members of a group (AGNEW 1961, LOOMAN 1963, RAMSEY & DE LEEUW 1964). The subjectivity of recognizing groups in a maze of links of different degrees is a basic limitation on the use of plexuses for classification.

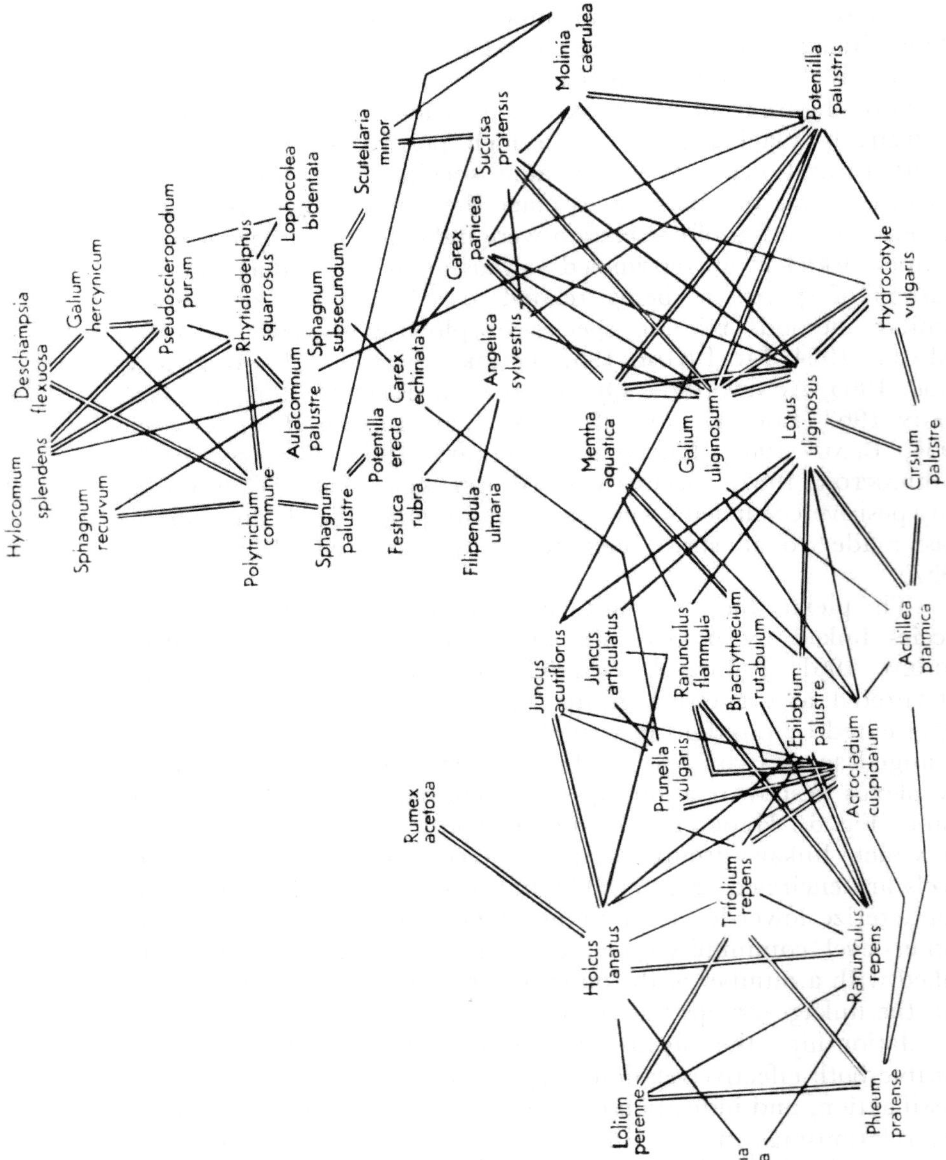

Fig. 5. A typical plexus diagram of positive correlations between species in pastures with *Juncus effusus* (AGNEW 1961, Fig. 1). Single line represents 5 % ≧ P, double line P ≧ 1 %.

This evident subjectivity may be a reason plexuses have not been more widely used as an aid to the classification of community samples, though such use is feasible in some cases. The Polish phytosociologists, who were among the early users of graphical methods,

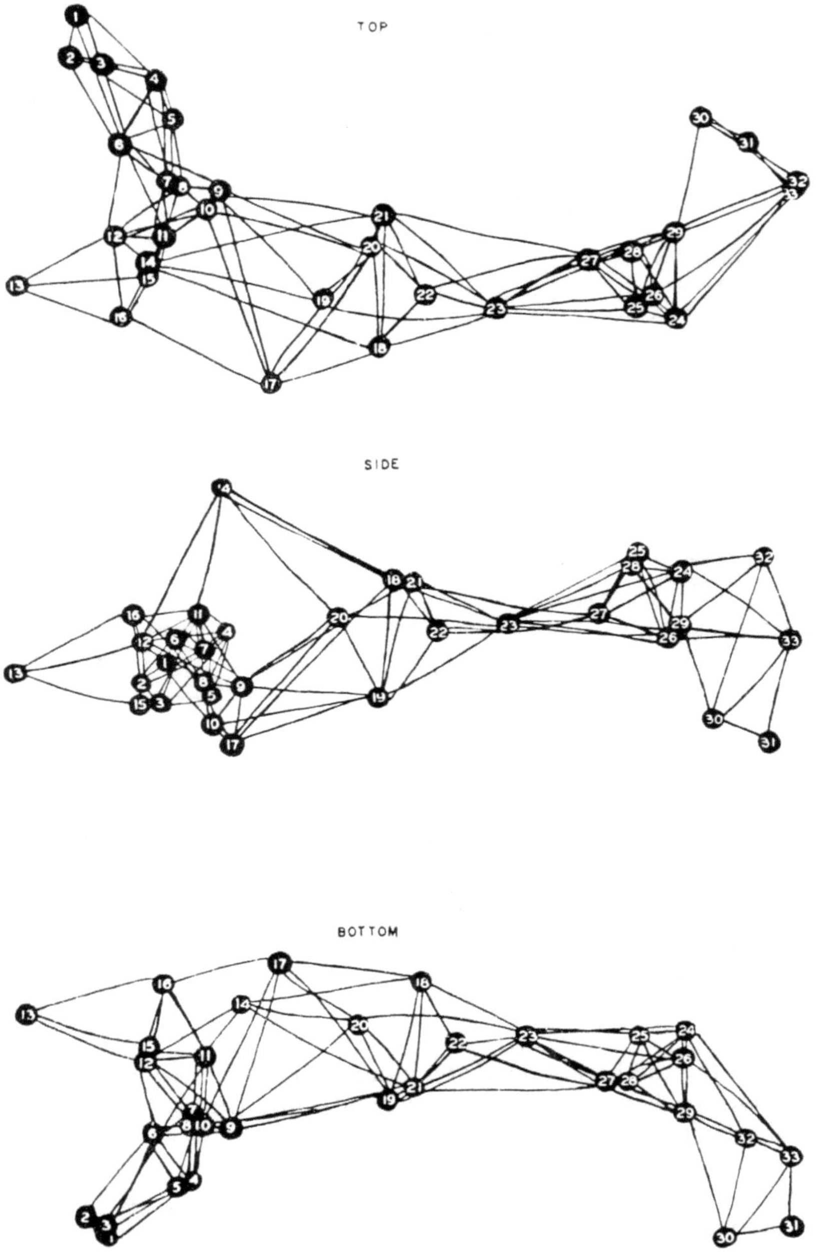

Fig. 6. A plexus or constellation of grassland species in Saskatchewan, a three dimension model photographed from three sides (LOOMAN 1963, Fig. 1).

Fig. 7. An arrangement of species by association measurements for Swiss mountain vegetation (Hegg 1965). Linkage-groups of species, by which community-types are characterized, are circled. A number of directions of community variation in relation to environment are indicated: acid soils (left), alpine vegetation (upper left), calcareous soils (right), wet soils (down), and nitrogen-rich soils (lower left).

used the method of dendrites along with the matrix method to show relations between phytosociological units (Fig. 8). Coefficients of similarity (CC) were shown on the links between units, although the graphical distance was not proportional to the magnitude of the coefficient (Faliński 1960, Izdebski 1963). Ramsey (1964) used the Sørensen coefficient (CC), with a minimum value of 70, to delimit initial group noda or central clusters of plots. These form the foci of groups in a plexus diagram and plots are added with progressively lower similarity values (*cf.* Ramsey & De Leeuw 1964).

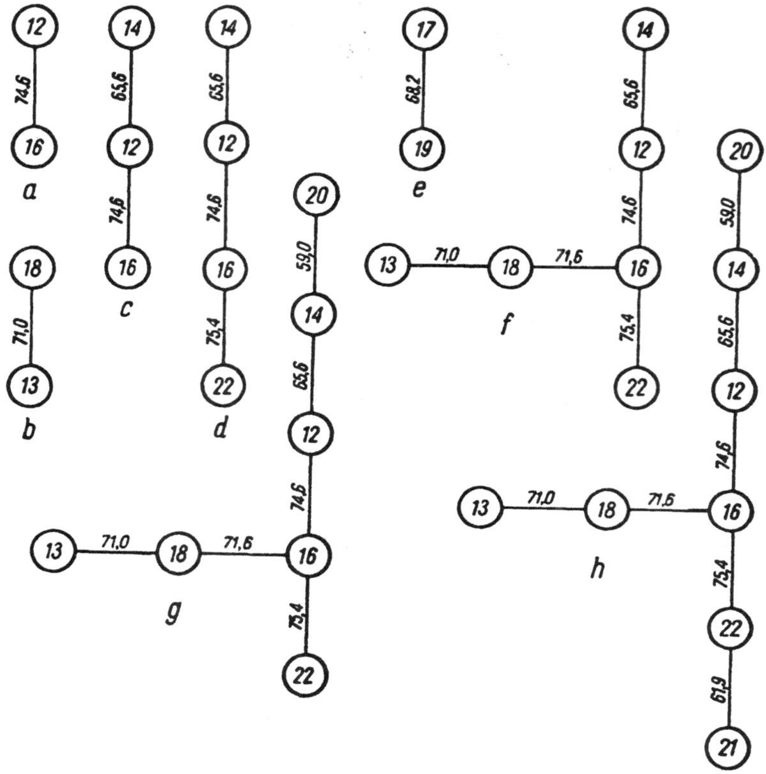

	12	13	14	16	17	18	19	20	21	22
12		61,3	**65,6**	**74,6**	55,2	64,7	51,8	58,5	61,0	65,7
13	62 19		62,1	59,0	65,4	**71,0**	58,3	50,8	52,6	50,0
14	64 21	58 18		57,1	51,8	62,5	48,0	**59,0**	56,4	63,6
16	67 25	61 18	63		59,6	**71,6**	56,6	46,9	59,2	**75,4**
17	58 16	52 17	54 14	57 17		62,1	**68,2**	50,9	50,0	43,3
18	68 22	62 22	64 20	67 24	58 18		55,5	55,4	58,5	62,8
19	54 14	48 14	50 12	53 15	44 15	54 15		47,1	47,0	50,0
20	65 19	59 15	61 18	64 15	55 14	65 18	51 12		48,1	47,8
21	82 25	76 20	78 22	81 24	72 18	82 24	68 16	79 19		**61,9**
22	70 23	64 16	66 21	69 26	60 13	70 22	56 14	67 16	84 26	

Fig. 8. A table of similarity coefficients and dendrite (FALIŃSKI 1960, Fig. 1,2). The matrix of KULCZYŃSKI (1928) coefficients (below) is the basis for a cumulative series of plexus diagrams *a* through *h* (above).

7.4.2 Definition and Representation of Groups

Groups or clusters may be objectively recognized when all pairs of the original matrix are at least as similar as some agreed-upon value of the similarity index, *i.e.* when the elements of the comparison matrix exceed a given value. All pairs of objects which are linked at the given level with any member of the group may be considered members of the group or it may be required, as Sørensen (1948) did, that all members be linked with each of the others above the selected similarity values.

The cohesiveness of a cluster may be expressed simply as a proportion of the possible links between the members of any group which is realized (*i.e.* 4 objects would have 6 possible links but could be linked with a minimum of 3). Wirth et al. (1966) proposed that cohesiveness of clusters may be expressed by the ratio:

$$\frac{\text{Total number of existing connections minus necessary connections.}}{\text{Total possible connections minus necessary connections.}}$$

If the components are simply linked linearly the connectedness is zero; when the ratio is one the cluster is maximally connected. Thus, patterns of clusters can be quantitatively compared.

Hopkins (1957) devised an objective method of producing 'basic units' or groups of significantly ($P > 0.001$ for SA) positively associated species within a site of a plant community. He suggested a method of identifying these as follows:

1. List all species with negative associations in order of decreasing number of these associations.
2. To any of these species add species which are positively associated with them.
3. When two or more groups contain a common species combine them.
4. Add extra groups of any positively related species to the list of groups.
5. Repeat 3.

The remaining groups are the basic units, and Hopkins shows a plexus diagram of such groups. Since by definition no member of a group has a significant association with any member of another group, there is no ready way of showing relations among groups. Hopkins secondarily assigned each of the sample areas to one or more basic units and estimated the proportion of the study area occupied by each basic unit. Araoz et al. (1971) form groups or noda by grouping sample quadrats according to their distances from randomly selected points. The interrelations between these are shown by linking noda with high association and low dissociation values to form a plexus diagram.

Most of the work using plexus diagrams has been concerned with identification of species groups or community-types. Numerical taxonomists have been doing much the same kind of thing (Wirth et al. 1966, Moss 1967, Gould & Garwood 1969). Wirth et al. show a series of successive plexus graphs formed as the strictness of the criterion for linking their objects (taxa) is sequentially relaxed. This is similar to Sørensen's method of forming clusters and is useful in showing the sequence of links formed. Gould & Garwood, similarly, form a series of increasingly encompassing (hierarchial) groups by cumulating patterns of correlation coefficients among dentition measurements of animals at decreasing levels of *r*. The representation of points (vertices) and lines (edges) in abstract space is the subject matter of graph theory which ecologists have been empirically, and largely unwittingly, dabbling in for four decades. Numerical taxonomists have been, more recently and self-consciously, doing the same thing, and have incorporated some of the more sophisticated aspects of graph theory into their analyses. Busacker & Saaty (1965) provide an introduction to graph theory which may help ecologists to improve their largely ad hoc graphic procedures. The plexus graphs commonly used by ecologists are usually 'undirected' — the lines connecting the points not having a direction. If these lines are given direction (by arrows) the graph is a 'directed' graph. This may be useful in graphic representations of a plexus to show nearest neighbor or 'most similar' relations not possible to show in the ordinary representation. Jackson & Crovello (1971) show a plexus imbedded in the first three dimensions or components of a factor analysis with the nearest neighbors or most similar points connected by directed lines. As ecologists have often noted, it is not usually possible to show the full matrix of similarity relations in two or three dimensions hence the plexus is always somewhat subjective and the actual relations distorted. In a directed graph the relations can be objectively shown by a factor analysis technique and the nearest neighbor relations superimposed on the graph illustrating where the full relationships are inadequately represented by the analysis.

7.4.3 Ordination

Plexus treatment may be effective also as an approach to ordination. Sukatschew (1928, 1932) and other Russian authors have,

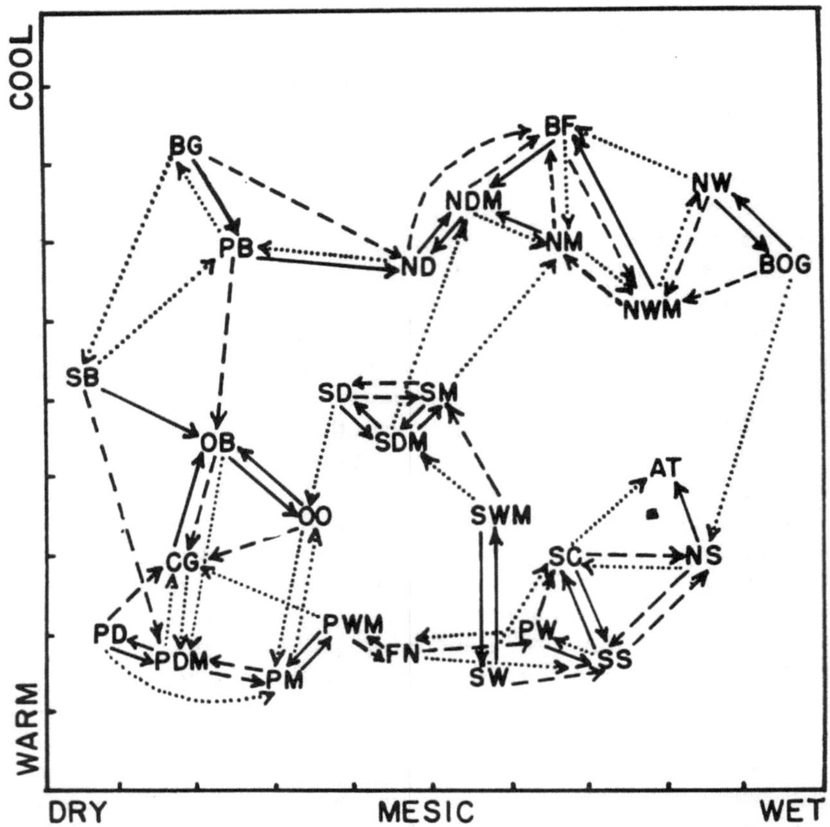

Fig. 9. A plexus ordination of community-types of Wisconsin in relation to moisture and temperature axes (Curtis 1959, Fig. 35). Solid arrows point to the most closely related community, dashed arrows to the second, and dotted arrows to the third most closely related community. Community-types ordinated: AT alder ticket, BF boreal forest, BG bracken-grassland, BOG open bog, CG cedar glade, FN fen, ND northern dry forest, NDM northern dry-mesic forest, NM northern mesic forest, NWM northern wet-mesic forest, NW northern wet forest, OB oak barrens, OO oak opening, PB pine barrens, PD dry prairie, PDM dry-mesic prairie, PM mesic prairie, PWM wet-mesic prairie, PW wet prairie, SB sand barrens, SC shrub-carr, SD southern dry forest, SDM southern dry-mesic forest, SM southern mesic forest, SS southern sedge meadow, SWM southern wet-mesic forest, SW southern wet forest.

without using similarity measurements, arranged community-types into ecological series intersecting one another and representing major directions of environmental and vegetational variation (see article 17, Fig. 1, and other figs. in articles 15 and 17). Wagner (1958) arranged middle European forest community-types into a plexus by their relations to elevation, moisture, and aridity. Comparable representations appear in the long tradition of succession diagrams in classical ecology showing the seral relationships of communities (or species) in successions. These diagrams were entirely subjective, and the essential difference of the plexus from them is the addition of quantitative methods of collecting data, the technique of condensing the primary data matrix into a matrix of similarity values, and the use of these values to identify clusters or directions of variation in the sample material.

CLAUSEN (1957a) described a graphic method of representing relationships by subjectively placing the elements of the matrix on a plane surface so that the distance from the nearest neighbors approximated the similarity values although they were not connected by lines. CURTIS (1959) arranged a wide range of different plant communities of Wisconsin into a plexus of community-types, comparing composite samples of each type by percentage presence of the more important undergrowth species and a PS calculation. A community-type central to the range of variation was placed in the center of a cubical space, and the other types were located with respect to it by triangulation procedures (Fig. 9). Two-dimensional views of this arrangement were used to plot environmental measurements for the community-types and to show how moisture, temperature, and internal light conditions in particular varied along the three axes of the plexus. The plexus was consequently an ordination of the community-types in relation to these three complex-gradients. The plexus was used also to show the distribution of species in relation to the community pattern and one another (Fig. 10), and the patterns of such broader community characteristics as the proportions in the vegetation of evergreen species, of shrubs and woody vines, and of major taxonomic groupings (Fig. 11).

In other studies the species or community plexus has been displayed on the main axes of environmental variables (WHITTAKER & FAIRBANKS 1958, OMURA & HOSOKAWA 1959, MARTIN 1960, AGNEW 1961, QUARTERMAN & KEEVER 1962, RAMSAY 1964, RAMSAY & DE LEEUW 1964). GIMINGHAM (1961, 1969) arranged a plexus diagram of heath stands from a broad geographic area of western Europe, considering SØRENSEN's CC values, to show major directions of variation (north-south and oceanic-continental climates). CURTIS' (1959) may be most significant among these applications of plexus technique for two reasons. First his plexus, rather than being a substitute for more formal ordination of a limited range of material, was the means of ordinating a range of samples so wide as to make difficult any other kind of ordination. Second his study, more effectively than any other, used the plexus to relate

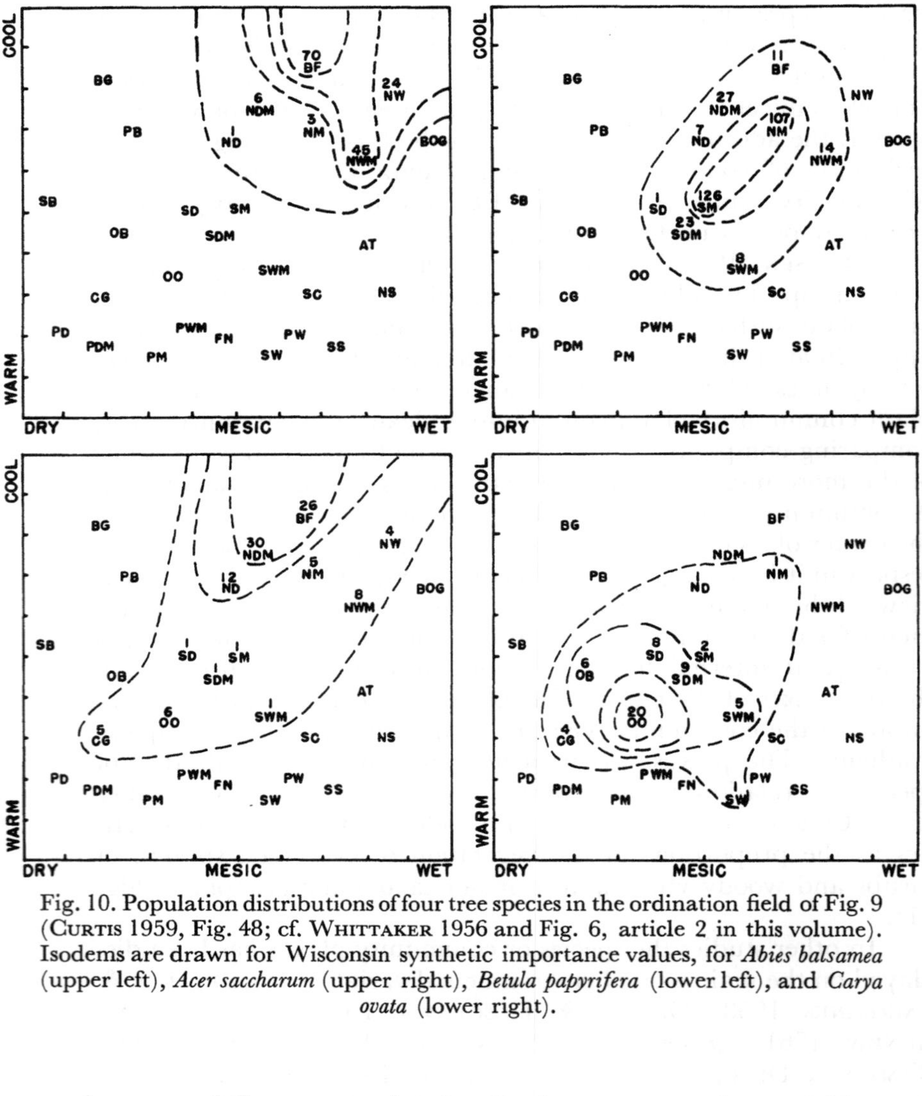

Fig. 10. Population distributions of four tree species in the ordination field of Fig. 9 (CURTIS 1959, Fig. 48; cf. WHITTAKER 1956 and Fig. 6, article 2 in this volume). Isodems are drawn for Wisconsin synthetic importance values, for *Abies balsamea* (upper left), *Acer saccharum* (upper right), *Betula papyrifera* (lower left), and *Carya ovata* (lower right).

environmental factors, species distributions, community variables, and community-types to one another.

In some cases species, or samples, or composite samples for community-types, can be arranged into a pattern showing their relationships to one another and recognized directions of environmental variation. In other cases the arrangement itself may reveal main directions of community variation that can be related to major environmental complex-gradients. The plexus may thus, in favorable cases, serve the purposes of direct and indirect gradient

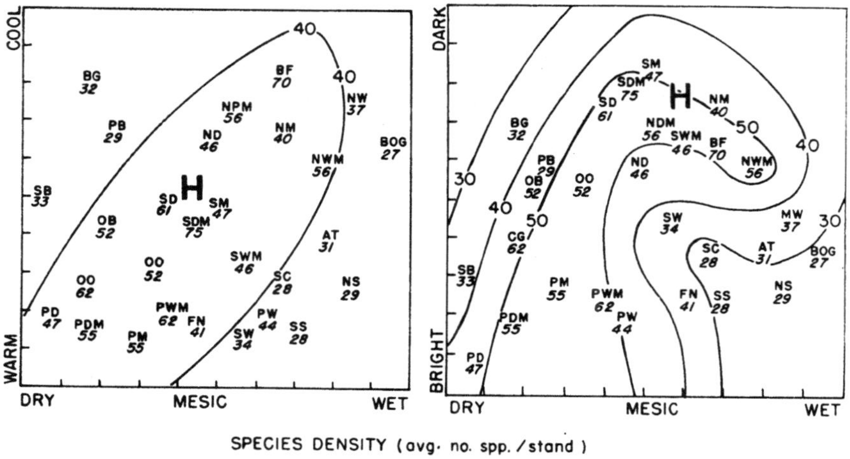

Fig. 11. Patterns of species diversity (average numbers of vascular plant species per sample) in the ordination field of Fig. 9, and the comparable ordination in relation to moisture and light axes (prepared from CURTIS 1959, Table IV—2 by R. H. WHITTAKER).

analysis, respectively. It is also possible to ordinate samples into a plexus by similarity measurements applied to environmental characteristics. HUGHES & LINDLEY (1955) calculated a matrix of similarity coefficients from a table of soil characteristics for four soil types and illustrated the relations between these as a plexus. HOLE & HIRONAKA (1960) calculated matrices of similarity values for Miami Family soil characteristics separately from laboratory and field data and a third matrix of representative soils of Great Soil Groups of the world. They represented the results as a three-dimensional model relating the soils types by perspective projection (Fig. 12).

There may be further interest in the construction of the different plexuses (of species, samples, and biotope measurements) possible from the same set of samples (or community types). No study was found comparing all three of these plexuses. Fig. 13 illustrates a case in which species distributions in samples (SA) and similarities (PS) of composite samples representing community-types were used to prepare plexuses which showed the same environmental axes. In this case the two plexuses independently revealed major directions of environmental effects on communities that had not been clearly recognized before the plexus treatment. BEALS & COPE (1964) calculated three separate matrices of similarity coefficients using the trees, spring herbs, and summer herbs found on different forest soil types and compared plexuses for the soil types based on

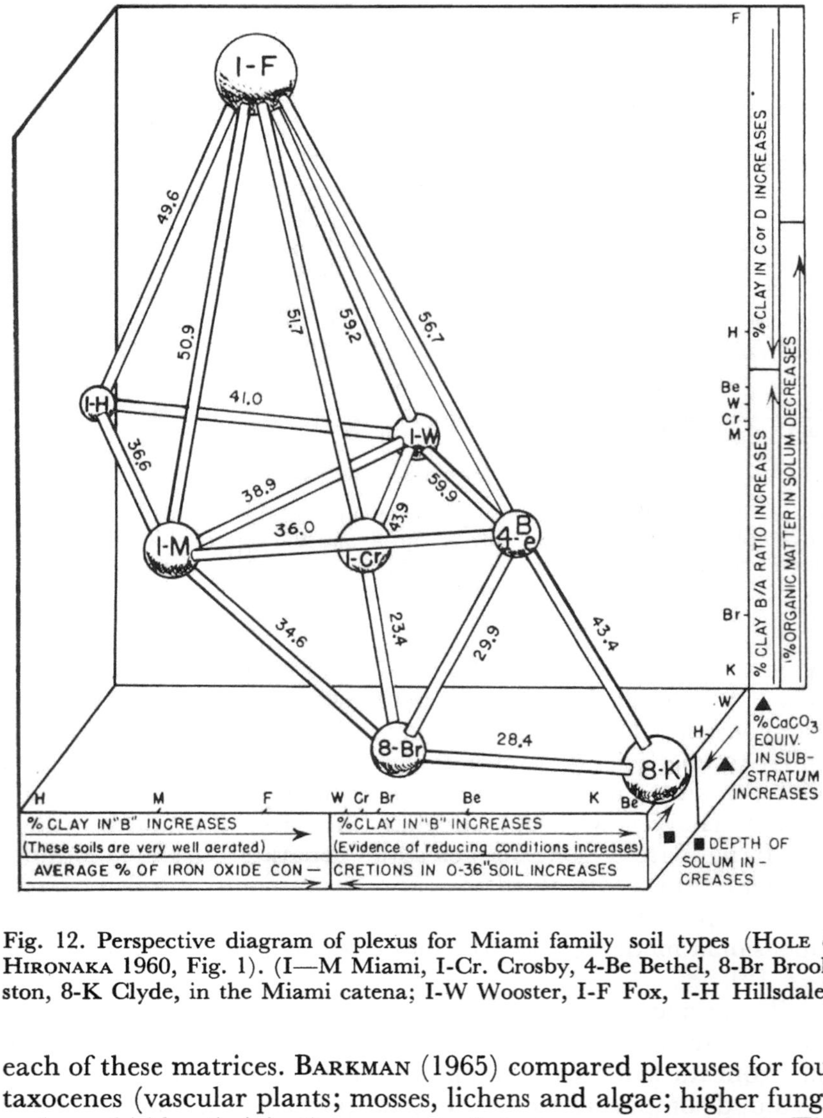

Fig. 12. Perspective diagram of plexus for Miami family soil types (HOLE & HIRONAKA 1960, Fig. 1). (I—M Miami, I-Cr. Crosby, 4-Be Bethel, 8-Br Brookston, 8-K Clyde, in the Miami catena; I-W Wooster, I-F Fox, I-H Hillsdale).

each of these matrices. BARKMAN (1965) compared plexuses for four taxocenes (vascular plants; mosses, lichens and algae; higher fungi; and carabid beetles) in the same set of nineteen communities in The Netherlands.

7.5 Conclusion

The versatility of the techniques we have been discussing deserves emphasis. The secondary matrix and the plexus may be used, separately or in combination, to seek or to display relationships of ob-

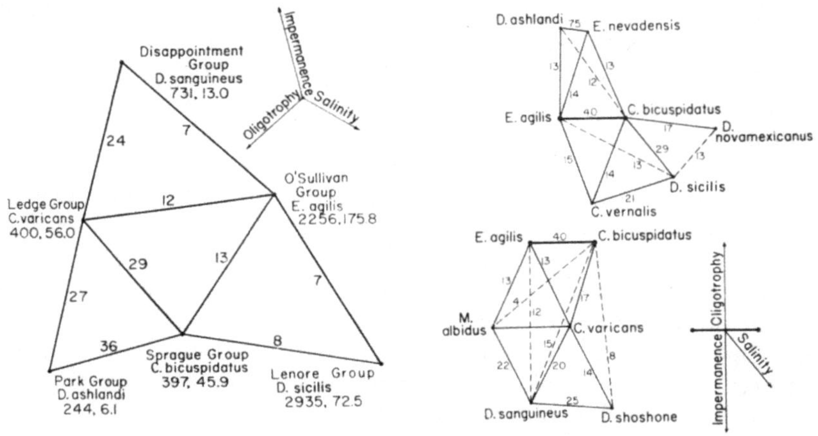

Fig. 13. A community-type plexus (left) and species-association plexus (right) for copepod samples from lakes and ponds in southeastern Washington (WHITTAKER & FAIRBANKS 1958, Figs. 3 and 5). Community-types are arranged by percentage similarities (numbers on lines) for composite samples; for each type are entered a characteristic species, mean total salt content of the water in parts per million (first number), and mean microsrustacean standing crop in individuals per liter (second number). Copepod species are arranged by percentage association values (numbers on lines); the solid lines are scaled for these values. The heavy line connecting *Eucyclops agilis* and *Cyclops bicuspidatus* is a center from which the remaining species radiate on the three environmental axes of oligotrophy (*Diaptomus ashlandi* and *Epischura nevadensis*), salinity (*C. vernalis*, *D. sicilis*, and *D. novemexicanus*), and seasonal impermance (*D. sanguineus* and *D. shoshone*).

jects to one another and environments. Species, community samples and biotopes may be the objects of such treatment, and there may be interest in comparing the results for these different objects in a given study. The levels on which the matrix and plexus can be applied range from use of replicate samples from a single stand to study species association and intra-community differentiation, to the use of individual samples from a narrower or wider range of community variation, to use of composite samples summarizing data for many individual samples representing a range of community-types. The techniques may be used to group or classify the objects or to ordinate them, or for both purposes; and the ordination makes possible relating to one another variables of environment, species distribution, and community characteristics.

The most general rewards of matrix and plexus approaches lie in the ordering of ecological material, and in revealing or representing relationships that are not otherwise evident. The plexus in particular may reveal relations of samples and species to one another in relation to environmental gradients. The plexus may then, like other ordinations, serve as a means of abstracting and understanding a community pattern in relation to a range of environmental variation. In some simple cases the matrix and plexus treatment may serve these purposes as well as any other, and there is only limited gain from the use of other, more formal techniques of ordination. Matrix and plexus techniques may become increasingly ineffective, however, with increase in number of objects, range of community variation, and number of directions of environmental and community interrelation. Though other means of ordination are affected by these same characteristics of the sample material, Wisconsin comparative ordination and factor analysis techniques will in some cases offer much more effective ordination than a plexus. Recent studies of distortions inherent in these other techniques imply that they too must be interpreted with caution (La France 1971, Austin & Noy-Meir 1972, article 11).

The matrix and plexus approach is thus not a generally sufficient answer to the problems of classification and ordination. However, the matrix and plexus techniques are more direct, and their implications are in some cases clearer to the investigator, than is the case with mathematically more complex procedures. These relatively simple approaches to community variation have continued value –as an alternative means of treating either small sample sets or very wide ranges of community variation, and as a preliminary step toward or a complement and aid to interpretation of other ordinations – that should not be overlooked.

7.6.1 Summary

Data from natural communities are often summarized as a community table or primary matrix, in which the columns represent samples and the rows species or environmental measurements. A variety of measurements can be applied to such tables to express the relative similarity to one another of samples, or of species, or of environmental variables. These similarity measurements of a given type may be compiled into a table (secondary matrix) in which the measurements indicate relationships of each sample (or species, or variable) to every other.

The secondary matrices can be rearranged that high similarity values are concentrated along the diagonal of the table. The ordered matrix may then be used either as: (i) an aid to classification of samples or grouping of species, when clusters of high similarity values are observed in the matrix, or (ii) an approach to ordination when the sequence of samples or species expresses their relation to a major environmental complex-gradient.

Values from the secondary matrix may also be used to arrange samples, or species, as points in a diagram or plexus in which the closeness of points to one another expresses their relative similarity. The plexus also may be used as an aid to classification or a means of ordination. With some sample sets, the plexus is an effective means of ordination in relation to two or three axes of environmental and community variation. The plexus may then be used to relate environmental measurements, species distributions, and community characteristics to one another, and as an abstract representation of a community pattern.

Though their effectiveness is more limited than that of some other ordination techniques, matrix and plexus approaches have continued value.

7.6.2 Zusammenfassung

Die Daten von natürlichen Pflanzengesellschaften werden oft in einer Gesellschaftstabelle oder 'Primär-Matrize' zusammengefasßt, in welcher die senkrechten Spalten die Aufnahmen und die Reihen die Arten oder Standorts-Messungen darstellen. Verschiedene Methoden können auf solche Tabellen angewandt werden, um die gegenseitige ähnlichkeit zwischen Aufnahmen, Arten oder Standortsvariabeln auszudrucken. Diese Ähnlichkeitswerte eines gegebenen Typs können in einer Tabelle (Sekundär-Matrize) zusammengestellt werden, in welcher die Messungen Beziehungen von jeder Aufnahme (oder Art, oder Variabeln) zu allen anderen anzeigen.

Die Sekundär -Matrizen können so umgeordnet werden, dasz hohe Ähnlichkeitswerte sich entlang der Tabellendiagonale konzentrieren. Die geordnete Matrize kann verwendet werden:

1. als eine Hilfe zur Klassifikation von Aufnahmen oder zur Gruppeirung von Arten, wenn Häufungen von hohen Ähnlichkeitswerten in der Matrize beobachtet werden, oder

2. als ein Weg zur Ordination, wann die Reihenfolge der Aufnahmen oder Arten ihre Beziehung zi einem Gradienten eines Haupt-Standortsfaktor ausdrückt.

Werte der Sekundär-Matrize können auch verwendet werden zur Ordnung von Aufnahmen oder Arten als Punkte in einem Diagramm oder Plexus, in welchem die Dichte der Punkte ihre relative Ähnlichkeit ausdrückt. Der Plexus kann auch verwendet werden als eine Hilfe zur Klassifikation oder ein Mittel zur Ordination.

Mit einigen Aufnahmengruppen ist der Plexus ein wirksames Mittel zur Ordination in Beziehung zu zwei oder drei Achsen von Standorts-oder Gesellschaftsvariation. In diesem Fall kann der Plexus verwendet werden um Standorts-Messungen, Arten-Verteilungen und Gesellschafts-Merkmale miteinander zu verknüpfen.

Obwohl die Wirksamkeit der Matrize- und Plexus -Verfahren beschränkter ist als diejenige einiger anderen Ordinationsverfahren, haben sie jedoch bleibenden Wert.

REFERENCES

AGNEW, A. D. Q., – 1961 – The ecology of *Juncus effusus* L. in North Wales. *J. Ecol.* 49: 83—102.

AGRELL, I., – 1963 – A sociological analysis of soil Collembola. *Oikos* 14: 237—247.

ARAOZ, J., G. SARMIENTO, & M. MONASTERIO, – 1971 – An essay in the use of association and dissociation measures in phytosociological classification. *J. Ecol.* 59: 39—50.

AUSTIN, M. P. & I. NOY-MEIR, – 1972 – The problem of non-linearity in ordination: experiments with two-gradient models. *J. Ecol.* 59: 763—773.

BALOGH, J., – 1958 – Lebensgemeinschaften der Landtiere. Akademie-Verlag, Berlin. 560 p.

BANNISTER, P., – 1968 – An evaluation of some procedures used in simple ordinations. *J. Ecol.* 56: 27—34.

BARKMAN, J. J., – 1958 – Phytosociology and Ecology of Cryptogamic Epiphytes. Van Gorcum, Assen. 628 pp.

BARKMAN, J. J., – 1965 – Die Kryptogamenflora einiger Vegetationstypen in Drente und ihr Zusammenhang mit Boden und Mikroklima. (Engl. summ.) In 'Biosoziologie,' ed R. TÜXEN, *Ber. Symp. Int. Ver. Vegetskunde.*, Stolzenau/Weser, 1960: 157—171.

BEALS, E. W., – 1965 – Species patterns in a Lebanese Poterietum. *Vegetatio* 13: 69—87.

BEALS, E. W. & J. B. COPE, – 1964 – Vegetation and soils in an eastern Indiana woods. *Ecology* 45: 777—792.

BEAMAN, J. H. & J. W. ANDRESEN, – 1966 – The vegetation, floristics and phytogeography of the summit of Cerro Potosi, Mexico. *Am. Midl. Nat.* 75: 1—33.

BECKING, R. W., – 1957 – The Zurich-Montpellier School of phytosociology. *Bot. Rev.* 23: 411—488.

BOYCE, A. J., – 1969 – Mapping diversity: A comparative study of some numerical methods. In 'Numerical Taxonomy', ed. A. J. COLE, pp. 1—30. Academic Press, New York.

BRAUN-BLANQUET, J., – 1951 – Pflanzensoziologie: Grundzüge der Vegetationskunde. 2nd ed., Wien, 631p. 3rd ed., 1964, 865p.

BRAY, J. R., – 1956 – A study of mutual occurrence of plant species. *Ecology* 37: 21—28.

BRODO, I. M., – 1961 – A study of lichen ecology in central Long Island, New York. *Am. Midl. Nat.* 65: 290—310.

BUSACKER, R. G. & T. L. SAATY, – 1965 – Finite Graphs and Networks. McGraw-Hill. New York. 294p.

CARMICHAEL, J. W. & P. H. A. SNEATH, – 1969 – Taxometric maps. *Syst. Zool.* 18: 402—415.

Češka, A., – 1968 – Application of association coefficients for estimating the mean similarity between sets of vegetational relevés. *Folia geobot. phytotax., bohemoslov.*, Praha. 3: 57—64.

Clausen, J. J., – 1957a – A comparison of some methods of establishing plant community patterns. *Bot. Tidsskr.* 53: 253—278.

Clausen, J. J., – 1957b – A phytosociological ordination of the conifer swamps of Wisconsin. *Ecology* 38: 638—646.

Cole, L.C., –1949 –The measure of interspecific association. *Ecology* 30: 411—424.

Cole, L. C., – 1957 – The measurement of partial interspecific association. *Ecology* 38: 226—233.

Cook, C. W., & R. Hurst, – 1963 – A quantitative measure of plant association on range in good and poor condition. *J. Range Mgmt.* 15: 266—273.

Crovello, T. J., – 1969 – Effects of change of characters and of number of characters in numerical taxonomy. *Am. Midl. Nat.* 81: 68—86.

Crovello, T. J., – 1970 – Analysis of character variation in ecology and systematics. *Ann. Rev. Ecol. Syst.* 1: 55—98.

Curtis, J. T., – 1959 – The Vegetation of Wisconsin: An Ordination of Plant Communities. Univ. Wisconsin, Madison. 657 pp.

Dagnelie, P., – 1960 – Contribution à l'étude des communautés végétale par l'analyse factorielle. (Engl. summ.) *Bull. Serv. Carte phytogéogr.* Sér. B, 5: 7—71, 93—195.

Dagnelie, P., – 1961 – L'application de l'analyse multivariable à l'étude des communautés végétales. *Bull. Inst. int. Statist.*, Rome, 33c Sess. p. 1—12.

Dagnelie, P., – 1962 – Étude statistique d'une pelouse à *Brachypodium ramosum.* III. Les liaisons interspécifiques. Presence et absence d'espèces. *Bull. Serv. Carte phytogéogr.* Sér. B, 7: 85—97.

Dahl, E., – 1957 – Rondane: Mountain vegetation in South Norway and its relation to the environment. *Skr. norske Vidensk-Akad.* Mat- naturv. Kl. 1956 (3): 1—374.

Davies, W., – 1971 – The phytopsammon of a sandy beach transect. *Am. Midl. Nat.* 86: 292—308.

Dix, R. L. & J. E. Butler, – 1960 – A phytosociological study of a small prairie in Wisconsin. *Ecology* 41: 316—327.

Ellenberg, H., – 1956 – Aufgaben und Methoden der Vegetationskunde. In 'Einführung in die Phytologie' by H. Walter, vol. IV. Grundlagen der Vegetationsgliederung, pt. I. Ulmer, Stuttgart. 136p.

Fager, E., – 1957 – Determination and analysis of recurrent groups. *Ecology* 38: 586—595.

Fager, F. W. & J. A. Mc Gowan, – 1963 – Zooplankton species groups in the North Pacific. *Science, N. Y.* 140: 453—460.

Faliński, J., – 1960 – Zastosowanie taksonomii wrocławskiej do fitosocjologii- (Germ. summ.: Anwendung der sog. 'Breslauer Taxonomie' in der Pflanzensoziologie). *Acta Soc. bot. Pol.* 29: 333—361.

Frydman, I. & R. H. Whittaker, – 1968 – Forest associations of southeast Lublin Province, Poland. (Germ. summ.) *Ecology* 49: 896—908.

Gams, H., – 1961 – Erfassung und Darstellung mehrdimensionaler Verwandtschaftsbeziehungen von Sippen und Lebensgemeinschaften. *Ber. geobot. Inst. ETH, Stiftg. Rübel*, Zurich, 1960, 32:96—115.

Ghent, A. W., – 1972 – A simplified computation procedure for Kendall's Tau suited to extensive species - density comparisons. *Am. Midl. Nat.* (in press).

Gilbert, M. & J. T. Curtis, – 1953 – Relation of the understory to the upland forest in the prairie-forest border region of Wisconsin. *Trans. Wis. Acad. Sci. Arts Lett.* 42: 183—195.

GIMINGHAM, C. H., – 1961 – North European heath communities: a 'network of variation'. *J. Ecol.* 49: 655—694.
GIMINGHAM, C. H., – 1969 – The interpretation of variation in north-European dwarf-shrub heath communities. (Germ. summ.) *Vegetatio* 17: 89—108.
GODRON, M., – 1967 – Les groupes écologiques imbriqués en éceailles. *Oecol. Plant.*, Paris 2: 217—226.
GOULD, S. J. & R. A. GARWOOD, – 1969 – Levels of integration in mammalian dentitions: An analysis of correlations in *Nesophontes micrus* (Insectivora) and *Oryzomys couesi* (Rodentia). *Evolution* 23: 276—300.
GREIG-SMITH, P., – 1952 – Ecological observations on degraded and secondary forest in Trinidad, British West Indies. II. Structure of the communities. *J. Ecol.* 40: 316—330.
GREIG-SMITH, P., – 1964 – Quantitative Plant Ecology. 2nd ed. Butterworths, London. 256 pp.
GUINOCHET, M., – 1955 – Logique et dynamique du peuplement végétale. Masson, Paris. 143 pp.
GUINOCHET, M. & P. CASAL, – 1957 – Sur l'analyse différentielle de Czekanowski et son application à la phytosociologie. *Bull. Serv. Carte phytogéogr.* Sér. B, 2: 25—33.
HAGEMEIER, E. M., – 1966 – A numerical analysis of the distributional patterns of North American mammals. II. Reevaluation of the provinces. *Syst. Zool.* 15: 279—299.
HALE, M. E., – 1955 – Phytosociology of corticolous cryptogams in the upland forests of southern Wisconsin. *Ecology* 36: 45—63.
HEGG, O., – 1965 – Untersuchungen zur Pflanzensoziologie und Ökologie im Naturschutzgebiet Hohgant (Berner Voralpen). (French & Engl. summs.) *Beitr. geobot, Landesaufn. Schweiz* 46: 1—188.
HOLE, F. D. & M. HIRONAKA, – 1960 – An experiment in ordination of some soil profiles. *Proc. Soil Sci. Soc. Am.* 24: 309—312.
HOPKINS, B., – 1957 – Pattern in the plant community. *J. Ecol.* 45: 451—463.
HOSOKAWA, T., M. OMURA & Y. NISHIHARA, – 1957 – Grading and integration of epiphyte communities. *Jap. J. Ecol.* 7: 93—98.
HUGHES, R. E. & D. V. LINDLEY, – 1955 – Application of biometric methods to problems of classification in ecology. *Nature, Lond.* 175: 806—807.
HURLBERT, S. H., – 1969 – A coefficient of interspecific association. *Ecology* 50: 1—9.
IZDEBSKI, K., – 1963 – Bory na Roztoczu Środkowym. (Russ. & Engl. summs: Pineto-Vaccinietum uliginosi, Pineto-Vaccinietum myrtilli and Abietum polonicum in Central Roztoze). *Annls Univ. Mariae Curie-Skłodowska*, Lublin, Sec. C. 1962, 17: 313—362.
JACKSON, R. C. & T. J. CROVELLO, – 1971 – A comparison of numerical and biosystematic studies of *Haplopappus*. *Brittonia* 23: 54—70.
JONES, E. W., – 1955 – Ecological studies on the rain forest of southern Nigeria. IV. The plateau forest of the Okumo Forest Reserve. *J. Ecol.* 43: 564—594.
JUHÁSZ-NAGY, P., – 1963 – Investigations on the Bulgarian vegetation. Some hygrophilous plant communities. (I-III) *Acta biol. debrecena* (*Acta Univ. debrecen*, Budapest, Ser. 2) 2: 47—70.
KERSHAW, K. A., – 1959 – An investigation of the structure of a grassland community. II. The pattern of *Dactylis glomerata*, *Lolium perenne* and *Trifolium repens*. *J. Ecol.* 47: 31—43.
KERSHAW, K. A., – 1960 – The detection of pattern and association. *J. Ecol.* 48: 233—242.

Kleiner, E. F. & K. T. Harper, – 1966 – An investigation of association patterns of prevalent grassland species in Red Butte Canyon, Salt Lake County, Utah. *Proc. Utah Acad. Sci.* 43 (2) : 29—36.

Kontkanen, P., – 1957 – On the delimitation of communities in research on animal biocoenotics. *Cold Spring Harb. Symp. quant. Biol.* 22: 373—378.

Kulczyński, S. M., – 1928 – Die Pflanzenassoziationen der Pieninen. *Bull. int. Acad. pol. Sci. Lett.*, Cl. Sci. Math. Nat., Sér. B., 1927 (Suppl 2): 57—203.

La France, C., – 1971 – Vegetation analysis and ordination: a modeling approach to the analysis of technique. *Ph. D. Thesis, Univ. Notre Dame.*

Lambert, J. M. & W. T. Williams, – 1962 – Multivariate methods in plant ecology. IV. Nodal analysis. *J. Ecol.* 50: 775—802.

Looman, J., – 1963 – Preliminary classification of grasslands in Saskatchewan. *Ecology* 44: 15—29.

Looman, J. & J. B. Campbell, – 1960 – Adaption of Sørensen's K. (1948) for estimating unit affinities in prairie vegetation. *Ecology* 41: 409—416.

Macan, T. T., – 1954 – A contribution to the study of the Corixidae (Hemipt). *J. Anim. Ecol.* 23: 115—141.

Mac fadyen, A., – 1954 – The invertebrate fauna of Jan Mayen Island (East Greenland). *J. Anim. Ecol.* 23: 261—297.

Mac fadyen, A., – 1963 – Animal Ecology, Aims and Methods. 2nd ed. Pitman, New York 344p.

Martin, R. H., – 1960 – The ecology of Groenvlei, a South African fen. *J. Ecol.* 48: 307—329.

Matuszkiewicz, W., – 1948 – Roślinność lasów okolik Lwowa (Engl. summ.: The vegetation of the forests of the environs of Lvov.) *Annls Univ. Mariae Curie-Skłodowska*, Lublin, C, 3: 119—193.

Mc Donough, W. T., – 1963 – Interspecific associations among desert plants. *Am. Midl. Nat.* 70: 291—299.

Mc Intosh, R. P., – 1957 – The York Woods, a case history of forest succession in southern Wisconsin. *Ecology* 38: 29—37.

Mc Intosh, R. P., – 1962 – Pattern in a forest community. *Ecology* 43: 25—33.

Mc Intosh, R. P., – 1967a – An index of diversity and the relation of certain concepts to diversity. *Ecology* 48: 392—404.

Mc Intosh, R. P., – 1967b – The continuum concept of vegetation. *Bot. Rev.* 33: 130—187.

Moore, G. W., W. S. Benninghoff & P. S. Dwyer, – 1967 – A computer method for the arrangement of phytosociological tables. In 'Proc. Ass. for Computing Machinery National Meeting, 1967,' pp. 297—299.

Morisita, M., – 1959 – Measuring of interspecific association and similarity between communities. *Mem. Fac. Sci. Kyushu Univ.*, Ser. E. (Biol.) 3: 65—80.

Moss, W. W., – 1967 – Some new analytic and graphic approaches to numerical taxonomy with an example from the Dermanyssidae (Acari). *Syst. Zool.* 16: 177—207.

Motyka, J., B. Dobrzański, & S. Zawadzki, – 1950 – Wstępne badania nad łąkami poludiniowo-wschodniej Lubelszczyzny. (Russ. & Engl. summs.: Preliminary studies on meadows in the southeast of the Province of Lublin). *Annls Univ. Mariae Curie-Skłodowska*, Lublin, Sec. E, 5: 367—447.

Newsome, R. D. & R. L. Dix, – 1968 – The forests of the Cypress Hills Alberta and Saskatchewan, Canada. *Am. Midl. Nat.* 80: 118—185.

Omura, M. & T. Hosokawa, – 1959 – On the detailed structure of a corticolous community analysed on the basis of interspecific association. *Mem. Fac. Sci. Kyushu Univ.*, Ser. E. (Biol.) , 3: 51—63.

Ono, Y., – 1961 – An ecological study of the brachyuran community on Tomoika

Bay, Amakusa, Kyushu. *Rec. oceanogr. Wks Japan,* spec. no. 5: 199—210.

ORLOCI, L. – 1972 – On objective measures of phytosociological resemblance. *Am. Midl. Nat.* (in press).

POORE, M. E. D., – 1955 – The use of phytosociological methods in ecological investigations. I. The Braun-Blanquet system. *J. Ecol.* 43: 226—244. II. Practical issues involved in an attempt to apply the Braun-Blanquet system. *J. Ecol.* 43: 245—269. III. Practical application. *J. Ecol.* 43: 606—651.

PRÉCSÉNYI, O., – 1958 – Über die interspezifische Korrelation. *Acta bot. hung.* 4: 155—158.

QUARTERMAN, E. & C. KEEVER, – 1962 – Southern mixed hardwood forest: Climax in the southeastern Coastal Plain: U. S. A. *Ecol. Monogr.* 32: 167—185.

RAMSAY, D. MC C., – 1964 – An analysis of Nigerian savanna II. An alternative method of analysis and its application to the Gombe sandstone vegetation. *J. Ecol.* 52: 457—466.

RAMSAY, D. MC C. & P. N. DE LEEUW, – 1964 – An analysis of Nigerian savanna I. The survey area and the vegetation developed over Bima sandstone. *J. Ecol.* 52: 233—254.

RENKONEN, O., – 1938 – Statistisch- ökologische Untersuchungen über die terrestriche Käferwelt der finnishchen Bruchmoore. (Finnish summ.) *Annls zool., Soc. zool. - bot. fenn. Vanamo (Suomal. eläin-ja kasvit Seur. van. eläin. Julk.)* 6: 1—231.

RENKONEN, O., – 1949 – Discussion on the ways of insect synecology. *Oikos* 1: 122—126.

RUŽIČKA, M., – 1961 – Flechten - Kieferwald auf den Flugsanden der Tiefebene Záhorská nížina (Cladonio - Pinetum zahoricum). (Russ. & Slovak summs.) *Biológia Bratisl.* 16: 881—894.

SANDERS, H. L. & R. R. HESSLER, – 1969 – Ecology of the deep-sea benthos. *Science, N. Y.* 163: 1419—1424.

SHANKS, R. E., – 1953 – Forest composition and species association in the beech-maple forest region of western Ohio. *Ecology* 34: 455—466.

SOKAL, R. R. & P. H. A. SNEATH, – 1963 – Principles of Numerical Taxonomy. Freeman, San Francisco. 359pp.

SØRENSEN, T. A., – 1948 – A method of establishing groups of equal amplitude in plant sociology, based on similarity of species content, and its application to analysis of the vegetation on Danish commons. *Biol. Skr.*, K. Danske Vidensk Selsk. 5 (4) : 1—34.

STRUIK, G. J. & J. T. CURTIS, – 1962 – Herb distribution in an *Acer saccharum* forest. *Am. Midl. Nat.* 68: 285—296.

SUKACHEV, V. N., – 1928 - Principles of classification of the spruce communities of European Russia. *J. Ecol.* 16: 1—18.

SUKATSCHEW, W. N., – 1932 – Die Untersuchung der Waldtypen des osteuropäischen Flachlandes. *Handb. biol. ArbMeth.* 11, 6: 191—250.

TAYLOR, R. J. & D. CAMPBELL, – 1969 – Biochemical systematics and phylogenetic interpretations in the genus *Aquilegia. Evolution* 23: 153—162.

VASILEVICH, V. I., – 1969 – Statistical Methods in Geobotany. (In Russian). USSR Academy of Sciences, Leningrad. 232pp.

VRIES, D. M. DE, – 1953 – Objective combinations of species. *Acta bot neerl.* 1: 497—499.

VRIES, D. M. DE, J. P. BARETTA & G. HAMMING, – 1954 – Constellation of frequent herbage plants, based on their correlation in occurrence. *Vegetatio*: 5/6: 105—111.

WAGNER, H., – 1958 – Grundfragen der Systematik der Waldgesellschaften. *Veröff. geobot. Inst. Rübel,* Zürich 33: 241—252.

WELCH, J. R., – 1960 – Observations on deciduous woodland in the eastern province of Tanganyika. *J. Ecol.* 48: 557—573.
WHITTAKER, R. H., – 1952 – A study of summer foliage insect communities in the Great Smoky Mountains. *Ecol. Monogr.* 22: 1—44.
WHITTAKER, R. H., – 1956 – Vegetation of the Great Smoky Mountains. *Ecol. Monogr.* 26: 1—80.
WHITTAKER, R. H., – 1962 – Classification of communities. *Bot. Rev.* 28: 1—239.
WHITTAKER, R. H., – 1967 – Gradient analysis of vegetation. *Biol. Rev.* 42: 207—264.
WHITTAKER, R. H. & C. W. FAIRBANKS, – 1958 – A study of plankton copepod communities in the Columbia Basin, southeastern Washington. *Ecology* 39: 46—65.
WIRTH, M., G. F. ESTABROOK & D. J. ROGERS, – 1966 – A graph theory model for systematic biology, with an example for the Oncidiinae (Orchidaceae) *Syst. Zool.* 15: 59—69.
YARRANTON, G. A., – 1967 – A quantitative study of the bryophyte and macrolichen vegetation of the Dartmoor granite. *Lichenologist* 3: 392—408.
YARRANTON, G. A. & W. J. BEASLEIGH, – 1968 – Towards a mathematical model of limestone pavement vegetation I. Vegetation and microtopography. *Can. J. Bot.* 46: 1591—1599.

8 WISCONSIN COMPARATIVE ORDINATION

Grant Cottam, F. Glenn Goff, and R. H. Whittaker

Contents

8 WISCONSIN COMPARATIVE ORDINATION

8.1 Introduction

Some of the earliest work in vegetational ordination was done at the University of Wisconsin more than twenty years ago. Since that time there has been a continuous increase in the use of various types of ordination procedures in vegetation studies. With wider usage of the approach have come better understanding and more formal mathematical techniques. The increasing capacity and availability of computers have promoted use of multivariate analytic methods in ecological research.

Because of the early development of certain methods of vegetation analysis at Wisconsin and the extensive application of these methods to the vegetation of the region, the University of Wisconsin has come to be known as one of the centers of 'ordination ecology.' Some authors (not, in general, those from Wisconsin) speak of a 'Wisconsin School' of plant ecology. While it is true that several techniques were independently developed by Wisconsin workers, most of these can be thought of as special cases of more general methods of multivariate analysis (e.g. SEAL 1964, KENDALL 1965) that are now coming into wider usage. The distinctive contribution from Wisconsin is discussed also by CURTIS (1959), McINTOSH (1958, 1967), GREIG-SMITH (1964), and WHITTAKER (1967).

The purpose of this paper is (i) to present a brief review of the development of ordination at the University of Wisconsin, (ii) to illustrate the technique of polar ordination and its effectiveness for vegetation research, and (iii) to indicate how the compositional gradient analysis and polar ordination techniques developed at Wisconsin relate to other techniques now in use.

8.2 Compositional Gradient Analysis

CURTIS & McINTOSH's (1951) compositional gradient analysis was based on use of species 'adaption values' for sample ordination. To obtain these values, the quantity of each species within each stand or sample was first expressed as a synthetic importance value — the sum of relative density, relative basal area, and relative frequency (CURTIS & McINTOSH 1951).

Assigning adaptation values to species was actually a two-stage process. Samples were first grouped into subsets, members of each of which were similar in their dominant species. These subsets were then arranged in a preliminary order along a compositional gradient by visual inspection of the patterns of species importance. An arrangement was sought which produced smooth patterns of change in the importance values of dominant species (either a monotonic taper from one extreme of the gradient toward the other, or tapers on each side of a peak within the gradient). This approach to ordination through the distributions of major species was termed the 'method of leading dominants.' Species distributions in this preliminary ordering of samples were observed, and each species was assigned an approximate value for relative position along the gradient on a scale from 1 to 10. These preliminary values were multiplied by the synthetic importance values of species to calculate a compositional index (cf. the weighted average technique, article 2) for each sample. Since the sum of the synthetic importance values for species in a given sample is 300 %, the compositional indices with species weighted 1 to 10 range between 300 and 3000. By plotting the importance values of a particular species in samples against the compositional indices of the samples, a scatter diagram such as that shown in Fig. 1 was produced.

The peak position of the species along the compositional gradient (x-axis) was determined by use of a moving-average curve

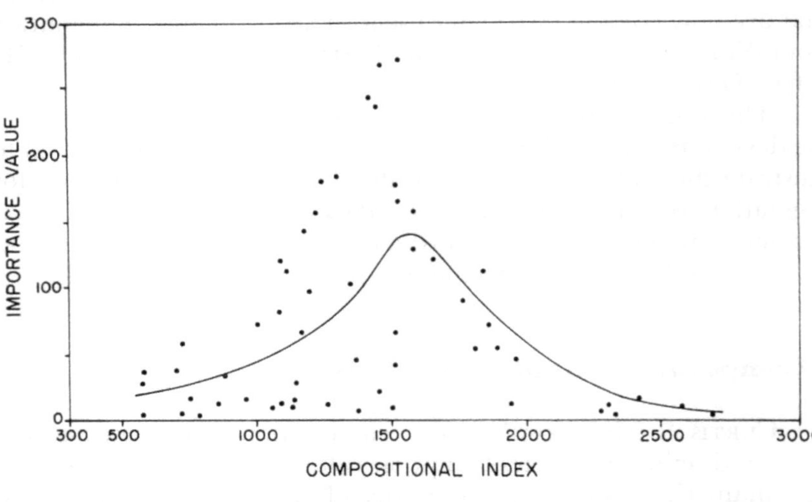

Fig. 1. A scatter diagram showing importance values of *Pinus strobus* in samples along the northern Wisconsin compositional gradient for forests. The solid line is a moving mean through the scatter of points; data are from Brown & Curtis (1952) records in the Wisconsin Plant Ecology laboratory files.

fitting technique. The adaptation value corresponding to this peak position (i.e., the peak position along the compositional gradient divided by 300) was the final adaptation value or weight assigned to the species. The final adaptation values were quite similar to the preliminary values. In some of the early studies this was taken as a validation of the preliminary ordering of samples. It is not, for the first assignment of values tends to fix the portion of the ordination transect in which a given species occurs. Studies independently relating species populations to environmental gradients are to be desired as a check on the assignment of species adaptation values.

The final adaptation values were used to arrange and group the samples into a composite transect, the data for which could be summarized in the form of Table I, article 2. Such transects effectively showed the rise and fall of species populations, and the change in community characteristics, along the compositional gradient. Characteristics of soils were found to vary along the gradient (Brown & Curtis 1952, Maycock & Curtis 1960), and it was shown that

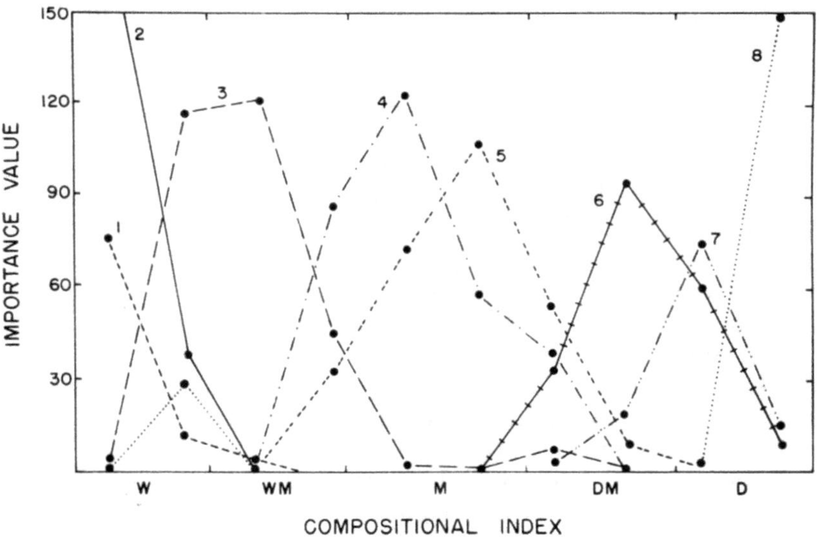

Fig. 2. Behavior of major tree species on the combined ordination of upland and lowland forests in northern Wisconsin: 1 tamarack (*Larix laricina*), 2 black spruce (*Picea mariana*), 3 white cedar (*Thuja occidentalis*), 4 eastern hemlock (*Tsuga canadensis*), 5 sugar maple (*Acer saccharum*), 6 white pine (*Pinus strobus*), 7 red pine (*P. resinosa*), and 8 jack pine (*P. banksiana*). Points are average Wisconsin synthetic importance values for species in sets of samples representing positions along the compositional gradient from wet (W, left), through mesic (M, middle), to dry (D, right) sites. From '*The Vegetation of Wisconsin*' by J. T. Curtis (1959) with permission of the University of Wisconsin Press.

other groups of organisms formed compositional gradients paralleling those for trees — undergrowth plants (GILBERT & CURTIS 1953), soil fungi (TRESNER et al. 1954, CHRISTENSEN et al. 1962), mosses and lichens on bark (HALE 1955, CULBERSON 1955), and birds (BOND 1957). One popular diagram resulting from these early studies showed the importance curves of several species superimposed as Fig. 2. Although studies of upland and lowland vegetation were conducted separately, the series of overlapping species curves, including both upland and lowland, was often shown on a single 'continuum' graph. Often in diagrams of this sort the curves were hand-smoothed.

GILBERT & CURTIS (1953) were the first workers at Wisconsin to use a matrix of interspecific association to order species. They used intensity of shading in the cells of the matrix to represent degree of interspecific association and simply rearranged the species to produce an ordering with heavily shaded cells along the main diagonal and lighter cells away from the diagonal (see article 7). Although they did not claim quantitative rigor, the ordering of species that resulted should be very similar if not identical to the ordering that would be produced by extracting the first eigenvector (principal component) from the same matrix of interspecific association. A similar approach has been applied by CLAUSEN (1957), SWINDALE & CURTIS (1957), and others to order species.

Following the work of CURTIS & MCINTOSH, several other studies used essentially the same techniques. BROWN & CURTIS (1952) applied the method of leading dominants to the forests of northern Wisconsin, CURTIS (1955) used a modification of the technique based on 'joint occurrence groups' to obtain gradient values for prairie species. Some studies applying the approach other than those mentioned include CHRISTENSEN et al. (1959), BRAY (1960), LINDSEY et al. (1961), ANDERSON (1963), LARSEN (1965), BUELL et al. (1966), ANDERSON et al. (1966), GEHLBACH 1967, and GEMBORYS & HODGKINS (1971).

All these studies had two characteristics in common. First, index values were assigned as weights for the species; values for samples could be found indirectly by use of the species values, but the samples were not ordered directly. Secondly, the species values were inferred from sociological data. While the patterns of community variation presumably resulted from the operation of underlying environmental variables, the indices were assigned primarily on the basis of species association or correlation and not by observation of species distributions along particular environmental gradients. It is in this respect that most of the Wisconsin work differs from the direct gradient analysis of WHITTAKER (1951, 1956, 1967), ROWE

(1956), Waring & Major (1964) and others who have assigned species indices along specified environmental gradients (article 2).

8.3 Polar Ordination

8.3.1 Technique

Wisconsin comparative or polar ordination was introduced by Bray & Curtis in 1957. The essential techniques of this, which may be the most distinctive contribution of Wisconsin plant ecology, can best be explained with a simple example. (The example and discussion relating to it were prepared by the third author on the basis of experience with the technique at Cornell University, and using 12 New Jersey saltmarsh samples from a larger set made available by the courtesy of L. F. M. Fresco.)

Table I

Primary matrix of species coverage percents in

New Jersey saltmarsh samples.

	Sample number	1	2	3	4	5	6	7	8	9	10	11	12
	Species												
1	Spartina alterniflora	75	30	5	20	5	1		10	1	2		
2	Salicornia europaea	5	10	2	1	1		2			2		
3	Atriplex patula var. hastata	1	10	2	1	1	2	5		1		5	2
4	Distichlis spicata		15	80	2	10	15	30	1	10	10	20	
5	Suaeda maritima				20	10							
6	Salicornia virginica				5	10							
7	Juncus gerardii			1			40	1					
8	Scirpus olneyi						5	20				1	
9	Spartina patens							20	10	50		2	5
10	Iva frutescens							5	1	2	1	20	10
11	Phargmites communis								1	10	20	5	30
12	Solidago sempervirens									1	5	1	2
	Species cover total	81	65	90	49	37	63	83	23	75	40	54	49

The samples are first compiled into a standard community-table or primary matrix, as illustrated in Table I. Species that occur in a single sample have been omitted. Further analysis is based on quantitative comparisons among these samples. A number of similarity measurements can be used (articles 6, 11) but many applications of Wisconsin comparative ordination since Bray & Curtis (1957) have used a double standardization before computing percentage similarities. The rows of the primary matrix are first standardized, with the importance values of each species converted into percentages of the maximum value in the row for that species. Second, these standardized percentages are summed in each column, and the values in each column are converted to percentages of the total for that column. Each sample is now represented by a column of relative importance values (totaling 100 % or 1.0). This double standardization is in no way essential to the Bray & Curtis technique, but it may be advantageous (see 8.3.3). Table II is the same primary matrix as Table I after such double standardization.

Sample similarities are now computed. When the columns in the table are relative importance values, percentage similarity

Table II

Primary matrix of species coverages in New Jersey saltmarsh samples after double standardization.

	Sample number	1	2	3	4	5	6	7	8	9	10	11	12
	Species												
1	Spartina alterniflora	62.5	15.5	4.5	13.4	3.5	0.8		31.1	0.7	1.3		
2	Salicornia europaea	31.3	38.6	13.4	5.0	5.3		7.3			9.7		
3	Atriplex patula var. hastata	6.2	38.6	13.4	5.0	5.3	12.1	18.2		5.3		22.7	9.1
4	Distichlis spicata		7.2	67.0	1.3	6.6	11.4	13.6	2.9	6.7	6.0	11.3	
5	Suaeda maritima				50.2	26.4							
6	Salicornia virginica				25.1	52.9							
7	Juncus gerardii			1.7			60.6	0.9					
8	Scirpus olneyi						15.1	36.4				2.3	
9	Spartina patens							14.5	46.6	53.4		1.8	4.5
10	Iva frutescens							9.1	11.7	5.3	2.4	45.3	22.7
11	Phragmites communis								7.8	17.8	32.2	7.6	45.5
12	Solidago sempervirens									10.7	48.3	9.1	18.2

computation simplifies to $PS = \Sigma \min(x,y)$, in which min (x,y) is the smaller of the two values for a given species in samples X and Y (article 6, formula 51). These values, computed in all possible directions between samples, are compiled into the secondary, sample-similarity matrix given as Table III. The problem is now to extract from these numbers major directions of variation in community composition, and to arrange the samples in relation to axes representing these major directions.

If the ordination is indirect, two samples which are most dissimilar may be chosen as the first end-points. The sample that has the lowest mean similarity with all other samples may be located first, in our example sample 6 (15.5 % bottom row of Table III). The sample with the lowest similarity value with this (sample 8, 3.7 %, Table III) may become the other end-point .These samples are now the poles of an axis along which all other samples are to be arranged, or ordinated.

Table III

Secondary matrix for sample similarities, New Jersey saltmarshes, percentage similarities of samples after double standardization of species coverage data.

Sample	1	2	3	4	5	6	7	8	9	10	11	12
1		53.0	24.1	23.4	14.1	7.1	13.5	31.1	6.1	11.0	6.2	6.2
2	53.0		38.5	24.7	20.7	20.2	32.7	18.4	12.7	17.0	29.9	9.1
3	24.1	38.5		15.8	20.7	26.0	35.2	7.4	12.7	17.0	24.7	9.1
4	23.4	24.7	15.8		66.4	7.1	11.3	14.6	7.0	7.6	6.3	5.0
5	14.1	20.7	20.7	66.4		12.7	17.2	6.4	12.6	12.6	11.9	5.3
6	7.1	20.2	26.0	7.1	12.7		39.5	3.7	12.7	6.9	25.7	9.1
7	13.5	32.7	35.2	11.3	17.2	39.5		26.5	31.9	15.7	42.7	22.7
8	31.1	18.4	7.4	14.6	6.4	3.7	26.5		63.3	14.4	23.9	24.0
9	6.1	12.7	12.7	7.0	12.6	12.7	31.9	63.3		37.7	35.8	43.7
10	11.0	17.0	17.0	7.6	12.6	6.9	15.7	14.4	37.7		25.1	52.8
11	6.2	29.9	24.7	6.3	11.9	25.7	42.7	23.9	35.8	25.1		50.2
12	6.2	9.1	9.1	5.0	5.3	9.1	22.7	24.0	43.7	52.8	50.2	
Mean	17.8	25.2	21.0	17.2	18.2	15.5	26.3	21.3	25.1	19.7	25.6	21.6

A distance between these end-point samples must be obtained; this distance becomes the length of the first axis. The complement of the similarity value for the end-point samples, their percentage difference $PD = 100\% - PS'$, is an expression of that distance, but not necessarily the best value. (i) If replicate samples from the same community are compared, their percentage similarities are not 100 % but some lower value; replicate similarities of 70 to 90 % are often observed (article 4). Distance between two samples is better expressed by this replicate similarity minus PS', than by 100 minus PS'. In many cases similarity measurements for replicate samples are not available ; the replicate similarity may be estimated as somewhat higher than the highest value in a secondary matrix of non-replicate samples, but usually less than 100 %. BRAY & CURTIS (1957) obtained a mean similarity (PS') of 82 % for seven replications; BEALS (1960) used 85 % as a reasonable value for ordinating a matrix in which no similarity exceeds 85 %. (ii) If the similarity value for the two end-points is zero, the distance between them may be not smaller than 100 but larger. Removal of one or both samples may be considered. It may also be possible to estimate the distance by summing the distances from an intermediate sample to the two end-points. There are no zero values in Table III. Since some, single-species saltmarshes in the larger set had 100 % similarities, 100 % has been used as the replicate similarity in the example; and the first axis length is $L = 100 - 3.7 = 96.3$. (The ordination itself is not much affected by the replicate similarity estimate, or by whether 100 % or such an estimate is used – see BANNISTER 1968.)

All other samples are now located along the axis between end-point samples 8 and 6. Fig. 3 illustrates a simple graphical technique by which sample 7 is located on the axis (BRAY & CURTIS 1957). The similarity values for sample 7 with samples 8 and 6 are each subtracted from 100 %, to obtain distances (D_1, D_2) of 73.5 and 60.5. A compass is set to these distances by the scale of the axis, and a line is dropped onto the axis from the point at which the arcs for the distances from samples 8 and 6 intersect. The distance of sample 7 along the axis from sample 8 (x) is 57.2; its vertical distance from the axis (e) is 46.1. These values are easily computed from the Pythagorean theorem (BEALS 1960, 1965):

$$x = (L^2 + D_1^2 - D_2^2)\ /2L, \text{ and } e = \sqrt{(D_1^2 - x_2)},$$

in which L is the distance between the end-points, and D_1 and D_2 are the distances of the sample being located from the first and second end-point respectively. All the samples of our example have been thus located in Fig. 3. In many such ordinations the samples

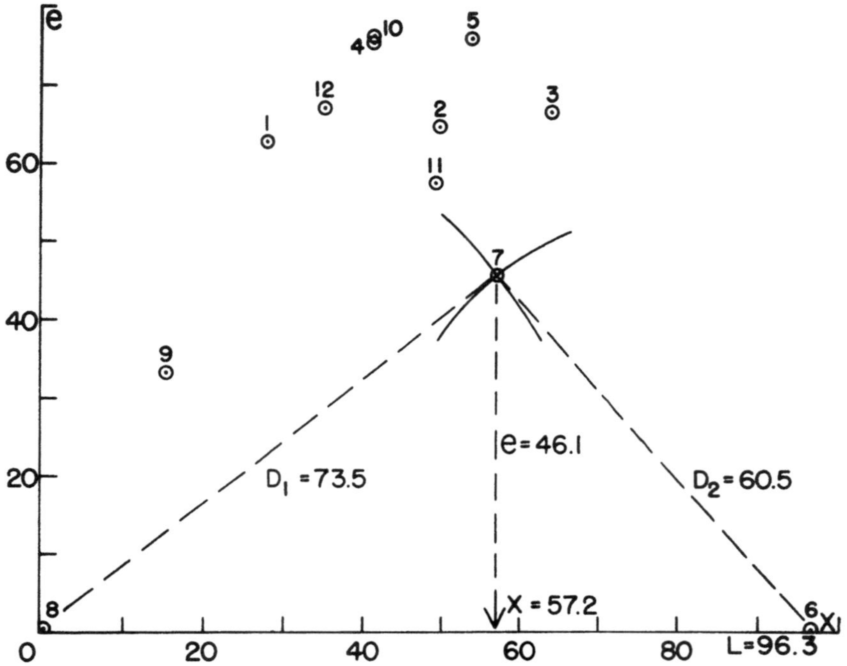

Fig. 3. The BRAY & CURTIS (1957) graphic technique for ordinating samples along an axis defined by end-point samples. For each sample its distances from the end-points intersect at a point defining relative location along the axis (x) and apparent departure from the axis (e) as shown for sample 7. The 12 samples of Tables I-III have been ordinated along an axis of length 96.3 between samples 8 and 6 by distance values (100 minus the percentage similarities in Table III) from end-point samples 8 and 6. Results from calculation by the formulas in the text are the same, as given in the first columns of Table IV.

form an arch over the axis (article 11). There may be little significance to the values for e, but the values for x should express relative position along a gradient of community composition between the end-point samples chosen.

A second axis of ordination is now desired. A second pair of end-point samples should be sought which are: (i) both among the samples in the middle part of the first axis, and (ii) close to one another in their positions along that axis, but (iii) most distant from one another among such pairs of samples. They should also (iv) have high values for e, but samples with the highest e values may not be the best choice of end-points. In our example samples 1 vs. 12 and 4 vs. 10 are likely choices of second end-point pairs; both of these

Table IV

New Jersey saltmarsh sample ordination, positions on (x) and distances from (e) two axes.

	First axis (8-6)		Second axis (1-12)	
Sample	x	e	x	e
1	28.0	63.0	0	0
2	49.6	64.8	14.6	44.7
3	64.2	66.7	33.5	68.1
4	41.1	74.8	30.0	70.4
5	54.0	76.4	38.4	76.8
6	96.3	0	48.9	79.1
7	57.2	46.1	54.9	66.8
8	0	0	41.4	55.1
9	15.6	33.2	77.1	53.7
10	41.1	75.1	77.3	44.2
11	49.5	57.7	80.5	48.0
12	35.2	67.4	93.7	0

pairs were tried but 1 and 12 both had the lower similarity (6.2) and gave an ordination more interpretable in terms of tide-levels and other relations. All the remaining samples are now located between samples 1 and 12, along an axis of length $L = 100 - 6.2 = 93.8$, by either the graphical technique or the computation given above. Results of ordination along both axes are given in Table IV. The two axes may now be used for a coordinate system, in which each sample is located by its positions on both axes (Fig. 4). The second axis is usually not strictly perpendicular to the first, but may be so treated as in Fig. 4.

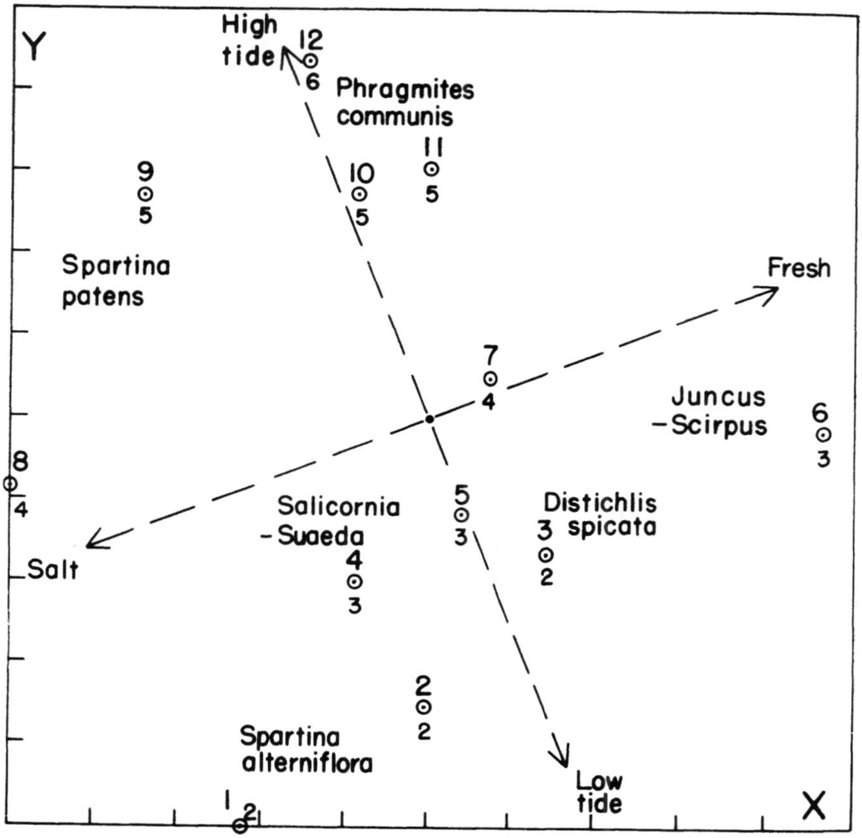

Fig. 4. A Wisconsin comparative ordination of 12 samples from New Jersey salt-marshes (unpublished data of L. F. M. FRESCO). Coverage values for species (Table I) were doubly standardized (Table II); percentage similarities of samples were computed (Table III), and distances (100 minus these similarities) were used to locate samples in relation to two pairs of end-points, samples 8 and 6, and 1 and 12 (Table IV). Sample numbers are given above their points, tide levels on a scale from 1 (lowest) to 7 (highest) below the points. The names are the dominant species by which the samples may be classified into types; these types relate to major environmental gradients as indicated by the broken lines.

The relationships of samples in the ordination field should now be inspected. If samples that are located close together are similar to one another in composition, the two-dimensional ordination may be sufficient, and no third axis may be needed. If samples close together are markedly different, this difference suggests the existence of a third direction of community variation and the need for a third ordination axis. The end-points for the third axis may best be chosen from among the samples in the central area of the two-dimensional ordination field, in the same way the second end-points were chosen. Our example will not be carried beyond the second axis. It is sometimes possible, but not usually profitable, to carry comparative ordination to a fourth or even a fifth axis.

8.3.2 Results

The axes and the community pattern are now to be interpreted. Environmental measurements may be plotted at the sample points in the ordination field. In Fig. 4 the numbers below the sample points are tide-levels on a scale from 1 (lowest, including pure *Spartina alterniflora* marshes not represented in our example) to 7 (highest, samples with *Phragmites* and *Iva* and additional species not given in Table I). The second axis is related to the tide-level gradient, but the sample relationships (especially samples 3 vs. 4 and 6 vs. 8) suggest that the gradient is obliquely related to the *Y* axis, as indicated. Field observations, and the presence of *Scirpus olneyi* and *Juncus gerardii* as indicators of fresh-water influence, imply that the first axis is related to salinity. The oblique sequence of samples 1, 2, 3, and 6 in particular appears to represent the gradient of increasingly strong fresh-water dilution. Names of dominance-types, for a preliminary classification of the samples, are entered in Fig. 4 (samples 7 and 8 are transitional to the lower and higher tide-level types).

The ordination thus unifies the samples into an understandable pattern of communities in relation to a range of environmental variation. Further characteristics of the pattern can be shown by plotting other data in the ordination field. In Fig. 5 coverage estimates for four major species are plotted and outlined by contour lines. Each species has a center of maximum coverage in one of the community-types, from which its coverage tapers into other types. Characteristics of the samples as wholes, estimated total coverage and richness in species, are plotted in the bottom panels of Fig. 5. The coverage data are irregular both in the example and the larger sample set; the suggested increase (apart from sample 9) toward lower tide levels is questionable. Species diversities, in contrast, show a clear increase from low to high tide levels. The ordination can thus relate to one another environmental factors, species populations, and community trends, in an intergrading pattern of community-types as recognized by the investigator (cf. Whittaker 1967).

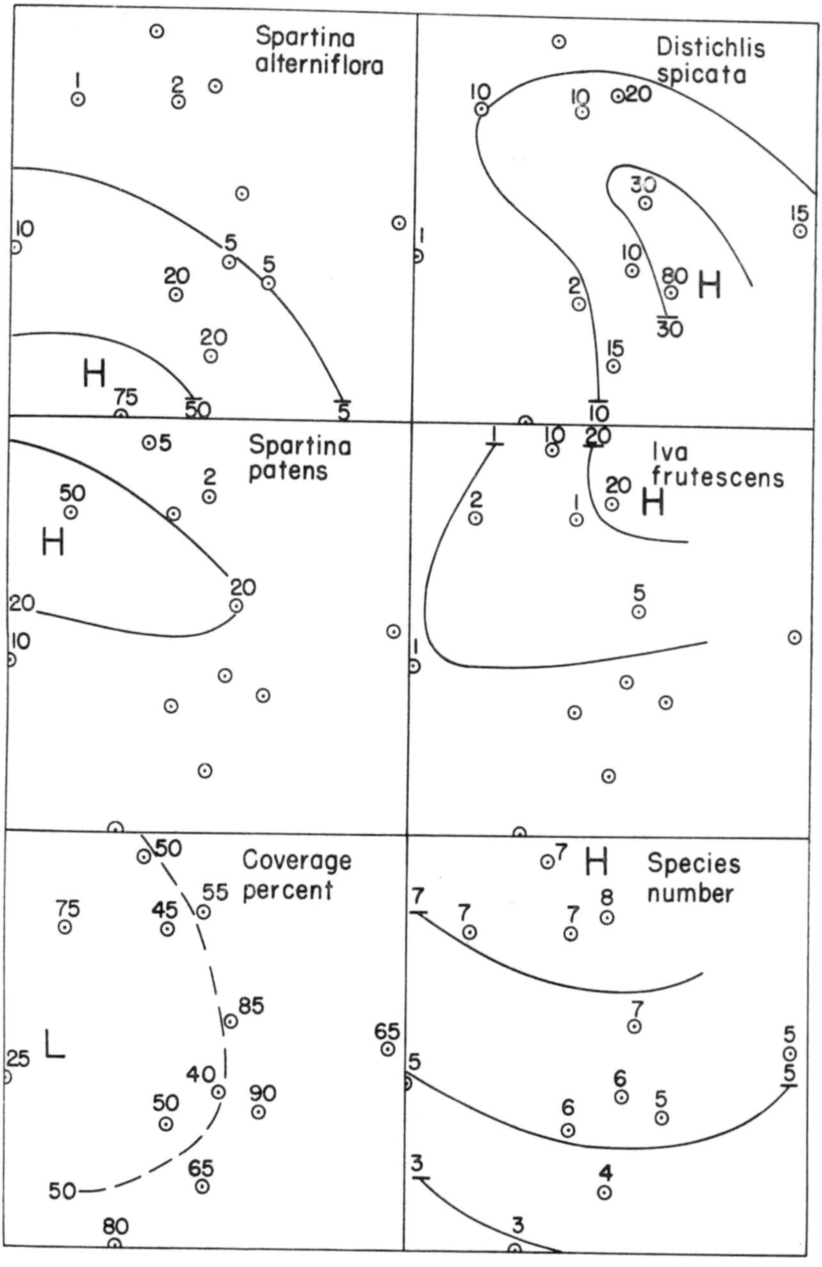

Fig. 5. Coverages of four major species, and sample coverages and species numbers, plotted for New Jersey saltmarsh samples as ordinated in Fig. 4.

Figs. 6–8 illustrate other results from comparative ordination. Fig. 6 plots symbols for community-types at positions for a larger number of samples in an ordination field (Ream 1963). The types occupy different parts of the field but are not sharply segregated or clustered. Fig. 7 indicates by sizes of circles the importance values for two species at sample positions in relation to three axes (Ayyad & Dix 1964). The sample positions may be conceived as a cloud of points in a three-dimensional space. Each species has its center of maximum importance in the space; in all directions away from this center its importance values decrease. It is consequently possible to ordinate species in relation to the same directions of community variation as those used for sample ordination. Fig. 8 represents an ordination of species by their centers in a three-dimensional space (Bray & Curtis 1957).

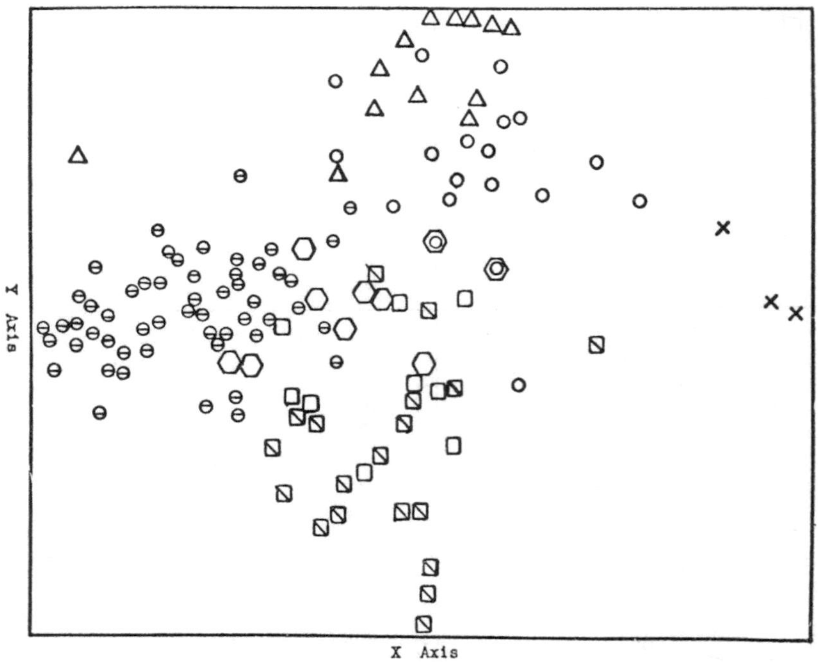

Fig. 6. A Wisconsin comparative ordination of forest samples from the Wasatch Mountains, Utah (Ream 1963). Samples have been classified into dominance-types that are represented by different symbols: triangles *Abies concolor*, circles *Pseudotsuga menziesii*, barred circles *Populus tremuloides*, squares *Pinus contorta*, barred squares *Picea engelmanni*, hexagons *Abies lasiocarpa*, x's *Pinus flexilis*.

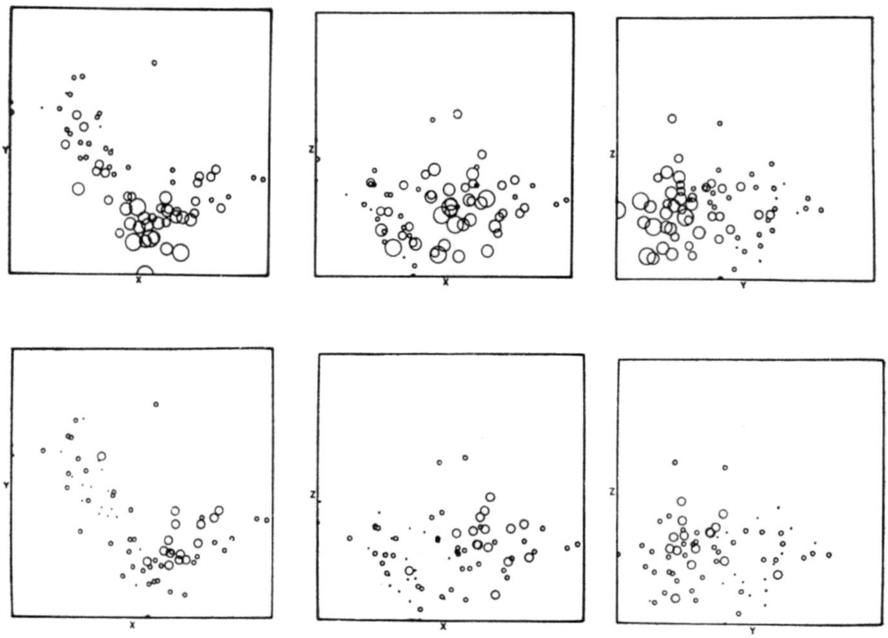

Fig. 7. Sample positions and patterns of species importance values in a three-dimensional ordination space for Saskatchewan prairies (Ayyad & Dix 1964, Fig. 2). Centers of circles locate samples in relation to the *x* and *y* axes on the left, the *x* and *z* axes in the middle, and the *y* and *z* axes on the right. Sizes of circles represent relative density values in these samples for two species—*Agropyron dasystachum* above, and *Koeleria cristata* below.

Ordinations thus provide useful graphical summaries of vegetational variation. A large amount of information can be conveyed in forms representing coherently and at a glance the interrelations of species, communities, and environments. An ordination may be colloquially referred to as a 'coat-rack for ideas' about vegetation, or a conceptual 'skeleton' by which the complex relationships in the field can be held together and given recognizable form. Once the ordination is prepared, the vegetation can be described by referring to segments or areas within the vegetational hyperspace, without the necessity for classification. On the other hand the vegetational hyperspace can be divided to give a classification that is based on the relationships revealed by ordination, or an ordination of samples that have been classified into units can reveal the relations of these units to one another and environment.

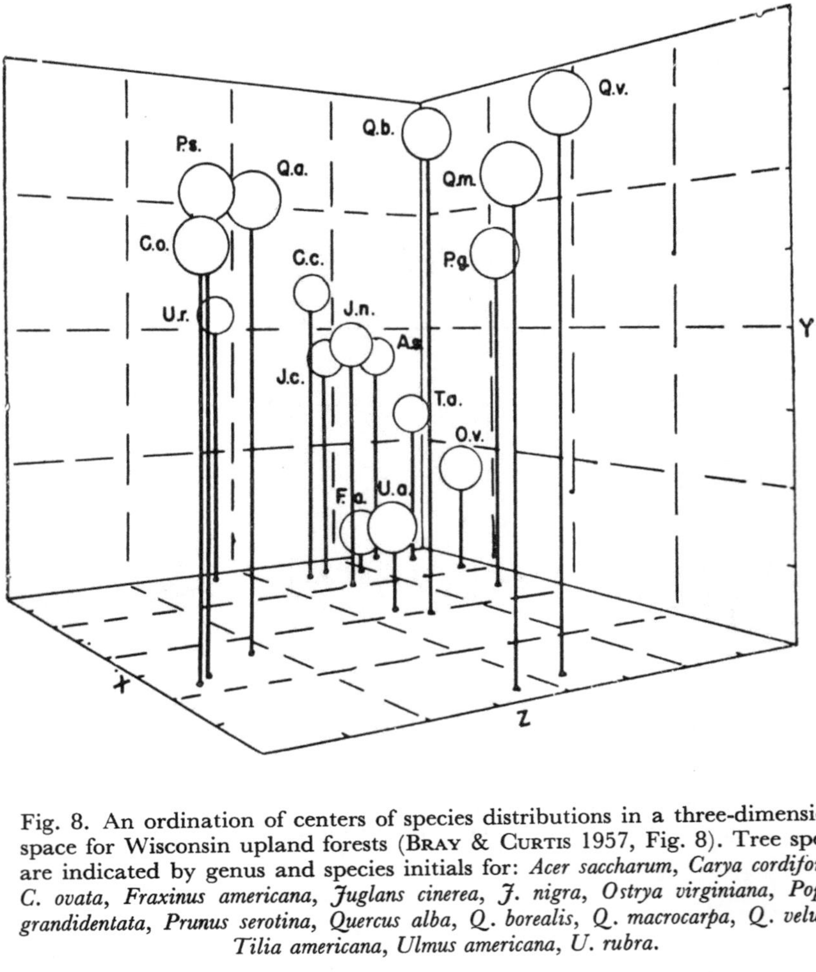

Fig. 8. An ordination of centers of species distributions in a three-dimensional space for Wisconsin upland forests (BRAY & CURTIS 1957, Fig. 8). Tree species are indicated by genus and species initials for: *Acer saccharum*, *Carya cordiformis*, *C. ovata*, *Fraxinus americana*, *Juglans cinerea*, *J. nigra*, *Ostrya virginiana*, *Populus grandidentata*, *Prunus serotina*, *Quercus alba*, *Q. borealis*, *Q. macrocarpa*, *Q. velutina*, *Tilia americana*, *Ulmus americana*, *U. rubra*.

8.3.3 PROBLEMS AND ALTERNATIVES

Ideally, a comparative ordination may yield a pattern with a major community-type of intermediate environments in the center, and other communities radiating from it along axes interpretable as complex-gradients. In some cases the central community-type is the prevailing climax for an area (FRYDMAN & WHITTAKER 1968). Most ordination techniques, however, under some circumstances produce distorted representations of community relationships to environment (article 11). Although the BRAY & CURTIS technique is less subject to such distortion than most others, curvilinear ordinations of communities forming a 'straight' ecological series along a gradient in the field are sometimes produced. Thus in the study by BEALS & COTTAM (1960) the central part of the ordination field was occupied by samples with successional species of *Populus* and *Betula*, while other samples representing a moisture gradient were curved around these with the end-points of the gradient, the wettest and driest samples, close together on one side of the field .The ordination is not erroneous in its arrangement of samples by their floristic relationships, but neither its axes nor diagonals to them can be simply interpreted as environmental gradients. It is not uncommon for some axes of a comparative ordination, particularly those extracted after the first two or three, to be difficult to interpret. In general the likelihood of curvilinear arrangement and/or uninterpretable axes is increased by, (i) increase in the range of communities sampled and degree of difference among them and (ii) increase in the number of environmental factors affecting these communities, particularly (iii) increase in the intensity and variety of disturbance factors.

These effects emphasize first, that Wisconsin comparative ordination is by no means a fool-proof technique and second, the importance of ecological understanding from other sources to interpret its results. These effects may have for some investigators a third implication: the desirability of choosing end-point samples on the basis of known ecological relationships. Choice of end-points by maximum dissimilarities may seem more objective. In favorable cases it results in a first axis that parallels the main direction of change in community composition in the sample set in response to environment. It may also, however, result in a first axis the direction of which is determined by chance inclusion of a sample that is extreme in relation to the rest of the set. When environmental relationships are reasonably understood, there may be advantage in choosing the pairs of end-points, or at least the first one or two pairs, to represent the poles of the environmental gradients judged to be most significant. Such choice may have as a welcome, secondary effect reduction in computation; if the end-points are chosen the whole secondary matrix is not needed, for only the similarities of end-points with other samples are used. When the investigator does not feel the environmental relationships should be thus assumed, end-points may be chosen by maximum dissimilarity and the axes interpreted later as in the example. There may also be advantage in trying both or balancing together dissimilarities and known or suspected environmental relations in the choice of end-points. Research purposes are often best served by experimenting with different end-point pairs and ordinations.

A variety of similarity measurements, values from which these are subtracted to obtain distances, and transformations of distance values have been experimented with (Bannister 1968, Loucks 1962, Beals 1960, Orloci 1966, Gimingham et al. 1966, Frydman & Whittaker 1968, Maarel 1969). Because of the simple geometry of sample location (Fig. 3), most of these have only modest effects on the resulting ordination; and they will not be discussed in detail (see articles 6.5 and 11.4.4). Probably the most significant choice is that of the similarity measurement itself. Choices include coefficient of community (*CC*), percentage similarity (*PS*) applied to the raw or relativized data, percentage similarity applied to the doubly standardized data (*PS'*), and Euclidean distance (*ED*). (If species rather than samples are to be ordinated, as is also possible, there are comparable choices among measures of species association and correlation, see article 6.5.) Some preliminary suggestions on choice of measurement (see also article 11) are: (i) *CC* may be preferred to the unstandardized *PS* for sample sets of high alpha and beta diversity, and for sets with diverse and unsystematic effects of disturbance on quantitative representation of species, effects that are not themselves to be the subject of ordination axes. (ii) The unstandardized *PS* may be preferred to *CC* for sample sets of low alpha and beta diversity, particularly if species populations are not irregular (differing in samples from similar biotopes in consequence of disturbance or fluctuation). (iii) In some cases use of both *CC* and *PS* should be part of the experimenting to see what ordination best clarifies community and species relationships (article 11). (iv) Experience with Wisconsin ordinations suggests, however, that the doubly standardized *PS'* often gives an optimum combination of the advantages of *CC* and *PS*. *PS'* retains the effect of quantitative difference in species representation between samples, while avoiding the over-weighting of a very few dominant species in the similarity value; and it permits the use for ordination of different kinds of importance values for different species and strata (e.g. density, or basal area, or the Wisconsin synthetic importance value for trees, frequencies or coverages for herbs). (v) It is uncertain that *ED* is ever to be preferred for polar ordination.

There remain problems of excessive numbers of species, numbers of samples, and range of community difference. These problems of sample-set dimensions may overtax either the computer, or the ordination capacity of the Bray-Curtis technique, or both. There is in general little or no loss in ordination effectiveness from reducing the number of species by eliminating those occurring in the fewest samples. Too wide a range of beta diversity, or community difference, is a more serious problem and may confront the investigator with a dilemma. If extreme samples are chosen as end-points, many other samples will have very low or zero similarities with the end-points and cannot be effectively ordinated by comparison with them. If less extreme samples, from within the range of community variation in the set, are chosen, there may be risk that more extreme samples will be reflected back to locations within the ordination field, and that ecological relationships may thereby be made obscure.

Three directions for solving problems of excessive number of samples or range in compositional difference are possible: (i) The number of samples may be reduced by random selection, or by a systematic selection of samples representing the range of different biotopes or community-types in the set. Such selection, that may be combined with reduction in number of species, may make a set of excessive size manageable for computer treatment, or may make a hand ordination possible if a computer is not available. It will not generally solve the problem of too wide a range of community difference. (ii) The samples may be classified into subsets representing community-types, composite samples (each averaging or summarizing the characteristics of a subset) may be obtained, and the composite samples may be ordinated. The procedure has the same advantages as (i) and may more effectively utilize the information available. Because the composite samples are richer in species than the individual samples their composition may overlap more, and the composite samples may be more effectively ordinated than individual samples when the range of community difference is wide. (iii) The sample set may be divided into two or more larger subsets, each comprising some range of environmental and community difference. Ordination may now be carried out within these manageable subsets to give results that are clearer and less subject to distortion. If the approach is direct, end-points for the first axes in each subset may be so chosen as to link together the subset ordinations; and end-points of the second (and higher) axes may be chosen to give parallel ordinations in relation to the second (and higher) axes. Such coordination of subsets is not likely to be easy but in principle, in favorable cases, it may permit comparative ordination of a sample set otherwise too diverse to handle by the Bray & Curtis or any other technique now known.

8.3.4 Modifications

Polar ordination differs from compositional gradient analysis in that: (i) Stands are ordered directly rather than by species indices, and (ii) multidimensional models are constructed rather than single-dimensional 'gradients.' It is worth noting that the gradient analysis approach can easily be extended to multiple directions. However, this was not understood at the time of the first polar ordination, and the extension from unidimensional to multidimensional models was one of the primary incentives for the new technique. Further examples of polar ordination include AYYAD & DIX (1964), McINTOSH & HURLEY (1964), ASHTON (1964), GITTINS (1965a, c), WHITE (1965), LARSEN (1965), SWAN & DIX (1966), FRYDMAN & WHITTAKER (1968), MAAREL (1969), and ROGERS (1970). A number of experiments with and modifications of the technique deserve mention.

MAYCOCK & CURTIS (1960) used species adaptation values and compositional indices to select the end-point samples for the first (direct) ordination axes, and sample dissimilarities to select the end-points for the second and third (indirect) axes. BRAY (1956) and SWINDALE & CURTIS (1957) used comparative ordination to ordinate species instead of samples and to derive adaptation values for compositional ordinations; GITTINS (1965a, b) has also used the technique to ordinate species. Over the past 4 or 5 years, students at Wisconsin have applied the polar ordination procedure to several different types of stand similarity matrices, including correlation, covariance, association, root mean square, Chi-square, and average absolute difference matrices. GOFF & COTTAM (1967) made use of residual similarity matrices for non-primary dimensions and based the selection of axis ends on standard deviation criteria rather than simply proximity on a former dimension. They also outlined an extension of ordination, termed 'index iteration,' which amounts essentially to characteristic vector analysis on an arbitrary numerical scale. This technique arose as an extension of the polar ordination technique.

ORLOCI (1966) and SWAN et al. (1969) have suggested ordinations employing *ED* instead of *PS'* to locate samples on strictly orthogonal axes. Later studies suggest that *ED* may give less effective ordination than *PS'*, *PS*, or *CC*, and that there may be no evident advantage in the use of strictly orthogonal axes and the additional computations necessary to produce these (BANNISTER 1968). SWAN (1970) suggested means of compensating for the effect of zero values on ordination. GAUCH (1972, article 4) shows that the relation of similarity values to separation along an environmental gradient fits a non-standardized complement of an error function, and that correction of distance values by the inverse of the error function gives improvement in ordination.

Polar ordination has been applied in several studies involving other than phytosociological data. For example, HOLE & HIRONAKA (1960) and MONK (1965) have used the technique to order samples by soil characteristics. BEALS (1960) applied parallel ordination procedures to birds and forest tree community data and then compared the patterns to infer relationships between these two biological strata. LOUCKS (1962) used a similar approach to compare the vegetation and environment of New Brunswick forests. His environmental model was constructed by numerically combining several environmental variables into three synthetic 'scalars' which were then plotted orthogonally to form the three-dimensional model. This was compared to a polar ordination of vegetation.

KNIGHT used polar ordination in his analysis of the 'structural-functional' aspects of Wisconsin prairies (1965) and forests (KNIGHT & LOUCKS 1969). In place of species he used such features as life-form, bark thickness, time of blooming, and drought resistance as the basis for stand comparison and ordination. After constructing the ordination he expressed stand positions in terms of new axes that seemed to account for the greatest amount of variation in the characteristics initially used to construct the model. These characteristics were then scaled according to the positions of their peaks along the new axes. This amounts to an axis rotation not greatly different from the type commonly used in factor analysis, but again it was independently derived and is mathematically less demanding than other possible methods.

In '*The Vegetation of Wisconsin*' CURTIS (1959) employed ordination to order entire community-types, rather than individual stands (article 7). All species that were 'prevalent' in any community were included in the analysis. Each of 28 terrestrial community-types was represented by the percentage presence of these prevalent species in samples of the type. Index of similarity values were calculated between these composite samples for community-types, but the model was constructed by means of direct triangulation, starting with the most central community and building outward, rather than by use of end-points. REAM (1963) likewise constructed a model of community-types, in this case those of the Wasatch mountain range in central Utah. He used polar ordination for both the initial stand model and the subsequent model of types. Polar ordination has also been applied to community-types by FRYDMAN & WHITTAKER (1968) and MAAREL (1969).

8.4 Conclusion

In conclusion, the principal contributions of the Wisconsin school appear to fall into three major categories. The first is in contributing to and stimulating research toward analysis of vegetation as a more or less continuous variable rather than as categorized types. The approach used at Wisconsin was based on the philosophy that the plants themselves are their own best indicators of environmental conditions, and that when ordered as described above the relationships of stands or species to the environmental complexes to which the plants were responding should be discernible. Much remains to be done in defining these relationships. It seems clear that the plants are seldom ordered along single-factor gradients, and that modeling of the operating of environment through time upon growth of individuals and populations will be necessary to explain in detail the correlations between species populations and environmental complexes. Yet the attempt to relate phytosociological ordinations and synthetic environmental variables, including stand structural variables containing information related to development of the included populations through time, represents a first-order attempt to define and deal quantitatively with ecological reality at the regional or landscape level.

The second major contribution is in the variety of techniques that were devised or adapted to refine the ordination of the various communities to which the methods were applied. No two of the major papers were based on identical techniques (see also Mc Intosh 1967, Whittaker 1967). It appeared at one time that principal components analysis or other more formal techniques might replace polar ordination as the preferred technique of vegetation analysis (Orloci 1966, Austin & Orloci 1966). Further experience with principal components analysis has, however, been disappointing because of the technique's vulnerability to distortion (Noy-Meir & Austin 1970, Jeglum et al. 1971, Austin & Noy-Meir 1972, article 11). Wisconsin comparative ordination may well be the sturdiest and most versatile of ordination techniques, equally adaptable to direct and indirect applications, as well as the least demanding of computational effort.

A third and most important contribution to date was the application of ordination methods to the entire vegetation of a major geographical unit, the state of Wisconsin. J. T. Curtis's book, '*The Vegetation of Wisconsin*' (1959, 1971), which is a synthesis of the work of many Wisconsin students, stands as a landmark in vegetational studies and provides a comprehensive treatment of the whole vegetation of an area.

8.5.1 Summary

Early ordination in Wisconsin was based on compositional gradient analysis. Vegetation samples were classified into dominance-types, and samples representing these types were arranged into an ecological series by the slopes of the populations of dominant species ('method of leading dominants'). Species were given numbers by their relative positions in the ecological series, and the numbers were used as weights to obtain compositional indices for samples expressing their positions along the gradient. The samples were then ordinated and compiled into transects by means of the compositional indices .

Later work has been based on the Wisconsin comparative or polar ordination of Bray & Curtis (1957) and variants thereof. Given a set of vegetation samples to be ordinated, similarity measures may be computed for each sample compared with every other sample. Two end-point samples are chosen either by maximum dissimilarity from the rest of the set and one another, or because they represent extremes of an environmental gradient. The similarity measures are used to arrange all other samples in sequence between the end-point pair along the axis. A second pair of end-point samples dissimilar from the first pair and one another are chosen, and all samples are arranged along the axis between the second pair. The two axes become coordinates of an ordination field in which all the samples may be located, and in which values for environmental variables, species populations, and community characteristics may be plotted at the points for samples. Samples may be ordinated in relation to third and higher axes when this is appropriate.

The technique is illustrated with a simple example. A variety of modifications that have been experimented with in the school of Wisconsin or suggested by others are mentioned. Experience and research comparing different ordinations suggest: (i) the continued value of Wisconsin comparative ordination as a most versatile and least vulnerable technique, (ii) its equal effectiveness for direct and indirect ordination, and (iii) its real value in research as a framework for investigating and understanding of vegetational relationships. Principal contributions of Wisconsin ecology include the development of the technique, the exploration of various alternatives, and the achievement by Curtis (1959) of an ordination and description of the vegetation of the state.

8.5.2 Zusammenfassung

Die frühe Ordination in Wisconsin berühte auf der vergleichenden Gradienten-Analyse. Vegetationsaufnahmen wurden in Dominanz-Typen klassifiziert und die Aufnahmen dieser Typen wurden nach dem Abfall der Populationen von dominierenden Arten zu einer ökologischen Reihe geordnet ('Methode der Leit-Dominanten'). Die Arten erhielten auf Grund ihrer relativen Lagen in den ökologischen Reihen eine Zahl und diese Zahlen wurden als Gewichte verwendet um Ordnungs-Indices für die Aufnahmen zu erhalten, die ihre Lage längs des Gradienten ausdrücken. Die Aufnahmen wurden dann mittels der Ordnungs-Indices in Transekten angeordnet und zusammengefaßt.

Spätere Arbeiten wurden auf der vergleichenden oder polaren Ordination der Wisconsin-Schule von Bray & Curtis (1957) und den Varianten dieser Methode gegründet. Für die Ordination einer Reihe von Vegetationsaufnahmen können für jede Aufnahme im Vergleich zu jeder anderen Ähnlichkeits-Werte errechnet werden. Zwei Endpunkt-Aufnahmen werden entweder auf Grund ihrer großten Unähnlichkeit mit den Übrigen Aufnahmen und unter einander oder weil Sie Extreme eines Standortsgradienten darstellen ausgewählt.

Die Verwandschaftswerte werden verwendet um alle übrigen Aufnahmen zwischen dem Paar der Endpunktaufnahmen längs der Achse anzuordnen. Ein zweites Paar von dem ersten Paar und unter einander unähnlichen Endpunkt-Aufnahmen werden ausgewählt und alle Aufnahmen werden längs des Achse zwischen dem zweiten Paar angeordnet. Die beiden Achsen werden Koördinaten eines Ordinationsfeldes, in welchem alle Aufnahmen geortet und in welchen Werte für Standortsvariabeln, Art-Populationen und Gesellschaftsmerkmale an den Aufnahme Punkten eingezeichnet werden können.

Die Technik wird durch ein einfaches Beispiel erläutert. Eine Mannigfaltigkeit von Modifikationen, mit denen in der Wisconsin-Schule experimentiert worden ist, oder die von Anderen vorgeschlagen worden sind, werden erwähnt. Erfahrung und vergleichende Untersuchung verschiedener Ordinationen ergeben: 1) den dauernden Wert der vergleichenden Ordination der Wisconsin-Schule als eine vielseitige wenigst empfindliche Technik, 2) ihre gleiche Brauchbarkeit für direkte und indirekte Ordination, und 3) ihre wirklichen Wert in der Forschung als Rahmen für die Untersuchung und daß Verständnis von Vegetationsbeziehungen. Haupt-Beiträge der Wisconsin-Ökologie umfassen die Entwicklung der Technik, die Untersuchung verschiedener Alternativen und die Vollendung einer Ordination und einer Beschreibung der Vegetation des Staates Wisconsin durch Curtis (1959).

REFERENCES

Anderson, D. J., – 1963 – The structure of some upland plant communities in

Caernarvonshire. III. The continuum analysis. *J. Ecol.* 51: 403–414.
ANDERSON, D. J., R. C. COOKE, T. T. ELKINGTON, & D. J. READ, – 1966 – Studies on structure in plant communities. II. The structure of some dwarf-heath and birch-copse communities in Skjaldfannardalur, north-west Iceland. *J. Ecol.* 54: 781–793.
ASHTON, P. S., – 1964 – Ecological studies in the mixed dipterocarp forests of Brunei State. *Oxford For. Mem.* 25: 1–75.
AUSTIN, M. P. & I. NOY-MEIR, – 1972 – The problem of non-linearity in ordination: experiments with two-gradient models. *J. Ecol.* 59: 763–774.
AUSTIN, M. P. & L. ORLOCI, – 1966 – Geometric models in ecology. II. An evaluation of some ordination techniques. *J. Ecol.* 54: 217–227.
AYYAD, M. A. G. & R. L. DIX, – 1964 – An analysis of a vegetation-microenvironmental complex on prairie slopes in Saskatchewan. *Ecol. Monogr.* 34: 421–442.
BANNISTER, P., – 1968 – An evaluation of some procedures used in simple ordinations. *J. Ecol.* 56: 27–34.
BEALS, E. W., – 1960 – Forest bird communities in the Apostle Islands of Wisconsin. *Wilson Bull.* 72: 156–181.
BEALS, E. W., – 1965 – Ordination of some corticolous cryptogamic communities in south-central Wisconsin. (Russ. summ.) *Oikos* 16: 1–8.
BEALS, E. W. & G. COTTAM, – 1960 – The forest vegetation of the Apostle Islands, Wisconsin. *Ecology* 41: 743–751.
BOND, R. R., – 1957 – Ecological distribution of breeding birds in the upland forests of southern Wisconsin. *Ecol. Monogr.* 27: 351–384.
BRAY, J. R., – 1956 – A study of mutual occurrence of plant species. *Ecology* 37: 21–28.
BRAY, J. R., – 1960 – The composition of savanna vegetation in Wisconsin. *Ecology* 41: 721–732.
BRAY, J. R. & J. T. CURTIS, – 1957 – An ordination of the upland forest communities of southern Wisconsin. *Ecol. Monogr.* 27: 325–349.
BROWN, R. T. & J. T. CURTIS, – 1952 – The upland conifer-hardwood forests of northern Wisconsin. *Ecol. Monogr.* 22: 217–234.
BUELL, M. F., A. N. LANGFORD, D. W. DAVIDSON, & L. F. OHMANN, – 1966 – The upland forest continuum in northern New Jersey. *Ecology* 47: 416–432.
CHRISTENSEN, E. M., J. J. CLAUSEN, & J. T. CURTIS, – 1959 – Phytosociology of the lowland forests of northern Wisconsin. *Am. Midl. Nat.* 62: 232–247.
CHRISTENSEN, M., W. F. WITTINGHAM, & R. O. NOVAK, – 1962 – The soil microfungi of wet-mesic forests in southern Wisconsin. *Mycologia* 54: 374–388.
CLAUSEN, J. J., – 1957 – A phytosociological ordination of the conifer swamps of Wisconsin. *Ecology* 38: 638–646.
CULBERSON, W. L., – 1955 – The corticolous communities of lichens and bryophytes in the upland forests of northern Wisconsin. *Ecol. Monogr.* 25: 215–231.
CURTIS, J. T., – 1955 – A prairie continuum in Wisconsin. *Ecology* 36: 558–566.
CURTIS, J. T. –1959 – The Vegetation of Wisconsin: An Ordination of Plant Communities. Univ. Wisconsin, Madison., Reprinted 1971. 657 pp.
CURTIS, J. T. & R. P. MCINTOSH, – 1951 – An upland forest continuum in the prairie-forest border region of Wisconsin. *Ecology* 32: 476–496.
FRYDMAN, I. & R. H. WHITTAKER, – 1968 – Forest associations of southeast Lublin Province, Poland. (Germ. summ.) *Ecology* 49: 896–908.
GAUCH, H. G., Jr., – 1972 – The relationship between sample similarity and ecological distance. (manuscript).
GEHLBACH, F. R., – 1967 – Vegetation of the Guadalupe Escarpment, New Mexico-Texas. *Ecology* 48: 404–419.

Gemborys, S. R. & E. J. Hodgkins, – 1971 – Forests of small stream bottoms in the Coastal Plain of southwestern Alabama. *Ecology* 52: 70–84.

Gilbert, M. L. & J. T. Curtis, – 1953 – Relation of the understory to the upland forest in the prairie-forest border region of Wisconsin. *Trans. Wis. Acad. Sci. Arts Lett.* 42: 183–195.

Gimingham, C. H., N. M. Pritchard, & R. M. Cormack, – 1966 – Interpretation of a vegetational mosaic on limestone in the island of Gotland. *J. Ecol.* 54: 481–502.

Gittins, R., – 1965a – Multivariate approaches to a limestone grassland community. I. A stand ordination. *J. Ecol.* 53: 385–401.

Gittins, R., – 1965b – Multivariate approaches to a limestone grassland community. II. A direct species ordination. *J. Ecol.* 53: 403–409.

Gittins, R., – 1965c – Multivariate approaches to a limestone grassland community. III. A comparative study of ordination and association-analysis. *J. Ecol.* 53: 411–425.

Goff, F. G. & G. Cottam, – 1967 – Gradient analysis: the use of species and synthetic indices. *Ecology* 48: 793–806.

Greig-Smith, P., – 1964 – Quantitative Plant Ecology. 2nd ed. Butterworths, London. 256 pp.

Hale, M. E., Jr., – 1955 – Phytosociology of corticolous cryptogams in the upland forests of southern Wisconsin. *Ecology* 36: 45–63.

Hole, F. D. & M. Hironaka, – 1960 – An experiment in ordination of some soil profiles. *Proc. Soil Sci. Soc. Am.* 24: 309–312.

Jeglum, J. K., C. F. Wehrhahn, & J. M. A. Swan, – 1971 – Comparisons of environmental ordinations with principal component vegetational ordinations for sets of data having different degrees of complexity. (French summ.) *Can. J. For. Res.* 1: 99–112.

Kendall, M. G., – 1965 – A Course in Multivariate Analysis. Griffins, London.

Knight, D. H., – 1965 – A gradient analysis of Wisconsin prairie vegetation on the basis of plant structure and function. *Ecology* 46: 744–747.

Knight, D. H. & O. L. Loucks, – 1969 – A quantitative analysis of Wisconsin forest vegetation on the basis of plant function and gross morphology. *Ecology* 50: 219–234.

Larsen, J. A., – 1965 – The vegetation of the Ennadai Lake Area, N. W. T.: studies in subarctic and arctic bioclimatology. *Ecol. Monogr.* 35: 37–59.

Lindsey, A. A., R. O. Petty, D. K. Sterling, & W. van Asdall, – 1961 – Vegetation and environment along the Wabash and Tippecanoe Rivers. *Ecol. Monogr.* 31: 105–156.

Loucks, O. L., – 1962 – Ordinating forest communities by means of environmental scalars and phytosociological indices. *Ecol. Monogr.* 32: 137–166.

Maarel, E. van der, – 1969 – On the use of ordination models in phytosociology. (Germ. summ.) *Vegetatio* 19: 21–46.

Maycock, P. F. & J. T. Curtis, – 1960 – The phytosociology of boreal conifer-hardwood forests of the Great Lakes region. *Ecol. Monogr.* 30: 1–35.

McIntosh, R. P., – 1958 – Plant communities. *Science, N.Y.* 128: 115–120.

McIntosh, R. P., – 1967 – The continuum concept of vegetation. *Bot. Rev.* 33: 130–187.

McIntosh, R. P. & R. T. Hurley, – 1964 – The spruce-fir forests of the Catskill Mountains. *Ecology* 45: 314–326.

Monk, C. D., – 1965 – Southern mixed hardwood forest of northcentral Florida. *Ecol. Monogr.* 35: 335–354.

Noy-Meir, I. & M. P. Austin, – 1970 – Principal component ordination and simulated vegetational data. *Ecology* 51: 551–552.

ORLOCI, L., – 1966 – Geometric models in ecology. I. The theory and application of some ordination methods. *J. Ecol.* 54: 193–215.
REAM, R. R., – 1963 – The vegetation of the Wasatch Mountains, Utah and Idaho. *Thesis, Univ. of Wisconsin.*
ROGERS, D. J., – 1970 – A preliminary ordination study of forest vegetation in the Kirchleerau area of the Swiss Midland. (Germ. summ.) *Ber. geobot. Inst. ETH, Stiftg. Rübel,* Zürich, 1969, 40: 28–78.
ROWE, J. S., – 1956 – Uses of undergrowth plant species in forestry. *Ecology* 37: 461–473.
SEAL, H. L., – 1964 – Multivariate Statistical Analysis for Biologists. John Wiley and Sons, Inc., New York.
SWAN, J. M. A., – 1970 – An examination of some ordination problems by use of simulated vegetational data. *Ecology* 51: 89–102.
SWAN, J. M. A. & R. L. DIX, – 1966 – The phytosociological structure of upland forest at Candle Lake, Saskatchewan. *J. Ecol.* 54: 13–40.
SWAN, J. M. A., R. L. DIX, & C. F. WEHRHAHN, – 1969 – An ordination technique based on the best possible stand-defined axes and its application to vegetational analyss. *Ecology* 50: 206–212.
SWINDALE, D. N. & J. T. CURTIS, – 1957 – Phytosociology of the larger submerged plants in Wisconsin lakes. *Ecology* 38: 397–407.
TRESNER, H. D., M. P. BACKUS, & J. T. CURTIS, – 1954 – Soil microfungi in relation to the hardwood forest continuum in southern Wisconsin. *Mycologia* 46: 314–333.
WARING, R. H. & J. MAJOR, – 1964 – Some vegetation of the California coastal redwood region in relation to gradients of moisture, nutrients, light, and temperature. *Ecol. Monogr.* 34: 167–215.
WHITE, K. L., – 1965 – Shrub-carrs of southeastern Wisconsin. *Ecology* 46: 286–304.
WHITTAKER, R. H., – 1951 – A criticism of the plant association and climatic climax concepts. *NW. Sci.* 25: 17–31.
WHITTAKER, R. H., – 1956 – Vegetation of the Great Smoky Mountains. *Ecol. Monogr.* 26: 1–80.
WHITTAKER, R. T., – 1967 – Gradient analysis of vegetation. *Biol. Rev.* 42: 207–264.

9 L'ANALYSE FACTORIELLE

PIERRE DAGNELIE

Sommaire

9 L'ANALYSE FACTORIELLE

9.1 Définition

Au départ, l'analyse factorielle réunissait un ensemble de méthodes mises au point en *psychologie expérimentale* et utilisées surtout dans l'analyse statistique des résultats de tests d'intelligence ou de comportement (Burt 1949, Harman 1960, Vincent 1953). Progressivement cependant, ces méthodes ont été intégrées dans le cadre des méthodes statistiques classiques (Kendall 1950, Kendall & Lawley 1956, Lawley 1955, Lawley & Maxwell 1963), et actuellement, l'analyse factorielle trouve sa place parmi les méthodes d'*analyse statistique à plusieurs variables*, en compagnie notamment de l'analyse des composantes, de l'analyse discriminante et de l'analyse de la dispersion (Anderson 1958, Dagnelie 1966, Kendall 1957, Seal 1964).

9.2 Un exemple simple

Le meilleur moyen de comprendre le sens et les principales notions de l'analyse factorielle est d'envisager un exemple simple, choisi dans le domaine psychologique.

Considérons *trois tests* auxquels peuvent être soumis un certain nombre de personnes, et désignons par z_1, z_2 et z_3, les variables observées qui correspondent aux résultats des trois tests. Nous supposerons que les cotes ('scores') z_j $(j=1,2,3)$ dépendent linéairement de deux *variables fondamentales* x_1 et x_2 communes à deux ou plusieurs tests et, dans chaque cas, d'une variable fondamentale v_j propre au test envisagé. Le *modèle mathématique* considéré est donc:

$$z_j = a_{j1}x_1+a_{j2}x_2+u_jv_j \qquad (j = 1,2,3),$$

les coefficients a_{j1}, a_{j2} et u_j, appelés *saturations* ('loadings'), indiquant dans quelle mesure les cotes z_j dépendent respectivement des variables x_1, x_2 et v_j.

Nous supposerons en outre que toutes les variables considérées sont *réduites*, c'est-à-dire de moyenne nulle et d'écart-type ou de variance unitaire, et que les variables fondamentales x_1, x_2 et v_j $(j=1,2,3)$ sont indépendantes les unes des autres. Dans ces conditions, les saturations sont en réalité les coefficients de corrélation des variables fondamentales avec les variables observées.

D'une manière plus concrète, les trois tests envisagés pourraient être constitués respectivement de *problèmes de raisonnement* (z_1), de *problèmes d'arithmétique* (z_2) et de *calculs d'addition* (z_3). Les aptitudes fondamentales susceptibles d'influencer les résultats obtenus par les différentes personnes pourraient être, dans ce cas, une aptitude au raisonnement (x_1) et une aptitude au calcul (x_2). La variable z_1 serait alors étroitement liée à x_1; de même, z_3 dépendrait surtout de x_2, tandis que z_2 serait largement influencé à la fois par x_1 et par x_2. On pourrait observer par exemple les relations suivantes:

$$z_1 = 0{,}70x_1+0{,}10x_2+0{,}71v_1$$
$$z_2 = 0{,}40x_1+0{,}70x_2+0{,}59v_2$$
$$z_3 = 0{,}10x_1+0{,}70x_2+0{,}71v_3,$$

dans lesquelles v_1, v_2 et v_3 sont les variables fondamentales propres à chacun des tests.

Les saturations des facteurs communs x_1 et x_2 peuvent être réunies en un tableau à double entrée (tableau I), auquel on ajoute généralement les *variances communes:*

$$h_j^2 = a_{j1}^2+a_{j2}^2 \qquad (j = 1,2,3),$$

et les *variances dues à chacun des facteurs:*

$$\lambda_i = a_{1i}^2+a_{2i}^2+a_{3i}^2 \qquad (i = 1,2).$$

Les *variances communes*, plus souvent appelées *communautés* ('communalities'), mesurent l'importance de l'ensemble des facteurs communs dans la détermination des résultats de chacun des tests; tandis que les variances dues aux facteurs mesurent l'importance relative des différents facteurs communs dans la détermination des résultats de l'ensemble des tests[1]).

TABLEAU I. Saturations, variances communes et variances dues aux facteurs communs x_1 et x_2, pour trois tests psychologiques.

j	a_{j1}	a_{j2}	h_j^2
1	0,70	0,10	0,50
2	0,40	0,70	0,65
3	0,10	0,70	0,50
λ_i	0,66	0,99	1,65

1) Pour éviter les confusions, nous utiliserons de préférence l'expression 'variance commune', laissant au mot 'communauté' son sens proprement phytosociologique.

Les saturations du tableau I peuvent également être présentées sous la forme d'une *matrice de corrélation*, réunissant les coefficients de corrélation entre les résultats des différents tests pris deux à deux. On démontre en effet facilement que, pour tout couple de tests (j,k):

$$r_{jk} = a_{j1}a_{k1} + a_{j2}a_{k2},$$

ce qui signifie simplement que les tests sont d'autant plus étroitement corrélés qu'ils dépendent dans une plus large mesure des mêmes aptitudes fondamentales. Les coefficients obtenus de cette manière pour l'exemple considéré ici sont réunis dans le tableau II.

TABLEAU II. Matrice de corrélation entre trois tests psychologiques.

j \ k	1	2	3
1	1	0,35	0,14
2	0,35	1	0,53
3	0,14	0,53	1

Mais en réalité, le problème de l'analyse factorielle se pose pratiquement dans le sens opposé à celui que nous avons envisagé jusqu' à présent: après avoir soumis un certain nombre de personnes à un certain nombre de tests, on calcule les coefficients de corrélation observés et on s'efforce d'en déduire un tableau de saturations. Dans ces conditions, l'analyse factorielle a notamment pour but, d'une part, d'*expliquer les corrélations observées* entre les tests en mettant en évidence certaines aptitudes fondamentales sous-jacentes et, d'autre part, d'identifier autant que possible ces aptitudes fondamentales.

En outre, à partir du moment où on considère des valeurs observées et non des valeurs théoriques, il faudrait ajouter au modèle mathématique un *terme aléatoire d'erreur*. Celui-ci est souvent englobé dans les facteurs spécifiques, qui pourraient alors être divisés pour chaque test en une composante spécifique proprement dite (systématique) et une composante d'erreur (aléatoire). En ce qui concerne notre exemple, qui n'est nullement artificiel, les *corrélations réellement observées* pour un ensemble de 145 personnes sont données par HARMAN (1967), et diffèrent à peine de celles du tableau II.

Cependant, posé de cette façon, le problème de l'analyse factorielle admet généralement une *infinité de solutions*. Les corrélations observées peuvent non seulement être expliquées en supposant l'existence des facteurs communs x_1 et x_2, intervenant respectivement dans les proportions a_{j1} et a_{j2} (tableau I), mais aussi en supposant par exemple l'existence des facteurs communs x'_1 et x'_2 et des saturations correspondantes données au tableau III. A partir de ce dernier tableau, nous pouvons en effet recalculer les coefficients de corrélation, comme précédemment, à l'aide de la relation:

$$r_{jk} = a'_{j1}a'_{k1} + a'_{j2}a'_{k2}.$$

Il est donc nécessaire de choisir, parmi cette infinité de solutions, la ou les solutions les plus intéressantes. Différentes méthodes de calcul des saturations permettent de réaliser ce choix.

TABLEAU III. Saturations, variances communes et variances dues aux facteurs communs x'_1 et x'_2, pour trois tests psychologiques.

j	a'_{j1}	a'_{j2}	h_j^2
1	0,50	−0,50	0,50
2	0,80	0,10	0,65
3	0,62	0,34	0,50
λ'_1	1,27	0,38	1,65

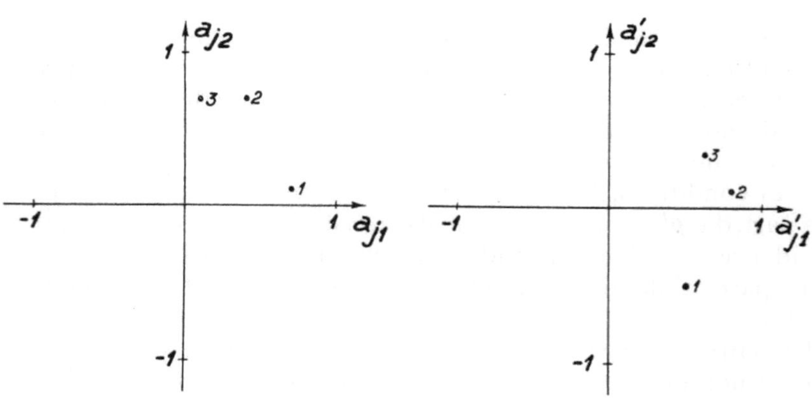

Fig. 1. Représentation graphique de trois tests psychologiques dans deux systèmes d'axes.

L'existence d'une infinité de solutions ne prête d'ailleurs pas à conséquence, comme le montre la *représentation graphique des tests*. En attribuant à chaque facteur un axe de coordonnées dans un système cartésien rectangulaire, il est en effet possible de faire correspondre à chaque test un point dont les coordonnées sont égales aux saturations. A titre d'exemple, la figure 1 donne la représentation des trois tests considérés jusqu'à présent (1, 2 et 3) dans deux systèmes d'axes, correspondant aux deux solutions envisagées (tableaux I et III).

La comparaison de ces deux graphiques montre que les positions relatives de l'origine et des trois points représentant les tests sont identiques: pour passer d'un système de facteurs à l'autre, il suffit en fait d'opérer une *rotation des axes* autour de l'origine. Inversement, les saturations obtenues en projetant les points-tests sur tout autre système d'axes orthogonaux issus de la même origine permettent de reconstruire comme ci-dessus la matrice de corrélation.

Enfin, les cotes initiales étant connues pour un certain nombre de personnes et les saturations étant calculées, on peut également se fixer comme but d'estimer, pour les différentes personnes, les valeurs des variables fondamentales, c'est-à-dire les cotes des différentes personnes en les différentes aptitudes fondamentales. Cette opération, appelée *estimation des facteurs*, permet d'obtenir une représentation graphique des personnes, en faisant correspondre à chacune d'elles un point dont les coordonnées sont les cotes fondamentales estimées.

Il est intéressant de comparer ces *deux types de représentation graphique*: celle des tests dans ce que nous appellerons l'*espace des saturations* et celle des personnes dans l'espace des facteurs communs ou *espace factoriel*. La première est telle que deux tests indiquant des aptitudes voisines sont représentés par des points situés à proximité l'un de l'autre. En outre, on montre aisément que les distances séparant les points-tests de l'origine sont égales aux racines carrées des variances communes, et sont donc le reflet de la précision avec laquelle les tests indiquent les aptitudes fondamentales communes. Dans la seconde représentation graphique, deux personnes possédant des aptitudes semblables sont également représentées par des points voisins, mais les distances séparant ces points de l'origine n'ont pas la même signification particulière.

Pour la facilité de l'interprétation, il est possible aussi de remplacer, dans les deux graphiques, les axes orthogonaux par des *axes obliques*.

9.3 Principe général

Tout ce qui vient d'être dit peut évidemment être généralisé au cas de *n tests dépendant de r facteurs* ($n \geqq r$). On a, pour chaque test ($j=1, \ldots, n$):

$$z_j = a_{j1}x_1 + a_{j2}x_2 + \ldots + a_{jr}x_r + u_j v_j = \sum_{i=1}^{r} a_{ji}x_i + u_j v_j$$

et
$$h_j^2 = a_{j1}^2 + a_{j2}^2 + \ldots + a_{jr}^2 = \sum_{i=1}^{r} a_{ji}^2;$$

pour chaque couple de tests (j et $k=1,\dots,n$):

$$r_{jk} = a_{j1}a_{k1}+a_{j2}a_{k2}+\dots+a_{jr}a_{kr} = \sum_{i=1}^{r} a_{ji}a_{ki};$$

et pour chaque facteur ($i=1,\dots,r$):

$$\lambda_i = a_{1i}^2+a_{2i}^2+\dots+a_{ni}^2 = \sum_{j=1}^{n} a_{ji}^2.$$

Pratiquement, l'analyse factorielle des cotes z_{jp}, obtenues en soumettant un certain nombre de personnes ($p=1,\dots, N$) à un certain nombre de tests ($j=1,\dots,n$), se réalise alors selon le schéma suivant.

A partir des cotes z_{jp}, on calcule le tableau ou la matrice des *corrélations observées* r_{jk}, ainsi qu'une première estimation des *variances communes* h_j^2. Cette première estimation peut notamment être obtenue à l'aide des coefficients de corrélation multiple, surtout pour de petites matrices de corrélation, ou à l'aide des valeurs les plus élevées des coefficients de corrélation observés, surtout pour de grandes matrices.

De ces coefficients de corrélation observés et de ces variances communes, on déduit ensuite les *saturations*, en utilisant le plus souvent la méthode des facteurs principaux ou la méthode centroïde. La première de ces deux méthodes a pour principe de faire intervenir le moins de facteurs possibles, en mettant en évidence les facteurs dont les variances λ_i sont les plus grandes. En termes matriciels, les variances λ_i dues aux facteurs sont d'ailleurs les valeurs propres de la matrice de corrélation dont les éléments diagonaux sont les variances communes, et les saturations a_{ji} sont les éléments des vecteurs propres correspondants[1]). La méthode centroïde ne possède pas les propriétés optimales de la méthode des facteurs principaux, mais elle donne cependant des résultats fort semblables, tout en étant d'un emploi beaucoup moins laborieux. Son emploi se justifie donc lorsque les moyens de calcul dont on dispose sont limités.

Au fur et à mesure du calcul des saturations, il y a lieu de tester autant que possible la *signification des facteurs* mis en évidence, de manière à savoir à quel moment le processus de calcul doit être interrompu. Il faut remarquer cependant que ce problème de test

1) Le calcul des valeurs et des vecteurs propres peut également être réalisé en maintenant les valeurs 1 dans la diagonale principale: cette façon de faire conduit à l'obtention des *composantes principales* ou des *axes principaux*, dans le cadre de ce qui est appelé *l'analyse des composantes*. Le but poursuivi dans ce cas est de condenser l'essentiel des informations apportées par un certain nombre de variables interdépendantes, en un nombre plus restreint de variables indépendantes (Dagnelie 1966, Kendall 1957, Kendall & Lawley 1956). Bien que fondamentalement différentes, l'analyse factorielle et l'analyse des composantes sont souvent confondues, les résultats obtenus dans les deux cas étant d'ailleurs fort semblables pour de grandes matrices de corrélation (Dagnelie 1960).

de signification n'a pas encore trouvé de solution entièrement satisfaisante dans le cadre de la méthode des facteurs principaux et de la méthode centroïde. En l'absence de tests tout à fait valables, une attention particulière doit être apportée au fait que les différents facteurs extraits peuvent ou ne peuvent pas trouver une interprétation logique.

Le nombre de facteurs à considérer étant fixé et les saturations ayant été estimées en première approximation, on peut *éventuellement* obtenir une meilleure estimation des variances communes et en déduire, comme il vient d'être dit, une meilleure estimation des saturations. Toutefois, l'utilisation de ce processus itératif ne s'avère pas toujours indispensable, surtout pour les grandes matrices de corrélation.

Enfin, il est possible d'effectuer l'*estimation des facteurs*, par régression multiple, et de passer à l'*interprétation des résultats*, à l'aide notamment des différents types de représentation graphique.

Des *compléments d'information* relatifs à ces différents problèmes peuvent être trouvés dans un travail publié antérieurement (Dagnelie 1960) ou dans certains ouvrages spécialisés, dont le plus intéressant, pour les non-psychologues, semble être à l'heure actuelle celui de Harman (1967). D'autres ouvrages, plus anciens ou plus spécialement orientés vers les problèmes psychologiques, sont dus à Burt (1940), Cattell (1952), Faverge (1954), Holzinger & Harman (1941), Horst (1965), Thomson (1956), Thurstone (1947) et Vernon (1952).

9.4 **Un exemple phytosociologique**

Nous résumerons ici, à titre d'exemple, l'étude que nous avons réalisée précédemment à partir de 80 *relevés de hêtraies* (Dagnelie 1956, 1960), établis selon la méthode de Braun-Blanquet (1951), dans un type de végétation déjà bien connu (Noirfalise 1956).

9.4.1 Corrélation entre espèces, saturations et variances communes

Dans le but de réduire autant que possible le volume des calculs, nous avons tout d'abord écarté de l'analyse les espèces les plus rares, qui ne peuvent apporter une information réellement intéressante. Pour les 38 espèces présentes au moins dans 10 % des relevés, nous avons ensuite calculé une *matrice de corrélation de point*, en tenant compte soit de la simple présence de chacune des espèces, soit pour les plus fréquentes d'entre elles, de leur plus ou moins grande abondance. Le tableau IV donne la liste des espèces considérées, et le tableau V présente une partie de la matrice de corrélation.

Les *saturations* des trois premiers facteurs ont été calculées par la méthode centroïde, en utilisant les coefficients de corrélation les plus élevés comme premières estimations des *variances communes*: le tableau IV donne les résultats ainsi obtenus. D'autre part, les tests de signification du genre du test χ^2 de Burt (1952) montrent que les deux premiers facteurs sont nettement significatifs, le troisième facteur étant juste significatif. Dans la suite, et pour simplifier notamment les représentations graphiques, nous considérerons essentiellement l'interprétation des *deux premiers facteurs*.

La première constatation a trait à l'existence d'une *relation étroite entre les variances communes des espèces et leur pouvoir différentiel*, tel qu'il est reconnu par les phytosociologues. Sur la base des trois facteurs extraits, les espèces se classent en effet comme suit dans l'ordre décroissant des variances communes:

0,40 et plus: *Festuca silvatica*, *Acer pseudoplatanus* (A), *Oxalis acetosella* et *Deschampsia caespitosa*;

de 0,39 à 0,30: *Deschampsia flexuosa*, *Anemone nemorosa*, *Fagus silvatica* (h), *Acer pseudoplatanus* (h), *Milium effusum*, *Juncus sp.*, *Agrostis vulgaris*, *Dicranum scoparium* et *Vaccinium myrtillus*;

de 0,29 à 0,20: *Quercus sessilis* (A), *Luzula nemorosa*, *Carex pilulifera*, *Picea excelsa* (h), *Polygonatum verticillatum*, *Carex remota*, *Leucobryum glaucum*, *Rubus idaeus*, *Epilobium spicatum*, *Fagus silvatica* (A), *Dryopteris spinulosa*, *Carex leporina*, *Dicranella heteromalla* et *Polytrichum formosum*;

de 0,19 à 0,10: *Quercus sessilis* (*h*), *Athyrium filix femina*, *Rubus sp.*, *Hypnum loreum*, *Fagus silvatica* (*a*), *Sorbus aucuparia* (*h*), *Thuidium tamariscinum*, *Luzula silvatica* et *Eurynchium myosuroides*;

moins de 0,10: *Mnium hornum* et *Hypnum cupressiforme*.

Les espèces généralement considérées comme 'compagnes' sont rassemblées au bas de l'échelle des variances communes. Seules viennent s'y joindre, dans les classes 'de 0,19 à 0,10' et 'moins de 0,10', *Athyrium filix femina*, *Luzula silvatica* et quelques espèces ligneuses présentes dans les strates inférieures (*Quercus sessilis*, *Fagus silvatica* et *Sorbus aucuparia*). Il faut cependant être attentif au fait que les variances communes doivent être considérées comme des indices de la valeur différentielle des espèces *au sein de l'ensemble des relevés considérés*, et non de la valeur des espèces en tant que caractéristiques ou, d'une manière plus générale, en tant qu'indicatrices écologiques: en effet, une espèce caractéristique, répartie d'une manière

TABLEAU IV. Liste des espèces observées, saturations et variances communes, pour 80 relevés de hêtraies.

j	Espèces (§)	Saturations a_{j1}	a_{j2}	a_{j3}	Variances communes (h_j^2)
1	Fagus silvatica (A)	-0,06	-0,35	-0,34	0,24
2	Acer pseudoplatanus (A)	0,39	-0,50	0,33	0,51
3	Quercus sessilis (A)	-0,48	0,18	0,17	0,29
4	Fagus silvatica (a)	-0,15	0,17	0,30	0,14
5	Fagus silvatica (h)	0,54	0,29	0,09	0,38
6	Acer pseudoplatanus (h)	0,23	-0,16	0,54	0,38
7	Quercus sessilis (h)	-0,09	0,41	0,13	0,19
8	Sorbus aucuparia (h)	-0,31	0,10	-0,12	0,12
9	Milium effusum	0,55	-0,21	0,11	0,36
10	Festuca silvatica	0,65	-0,26	0,21	0,54
11	Luzula nemorosa	0,50	0,15	-0,11	0,29
12	Oxalis acetosella	0,59	-0,11	0,36	0,48
13	Deschampsia caespitosa	0,65	-0,08	0,17	0,46
14	Juncus sp.	0,51	0,30	0,08	0,36
15	Carex remota	0,34	0,29	-0,25	0,26
16	Carex leporina	0,44	0,10	0,15	0,23
17	Vaccinium myrtillus	-0,47	0,27	0,18	0,32
18	Deschampsia flexuosa	-0,53	0,30	-0,11	0,39
19	Carex pilulifera	0,14	0,24	-0,46	0,29
20	Polytrichum formosum	0,09	0,35	0,25	0,20
21	Dicranum scoparium	-0,33	0,33	0,35	0,34
22	Dicranella heteromalla	0,17	-0,31	-0,29	0,21
23	Agrostis vulgaris	0,36	0,42	-0,23	0,35
24	Rubus idaeus	0,41	0,26	-0,12	0,25
25	Rubus sp.	0,31	0,25	-0,03	0,16
26	Dryopteris spinulosa	0,33	0,16	0,32	0,24
27	Mnium hornum	-0,15	-0,22	0,13	0,09
28	Thuidium tamariscinum	-0,13	0,13	0,29	0,12
29	Eurynchium myosuroides	-0,26	-0,18	-0,11	0,11
30	Hypnum cupressiforme	-0,13	0,08	0,19	0,06
31	Hypnum loreum	-0,22	0,16	0,29	0,16
32	Picea excelsa (h)	0,16	0,43	-0,24	0,27
33	Anemone nemorosa	0,51	-0,30	0,19	0,39
34	Polygonatum verticillatum	0,47	-0,16	0,16	0,27
35	Athyrium filix femina	0,38	-0,08	0,21	0,19
36	Luzula silvatica	0,18	0,27	0,11	0,12
37	Leucobryum glaucum	-0,38	0,11	0,32	0,26
38	Epilobium spicatum	0,47	0,13	0,13	0,25
	Variances (λ_i)	5,56	2,48	2,23	10,27

(§) A = strate arborescente; a = strate arbustive; h = strate herbacée.

aléatoire dans tout le groupement végétal considéré, n'est théoriquement liée, au sein du groupement, à aucune autre espèce. Sa variance commune doit théoriquement y être nulle, et il serait donc faux d'affirmer que les variances communes faibles sont l'apanage d'espèces dont le pouvoir indicateur est faible.

TABLEAU V. Extrait de la matrice de corrélation relative à 38 espèces étudiées dans 80 relevés de hêtraies (seules les dix premières espèces sont considérées).

Espèces (j)	1	2	3	4	5	6	7	8	9	10
1	1									
2	0,059	1								
3	—0,285	—0,261	1							
4	—0,234	—0,040	0,175	1						
5	—0,183	0,156	—0,193	0,096	1					
6	—0,150	—0,515	—0,020	0,033	0,227	1				
7	—0,058	—0,088	0,082	0,282	0,275	0,255	1			
8	0,063	—0,109	0,152	0,089	—0,128	—0,194	0,181	1		
9	0,008	0,389	—0,319	0,015	0,200	0,119	—0,126	—0,148	1	
10	0,029	0,427	—0,302	—0,041	0,358	0,338	—0,164	—0,334	0,444	1
⋮	⋮	⋮	⋮	⋮	⋮	⋮	⋮	⋮	⋮	⋮

9.4.2 GROUPES ÉCOLOGIQUES ET SOCIOLOGIQUES

La figure 2, établie à partir des saturations du tableau IV, montre comment les espèces considérées se répartissent dans l'espace factoriel relatif aux deux premiers facteurs. Sur la base de ce graphique, nous avons pu définir *quatre groupes d'espèces*.

Un premier groupe (*A*) réunit les espèces différentielles des sols frais et riches: *Anemone nemorosa*, *Festuca silvatica*, *Milium effusum*, *Polygonatum verticillatum*, *Athyrium filix femina*, *Oxalis acetosella* et *Deschampsia caespitosa*.

Le deuxième groupe d'espèces (*B*) comprend essentiellement les différentielles des sols frais à humides, mais pas spécialement riches: *Carex leporina*, *Epilobium spicatum*, *Luzula nemorosa*, *Dryopteris spinulosa*, *Rubus idaeus*, *Rubus sp.*, *Fagus silvatica* (*h*), *Juncus sp.* et *Carex remota*. Quant au groupe *C*, il englobe les espèces des sols pauvres et fréquemment secs: *Dicranum scoparium*, *Deschampsia flexuosa*, *Vaccinium myrtillus*, *Quercus sessilis* (*A*) et *Leucobryum glaucum*.

Enfin, un dernier groupe beaucoup moins bien défini (*D*) réunit les 'compagnes', qui se trouvent à proximité de l'origine du système d'axes: la distance de l'origine au point qui représente une espèce est en effet égale à la racine carrée de la variance commune de cette espèce. La circonférence qui a été tracée en prenant l'origine comme centre correspond à une variance commune de 0,10: son rayon est égal à $\sqrt{0,10}$. Dans le cercle ainsi défini et aux alentours immédiats, se trouvent localisées les espèces suivantes: *Carex pilulifera*, *Luzula*

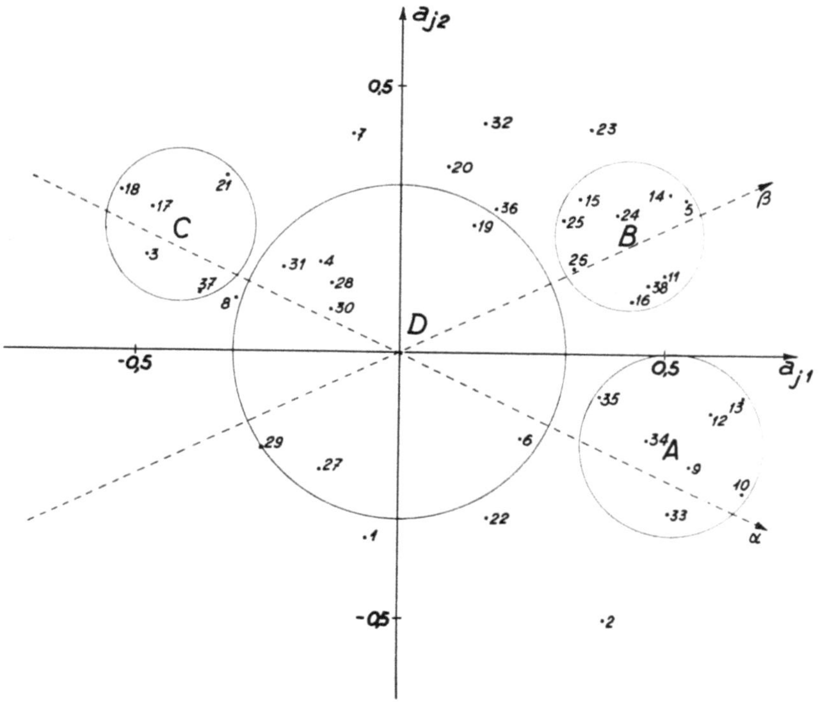

Fig. 2. Répartition des 38 espèces en fonction des saturations des deux premiers facteurs.

silvatica, Polytrichum formosum, Fagus silvatica (*a*), *Thuidium tamariscinum, Hypnum cupressiforme, Hypnum loreum, Sorbus aucuparia* (*h*), *Eurynchium myosuroides, Mnium hornum, Fagus silvatica* (*A*), *Dicranella heteromalla* et *Acer pseudoplatanus* (*h*).

Les trois premiers groupes contiennent chacun des espèces qui sont assez étroitement liées entre elles, et vraisemblablement liées aussi à des conditions écologiques particulières: ils peuvent être considérés comme des *groupes écologiques ou sociologiques*. Les espèces du dernier groupe, par contre, n'ont vraisemblablement pas de signification écologique précise, à l'intérieur du type de végétation considéré, sauf peut-être celles qui sont les plus éloignées de l'origine, notamment *Luzula silvatica* et *Sorbus aucuparia*. Enfin, quelques espèces n'ont pas encore été citées: *Agrostis vulgaris, Picea excelsa* (*h*), *Quercus sessilis* (*h*) et *Acer pseudoplatanus* (*A*). Certaines d'entre elles peuvent éventuellement être rattachées aux groupes sociologiques qui viennent d'être définis, notamment *Acer pseudoplatanus* (groupe *A*) et *Agrostis vulgaris* (groupe *B*).

La signification écologique donnée aux espèces est, elle aussi, valable uniquement pour l'ensemble des relevés étudiés. C'est ainsi qu'*Epilobium spicatum*, généralement considérée comme une 'espèce de coupes', est associée *dans les hêtraies étudiées* aux différentielles de fraîcheur.

Les facteurs orthogonaux mis en évidence par l'analyse factorielle ne peuvent cependant pas être interprétés tels quels: dans le cas présent, ils peuvent avantageusement être remplacés par deux autres facteurs, qui ne sont plus indépendants l'un de l'autre. Ces *deux facteurs obliques* sont matérialisés sur la figure 2 par les axes α et β, qui font chacun un angle de 25° avec l'axe du premier facteur initial (a_{j1}). L'un (α) permet d'opposer directement les espèces des sols pauvres et généralement secs (groupe *C*) aux espèces des sols frais et riches (groupe *A*); l'autre (β) est dirigé vers le groupe des différentielles des sols frais à humides (groupe *B*).

9.4.3 Classification des relevés

D'autre part, des groupes sociologiques étant ainsi définis, il est possible d' établir une *classification des relevés* en fonction de l'importance relative de ces différents groupes. Quatre types de relevés peuvent ainsi être distingués: les premiers réunissent à la fois des espèces des sols riches et des sols frais à humides (groupes *A* et *B*); les seconds contiennent surtout des espèces différentielles des sols frais à humides (groupe *B*); les troisièmes réunissent ces dernières espèces et celles des sols pauvres et généralement secs (groupes *B* et *C*); tandis que les derniers relevés comprennent essentiellement des espèces de ce dernier groupe (*C*). De plus, des 'compagnes' (*D*) sont présentes dans tous les relevés.

Le tableau VI résume les caractéristiques essentielles de cette classification, en donnant la fréquence moyenne (en %) et la 'présence' moyenne des espèces de chaque groupe dans chaque type de relevés. Cette succession de quatre types de hêtraies est d'ailleurs très comparable à celle donnée par Noirfalise (1956): Luzulo-Fagetum festucetosum, Luzulo-Fagetum caricetosum, Luzulo-Fagetum typicum et Luzulo-Fagetum vaccinietosum, sans oublier les 'hêtraies' du Quercetum sessiliflorae-Betuletum.

En outre, l'*estimation des facteurs* peut être réalisée, et les résultats obtenus permettent d'établir un diagramme de dispersion, dans lequel chaque relevé est représenté par un point (figure 3). L'examen de ce diagramme montre tout d'abord que la répartition des relevés est très irrégulière et, en particulier, que les relevés ne

Tableau VI. Fréquence moyenne (en %) et 'présence' moyenne des espèces de quatre groupes sociologiques (*A* à *D*) dans quatre types de relevés (1 à 4)[1].

	1	2	3	4
A	53 III	9 I	5 I	1 I
B	43 III	45 III	31 II	1 I
C	10 I	14 I	55 III	70 IV
D	30 II	30 II	35 II	30 II

sont pas réunis en noyaux ou en groupes nettement séparés les uns des autres. Ceci semble indiquer que, sur la base des données étudiées, il n'est pas possible de définir un petit nombre de types phytosociologiques *nettement* distincts les uns des autres.

L'estimation des facteurs permet aussi de vérifier la *valeur d'une classification préalablement établie* et de comparer deux ou plusieurs classifications. C'est dans ce but qu'a été conçue la figure 4, qui représente par des symboles différents les relevés appartenant, d'une part, aux différents types phytosociologiques définis précédemment (Dagnelie 1956) et, d'autre part, aux différents types de végétation qui viennent d'être présentés.

En ce qui concerne la *première classification*, on constate une certaine localisation de chaque type dans un secteur donné de l'espace factoriel floristique, mais cette localisation est loin d'être parfaite. Pour la *nouvelle classification* qui vient d'être présentée, la correspondance entre les types de végétation et la répartition des relevés dans l'espace factoriel est légèrement meilleure, ce qui n'a rien d'étonnant puisque ces types ont précisément été définis à l'aide des groupes sociologiques mis en évidence par l'analyse factorielle. Le diagramme relatif à cette dernière classification montre également dans quels secteurs ces différents groupes d'espèces ont une importance prépondérante: le groupe *A* est localisé dans la partie inférieure droite (relevés désignés par le symbole o); le groupe *C* dans la partie gauche du graphique (symboles + et /); tandis que le groupe *B* se rencontre partout sauf à l'extrême gauche (symboles o, — et +).

1) Les chiffres romains indiquent les 'classes de présence', au sens de Braun-Blanquet (1951).

On définit ainsi pour chaque groupe d'espèces une zone de *prépondérance dans l'espace factoriel floristique*, et il est possible de voir quelles sont les étendues relatives des différentes zones.

9.4.4 Répartition des espèces dans l'espace factoriel

Enfin, cette dernière notion peut également être appliquée aux espèces, en définissant pour chacune d'elle une *zone de répartition dans l'espace factoriel*. Comme il fallait s'y attendre, on observe des zones d'autant mieux localisées que les espèces sont de meilleures différentielles, ou encore que leur variance commune est plus élevée. A titre d'*exemples*, nous donnons ici les diagrammes de répartition de deux espèces (figure 5). Le diagramme relatif à *Festuca silvatica* montre l'existence d'une zone relativement étroite, qui correspond d'ailleurs approximativement à la zone de prépondérance du groupe

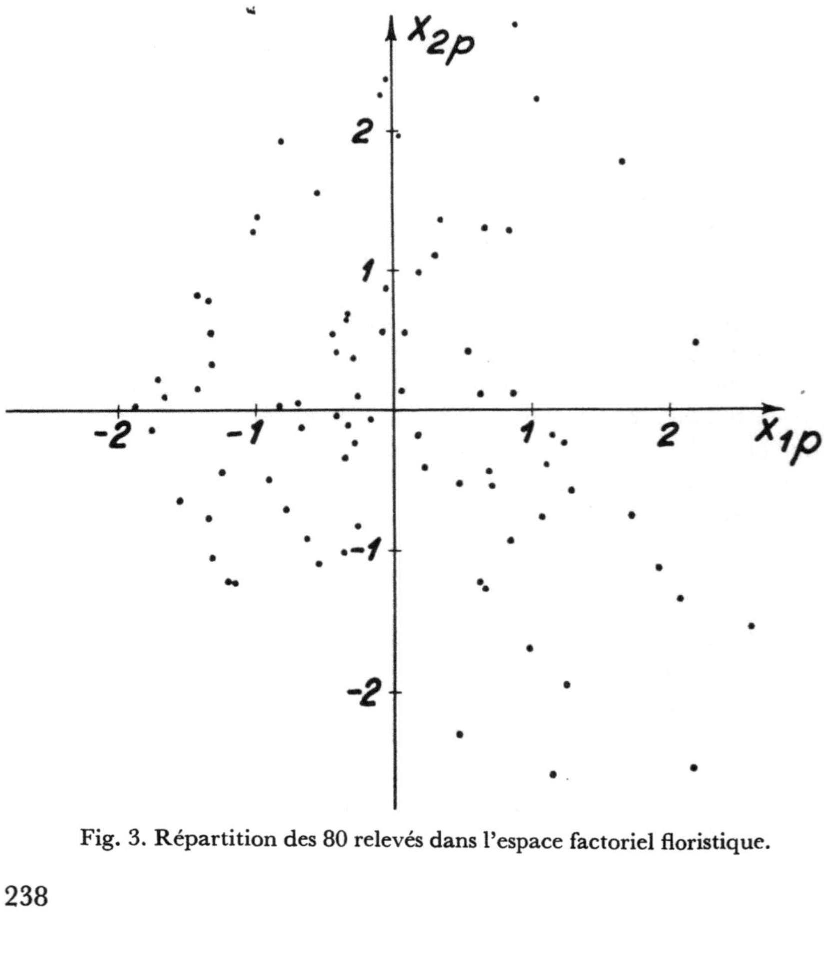

Fig. 3. Répartition des 80 relevés dans l'espace factoriel floristique.

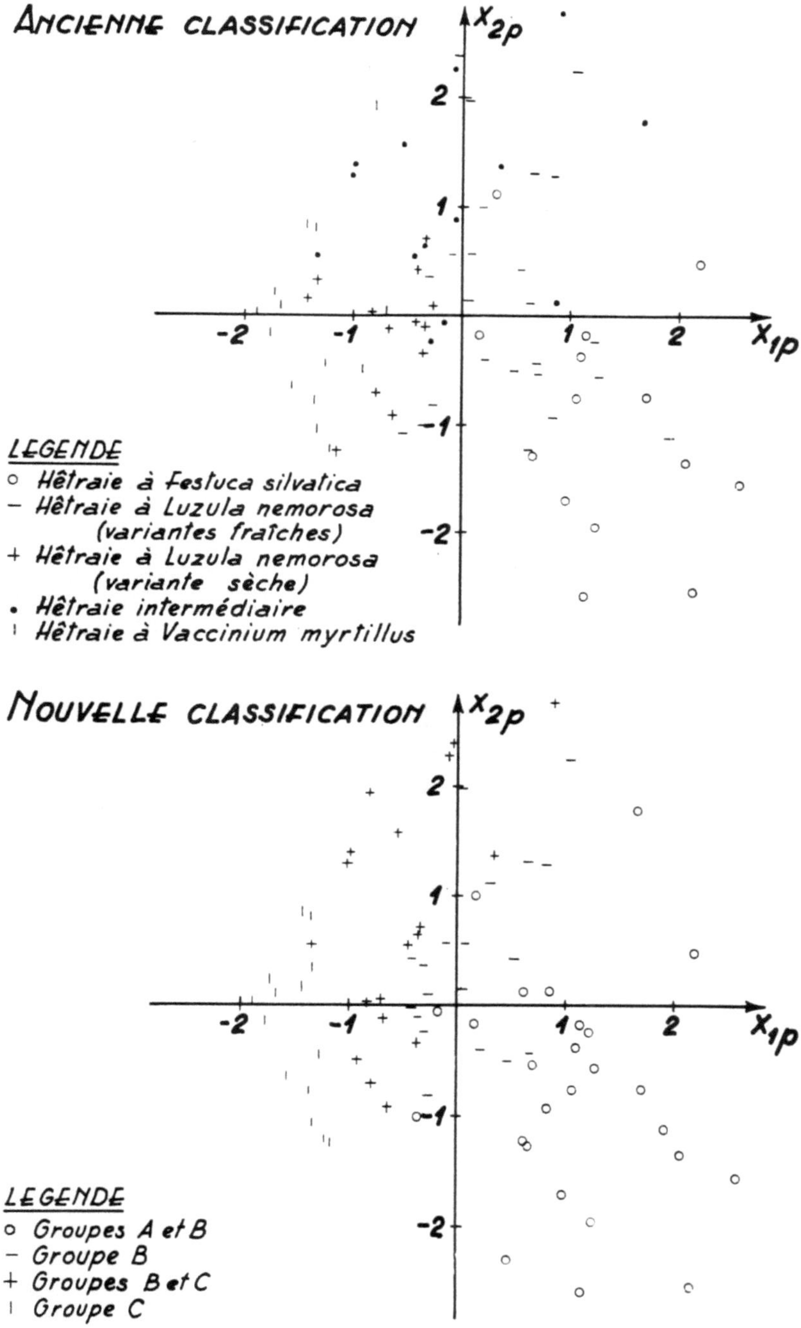

Fig. 4. Répartition des principaux types de relevés dans l'espace factoriel floristique, pour deux classifications différentes.

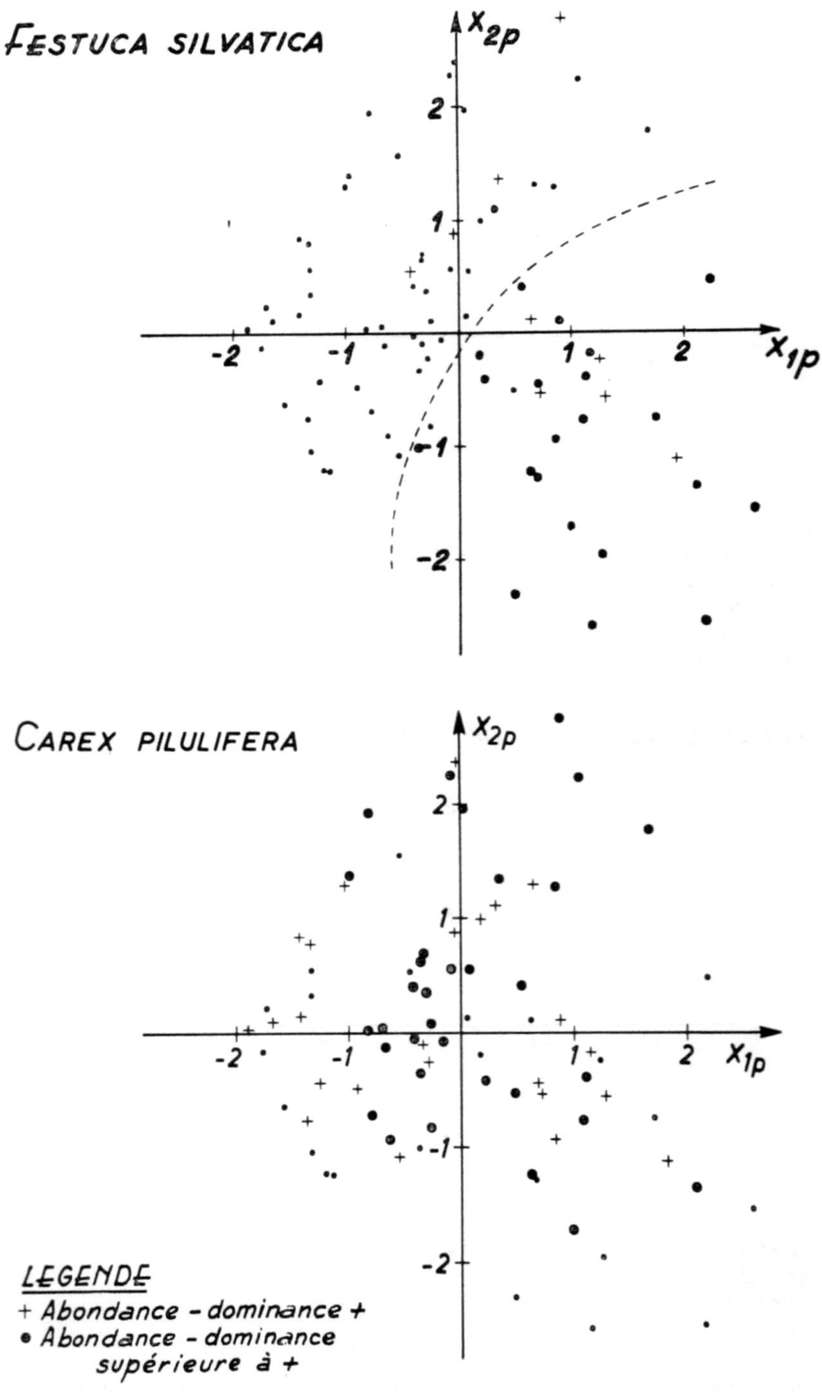

Fig. 5. Répartition de *Festuca silvatica* et de *Carex pilulifera* dans l'espace factoriel floristique.

A (figure 4). A l'intérieur de cette aire, la fétuque est présente dans 93 % des relevés (26 sur 28); à l'extérieur de cette zone, dans 10 % des relevés seulement (5 sur 52). Par contre, une espèce comme *Carex pilulifera*, qui n'est que faiblement liée aux deux premiers facteurs mis en évidence ($h_j^2 = 0{,}08$ pour les deux premiers facteurs), se présente dans la quasi-totalité de l'espace factoriel. Néanmoins, il existe une certaine localisation des relevés où cette espèce est assez abondante.

9.5 Discussion et conclusions

Dans l'exemple traité ci-dessus, l'analyse factorielle nous a permis de résoudre des questions telles que:
— l'étude de l'*homogénéité floristique* d'un ensemble de relevés;
— l'estimation du *pouvoir différentiel des espèces* au sein d'un ensemble de relevés;
— la définition de *groupes sociologiques d'espèces* étroitement liées entre elles;
— l'établissement d'une *classification floristique* des relevés et la comparaison de deux ou plusieurs classifications;
— l'étude de la *répartition des espèces et des groupes d'espèces* dans l'ensemble des relevés étudiés.

L'analyse factorielle peut aussi intervenir dans l'interprétation *d'observations écologiques*, les différentes caractéristiques écologiques observées prenant alors la place des espèces (Dagnelie 1960, 1965b). Dans ce cas, les problèmes qui peuvent être abordés sont essentiellement:
— l'étude de *l'homogénéité écologique* d'un ensemble de relevés;
— l'identification des *facteurs écologiques prépondérants* pour un ensemble de relevés;
— la détermination du *pouvoir indicateur* et de *l'amplitude écologique des espèces*, et d'une façon plus générale, l'étude des relations existant entre les facteurs écologiques et la répartition des espèces.

A plus petite échelle, l'analyse factorielle peut rendre des services dans *l'étude individuelle d'une communauté végétale* (Dagnelie 1962b), en ce qui concerne notamment:
— l'analyse de l'homogénéité floristique ou écologique, et de la *structure* d'une communauté végétale;
— la *comparaison de différents types de relevés*.

Enfin, l'analyse factorielle peut également intervenir dans l'étude des *similitudes entre relevés* ou groupes de relevés, la matrice de corrélation entre espèces étant alors remplacée par une matrice de corrélation ou de similitude entre relevés ou groupes de relevés (DAGNELIE 1960)[1]).

De toutes les méthodes d'analyse statistique à plusieurs variables, l'analyse factorielle pourrait ainsi être appelée à jouer le rôle le plus important en matière phytosociologique et écologique. Elle occupe certes une place de choix dans l'ensemble des méthodes d'analyse des liaisons entre espèces et entre variables écologiques (DAGNELIE 1962a, 1965a, GOUNOT 1961, GREIG-SMITH 1964), et elle s'intègre parfaitement dans le cadre des méthodes *d'analyse des gradients*, au sens de WHITTAKER (1967). Bien qu'étant surtout utile dans ce dernier domaine, l'analyse factorielle, ou l'analyse des composantes, peut avantageusement être associée aussi à certains processus de *classification* (GROENEWOUD 1965).

Divers facteurs doivent cependant freiner *l'utilisation de l'analyse factorielle en phytosociologie et en écologie*. D'une part, il s'agit d'un ensemble de méthodes souvent mal connues et fort complexes pour de nombreux biologistes. D'autre part, toute utilisation de l'analyse factorielle soulève une série de problèmes, relatifs notamment à la codification des données et au choix de différentes méthodes de calcul: choix d'un paramètre d'association ou de corrélation, calcul des variances communes, calcul des saturations, test de signification, etc. (DAGNELIE 1960).

Ces différents problèmes peuvent cependant être résolus sans trop de difficultés par un travail d'équipe ou de collaboration entre chercheurs de disciplines différentes (botanistes et statisticiens par exemple). De même, les difficultés soulevées par la *réalisation pratique des calculs* deviennent chaque jour moins importantes, en raison de l'utilisation de plus en plus fréquente des moyens modernes de traitement de l'information (machines mécanographiques et ordinateurs). Cependant, la nécessité d'un travail d'équipe ou de collaboration, et d'un recours fréquent à de puissants moyens de calcul, implique évidemment que seuls des problèmes d'une certaine ampleur peuvent justifier l'emploi de l'analyse factorielle.

Enfin, une limitation impérieuse, qui est apparue dès les premiers travaux (GOODALL 1954), a trait au *caractère linéaire du modèle utilisé*. Il s'agit là d'un point important, car la répartition des espèces est rarement liée de manière linéaire, ou même approximativement linéaire, aux facteurs écologiques qui la régissent. Ce point mériterait certainement d'être mieux étudié, tant au niveau des applications que sur le plan théorique.

1) L'opposition entre l'étude des corrélations entre espèces et l'étude des similitudes ou des corrélations entre relevés est souvent marquée par l'utilisation des lettres Q et R ('*Q-techniques*' et '*R-techniques*'). Cette écriture conventionnelle peut cependant prêter à confusion (DAGNELIE 1968).

9.6 **Addendum**

Depuis la rédaction de ce texte, en 1967, plusieurs ouvrages ont été consacrés à l'analyse statistique à plusieurs variables, et en particulier à l'analyse factorielle, notamment par MORRISON (1967) et ÜBERLA (1968).

De même, de nombreux travaux ont été publiés, entre autres dans le *Journal of Ecology* et le *Canadian Journal of Botany*, en ce qui concerne l'emploi de l'analyse factorielle et de l'analyse des composantes en phytosociologie et en écologie. Nous citerons seulement ici le travail de AUSTIN (1968), qui fournit une discussion générale du problème et un exemple d'utilisation, celui de ORLOCI (1968), qui fait de nombreuses allusions à l'emploi de l'analyse des composantes, le livre de GOUNOT (1969), qui replace également l'analyse factorielle dans son cadre général, et les comptes-rendus de *l'International Symposium on Statistical Ecology* de New Haven (PATIL, PIELOU et WATERS 1971).

9.7.1 RÉSUMÉ

L'analyse factorielle est constituée d'un ensemble de méthodes mises au point en psychologie expérimentale et utilisées surtout dans l'interprétation statistique des résultats de tests psychologiques. Avec l'analyse des composantes notamment, l'analyse factorielle figure en bonne place parmi les méthodes d'analyse statistique à plusieurs variables (paragraphe 9.1).

Les principales notions d'analyse factorielle sont introduites en partant d'un exemple simple, emprunté au domaine psychologique: les résultats de trois tests (problèmes de raisonnement, problèmes d'arithmétique et calculs d'addition), auxquels peuvent être soumises différentes personnes, sont mis en rapport avec deux aptitudes fondamentales de ces personnes (aptitude au raisonnement et aptitude au calcul). Dans une telle situation, le but poursuivi est d'expliquer les corrélations observées entre les résultats des tests, en mettant en évidence différentes aptitudes fondamentales sous-jacentes et en identifiant autant que possible ces aptitudes (paragraphe 9.2).

Le modèle général de l'analyse factorielle est ensuite présenté et une série de références bibliographiques sont données (paragraphe 9.3).

La transposition de l'analyse factorielle en phytosociologie peut se faire en considérant que les espèces végétales tiennent lieu de tests, que les relevés de végétation remplacent les personnes et que des facteurs écologiques se substituent aux aptitudes fondamentales.

L'objectif est alors, à partir des corrélations observées entre espèces, de mettre en évidence et d'identifier autant que possible les principaux facteurs écologiques sous-jacents. Subsidiairement, d'autres problèmes peuvent être abordés, tels l'étude du pouvoir différentiel des espèces, la définition de groupes sociologiques d'espèces étroitement liées entre elles, l'établissement d'une classification floristique des relevés, la comparaison de deux ou plusieurs classifications des relevés et l'étude de la répartition des espèces et des groupes d'espèces dans l'ensemble des relevés étudiés. Un exemple relatif à 38 espèces et 80 relevés de végétation de hêtraies est présenté de façon détaillée (paragraphe 9.4).

En conclusion, d'autres possibilités d'utilisation de l'analyse factorielle sont esquissées, en ce qui concerne notamment l'interprétation d'observations écologiques et l'étude de la structure de la végétation. Enfin, les principaux facteurs susceptibles de freiner l'utilisation de l'analyse factorielle en phytosociologie et en écologie sont énumérés et discutés (paragraphe 9.5).

Un addendum donne quelques références bibliographiques postérieures à la rédaction du texte principal (paragraphe 9.6).

9.7.2 Zusammenfassung

Die Faktoren-Analyse besteht aus einer Gruppe von Methoden, die in der experimentellen Psychologie entwickelt und besonders in der statistischen Deutung der Ergebnisse von psychologischen Prüfungen verwendet worden sind. Besonders mit der Hauptkomponenten-Analyse nimmt die Faktoren-Analyse einen Hauptplatz zwischen den multivariabelen statistischen Analysen ein (Par. 9.1).

Die Haupt-Begriffe der Faktoren-Analyse werden an Hand eines aus dem Gebiet der Psychologie stammenden einfachen Beispiels eingeführt: die Ergebnisse von drei Prüfungen (Probleme der Urteilskraft, Probleme der Arithmetik und der Additions-Rechnung) denen verschiedene Personen unterworfen werden können, sind mit zwei grundlegenden Fähigkeiten dieser Personen (Fähigkeit zur Urteilskraft und Fähigkeit zur Berechnung (Kalkül)) in Beziehung gesetzt worden. In einer solchen Lage ist das angestrebte Ziel, die zwischen den Prüfungs-Ergebnissen beobachteten Wechselbeziehungen (Korrelationen) zu erklären, zu dem verschiedene zugrundeliegende Grund-Fähigkeiten hervorgehoben und diese so weit wie möglich erkannt werden (Par. 9.2).

Das allgemeine Modell der Faktoren-Analyse wird anschliessend dargestellt und eine Reihe bibliographische Hinweise werden angegeben (Par. 9.3).

Die Übertragung der Faktoren-Analyse in die Pflanzensoziologie kann durchgeführt werden in dem man annimmt, daß die Pflanzen-Arten der Prüfung unterzogen werden, daß die Vegetations-Aufnahmen die Personen ersetzen und daß Standortsfaktoren an die Stelle der Grund-Fähigkeiten treten.

Das Ziel ist nun, ausgehend von den die zwischen Arten beobachteten Korrelationen, die zugrundeliegenden ökologischen Faktoren soweit wie möglich hervorzuheben und zu bestimmen. Anschließend können andere Probleme erörtert werden, wie das Studium der Unterscheidungs-Fähigkeit der Arten, die Bestimmung soziologischer Gruppen von eng untereinander verbundenen Arten, die Aufstellung einer floristischen Einteilung (Klassifikation) der Aufnahmen, der Vergleich von zwei oder mehr Klassifikationen von Aufnahmen und das Studium der Verteilung der Arten und Artengruppen in der Gesamtheit der untersuchten Aufnahmen. Ein Beispiel mit 38 Arten und 80 Vegetationsaufnahmen von Buchenwäldern wird in eingehender Weise gegeben. (Par. 9.4).

Zum Schluß werden andere Möglichkeiten zur Anwendung der Faktoren-Analyse beschrieben, besonders die Deutung von ökologischen Beobachtungen und das Studium der Vegetationsstruktur. Schließlich werden die wichtigsten Faktoren, die geeignet sind die Anwendung der Faktoren-Analysen in der Pflanzensoziologie und Ökologie zu erschweren, aufgezählt und diskutiert (Par. 9.5).

Ein Nachtrag gibt einige bibliographische Hinweise nach der Abfassung des Haupttextes. (Par. 9.6).

9.7.3 Summary

Factor analysis consists of a set of methods developed in experimental psychology and used especially in the interpretation of results of psychological tests. Together with principal components analysis, factor analysis is one of the most effective methods of multivariate analysis (paragraph 9.1).

The principal concepts of factor analysis are introduced with a simple example taken from psychology: results of three tests (problem reasoning, arithmetic problems, and addition), that can be applied to various persons, are related to two fundamental abilities of these persons (reasoning ability and calculation ability). In such a situation the aim is an explanation of the observed correlations between the test results, making evident various underlying fundamental abilities and identifying those abilities as far as possible (paragraph 9.2).

The general model of factor analysis is then presented, and a list of references is given (paragraph 9.3).

Factor analysis can be applied in phytosociology, by considering that the plant species take the place of the tests, the relevés of vegetation replace the persons, and the ecological factors correspond to the fundamental abilities. The objective is then, on the basis of observed correlations between species, to make evident and to identify as far as possible the underlying ecological factors. In addition, other problems can be approached such as the study of the indicator value of species, the definition of sociological groups of closely associated species, the erection of a floristic classification of the relevés, the comparison of two or more classifications of the relevés, and the study of the distribution of species and species groups in the set of relevés studied. An example involving 38 species and 80 relevés of beech forests is presented in detail (paragraph 9.4).

In conclusion, other possible uses of factor analysis are outlined, notably those concerning the interpretation of ecological observations and the study of the structure of vegetation. Finally the main factors limiting the utilization of factor analysis in phytosociology and ecology are listed and discussed (paragraph 9.5).

An addendum gives some bibliographic references that appeared after the preparation of the main text (paragraph 9.6).

RÉFÉRENCES

ANDERSON, T. W., – 1958 – An Introduction to Multivariate Statistical Analysis. Wiley, New York, 374 p.

AUSTIN, M. P., – 1968 – An ordination study of a chalk grassland community. *J. Ecol.* 56: 739—757.

BRAUN-BLANQUET, J., – 1951 – Pflanzensoziologie. Grundzüge der Vegetationskunde. Springer, Wien, 631 p.

BURT, C., – 1940 – The Factors of the Mind. Univ. Press, London, 509 p.

BURT, C., – 1949 – Alternative methods of factor analysis and their relations to Pearson's method of principal axes. *Br. J. Psychol., Statist. Sect.* 2: 98—121.

BURT, C., – 1952 – Tests of significance in factor analysis. *Br. J. Psychol., Statist. Sect.* 5: 109—133.

CATTELL, R. B., – 1952 – Factor Analysis. Harper, New York, 462 p.

DAGNELIE, P., – 1956 – Recherches sur la productivité des hêtraies d'Ardenne en relation avec les types phytosociologiques et les facteurs écologiques. *Bull. Inst. agron. Stat. Rech. Gembloux* 24: 369—410.

DAGNELIE, P., – 1960 – Contribution à l'étude des communautés végétales par l'analyse factorielle. *Bull. Serv. Carte phytogéogr., Sér. B,* 5: 7—71 et 93—195.

DAGNELIE, P., – 1962a – L'application de l'analyse multi-variable à l'étude des communautés végétales. *Bull. Inst. int. Statist.* 39(2): 265—275.

DAGNELIE, P., – 1962b – Etude statistique d'une pelouse à *Brachypodium ramosum*: les liaisons interspécifiques. *Bull. Serv. Carte phytogéogr., Sér. B,* 7: 85—97 et 149—160.

DAGNELIE, P., – 1965a – L'étude des communautés végétales par l'analyse statistique des liaisons entre les espèces et les variables écologiques: principes fondamentaux. *Biometrics* 21: 345—361.

DAGNELIE, P., – 1965b – L'étude des communautés végétales par l'analyse statistique des liaisons entre les espèces et les variables écologiques: un exemple. *Biometrics* 21: 890—907.

DAGNELIE, P., – 1966 – Introduction à l'analyse statistique à plusieurs variables. *Biom.-Praxim.* 7: 43—66.

DAGNELIE, P., – 1968 – Introduction aux problèmes et aux méthodes de classification numérique. *Biom.-Praxim.* 9: 87—111.

FAVERGE, J. M., – 1954 – Méthodes statistiques en psychologie (2 vol.). Presses Univ. de France, Paris.

GOODALL, D. W., – 1954 – Objective methods for the classification of vegetation. III. An essay in the use of factor analysis. *Austral. J. Bot.* 2: 304—324.

GOUNOT, M., – 1961 – Les méthodes d'inventaire de la végétation. *Bull. Serv. Carte phytogéogr., Sér. B*, 6: 7—73.

GOUNOT, M., – 1969 – Méthodes d'étude quantitative de la végétation. Masson, Paris, 314 p.

GREIG-SMITH, P., – 1964 – Quantitative Plant Ecology. Butterworths, London, 2nd ed. 256 p.

GROENEWOUD, H. VAN, – 1965 – Ordination and classification of Swiss and Canadian coniferous forests by various biometric and other methods. (Germ. summ.) *Ber. geobot. Inst. ETH, Stiftg. Rübel*, Zürich, 36: 28—102.

HARMAN, H. H., – 1967 – Modern Factor Analysis. Chicago Univ. Press, 474 p.

HOLZINGER, K. J. & HARMAN, H. H., – 1941 – Factor Analysis. A Synthesis of Factorial Methods. Chicago Univ. Press, 417 p.

HORST, P., – 1965 – Factor Analysis of Data Matrices. Holt, Rinehart et Winston, New York, 730 p.

KENDALL, M. G., – 1950 – Factor analysis as a statistical technique. *Jl. R. statist. Soc., Ser. B*, 12: 60—73.

KENDALL, M. G., – 1957 – A Course in Multivariate Analysis. Griffin, London, 185 p.

KENDALL, M. G. & LAWLEY, D. N., – 1956 – The principles of factor analysis. *Jl. R. statist. Soc., Ser. A*, 119: 83—84.

LAWLEY, D. N., – 1955 – A statistical examination of the centroid method. *Proc. R. Soc. Edinb., Sect. A*, 64: 175—189.

LAWLEY, D. N. & MAXWELL, A. E., – 1963 – Factor Analysis as a Statistical Method. Butterworths, London, 117 p.

MORRISON, D. F., – 1967 – Multivariate Statistical Methods. Mc Graw Hill, New York, 338 p.

NOIRFALISE, A., – 1956 – La hêtraie ardennaise. *Bull. Inst. agron. Stat. Rech. Gembloux* 24: 208—239.

ORLOCI, L., – 1968 – A model for the analysis of structure in taxonomic collections. *Can. J. Bot.* 46: 1093—1097.

PATIL, G. P., PIELOU, E. C. & WATERS, W. E., – 1971 – Statistical Ecology, Vol. 3: Many Species Populations, Ecosystems, and Systems Analysis. Pennsylvania State Univ. Press, 462 p.

SEAL, H. L., – 1964 – Multivariate Statistical Analysis for Biologists. Methuen, London, 207 p.

THOMSON, G. H., – 1956 – The Factorial Analysis of Human Ability. London Univ. Press, 383 p.

THURSTONE, L. L., – 1947 – Multiple-Factor Analysis. Chicago Univ. Press, 535 p.

ÜBERLA, K., – 1968 – Faktorenanalyse. Springer, Berlin, 399 p.
VERNON, P. E., – 1952 – La structure des aptitudes humaines (Trad.: M. REUCHLIN). Presses Univ. de France, Paris, 195 p.
VINCENT, D. F., – 1953 – The origin and development of factor analysis. *Appl. Statist.* 2: 107—117.
WHITTAKER, R. H., – 1967 – Gradient analysis of vegetation. *Biol. Rev.* 42: 207—264.

10 ORDINATION BY RESEMBLANCE MATRICES

Laszlo Orloci

Contents

10 ORDINATION BY RESEMBLANCE MATRICES

10.1 Introduction

The term *ordination* seems to be used by phytosociologists in a rather loosely defined sense. It may describe any method in which given phytosociological entities, as abstract spatial points, are ordered according to one or several of their properties in such a manner that their arrangement will reveal some useful information about their relationships (cf. GREIG-SMITH 1964). We would be justified on the basis of this definition to include in the general class of ordinations methods as varied as continuum analysis, gradient analysis, component analysis, factor analysis, and many more, which differ greatly both in underlying theory and also in practical effectiveness to achieve specific goals (cf. DAGNELIE 1960, WHITTAKER 1967, MC INTOSH 1967, PRÉCSÉNYI & SZÖCS 1969, and references therein).

The concept of ordination as a structuring procedure (cf. LAMBERT & DALE 1964) in which the total variability in a sample is apportioned between new axes fits very closely the ordination concept as it is understood in the present paper. Efficient structuring, of course, is a concept closely related to the notion of parsimonious summarization (SEAL 1964) in which an abstract reference system is extracted from a set of resemblance values under strictly defined optimality conditions.

The limitation of the ordination concept to the notion of parsimonious summarization would however be somewhat too narrow for our purpose. We want to define a class which will include not only summarization techniques, such as for instance component analysis, but also ordinations of some other types, such as factor analysis, in which it is possible to test specific hypotheses regarding the efficiency of the ordination model in accounting for the common variation in a given population.

The ordination methods in this broader but still manageable class include the open-model of factor analysis, component analysis, position vectors ordination, polar ordination (several methods), principal axes analysis, and other techniques, commonly referred to as multidimensional scaling. A common property of the ordinations we have in mind is an ability to extract metric information from a given resemblance matrix. This is new information on the basis of which specific phytosociological relationships, or trends in these relationships, can more easily be recognized.

We shall consider some general concepts first, before turning to a discussion of the admissible resemblance functions and methods for ordination.

10.2 The Purpose of Ordination

The actual purpose of ordinations may differ considerably. Factor analysis, for instance, is aimed at the isolation of the common variation from the specifics in a population (Lawley & Maxwell 1963, Seal 1964, Cattell 1965), while another commonly used technique, component analysis, is intended to summarize the total variation (Seal 1964). Whereas these are often recognized as the direct goals in ordinations, the primary results of an ordination, such as the component scores for instance, may serve as input data in classifications (e.g. Goodall 1954, Ivimey-Cook 1968), or in further ordinations (e.g. Quadling 1967). Ordinations may also provide a basis for other specialized analyses such as the quantification of successional trends (Maarel 1969), group identification (Dagnelie 1960, 1965), cluster recognition (e.g. Gittins 1965a,b, Greig-Smith et al. 1967), or graphical discriminant analysis (Orloci 1968). Most often, however, ordinations are concerned with the identification of trends and patterns of variation, the explanation of these, or their generalization as hypotheses in terms of correlations with external variables (see Whittaker 1967, Gittins 1969, and references therein). Other examples include Bray & Curtis (1957), Curtis (1959), Maycock & Curtis (1960), Maycock (1963), Mc Intosh & Hurley (1964), Ashton (1964), Gittins (1965b), van Groenewoud (1965), Ivimey-Cook & Proctor (1967), Yarranton (1967a,b,c), Kershaw (1968), Austin & Greig-Smith (1968), Chandapillai (1970), Grigal & Goldstein (1970), Moore et al. (1970), Phipps (1970), and Austin et al. (1971), among others. The secondary analyses may themselves represent complex analytical techniques such as trend surface analysis (Sneath 1967, Gittins 1968), canonical correlation analysis (e.g. Austin 1968), curve seeking (Sneath 1966), or axis rotation (e. g. Sokal 1958, Sokal & Rohlf 1962).

10.3 Stochastic vs. Descriptive Ordinations

It is most desirable to consider ordinations as stochastic (probabilistic) models. If, for instance, the assumption of an underlying multivariate normal distribution is acceptable then it is possible in component analysis to test the hypothesis that the last k ordination axes, corresponding to the k smallest eigenvalues of the covariance matrix, are statistically indistinguishable. Factor analysis, under the same assumptions regarding the nature of the underlying population, will provide a test for the hypothesis that a set of m common-factors completely describes the covariance structure of a given population.

A stochastic use of ordinations, while most desired for its hypothesis-testing aspects, may however be inappropriate in many practical situations when the specific assumptions that are implicit in the model are unwarranted .The ordination, in such situations, reduces to a deterministic use of the mathematical model, aimed at precise descriptions and hypothesis generation.

10.4 Hypothesis Testing in Ordinations

Different methods may be used by phytosociologists for testing given hypotheses .These may be based on (i) consistency of the observed ordination results with known facts, (ii) consistency of ordination results in repeated field surveys or experiments, and (iii) the use of standard statistical indices with known probability distributions. The first of these methods may not seem too rigorous. It can, nevertheless, prevent the phytosociologist from attributing much significance to observed results which, according to past scientific experience, could have arisen with a relatively high probability as a consequence of sampling or experimental error. Repeated ordinations of materials derived from successive surveys in the same area may provide a most reliable basis of significance tests. This method, however, can be very uneconomical in most types of practical applications. The most common methods of hypothesis testing are based on the use of standard statistical indices, which presumably possess specific distributional properties. The use of these indices, however, can be very hazardous in the hands of the practicing phytosociologist who is required to accept specific assumptions purely on an intuitive basis without adequate facilities required for a reliable theoretical or experimental verification.

10.5 Ordination Efficiency

Different criteria are suggested by the various authors to evaluate the relative efficiency of an ordination. VAN GROENEWOUD

(1965) mentions (i) the need to account for the largest possible portion of variation in the sample by the simplest possible ordering of individuals, and (ii) meaningfulness in ecological terms (see also AUSTIN & NOY-MEIR 1971). He further considers (iii) the nature of the resemblance function used, and (iv) the existence of facilities whereby new individuals can be placed in a completed ordination.

We may add at this point that component analysis is expected to be the most efficient method for the summarisation of linear variation, using the smallest possible number of dimensions. The question of ecological meaningfulness can be answered on the basis of the potentials of the model itself, whether or not it is capable of detecting the kind of variability which would interest the phytosociologist, and on the basis of the correlation of the ordination results with given ecological factors. It seems that the nature of the resemblance function is also very important for the simple reason that it must qualify as a valid spatial parameter in the assumed sample space of operations of a given technique. And further, the results of the ordination may greatly depend on the resemblance function. The placement of newly-acquired individuals in a completed ordination is usually accomplished by the transformation $\mathbf{Y} = \boldsymbol{\alpha}'\mathbf{A}$ which produces coordinates for an individual (vector) $\mathbf{A}$ given $\boldsymbol{\alpha}$, the matrix of transformation coefficients. Such a transformation is readily available in component analysis, and can also be derived in other ordination methods.

There are further criteria to be considered regarding an evaluation of the relative efficiency of given ordination techniques. Among these we may mention (v) the stochastic properties of the model and (vi) the type of variation, whether linear or curvilinear, that can be detected by the model. The conventional ordination methods are deterministic in nature with the exception of factor analysis and perhaps component analysis, which can be used in association with standard statistical tests based on the postulated normal distribution of joint frequencies in the underlying population. The conventional methods, however, appear to have one really serious defect regarding criterion (vi): they treat all variation as if it were linear. This implies that ecological properties with curvilinear variation cannot be handled by these ordination methods without distortion (see GOODALL 1954, VAN GROENEWOUD 1965, SWAN 1970, AUSTIN 1968, NOY-MEIR & AUSTIN 1970, AUSTIN & NOY-MEIR 1971, GAUCH & WHITTAKER 1971, 1972). What is meant by the inability of ordinations to account for non-linear variation can be illustrated by the ordination axes of component analysis. These axes are linearly orthogonal (in terms of zero covariances) but they may be curvilinearly related. The position of these axes in no way is affected by the existence of higher order variation in the experimental data. There seems to be no satisfactory solution to the problem of non-linear ordinations, although many attempts have been made to respond to this problem. AUSTIN & NOY-MEIR (1971) gives a summary of several possibilities. They suggest to limit the scope of ordinations to situations of narrow vegetational and environmental variation where the assumption of linear variation is more acceptable, and to always take into account the linear nature of the model when interpreting the ordination results in ecological terms. Another approach may include manipulations of the ordination results, aimed at the extraction of the non-linear component of variation, by techniques such as trend surface analysis (e. g. GITTINS 1968) and other regression methods concerned with curve-seeking (e.g. SNEATH 1966). A further possibility is the standardization, or transformation of the data (e. g. SWAN 1970), or the use of some specialized multivariate technique such as for instance MC DONALD's (1962) non-linear factor analysis and KRUSKAL's (1964a,b) method of multidimensional scaling. The method of SHEPARD & CARROLL (1965) has also been mentioned as another possibility (see AUSTIN & NOY-MEIR 1971).

It seems that, in need of a better method, the practicing phytosociologist may continue using the linear models of ordination, and when it appears necessary, he may further scrutinize the ordination results for non-linear variation on the basis of the use of some other method. We must, of course, assume that he will use ordinations, or as a matter of fact any other method, after careful consideration of their relevance to the problems on hand. The ordination methods described in the present paper should not be regarded as general strategies. They should rather be considered as highly specialized techniques intended to answer specific questions. The failure of the phytosociologist to recognize this will inevitably result in a wasteful analysis.

10.6 A Model For Ordination

We shall define ordination *entities* in a general sense, as *individuals* or characters. An individual, the basic distinguishable unit for observation, will be identified with a site (quadrat, sample plot), and the characters will be designated as entities on the basis of which the individuals are described. Among the characters we shall consider only those which vary among the individuals. The *variates* may represent the quantity or presence of different species within the sites, or any other properties, vegetational or environmental, on the basis of which the characteristics of the sites can be effectively described. It should be noted that although we designate the sites as individuals whose ordination is contemplated, there should be no conceptual difficulty in identifying entities of some other kind as individuals and using them accordingly in an ordination.

We may picture an individual as a fixed point in space whose position is determined by the size or state of the characters which it possesses. It is implicit in the foregoing analogy that the characters are regarded as axes of a reference system. This system with all of its properties constitutes the sample space for ordination. It may be noted that the *sample space*, as we define it, is an abstraction.

Our ordination model consists of (i) a collection (sample, population) **X** of c individuals described on the basis of r variates, (ii) a resemblance function f which generates the resemblance matrix **R** from **X**, and (iii) statements in a formal algebraic language specifying the steps in the analysis. The c individuals (column vectors of **X**) identify the entities to be ordinated, and the r variates (row vectors of **X**) are those properties on the basis of which the pairwise resemblance of individuals is determined. We operate this model on the assumption that the variates are commensurable (see Pielou 1969), and with purpose to extract metric information from **R**. In some instances our model will have stochastic properties. More often, however, we shall use the model in a deterministic context.

10.7 Ordination by Euclidean Matrices

10.7.1 The Sample Space

10.7.1.1 *Definition of a Metric Space*

Firstly, we shall describe the properties of a general metric space. The metric properties are quite relevant in developing our idea of ordinations in an Euclidean sample space which completely possesses these properties. A metric space is defined as a collection of objects **X** and a metric (distance) function d. We note in this connection that (i) a metric function cannot be negative, (ii) it must be symmetric so that it will not be affected by the sequence in which the individual objects are compared, and (iii) it must possess the triangle inequality property, implying that given three individuals as points of a sample space the sum length of any two sides of the triangle which they define must not be less than the length of the third side. In other words, any resemblance function which can produce negative values is automatically disqualified as a metric. And further, a resemblance function which gives one result when the comparison is from A to B, and another when the comparison is from B to A is useless as a metric function. A function which does not conform with the triangle inequality axiom will also be rejected as a metric measure. Although we must reject any function as a metric if it fails on any one of the three metric space axioms, we must also note that the metric use of a non-metric function does not necessarily result in serious distortion in some ordinations. This problem will be considered later in the paper when we discuss the concept of distortion which results from the Euclidean use of given non-Euclidean measures of resemblance in phytosociological ordinations.

10.7.1.2 *Some Notes on Resemblance Functions*

Several measures of resemblance which can be used in ordinations are discussed in article 6 by GOODALL. With this point in mind we shall limit our discussion of the resemblance measures to those properties which must be considered by the phytosociologist when selecting an appropriate measure for ordination. The resemblance functions of the present chapter, with few exceptions, are meaningful in an Euclidean sample space. This space is distinguished from other metric spaces by the fact that it is only the Euclidean space where a scalar product of vectors exists.

A commonly used resemblance function is the Euclidean distance, or common distance, which we know from every day experience. The simplest definition of such a distance between two quadrats of vegetation is given as the square root of the sum of squared differences between their corresponding species. It, however, should be noted at this point that (i) when we decide to define a distance in this way we also implicitly assume that the individual species, whose squared differences are summed, are statistically independent, and (ii) when the original data are not adjusted or standardized in some appropriate manner (see ORLOCI 1967a) then this distance will not have a fixed upper limit — that is to say that it will not be a relative measure on the basis of which we can make direct comparisons between ordinations.

Our initial definition of Euclidean distance is seriously weakened by its total reliance on absolute quantity, without any regard to the relative importance or weight of the different species. In this connection we should mention that different weighting procedures were suggested which used the total species correlations (see MAC NAUGHTON-SMITH et al. 1964) or the total chi square (see WILLIAMS et al. 1964). We may note that (i) these weighting techniques will both have the same effect when the data represents species presence, and (ii) an Euclidean distance so weighted will not evenly retain its metric properties in sets of quadrats in which the weight of a species does not remain constant. In more specific terms of our ordination problem, such a 'distance' measure will not be allowed to serve as a definition of resemblance if the goal is a simultaneous analysis of several sets of quadrats which associate with greatly different weights of the species.

It is a peculiarity of the Euclidean distance that it may show a low degree of resemblance between two quadrats in which all species are identical, if the species values are greatly different, and conversely, it may indicate a high degree of resemblance between two quadrats which have no species in common, if the species values are generally low. A procedure which will help overcoming this weakness of the Euclidean distance involves the standardization of the data in such a manner that each observation is divided by the square root of the sum of squared observations in the quadrat (column vector of **X**) of its occurrence. This kind of standardization has a definite geometric meaning (see Orloci 1967a, Pielou 1969), and it also has the advantage of being independent from the set in which a quadrat occurred. For these reasons when the Euclidean distance is computed by this method it will remain a valid Euclidean spatial parameter, and it will continue to be admissible as a resemblance function in ordinations.

In some statistical applications distance is defined in the manner of Hotelling's T^2 statistic, or as a fraction of this statistic (see Morrison 1967). This definition is equivalent to our first definition of Euclidean distance if the species are statistically independent and have unit variance. Due to standardization and transformation implicit in the T^2 statistic, its application as a resemblance function in Euclidean ordinations is restricted to sets of quadrats in which the pattern of covariance relationships of species remains constant. This statistic is further limited to sets of quadrats for which the assumption of multivariate normality holds.

Throughout the preceding discussions, regarding the Euclidean distance, we implicitly assumed that the species can be represented by perpendicular axes. This assumption, however, is unnecessary, and in some instances, it may not even be desired. As an alternative assumption we may stipulate that the species are correlated, and as such, they can be represented by oblique axes. The obliqueness can be measured by the product-moment correlation coefficient. A definition of Euclidean distance on the basis of the assumption that the reference axes are oblique has been given by Gengerelli (1963) and its use has also been demonstrated by CARLSON (1970). The principal advantage of their definition of distance is that it removes the effect of linear species correlations from the value of absolute distance. In a sense this property makes this definition quite similar, though not identical in general, to a fraction of the T^2 statistic, also known as generalized distance. Since in this distance we use the correlation coefficient as an element, it will have similar locally metric properties as T^2. This property should not be ignored when the application of these measures is considered in a simultaneous ordination of sets of quadrats with unequal correlation matrices.

The Euclidean distance will simplify to $\sqrt{(b+c)}$ when the data represent presence scores of species. In this and following expressions for comparison of two quadrats, the symbol a is the number of species present in both quadrats, b the number in the first quadrat only, and c the number in the second quadrat only.

The $\sqrt{(s)}$ fraction of this distance, where s is the total number of species in the sample, represents the one-complement of the simple matching coefficient (SOKAL AND MICHENER 1958). The sampling distribution of this coefficient was derived by GOODALL (1967). While division by s will not interfere with the admissibility of this distance as a resemblance function in an Euclidean ordination, division by $(a+b+c)$ definitely will. Such a division is implicit in the JACCARD (1901) coefficient (39). (Numbers in parentheses are those of formulas in article 6.) For some interesting properties of this coefficient the reader should consult a paper by HOLGATE (1971). The one complement of JACCARD's coefficient is admissible as a resemblance function in Euclidean ordinations in sets of quadrats in which no species is missing in more than one quadrat (see ORLOCI 1971).

A variant of the JACCARD coefficient which has been used in ordination is the SØRENSEN (1948) index (40), $CC=2a/(2a+b+c)$. It is interesting to point out that the one complement of this index, given by $CD=1-CC=(b+c)/(2a+b+c)$, can be regarded as an Euclidean distance in sets of quadrats for which $(2a+b+c)$ is a constant quantity. This function and a related function $2CD$, which was treated by VAN DER MAAREL (1969) as a metric essentially in the same context as discussed earlier by GOWER (1966), have been used in ordinations.

A function, $PD=1-PS$, has been used for quantitative data (e.g. BRAY & CURTIS 1957) where $PS=2\sum\min(x_i,y_i)/\sum(x_i+y_i)$. The 100 multiple of PS is a measure of percentage similarity (50); x_i and y_i represent the quantities of the ith species in the two quadrats (x and y), and the summations are taken over all the species from $i=1$ to r. It can be shown that when the data represent presence scores of species PD is the same as CC, related to the Euclidean distance. When the data represent quantities other than presence scores the function PD will be a metric function, equivalent to the absolute value function and not to the Euclidean distance, in sets of quadrats for which $\sum(x_i+y_i)$ is a constant quantity for all pairs of quadrats. To show this relationship we rewrite PD in the form $PD = \sum|x_i-y_i|/\sum(x_i+y_i)$, where the quantity in the numerator is the familiar absolute value function. We conclude that PD, when computed from quantitative data, will be a fraction of the absolute value function in sets where the sum of species values is constant for any pair of quadrats. A variant of PD (51) has been suggested by WHITTAKER (1952) in which the sum of relativized species values is a constant at 1.0, so that for any pair of quadrats $PD = \sum|x_i-y_i|/2. = 1-\sum\min(x_i,y_i)$.

In sets where $(2a+b+c)$, or $\sum(x_i+y_i)$, is not a constant, the use of CD, or PD, as a metric ordination parameter is not strictly justified. It seems though that CD, and PD, will retain traces of the metric properties, and even when the sums of species values differ considerably between the quadrat pairs, the distortion in the ordination resulting from the metric use of CD, or PD, may not be too extensive. A frequently used variant of these indices is a quantity defined by $D=\max L-L$ in which L is either CC or PS, and the highest value observed (ideally a value for replicate samples) becomes max L. This definition has been used as a substitute for Euclidean distance (e.g. BEALS 1960). Such a use of D, however, is not quite appropriate (see AUSTIN & ORLOCI 1966) for the following reasons: If we put $D=1-L+e$, where $e=\max L-1$, we can clearly see that an adjustment is implicit in D which takes the form of adding a constant quantity e to the function $1-L$. A consequence of such an adjustment in sets of quadrats where $1-L$ is a metric function, but e is not a zero value, is that D may fail as a metric measure of resemblance. This failure will distort the ordination in those circumstances in which the metric nature of $1-L$ is assumed. In sets where e is zero and $2a+b+c$, or $\sum(x_i+y_i)$, is a constant, the quantity $\max L-L=1-L$ can be used as an Euclidean distance in the case of presence data, or as an absolute value function in the case of quantitative data.

Ordinations often are based on resemblance measures which can be derived from Euclidean distance. An especially important group of such measures is represented by the scalar products. A scalar product of two vectors (species or quadrats) is defined as the product of their lengths and the cosine of their subtending angle. The relevant expressions are given by ORLOCI (1967b), including a detailed discussion of their effect upon principal component analysis.

Regarding the general topic of scalar products we may comment further on the similarity coefficients which can be represented in terms of the general function $\rho = a/m$. The similarity coefficient of RUSSELL & RAO (1940) is of this type in which $m = a+b+c+d$ (symbols as above, with $d =$ number of species in neither quadrat). We can similarily express the JACCARD coefficient (39) in which $m = a+b+c$, and the coefficient (43) of OCHIAI (1957) in which $m = \sqrt{((a+b)(a+c))}$. The coefficient of RUSSELL & RAO represents an average scalar product of vectors of binary elements. The coefficient of JACCARD, however, qualifies as a scalar product only in sets of quadrats in which m is a constant quantity. The coefficient of OCHIAI, on the other hand, is a true scalar product between vectors of unit length. Note that these coefficients can be converted into a standard (Euclidean distance) measure of dissimilarity in the manner of $ED = \sqrt{(2(1-\rho))}$ with values occurring between zero and $\sqrt{(2)}$.

None of the a/m coefficients so far mentioned could be rightly labelled as a product-moment correlation coefficient since none of them achieves a translation of the co-ordinate origin of the reference system to the common centroid (center of gravity). A peculiarity of the product-moment correlation coefficient is that it uses the centroid of the system about which the scalar product of vectors is computed. This property will qualify the correlation coefficient as an admissible function in, for instance, component analysis or position vectors ordination. An Euclidean distance corresponding to the product-moment correlation coefficient is given by $\sqrt{(2(1-\rho))}$ where ρ represents the product-moment correlation coefficient. This distance has values between 0 and 2.

10.7.1.3 *A Classification of Resemblance Functions*

The resemblance functions which we have so far discussed can be associated with different ordination methods. We summarize some of their important properties in Table I. Shown in this table is the classification of the different coefficients according to their distance or scalar product properties, and according to admissibility as resemblance functions in the different ordination methods .The idea of local admissibility, indicated by an L between parentheses, is introduced in the table to indicate that in some sets of quadrats these functions may not be acceptable as valid definitions of resemblance for Euclidean ordination. Because the metric properties of some coefficients change depending on the data, we used the symbol P

TABLE I. Distance and scalar product properties of coefficients, and their admissibility in ordinations.

Coefficient	Classification: Euclidean distance	Classification: Scalar product centered	Classification: Scalar product not centered	Classification: Absolute value function	Ordination method of admissibility
Simple matching			X		
One-complement of simple matching	X				C, Po
JACCARD			X (L)		
One-complement of JACCARD	X (L)				C, Po
SØRENSEN			X (L,P)		
One-complement of SØRENSEN	X (L,P)			X (L)	C,Po
RUSSELL and RAO			X		
One-complement of RUSSELL and RAO	X				C,Po
OCHIAI			X		
One-complement of OCHIAI	X				C,Po
Simple product moment correlation		X			C,P
One-complement of simple product moment correlation	X				C,Po
Variance/covariance		X			C,P,F

Legends to symbols: (L) - locally, (P) - presence data, C - component analysis, P - position vectors ordination, Po - polar ordination, F - factor analysis.

between parentheses to signify that the use of presence data is assumed.

10.7.1.4 *Resemblance Functions and Sample Structure*

The phytosociologist often thinks of sample structure as a covariance structure which he sees in the raw data. He may not realize that such a structure can completely be destroyed by the different adjustments or standardizations usually implicit in the resemblance function. For example, when he uses the product-moment correlation coefficient as a measure of species resemblance, he is in fact introducing a standardization which makes the variance of every species equal to unity. Thus after applying the correlation coefficient to the data, it is no longer valid to think of structure as it may have existed in the original data, since then that type of structure will have been completely destroyed by the correlation coefficient. This effect of the resemblance function on the sample structure may be visualized as a scrambling of the original spatial configuration of individuals, shifting their position relative to one another, in a manner depending on the nature of the resemblance function.

10.7.2 Factor Analysis

Factor analysis is discussed by Dagnelie in article 9. It, however, is necessary to briefly describe some aspects of this method on the basis of which we will be able to make a comparison between factor analysis and component analysis. We characterize factor analysis as a statistical method aimed at the estimation of the underlying common covariance structure of a multivariate normal population. In this respect factor analysis differs from component analysis which is a variance-oriented method of summarization. Both factor analysis and component analysis, when used as stochastic models, require the assumption that the underlying joint probability distribution is multivariate normal, and that the variates are commensurable.

The basic model in factor analysis can be given by the expression $x_i = \sum a_{ij} Y_j + e_i$ which implies that each observed variate x_i is considered to be a composite of the effect of m common factors Y_j, $j=1$ ..., m, and a specific factor e_i. The symbol a_{ij} represents factor loadings. It follows from this definition that the basic computational problem actually amounts to finding the factor loadings and the variances of the specific factors. In contrast with the extraction of component scores in component analysis, the factor scores (coordinates) cannot be derived by a simple linear transformation in factor analysis, but they must be estimated by one of several possible methods (see Lawley & Maxwell 1963). The factor analysis model, like the component analysis model, is linear.

There is another important difference between factor analysis and component analysis. regarding the covariance structure. This structure is completely destroyed in the course of factor analysis. The common factor space thus cannot be regarded as an orthogonal projection of the original r dimensional configuration (as depicted by the covariance matrix) within the ordination space of the m common factors.

The assumption that the underlying multivariate distribution is normal, and the fact that the phytosociologist must specify the number of common factors before commencing with the analysis, will impose limitations upon the applicability of factors analysis in phytosociology. The assumption of normality may require data transformations which may or may not be particularly helpful. The choice of an initial value for m, the number of postulated common factors, is a matter of personal judgment. Several trials may be required before the postulated m-factor model will give a sufficiently close fit to the data. The application of factor analysis to problems in phytosociology has apparently been limited to the use of centroid (principal) factor analysis (Dagnelie 1960, 1965, Szöcs 1971). The phytosociological value of the maximum-likelihood solution in association with factor rotation, however, has yet to be clearly demonstrated.

10.7.3 Component Analysis

10.7.3.1 *General Characteristics of the Method*

While factor analysis attempts to elicit the underlying common covariance structure of a given population in terms of m common factors, component analysis is aimed at the efficient summarization of the total variation in the sample. The two techniques are further distinguished by other properties in addition to those which we already mentioned: (i) The principal components are variant under changes in the scale of the variates. This implies, for example, that the ordination axes (components) extracted from a correlation matrix are expected to differ from those which are extracted from a covariance matrix. A peculiar effect of the scale used is that after standardization of each variate by a factor $1/S_i$ the component scores are expected to differ, by a factor other than $1/S_i$, from the component scores extracted from the unstandardized variates. There is no direct transference between the two sets of component scores in the manner of a simple scaling of the results. This is not so in factor analysis where such a transference is readily available. (ii) There is a simple linear relationship between the components and the original variates in component analysis. This is not so with the factor scores in factor analysis. (iii) The concepts of common and specific factors do not apply to component analysis. (iv) The variates are assumed to be simple transforms $x_i = \sum a_{ij} Y_j$ of r components, $j = 1,...,r$. Apparently no criteria are available which would specify the number of components needed to account for a significant portion of the total variance.

10.7.3.2 *The Basic (R-type) Algorithm*

The technique of component analysis is illustrated by an example at the end of this chapter. Assuming that the sample consists of r species and c quadrats we can outline an algorithm for component analysis in the following manner: (i) Adjust or standardize the original data to derive an rxc matrix **A** with elements $a_{ij}=(x_{ij}-\bar{x}_i)/S_i$ where x_{ij} is the original value of species i in quadrat j, $\bar{x}_i$ is the mean value of species i in the sample set, and S_i is a factor of standardization chosen by the phytosociologist (set ORLOCI 1967b). (ii) Generate an rxr matrix **R** of scalar products in accordance with $\mathbf{R}=\mathbf{AA}'$; the elements of **R** relate the species in pairs. (iii) Find the eigenvalues (λ_1, ..., λ_r) and the corresponding eigenvectors ($\boldsymbol{\alpha}_1,...,\boldsymbol{\alpha}_r$) of **R**. It is assumed that the eigenvectors are so standardized that their lengths will be equal to unity, i.e. $\boldsymbol{\alpha}_i' \boldsymbol{\alpha}_i=1$. Note that any element a_{hi} is regarded as a direction cosine of species h and component i. (iv) The component scores are computed for quadrats by the transformation $\mathbf{Y}=\boldsymbol{\alpha}'\mathbf{A}$. A characteristic element y_{ij} of **Y** represents the component score (co-ordinate) of the jth quadrat on the ith component. The components are the new ordination axes subject to the condition of mutual linear independence. The dispersion on the ith component is defined by the ith eigenvalue λ_i, and the efficiency of the ith component in accounting for variation in the sample is given by $\lambda_i/\sum\lambda_h$, the summation taken from $h=1$ to r. The correlation of species h and component i is defined by $a_{hi}\sqrt{(\lambda_i/\rho(h,h))}$ where $\rho(h,h)$ is the hth element in the principal diagonal of **R**. The basic computational problem in component analysis is concerned with the extraction of the eigenvalues and eigenvectors of the scalar product matrix. For computer use we recommend the JACOBI (NEUMAN) iterative method for which an ALGOL 60 algorithm is described by GRENCH & THACHER (1965), or IBM's EIGEN subroutine in Fortran IV (IBM 1968). The underlying concepts are discussed by GREENSTADT (1960).

10.7.3.3 *A Simplified (Q-type) Algorithm Based on Scalar Products*

The basic algorithm, which has just been described, is not a particularly practical one for deriving component scores when the number of species is relatively large as compared to the number of quadrats. When the number of quadrats is relatively small we can obtain the component scores more simply by analysis of a scalar product matrix of quadrats, **Q**=**A'A**, in which **A** matrix is identical to the **A** matrix from which the **R** matrix is computed. The analysis may be called in this case as the *Q-type component analysis*, signifying that, in contrast with the *R-type component analysis*, here we are analysing a **Q** matrix .This analysis has been described in detail by ORLOCI (1966, 1967b), GOWER (1966), and PIELOU (1969). These authors point out that the two analyses can produce identical component scores. This duality is a consequence of the relationship $\mathbf{Y}=\beta=\alpha'\mathbf{A}$, where the column vectors of β are the eigenvectors of **Q** so adjusted that their squared lengths are equal to the corresponding eigenvalues, i.e. $\beta_j'\beta_j=\lambda_j$.

10.7.3.4 *A Simplified (D-type) Algorithm Based on Distance*

In both the R-type and Q-type component analysis we have used the scalar product function as the basic definition of phytosociological resemblance. It, however, has been showed by GOWER (1966) that the same results can be obtained from the analysis of Euclidean distance. The distance matrix which we need to define is $\mathbf{D}^1$ with elements given by $-0.5d_{jk}^2$ where the d_{jk}^2 is the squared Euclidean distance computed between quadrat j and quadrat k. The matrix $\mathbf{D}^1$ has cxc elements, its non-negative eigenvalues and the associated eigenvectors are identical to those of the **Q** matrix. It directly follows from these identities that the analyses of $\mathbf{D}^1$, **Q**, or **R** will produce identical components scores for the quadtats. The user may find that in some ordinations $\mathbf{D}^1$ is much simpler for computation than either the **R** or the **Q** matrix.

10.7.3.5 *Mapping new Individuals in Scatter Diagrams*

It is possible to map new individuals in a scatter diagram on the basis of a transformation $y_{jk}=\alpha_j'\mathbf{A}_k$ where y_{jk} is the co-ordinate of the individual $\mathbf{A}_k$ on the jth ordination axis, and the elements of α_j represent direction cosines. We note that the species values in $\mathbf{A}_k$ must be adjusted similarily as in the original data which formed the basis of the ordination. The direction cosines can be obtained directly in a component analysis of an **R** matrix, or computed by $a_{hi}=(\sum a_{hj}y_{ij})/\lambda_i$ where a_{hi} is the direction cosine of the hth species relative to the ith ordination axis, a_{hj} is the value of species h in the jth quadrat, y_{ij} is the co-ordinate of the jth quadrat on the ith ordination axis, and λ_i is a measure of dispersion on the ith ordination axis. The summation is taken from $j=1$ to c.

10.7.3.6 *Number of Component Axes Required*

The phytosociologist may want to limit the analysis only to those components which correspond to the first *m* largest eigenvalues. Alternatively he may discard all components from the end of the list that are statistically indistinguishable (see Seal 1964). He may also consider the ecological meaningfulness of the components and reject those to which such a meaning cannot be attached (see Goodall 1954, Greig-Smith 1964, 1971).

10.7.3.7 *Remarks on Scatter Diagrams*

The joint distribution of the principal components is plotted in scatter diagrams with the purpose of graphical illustration of some ordination properties or the relationship of ordination results to external factors. It must, however, be noted that a scatter diagram represents an orthogonal projection in two dimensions of a potentially very complex multidimensional configuration of points (quadrats). A degree of distortion may thus occur in the projection depending on the original dimensions of the configuration. The distortion will occur in the sense that points, separated by great distances in the multidimensional configuration, may occupy a common position in the projection. It follows, therefore, that gaps between points in a scatter diagram will always signify separations at least as large as the real separations in the multidimensional configuration, whereas the lack of such gaps will not necessarily mean a similar lack of real separations.

10.7.3.8 *An Example*

The extraction of component scores is illustrated on an example so devised that the computational steps can be easily reconstructed. Consider a data matrix of 5 species (rows) and 4 quadrats (columns):

$$\mathbf{X} = \begin{matrix} x_{11} x_{12} x_{13} x_{14} \\ x_{21} x_{22} x_{23} x_{24} \\ x_{31} x_{32} x_{33} x_{34} \\ x_{41} x_{42} x_{43} x_{44} \\ x_{51} x_{52} x_{53} x_{54} \end{matrix} = \begin{matrix} 1 & 3 & 4 & 4 \\ 3 & 3 & 1 & 1 \\ 5 & 3 & 5 & 3 \\ 1 & 1 & 3 & 3 \\ 4 & 2 & 1 & 1 \end{matrix} .$$

Each entry x_{ij} in this matrix represents the quantity of a species in a given quadrat. The sample means for species are given by

$$\bar{\mathbf{X}} = \begin{matrix} \bar{x}_1 \\ \bar{x}_2 \\ \bar{x}_3 \\ \bar{x}_4 \\ \bar{x}_5 \end{matrix} = \begin{matrix} 3 \\ 2 \\ 4 \\ 2 \\ 2. \end{matrix}$$

From $\mathbf{X}$ we derive $\mathbf{A}$ according to $a_{ij} = (x_{ij} - \bar{x}_i)/S_i$ in which for reasons of simplicity we set S_i equal to 1. The $\mathbf{A}$ matrix is given by

$$\mathbf{A} = \begin{matrix} a_{11}\, a_{12}\, a_{13}\, a_{14} \\ a_{21}\, a_{22}\, a_{23}\, a_{24} \\ a_{31}\, a_{32}\, a_{33}\, a_{34} \\ a_{41}\, a_{42}\, a_{43}\, a_{44} \\ a_{51}\, a_{52}\, a_{53}\, a_{54} \end{matrix} = \begin{matrix} -2 & 0 & 1 & 1 \\ 1 & 1 & -1 & -1 \\ 1 & -1 & 1 & -1 \\ -1 & -1 & 1 & -1 \\ 2 & 0 & -1 & -1. \end{matrix}$$

The $\mathbf{R}$ matrix of scalar products for species is computed according to

$$\mathbf{R} = \mathbf{AA}' = \begin{matrix} 6 & -4 & -2 & 4 & -6 \\ -4 & 4 & 0 & -4 & 4 \\ -2 & 0 & 4 & 0 & 2 \\ 4 & -4 & 0 & 4 & -4 \\ -6 & 4 & 2 & -4 & 6, \end{matrix}$$

and the $\mathbf{Q}$ matrix of scalar products for quadrats according to

$$\mathbf{Q} = \mathbf{A'A} = \begin{matrix} 11 & 1 & -5 & -7 \\ 1 & 3 & -3 & -1 \\ -5 & -3 & 5 & 3 \\ -7 & -1 & 3 & 5. \end{matrix}$$

Note the equality of the sum of elements in the diagonal cells of $\mathbf{R}$ and $\mathbf{Q}$. The distance matrix of quadrats is given by

$$\mathbf{D}^1 = \begin{matrix} 0 & -6 & -13 & -15 \\ -6 & 0 & -7 & -5 \\ -13 & -7 & 0 & -2 \\ -15 & -5 & -2 & 0 \end{matrix}$$

where the elements represent $-0.5d_{jk}^2$ values.

The eigenvalues of $\mathbf{R}$, $\mathbf{Q}$, and $\mathbf{D}^1$ are identical (within rounding errors) with the exception of the fourth eigenvalue of $\mathbf{D}^1$ which is a negative value and should be ignored. The non-zero eigenvalues include $\lambda_1 = 18.591$, $\lambda_2 = 4.672$, and $\lambda_3 = 0.737$. Two more eigenvalues of $\mathbf{R}$ and one more of $\mathbf{Q}$ are zeros. These indicate that the four-point configuration is embedded in a three dimensional space. A relatively large λ_1 clearly is an indication that regarding the linear variation in **X** the summarization is quite successful. The eigenvectors (containing the transformation coefficients) of **R**, adjusted to unit length, are given by

$$\alpha' = \begin{matrix} 0.558 & -0.421 & -0.153 & 0.421 & -0.558 \\ -0.143 & -0.343 & 0.850 & 0.343 & 0.143 \\ 0.411 & 0.452 & 0.503 & -0.452 & -0.411 \end{matrix} .$$

The elements in each row of this matrix represent the direction cosines of species relative to a given principal component axis. The component scores are obtained according to

$$\mathbf{Y} = \beta = \alpha'\mathbf{A} = \begin{matrix} -3.226 & 0.735 & -0.235 \\ -0.690 & -1.537 & 0.401 \\ 1.805 & 1.252 & 0.420 \\ 2.110 & -0.449 & -0.587 \end{matrix}$$

The elements in the columns of **Y** are the component scores (coordinates) of quadrats on the new ordination axes. These scores are identical for the **R**, **Q**, and $\mathbf{D}^1$ matrices. In the present example the component scores can be produced least tediously on the basis of the $\mathbf{D}^1$ matrix.

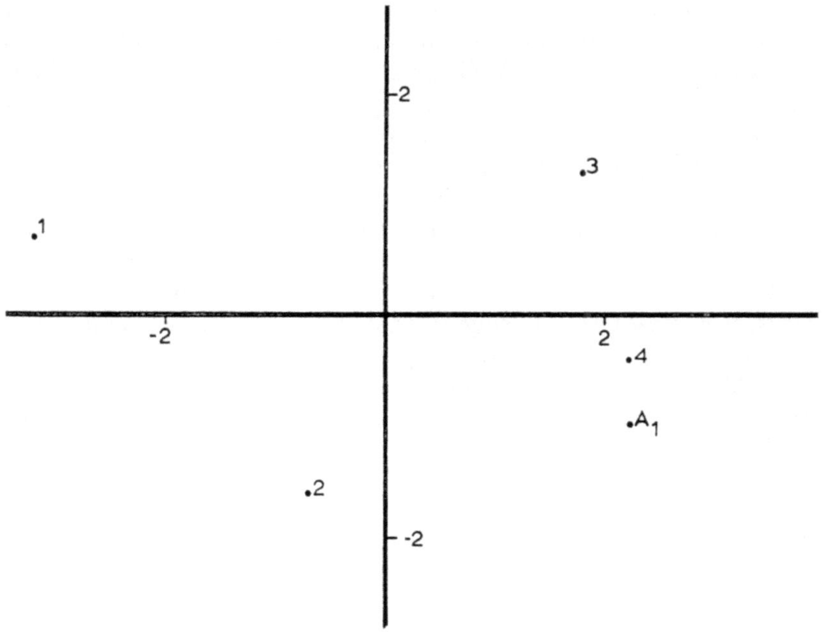

Fig. 1. Scatter diagram produced by component analysis from data given in the main text.

The scatter diagram of the first and second ordination axis is given in Fig. 1. Point $\mathbf{A}_1$ in this diagram represents an external quadrat in which the species values are the elements in $\mathbf{X}_1 = (5\ 2\ 3\ 3\ 1)$, so adjusted that $\mathbf{A}_1 = \mathbf{X}_1 - \overline{\mathbf{X}} = (2\ 0\ -1\ 1\ -1)$. The component scores for individual $\mathbf{A}_1$ are given by $y_{11} = \alpha_1' \mathbf{A}_1 = 2(0.558) + 0(-0.421) + (-1)(-0.153) + 1(0.421) + (-1)(-0.558) = 2.248$ and $y_{21} = \alpha_2' \mathbf{A}_1 = 2(-0.143) + 0(-0.343) + (-1)(0.850) + 1(0.343) + (-1)(0.143) = -0.936$.

10.7.3.9 *Summary of Component Analysis*

Component analysis as a mathematical method of data analysis is aimed at the extraction of a new set of co-ordinates (component scores) from a resemblance matrix whose elements qualify as Euclidean spatial parameters. These components are linearly independent, and as such, they provide the most efficient means for a parsimonious summarization of linear vegetational, or environmental, variation in a given sample. The field of phytosociological applications of component analysis includes trend seeking, search for clusters, discriminant-type assignments, and production of new data for input in further analyses.

We have outlined different algorithms on the basis of which phytosociological entities can be ordinated. While these algorithms are associated with different resemblance functions, which generate the coefficient matrices ($\mathbf{R}$, $\mathbf{Q}$, or $\mathbf{D}^1$) for analysis, they nevertheless represent alternative avenues of approach leading to the same set of component scores in their final results.

It may be pointed out that the application of component analysis as a general strategy is seriously weakened by its linear nature. Because the assumption of linearity can be expected to be a relatively reasonable assumption in the description of species and environmental correlations within relatively short segments of their gradients, it is reasonable to suggest that the application of component analysis should be limited to sample sets of relatively high homogeneity. Alternatively, component analysis may be followed by secondary analyses aimed at the extraction of higher-order variation, not detectable by the linear ordination model.

10.7.4 MISCELLANEOUS METHODS

10.7.4.1 *Polar Ordinations*

Component analysis, or factor analysis, is totally unsuited for longhand calculations, even in the case of relatively small sets of data.

It is the computational difficulty in these methods which probably stands out as one of the main reasons for the introduction of simplified methods, such as for instance the BRAY & CURTIS (1957) ordination and its many variants, also including those introduced to improve upon these simplified methods (see BEALS 1960, AUSTIN & ORLOCI 1966, SWAN et al. 1969, and VAN DER MAAREL 1969, among others). The improvements were concerned with (i) the nature of the resemblance function, (ii) the perpendicularity and common origin of the ordination axes, and (iii) the choice of end stands, i.e. ordination poles — hence the term *polar* ordination, to mark out the direction of the ordination axes. In the method of AUSTIN and ORLOCI the resemblance functions $1-L$, or max $L-L$, have been replaced by Euclidean distance, and the axes were selected in perpendicular directions. SWAN and his collaborators defined a first axis as that line, connecting two strategically chosen stands (quadrats) in sample space, on which the sum of squared projections of the inter-stand distances is a possible maximum. They extracted the subsequent ordination axes on the basis of the same principle from the residual distances. A computationally less tedious method was proposed by VAN DER MAAREL to achieve similar goals. In his method the first ordination axis is comparable to the first axis in the SWAN ordination, but the subsequent axes are produced from the original distances rather than the residuals.

Polar ordinations, similarly to component analysis, may represent linear models, although when used with certain types of resemblance functions they may be non-linear (see GAUCH & WHITTAKER 1972). However, they are less precise mathematically, and also, they are less efficient as methods of summarization than component analysis. These ordinations, nevertheless, are computationally simple, or relatively so, and can be handled with relative ease by the less mathematically minded phytosociologist. These properties, and the fact that they have produced meaningful ecological results, seem to justify the continued use of simple ordinations in phytosociology.

10.7.4.2 *Position Vectors Ordination*

The SWAN ordination as well as the method of VAN DER MAAREL are potentially less efficient than the position vectors ordination (ORLOCI 1966, FIELD & ROBB 1970, MARTIN 1969) in accounting for linear variability in a given sample. This is a consequence of the fact that the axes in position vectors ordination, similarily to the component axes, originate at the centroid of the sample space, and that each axis extracted by position vectors ordination approximates a component axis as closely as possible, under the condition that each ordination axis must go through an existing point (quadrat in our case) in sample space.

We shall illustrate position vectors ordination on the data given by **X** in paragraph 10.7.3.8. The analysis begins with computation of a **Q** matrix:

$$\mathbf{Q} = \mathbf{A}'\mathbf{A} = \begin{matrix} q_{11}q_{12}q_{13}q_{14} \\ q_{21}q_{22}q_{23}q_{24} \\ q_{31}q_{32}q_{33}q_{34} \\ q_{41}q_{42}q_{43}q_{44} \end{matrix} = \begin{matrix} 11 & 1 & -5 & -7 \\ 1 & 3 & -3 & -1 \\ -5 & -3 & 5 & 3 \\ -7 & -1 & 3 & 5. \end{matrix}$$

A characteristic element q_{jk} in this matrix defines the Q-type scalar product of quadrat j and k in a manner described in detail in paragraph 10.7.3.8.

Position vectors ordination derives its name from the use of position vectors as ordination axes. The position vector of a quadrat is defined as the line which connects the centroid of the sample space with the quadrat as a spatial point The initial goal is to identify the position vector which lies nearest to the first principal component axis. The criterion to be used is the sum of squared projections of all position vectors on the line of a given position vector. This sum for the jth position vector (quadrat) is given by $S_j = \sum q^2_{jk}/q_{jj}$ where the summation is from $k=1$ to c. The numerical values for the first and subsequent quadrats are given by $S_1 = 11^2/11 + 1^2/11 + (-5)^2/11 + (-7)^2/11 = 17.818$, $S_2 = 6.667$, $S_3 = 13.600$, and $S_4 = 16.8$ respectively. The maximum occurs on the line of the position vector of the first quadrat. We thus chose the first position vector as the first ordination axis. The co-ordinates on the first axis are computed according to $y_{1k} = q_{1k}/\sqrt{(q_{11})}$ for which the numerical values are given by $y_{11} = 11/\sqrt{(11)} = 3.317$, $y_{12} = 1/\sqrt{(11)} = 0.302$, $y_{13} = -5/\sqrt{(11)} = -1.508$, and $y_{14} = -7/\sqrt{(11)} = -2.111$. Note the closeness of these co-ordinate values to the component scores in the first column of **Y**, given in paragraph 10.7.3.8. The dispersion on the first ordination axis is 17.818, nearly equal to the first eigenvalue (18.591) of the **Q**, **R**, or $\mathbf{D}^1$ matrix After computation of the first axis co-ordinates follows the computation of the first residual scalar product matrix **Q**′ with elements defined by $q'_{jk} = q_{jk} - y_{1j}y_{1k}$. The value for instance in the second cell of the third row of **Q**′ is $q'_{32} = q_{32} - y_{13}y_{12} = -3 - (-1.508)(0.302) = -2.545$. The complete residual matrix is

$$\mathbf{Q}' = \begin{matrix} 0.000 & 0.000 & 0.000 & 0.000 \\ 0.000 & 2.909 & -2.545 & -0.364 \\ 0.000 & -2.545 & 2.727 & -0.183 \\ 0.000 & 0.364 & -0.183 & 0.545 \end{matrix}$$

within rounding errors.

We now repeat the procedure on $\mathbf{Q}'$. The sum of squared projections is a maximum on the second position vectoer ($S_2=5.182$), and the co-ordinates are $y_{21}=0, y_{22}=1.706, y_{23}=-1.492$, and $y_{24}=-0.213$. These results are directly comparable to the second largest eigenvalue (4.672) and corresponding component scores given in paragraph 10.7.3.8. The discrepancy between the two sets of results is explainable by the fact that position vectors ordination extracted a first axis which slightly deviates from the first component axis. This has a cumulative effect resulting in an increasing departure of the subsequent ordination axes from the corresponding components.

The second residual matrix $\mathbf{Q}''$ is derived from the first residual $\mathbf{Q}'$ according to the expression $q''_{jk}=q'_{jk}-y_{2j}y_{2k}$ for which the computed values are given by

$$\mathbf{Q}'' = \begin{matrix} 0.000 & 0.000 & 0.000 & 0.000 \\ 0.000 & 0.000 & 0.000 & 0.000 \\ 0.000 & 0.000 & 0.500 & -0.500 \\ 0.000 & 0.000 & -0.500 & 0.500 \end{matrix}$$

The sum of squared projections is equal on both non-zero position vectors, $S_3=S_4=1.000$. The corresponding co-ordinates then are $y_{31}=0, y_{32}=0, y_{33}=-0.707$, and $y_{34}=0.707$. These results are directly comparable to the third largest eigenvalue (0.737) and associated component scores. It may be useful at this point if we noted that the above example is rather peculiar in that respect, that the sequence of position vectors happens to be identical with the sequence in which they are extracted as ordination axes. This is not generally the rule, and it would not have happened if the quadrats were listed in a different sequence.

Note that the third residual matrix $\mathbf{Q}'''$ is a zero matrix, indicating that all variation in $\mathbf{Q}$ has been accounted for by the first three ordination axes. The total variation, i.e. the sum of all diagonal elements in $\mathbf{Q}$ is equal to 24. The total dispersion on the three ordination axes (17.818+5.182+1.000) is also equal to 24.

Position vectors ordination will be a reliable approximation to component analysis when two specific circumstances happen to occur simultaneously: (i) a relatively uniform phytosociological sample, and (ii) a large number of points (quadrats in our case). We can anticipate position vectors ordination to produce an accurate approximation to component analysis under these circumstances, because when there are many points, and if these points are more-or-less evenly dispersed in the relevant part of sample space, then it is quite probable that there will be at least one point at or very near each of the principal component axes. These points will be found by position vectors ordination. The ordination axes will go through these points and the centroid (average quadrat) of the system.

10.8 Ordination by non-Euclidean matrices

10.8.1 The Sample Space

The sample space, in this case, can be derived by negation of the properties of an Euclidean space. It will be sufficient here if we simply stated that the space we have in mind is one in which the spatial parameters are functions other than an Euclidean distance or a scalar product. Examples of such functions have already been discussed in chapter 10.7.1. The present chapter will further elaborate on the use of non-Euclidean resemblance functions in phytosociological ordinations. Two methods are discussed in this connection, including Gower's principal axis analysis and Kruskal's method of multidimensional scaling. It should be mentioned, however, that the use of these methods is not really limited to non-Euclidean functions.

10.8.2 Principal Axes Analysis

The underlying mathematical model in principal axes analysis is the same as in component analysis (see Gower 1966, Pielou 1969). In both of these methods ordination axes are obtained on the basis of the use of an eigenvalue and vector procedure. Principal axes analysis, however, differs from component analysis in one important respect — it is not limited to the use of Euclidean distance or scalar products.

We shall define principal axes analysis in the broadest sense as a method aimed at the extraction of Cartesian co-ordinates from any given resemblance matrix. This does not need to be a metric matrix. The use of a non-metric matrix, however, will lead to distortion related to the inefficiency of ordination distances **M** to account for the observed values in the resemblance matrix **D**. Ordination distances are defined by $m_{jk}^2 = \sum(y_{ij} - y_{ik})^2$ where the y_{ij} and y_{ik} values are the co-ordinates of quadrat j and quadrat k on the ith ordination axis. The distortion so-defined can be measured by the function $\sigma_{MD} = 100(1 - \rho_{MD}^2)$, where ρ_{MD}^2 is the squared product-moment correlation coefficient, computed between the elements of **M** and the corresponding elements of **D**.

We can suggest that degree of distortion in the ordination will be dependent on the metric properties of the resemblance function which generates the **D** matrix. To illustrate the point in question, we shall consider the following functions:

(1) $d_{jk}=2(1-2\sum \min(x_{ij},x_{ik}) / \sum (x_{ij}+x_{ik}))$

(2) $d_{jk}=2(\max L-2\sum \min(x_{ij},x_{ik}) / \sum (x_{ij}+x_{ik}))$

(3) $d_{jk}=2(\sum(x_{ij}+x_{ik})-2\sum \min(x_{ij},x_{ik}))=\sum |x_{ij}-x_{ik}|$, where x_{ij} and x_{ik} represent species values in the quadrats, $\sum \min(x_{ij},x_{ik})$ is the sum of lesser values in quadrats j and k, and $\max L$ is a maximum value of $2\sum \min(x_{ij},x_{jk})/\sum(x_{ij}+x_{ik})$.

Similar expressions have already been discussed in some respect in chapter 10.7.1. Function (3) is the absolute value function. This metric function is degenerated into a dissimilarity function in (1) and (2) with properties that fail on at least one, but possibly more, of the metric space axioms.

We have generated successive **D** matrices on the basis of functions (1), (2), and (3). The eigenvalues and corresponding eigenvectors were extracted from each **D** matrix after multiplication of its elements by the scalar −0.5. The co-ordinates of quadrats then were obtained by adjustment of the eigenvalues in the same manner as in component analysis of a **Q** or $\mathbf{D}^1$ matrix (see chapter 10.7.3). The data used in the examples are given in Table II.

II. Tree counts in a Wisconsin maple forest [1]).

t number	1	2	3	4	5	6	7	8	9	10	11	12	13	14	15	16	17	18	19	20	21	22	23	24	25[1])
Maple	8	4	3	7	0	0	4	5	8	1	1	5	8	9	9	10	10	5	6	8	3	2	2	6	4
od	0	7	4	1	1	6	2	0	0	3	0	4	1	0	0	3	0	0	0	0	4	6	0	0	0
y Elm	0	1	0	0	5	4	7	0	0	6	5	3	1	0	0	0	0	0	0	0	2	5	5	2	0
od	0	0	1	0	4	0	0	0	0	1	0	0	0	0	0	0	0	0	0	0	0	0	0	0	0
k	0	0	0	0	0	0	0	0	0	0	0	0	3	1	0	0	0	2	1	0	0	0	0	0	1
oud	0	1	1	0	0	0	0	0	0	0	0	0	0	0	0	0	0	0	0	0	0	0	0	0	0
ut	1	0	0	0	0	0	0	0	0	0	0	0	0	0	0	0	0	0	0	0	0	0	0	0	0
herry	1	0	0	0	0	0	0	0	0	0	1	0	0	0	0	0	0	0	0	0	0	0	0	0	0
rry	0	0	0	1	0	0	0	0	0	0	0	0	0	0	0	0	0	0	0	0	0	0	0	0	0
Oak	0	0	0	0	0	0	0	0	0	0	0	0	0	0	0	0	0	0	0	0	0	0	0	0	1

obtained by sampling on a vegetation map given by Curtis (1956). Sampling map units are equivalent et square quadrats in natural size.

No distortion occurred in the ordination (in terms of σ_{MD}) when **D** was generated by function (3). For these results the scatter diagram of the first two ordination axes is plotted in Fig. 2. The lack of

Fig. 2. Scatter diagram produced by principal axes analysis of the absolute value function based on data in Table II. See the main text for explanations.

distortion indicates that the information in **D** can be completely accounted for by the Cartesian co-ordinates of principal axes analysis. This implies an identity of the original configuration, as depicted by **D** (or -0.5**D**), and the derived configuration, as defined by the ordination distances **M**. The analysis of the same data on the basis of function (1), however, resulted in about 17 % distortion. An even larger amount of distortion (about 18 %) occurred when we used function (2) to generate the **D** matrix. The value of max$L$ was set at 0.5 in this example. The scatter diagrams for the ordination of functions (1) and (2) are given in Figs. 3—4.

"The generality of these results, however, remains to be seen."

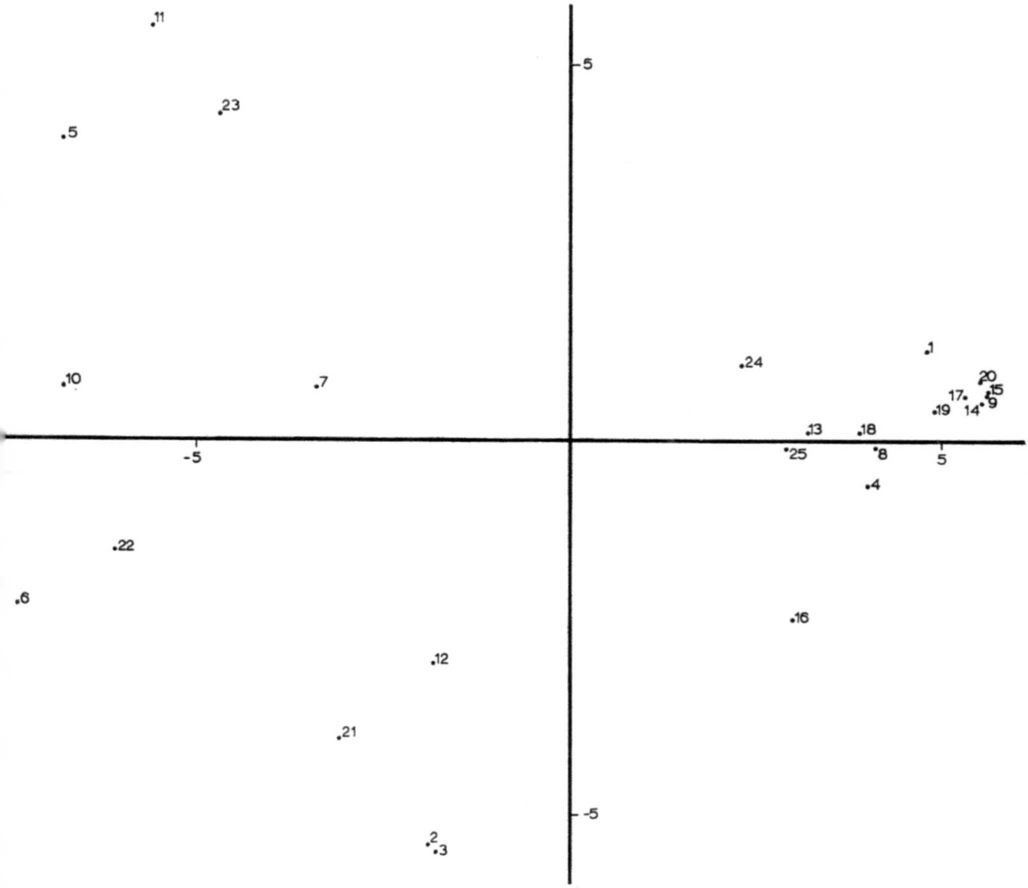

. Scatter diagram produced by principal axes analysis of a variant of the $CD = 1 - CC$ type SØRENSEN index based on data in Table II. See the main text for explanations.

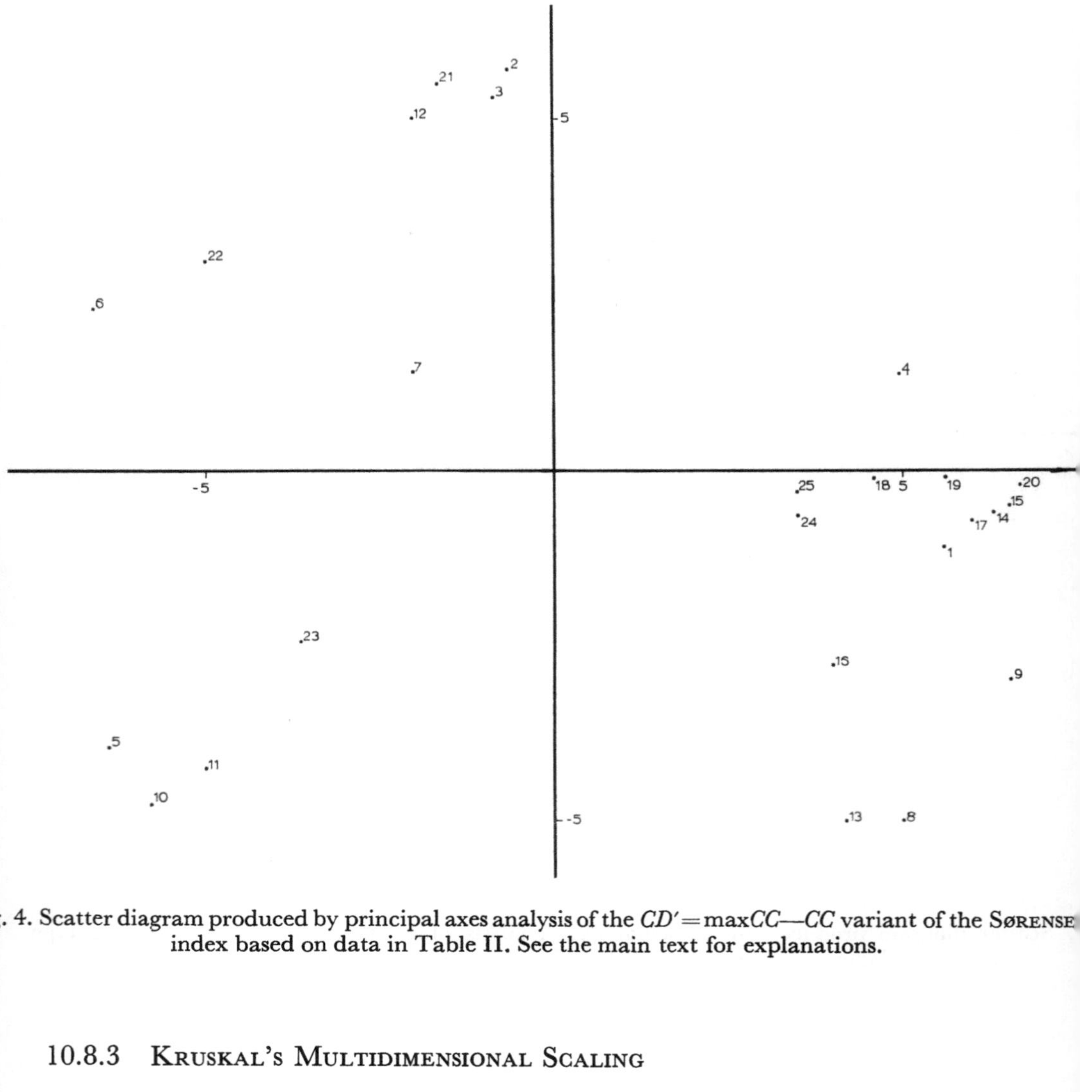

g. 4. Scatter diagram produced by principal axes analysis of the $CD' = \max CC$—CC variant of the SØRENSE index based on data in Table II. See the main text for explanations.

10.8.3 KRUSKAL'S MULTIDIMENSIONAL SCALING

While the metric configuration **M** is derived by an eigenvalue and vector procedure in principal axes analysis, this type of configuration, or the co-ordinates which describe it, can also be derived by other methods from a given set of experimental (observed) resemblance values. A method particularly well-suited to perform such a task has been proposed by KRUSKAL (1964a,b).

The KRUSKAL method begins by specifying t the number of ordination axes to be extracted. Then an arbitrary set of c points is placed in the t-dimensional ordination space, and the initial ordination resemblance values, often as distances or dissimilarities, are computed. We shall designate these dissimilarities by (d_{jk}). This is followed by fitting the regression line $\hat{d}_{jk}=f(\delta_{jk})$ to the points in the scatter diagram of the ordination dissimilarities (d_{jk}) and the experimental resemblance values (δ_{jk}). We note that the KRUSKAL method is little restricted as far as the definition of d_{jk} or δ_{jk} is concerned, and that this method can handle different types of regression models whose choice is largely a matter of personal preference (see KRUSKAL & CARMONE 1971). Different types of regression models, however, yield different regression estimates $\hat{d}_{jk}$, and consequently, different sets of ordination co-ordinates.

The stress, or distortion, in the ordination configuration, which we seek to minimize in the KRUSKAL method, is measured by the residual sum of squares in (d_{jk}) with respect to the regression estimates $(\hat{d}_{jk})$, and not by the fit of the (d_{jk}) to the experimental resemblance values (δ_{jk}). KRUSKAL & CARMONE have used as a measure of stress $S d\hat{d}=\sqrt{(\sum\sum(d_{jk}-\hat{d}_{jk})^2/\sum\sum(d_{jk}-\bar{d})^2)}$. In this expression d_{jk} is an ordination disssimilarity, $\bar{d}$ is the mean of all d_{jk}, $\hat{d}_{jk}$ is a regression estimate of d_{jk} at δ_{jk}, and the summations are taken from $j=1$ to $c-1$ and $k=j+1$ to c. Following the calculation of stress, all points in the initial ordination configuration are moved a bit, i.e. their co-ordinates are changed a little to decrease the value of stress. The procedure is iterated, starting with a new regression analysis, until the value of stress drops below a specified threshold limit. The output includes t sets of c co-ordinates which can be plotted jointly in two dimensional scatter diagrams. The accuracy of the scatter diagrams in representing the ordination configuration (d_{jk}) can be measured by $\sigma_{d*d}=\sqrt{(1-\rho^2_{d*d})}$ where ρ^2_{d*d} is the squared correlation of the ordination dissimilarities (d_{jk}) and the distances in the scatter diagram (d^*_{jk}).

There seems to be no example in phytosociology which used the KRUSKAL method for ordination. This method, however, appears to have many attractive properties for the phytosociologist. The method's extreme flexibility to select from different resemblance functions on the basis of which the δ_{jk} and d_{jk} values are computed, and the reliance on regression analysis, the model of which to be chosen by the user himself, can be mentioned in this connection. By using the KRUSKAL method it is possible to derive an ordination configuration (d_{jk}) from a matrix of observed resemblances (δ_{jk}) in such a manner that the former will be some specified non-linear function of the latter. This and the other properties which we already mentioned definitely indicate a good potential of the KRUSKAL method in phytosociological ordinations.

10.9 **Comparison of Methods**

The ordination methods which we discussed differ greatly on the basis of conceptual grounds and also in terms of the goals which they may attempt to achieve. Among these methods factor analysis stands out as an ordination method principally concerned with estimation of population properties. This method is covariance -oriented and it is burdened by several restrictive assumptions. These, and probably also the lack of sufficient examples, may be reasons why factor analysis, despite its effective use by Dagnelie (1960, article 9) has not aroused wider interest among phytosociologists.

Component analysis has mainly been used as a descriptive (deterministic) model without much regard of statistical considerations. It has been proved to be a useful method in handling continuous linear variability in phytosociological samples in connection with different practical problems which range from trend seeking to classifications. The classification function of component analysis, similarly as in all the ordination techniques, has however been reduced in importance by the availability of more efficient classification techniques.

The principal justification for the use of polar ordinations and the position vectors ordination is that they are simple conceptually and easily executed computationally, or relatively so, as compared to component analysis. This advantage, nevertheless, is being fast eroded away as the phytosociologist gains more ready access to the use of digital computers.

The method of principal axes analysis is computionally similar to component analysis. It, nevertheless, permits the use of almost any resemblance function as an ordination parameter. This property is in sharp contrast with component analysis where the admissible resemblance functions must qualify as Euclidean spatial parameters. Kruskal's method also can handle the ordination of different non-Euclidean resemblance matrices without much restrictions on the function which generates them. This method has the added advantage of a regression device whereby the ordination configuration can be made to be a non-linear function of the observed resemblances. Such a non-linear derivation of an ordination may very much interest the phytosociologist in his search for an ecologically more meaningful ordination method.

10.10.1 Summary

Ordination is used as a collective term in the present paper to describe the different methods which can extract metric information, such as for instance a new set of co-ordinates, from a given resemblance matrix. While the elements in such a matrix define the pairwise similarity, or dissimilarity, between phytosociological entities, the entities themselves may represent vegetation units (quadrats, sample plots), individual species, or other kinds of objects, vegetational or environmental, that can be recognized without ambiguity. The present paper is aimed at a comparative review of the different ordination methods which have been used, or have a good potential, as a basis for the mathematical analysis of phytosociological data.

Two general groups of ordination methods are discussed. Those of the first group are restricted to the analysis of resemblance matrices in which every element will qualify as either a common (Euclidean) distance or a centered scalar product. The methods of this group include CATTELL's open model factor analysis, component analysis, several variants of polar ordination, and position vectors ordination.

The ordinations methods of the second group, in contrast with those of the first group, are not limited to the use of distances or scalar products. As a matter of fact they can handle resemblance matrices of almost any kind. The different definitions of resemblance, however, will potentially lead to different ordination results. Described in this group are GOWER's method of principal axes analysis, which we broadened in our interpretations of the admissible definitions of resemblance, and KRUSKAL's method of multidimensional scaling.

The first parts of the paper are devoted to a discussion of the general concepts about ordinations, regarding their purpose, the underlying common model, hypothesis testing, and the important efficiency criteria. The discussion of the general concepts is followed by a systematic description of ordination methods with a view at their instrinsic properties and generic relationships as they relate to phytosociological applications. The different methods are illustrated by several numerical examples.

10.10.2 ZUSAMMENFASSUNG

Ordination wird in diesem Beitrag als ein Sammel-Begriff verwendet um verschiedene Methoden zu beschreiben mit denen metrische Information gewonnen werden kann, wie z.B. ein neues System von Koordinaten, aus einer gegebenen Ähnlichkeits-Matrix. Während die Elemente in einer solchen Matrix die paarweise Ähnlichkeit oder Unähnlichkeit zwischen pflanzensoziologischen

Einheiten bestimmen, können diese selbst Vegetationseinheiten (Quadrate, Probeflächen), einzelne Spezies, oder andere Arten von Objekten, sowohl der Vegetation als auch des Standorts darstellen, die nur zweideutig erkannt werden können.

Dieser Beitrag bezweckt eine vergleichende Übersicht über die verschiedenen Ordinationsmethoden, die gebraucht worden sind oder die eine gute Möglichkeit als Grundlage für die mathematische Analyse von pflanzensoziologischen Daten enthalten.

Zwei allgemeine Gruppen von Ordinationsmethoden werden besprochen: Diejenigen der ersten Gruppe sind auf die Analyse von Ähnlichkeits-Matrizen beschränkt, in welchen jedes Element entweder als ein gewöhnlicher (euklidischer) Abstand oder als 'zentriertes Skalar-Produkt' dargestellt sind. Die Methoden dieser Gruppe umfassen u.a. CATTEL's offene Model-Faktoren-Analyse, Komponenten-Analyse, mehrere Variaten der polaren Ordination und Positionsvectoren-Ordination.

Die Ordinations-Methoden der zweiten Gruppe sind, im Gegensatz zu denjenigen der ersten Gruppe, nicht beschränkt auf die Anwendung von Abständen oder Skalarprodukten. Tatsächlich können sie fast alle Arten von Ähnlichkeits-Matrizen behandeln. Die verschiedenen Definitionen von Ähnlichkeit können jedoch zu verschiedenen Ordinations-Ergebnisesn führen. In dies Gruppe werden GOWER's Methode der Hauptachsen-Analyse beschrieben, die in unserer der Interpretation der zulässigen Definitionen der Ähnlichkeit etwas erweitert wird, und KRUSKAL's Methode der 'multidimensional scaling' beschrieben.

Die ersten Teile des Beitrags sind einer Diskussion der allgemeinen Begriffe über Ordinationen in Hinblick auf ihre Ziel, das zugrundeliegende gemeinsame Modell, die Prüfung von Hypothesen und die wichtigen Kriterien der Wirksamkeit, gewidmet. Der Diskussion der allgemeinen Begriffe folgen eine systematische Beschreibung der Ordinations-Methoden in Hinblick auf ihre eigentlichen Eigenschaften und ihre Verwandtschaft, soweit sie sich auf pflanzensoziologischen Anwendungen beziehen. Die verschiedenen Methoden werden durch mehrere numerischen Beispiele erläutert.

RERERENCES

ASHTON, P. S., – 1964 – Ecological studies in the mixed dipterocarp forests of Brunei State. *Oxf. For. Mem.* 25: 1—75.

AUSTIN, M. P., – 1968 – An ordination study of a chalk grassland community. *J. Ecol.* 56: 739—757.

AUSTIN, M. P., P. S. ASHTON, & P. GREIG-SMITH, – 1971 – The application of quantitative methods to vegetation survey. III. A re-examination of rain forest data from Brunei. *J. Ecol.* (In press.)

AUSTIN, M. P. & P. GREIG-SMITH, – 1968 – The application of quantitative methods to vegetation survey. II. Some methodological problems of data from rain forests. *J. Ecol.* 56: 827—844.

AUSTIN, M. P. & I. NOY-MEIR, – 1971 – The problem of non-linearity in ordination: experiments with two gradient models. *J. Ecol.* 59: 763—773.

AUSTIN, M. P. & L. ORLOCI, – 1966 – Geometric models in ecology. II. An evaluation of some ordination techniques. *J. Ecol.* 54: 217—227.

BEALS, E., – 1960 – Forest bird communities in the Apostle Islands of Wisconsin. *Wilson Bull.* 72: 156—181.

BRAY, R. J. & J. T. CURTIS, – 1957 – An ordination of the upland forest communities of southern Wisconsin. *Ecol. Monog.* 27: 325—349.

CARLSON, K. A., – 1970 – A multivariate classification of reformatory inmates. *Ph. D. Thesis.* University of Western Ontario, London, Canada.

CATTELL, R. B., – 1965 – Factor analysis: an introduction to essentials. *Biometrics* 21: 190—215, 405, 435.

CHANDAPILLAI, M. M., – 1970 – Variation in fixed dune vegetation at Newborough Warren, Anglesey, *J. Ecol.* 58: 193—201.

CURTIS, J. T., – 1956 – Plant Ecology Workbook. Burgess, Minneapolis.

CURTIS, J. T., – 1959 – The vegetation of Wisconsin: An Ordination of Plant Communities. University of Wisconsin, Madison. 657 pp.

DAGNELIE, P., – 1960 – Contribution à l'étude des communautés végétales par l'analyse factorielle. *Bull. Serv. Carte phytogéogr.*, Paris, Sér. B. 5: 7—71, 93—195.

DAGNELIE, P., – 1965 – L'étude des communautés végétales par l'analyse statistique des liaisons entre les espèces et les variables écologiques: un exemple. *Biometrics* 21: 890—907.

FIELD, J. G. & F. T. ROBB, – 1970 – Numerical methods in marine ecology. 2. Gradient analysis of rock shore samples from False Bay. *Zoologica africana.* 5: 191—210.

GAUCH, H. G. JR. & R. H. WHITTAKER, – 1971 – Coenocline simulation. *Ecology* 53: 446—451.

GAUCH, H. G. JR. & R. H. WHITTAKER, – 1972 – Comparison of ordination techniques. *Ecology.* (In press).

GENGERELLI, J. A., – 1963 – A method for detecting subgroups in a population and specifying their membership. *J. Psychol.* 55: 457—468.

GITTINS, R., – 1965a – Multivariate approaches to a limestone grassland community. I. A stand ordination. *J. Ecol.* 53: 385—401.

GITTINS, R., – 1965b – Multivariate approaches to a limestone grassland community. III. A comparative study of ordination and association analysis. *J. Ecol.* 53: 411—425.

GITTINS, R., – 1968 – Trend -surface analysis of ecological data. *J. Ecol.* 56: 845—869.

GITTINS, R., – 1969 – The application of ordination techniques. In 'Ecological Aspects of the Mineral Nutrition of Plants,' R. H. RORISON, ed. pp. 37—66. *Symp. Brit. Ecol. Soc.* 1968, Blackwell, London.

GOODALL, D. W., – 1954 – Objective methods for the classification of vegetation. III. An essay in the use of factor analysis. *Aust. J. Bot.* 2: 304—324.

GOODALL, D. W., – 1967 – The distribution of the matching coefficient. *Biometrics* 23: 647—656.

GOWER, J. C., – 1966 – Some distance properties of latent root and vector methods used in multivariate analysis. *Biometrika* 53: 325—338.

GREENSTADT, J., – 1960 – The determination of the characteristic roots of a matrix by the Jacobi method. In 'Mathematical Methods for Digital Computers,' A. RALSTON & H. S. WILF, eds., pp. 84—91. Wiley, New York.

GREIG-SMITH, P., – 1964 – Quantitative Plant Ecology. 2nd ed. Butterworths, London. 256 pp.

GREIG-SMITH, P., – 1971 – Analysis of vegetation data: the user viewpoint. In 'Statistical Ecology,' C. P. Patil ed., 3: 149—166. Pennsylvania State University Press, London.

GREIG-SMITH, P., M. P. AUSTIN, & T. C. WHITMORE, – 1967 – The application of quantitative methods to vegetation survey. I. Association analysis and principal component ordination of rain forest. *J. Ecol.* 55: 483—503.

GRENCH, R. E. & H. C. THACHER, – 1965 – Collected Algorithms 1960—1963 from the Communications of the Association for Computing Machinery. Clearing House for Federal Scientific and Technical Information, U. S. Department of Commerce, Springfield, Virginia.

GRIGAL, D. F. & R. A. GOLDSTEIN, – 1970 – Classification and ordination of an intensively sampled oak-hickory forest. Walker Branch Watershed. Oak Ridge National Laboratory, Tennessee. (Mimeographed).

GROENEWOUD, H. VAN, – 1965 – Ordination and classification of Swiss and Canadian coniferous forests by various biometric and other methods. (Germ. summ.) *Ber. geobot. Inst. ETH, Stiftg Rübel*, Zürich, 1964, 36: 28—102.

HOLGATE, P. – 1971 – Notes on the Marcyewski-Steinhaus coefficient of similarity In 'Statistical Ecology,' G. P. Patil, ed., 3: 181—193. Pennsylvania State University Press, London.

IBM, 1968 — System/360 Scientific Subroutine Package (360A-CM-03X) Version III Programmer's Manual Docoment H20-0205-3, IBM, White Plains, N.Y. 454 pp.

IVIMEY-COOK, R. B., – 1968 – Investigations into the phenetic relationships between species of Ononis *L. Watsonia* 7: 1—23.

IVIMEY-COOK, R. B. & M. C. F. PROCTOR, – 1967 – Factor analysis of data from an east Devon heath: comparison of principal component and rotated solutions. *J. Ecol.* 55: 405—413.

JACCARD, P., – 1901 – Distribution de la flore alpine dans le Bassin des Dranses et dans quelques régions voisines. *Bull. Soc. vaud. Sci. nat.* 37: 241—272.

KERSHAW, K. A., – 1968 – Classification and ordination of Nigerian Savanna vegetation. *J. Ecol.* 56: 467—482.

KRUSKAL, J. B., – 1964a – Multidimensional scaling by optimizing goodness of fit to a nonmetric hypothesis. *Psychometrika* 29: 1—27.

KRUSKAL, J. B., – 1964b – Nonmetric multidimensional scaling: a numerical method. *Psychometrika* 29: 115—129.

KRUSKAL, J. B. & F. CARMONE, – 1971 – How to use M-D-SCAL (Version 5M) and other useful information. Bell Telephone Laboratories, Murray Hill, New Jersey, U. S. A., and University of Waterloo, Waterloo, Ontario, Canada. (Mimeographed.)

LAMBERT, J. M. & M. B. DALE, – 1964 – The use of statistics in phytosociology. *Adv. ecol. Res.* 2: 59—99.

LAWLEY, D. N. & A. E. MAXWELL, – 1963 – Factor Analysis as a Statistical Method. Butterworths, London. 117 pp.

MAAREL, E. VAN DER, – 1969 – On the use of ordination models in phytosociology. (Germ. summ.) *Vegetatio* 19: 21—46.

MACNAUGHTON-SMITH, P., W. T. WILLIAMS, M. B. DALE, & L. G. MOCKETT, – 1964 – Dissimilarity analysis: a new technique of hierarchical sub-division. *Nature, Lond.* 202: 1034—1035.

MARTIN, J., – 1969 – Ordination methods in lichenocoenology. In: 'Quantitative Methods of Vegetation Study,' p. 86. Collected Conference Reports,

Institute of Zoology and Botany, Academy of Sciences, Tartu State University, Estonian S. S. R.

MAYCOCK, P. F., – 1963 – The phytosociology of the deciduous forests of extreme southern Ontario. *Can. J. Bot.* 41: 379—438.

MAYCOCK, P. F. & J. T. CURTIS, – 1960 – The phytosociology of Boreal conifer-hardwood forests of the Great Lakes region. *Ecol. Monogr.* 30: 1—35.

MC DONALD, R. P., – 1962 – A general approach to non-linear factor analysis. *Psychometrika* 27: 397—415.

MC INTOSH, R. P., – 1967 – The continuum concept of vegetation. *Bot. Rev.* 33: 130—187.

MC INTOSH, R. P. & R. T. HURLEY, – 1964 – The spruce-fir forests of the Catskill Mountains. *Ecology* 45: 314—326.

MOORE, J. H., P. FITZSIMONS, E. LAMBE, & J. WHITE, – 1970 – A comparison and evaluation of some phytosociological techniques. (Germ. summ.) *Vegetatio* 20: 1—20.

MORRISON, D. F., – 1967 – Multivariate Statistical Methods. Mc Graw-Hill, New York. 338 pp.

NOY-MEIR, I. & M. P. AUSTIN, – 1970 – Principal component ordination and simulated vegetational data. *Ecology* 51: 551—552.

OCHIAI, A., – 1957 – Zoogeographic studies on the soleoid fishes found in Japan and its neighbouring regions. *Bull. Jap. Soc. scient. Fish.* 22: 526—530.

ORLOCI, L., – 1966 – Geometric models in ecology. I. The theory and application of some ordination methods. *J. Ecol.* 54: 193—215.

ORLOCI, L., – 1967a – An agglomerative method for classification of plant communities. *J. Ecol.* 55: 193—205.

ORLOCI, L., – 1967b – Data centering: a review and evaluation with reference to component analysis. *Syst. Zool.* 16: 208—212.

ORLOCI, L., – 1968 – A model for the analysis of structure in taxonomic collections. *Can. J. Bot.* 46: 1093—1097.

ORLOCI, L., – 1971 – Notes on the 'Notes on the Marcyewski-Steinhaus coefficient of similarity' - a paper by P. Holgate. In 'Statistical Ecology,' G. P. Patil, ed., 3: 192—193. Pennsylvania State University Press, London.

PHIPPS, J. B., – 1970 – Studies in the Arundinelleae (Gramineae). X Preliminary taximetrics. *Can. J. Bot.* 48: 2333—2356.

PIELOU, E. C., – 1969 – An Introduction to Mathematical Ecology. Wiley-Interscience, New York. 286 pp.

PRÉCSÉNYI, I. & Z. SZÖCS, – 1969 – Javaslat a kontinuum-vizsgálat egy módszerének módosítására. (Magyar with Engl.summ.: Modification of a method of continuum studies). *Bot. Közl.* 56: 189—196.

QUADLING, C., – 1967 – Evaluation of tests and grouping of cultures by a two-stage principal component method. *Can. J. Microbiol.* 13: 1379—1400.

RUSSELL, P. F. & T. R. RAO, – 1940 – On habitat and association of species of anopheline larvae in south-eastern Madras. *J. Malar. Inst. India* 3: 153—178.

SEAL, H. L., – 1964 – Multivariate Statistical Analysis for Biologists. Methuen, London. 207 pp.

SHEPARD, R. N. & J. D. CARROLL, – 1965 – Parametric representation of non-linear data structures. In 'Multivariate Analysis,' P. R. Krishnaiah, ed. pp. 561—592. Academic Press, London.

SNEATH, P. H. A., – 1966 – A method for curve seeking from scattered points. *Comput. J.* 8: 383—391.

SNEATH, P. H. A., – 1967 – Trend-surface analysis of transformation grids. *J. Zool.* 151: 65—122.

SOKAL, R. R., – 1958 – Quantification of systematic relationships and of phylogenetic trends. *Proc. Tenth Int. Congr. Ent.*, Montreal 1956, 1: 409—415.

SOKAL, R. R. & C. D. MICHENER, – 1958 – A statistical method for evaluating systematic relationships. *Kans. Univ. Sci. Bull.* 38: 1409—1438.
SOKAL, R. R. & F. J. ROHLF, – 1962 – The comparison of dendrograms by objective methods. *Taxon* 11: 33—39.
SØRENSEN, T., – 1948 – A method of establishing groups of equal amplitude in plant sociology based on similarity of species content, and its application to analyses of the vegetation on Danish Commons. *Biol. Skr. K. danske Vidensk. Selsk.* 5 (4): 1—34.
SWAN, J. M. A., – 1970 – An examination of some ordination problems by use of simulated vegetation data. *Ecology* 51: 89—102.
SWAN, J. M. A., R. L. DIX, & C. F. WEHRHAN, – 1969 – An ordination technique based on the best possible stand-defined axes and its application to vegetation analysis. *Ecology* 50: 206—212.
SZÖCS, Z., – 1971 – A Vértes-Hegység bükkösei: II. Interspecifikus korrelációvizsgálatok. (Magyar with Engl. summ.: The beech forests of the Vértes-Mountain: II. An investigation of interspecific correlations.) *Bot. Közl.* 58: 47—52.
WHITTAKER, R. H., – 1952 – A study of summer foliage insect communities in the Great Smoky Mountains. *Ecol. Monogr.* 22: 1—44.
WHITTAKER, R. H., – 1967 – Gradient analysis of vegetation. *Biol. Rev.* 49: 207—264.
WILLIAMS, W. T., M. B. DALE, & P. MACNAUGHTON-SMITH, – 1964 – An objective method of weighting in similarity analysis. *Nature, Lond.* 201: 426.
YARRANTON, G. A., – 1967a – Principal components analysis of data from saxicolous bryophyte vegetation at Steps Bridge, Devon. I. A quantitative assessment of variation in the vegetation. *Can. J. Bot.* 45: 93—115.
YARRANTON, G. A., – 1967b – Principal components analysis of data from saxicolous bryophyte vegetation at Steps Bridge, Devon. II. An experiment with heterogeneity. *Can. J. Bot.* 45: 229—247.
YARRANTON, G. A., – 1967c – Principal components analysis of data from saxicolous bryophyte vegetation at Steps Bridge, Devon. III. Correlation of variation in vegetation with environmental variables. *Can. J. Bot.* 45: 249—258.

11 EVALUATION OF ORDINATION TECHNIQUES

ROBERT H. WHITTAKER AND HUGH G. GAUCH, JR.

Contents

11 EVALUATION OF ORDINATION TECHNIQUES

11.1 Introduction

Historically, ordination techniques have had a number of sources (WHITTAKER 1967): (i) their first appearance in the Russian school of RAMENSKY (1930, see article 5), (ii) use of similarity measurements in the Polish school (KULCZYŃSKI 1928, MOTYKA 1947, MATUSZKIEWICZ 1948, FALIŃSKI 1960, etc.), (iii) three independent developments of direct gradient analysis (articles 2 and 3) – for indicator use in applied phytosociology in Germany (ELLENBERG 1948, 1950, 1952), for analysis of the complex vegetation patterns of American mountains (WHITTAKER 1948, 1951, 1956, 1960), and for study of the vegetation of Wisconsin (CURTIS & MCINTOSH 1951, BROWN & CURTIS 1952, CURTIS 1955, 1959); (iv) exploration of comparative, indirect techniques by Americans (CULBERSON 1955, BRAY 1956, BOND 1957, CLAUSEN 1957, WHITTAKER & FAIRBANKS 1958), and especially the development of Wisconsin comparative ordination by BRAY & CURTIS (1957, article 8); (v) the introduction of the more formal techniques of factor analysis (DAGNELIE 1960, article 9) and principal components analysis (GOODALL 1954, GROENEWOUD 1965, ORLOCI 1966, article 10); and (vi) the recent proliferation of techniques related to principal components analysis and Wisconsin comparative ordination (articles 8 and 10).

Among the latter the different combinations of ordination principles, similarity coefficients, and axis determinations provide an almost uncountable number of possible techniques. There is need for evaluation of these techniques in terms of freedom from distortion, lucidity of data treatment, effort and expense of computation, and over-all research effectiveness. Ordination procedures have been more rapidly proposed than tested, but a few studies comparing and evaluating different techniques are now available. In an effort at preliminary clarification of a now confusing field, this article will attempt a classification of ordination techniques and an ordination of some of these by relative usefulness.

11.2 Kinds of Ordination Techniques

Table I gives a classification of techniques of ordination as a means of summarizing major possibilities. The techniques can be classified by a number of criteria: (i) the model or principle by which samples, or species, are arranged, (ii) whether arrangement is direct, using environmental gradients accepted as given as bases of ordination, or indirect, using directions of variation extracted from measurements applied to samples, (iii) whether samples or species are ordinated, hence for comparative techniques, whether a Q-matrix of sample similarities or an R-matrix of species correlations is employed, (iv) in these, which similarity measurement is used (for a Q-matrix CC, PS, or ED, for an R-matrix SA, SC, etc., see article 6), (v) whether axes are required to be perpendicular, or may be treated as oblique in relation to one another. Additional possibilities include (vi) R- and Q-matrixes for environmental variables or for growth-forms or other species-groups, rather than importance values for species. The first three of these criteria have been incorporated into Table I. These techniques are almost as difficult to classify satisfactorily as communities, and the reader may observe omissions from and overlaps in Table I.

The first two techniques in the table, ordination by an environmental index or by a weighted average, represent direct gradient analysis as this developed in American ecology especially (articles 2 and 3). Environmental indices include such measurements as elevation and soil pH that (though they are not assumed to be the critical factors affecting plant populations) give effective indication of relative position along complex-gradients involving a number of environmental variables. Samples may be ordinated by such measurements of their biotopes; species, in contrast, are ordinated by their distributions along a gradient. Species may, in particular, be arranged in sequence by positions of their population centers or modes along a gradient. Sets of species centered in different parts of a given gradient are ecological groups; and these may in turn be the basis of sample ordination by the second technique, that of weighted averages (article 2). Weighted averages are normally used to ordinate samples (2a) by weights applied to species, but it is also possible to use weights for sample positions along a gradient to ordinate species (2b); weighted averages are in the latter case computed for each species with the weights applied to its importance values in different samples.

Most of the remaining techniques are 'comparative' in the sense that they are based on measurements comparing samples by similarity of composition, or species by similarity of distribution (article 6). Matrix and plexus techniques (article 7) are informal techniques of ordinating by rearranging samples (or species) in the matrices of similarity values, or arranging samples (or species) into graphical representations (plexuses) of their relative similarities. These techni-

TABLE I

A Classification of Ordination Techniques

Direct	Indirect
1. Environmental index	
a. Sample ordination	
b. Species ordination	
2. Weighted average	
a. Sample ordination	
b. Species ordination	
3. Matrix and plexus (comparative, non-axial)	
a. Sample (Q-matrix and sample plexus)	
b. Species (R-matrix and species plexus)	
4. Comparative, axial ordination	
a. Sample ordination from a Q-matrix	
Two end-point samples chosen for environmental relations	Wisconsin comparative or polar ordination (two most dissimilar end-point samples)
Single end-point sample	
Multiple reference samples	
b. Species ordination from an R-matrix	
Chosen end-point species	Wisconsin inverse
5. Projective	
a. Sample ordination	
Importance projection	Principal components analysis
Similarity projection	Principal coordinates, etc.
b. Species ordination	
6. Correlative	
a. Sample correlation	
Sample composition on environmental factors or scalars	Inverse factor analysis
b. Species correlation	
Species occurrence on environmental factors or scalars	Factor analysis, s.s. (both species and sample ordinations) Canonical correlation analysis

ques have been termed here 'non-axial' in the sense that a coordinate system need not be assumed as the basis of the arrangement (though one may be suggested by the results of that arrangement). The techniques from 4 on in Table I employ axes forming coordinate systems, though these axes are not in all cases perpendicular.

The most familiar axial technique is that of the Wisconsin comparative, or BRAY & CURTIS (1957), ordination (article 8). Uses of the technique in the Wisconsin school were in most cases indirect, in the sense that most-dissimilar samples were chosen, often without regard to known environmental relations, to serve as end points and define the axes of ordination. The technique can serve equally well for direct ordination, with end-point samples chosen to represent the extremes of environmental gradients. Though Wisconsin comparative ordination has normally used a Q-matrix (of percentage similarity values), it can also be applied to a R-matrix to ordinate species (GITTINS 1965, BEALS 1965). Wisconsin ordination has used two samples (or species) as end points for axes; but for direct comparative ordination particularly it is possible to use one, or more than two, reference samples. Thus measurement of degrees of retrogression (article 4) may arrange samples by relative similarity to a single (undisturbed, control) reference sample. WHITTAKER (1960, article 2, Fig. 3) used three reference samples (or average composite samples) representing the extremes and the middle of a gradient, and the RAMENSKY approach (article 5) may use five or more reference samples along a given complex-gradient. Since in comparative ordination location of samples may be least effective near the middle of an axis represented only by end points, use of larger numbers of reference points deserves further study. A number of other variants on comparative ordination have been developed by LOUCKS (1962), KNIGHT (1965), MONK (1965), ORLOCI (1966), GOFF & COTTAM (1967), SWAN et al. (1969), and MAAREL (1969), see also article 8.

These techniques grouped as comparative, axial ordinations locate samples by a two-dimensional, pythagorean projection onto each axis. Principal components analysis, in contrast, projects samples from a multidimensional space onto axes that minimize variance of sample distances from a given axis, hence maximizing variance along the axis (ORLOCI 1966, GITTINS 1969, article 10.7.3). Successive axes are so chosen that each minimizes variance in a direction perpendicular to the preceding axes. Principal components analysis thus represents indirect, projective ordination in a mathematically ideal form (article 10), but variants again are possible. Principal coordinates or principal axes ordinations (GOWER 1966, 1969, PIELOU 1969, article 10.8.2) and multidimensional scaling (KRUSKAL 1964a, b, article 10.8.3) may employ similarity measurements other than Euclidean distance for indirect ordination. Position vectors ordination selects axes through actual samples, those samples the position vectors of which lie closest to principal component axes (ORLOCI 1966, article 10.7.4.2). Similarity projection (GAUCH & WHITTAKER 1972b) is a direct ordination that can use any sample similarity measurement to project sample positions onto axes defined by end-point samples. Importance projection is a direct technique applied to the primary matrix of importance values in samples, rather than to a secondary matrix (GAUCH & WHITTAKER 1972b). It is identical in function to one of the variants in the preceding group – Wisconsin comparative ordination using a Q-matrix of Euclidean distance values and the same end-point samples. Principal components analysis gives the same sample ordination whether based on a Q- or an R-matrix (article 10), but similarity projection can be applied to ordinate species from an R-matrix of species correlation values.

A final group of techniques are based on correlation measurements. Factor analysis and principal components analysis are related techniques (see article 10); the latter is sometimes regarded as a reduced version of factor analysis in the broad sense (CATTELL 1965). Factor analysis (DAGNELIE 1960, 1962, GREIG-SMITH 1964, WHITTAKER 1967, article 9), however, uses an R-matrix of SC or SA values and derives extracted factors as principal directions of species correlation. These extracted factors become axes (that need not be perpendicular) for ordination both of species in a loading space and of samples in a factor-value space. Factor analysis can be applied also to an R-matrix of correlations of environmental measures for samples, or to a mixed R-matrix of correlations of species distributions and environmental measurements (DAGNELIE 1960). Inverse factor analysis employing a Q-matrix is possible (DAGNELIE 1960) but of more uncertain meaning. Canonical correlation analysis (AUSTIN 1968, PIELOU 1969) seeks joint relationships in matrices of two types of variables, such as environmental and community measurements; efforts at ecological application to date seem not promising (AUSTIN 1968). For direct ordinations more conventional multiple correlation techniques could, in principle, be applied. Correlations of species importance values with environmental measurements (or scalars based on combining these) could be used for direct ordination of species in terms of relative response to the independent variables as axes. Correlation of community measurements (weighted averages, life-form representation, productivity, etc.) with environmental measurements as independent variables, could be used to obtain regression equations predictive of the community variables, and these equations used to ordinate samples. The utility of such ordinations may be limited by the complex response surfaces formed by both species populations and community variables in relation to environmental variables (GITTINS 1968, AUSTIN 1972, article 3); these limitations may account for the fact that such correlative techniques have not been used for direct ordination.

11.3 Bases of Evaluation

11.3.1 ASSUMPTIONS ABOUT SYNECOLOGICAL DATA

If ordinations are to be tested and the tests evaluated, the tests must be based on assumptions about the characteristics of community data. The characteristics include the ways species and communities relate to environmental gradients and one another, and properties of community samples. Some of these relationships (see article 3) are:

(i) Correlated variation through space of environmental factor-gradients. These factor-gradients may be grouped, or treated in terms of complex-gradients or major directions of variation in environment to which communities respond. Indices, based on measurements of either environments or communities, and expressing relative positions along complex-gradients are desired.

(ii) Individuality of species distributional response to environmental factors and complex-gradients.

(iii) Bell-shaped, apparently Gaussian form of most species populations in response to most environmental complex-gradients. In relation to more than one gradient the distributions become Gaussian solids or atmospheric distributions.

(iv) Continuous intergradation of communities, and of samples representing them, along most complex-gradients.

(v) The consequent formation by the communities of a landscape of a complex, multidimensional population pattern.

11.3.2 Criteria of Evaluation

Among the properties of ordination procedures which are desired, and by which they may be evaluated are:

(i) Relative freedom from distortion — relative success of the ordination in representing the relations of samples and species to an environmental complex-gradient or pattern, vs. representation that is distorted.

(ii) Range — the relative extent of community or environmental variation that the ordination technique can treat without undue distortion.

(iii) Lucidity — clarity to the investigator and others of the basis of the calculations, of their relation to the characteristics of his data, and hence of both the computational and ecological meaning of the ordination results.

(iv) Efficiency — effective computation and use of information, vs. either over-elaborate computation or sacrifice of information that might help to clarify relationships.

(v) Heuristic function — success in utilizing information to reveal otherwise unrecognized relations of species and communities to environment, relations on which ecological interpretation and possible further research may be based.

(vi) Computational expense and dependence on access to computers.

11.3.3 SIMULATED COENOCLINES

Of these criteria it is freedom from distortion (i) that can be most effectively tested and may in turn affect other criteria. Distortion is most easily tested by submitting synthetic or simulated coenoclines to ordination. When samples of known relations to one another along a linear coenocline axis are ordinated, distortion appears as altered spacing of the samples and curvature of the axis with displacement of samples into more dimensions or axes then are present in the simulated data. A simple, simulated coenocline of bell-shaped curves, evenly spaced along the axis, was first used for testing ordination by SWAN (1970), and NOY-MEIR & AUSTIN (1970). A two-dimensional model of binomial solids was tested by AUSTIN & NOY-MEIR (1972). Results from direct gradient analysis were studied by GAUCH & WHITTAKER (1972a) to develop a more realistic coenocline. The field data studied suggested these assumptions beyond those stated in section 11.3.1:

i) The modes, or peak importance values, of minor species are distributed at random along the coenocline axis. The modes of major species may, however, be distributed more evenly because of selection toward the reduction of competition by the centering of their populations in different parts of a gradient (WHITTAKER 1967).

ii) The heights of these modes, for the species of a coenocline, form either lograndom or lognormal distributions. When maximum importance values of tree species in forest coenoclines were grouped into octaves or doubling-units, the numbers of species in octaves were approximately even, implying lograndom distribution. (In this case a limit is imposed on the upper end of this distribution because maximum importance values for the dominants of different parts of the coenocline cannot exceed a certain range of values, and a lower limit is imposed by sample adequacy for rare species.) When plant species not subject to an importance-value ceiling (forest undergrowth species, and desert species) were treated the same way, their importance values formed lognormal distributions. Lograndom and lognormal distributions were used as alternative assumptions for the coenocline program; lograndom distribution was used for the tests that follow.

iii) Relative widths, or dispersions, of species populations in coenoclines were observed. Using as units the steps or intervals of composite transects, species whose distributions reached neither end of a transect were arranged into frequency distributions by the numbers of steps through which their distributions extended. The results suggest a normal distribution for these population amplitudes in relation to a complex-gradient. The standard deviation for species amplitudes was around 0.3 times its average value (i.e. these amplitudes have a coefficient of variation of about 0.3). It is assumed that this coefficient of variation, computed for total numbers of transect steps, should apply also to amplitudes expressed as standard deviations.

iv) Correlations between modal positions, peak importance values, and amplitudes for species populations in coenoclines were considered. No correlation between peak importance value and amplitude was found, despite the suggestion of Mc Naughton & Wolf (1970) that a weak correlation might exist. When species diversity changes markedly along a complex-gradient, correlations of peak importance value and amplitude with modal position might be expected, but no such correlations were found in the forest and desert coenoclines tested. Non-correlation of the three parameters of species populations was assumed in the program.

v) Coenoclines must differ from one another, however, in richness in species of particular communities or samples (alpha diversity), in degree of compositional change or species turnover along the coenocline (beta diversity), and in total richness in species (gamma diversity). Field samples used for ordination must differ further in the degrees of sample error or population irregularity affecting them. The simulation program[1]) was designed to produce coenoclines in which alpha diversity, beta diversity, and sample error could be independently varied. Representative simulated coenoclines are illustrated in Fig. 1.

11.3.4 Tests of Ordination

Tests of direct ordinations, based primarily on relative closeness of observed species distributions in transects to the ideal bell-shaped curve, have been described (article 2.2.4). Tests of indirect ordination are based primarily on detecting relative distortion when a coenocline that represents a linear axis is ordinated. (Certain tests can also be applied to a two-dimensional pattern of communities as simulated by Austin & Noy-Meir, 1972.) Possible tests include:

i) Observation of the arrangement of samples in an ordination space, as compared with known relationships of these samples along a simulated or actual coenocline (or in a simulated or actual community pattern).

[1]) The program, termed 'Coenocline Simulation' developed in IBM FORTRAN IV and cataloged as CEP-1, is further described by Gauch & Whittaker (1972a) and will be made available at cost on request to the authors.

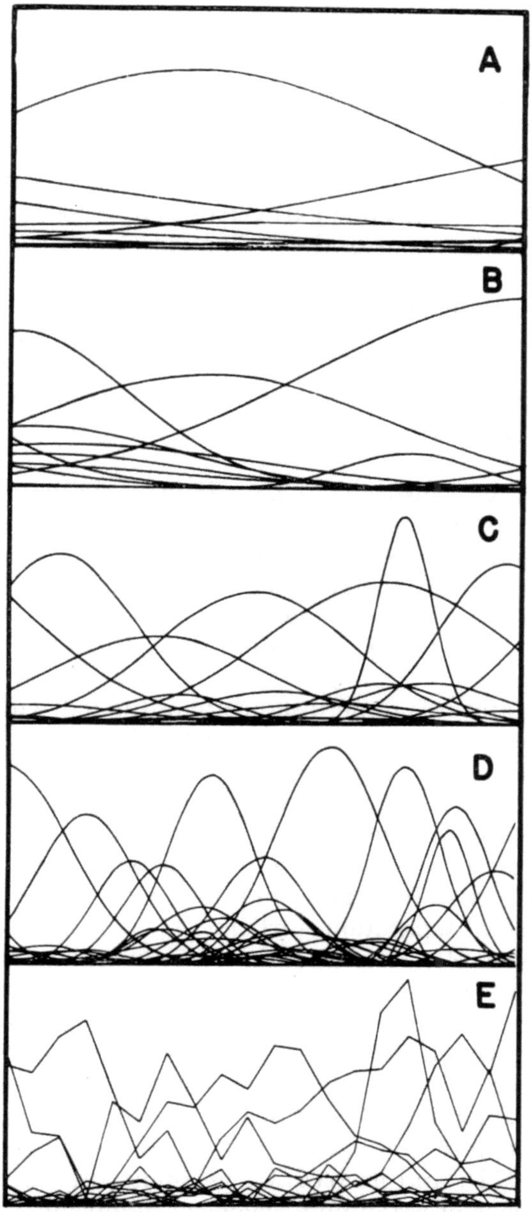

Fig. 1. Five simulated coenoclines of the type used for testing of ordination performance (GAUCH & WHITTAKER 1972a). Characteristics of the species distributions are designed to conform to those of natural communities along environmental gradients (cf. article 2, Fig. 1). Each curve represents a species population, with importance values on the vertical axis, and a gradient in relation to which the species is distributed on the horizontal axis. Alpha and beta species diversities double from A to B, B to C, and C to D. Coenocline C has a beta diversity of $\mathcal{Z}=6.0$ (4.45 half-changes) and total species number of 52. E differs from C in the introduction of sampling error. The four diversity levels are intended to resemble conditions in actual forest coenoclines: A should resemble a taiga coenocline with only two dominant tree species, B and C warm-temperate coenoclines of lower and higher diversities, and D a subtropical forest coenocline.

ii) Correlation of positions of samples as ordinated, with known positions in a coenocline or pattern. For a coenocline PEARSON'S coefficient of correlation (R) may be computed for sample position on the simulated coenocline axis versus sample location when projected onto the first axis of the ordination (GAUCH & WHITTAKER 1972b).

iii) Average distance (P) of samples from the ordination axis, divided by the length of the ordination axis (GAUCH & WHITTAKER 1972b). Since the coenocline has a single axis, a zero value of P is expected; values of P above zero indicate that sample positions have been displaced into non-existent higher axes, producing spurious information.

iv) Relative clustering (CLS). If in a simulated coenocline samples are placed at uniform intervals along the axis, their average distance from the midpoint is 0.5 times half the axis length. A value of CLS below 0.5 indicates the points have been clustered in toward the center, a value above 0.5 that they have been displaced outward toward the ends of the axis. CLS has proved less useful than measures ii) and iii).

v) Correlation of sample distances as ordinated, with known sample separation. This test may be ineffective because it fails to detect curvilinear distortion. It is not valid as used by AUSTIN & ORLOCI (1966) to test BRAY & CURTIS ordination against principal components, accepting as true sample separation Euclidean distance. (Euclidean distance is a premise of principal components but not of BRAY & CURTIS ordination, and does not bear a linear relation to environmental separation—article 4.)

vi) Per cent of variance accounted for by a given number of axes is a mathematically attractive test of ordination efficacy (ORLOCI 1966, BANNISTER 1968). This test also is prejudiced by its assumption that variance, which the axes of principal components analysis are defined to maximize, is a fair test applied to other ordinations. There is no demonstrated relationship between per cent extraction and ecological interpretability (JEGLUM & SWAN 1972, AUSTIN & NOY-MEIR 1972), and no basis for assuming that an axis of maximum variance is of maximum ecological significance.

11.3.5 ORDINATIONS TESTED

SWAN (1970) used a simple simulated coenocline to test the effect on BRAY & CURTIS ordination of zero importance values for species and curvilinearity of species distributions. NOY-MEIR & AUSTIN (1970) observed the distortion of the same coenocline by principal components analysis. AUSTIN & NOY-MEIR (1972) showed distortion by principal components and BRAY & CURTIS ordinations applied to a two-dimensional simulated community pattern. WESTMAN (1971) studied effects of several different techniques (including BRAY & CURTIS, principal components, and factor analysis) on actual data of a redwood to pygmy forest coenocline. Other studies testing or comparing results of ordinations include DAGNELIE (1960), GROENEWOUD (1965), ORLOCI (1966), AUSTIN & ORLOCI (1966), BANNISTER (1968), MOORE et al. (1970), JEGLUM et al. (1971), JEGLUM & SWAN (1972), and LAFRANCE (1972).

In the study of GAUCH & WHITTAKER (1972b) simulated coenoclines (as described above) varying in alpha diversity, beta diversity, and degree of noise or sample error were subjected to indirect ordination by eight combinations of technique and sample similarity measurement. The similarity measurements employed were (i) coefficient of community (CC, SØRENSEN variant, article 6 formula 40, subtracted from a maximum similarity of 100 % for noiseless coenoclines, or from lower values for coenoclines with sample error or noise to give a distance measure, CD), (ii) percentage similaritity (PS, article 6 formula 50, also subtracted from a maximum value of 100 % or a lower value with noise to give a distance, PD), and (iii) Euclidean distance (ED, article 6 formula 62). The combinations of technique and measurement were:

i) Wisconsin comparative (BRAY & CURTIS) ordination with CD
ii) Similarity projection with CD
iii) Wisconsin comparative ordination with PD
iv) Similarity projection with PD
v) Wisconsin comparative ordination with ED
vi) Similarity projection with ED
vii) Importance projection
viii) Principal components analysis.

11.4 **Results**

11.4.1 CHARACTER OF DISTORTIONS

As shown by NOY-MEIR & AUSTIN (1970), principal components analysis produces curvilinear distortions of simulated coenoclines. In Fig. 2 (from GAUCH & WHITTAKER 1972b) the dashed lines ending in arrows represent the expected axis of the coenocline if the relations of samples along it were correctly reproduced by the ordination. As shown in the lower left panel of Fig. 2, principal

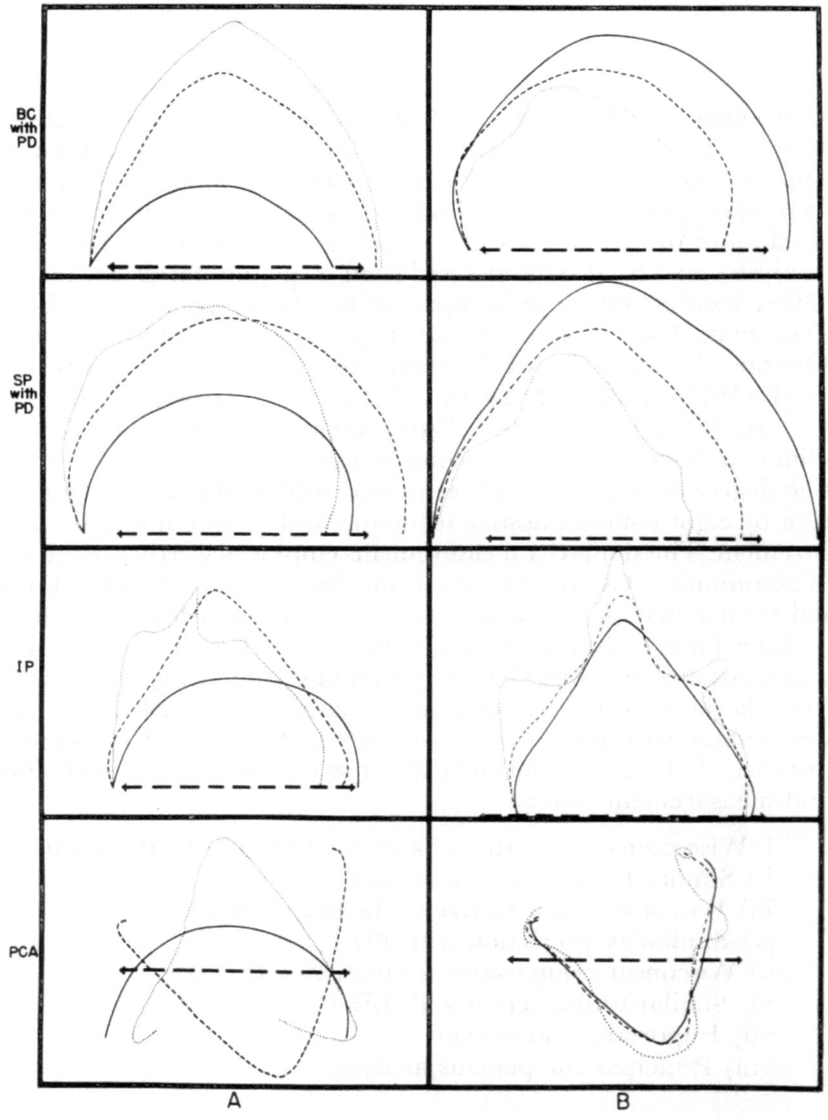

Fig. 2. Results of ordinating simulated coenoclines (see Fig. 1) by different techniques, with varying levels of beta diversity and sampling error (GAUCH & WHITTAKER 1972b). The dashed arrows represent the expected or 'correct' ordination, on the basis that the simulated coenocline is a linear community gradient in response to an environmental gradient. The solid, broken, and dotted lines show the manner in which the samples were arranged by the ordinations, in relation to the first two axes of ordination only. The ordinations on the left are for coenoclines without sampling error, differing by three levels of beta diversity with Z equalling 3 (solid lines), 6 (broken lines), and 12 (dotted lines); the coenoclines thus resemble those of Fig. 1 B, C, and D. The ordinations on the right are for coenoclines of the same beta diversity ($Z=6.0$), differing by three levels of sampling error with replicate sample similarity of 100 % (solid lines), 70 % (broken lines), and 55 % (dotted lines). The four ordination techniques compared are: top, Wisconsin comparative or BRAY & CURTIS using percentage difference (BC with PD); similarity projection using percentage difference (SP with PD), importance projection (IP) and, bottom, principal components analysis (PCA).

components ordination produces from coenoclines curves of varying forms and degrees of departure from the true axis. The full complexity of these curves is not represented by the two-dimensional projection, for they extend into higher dimensions. In extreme cases principal components can produce from a coenocline a multidimensional doodle. Figs. 3 and 4 represent the looping of one and two axes as recognized in environmental ordinations when the samples were subjected to principal components analysis by WESTMAN (1971) and JEGLUM et al. (1971).

Principal components, principal coordinates, and factor analysis are all non-polar, indirect ordinations. No end points for the axes are specified, but the computations extract axes as principal direc-

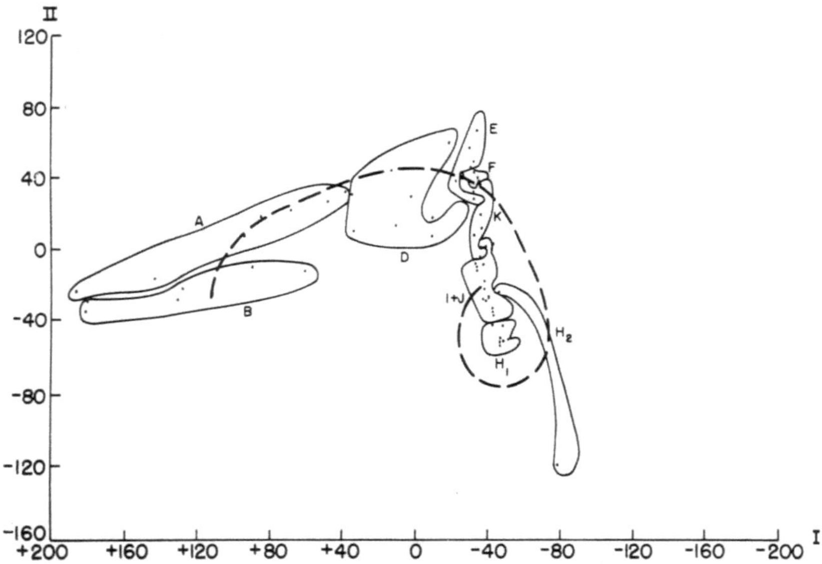

Fig. 3. Results of principal components ordination of field data representing an actual coenocline (WESTMAN 1971). The coenocline connects *Sequoia sempervirens* forests (A, B) through *Pinus muricata* heath (E, F) and other intermediate types, to pygmy forests of *Cupressus pygmaea* and *Pinus contorta* ssp. *bolanderi* (H2, I), on marine terraces of Mendocino County, California. The principal environmental factor is believed to be soil nutrient conditions, from the normal soils for the region under *Sequoia* forest to the intensely leached and sterile podsols under the pygmy forests. In principal components ordination the nutrient axis becomes a spiral as indicated by the broken line.

tions of variation in the samples submitted. The computations interpret directions of variation on the basis of assumptions of linear and monotonic relations of species importance-values to environmental

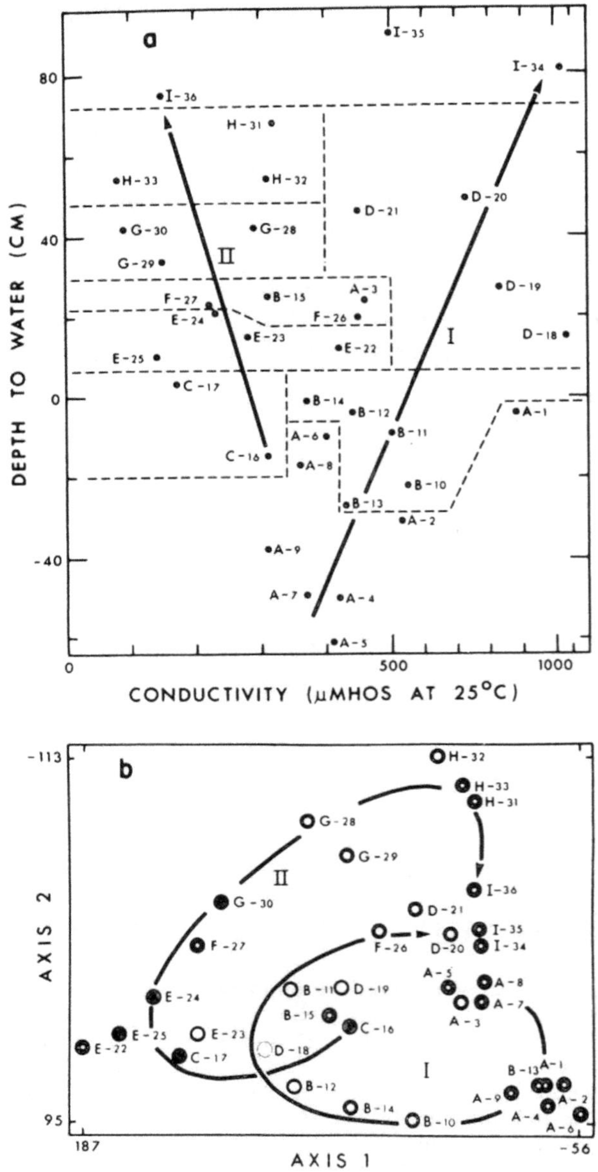

Fig. 4. Results of principal components analysis of field data representing a two-dimensional community pattern (JEGLUM et al. 1971, Fig. 1). Above is a direct, environmental ordination using gradients of water level and water conductivity. Two coenoclines or community-gradients are indicated by the arrows. Below, a principal components ordination projected onto the plane of the first two axes; the community-gradients have become overlapping, open loops.

gradients and one another. The actual distributions of species importance values are curvilinear and often unimodal or 'ditonic,' or sloping in two directions from a peak value; less frequently they are bimodal or 'polytonic,' with two or more reversals of slope. These curvilinear relationships can only be misinterpreted by the computations as rectilinear species responses along a coenocline axis that is curvilinear in ordination space. The distortions of actual ecological data by these techniques are likely to be uninterpretable unless (as in Figs. 3 and 4) the ecological relationships are known apart from the indirect ordination. When distortion is severe, the extracted axes are devoid of recognizable ecological meaning, and the ordination cannot serve as a heuristic technique for revealing significant directions of community variation in response to environment.

Polar ordinations (Wisconsin comparative, similarity projection, importance projection, etc.) produce distortions of different character—coenoclines become arches of forms from segmental and Roman to Gothic and Saracenic (Fig. 2). Since the poles of the first axis are anchored by specific end-point samples, the arrangement of samples as projected onto the first axis is not ecologically meaningless, though their spacing along the axis is distorted. In some cases samples are projected onto the axis outside the end-points as extremes of the true axis. When the ordination forms an arch, sample distances (*e*) from the first axis are largely determined by position along the arch, and hence may be ecologically meaningless.

Austin & Noy-Meir (1972) have applied principal components analysis to a two-dimensional simulated pattern of communities, as represented in a grid of equidistant samples. The ordination produces from such grids complex baskets and other curvilinear planes in three and higher dimensions (Fig. 5). As in two-dimensional ordination of a coenocline, the end samples are turned inside other samples when projected onto the first axis, so in three-dimensional ordination of a pattern, the corner samples are turned inside other samples when projected onto the plane of the first two axes. Wisconsin comparative ordination applied to the same grid of samples produces distortion into a third dimension, with bulging or constriction of the pattern as projected onto a two-dimensional plane (Fig. 6). Wisconsin comparative ordination did not, however, produce as severe distortions as principal components analysis, or displace the corner samples inside other samples in Austin & Noy-Meir's (1972) test.

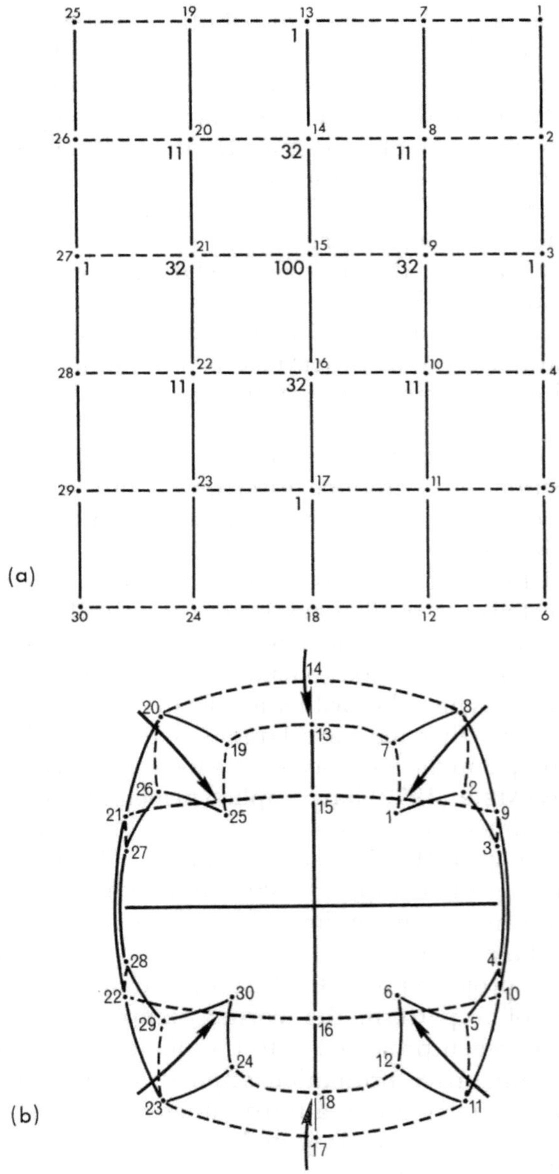

Fig. 5. Results of principal components ordination of a two-dimensional pattern of vegetation (Austin & Noy-Meir 1972). Above (a) is the simulated community pattern with a grid of samples as indicated by the small numbers. The larger numbers are importance values for one of the species, and these form a bell-shaped solid centered at sample 15. Below (b) is the result of principal components ordination of the samples, as projected onto a two-dimensional surface. Principal components ordination produces a complex curvilinear surface from the data, with the edges and corner samples turned within other samples.

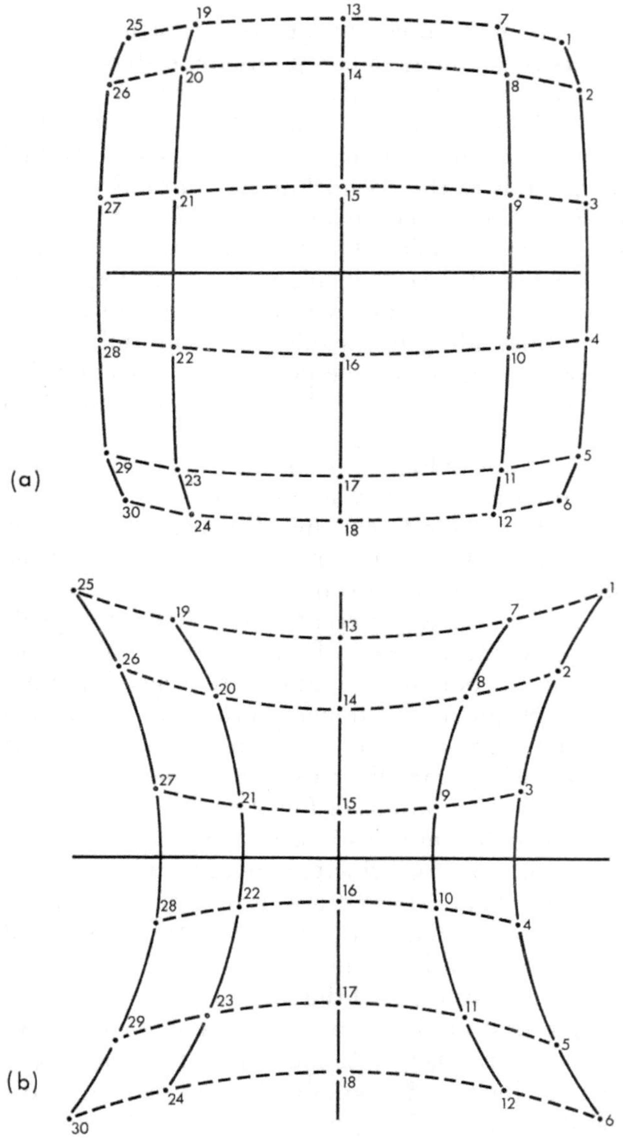

Fig. 6. Results of Wisconsin comparative (Bray & Curtis) ordination of a two-dimensional pattern of vegetation (Austin & Noy-Meir 1972). The pattern is represented by a grid of samples, as illustrated in Fig. 5 (a). The Bray & Curtis ordination produces distortion of the community pattern into a third dimension and curvilinearity in its projection onto a two-dimensional surface as illustrated; but the distortion is less severe than that of principal components analysis. The pattern used for the lower ordination (b) differed from that of (a) in the addition of species in the center of the pattern.

11.4.2 Effect of Beta Diversity

The curves of Fig. 2 represent degrees of distortion with three levels of beta diversity, or range in compositional difference of samples along the coenocline. (Beta diversity doubles from $Z=3$, solid lines, to $Z=6$, dashed lines, and $Z=12$, dotted lines; for PS and noiseless data $1.35Z$ equals one half-change (Whittaker & Gauch 1972b). When $Z=5$ the ends of the coenocline have practically no species in common.) For principal components analysis distortion increases with increasing beta diversity. The first axis, at least, of the lowest beta diversity in Fig. 2 lower left might retain ecological meaning; for the higher levels of beta diversity any ecological interpretation of axes may be misleading or meaningless. The polar ordinations show increasing arching of sample positions in a second dimension with increasing beta diversity, illustrated most clearly in Fig. 2 upper left.

The left hand side of Fig. 7 shows the effect of increasing beta diversity on ordination efficacy, as measured by *R* (coefficient of correlation for sample positions) and *P* (mean distance of points from the first axis). All ordination techniques lose efficacy with increasing beta diversity, as shown by decrease in *R* and increase in *P*, but the loss occurs at different rates. Generally speaking the loss and accompanying increase of distortion is more rapid in the sequence: Wisconsin comparative, similarity projection, importance projection, and principal components ordination. Less consistent reduction of *CLS* values below 0.5 (implying sample clustering) occurs with increasing beta diversity. The other measurements in Fig. 7 reinforce the observations of Fig. 2 on the relative vulnerability of the ordination techniques to distortion, from least for Wisconsin comparative to most for principal components.

Jeglum et al. (1971) showed effects of beta diversity on direct or environmental ordination and principal components analysis for a set of sample representing communities ranging from marsh to forest. Separate ordinations used the set as a whole, half sets (woody and nonwoody types), and quarter sets (marshes, meadows, shrub fens, and forests and muskegs). As the range of community diversity in the sample set increased, the effectiveness of principal components analysis decreased (Fig. 4) whereas that of the environmental ordination increased.

11.4.3 Effects of Noise and Alpha Diversity

Effects of noise or sample error are shown by the right, contrasted with the left, side of Fig. 2, and by the right side of Fig. 7.

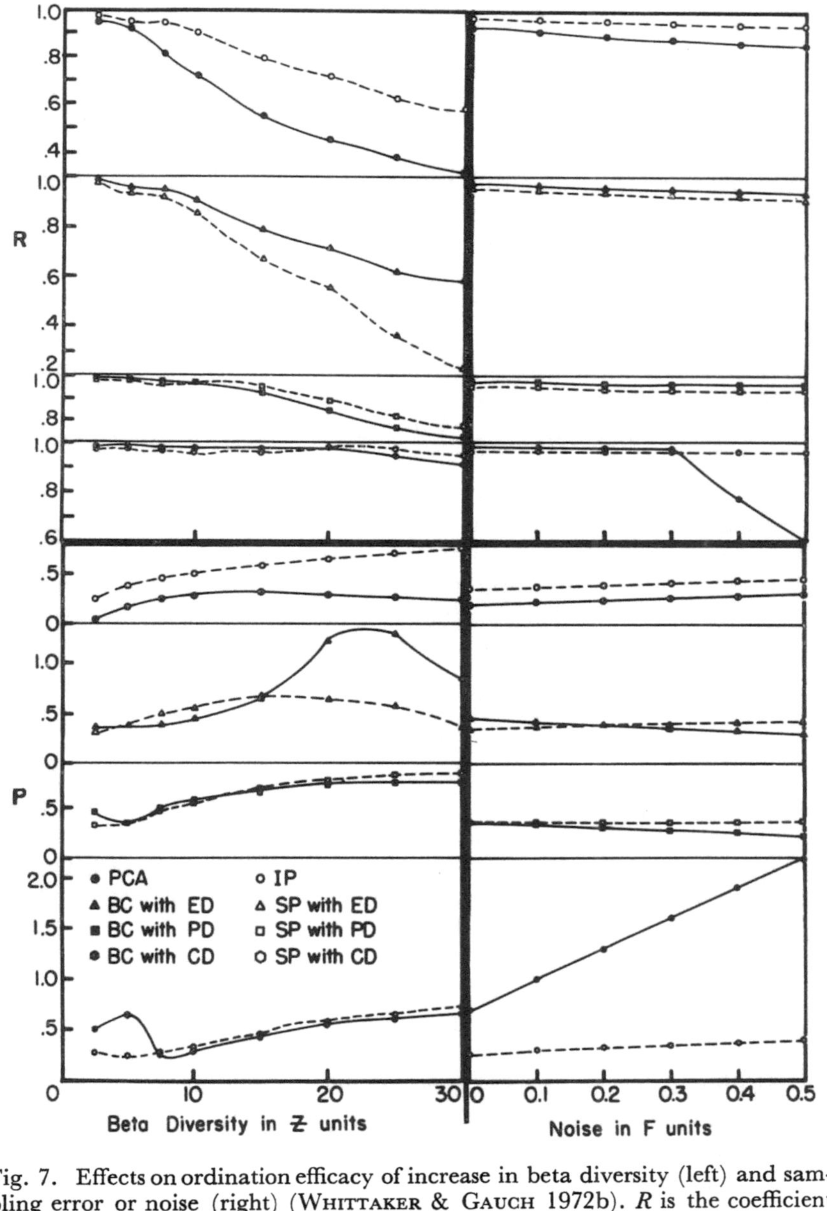

Fig. 7. Effects on ordination efficacy of increase in beta diversity (left) and sampling error or noise (right) (WHITTAKER & GAUCH 1972b). *R* is the coefficient of correlation of sample position on the simulated coenocline axis versus sample location when projected onto the first axis of the ordination; decreasing *R* indicates loss of ordination efficacy. P is average distance of samples from the axis after ordination, divided by the length of the axis; increasing *P* implies loss of ordination efficacy. Beta diversity is measured in '$\mathcal{Z}$' units (one half-change = 1.349 $\mathcal{Z}$); sampling error in 'F' units (0.2 F = replicate percentage similarity of 76 %, 0.5F = 55 %). In each of the four quarters of the figure the same four ordination techniques are tested, from top to bottom: Wisconsin comparative, similarity projection, importance projection, and principal components analysis (see also Fig. 2).

The general effect of noise is to scatter points from their original positions, as shown by the examples of Fig. 2B. Fig. 7 shows behavior of the performance statistics *R* and *P* as affected by increasing levels of noise imposed on the data, for coenoclines having approximately 50 species and a beta diversity of $Z=6$. The over-all sequence of increasing vulnerability to deterioration with noise is, again, from Wisconsin comparative to similarity projection and importance projection to principal components ordination. This sequence is somewhat affected by the similarity measurements used; those using ED deteriorate more rapidly than those using PD, and those using CD deteriorate least rapidly. Wisconsin comparative ordination employing CD performs best of the techniques, given slight or moderate noise, but also deteriorates rapidly at high levels of noise.

Increase in alpha diversity in coenoclines, in the range from 20 to 50 species, improves ordination efficacy only slightly. A corollary is that, for vegetation fairly rich in species, reduction in the number of species used for ordination only slightly reduces ordination efficacy (AUSTIN & GREIG-SMITH 1968). Reproducibility or consistency of ordinations (for replicate coenoclines of the same alpha and beta diversity) increases, however, as alpha diversity increases up to a species number of about 35. Although increasing alpha diversity has a generally minor effect on these ordinations, it tends to counteract in some degree one of the concommitants of distortion: As distortion and *P* increase, and *R* decreases, the consistency of replicate ordinations for coenoclines of the same characteristics decreases. The greater the distortion, the less predictable is the ordination, the character of the distortion, and the way the axes relate to hypothetical or actual environment.

11.4.4 PERFORMANCE OF SIMILARITY MEASUREMENTS

Similarity measurements are reviewed in article 6, and mathematical properties of some of those used for ordination are discussed in article 10. Two ordination techniques – Wisconsin comparative and similarity projection – were tested by GAUCH & WHITTAKER (1972b) using the three major measures of sample similarity – coefficient of community (CC, as CD), percentage similarity (PS, as PD), and Euclidean distance (ED). There were some variations in the performances of these with different levels of beta diversity and sample error; but the over-all result was that ordination efficacy decreased from coefficient of community (best) through percentage similarity, to Euclidean distance (worst). It may seem surprising that CC, which sacrifices the quantitative information incorporated in PS, should give better ordination performance in many cases (see also AUSTIN & GREIG-SMITH 1968, AUSTIN & NOY-MEIR 1972). CC is affected less, however, by the effects of complex and ditonic species distributions curves on the similarity measurement. PS bears a more strongly curvilinear relation to separation along environmental gradients than CC (GAUCH 1972). Thus in some ordination circumstances more is gained by the use of the less strongly curvilinear measure, CC, than is lost by sacrifice of information. For a sample set representing a limited range of communities, in which most species have monotonic distributions along a given axis, the advantage is probably reversed. Also, in sample sets with contrasting alpha diversities of samples, use of CC tends to spread apart species-rich samples and to compress together species-poor samples in ordination space (ORLOCI 1966). That ED should perform worst is understandable on the basis that its squaring of importance values exaggerates the weighting on major species, and the curvilinear relations and sample errors affecting their importance values, beyond the corresponding effects on percentage similarity.

Use of relativized, vs. unrelativized importance values had little effect on the ordinations compared by GAUCH & WHITTAKER (1972b). The coenocline samples being ordinated did not, however, differ widely in total importance values. When samples of different sizes are being ordinated by percentage similarity, relative importance values are to be preferred; and there is probably advantage in normal use of the relativized form of percentage similarity and distance (article 6 formula 51) in ordination programs. Contrast in sample size can have stronger distorting effects on ordination with ED than with PS, and can result in the anomaly that two samples sharing all or many of their species can have a larger ED than two samples with no species in common (ORLOCI 1967, BANNISTER 1968, article 10). ORLOCI (1967, article 10.7.1.2) suggests a standardized Euclidean distance as a solution. The present authors have used also a Euclidean distance relativized between 0 and 1.0: $ED' = \sqrt{\sum_i (x_i - y_i)^2/2}$, in which x_i and y_i are relative importance values (totaling 1.0 for each sample) for a given species in samples X and Y). The Wisconsin school has used a double standardization (article 8), and AUSTIN & GREIG-SMITH (1968) and AUSTIN & NOY-MEIR (1972) have found that some ordinations are improved by such treatment of the data.

The ordination properties of percentage similarity computed from doubly standardized data have not been compared with coefficient of community and percentage similarity, but should lie between these two. The same should be true of importance values that damp the contrasts in actual species measurements, such as the 5-point BRAUN-BLANQUET (1964) and 10-point Domin (CAIN & OLIVEIRA 1959, BANNISTER 1966) scales. In ordinations other than principal components analysis, use of percentage similarities based on these or other scale values may well be considered, for such similarities retain some quantitative information without the over-weighting of dominants that affects percentage similarity and Euclidean distance.

Efforts have also been made to improve ordination by use of transforms of percentage similarity–the arcsin (LOUCKS 1962), reciprocal (GIMINGHAM et al. 1966), and logarithm (FRYDMAN & WHITTAKER 1968). The arcsin transformation seems to produce little or no improvement in ordination (ORLOCI 1966). None of these transforms fits the relationship of percentage similarity to environmental separation, which is a non-standardized complement of an error function (GAUCH 1972, article 4). The inverse of the error function provides a suitable transform that improves ordination efficacy within the limits of moderate beta diversity (GAUCH 1972). SWAN (1970, JEGLUM & SWAN 1972) has commented on the distorting effect of zero importance values and possible corrections. In sample sets that are not of low species diversity minor species (i.e. those occuring in few samples) can be deleted to simplify calculations and reduce the distorting effect of zeros, and such deletion does not significantly reduce ordination efficacy (AUSTIN & GREIG-SMITH 1968, JEGLUM & SWAN 1972). For sample sets of high beta diversity no transformation is known that can solve the various problems for ordination caused by numerous zero importance values, ditonic species curves, the curvilinearity of similarity indices, and the insensitive expression of sample separation by these at low and zero values.

11.4.5 RANKING OF ORDINATION TECHNIQUES

The over-all ranking of the eight ordination techniques tested by GAUCH & WHITTAKER (1972b) may be suggested as: (i) best, Wisconsin comparative and similarity projection with coefficient of community, (ii) the same with percentage similarity, (iii) the same with Euclidean distance (and importance projection), and (iv) worst, principal components analysis. WESTMAN (1971) found similarity projection best, Wisconsin comparative nearly as good, followed by principal components analysis and importance projection, with factor analysis least successful. Some variation in the sequence of relative success of techniques must be expected for sample sets of different characteristics. We suggest as a general preference use of the Wisconsin comparative ordination of Bray & Curtis (1957). Use in it of both coefficient of community and relativized percentage similarity is suggested when parallel computations are feasible; but when a single ordination is to be carried out there may be advantage in an index intermediate in function to CC and PS (i.e. PS calculated for doubly standardized data or for Braun-Blanquet or other scale values). Orloci (article 10) discusses, and Table I includes, ordination techniques that cannot be ranked here. It seems likely that other indirect, projective techniques (e.g. principal coordinates, multidimensional scaling) will give ordination results broadly comparable to those of principal components; some improvement may, however, result from use in them of PD or CD in place of ED. Coenocline simulation should make possible testing the properties of these, and should make possible also a trial assessment of relative efficacy of various ordination techniques and similarity measures for field data the essential characteristics of which can be incorporated in a simulated coenocline.

11.5 Conclusion

There is a natural attraction, in an area of quantitative techniques, toward the use of mathematically more formal and apparently more elegant procedures. The attraction has led in some cases to uncritical acceptance of principal components analysis as a mathematically gratifying technique without reference to implications of ecological data for its performance. The results reviewed here show that the mathematically based preference for principal components analysis as a technique and Euclidean distance as a measure is in direct opposition to the relative ineffectiveness of these in the ordinations tested. Under a variety of situations the simpler procedures produce more satisfactory ordinations. The result relates to an earlier observation that, 'the character of species distributions and compositional gradients imply diminishing returns from complex mathematical treatment of vegetation. Advances in theory of vegetation structure and classification have come predominantly from simpler techniques, from the results of which the basis of complex and indirect techniques may be understood. This is by no means to oppose mathematical treatment, but to balance against its interest and value the fact that for some kinds of vegetation research simpler and more direct methods, carried out with skill and perceptiveness, are more effective.' (Whittaker 1967). It may be stated as a canon of method in synecology that simplicity of technique, and clarity to the investigator of its operation with his data, is to be valued above mathematical elegance.

This principle applies to choice between Wisconsin comparative and principal components ordinations: the former is superior by the first three criteria given above (11.3.2) – relative freedom from distortion, range of community variation that can be dealt with, and clarity of technique. Because of the distortion principal components does not use information more efficiently (in the sense of producing interpretable results, at least), or have wide heuristic usefulness. It is an attractive idea that principal components analysis might objectively discover the major directions of variation in a set of community samples, and thereby point out relationships otherwise unobserved but deserving of study. However, the axes principal components analysis offers are often meaningless. It is (regrettably or not) better in many cases for the investigator to define his axes, or at least the first axis, by choice of end-point samples. Wisconsin comparative ordination permits such axis choice, as a basis for computations that are far more lucid to the investigator than those of principal components. In the latter the arithmetic relationships determining axis direction are generally hidden, even for an investigator who understands the principle of the technique, in an almost impenetrable thicket of computation.

There has been a progression of preference in the history of gradient analysis from direct techniques accepting environmental axes, to Wisconsin comparative ordination with axes indirectly defined by most dissimilar samples, to principal components and factor analysis with axes defined by maximum variance or covariance for the whole sample set. The progression has moved toward increase in apparent objectivity, but the objectivity thus gained is largely illusory. The extreme stands defining the axes of an indirect Wisconsin comparative ordination may be included in the sample set by subjective choice, or by accident of sample availability. Even without distortion, the direction of the first axis of a principal components analysis is determined by the range of samples the investigator happens to choose, or has available, for the sample set. With distortion the first and further axes may be more purely artifacts of calculation. The circumstances of ordination research do not cooperate with the preference for objectivity and independently repeatable results that is instinctive to scientists. There is scarcely a real basis for preferring indirect axis determination to direct; no prejudice should attach to a self-conscious subjectivity and informed choice of axes. One of the results of research into ordination techniques is likely to be partial reversal of the progression of preference, toward increased emphasis of direct gradient analysis.

We suggest the following as a first axis of ordination of the techniques reviewed (with decreasing simplicity, clarity, and applicability, and increasing reliance on mathematical treatment):

i) Direct ordination with samples arranged into transects, and patterns or mosaics, can in favorable circumstances give particularly clear results. Direct ordination can also handle longer axes of community variation (by either environmental index or weighted-average ordination) without distortion than other techniques. It is thus of widest applicability, for sample sets ranging from small to very large. For problems of small scale involving a single environmental complex-gradient it is a simple and lucid technique. For such larger projects as transect analysis of the vegetation of mountain ranges (WHITTAKER 1956, 1960, WHITTAKER & NIERING 1965, 1968b) it is demanding of a rather large effort in field sampling and data treatment. Given such effort large sample sets of wide community variation, that would be difficult to handle with any other known ordination technique, can be treated to give results including detailed knowledge of species distributions (WHITTAKER 1960, WHITTAKER & NIERING 1964, 1968a) scarcely obtainable by other techniques.

ii) Wisconsin comparative and other polar techniques have for many applications the optimum combination of computational effort and clarity. As advantages over the preceding: (a) They do not require large numbers of samples that can be arranged into composite transects, and they do not require that the samples represent positions along recognized environmental gradients. (b) Both direct and indirect ordinations are feasible. The former are often to be preferred, but the latter may be the choice when directions of community variation are obscure, or when the object is only a sample arrangement, the relation of which to environmental variation may be later determined. (c) For such arrangement polar ordinations may utilize limited sample sets representing more than one direction of variation more efficiently than the preceding, and like the preceding may effectively reveal species and community relationships. For such arrangement polar ordinations are subject to less distortion, and can handle larger ranges of community variation, than the following.

iii) Principal components analysis, factor analysis, and related techniques are best regarded as specialty techniques, likely to be most useful for sample sets that have low beta diversity and can be interpreted by detailed knowledge of species relationships from other sources (WHITTAKER 1967). Such circumstances characterize the most successful applications to date (e.g. DAGNELIE 1960, 1962, article 9, and JEGLUM et al. 1971). Despite the limitations of these techniques, their mathematical power and the fact they produce both species and sample ordinations as well as relative variance along axes, will give them continued attraction for research circumstances that minimize their limitations. In some cases they may be more effectively applied to measurements of environmental variables or community characteristics (JAMES 1971) that are monotonic across the range of a sample set, than to measurements of species populations.

The authors and others are working toward a formal ordination technique more appropriate to ecological data than principal components and factor analysis. Such a technique is likely to be based on derivation of a first axis by end-point samples (or principal components), followed by iterative improvement of normal or other curvilinear fits to species importance values and thereby of sample arrangement along the axis. It may also be complex and demanding of computational effort and subject to limitations of range of material that can be successfully handled. It is unlikely to offer a single solution for all ordination problems or a general replacement of other techniques.

We suggest also a second canon of method for synecological material, that research rewards are likely to be directly proportional to the investigator's willingness to explore alternative techniques and investment of effort, interest, and understanding. The rewards may well be inversely proportional to the degree of reliance on a single, mathematically advanced, technique. Understanding may gain when direct and indirect techniques are used in conjunction with one another. More complex techniques generally gain in value when such simpler techniques as direct ordination, matrix and plexus treatment, and polar ordination can aid in interpreting their results.

The fact that ordinations offer no solution to all problems suggests also the continued value of community classification. It is difficult to frame a general statement on when research should be based on ordination and when on classification. Classification is (in some forms at least) the less demanding approach, and it may be the more natural basis for both preliminary work and for mapping and other practical applications. Ordinations are in many cases more demanding of research effort, and for this effort may more effectively reveal relationships of communities and species with one another and environment. Penetrating study of some of these relations through classification is possible, but gradient relationships and continuities are often lost sight of when vegetation is divided into classes. Choice may have little to do with the character of vegetation: Classification may seem the more natural approach to vegetation made mosaic-like by human disturbance, yet ordination is by no means inappropriate to such vegetation; ordination may seem most appropriate to extensively intergrading, undisturbed communities, yet classification of such vegetation is also feasible and usually also desired. Choice of ordination or classification is likely to be determined by the experience and interests of the investigator. Thus the appropriate suggestion may be that an investigator making either choice of primary method use the other in combination with it. Most gradient analyses use classification as one means of summarizing relationships; and many studies based on classification can gain by using ordination to interrelate, as parts of the vegetation pattern of a landscape, the community-types derived from one of the varied approaches to classification that concern the second half of this book.

11.6.1 Summary

A rather large number of ordination techniques have now been proposed. These may be classified by their direct or indirect approach to environmental gradients, the model or principle on which ordination is based, and whether samples or species are ordinated. These ordination procedures need evaluation in terms of freedom from distortion of sample positions, ranges of sample variation that can be handled, clarity of data treatment and results, computational expense, and general effectiveness for research.

Performance of ordination techniques is best evaluated using simulated coenoclines, or artificial community-gradients. 'Samples' in these have known relationships, permitting different ordinations of these samples to be compared by (i) observation of the way the samples are arranged, (ii) coefficients of correlation for actual positions of samples in a simulated coenocline, vs. apparent position as projected onto an ordination axis, and (iii) average distance of samples from the ordination axis. A number of ordination techniques have been compared by these means.

All techniques tested produce some degree of curvilinear distortion of an originally straight coenocline axis. Increasing beta diversity (range of community variation) has a strong effect in increasing degree of distortion. Increasing degree of sample error and decreasing alpha diversity (of samples) have generally smaller effects. Vulnerability of ordination techniques to distortion increases in the approximate sequence: Wisconsin comparative ordination, similarity projection, importance projection, principal components analysis, and factor analysis. In the first two the different similarity measurements tend to give increasing distortion in the sequence, coefficient of community, percentage similarity, Euclidean distance (worst). Thus there is often inverse relationship between mathematical complexity or elegance, and research effectiveness, of ordination procedures.

Wisconsin comparative (Bray & Curtis) ordination is recommended as the most generally useful of these techniques. Specification of end points with known ecological meaning, and use of both percentage similarity and coefficient of community in a given study, are suggested. For some studies direct gradient analysis has advantage over comparative ordination because of the clarity of its results and the wide ranges of community variation that can be analyzed without distortion. For a moderate or wide range of community variation, principal components analysis does not produce ecologically interpretable axes. Principal components and factor analysis are best regarded as specialty techniques most useful for narrow ranges of community variation.

Research rewards from ordination tend to increase when different techniques are explored in a given study, and when simpler and more direct techniques are used, along with ecological understanding from other sources, to interpret results from more complex and indirect techniques.

11.6.2 Zusammenfassung

Eine ziemliche große Zahl von Ordinationsverfahren ist bis jetzt vorgeschlagen worden. Diese können auf Grund ihrer direkten oder indirekten Annäherung an die Standorts-gradienten, an das Modell oder den Grundgedanken, auf denen die Ordination berüht und ob die Aufnahmen oder Arten angeordnet sind, klassifiziert werden. Diese Ordinationsverfahren brauchen eine Bewertung im Sinne der Verzerrungs-Freiheit der Aufnahme-Lagen, der Breite der Aufnahme-Variation die gefaßt werden kann, der Klarheit der Daten-Behandlung und der Resultanten, Berechnungskosten und allgemeine Wirksamkeit für die Forschung.

Die Leistung vor Ordinationstechniken kann am besten an Hand simulierter Coenoclines oder künstlicher Gesellschafts-Gradienten bewertet werden. 'Aufnahmen' in diesen haben bekannte Beziehungen was die Vergleichung von verschiedenen Ordinationen dieser Aufnahmen ermöglicht und zwar durch:

1) Beobachtung der Anordnungsweise der Aufnahmen,

2) Korrelationskoeffizienten für die tatsächlichen Lagen der Aufnahmen in einer simulierten Coenocline und die augenschein-Lage nach der Projektion auf eine Ordinationsachse, und

3) mittlerer Abstand der Aufnahmen von der Ordinationsachse. Eine Anzahl von Ordinationstechniken sind nach diesen Gesichtspunkten verglichen worden.

Alle geprüfte Techniken liefern eine gewisse kurvilinieare Verzerrung einer urspränglich geraden Coenoclinen-Axe. Eine zunehmende beta-Diversität (Breite der Gesellschafts-Variation) hat einen starken Einfluß auf einem höheren Verzerrungsgrad. Ein Ein zunehmender Grad von Aufnahmefehlern und eine abnehmende alpha-Diversität (von Aufnahmen) haben im allgemeinen kleinere Wirkungen. Die Empfindlichkeit der Ordinationstechniken für Verzerrung nimmt etwa zu in der Reihenfolge: Vergleichende Ordination der Wisconsin-Schule, Ähnlichkeits-Projektion, Wichtigkeits-Projektion Haupt-Komponenten-Analyse, und Faktoren-Analyse. Bei den beiden ersten Techniken neigen die verschiedenen Ähnlichkeits-Messungen, in der Folge der Gemeinschaftskoeffizienten der prozentualen Ähnlichkeit und des Euklidisschen Abstandes meist in zunehmendem Maß zur Verzerrung. So gibt es bei den Ordinationsverfahren oft eine umgekehrte Beziehung zwischen mathematischer Komplexität oder Eleganz und der Wirksamkeit der Forschung.

Die vergleichende Wisconsin-Ordination (Bray & Curtis) wird als die um allgemeinsten brauchbare dieser Techniken empfohlen. Eine genaue Aufstellung von End-Punkten mit bekannter ökologischen Bedeutung und der Gebrauch sowohl der prozentualen Ähnlichkeit und der Gemeinschaftskoeffizienten in einer bestimmten Untersuchung, wird vorgeschlagen.

Für einige Untersuchungen hat die direkte Gradienten-Analyse Vorteile vor der vergleichende Ordination wegen der Klarheit ihrer Ergebnisse und der große Breite der Gemeinsshaftsvariation die ohne Verrzerung analysiert werden kann.

Für eine mäßige oder große Breite der Gemeinschaftsvariation gibt eine Haupt-Komponenten-Analyse keine ökologisch interpretierbaren Achsen. Hauptkomponenten- und Faktoren-Analyse können am besten als brauchbare Spezial-Techniken für geringe Variations-Breiten der Gesellschaften betrachtet werden.

Der wissenschaftliche Gewinn der Ordination neigt zur Zunahme wann in einer gegebenen Untersuchung verschiedene Techniken verwendet werden und wenn einfachere und unmittelbarere Techniken zusammen mit ökologischen Verstandnis aus anderen Quellen zur Deutung der Ergebnissen von Komplexeren und indirekteren Techniken gebraucht werden.

REFERENCES

AUSTIN, M. P., – 1968 – An ordination study of a chalk grassland community. *J. Ecol.* 56: 739—757.

AUSTIN, M. P., – 1972 – Models and analysis of descriptive vegetation data. (manuscript).

AUSTIN, M. P. & P. GREIG-SMITH, – 1968 – The application of quantitative methods to vegetation survey. II. Some methodological problems of data from rain forest. *J. Ecol.* 56: 827—844.

AUSTIN, M. P. & I. NOY-MEIR, – 1972 – The problem of non-linearity in ordination: experiments with two-gradient models. *J. Ecol.* 59: 763—774.

AUSTIN, M. P., & L. ORLOCI, – 1966 – Geometric models in ecology. II. An evaluation of some ordination techniques. *J. Ecol.* 54: 217—227.

BANNISTER, P., – 1966 – The use of subjective estimates of cover -abundance as the basis for ordination. *J. Ecol.* 54: 665—674.

BANNISTER, P., – 1968 – An evaluation of some procedures used in simple ordinations. *J. Ecol.* 56: 27—34.

BEALS, E. W.,–1965–Species patterns in a Lebanese Poterietum. *Vegetatio* 13: 69—87.

BOND, R. R., – 1957 – Ecological distribution of breeding birds in the upland forests of southern Wisconsin. *Ecol. Monogr.* 27: 351—384.

BRAUN-BLANQUET, J., – 1964 – Pflanzensoziologie: Grundzüge der Vegetationskunde. 3rd ed. Springer, Wien. 865 pp.

BRAY, J. R., – 1956 – A study of mutual occurrence of plant species. *Ecology* 37: 21—28.

BRAY, J. R. & J. T. CURTIS, – 1957 – An ordination of the upland forest communities of southern Wisconsin. *Ecol. Monogr.* 27: 325—349.

BROWN, R. T. & J. T. CURTIS, – 1952 – The upland conifer-hardwood forests of northern Wisconsin. *Ecol. Monogr.* 22: 217—234.

CAIN, S. A. & G. M. DE OLIVEIRA CASTRO, – 1959 – Manual of Vegetation Analysis. Harper, New York. 325 pp.

CATTELL, R. B., – 1965 – Factor analysis: an introduction to essentials. I. The purpose and underlying models. *Biometrics* 21: 190—215.

CLAUSEN, J. J., – 1957 – A comparison of some methods of establishing plant community patterns. *Bot. Tidsskr.* 53: 253—278.

CULBERSON, W. L., – 1955 – The corticolous communities of lichens and bryophytes in the upland forests of northern Wisconsin. *Ecol. Monogr.* 25: 215—231.

CURTIS, J. T., – 1955 – A prairie continuum in Wisconsin. *Ecology* 36: 558—566.

CURTIS, J. T., – 1959 – The Vegetation of Wisconsin: An Ordination of Plant Communities. University of Wisconsin, Madison. 657 pp.

CURTIS, J. T. & R. P. MCINTOSH, – 1951 – An upland forest continuum in the prairie-forest border region of Wisconsin. Ecology 32: 476—496.

DAGNELIE, P., – 1960 – Contribution à l'étude des communautés végétales par l'analyse factorielle. (Engl. summ.) *Bull. Serv. Carte phytogéogr.*, Paris, Sér. B, 5: 7—71, 93—195.

DAGNELIE, P., – 1962 – L'étude des communautés végétales par l'analyse des liaisons entre les espèces et les variables écologiques. Inst. Agron. de l'État, Gembloux. 135 pp.

ELLENBERG, H., – 1948 – Unkrautgesellschaften als Mass für den Säuregrad, die Verdichtung und andere Eigenschaften des Ackerbodens. *Ber. Landtech.* 4: 130—146.

ELLENBERG, H., – 1950 – Landwirtschaftliche Pflanzensoziologie. I. Unkrautgemeinschaften als Zeiger für Klima und Boden. Ulmer, Stuttgart. 141 pp.

ELLENBERG, H., – 1952 – Landwirtschaftliche Pflanzensoziologie. II. Wiesen und Weiden und ihre standörtliche Bewertung. Ulmer, Stuttgart. 143 pp.

FALIŃSKI, J. B., – 1960 – Zastosowanie taksonomii wrokławskiej do fitosocjologii. (Germ. summ.: Anwendung der sog. 'Breslauer Taxonomie' in der Pflanzensoziologie). *Acta Soc. bot. pol.* 29: 333—361.

FRYDMAN, I. & R. H. WHITTAKER, – 1968 – Forest associations of southeast Lublin Province, Poland. (Germ. summ.) *Ecology* 49: 896—908.

GAUCH, H. G. Jr., – 1972 – The relationship between sample similarity and ecological distance. *Ecology* (in press).

GAUCH, H. G. Jr. & R. H. WHITTAKER, – 1972a – Coenocline simulation. *Ecology* 53: 446—451.

GAUCH, H. G. Jr. & R. H. WHITTAKER, – 1972b – Comparison of ordination techniques. *Ecology* (in press).

GIMINGHAM, C. H., N. M. PRITCHARD & R. M. CORMACK, – 1966 – Interpretation of a vegetational mosaic on limestone in the island of Gotland. *J. Ecol.* 54: 481—502.

GITTINS, R., – 1965 – Multivariate approaches to a limestone grassland community. II. A direct species ordination. *J. Ecol.* 53: 403—409.

GITTINS, R., – 1968 – Trend-surface analysis of ecological data. *J. Ecol.* 56: 845—869.

GITTINS, R., – 1969 – The application of ordination techniques. In, 'Ecological Aspects of the Mineral Nutrition of Plants,' ed. I. H. RORISON. *Symp. Brit. Ecol. Soc.*, 1968, 9: 37—66.

GOFF, F. G. & G. COTTAM, – 1967 – Gradient analysis: the use of species and synthetic indices. *Ecology* 48: 793—806.

GOODALL, D. W., – 1954 – Objective methods for the classification of vegetation. III. An essay in the use of factor analysis. *Aust. J. Bot.* 2: 304—324.

GOWER, J. C., – 1966 – Some distance properties of latent root and vector methods used in multivariate analysis. *Biometrika* 53: 325—338.

GOWER, J. C., – 1969 – Multivariate analysis and multidimensional geometry. *Statistician* 17: 13—28.

GREIG-SMITH, P., – 1964 – Quantitative Plant Ecology. 2nd ed. Butterworths, London. 256 pp.

GROENEWOUD, H. VAN, – 1965 – Ordination and classification of Swiss and Canadian coniferous forests by various biometric and other methods. (Germ. summ.) *Ber. geobot. Inst. ETH, Stiftg. Rübel*, Zürich, 1964, 36: 28—102.

JAMES, F. C., – 1971 – Ordinations of habitat relationships among breeding birds. *Wilson Bull.* 83: 215—236.

JEGLUM, J. K. & J. M. A. SWAN, – 1972 – Comments on species selection criteria and the adjustment of vegetational data for principal component analysis. (manuscript).

JEGLUM, J. K., C. F. WEHRHAHN, & J. M. A. SWAN, – 1971 – Comparisons of environmental ordinations with principal component vegetational ordinations for sets of data having different degrees of complexity. (French summ.) *Can. J. For. Res.* 1: 99—112.

KNIGHT, D. H., – 1965 – A gradient analysis of Wisconsin prairie vegetation on the basis of plant structure and function. *Ecology* 46: 744—747.

KRUSKAL, J. B., – 1964a – Multidimensional scaling by optimizing goodness of fit to a nonmetric hypothesis. *Psychometrika* 29: 1—27.

KRUSKAL, J. B., – 1964b – Nonmetric multidimensional scaling: a numerical method. *Psychometrika* 29: 115—129.

KULCZYŃSKI, S., – 1928 – Die Pflanzenassoziationen der Pieninen. *Bull. int. Acad. pol. Sci. Lett.*, Cl. Sci. Math. Nat., sér B, 1927 (Suppl. 2): 57—203.

LAFRANCE, C. R., – 1972 – Sampling and ordination characteristics of computer simulated individualistic communities. *Ecology* (in press).

LOUCKS, O. L., – 1962 – Ordinating forest communities by means of environmental scalars and phytosociological indices. *Ecol. Monogr.* 32: 137—166.

MAAREL, E. VAN DER., – 1969 – On the use of ordination models in phytosociology. (Germ. summ.) *Vegetatio* 19: 21—46.

MATUSZKIEWICZ, W., – 1948 – Roślinność lasów okolik Lwowa (Engl. summ.: The vegetation of the forests of the environs of Lvov). *Annls Univ. Mariae Curie-Sklodowska*, Sect. C, 3: 119—193.

McNAUGHTON, S. J. & L. L. WOLF, – 1970 – Dominance and the niche in ecological systems. *Science, N. Y.* 167: 131—139.

MONK, C. D., – 1965 – Southern mixed hardwood forest of northcentral Florida. *Ecol. Monogr.* 35: 335—354.

MOORE, J. J., P. FITZSIMONS, E. LAMBE, & J. WHITE,–1970–A comparison and evaluation of some phytosociological techniques. (Germ. summ.) *Vegetatio* 20: 1—20.

MOTYKA, J., – 1947 – O zadaniach i metodach badań geobotanicznych. (French summ.: Sur les buts et les méthodes des recherches géobotaniques). *Annls Univ. Mariae Curie-Sklodowska*, Sect. C (suppl); 1: 1—168.

NOY-MEIR, I. & M. P. AUSTIN, – 1970 – Principal component ordination and simulated vegetational data. *Ecology* 61: 551—552.

ORLOCI, L., – 1966 – Geometric models in ecology. I. The theory and application of some ordination methods. *J. Ecol.* 54: 193—215.

ORLOCI, L., – 1967 – An agglomerative method for classification of plant communities. *J. Ecol.* 55: 193—206.

PIELOU, E. C., – 1969 – An Introduction to Mathematical Ecology. Wiley-Interscience, New York. 286 pp.

RAMENSKY, L. G., – 1930 – Zur Methodik der vergleichenden Bearbeitung und Ordnung von Pflanzenlisten und anderen Objekten, die durch mehrere, verschiedenartig wirkende Faktoren bestimmt werden. *Beitr. Biol. Pfl.* 18: 269—304.

SWAN, J. M. A., – 1970 – An examination of some ordination problems by use of simulated vegetational data. *Ecology* 51: 89—102.

SWAN, J. M. A., R. L. DIX, & C. F. WEHRHAHN, – 1969 – An ordination technique based on the best possible stand-defined axes and its application to vegetational analysis. *Ecology* 50: 206—212.

WESTMAN, W. E., – 1971 – Production, nutrient circulation, and vegetation-soil relations of the pygmy forest region of Northern California *Thesis, Cornell Univ., Ithaca, N. Y.*

WHITTAKER, R. H., – 1948 – A vegetation analysis of the Great Smoky Mountains. *Thesis, Univ. Illinois, Urbana.*

WHITTAKER, R. H., – 1951 – A criticism of the plant association and climatic climax concepts. *NW. Sci.* 25: 17—31.

WHITTAKER, R. H., – 1956 – Vegetation of the Great Smoky Mountains. *Ecol. Monogr.* 26: 1—80.

WHITTAKER, R. H., – 1960 – Vegetation of the Siskiyou Mountains, Oregon and California. *Ecol. Monogr.* 30: 279—338.

WHITTAKER, R. H., – 1967 – Gradient analysis of vegetation. *Biol. Rev.* 42: 207—264.

WHITTAKER, R. H. & C. W. FAIRBANKS, – 1958 – A study of plankton copepod communities in the Columbia Basin, southeastern Washington. *Ecology* 39: 46—65.

WHITTAKER, R. H. & W. A. NIERING, – 1964 – Vegetation of the Santa Catalina Mountains, Arizona. I. Ecological classification and distribution of species. *J. Ariz. Acad. Sci.* 3: 9—34.

WHITTAKER, R. H. & W. A. NIERING, – 1965 – Vegetation of the Santa Catalina

Mountains, Arizona (II). A gradient analysis of the south slope. *Ecology* 46: 429—452.

WHITTAKER, R. H. & W. A. NIERING, – 1968a – Vegetation of the Santa Catalina Mountains, Arizona. III. Species distribution and floristic relations on the north slope. *J. Ariz. Acad. Sci.* 5: 3—21.

WHITTAKER, R. H. & W. A. NIERING, – 1968b – Vegetation of the Santa Catalina Mountains, Arizona. IV. Limestone and acid soils. *J. Ecol.* 56: 523—544.

12 APPROACHES TO CLASSIFYING VEGETATION

Robert H. Whittaker

Contents

12 APPROACHES TO CLASSIFYING VEGETATION

12.1 The Basis of Classification

12.1.1 Class-Concepts

Plant communities are classified through a process of interaction between a phytosociologist and vegetation (Whittaker 1962). The phytosociologist moves through the landscape and observes that certain kinds of communities repeat themselves: similar combinations of species appear in similar biotopes at a number of points in the landscape. Similar communities, once observed, are grouped together in the phytosociologist's memory as an informal, preliminary conception of a community-type. The manner in which the phytosociologist groups communities, and frames his conception of a type, is likely to be influenced by what he thinks is most important and interesting about these communities. He may also change his conception of a type as further observations influence his interpretation of it. In time, however, the phytosociologist wishes to abstract from his conception a more formal definition of a community-type which can be communicated to others. This definition will describe a *class*, a grouping of individuals or observations by their shared characteristics. In this case the class unites a number of stands, or particular communities in the field (or samples representing these), that are alike in ways the phytosociologist regards as significant.

The formal definition is a *class-concept* which should, preferably, be simply and clearly stated. The phytosociologist will state the class-concept according to his interests and view of what is important for classification. If he is most interested in vegetation stucture, he may state a definition of a formation: Forests dominated by needle-leaved trees in subarctic and subalpine climates are members of community-type (formation) *A*. If his interests center on floristic composition, he may define an association: Plant communities in which several of the following character-species ... occur together belong to community-type (association) *B*. If he is concerned mainly with dominant species, he may define a dominance-type: Grasslands in which the two species *a* and *b* are more important than any others represent community-type (dominance-type) *C*. Such statements of class-concepts are also intensional definitions of community-types. The particular communities in the field for which a defining statement becomes true, which therefore conform to the class-concept, constitute its extensional definition. A phytosociological table giving the species composition of a number of observed communities that fit the definition of Association *B* is thus an approach toward extensional definition of that association. The intensional definition is an abstraction narrowly limited to some most essential characteristics of the association. A community table as an aproach to extensional definition is a much less selective abstraction that communicates many of the characteristics of the community-type as these are shared by, or differ between, the vegetation samples compiled in the table.

12.1.2 Natural vs. Artificial Units

At an earlier period ecologists and phytosociologists were agitated by debate on whether associations and other community-types were real or abstract, artificial or natural units; some of the history of such debate has been reviewed elsewhere (Whittaker 1962). The first issue, of the community-type as a reality or abstraction, is the more easily resolved. A community-type, as a class-concept, is inescapably an abstraction. This abstraction is a reality as such in its psychological and cultural context. As such it is to be clearly distinguished from the plant communities in the field, that are also accepted as realities, as phenomena subject to observation and abstraction. The second issue, of the community-type as a natural or artificial unit may require somewhat more consideration, for it leads to the very basis of the problems of classifying vegetation (Whittaker 1956, 1962).

The question must be stated in terms of relationships between species and plant communities being classified before it can be answered. To consider the kinds of relationships that characterize a 'natural' unit, we may take an example from a different area of study — the 'good' species in the taxonomy of individual organisms. A large number of characteristics are shared by members of the species and distinguish those members from the members of other species. Individuals with characteristics intermediate between species *A* and other species (i.e. hybrids) are absent or rare. The species as a class is thus relatively homogeneous within itself, and discontinuous with other, comparable classes. Genetic exchange among its members by sexual reproduction, and genetic or other barriers to hybridization, maintain its internal homogeneity and external discontinuity. Because many of the characteristics of its members are correlated with one another, different taxonomists, chosing different characteristics for their class-concepts of the species, might come to the same extensional definitions — they might group the same individual specimens into the species by different characteristics. The species as a unit is then 'natural,' for the characteristics of its members have a relatively strong effect in determining the membership and boundary of the class in extension, in relative independence of the characteristics that particular taxonomists choose for emphasis in their intensional definitions.

With plant communities, it is not thus. We have commented (articles 3,5) on the two principles stated by RAMENSKY (1926) and GLEASON (1926). (i) Species are distributed 'individualistically,' each according to its own way of relating to environment, hence no two alike. Species do not fit naturally into groupings that correspond to community-types and are discontinuous with other such groupings. Ecological groups and character-species groups are primarily arbitrary groupings of species by similarity of distributional relationships; the limits of such a grouping and the number of species it is to comprise must be decided by the ecologist. Because individualistic species are combined with one another in varied combinations and proportions in communities, the phytosociologist must choose what combinations he is to recognize as associations or other community-types. (ii) Plant communities often (usually, if not affected by different disturbances or by environmental discontinuity) intergrade continuously. Boundaries of community-types are in consequence often arbitrary. The number of community-types into which a vegetational continuum is divided, and the locations of the boundaries of these types, are determined by the phytosociologist's choice of the characteristics he uses to define those types. In these respects plant communities are widely different from 'good' species. Community-types are not natural but arbitrary units in the sense that their extensional definitions are strongly influenced if not wholly determined by phytosociologists' choices of the characteristics by which communities are to be classified.

Classifications are further affected by the relative independence of these different characteristics (see article 3). Not only species but also groupings of species by growth-forms, life-forms, or strata show relative independence of one another, and may be differently combined into particular communities. Each growth-form has a different relation to environments, and a different area of those environments in which it is important or dominant, from any other growth-form (WHITTAKER & NIERING 1965). Over-all community characteristics such as total coverage, biomass, productivity, and species-diversity are differently related to environmental gradients. Plant communities are complex, many-faceted phenomena whose many characteristics are in some respect functionally interrelated, and yet are not simply correlated with one another.

12.1.3 Implications

The continuity of communities and the relative independence of their characteristics have major implications for principles in the classification of communities (Whittaker 1962). Among these implications:

i) There are many possible approaches to classifying communities, emphasizing different kinds of characteristics as bases of classification.

ii) The community-types, or units, that result from different approaches may not correspond in their boundaries and memberships. Units of one system may bear no simple relation to those of another. This fact makes interesting the study of comparative classifications — application of several methods to a given problem, so that the effects of these approaches may be compared (Miller 1951, Ellenberg 1967).

iii) No one approach to classification can claim exclusive merit. Different approaches, based on different characteristics of communities, may reveal different relationships of interest. Different approaches to classification are to be judged by their efficiency, utility, and productiveness for the study of different kinds of vegetation by ecologists or phytosociologists of different interests.

iv) There may thus be merit, as English-language ecologists have often felt (Whittaker 1962, article 14), in an open-minded and informal approach to classification, seeking for each vegetation area and research purpose a most useful classification.

v) There is also value, as many phytosociologists would stress, in a standardization of the approach to classification by many researchers (Braun-Blanquet 1951, article 20). Such standardization favors the efficient collection, organization, and communication of information about plant communities and makes possible the effective relation of the research of each individual to that of many others.

12.2 **History: the Traditions**

Because of the varied possibilities for classifying vegetation many schools have developed, distinctive in their responses to the vegetation problems of different areas but complexly interrelated by sharing of ideas, forming a kind of braided stream of phytosociological history. The history of schools has been reviewed elsewhere (WHITTAKER 1962) and is to be treated in another part of the *Handbook*; it cannot be given in detail here. We shall outline only briefly some aspects of that history involving classification, in terms of the major traditions of ecology and phytosociology (WHITTAKER 1962, cf. ALEKSANDROVA 1969, SHIMWELL 1971).

12.2.1 THE PHYSIOGNOMIC TRADITION

The origins of scientific classification of vegetation appear in the work of early students of plant geography, notably HUMBOLDT (1807) and GRISEBACH (1838). From HUMBOLDT may be traced the idea of growth-forms as major types of plants characterizing communities and of associations of plants, from GRISEBACH the concept of the formation as a distinctive kind of plant community, characterized by its growth-forms. Growth-forms determine the visible structure, or physiognomy, of plant communities. Vegetation study in which physiognomy and its relations to climate are central may be considered part of the Physiognomic Tradition. From its origins in Europe the physiognomic approach spread to students on all continents and a wide range of work in plant ecology and phytosociology, geography, and climatology. The Physiognomic Tradition thus both was parental to the five regional traditions, and is still a major approach to vegetation in its own right (article 13).

12.2.2 THE NORTHERN TRADITION

Emphasis of physiognomy was a guiding influence in the early development of the Northern Tradition of phytosociology in the Scandinavian and Baltic area. The northern authors were dealing with vegetation mostly rather poor in species, but in which the layers or strata (e.g. the tree, shrub, herb, and moss strata of a forest) were conspicuous. As schools of more intensive vegetation research developed in the North, concern shifted from broad physiognomic relations to study of communities through their strata and the major or dominant species of these. Three directions characterized different major schools: (i) Vegetation was classified by the dominant species of strata (and the sociation as a unit) in the school of Uppsala (HULT 1881, SERNANDER 1898, FRIES 1913, DU RIETZ 1921, article 18.) (ii) Treatment of the strata themselves as units (unions) was developed by GAMS (1918) in Austria, LIPPMAA (1939) in Estonia, and others (article 16). (iii) The undergrowth strata of forests were used as the basis of classification and indication of site quality by the Finnish School of CAJANDER (1909, 1949, article 15). The Northern Tradition included several other schools, among them MARKUS' (1930) study of nature-complexes in Estonia, the Danish schools of RAUNKIAER (1934) and of BÖCHER (1933, 1954), Icelandic work relating narrowly defined formations to biotopes and environmental gradients (HANSEN 1930, STEINDÓRSSON 1954), and the Dutch grassland research of DE VRIES (1948). Vegetation research in the Northern Tradition has been both distinctive and diverse in its response to the properties of northern vegetation (article 18), but many recent authors replace the northern approaches by, or fit them into, those of the Southern Tradition's school of BRAUN-BLANQUET.

12.2.3 THE SOUTHERN TRADITION

In its earlier period the Southern Tradition was centered in the cities of Zürich and Montpellier under the leadership of SCHRÖTER (SCHRÖTER & KIRCHNER 1902), and FLAHAULT (1893, FLAHAULT & SCHRÖTER 1910). The largest school in the tradition, that of BRAUN-BLANQUET, is often termed the Zürich-Montpellier or Swiss-French School; but other schools in those cities have included that of BROCKMAN-JEROSCH & RÜBEL (1912, RÜBEL 1930) in Zürich, concerned with physiognomic relations of vegetation, SCHMID's (1941, 1952, 1961) approach in Zürich through regional floristic units, and the emphasis of succession and habitat relations of LÜDI (1921, 1948) in Zürich, and of KUHNHOLZ-LORDAT (1952) in Montpellier. The Southern Tradition in a broader sense includes other distinctive work such as that of MEUSEL (1939a, 1954) in Germany, concerned with landscape interpretation and chorological relations, AICHINGER's (1952, 1954) system for treating vegetational dynamics in Austria, GAUSSEN's (1933, 1951) approach to vegetation mapping in France, and the work of NEGRI (1927, 1954) and others in Italy and of VILLAR (1929) in Spain. The center of the greatest single school of phytosociology, that of BRAUN-BLANQUET (1913, 1921, 1951) has been at Montpellier, at the 'Station Internationale de Géobotanique Mediterranéene et Alpine'; but the 'Zentralstelle für Vegetationskartierung' at Stolzenau, Germany, became under TÜXEN (1937, 1968b, 1970) a second center. The school of BRAUN-BLANQUET has been characterized by concern with full floristic composition of vegetation, and by the creation for community taxonomy of a formal hierarchy of which the association is the basic unit (article 20). Approaches concerned with gradient relations and continuities of communities, related to American gradient analysis, have been developed by Ellenberg (1950, 1952), Dagnelie (1960), Maarel (1969, Maarel & Leertouwer 1967), and Moore et al. (1970). The school has produced an enormous literature on the vegetation of Europe and has established beachheads of application by investigators on other continents.

12.2.4 The Russian Tradition

Early work on vegetation in Russia was physiognomic. As more intensive study of vegetation developed, the concept of the sociation was imported from the school of Uppsala; research was based on this unit (which the Russians term 'association') by the schools of Sukachev (1928, 1932) at Leningrad and of Alekhin (1926, 1932) and Katz (1929, 1933) at Moscow. Much Russian work following that of Keller (Dimo & Keller 1907, Keller 1925-6) arranged communities in sequences or ecological series along environmental gradients. Studies of forests by Sukachev (1928, 1932) and others, employing sociations as units arranged in ecological series, relates to the Finnish school though the units differ from Cajander's site-types. Sukachev (1954), like his predecessors Morosov (1928) and Kruedener (1926), conceives of forest communities as landscape units or biogeocoenoses, interrelating the various climatic, topographic, edaphic, biological, and communal factors of forests and their environments. Russian phytocoenology has included also the stratal or synusial approaches developed by Keller (1932) and Grossheim (1930) and Ramensky's (1926, 1930, article 5) emphasis of species individuality and vegetational continua. Other developments in Russia are discussed by Aleksandrova (1969, article 17).

12.2.5 The British Tradition

Early work of Moss (1910) and Tansley (1911, 1920) used the formation as a unit but interpreted this in terms of successional processes, not simply as a regional climax unit. Unlike the American author Clements, Tansley (1920, 1939) and most other British ecologists accepted the 'polyclimax' view that there might be a number of stable or climax communities in a given area in addition to the prevailing or climatic climax. Tansley (1939) consequently recognizes formations in British vegetation that would be treated as successional by the School of Clements. Formations were divided into broadly defined 'associations' characterized by their dominant species (TANSLEY 1939, BEARD 1955). British Commonwealth authors have applied the ideas of TANSLEY and CLEMENTS in other parts of the world; but Australian ecology developed indigenous approaches to classification of some interest (WOOD 1939, CROCKER & WOOD 1947, BEADLE 1948, BEADLE & COSTIN 1952). Much Commonwealth ecology has been physiognomic; to the use of formations BEARD (1955, article 13) has added the concept of formation series, an ecological series of formations along a major environmental gradient. Recent developments in British ecology include the work of POORE (1955-6, 1962) approaching vegetation through *noda* conceived as points of reference in vegetation that often intergrades continuously, GIMINGHAM'S (1961, 1969) treatment of heath vegetation as a network of variation, the numerical classification through 'association-analysis' and related techniques of WILLIAMS & LAMBERT (1959, LAMBERT & DALE 1964, article 19), and interest in indirect ordination through principal components analysis (GOODALL 1954, AUSTIN & ORLOCI 1966, GREIG-SMITH et al. 1967, YARRANTON 1967, GITTINS 1969, article 10).

12.2.6 THE AMERICAN TRADITION

Concern with vegetation development characterized the early work of COWLES (1899, 1901), CLEMENTS (1905, 1916) and COOPER (1913). In the hands of CLEMENTS (1916, 1936, WEAVER & CLEMENTS 1938, article 14) emphasis of succession and climax became the basis of a distinctive school and system of vegetation study. The basic units were formations and associations as climax communities adapted to the climates of geographic regions (article 14). COUPLAND (1961) applied CLEMENTS' classification to grasslands, and BRAUN (1938, 1947, 1950, 1956) studied the eastern forests with some of the concepts of CLEMENTS employed in a different perspective. Other approaches include the physiognomic systems of DANSEREAU (1951, 1957) and KÜCHLER (1949), article 13, the use of stratal units by CAIN (1936) and BILLINGS (1945) and of the sociation by DAUBENMIRE (1952) and HANSON (1953), use of concepts of BRAUN-BLANQUET by CONARD (1935, 1952), DANSEREAU (1943, 1946), BECKING (1957), and JANSSEN (1967), and application of the Finnish site-types by HEIMBERGER (1934). GLEASON (1926, 1939) dissented from the school of CLEMENTS to assert (like RAMENSKY) the principles of species individuality and vegetational continuity. Research in gradient analysis, approaching vegetation through gradient relations of environment, species populations, and communities was developed by WHITTAKER (1951, 1956, 1960, 1967, articles 2 and 3) and the Wisconsin school (CURTIS & MCINTOSH 1951, CURTIS 1959, BRAY & CURTIS 1957, MCINTOSH 1967, article 8).

12.3 **Major Approaches**

Though the schools of ecology and phytosociology are many, the basic major approaches to classification are fewer. We shall distinguish twelve for a brief review of different possibilities in classification, with reference to a few publications representing each. Some of these are given more extended discussion in the articles that follow:

12.3.1 PHYSIOGNOMIC UNITS

The first classifications of vegetation, developed in classic works on plant geography, were based on the physiognomy, or structure, of vegetation. Physiognomy is defined by the structural types of plants, or growth-forms (e.g. grasses, or broadleaved deciduous trees) that dominate or are most conspicuous in communities. A community-type on a given continent defined by growth-form dominance (and major features of environment) is a *formation*. A grouping of similar formations occurring in similar climates of the different continents is a *formation-type*. These concepts have been the basis of extensive investigations in biogeography. Of the rich literature in the area we may mention only a few classic or recent works (GRISEBACH 1872, WARMING 1909, RÜBEL 1930, SCHIMPER & FABER 1935, KÜCHLER 1949, BEARD 1955, DANSEREAU 1957, SCHMITHÜSEN 1959, FOSBERG 1961, and ELLENBERG & MUELLER-DOMBOIS 1967). Ecologists concerned with animals as well as plants have often used as corresponding units the *biome* and *biome-type* (CLEMENTS & SHELFORD 1939, ALLEE et al., 1949, ODUM 1953, KENDEIGH 1954, 1961, TISCHLER 1955). The physiognomic approach is discussed in the following article (13) by BEARD.

12.3.2 ENVIRONMENTAL UNITS

Physiognomic units classify vegetation by structure as an expression of environment. It is quite possible to turn this approach around, and to classify the environments themselves, with the classification guided by the intention that the environmental units will also be characterized by difference of structure. Some systems for the classification of climates (Köppen 1923, Thornthwaite 1948) seek to define climatic units that will correspond to major formation-types. A number of authors (e.g. Merriam 1894, Emberger 1936, 1942) have sought to define altitudinal zones primarily on the basis of climate; the system of Holdridge (1947, 1967, Tosi 1960) employs a more complex classification of zones by both temperature and moisture conditions and seeks to assign formation-types of the world to these. Some authors have designed classifications including *biotope-types* or *habitat-types* based on characteristics of local environments as well as climates; among these may be mentioned Villar (1929), Dansereau (1957: 130), Pearse (1926), and Elton & Miller (1954).

12.3.3 Landscape Units

Because of the many-faceted character of communities and the functional interrelation of their various characteristics, it is natural to think that all characteristics of communities and environments should be considered together in classifying them. The ideal becomes difficult in practice. Confronted with an almost unmanageable number of characteristics, the phytosociologist must choose among them that which is to be emphasized in a given classification. There is no very clear basis for the choice, and different choices will produce different classifications. Most plant geographers and phytosociologists have preferred to specify one kind of community characteristic as a primary basis for a clearer and more consistent classification. The many-factor approach has, however, been applied to units on two levels. (i) Landscapes. All characteristics of geographic areas – climate, geological substrate and topography, flora and fauna, kinds of natural communities, and cultural effects of man – may be considered together to define geographic or landscape units (Passarge 1921, 1927, 1929, Regel 1939, Tüxen 1968a). When landscapes as wholes are treated as units or grouped into classes by considering their many characteristics, the results are *landscape-types* (Regel 1949, Troll 1950). (ii) Microlandscapes. Smaller, homogeneous parts of the earth's surface may be treated as ecosystems, considering the community and its environment together, and classified into microlandscape-types. Markus designated a local landscape unit as a *nature-complex*, and classified these (Markus 1926, 1929, 1930, Regel 1939). In Russia the local landscape unit has been designated a *biogeocoenose*, and the concept has had extensive influence on the classification of forests as biogeocenose-types

(Morosov 1928, Kruedener 1926, Paczoski 1930, Sukachev 1944, 1954, 1960) and on other aspects of ecology. Among other West-European approaches considering environment and community together may be mentioned Negri (1927), Gams (1927), Hansen (1930), Scharfetter (1932), Sørensen (1937), Däniker (1939a,b), Gradmann (1941), Lüdi (1948), Kuhnholtz-Lordat (1952), Etter (1954), Ehrendorfer (1954), and Sjörs (1955).

12.3.4 Biotic Areas

In some approaches the geographic areas of species are emphasized as a basis of dividing of the earth's surface into biotic units. Schmid (1922, 1941, 1942, 1955, 1956b, 1961, 1963) has developed a distinctive approach to vegetation classification and interpretation. Geographic or regional units, for the recognition of which areas of species distributions are emphasized, are designated *Vegetationsgürtel*. From the flora of a regional unit, its *Artengarnitur*, species are variously distributed in different combinations into intergrading, local *Biocoenosen* that are not classifiable in any real sense, but among which abstract types or models may be recognized (Schmid 1940, 1942, 1950, 1952, Ehrendorfer 1954). Among the species themselves structural types or growth-forms and ecological response types (representation-forms) are distinguished (Schmid 1950, 1952, 1956a, 1963, Heuer 1949, Schwartz 1955, Saxer 1967); these many-sided ecological types are preferred to the traditional life-forms. The spectrum of growth-forms contributes to the characterization of the biocenose-type and permits its comparison with other biocoenoses. Major phytocoenoses of normal climates and soils in a vegetation-girdle are termed *regional* phytocoenoses, as distinguished from *local* phytocoenoses of distinctive biotopes (Schmid 1961). Although definition of the vegetation-girdles is in principle floristic, they correspond in general to regional vegetation types that may be recognized by physiognomy and dominance. Schmid's system has been applied to both intensive analysis of forest communities (Schmid 1936b, Heuer 1949, Schwartz 1955, Saxer 1955, 1967), and to broad geographic interpretation of European vegetation in terms of vegetation-girdles and their historic relations (Schmid 1949, 1961). The work of Meusel (1939a, 1939b, 1940, 1943, 1951, 1954) is distinct from that of Schmid but related in its concern with many-sided landscape relationships and with species distributions. Meusel (Meusel et al. 1965, cf. Böcher 1954) develops the concept of areal-types or distributional classes of species as an approach to vegetation interpretation and classification. In the United States DICE (1938, 1943, 1952, GOLDMAN & MOORE 1946, KENDEIGH 1961) recognizes as *biotic provinces* geographic units characterized by climate, physiography, soils, and major natural communities that develop in relation to these. Biotic provinces are on this basis landscape units, in the definition of which characteristic communities are emphasized. Their principal use, however, has been as faunal or biogeographic units appropriate for interpretation of species distributions and the evolutionary differentiation of species and subspecies.

12.3.5 ZONES AND SERIES

Some approaches to classification are based on dividing a vegetational gradient into segments, or *zones*. MERRIAM (1890, 1898, 1899, MUESEBECK & KROMBEIN 1952, KENDEIGH 1954) recognized altitudinal zones in mountains of the United States and sought to relate the zones to temperature in transcontinental belts. In many areas the zones were defined by dominant plant species, but they were also biotic units in the sense of the preceding paragraph, units that were thought to be definable by the distributions of animal species. Many authors have treated as zones (which together form successional series) the bands of vegetation surrounding water bodies. Treatment of vegetation in terms of zones has had a longer history in Europe (BRAUN-BLANQUET 1964). More recently it has been applied to relate the vegetations of tropical mountains to one another (TROLL 1961, HEDBERG 1951) and to classify communities of the marine littoral (DOTY 1957, STEPHENSON & STEPHENSON 1961, SOUTHWARD 1958, DAHL 1952-3). BEARD (1944, 1955, article 13) has related formations in the American Tropics to one another as *formation-series* – sequences of formations along major environmental gradients. Finnish (CAJANDER 1903, CAJANDER & ILVESSALO 1921, KUJALA 1945, article 15) and Russian (KELLER 1925-6, SUKACHEV 1928, 1932, article 17) work has related smaller vegetation units to one another in sequences along environmental gradients; such a sequence is an *ecological series*. All these approaches have in common a combination of classification (dividing community gradients into units, segments, or zones) and ordination (relating these units to one another along environmental axes). In many cases the units are dominance-types; but in other treatments they have been defined by climate, or as formations, associations, or sociations (see below). Ecological series can be used in any circumstances in which relations of community-types to environmental gradients can be recognized. Use of *zones* usually implies, however, that one environmental gradient has so great an effect on communities that other environmental variables may be subordinated to it. The communities are then treated as belts or zones in relation to that complex-gradient. Zones as dominance-types seem more clearly defined in communities of more severe environments, such as those of mountain ranges in dry climates, and the seacoast and shores of inland water bodies.

12.3.6 Species Dominance

It is natural to classify plant communities by what is, after physiognomy, their most obvious characteristic: dominant species. Classification by dominance-types has a number of limitations, but is often convenient and adequate to the purpose of a given study. Discussions of classification by dominance include Du Rietz (1932), Gams (1933), Crocker & Wood (1947), Beadle & Costin (1952), Tansley & Chipp (1926), Tansley (1939), Beard (1955), Poore (1955-6), Munz & Keck (1949-50), and Whittaker (1956, 1960). Many authors have combined dominance-types for lower-level with formations for higher-level classification. This approach, widespread among English-language ecologists particularly, is further discussed in a review by the author (Whittaker 1962) and in articles 13, 14, and 17 below. Many authors have used the term 'association' for dominance-types, but there may be advantage in applying 'association' only to the floristically conceived units of the School of Braun-Blanquet.

12.3.7 Vegetation Dynamics

A successional sequence is an 'ecological series' in time (or considered in movement along an environmental gradient in the present, as in the case of a pond margin). Successions in different habitats tend to converge toward climax communities that are more similar than the early stages of the successions. It is consequently possible to group the diverse successional communities of an area around a kind of climax toward which they lead, as a basis of classification. This combination of successional relationships with definition of the climax community-types as formations or dominance-types has been applied in English-language ecology (Moss 1910, Tansley

1911, 1920, 1939, ADAMSON & OSBORN 1924, STAMP 1925, PHILLIPS 1930, WOOD 1937, ROSAYRO 1950, COUPLAND 1961). The approach of most British ecologists to classification is informal, but a distinctive approach to formal classification developed in the United States under the leadership of CLEMENTS (1916, 1928, 1936, WEAVER & CLEMENTS 1929, CLEMENTS & SHELFORD 1939). CLEMENTS considered that there was a single, climatic climax community for a given area, toward which all successions converged. Climax communities were conceived as formations, and geographic subdivisions of formations were recognized as 'associations.' Successional or developmental types were termed 'associes.' A quite different system for intensive study of local vegetation by relationships of narrowly defined vegetation-development -types (*Vegetationsentwicklungstypen*) was developed in the Austrian mountains by AICHINGER (1952, 1967, WENDELBERGER 1951).

12.3.8 STRATAL UNITS

It is posssible to classify fractions of plant communities, and in particular to use different units of classification for the different strata or life-forms. Plants of the tree, shrub, herb, and thallophyte strata of a given community may in this case belong to different units, and there may be no simple distributional correspondence of these units when they are observed over a wider range of environments. This synusial treatment of vegetation has a long history from formulation of the approach by GAMS (1918, 1927) through the Estonian school of LIPPMAA (1933, 1935, 1939) to the present. The units of such classification are often termed *unions*. Synusial approaches through stratal or life-form units have had most extensive use for communities of epiphytic (BARKMAN 1958, HOSOKAWA 1954, WILMANNS 1962), and aquatic (e.g. VAARAMA 1938, JØRGENSEN 1948, KORNAŚ & MEDWECKA-KORNAŚ 1950, HÄYRÉN 1956, HARTOG & SEGAL 1964, SCHMITZ 1965) plants. The approach is reviewed in more detail by BARKMAN in article 16.

12.3.9 SOCIATIONS

It is possible also to classify communities by combinations of unions (or of dominant species of strata) as an approach related to the preceding one. By this approach two plant communities, one combining unions *A*, *B*, and *C*, and the other unions *A*, *D*, and *C* belong to different community-types. The classification of vegetation by combinations of stratal dominants developed in Sweden as a characteristic of the Uppsala school (FRIES 1913, DU RIETZ 1921, 1932, 1936) and has been applied to other northern and alpine vegetation (BOLLETER 1921, GAMS 1927, NORDHAGEN 1928, 1937, 1943, BÖCHER 1933, 1954, HANSON 1953, DAUBENMIRE 1952, 1954), and in Russia (article 17). Several terms have been used, but *sociation* is now accepted for such units. The most extensive use of sociations has been in Scandinavia, where much of the recent work employing these units fits them into the system of Braun-Blanquet (NORDHAGEN 1937, 1954, DAHL 1957). Article 18 by TRASS and MALMER discusses Scandanavian phytosociology further.

12.3.10 FOREST SITE-TYPES

A somewhat different approach with a stratal emphasis was developed in Finland by CAJANDER (1909, 1949, CAJANDER & ILVESSALO 1921). CAJANDER considered that the undergrowth of northern forests was relatively independent of the species of the canopy dominant. Forest site-types could be defined by composition of the undergrowth and used to indicate environment and its potential for growth of canopy trees. The site-types were grouped into classes by moisture conditions, and were arranged into ecological series in relation to major environmental gradients. Site-types have been widely used in northern Europe and have had more limited application in North America (HEIMBURGER 1934, KUJALA 1945). In many of the applications outside Finland the site-type concept departs from that of CAJANDER to converge with that of the sociation (ARNBORG 1953, LINDQUIST 1954, SUKATSCHEW 1932, GAMS 1933) or the association of BRAUN-BLANQUET. The site-type approach is considered in article 15 by FREY.

12.3.11 NUMERICAL CLASSIFICATION

Numerical classifications are those based directly on measurements of relative similarity of either the distributions of species or the compositions of samples. TUOMIKOSKI (1942), DAGNELIE (1960, 1962), and HEGG (1965) have considered groups of correlated species as a basis of classification. GOODALL (1953) used absence of species correlations as a criterion for homogeneity of lower-level vegetation units; WILLIAMS & LAMBERT (1959, 1961, LAMBERT & DALE 1964, LAMBERT & WILLIAMS 1962, GREIG-SMITH 1964) used correlations among species to classify samples in a dichotomous hierarchy. Measures of similarity of plant community samples (KULCZYŃSKI 1928) were used by MOTYKA (1947, MOTYKA et al. 1950) and MATUSZKIEWICZ (1948, 1950, 1952) to arrange the samples in an ordered matrix and classify the vegetation into units; SØRENSEN (1948), GUINOCHET (1954), ELLENBERG (1956), DAHL (1957), BARKMAN (1958) and others have used sample similarities as an aid in classification. More formal techniques of classifying samples by quantitative comparisons are described by EDWARDS & CAVALLI-SFORZA (1965), MAC NAUGHTON-SMITH et al. (1964), and ORLOCI (1967). POORE (1955-6, 1962) and GIMINGHAM (1961) used similarity measurements along with dominant and constant species to derive vegetation units termed *noda*. Vegetation is conceived as a multidimensional pattern or network of variation, in which a nodum is an abstract reference point. The term *nodum* might well be used for community-types on any level that are derived from numerical techniques. Such community-types may or may not correspond to any other unit of classification described here. Numerical techniques cannot solve the fundamental problems of classifying continuously intergrading communities (see also ASHBY 1936, 1948, DAHL 1957, GREIG-SMITH 1964, WHITTAKER 1962, 1967) but can provide results of interest. Numerical approaches are developed further in article 19 by GOODALL, see also article 6 by MCINTOSH.

12.3.12 FLORISTIC UNITS

It is possible, finally, to base classification on the full floristic composition of communities, without reliance on numerical techniques. In the school of BRAUN-BLANQUET (1913, 1921, 1932, 1951, 1964, ELLENBERG 1956, BECKING 1957, TÜXEN 1968b, 1970) vegetation samples (relevés) are grouped into community-types by similarity of composition, especially by representation of character-species. Character-species are species whose distributions are centered in, or largely limited to, a given community-type. The basic unit of the system is the *association*. Associations are grouped into higher units termed alliances, alliances into orders, and orders into classes to produce a formal hierarchy of community classification. A community-type or unit, on any level, in this system may be termed a *syntaxon* (BARKMAN et al. 1958). The higher units also may be defined by character-species. Associations may be divided into subassociations, and these into variants and facies as lower-level units. Subassociations and variants may be defined by differential-species – species that are (whatever their over-all distributions) largely limited to one or the other of two lower units being compared. The system of BRAUN-BLANQUET is the most widely applied and most effectively standardized of all approaches to classification, and has been adapted to diverse kinds of communities. The range of units from narrow to broad has made the system effective for such varied purposes as intensive study of local communities for environmental indication and land management on the one hand, and classification into higher units of the vegetation of all western Europe on the other. The system has limitations that have been commented on (POORE 1955-6, WHITTAKER 1962), but so do all approaches to vegetation classification. The final article (20) by WESTHOFF and MAAREL, among those following, will review this most successful of approaches to formal classification of plant communities.

12.4.1 SUMMARY

Plant communities are complex phenomena which can be variously classified. Because of the individualistic distributions of species and the continuity of communities, there is no single, natural unit of classification. Different choices of ways of defining community-types imply different classifications of the same vegetation.

Many schools have developed different systems of classification adapted to different kinds of landscapes and research interests. The earliest means of classification was by vegetation structure or physiognomy, but divergent later approaches are briefly described in terms of five regional traditions, each including several schools – the Northern or Scandinavian and Baltic, the Southern European, the Russian, the British, and the American.

Concepts of classification are shared by some of the schools, and the number of major ways of classifying communities is smaller than the number of schools. Twelve basic approaches to classification (and the units they employ) are distinguished: (1) physiognomic or structural (formation, formation-type), (2) environmental (biotope-type, etc.), (3) many-factor or landscape (landscape-type, micro-landscape-type or biogeocoenose-type), (4) biotic areas (vegetation-girdle, biotic province), (5) segments of community-gradients (life-zone, ecological series), (6) dominant species (dominance-type), (7) vegetation dynamics (American formation and association as regional vegetation types, associes and vegetation-development-type), (8) stratal or life-form divisions (union), (9) stratal combinations (sociation), (10) forest undergrowth types (site-type), (11) numerical comparisons (nodum), and (12) floristic units of BRAUN-BLANQUET (association and other syntaxa). Of these numbers 1, 6, 8, 9, 10, 11, and 12 are further discussed in the articles that follow.

12.4.2 Zusammenfassung

Pflanzengesellschaften sind komplexe Erscheinungen die auf verschieden Weisen klassifiziert werden können. Wegen der besonderen Verbreitung der Arten und dem ununterbrochenen Zusammenhang der Gesellschaften, gibt es keine einzige, natürliche Klassifikationseinheit. Verschiedene Möglichkeiten zur Definition von Gesellsshaftstypen schließen verschiedene Klassifikationen derselben Vegetation ein.

Viele Schulen haben verschiedene Klassifikationssysteme entwickelt die an verschiedene Arten von Landschaft und Forschungsinteressen angepaßt sind. Die frühesten Klassifikationsverfahren gingen von der Vegetationsstruktur oder der Physiognomie aus; abweichende spätere Versuche werden kurz beschrieben in Ausdrücken von fünf regionalen Traditionen, die jede mehrere Schulen umfassen: die Nordische oder Skandinavisch-Baltische, die Südeuropäische, die Russische, die Britische und die Amerikanische.

Einige dieser Schulen haben gemeinsame Klassifikationsbegriffe und die Zahl der Hauptrichtungen der Gesellschaftsklassifikation ist niedriger als die Zahl der Schulen. Zwölf Grund-Züge der Klassifikation (und der Einheiten, die sie verwenden) werden unterschieden:

1) physiognomisch oder strukturell (Formation, Formationstyp),

2) standortlich (Biotop-Typ usw.),

3) multifaktoriell oder landschaftlich (Landschaftstyp, Mikro-Landschaftstyp oder Biogeozönose-Typ),

4) biotische Gebiete (Vegetationsgürtel, biotische Provinzen),

5) Segmente von Gesellschaftsgradienten (Lebenszonen, ökologische Reihen),

6) herrschende Arten (Dominanztyp),

7) Vegetationsdynamik (Amerikanische Formation und Assoziation als regionale Vegetationstypen, Associes und Vegetationsentwicklungstyp),

8) Einteilungen nach Schichten (Union oder Verein),

9) Kombination von Schichten (Soziation),

10) Wald-Unterwuchstypen (Standortstyp),

11) zahlenmäßige (numerische) Vergleiche (Nodum),

12) floristische Einheiten von Braun-Blanqet (Assoziation und andere Syntaxa).

Von diesen werden 1, 6, 8, 9, 10, 11 und 12 in den folgenden Beiträgen weiter besprochen.

REFERENCES

ADAMSON, R. S. & T. G. B. OSBORN, – 1922 – On the ecology of the Ooldea District. *Trans. R. Soc. S. Aust.* 46: 539—564.

AICHINGER, E., – 1952 – Fichtenwälder und Fichtenforste als Waldentwicklungstypen. Ein forstwirtschaftlicher Beitrag zur Beurteilung der Fichtenwälder and Fichtenforste. *Angew. PflSoziol.*, Wien 7: 1—179.

AICHINGER, E., – 1954 – Statische und dynamische Betrachtung in der pflanzensoziologischen Forschung. *Veröff. geobot. Inst. Rübel*, Zürich 29: 9—28.

AICHINGER, E., – 1967 – Die Waldentwicklungstypen im Raume von Kirchleerau. (Engl. and French summs.) *Veröff. geobot. Inst. ETH, Stftg. Rübel*, Zürich 39: 187—270, 283, 293.

ALECHIN, W. W., – 1926 – Was ist eine Pflanzengesellschaft? Ihr Wesen und ihr Wert als Ausdruck des sozialen Lebens der Pflanzen. *Beih. Repert. Spec. nov. Regni veg.*, Beih. 37: 1—50.

ALECHIN, W. W., – 1932 – Die vegetationsanalytischen Methoden der Moskauer Steppenforscher. *Handb. biol. ArbMeth.* 11, 6: 335—373.

ALEKSANDROVA, VERA, D., – 1969 – Classification of Vegetation. Principles of Classification and Classification Systems of Various Phytocoenological Schools. (Russian). Nauka, Leningrad. 275 pp.

ALLEE, W. C., A. E. EMERSON, O. PARK, T. PARK & K. P. SCHMIDT, – 1949 – Principles of Animal Ecology. Saunders, Philadelphia. 837 pp.

ARNBORG, T., – 1953 – Det nordsvenska skogstypsschemat. 3rd ed. Svenska Skogsvardsför. Förlag, Stockholm. 20 pp.

ASHBY, E., – 1936 – Statistical ecology. *Bot. Rev.* 2: 221—235.

ASHBY, E., – 1948 – Statistical ecology. II. —A reassessment. *Bot. Rev.* 14: 222—234.

AUSTIN, M. P. & L. ORLOCI, – 1966 – Geometric models in ecology. II. An evaluation of some ordination techniques. *J. Ecol.* 54: 217—227.

BARKMAN, J. J., – 1958 – Phytosociology and Ecology of Cryptogamic Epiphytes, Including a Taxonomic Survey and Description of their Vegetation Units in Europe. Van Gorcum, Assen. 628 pp.

BARKMAN, J. J., H. DOING KRAFT, C. G. VAN LEEUWEN, & V. WESTHOFF, – 1958 – Enige opmerkingen over de terminologie in de vegetatiekunde. *Corresp Bl. Florist. Veget. Onderz. Neder.* 8: 87—93.

BEADLE, N. C. W., – 1948 – The Vegetation and Pastures of Western New South Wales, with Special Reference to Soil Erosion. Tennant, Govt. Printer, Sidney. 281 pp.

BEADLE, N. C. W. & A. B. COSTIN, – 1952 – Ecological classification and nomenclature, with a note on pasture classification by C. W. E. MOORE. *Proc. Linn. Soc. N. S. W.* 77: 61—82.

BEARD, J. S., – 1944 – Climax vegetation in tropical America. *Ecology* 25: 127—158.

BEARD, J. S., – 1955 – The classification of tropical American vegetation-types. *Ecology* 36: 89—100.

BECKING, R. W., – 1957 – The Zürich-Montpellier school of phytosociology. *Bot. Rev.* 23: 411—488.

BILLINGS, W. D., – 1945 – The plant associations of the Carson Desert Region, western Nevada. *Butler Univ. bot. Studs.* 7: 89—123.

BÖCHER, T. W., – 1933 – Studies on the vegetation of the east coast of Greenland between Scoresby Sound and Angmagssalik (Christian IX.s Land). *Meddr. Grønland* 104 (4): 1—132.

BÖCHER, T. W., – 1954 – Oceanic and continental vegetational complexes in Southwest Greenland. *Meddr. Grønland* 148 (1): 1—336.

BOLLETER, R., – 1921 – Vegetationsstudien aus dem Weisstannental. *Jb. St. Gall. naturw. Ges.* 57 (2): 1—140.

BRAUN, E. LUCY, – 1938 – Deciduous forest climaxes. *Ecology* 19: 515—522.

BRAUN, E. LUCY, – 1947 – Development of the deciduous forests of eastern North America. *Ecol. Monogr.* 17: 211—219.

BRAUN, E. LUCY, – 1950 – Deciduous Forests of Eastern North America. Blakiston, Philadelphia. 596 pp.

BRAUN, E. LUCY, – 1956 – The development of association and climax concepts: their use in interpretation of the deciduous forest. *Am. J. Bot.* 43: 906—911.

BRAUN-BLANQUET, J., – 1913 – Die Vegetationsverhältnisse der Schneestufe in den Rätisch-Lepontischen Alpen: Ein Bild des Pflanzenlebens an seinen äussersten Grenzen. *Neue Denkschr. Schweiz. naturf. Ges.* 48: 1—347.

BRAUN-BLANQUET, J., – 1921 – Prinzipien einer Systematik der Pflanzengesellschaften auf floristischer Grundlage. *Jb. St. Gall. naturw. Ges.* 57 (2): 305—351.

BRAUN-BLANQUET, J., – 1932 – Plant Sociology, the Study of Plant Communities. Transl. by G. D. FULLER and H. S. CONARD. Mc Graw-Hill, New York. 439 pp.

BRAUN-BLANQUET, J., – 1951 – Pflanzensoziologie: Grundzüge der Vegetationskunde. 2nd ed. 631 pp. 3rd ed. 1964. 865 pp. Springer, Wien.

BRAY, J. R. & J. T. CURTIS, – 1957 – An ordination of the upland forest communities of southern Wisconsin. *Ecol. Monogr.* 27: 325—349.

BROCKMAN-JEROSCH, H. & E. RÜBEL, – 1912 – Die Einteilung der Pflanzengesellschaften nach ökologisch-physiognomischen Gesichtspunkten. Engelmann, Leipzig. 72 pp.

CAIN, S. A., – 1936 – Synusiae as a basis for plant sociological field work. *Am. Midl. Nat.* 17: 665—672.

CAJANDER, A. K., – 1903 – (1906) – Beiträge zur Kenntniss der Vegetation der Alluvionen des nördlichen Eurasiens. I. Die Alluvionen des unteren Lena-Thales. *Acta Soc. Sci. fenn.* 32 (1): 1—182.

CAJANDER, A. K., – 1909 – Ueber Waldtypen. *Acta for. fenn.* 1 (1): 1—175.

CAJANDER, A. K., – 1949 – Forest types and their significance. *Acta for. fenn.* 56 (4): 1—71.

CAJANDER, A. K. & Y. ILVESSALO, – 1921 – Über Waldtypen II. *Acta for. fenn.* 20 (1): 1—77.

CLEMENTS, F. E., – 1905 – Research Methods in Ecology. Univ. Publ. Co., Lincoln, Nebr. 334 pp.

CLEMENTS, F. E., – 1916 – Plant succession: an analysis of the development of vegetation. *Publs Carnegie Instn*, Washington 242: 1—512.

CLEMENTS, F. E., – 1928 – Plant Succession and Indicators. Wilson, New York. 453 pp.

CLEMENTS, F. E., – 1936 – Nature and structure of the climax. *J. Ecol.* 24: 252—284.

CLEMENTS, F. E. & V. E. SHELFORD, – 1939 – Bio-ecology. Wiley, New York. 425 pp.

CONARD, H. S., – 1935 – The plant associations of central Long Island; a study in descriptive plant sociology. *Am. Midl. Nat.* 16: 433—516.

CONARD, H. S., – 1952 – The vegetation of Iowa: an approach toward a phytosociologic account. *Stud. nat. Hist. Iowa Univ.* 19 (4): 1—166.

COOPER, W. S., – 1913 – The climax forest of Isle Royale, Lake Superior, and its development. *Bot. Gaz.* 55: 1—44, 115—140, 189—235.

COUPLAND, R. T., – 1961 – A reconsideration of grassland classification in the northern Great Plains of North America. *J. Ecol.* 49: 135—167.

COWLES, H. C., – 1899 – The ecological relations of the vegetation on the sand dunes of Lake Michigan, I. Geographical relations of the dune floras. *Bot. Gaz.* 27: 95—117, 167—202, 281—308, 361—391.

COWLES, H. C., – 1901 – The physiographic ecology of Chicago and vicinity; a study of the origin, development, and classification of plant societies. *Bot. Gaz.* 31: 73—108, 144—182.

CROCKER, R. L. & J. G. WOOD, – 1947 – Some historical influences on the development of the South Australian vegetation communities and their bearing on concepts and classification in ecology. *Trans. R. Soc. S. Aust.* 71: 91—136.

CURTIS, J. T., – 1959 – The Vegetation of Wisconsin; An Ordination of Plant Communities. Univ. Wisconsin, Madison. 657 pp.

CURTIS, J. T. & R. P. MC INTOSH, – 1951 – An upland forest continuum in the prairie-forest border region of Wisconsin. *Ecology* 32: 476—496.

DAGNELIE, P., – 1960 – Contribution à l'étude des communautés végétales par l'analyse factorielle. (Engl. summ.) *Bull. Serv. Carte phytogéogr.*, Paris, Sér. B, 5: 7—71, 93—195.

DAGNELIE, P., – 1962 – L'étude des communautés végétales par l'analyse des liaisons entre les espèces et les variables écologiques. Inst. Agron. de l'État, Gembloux. 135 pp.

DAHL, E., – 1952-3 – Some aspects of the ecology and zonation of the fauna on sandy beaches. *Oikos* 4: 1—27.

DAHL, E., – 1957 – Rondane: mountain vegetation in South Norway and its relation to the environment. *Skr. norske Vidensk-Akad.* Mat-naturv. Kl. 1956 (3): 1—374.

DÄNIKER, A. U., – 1939a – Die Pflanzengesellschaft, ihre Struktur und ihr Standort. *Ber. schweiz. bot. Ges.* 49: 522—540.

DÄNIKER, A. U. – 1939b – Die Biozönose als Einheit der Vegetation. *Verhandl. schweiz. naturf. Ges.* 1939: 65—67.

DANSEREAU, P., – 1943 – L'érablière laurentienne. I. Valeur d'indice des espèces. *Can. J. Res.*, Sect. C, Bot. 21: 66—93.

DANSEREAU, P., – 1946 – L'érablière laurentienne. II. Les successions et leurs indicateurs. *Can. J. Res.*, Sect. C, Bot. 24: 235—291.

DANSEREAU, P., – 1951 – Description and recording of vegetation upon a structural basis. *Ecology* 32: 172—229.

DANSEREAU, P., – 1957 – Biogeography: An Ecological Perspective. Ronald, New York. 394 pp.

DAUBENMIRE, R., – 1952 – Forest vegetation of northern Idaho and adjacent Washington, and its bearing on concepts of vegetation classification. *Ecol. Monogr.* 22: 301—330.

DAUBENMIRE, R., – 1954 – Vegetation classification. *Veröff. geobot. Inst. Rübel*, Zürich 29: 29—34.

DICE, L. R., – 1938 – The Canadian biotic province with special ıeference to the mammals. *Ecology* 19: 503—514.

DICE, L. R., – 1943 – The Biotic Provinces of North America. Univ. Michigan, Ann Arbor. 78 pp.

DICE, L. R., – 1952 – Natural Communities. Univ. Michigan, Ann Arbor. 547 pp.

DIMO, N. A. & B. A. KELLER, – 1907 – Im Gebiet der Halbwüste. Bodenkundliche und botanische Untersuchungen in Süden des Zarizyner Kreises im Saratowschen Gouvernement. (In Russian) Saratow.

DOTY, M. S., – 1957 – Rocky intertidal surfaces. In 'Treatise on Marine Ecology and Paleoecology,' ed. J. W. HEDGPETH, vol. 1, Ecology. *Mem. geol. Soc. Am.* 67 (1):535—585.

Du Rietz, G. E., – 1921 – Zur methodologischen Grundlage der modernen Pflanzensoziologie. Holzhausen, Wien. 267 pp.

Du Rietz, G. E., – 1932 (1930) – Vegetationsforschung auf soziationsanalytischer Grundlage. *Handb. biol. ArbMeth.* 11, 5: 293—480.

Du Rietz, G. E., – 1936 – Classification and nomenclature of vegetation units 1930—1935. *Svensk bot. Tidskr.* 30: 580—589.

Edwards, A. W. F. & L. L. Cavalli-Sforza, – 1965 – A method for cluster analysis. *Biometrics* 21: 362—375.

Ehrendorfer, F., – 1954 – Gedankungen zur Frage der Struktur und Anordnung der Lebensgemeinschaften. *Angew. PflSoziol.*, Wien, Festschr. Aichinger 1: 151—167.

Ellenberg, H., – 1950 – Landwirtschaftliche Pflanzensoziologie. I. Unkrautgemeinschaften als Zeiger für Klima und Boden. Ulmer, Stuttgart. 141 pp.

Ellenberg, H., – 1952 – Landwirtschaftliche Pflanzensoziologie. II. Wiesen und Weiden und ihre standörtliche Bewertung. Ulmer, Stuttgart. 143 pp.

Ellenberg, H., – 1956 – Aufgaben und Methoden der Vegetationskunde. In 'Einführung in die Phytologie' by H. Walter, Vol. IV, Pt. 1. Ulmer, Stuttgart 136 pp.

Ellenberg, H. ed., – 1967 – Vegetations- und Bodenkundliche Methoden der forstlichen Standortskartierung: Ergebnisse eines internationalen Methodenvergleichs im Schweizer Mittelland. *Veröff. geobot. Inst. ETH, Stiftg Rübel*, Zürich 39: 1—296.

Ellenberg, H. & D. Mueller-Dombois, – 1967 – Tentative physiognomic-ecological classification of plant formations of the earth. *Ber. geobot. Inst. ETH, Stiftg. Rübel*, Zürich 1965/6, 37: 21—55.

Elton, C. S. & R. S. Miller, – 1954 – The ecological survey of animal communities; with a practical system of classifying habitats by structural characters. *J. Ecol.* 42: 460—496.

Emberger, L., – 1936 – Remarques critiques sur les étages de végétation dans les montagnes marocaines. *Ber. schweiz. bot. Ges.* 46: 614—631.

Emberger, L., – 1942 – Un projet d'une classification des climats du point de vue phytogéographique. *Bull. Soc. Hist. nat. Toulouse* 77: 97—124.

Etter, H., – 1954 – Grundsätzliche Betrachtungen zur Beschreibung und Kennzeichnung der Biochore. (French summ.) *Vegetatio* 5/6: 500—510.

Flahault, C., – 1893 – Les zones botaniques dans le Bas-Languedoc et les pays voisins. *Bull. Soc. bot. Fr.* 40 (Sess. extraord.): xxxvi—lxii.

Flahault, C. & C. Schröter, – 1910 – Rapport sur la nomenclature phytogéographique. *Actes 3me Int. bot. Congr.*, Bruxelles 1910, 1: 131—164.

Fosberg, F. R., – 1961 – A classification of vegetation for general purposes. *Trop. Ecol.* 2: 1—28. Also in 'Guide to the check sheet for IBP areas,' by G. F. Peterken. *IBP Handbook* 4: 73—120. Blackwell, Oxford and Edinburgh.

Fries, T. C. E., – 1913 – Botanische Untersuchungen im nördlichsten Schweden: Ein Beitrag zur Kenntnis der alpinen und subalpinen Vegetation in Torne Lappmark. Vetensk. och prakt. unders. i Lappland, anordn af Luossavaara-Kiirunavaara Aktiebolag. Flora och Fauna 2: 1—361. Almqvist & Wiksells, Stockholm.

Gams, H., – 1918 – Prinzipienfragen der Vegetationsforschung: Ein Beitrag zur Begriffsklärung und Methodik der Biocoenologie. *Vjschr. naturf. Ges. Zürich* 63: 293—493.

Gams, H., – 1927 – Von den Follatères zur Dent de Morcles: Vegetationsmonographie aus dem Wallis. *Beitr. geobot. Landesaufn. Schweiz* 15: 1—760.

Gams, H., – 1933 – Die Stellung der Waldtypen im Vegetationssystem. *Forstarchiv* 9: 53—59.

GAUSSEN, H., – 1933 – Géographie des plantes. Colin, Paris. 222 pp. 2nd ed., 1954, 223 pp.
GAUSSEN, H., – 1951 – Le dynamisme des biocénoses végétales. *Année biol.*, 3e Sér., 27: 89—101.
GIMINGHAM, C. H., – 1961 – North European heath communities; a 'network of variation' *J. Ecol.* 49: 655—694.
GIMINGHAM, C. H., – 1969 – The interpretation of variation in north-European dwarf-shrub heath communities. (Germ. summ.) *Vegetatio* 17: 89—108.
GITTINS, R., – 1969 – The application of ordination techniques. In 'Ecological Aspects of the Mineral Nutrition of Plants,' ed. I. H. RORISON, *Symp. Brit. Ecol. Soc.* 1968, 9: 37—66.
GLEASON, H. A., – 1926 – The individualistic concept of the plant association. *Bull. Torrey bot. Club* 53: 7—26.
GLEASON, H. A., – 1939 – The individualistic concept of the plant association. *Am. Midl. Nat.* 21: 92—110.
GOLDMAN, E. A. & R. T. MOORE, – 1946 – The biotic provinces of Mexico. *J. Mammal.* 26: 347—360.
GOODALL, D. W., – 1953 – Objective methods for the classification of vegetation. I. The use of positive interspecific correlation. *Aust. J. Bot.* 1: 39—63.
GOODALL, D. W., – 1954 – Objective methods for the classification of vegetation. III. An essay in the use of factor analysis. *Aust. J. Bot.* 2: 304—324.
GRADMANN, R., – 1941 – Methodische Grundfragen und Richtungen der Pflanzensoziologie. *Beih. Repert. Spec. nov. Regni Veg.* 131: 1—41.
GREIG-SMITH, P., – 1964 – Quantitative Plant Ecology. 2nd ed. Butterworths, London. 256 pp.
GREIG-SMITH, P., M. P. AUSTIN, & T. C. WHITMORE, – 1967 – The application of quantitative methods to vegetation survey. I. Association-analysis and principal component ordination of rain forest. *J. Ecol.* 55: 483—503.
GRISEBACH, A., – 1838 – Ueber den Einfluss des Climas auf die Begränzung der natürlichen Floren. *Linnaea* 12: 159—200.
GRISEBACH, A., – 1872 – Die Vegetation der Erde nach ihrer klimatischen Anordnung. Ein Abriss der vergleichenden Geographie der Pflanzen. Engelmann, Leipzig. 2 vols., 603 and 709 pp.
GROSSHEIM, A. A., – 1930 – Zur Frage nach dem Zustandekommen der Pflanzendecke. *Beitr. Biol. Pfl.* 18: 225—286.
GUINOCHET, M., – 1954 – Sur les fondements statistiques de la phytosociologie et quelques unes de leurs conséquences. *Veröff. geobot. Inst. Rübel*, Zürich 29: 41—67.
HANSEN, H. MØLHOLM, – 1930 – Studies on the vegetation of Iceland. In 'The Botany of Iceland,' ed. J. L. A. KOLDERUP-ROSENVINGE & E. WARMING, vol. 3, pt. 1, no. 10: 1—186. Frimodt, Copenhagen.
HANSON, H. C., – 1953 – Vegetation types in northwestern Alaska and comparisons with communities in other Arctic regions. *Ecology* 34: 111—140.
HARTOG, C. DEN, & S. SEGAL, – 1964 – A new classification of the water-plant communities. *Acta bot. neerl.* 13: 367—393.
HÄYRÉN, E., – 1956 – Über die Algenvegetation des sandigen Geolitorals am Meere in Schweden und in Finnland. *Svensk bot. Tidskr.* 50: 257—269.
HEDBERG, O., – 1951 – Vegetation belts of the East African mountains. *Svensk. bot. Tidskr.* 45: 141—202.
HEGG, O., – 1965 – Untersuchungen zur Pflanzensoziologie und Ökologie im Naturschutzgebiet Hohgant (Berner Voralpen), mit einem Beitrag zur Methodik der floristisch-statistischen Erfassung pflanzensoziologischer Zusammenhänge. (French and Engl. summs.) *Beitr. geobot. Landesaufn. Schweiz* 46: 1—188.

Heimburger, C. C., – 1934 – Forest-type studies in the Adirondack region. *Mem. Cornell Univ.* (*N.Y.*) *agric. Exp. Stn.* 165: 1—122.

Heuer, Ilse, – 1949 – Vergleichende Untersuchungen an der Föhrenbeständen des Pfynwaldes (Wallis); Versuch einer biocoenologischen Analyse. *Beitr. geobot. Landesaufn. Schweiz* 28: 1—185.

Holdridge, L. R., – 1947 – Determination of world plant formations from simple climatic data. *Science, N. Y.* 105: 367—368.

Holdridge, L. R., – 1967 – Life Zone Ecology. Rev. ed. Tropical Science Center, San Jose. 206 pp.

Hosokawa, T., – 1954 – On the *Campnosperma* forests of Kusaie in Micronesia, with special reference to the community units of epiphytes. *Vegetatio* 5/6: 351—360.

Hult, R., – 1881 – Försök till analytisk behandling af växtformationerna. *Meddn. Soc. Fauna Flora fenn.* 8: 1—155.

Humboldt, A. von, – 1807 – Ideen zu einer Geographie der Pflanzen nebst einem Naturgemälde der Tropenländer. Cotta, Tübingen. 182 pp.

Janssen, C. R., – 1967 – A floristic study of forests and bog vegetation, northwestern Minnesota. *Ecology* 48: 751—765.

Jørgensen, E. G., – 1948 – Diatom communities in some Danish lakes and ponds. *Biol. Skr.*, K. Danske Vidensk. Selsk. 5 (2): 1—140.

Katz, N. J., – 1929 – Die Zwillingsassoziationen und die homologen Reihen in der Phytosoziologie. *Ber. dt. bot. Ges.* 47: 154—164.

Katz, N. J., – 1933 – Die Grundprobleme und die neue Richtung der Phytosoziologie. *Beitr. Biol. Pfl.* 21: 133—166.

Keller, B. A., – 1925-6 – Die Vegetation auf den Salzböden der russischen Halbwüsten und Wüsten. *Z. Bot.* 18: 113—137.

Keller, B. A., – 1932 – Die Methoden zur Erforschung der Ökologie der Steppen- und Wüstenpflanzen. *Handb. Biol. ArbMeth.* 11, 6: 1—128.

Kendeigh, S. C., – 1954 – History and evaluation of various concepts of plant and animal communities in North America. *Ecology* 35: 152—171.

Kendeigh, S. C., – 1961 – Animal Ecology. Prentice-Hall, Englewood Cliffs. 468 pp.

Köppen, W., – 1923 – Die Klimate der Erde: Grundriss der Klimakunde. de Gruyter, Berlin & Leipzig. 369 pp.

Kornaś, J. & A. Medwecka-Kornaś, – 1950 – Associations végétales sous-marines dans le Golfe de Gdańsk (Baltique polonaise). (Engl. summ.) *Vegetatio* 2: 120—127.

Kruedener, A. von, – 1926 – Waldtypen als kleinste natürliche Landschaftseinheiten bzw. Mikrolandschaftstypen. *Petermanns geogr. Mitt.* 72: 150—158.

Küchler, A. W., – 1949 – A physiognomic classification of vegetation. *Ann. Ass. Am. Geogr.* 39: 201—210.

Kuhnholtz-Lordat, G., – 1952 – Le tapis végétale dans ses rapports avec les phénomènes actuels de surface en Basse-Provènce (de Cassis à Bandol). *Encyclopéd. Beogéogr. et Écol.* 9: 1—208. Lechevalier, Paris.

Kujala, V., – 1945 – Waldvegetationsuntersuchungen in Kanada mit besonderer Berücksichtigung der Anbaumöglichkeiten kanadischer Holzarten auf natürlichen Waldböden in Finnland. *Ann. Acad. Scient. fenn.* (*Suomal. Tiedeakat. Toim.*) Ser. A, 4 Biol. 7: 1—434.

Kulczyński, S., – 1928 – Die Pflanzenassoziationen der Pieninen. *Bull. int. Acad. pol. Sci. Lett.*, Cl. Sci. Math. Nat., Sér. B 1927 (Suppl. 2) : 57—203.

Lambert, Jean M. & M. B. Dale, – 1964 – The use of statistics in phytosociology. *Adv. ecol. Res.* 2: 59—99.

Lambert, Jean M. & W. T. Williams, – 1962 – Multivariate methods in plant ecology. IV. Nodal analysis. *J. Ecol.* 50: 775—802.

LINDQUIST, B., – 1954 – Ein Waldtypenschema für die skandinavischen Buchenwälder. *Angew. PflSoziol.* Wien, Festschr. Aichinger 2: 965—970.

LIPPMAA, T., – 1933 – Aperçu général sur la végétation autochtone du Lautaret (Hautes - Alpes) avec des remarques critiques sur quelques notions phytosociologiques. (Eston. summ.) *Acta Inst. Horti. bot. tartu.* 3 (3): 1—108.

LIPPMAA, T., – 1935 – Une analyse des forêts de l'île estonienne d'Abruka (Abro) sur la base des associations unistrates. *Acta Inst. Horti bot. tartu.* 4 (no. 1/2, art. 5): 1—97.

LIPPMAA, T., – 1939 – The unistratal concept of plant communities (the unions). *Am. Midl. Nat.* 21: 111—145.

LÜDI, W., – 1921 – Die Pflanzengesellschaften des Lauterbrunnentales und ihre Sukzession: Versuch zur Gliederung der Vegetation eines Alpentales nach genetisch-dynamischen Gesichtspunkten. *Beitr. geobot. Landesaufn. Schweiz* 9: 1—364.

LÜDI, W., – 1948 – Die Pflanzengesellschaften der Schinigeplatte bei Interlaken und ihre Beziehungen zur Umwelt. Eine vergleichend ökologische Untersuchung. *Veröff. geobot. Inst. Rübel*, Zürich, 23: 1—400.

MAAREL, E. VAN DER, – 1969 – On the use of ordination models in phytosociology. (Germ. summ.) *Vegetatio* 19: 21—46.

MAAREL, E. VAN DER & J. LEERTOUWER, – 1967 – Variation in vegetation and species diversity along a local environmental gradient. *Acta bot. neerl.* 16: 211—221.

MACNAUGHTON-SMITH, P., W. T. WILLIAMS, M. B. DALE, & L. G. MOCKETT, — 1964 – Dissimilarity analysis: a new technique of hierarchial sub-division. *Nature, Lond.* 202: 1034—1035.

MARKUS, E., – 1926 – Verschiebung der Naturkomplexe in Europa. *Geogr. Z.* 32: 516—541.

MARKUS, E., – 1929 – Die Grenzverschiebung des Waldes und des Moores in Alatskivi. *Acta Comment., Univ. tartu.*, Ser. A, 14 (3): 1—157.

MARKUS, E., – 1930 – Naturkomplexe von Alatskivi. *Acta Comment., Univ. tartu.*, Ser. A, 18 (8): 1—13.

MATUSZKIEWICZ, W., – 1948 – Roślinność lasów okolik Lwowa. (Engl. summ.: The vegetation of the forests of the environs of Lvov). *Annls. Univ. Mariae Curie-Skłodowska*, Lublin, Sect C, 3: 119—193.

MATUSZKIEWICZ, W., – 1950 – Badania fitosocjologiczne nad lasami bukowymi w Sudetach. (Russ. & Engl. summs.: Phytosociological researches on the beech-forests in the Sudetts-Mtns). *Annls. Univ. Mariae Curie-Skłodowska*, Lublin, Sect. C, Suppl. 5: 1—196.

MATUSZKIEWICZ, W., – 1952 – Zespoły leśne Białowskiego Parky Narodowego. (Russ, & Germ. summs.: Die Waldassoziationen von Białowieza-Nationalpark). *Annls. Univ. Mariae Curie-Skłodowska*, Lublin, Sect. C, Suppl. 6: 1—218.

MCINTOSH, R. P., – 1967 – The continuum concept of vegetation. *Bot. Rev.* 33: 130—187. Responses, *ibid.* 34: 253—332, 1968.

MERRIAM, C. H., – 1890 – Results of a biological survey of the San Francisco Mountain region and desert of the Little Colorado, Arizona. *N. Am. Fauna* 3: 1—136.

MERRIAM, C. H., – 1894 – Laws of temperature control of the geographic distribution of terrestrial animals and plants. *Natn. geogr. Mag.* 6: 229—238.

MERRIAM, C. H., – 1898 – Life zones and crop zones of the United States. *Bull. Bur. biol. Surv. U. S. Dep. Agric.* 10: 1—79.

MERRIAM, C. H., – 1899 – Results of a biological survey of Mount Shasta, California. *N. Am. Fauna* 16: 1—179.

MEUSEL, H., – 1939a – Die Vegetationsverhältnisse der Gipsberge im Kyffhäuser und im südlichen Harzvorland. Ein Beitrag zur Steppenheidefrage. *Hercynia* 2: 1—372.

MEUSEL, H., – 1939b – Pflanzensoziologische Systematik. *Lotos* 4: 393—401.

MEUSEL, H., – 1940 – Die Grasheiden Mitteleuropas: Versuch einer vergleichenden-pflanzensoziologischen Gliederung. *Bot. Arch.*, Leipzig 41: 357—519.

MEUSEL, H., – 1943 – Über die Grundlagen der Vegetationsgliederung. *Forschn. Fortschr.* 19: 34—36.

MEUSEL, H., – 1951 – Über Pflanzengemeinschaften: Probleme der Vegetationskunde, behandelt en einigen Pflanzenvereinen der Heimat. *Urania*, Jena 14: 95—106, 178—188.

MEUSEL, H., – 1954 – Über die umfassende Aufgabe der Pflanzengeographie. *Veröff. geobot. Inst. Rübel*, Zürich 29: 68—80.

MEUSEL H., E. JÄGER & E. WEINERT, – 1965 – Vergleichende Chorologie der zentraleuropäischen Flora. Fischer, Jena. 583+258 pp.

MILLER, A. H., – 1951 – An analysis of the distribution of the birds of California. *Univ. Calif. Publs. Zool.* 50: 531—644.

MOORE, J. J., P. FITZSIMONS, E. LAMBE, & J. WHITE, – 1970 – A comparison of some phytosociological techniques (Germ. summ.) *Vegetatio* 20: 1—20.

MOROSOW, G. F., – 1928 – Die Lehre vom Walde. Neumann, Neudamm. 375 pp.

MOSS, C. E., – 1910 – The fundamental units of vegetation. Historical development of the concepts of the plant association and plant formation. *New Phytol.* 9: 18—53.

MOTYKA, J., – 1947 – O zadaniach i metodach badán géobotanicznych. (French summ.: Sur les buts et les méthodes des recherches geobotaniques). *Annls. Univ. Mariae Curie-Skłodowska*, Lublin, Sect. C, Suppl. 1: 1—168.

MOTYKA, J., B. DOBRZAŃSKI, & S. ZAWADZKI, – 1950 – Wstępne badania nad łąkami południowo-wschodniej Lubelszczyzny. (Russ. & Engl. summs.: Preliminary studies on meadows in the south-east of the province of Lublin). *Annls. Univ. Mariae Curie-Skłodowska*, Lublin, Sect. E, 5: 367—447.

MUESEBECK, C. F. W. & K. V. KROMBEIN, – 1952 – Life zone map. *Syst. Zool.* 1: 24—25.

MUNZ, P. A. & D. D. KECK, – 1949-50 – California plant communities. *Aliso* 2: 87—105, 199—202. Also in 'A California Flora,' by P. A. MUNZ, 1959, pp. 10—20. Univ. California, Berkeley & Los Angeles.

NEGRI, G., – 1927 – Recenti contributi alla concezione sinecologica dei consorzi vegetali. *Nuovo G. bot. ital.*, N. S. 34: 872—885. (Abstr. in *Atti Soc. ital. Progr. Sci.* 16: 578, 1928).

NEGRI, G., – 1954 – Interpretazione individualistica del paesaggio vegetale. *Nuovo G. bot. ital.*, N. S. 61: 579—694.

NORDHAGEN, R., – 1928 – Die Vegetation und Flora des Sylenegebietes. I. Die Vegetation. *Skr. norske Vidensk-Akad.*, Math-naturv. Kl. 1927 (1): 1—612.

NORDHAGEN, R., – 1937 – Versuch einer neuen Einteilung der subalpinen-alpinen Vegetation Norwegens. *Bergens Mus. Årbok*, Naturv. rekke 1936 (7):1—88.

NORDHAGEN, R., – 1943 – Sikilsdalen og Norges Fjellbeiter: En plantesosiologisk monografi. *Bergens Mus. Skr.* 22: 1—607.

NORDHAGEN, R., – 1954 – Vegetation units in the mountain areas of Scandinavia. *Veröff. geobot. Inst. Rübel*, Zürich 29: 81—95.

ODUM, E. P., – 1953 – Fundamentals of Ecology. Saunders, Philadelphia. 384 pp. 2nd ed., 1959, 546 pp. 3rd ed., 1971, 574 pp.

ORLOCI, L., – 1967 – An agglomerative method for classification of plant communities. *J. Ecol.* 55: 193—206.

PACZOSKI, J. K., – 1930 – Lasy Białowieży. (Germ. summ.: Die Waldtypen von

Białowieza). *Państwowa Rada Ochrony Przyrody*, Monogr. Nauk 1: 1—575. Poznań.
PASSARGE, S., – 1921 – Vergleichende Landschaftskunde. I. Aufgaben und Methoden der vergleichenden Landschaftskunde. Reimer/Vohsen, Berlin. 71 pp.
PASSARGE, S., – 1927 – Aufgaben und Methoden der Landschaftskunde, erläutert an den Elementen der nordwestdeutschen Landschaft. *Geogr. Anz.* 28: 44—49.
PASSARGE, S., – 1929 – Botanische und geographische Pflanzenvereine. *Naturwissenschaften* 17: 565—566.
PEARSE, A. S., – 1926 – Animal Ecology. McGraw-Hill, New York. 2nd ed., 1939, 642 pp.
PHILLIPS, J. F. V., – 1930 – Some important vegetation communities in the Central Province of Tanganyika Territory (formerly German East Africa): A preliminary account. *J. Ecol.* 18: 193—234.
POORE, M. E. D., – 1955a – The use of phytosociological methods in ecological investigations. I. The Braun-Blanquet system. *J. Ecol.* 43: 226—244.
POORE, M. E. D., – 1955b – The use of phytosociological methods in ecological investigations. II. Practical issues involved in an attempt to apply the Braun-Blanquet system. *J. Ecol.* 43: 245—269.
POORE, M. E. D., – 1955c – The use of phytosociological methods in ecological investigations. III. Practical application. *J. Ecol.* 43: 606—631.
POOORE, M. E. D., – 1956 – The use of phytosociological methods in ecological investigations. IV. General discussion of phytosociological problems. *J. Ecol.* 44: 28—50.
POORE, M. E. D., – 1962 – The method of successive approximation in descriptive ecology. *Adv. ecol. Res.* 1: 35—68.
RAMENSKY, L. G., – 1926 – Die Grundgesetzmässigkeiten im Aufbau der Vegetationsdecke. (Abstr. from *Vêstn. opytn. Dêla, Voronezh* 1924: 37—73). *Bot. Zbl.* N. F. 7: 453—455.
RAMENSKY, L. G., – 1930 – Zur Methodik der vergleichenden Bearbeitung und Ordnung von Pflanzenlisten und anderen Objekten, die durch mehrere, verschiedenartig wirkende Faktoren bestimmt werden. *Beitr. Biol. Pfl.* 18: 269—304.
RAUNKIAER, C., – 1934 – The Life Forms of Plants and Statistical Plant Geography. Clarendon, Oxford. 632 pp.
REGEL, C., – 1939 – Komplexe, Landschaft, Vegetationsprovinz. *Verh. schweiz. naturf. Ges.* 1939: 68—70.
REGEL, C., – 1949 – Landschaft und Pflanzenverein, mit besonderer Berücksichtigung russischer Forschungen. (French and Ital. summs.) *Geographica helv.* 4: 243—254.
ROSAYRO, R. A. DE, – 1950 – Ecological conceptions and vegetational types with special reference to Ceylon. *Trop. Agric., Ceylon* 106: 108—121.
RÜBEL, E., – 1930 – Pflanzengesellschaften der Erde. Huber, Bern-Berlin. 464 pp.
SAXER, A., – 1955 – Die Fagus-Abies und Piceagürtelarten in der Kontaktzone der Tannen und Fichtenwälder der Schweiz. *Beitr. geobot. Landesaufn. Schweiz.* 36: 1—198.
SAXER, A., – 1967 – Eine Waldkartierung im aargauischen Suhrental nach der Methode von E. Schmid. (Engl. and French summs.) *Veröff. Geobot. Inst. ETH, Stftg. Rübel*, Zürich 39: 149—185, 282, 292—3.
SCHARFETTER, R., – 1932 (1928) – Die kartographische Darstellung der Pflanzengesellschaften. *Handb. Biol. ArbMeth.* 11, 5: 77—164.
SCHIMPER, A. F. W. & F. C. VON FABER, – 1935 – Pflanzengeographie auf physiologischer Grundlage. 3rd ed. Fischer, Jena. 2 vols, 588 & 1612 pp.
SCHMID, E., – 1922 – Biozönologie und Soziologie. *Naturw. Wschr.*, N. F. 21: 518—523.

SCHMID, E., – 1936 – Die Reliktföhrenwälder der Alpen. *Beitr. geobot. Landesaufn. Schweiz* 21: 1—190.
SCHMID, E., – 1940 – Die Vegetationskartierung der Schweiz im Masstab 1: 200,000. *Ber. geobot. Inst. Rübel*, Zürich 1939: 76—85.
SCHMID, E., – 1941 – Vegetationsgürtel und Biocoenose. *Ber. schweiz. bot. Ges.* 51: 461—474.
SCHMID, E., – 1942 – Über einige Grundbegriffe der Biocoenologie. *Ber. geobot. Inst. Rübel*, Zürich 1941: 12—26.
SCHMID, E., – 1949 – Prinzipien der natürlichen Gliederung der Vegetation des Mediterrangebietes (mit einer halb schematischenKarte). *Ber. schweiz. bot. Ges.* 59: 169—200.
SCHMID, E., – 1950 – Zur Vegetationsanalyse numidischer Eichenwälder. *Ber. geobot. Inst. Rübel*, Zürich 1949: 23—39.
SCHMID, E., – 1952 – Natürliche Vegetationsgliederung am Beispiel des Spanischen Rif. *Ber. geobot. Inst. Rübel*, Zürich 1951: 55—79.
SCHMID, E., – 1955 – Der Ganzheitsbegriff in der Biocoenologie und in der Landschaftskunde. *Geographica helv.* 10: 153—162.
SCHMID, E., – 1956a – Die Wuchsformen der Dikotyledonen. *Ber. geobot. Inst. Rübel*, Zürich 1955: 38—50.
SCHMID, E., – 1956b – Die Vegetationsgürtel der Iberisch-Berberischen Gebirge. *Veröff. geobot. Inst. Rübel*, Zürich 31: 124—163.
SCHMID, E., – 1961 – Erläuterung zur Vegetationskarte der Schweiz (French & Engl. summs.) *Beitr. geobot. Landesaufn. Schweiz* 39: 1—52.
SCHMID, E., – 1963 – Die Erfassung der Vegetationseinheiten mit floristischen und epimorphologischen Analysen. *Ber. schweiz. bot. Ges.* 73: 276—324.
SCHMITHÜSEN, J., – 1959 – Allgemeine Vegetationsgeographie. In, 'Lehrbuch der allgemeinen Geographie,' ed. E. OBST, vol. 4. de Gruyter, Berlin. 261 pp.
SCHMITZ, W., – 1965 – Die Soziologie aquatischer Mikrophyten. (Engl. summ.) In, 'Biosoziologie,' ed. R. TÜXEN, *Ber. Symp. int. Vereinig. Vegetationskunde*, Stolzenau/Weser 1960, 4: 120—139.
SCHRÖTER, C. & O. KIRCHNER, – 1902 – Die Vegetation des Bodensees. 2. Teil von C. SCHRÖTER. *Bodensee-Forschungen* 9 (2): 1—86. Komm.-Verlag Schr. Ver. Gesch. Bodensees und seiner Umgebung von J. T. Stettner, Lindau.
SCHWARZ, U., – 1955 – Die natürlichen Fichtenwälder des Juras. *Beitr. geobot. Landesaufn. Schweiz* 35: 1—143.
SERNANDER, R., – 1898 – Studier öfver vegetationen i mellersta Skandinaviens fjälltrakter. 1. Om tundraformationer i svenska fjälltrakter. *Ofvers. K. svenska Vetensk. Akad. Förhandl.* 6: 325—367.
SHIMWELL, D. W., – 1971 – Description and Classification of Vegetation. Sidgwick & Jackson, London. 322 pp.
SJÖRS, H., – 1955 – Remarks on ecosystems. *Svensk. bot. Tidskr.* 49: 155—169.
SØRENSEN, T., – 1937 – Remarks on the flora and vegetation of Northeast Greenland 74 ° 30' — 79 ° N. lat. *Meddr. Grønland* 101(4): 108—140.
SØRENSEN, T., – 1948 – A method of establishing groups of equal amplitude in plant sociology based on similarity of species content and its application to analysis of the vegetation on Danish commons. *Biol. Skr.*, K. danske Vidensk. Selsk. 5 (4): 1—34.
SOUTHWARD, A. J., – 1958 – The zonation of plants and animals on rocky sea shores. *Biol. Rev.* 33: 137—177.
STAMP, L. D., – 1925 – The vegetation of Burma from an ecological standpoint. *Res. Monogr. Univ. Rangoon* 1: 1—58.
STEINDÓRSSON, S., – 1954 – The coastline vegetation at Gásar in Eyjafjörður in the North of Iceland. *Nytt. Mag. Bot.* 3: 203—212.

STEPHENSON, T. A. & ANNE STEPHENSON, – 1961 – Life between tide-marks in North America. IVA. Vancouver Island. I. *J. Ecol.* 49: 1—29.
SUKACHEV, V. N., – 1928 – Principles of classification of the spruce communities of European Russia. *J. Ecol.* 16: 1—18.
SUKATSCHEW, W., – 1932 – Die Untersuchung der Waldtypen des osteuropäischen Flachlandes. *Handb. biol. ArbMeth.* 11, 6: 191—250.
SUKATSCHEW, W., – 1944 – On principles of genetic classification in biocenology. (Russ. with Engl. summ.) *Zhur. Obschch. Biol.* 5: 213—227.
SUKATSCHEW, W., – 1954 – Die Grundlagen der Waldtypen. *Angew. PflSoziol.*, Wien, Festschr. Aichinger 2: 956—964.
SUKATSCHEW, W., – 1960 – The correlation between the concept of 'forest ecosystem' and 'forest biogeocenose' and their importance for the classification of forests. *Silva fenn.* 105: 94—97.
TANSLEY, A. G., ed., – 1911 – Types of British Vegetation. Cambridge Univ., Cambridge. 416 pp.
TANSLEY, A. G., – 1920 – The classification of vegetation and the concept of development. *J. Ecol.* 8: 118—149.
TANSLEY, A. G., – 1939 – The British Islands and their Vegetation. Cambridge Univ., Cambridge. 930 pp.
TANSLEY, A. G. & T. F. CHIPP, eds., – 1926 – Aims and Methods in the Study of Vegetation. Brit. Emp. Veget. Comm. & Crown Agents for Colonies, London. 383 pp.
THORNTHWAITE, C. W., – 1948 – An approach toward a rational classification of climate. *Geogrl. Rev.* 38: 55—94.
TISCHLER, W., – 1955 – Synökologie der Landtiere. Fischer, Stuttgart. 414 pp.
TOSI, J. A., Jr., – 1960 – Zonas de vida natural en el Perú. Memoria explicativa sobre el mappa ecológico del Perú. (Engl., French, Germ. summs.) *Bol. Técn. Inst. interamer. cienc. agríc.* OEA Zona Andina, 5: 1—271.
TROLL, C., – 1950 – Die geographische Landschaft und ihre Erforschung. *Studium gen.* 3: 163—181.
TROLL, C., – 1961 – Klima und Pflanzenkleid der Erde in dreidimensionaler Sicht. *Naturwissenschaften* 48: 332—348.
TUOMIKOSKI, R., – 1942 – Untersuchungen über die Vegetation der Bruchmoore in Ostfinnland. I. Zur Methodik der pflanzensoziologischen Systematik. (Finn. summ.) *Ann. Bot., Soc. zool. -bot. fenn. Vanamo* (*Suomal. eläin-ja kasvit. Seur. van. kasvit. Julk.*) 17(1):1-203.
TÜXEN, R., – 1937 – Die Pflanzengesellschaften Nordwestdeutschlands. *Jber. naturh. Ges. Hannover* 1929/30—1935/36, 81—87: 1—170. *Mitt. flor.-soz. Arb Gemein Neidersachsen* 3: 1—170.
TÜXEN, R., ed., — 1968a — Pflanzensoziologie und Landschaftsökologie. *Ber. Symp. int. Vereinig. Vegetationskunde*, Stolzenau/Weser 1963, 7 : 1—426.
TÜXEN R., ed., — 1968b — Pflanzensoziologische Systematik. *Ber. Symp. int. Vereinig. Vegetationskunde*, Stolzenau/Weser 1964, 8 : 1—347.
TÜXEN, R., – 1970 – Pflanzensoziologie als synthetische Wissenschaft. In 'Vegetatiekunde als synthetische wetenschap.' ed. H. J. VENEMA, H. DOING, and I. S. ZONNEVELD. *Misc. Pap. Landbouwhogeschool Wageningen*, 5: 141—159.
VAARAMA, A., – 1938 – Wasservegetationstudien am Grosssee Kallavesi. (Finn. summ.) *Ann. Bot., Soc. zool. -bot. fenn. Vanamo* (*Suomal. eläin-ja kasvit. Seur. van. kasvit. Julk.*) 13 (1): 1—318.
VILLAR, E. H. DEL, – 1929 – Geobotánica. Editorial Labor, Barcelona-Buenos Aires. 339 pp.
VRIES, D. M. DE, – 1948 – Method and survey of the characterization of Dutch grasslands. (Esperanto summ.) *Vegetatio* 1: 51—57.

WARMING, E., – 1909 – Oecology of Plants: An Introduction to the Study of Plant-Communities. Oxford Univ., Oxford. 422 pp.
WEAVER, J. E. & F. E. CLEMENTS, – 1929 – Plant Ecology. McGraw-Hill, New York. 520 pp. 2nd ed., 1938, 601 pp.
WENDELBERGER, G., – 1951 – Das vegetationskundliche System Erwin Aichingers und seine Stellung im pflanzensoziologischen Lehrgebäude Braun-Blanquets. *Angew. PflSoziol.*, Wien 1: 69—92.
WHITTAKER, R. H., – 1951 – A criticism of the plant association and climatic climax concepts. *NW. Sci.* 25: 17—31.
WHITTAKER, R. H., – 1956 – Vegetation of the Great Smoky Mountains. *Ecol. Monogr.* 26: 1—80.
WHITTAKER, R. H., – 1960 – Vegetation of the Siskiyou Mountains, Oregon and California. *Ecol. Monogr.* 30: 279—338.
WHITTAKER, R. H., – 1962 – Classification of natural communities. *Bot. Rev.* 28: 1—239.
WHITTAKER, R. H., – 1967 – Gradient analysis of vegetation. *Biol. Rev.* 42: 207—264.
WHITTAKER, R. H. & W. A. NIERING, – 1965 – Vegetation of the Santa Catalina Mountains, Arizona. (II) A gradient analysis of the south slope. *Ecology* 46: 429—452.
WILLIAMS, W. J. & JEAN M. LAMBERT, – 1959 – Multivariate methods in plant ecology. I. Association-analysis in plant communities. *J. Ecol.* 47: 83—101.
WILLIAMS, W. J. & JEAN M. LAMBERT, – 1961 – Multivariate methods in plant ecology. III. Inverse association-analysis. *J. Ecol.* 49: 717—729.
WILMANNS, O., – 1962 – Rindenbewohnende Epiphytengemeinschaften in Südwestdeutschland. *Beitr. naturk. Forsch. Südwestdeutschl.* 21 (2): 87—164.
WOOD, J. G., – 1937 – The Vegetation of South Australia. Trigg, Govt. Printer, Adelaide. 164 pp.
WOOD, J. G., – 1939 – Ecological concepts and nomenclature. *Trans. R. Soc. S. Aust.* 63: 215—223.
YARRANTON, G. A., – 1967 – Principal components analysis of data from saxicolous bryophyte vegetation at Steps Bridge, Devon. III. Correlation of variation in vegetation with environmental variables. *Can. J. Bot.* 45: 249—258.

13 THE PHYSIOGNOMIC APPROACH

John S. Beard

Contents

13 THE PHYSIOGNOMIC APPROACH

13.1 Introduction

Vegetation possesses two principal properties, floristic composition and physiognomy. Different workers have tended to be more interested in the one than the other, but in general in any locality it has been the physiognomy which has received first attention with a later progression to more detailed floristic studies. This happened for example in Europe, where in the later stage interest in physiognomy declined among phytosociologists but remained active among biogeographers working on vegetation on a world scale. At the same time there are countries, in the tropics especially, where local vegetation studies are still in an early stage and where physiognomy is consequently the primary concern. There are thus two developments to be discussed further – the physiognomic approach in biogeography, and physiognomic approaches to more local areas, particularly in British Commonwealth countries.

13.2 History and Major Concepts

Many of the great early biogeographers were primarily interested in the floristics of vegetation, e.g. DE CANDOLLE (1855), DRUDE (1890), and SCHIMPER (1898). GRISEBACH (1838, 1872) was the first great physiognomist and it was he who at a very early date (1838) introduced the term *formation* in its physiognomic connotation: 'I give the name phytogeographical formation to a group of plants, such as a meadow or a forest, that has a fixed physiognomic character.' The term formation was used in this sense by some later authors of the 19th Century but others used it for floristic units, that is as a synonym for association.

WARMING was the next great physiognomist and his papers appearing from 1895 attracted great attention culminating in the publication of his book '*Oecology of Plants*' (1909) specially written for English translation. For WARMING (p. 145) 'an association is a community of definite floristic composition within a formation', and (p. 140), 'a formation is an expression of certain defined conditions of life and is not concerned with floristic differences.' Whatever other influences later affected thought in the physiognomic field, these two propositions of WARMING's have remained basic. WARMING's formations were mainly characterised by dominant growth-form, and he was relatively little concerned with further details of vegetation structure, as these were studied later by the British school.

The other great Danish physiognomist, RAUNKIAER, whose collected papers from 1904 also found their way into a special English edition (1934), influenced the main stream of thought rather less than WARMING. RAUNKIAER's system is based purely on *life-forms*, a classification of plant forms according to the position and character of their perennating buds. His system is more floristic than physiognomic; the *biological spectra* (percentages of the species in a flora that belong to the various life-forms) facilitate comparisons of the floras of different geographic areas and communities but do not effectively characterize vegetation structure. Biological spectra have been employed by many workers, but this approach is of little value in, for instance, the recognition of forest types in the tropics. Just one element of RAUNKIAER's work, the *leaf-size classes*, was picked out as of practical value by the British physiognomic school and incorporated as a standard criterion of classification. The RAUNKIAER approach is further discussed by BRAUN-BLANQUET (1951, 1964), CAIN (1950, CAIN & OLIVEIRA 1959) and DANSEREAU (1957) and in Part II of this *Handbook*.

From the work of GRISEBACH, WARMING and others the plant *formation* has emerged as a key concept, equivalent to the *biome* of animal ecologists. The formation can be more broadly or more narrowly defined but the essential concept is a classification of vegetation by its physiognomy or structure, particularly growth-form dominance. By structure we understand here the organisation of the height and degree of density of these layers. Environment must usually also be part of the definition: A formation is a major kind of plant community on a given continent, characterized by physiognomy and a range of environments to which that physiognomy is a response. Thus, tropical grassland (savanna), temperate grassland (steppe), alpine meadow, and saltmarsh are regarded as different formations though all are dominated by the same growth-form (grasses and grass-like plants). Table I (WHITTAKER 1970) gives a simple outline of major growth-forms; for more detailed treatments see DU RIETZ (1931), ELLENBERG (1956), and SCHMITHÜSEN (1968). Many authors, especially RICHARDS, TANSLEY and WATT who will be mentioned below, have used the term 'life-form' in this sense but, in order to avoid confusion, the latter should be restricted to RAUNKIAER's life-forms which are not at all the same thing.

Table I

Major plant growth-forms on land

Trees, larger woody plants, mostly well above three meters tall
- Needle-leaved (mainly conifers – pine, spruce, larch, redwood, and so on)
- Broad-leaved evergreen (many tropical and subtropical trees, mostly with medium -sized leaves)
- Evergreen-sclerophyll (with smaller, tough, evergreen leaves)
- Broad-leaved deciduous (leaves shed in the Temperate Zone winter, or in the tropical dry season)
- Thorn-trees (armed with spines, in many cases with compound, deciduous leaves)
- Rosette trees (unbranched, with a crown of large leaves – palms and tree-ferns)

Lianas (woody climbers or vines)

Shrubs, smaller woody plants, mostly below three meters in height
- Needle-leaved
- Broad-leaved evergreen
- Broad-leaved deciduous
- Evergreen-sclerophyll
- Rosette shrubs (yucca, agave, aloe, palmetto, and so on)
- Stem succulents (cacti, certain euphorbias, and so on)
- Thorn-shrubs
- Semishrubs (suffrutescent, that is, with the upper parts of the stems and branches dying back in unfavorable seasons)
- Subshrubs or dwarf-shrubs (low shrubs spreading near the ground surface, less than 25 cm high)

Epiphytes (plants growing wholly above the ground surface, on other plants)

Herbs, plants without perennial aboveground woody stems
- Ferns
- Graminoids (grasses, sedges, and other grasslike plants)
- Forbs (herbs other than ferns and graminoids)

Thallophytes
- Lichens
- Mosses
- Liverworts

The matter of scale also materially influences formation concepts. The essential definition of formation offers much play for individual choice of how broadly or narrowly formations are to be defined. Definition by only the major growth-form of the uppermost stratum (or the stratum of greatest coverage) produces relatively few broadly defined formations; distinction by different combinations of growth-forms of the same or different strata produces a larger number of more narrowly defined formations. For treatment of continental or world vegetation few, broad formations can be conveniently dealt with, whereas detailed local studies may require recognition of subdivisions and minor types as formations or subformations.

Within this framework it is now generally recognised that physiognomy expresses biotope conditions and that similar biotopes in different parts of the world will exhibit and can be recognised by vegetation of similar physiognomy. Local influences are sometimes pronounced because different growth-forms are available in different geographic areas. For example northern coniferous forests (taiga) and the Australian mallee (of giant shrubs) have physiognomies all their own, while Southern Africa, Australia and middle America have all produced growth-forms adapted to desert conditions, some strikingly convergent but others quite different. Broadly, however, there is a convergence in physiognomy in corresponding major environments on a worldwide scale, and world-wide systems of vegetation classification must inevitably be physiognomic in character. The evolutionary convergence of formations of similar environments on different continents has emerged as a major principle in biogeography. These generalised, world-scale units that group together similar formations from different continents are best termed *formation-types* (or *biome-types*).

13.3 **Physiognomic Systems**

We need to discuss (before going on to British applications to local areas) four major ways of approaching vegetation through physiognomy. These are: (i) Outline classifications of formation-types of the world, (ii) Interrelation of formations or formation-types with types of climates, (iii) Direct descriptive systems that need not rely on formations or formation-types, and (iv) Treatment in terms of sequences of formations along environmental gradients (formation-series) or of physiognomic continua along environmental gradients (ecoclines). Much of the difficulty of physiognomic research results from the fact that many formations intergrade continuously. We are dealing with merging phenomena which are not inherently suitable for classification, though they can be profitably classified. Given this difficulty, our four approaches entail: (i) Derivation of a classification by typification – recognition of major types of communities by physiognomy, without attempting to provide in the system for intergradations and minor physiognomic types. (ii) Classification that gives joint emphasis to climate and physiognomy, and must again limit itself to broad relationships. (iii) Use of descriptive characterization of physiognomy without formal classification. (iv) Concern with relations to environmental gradients and intergradations, with or without classification.

13.3.1 Formation-Types

There are a number of systems of formation-types or biome-types in the literature; but the systems that have had widest influence in

TABLE II

The formation-types of SCHIMPER and VON FABER (1935)

1. *Tropical Rain Forest* occupies regions of high and constant rainfall and temperature. The forest is many-layered, leaves are mainly evergreen, large, entire; trees tall and buttressed; epiphytes and lianas very common. The flora is very rich. Amazonia, Congo Basin, Malaysia.
2. *Subtropical rainforest* is found in humid subtropical regions with some seasonal variation in temperature and rainfall. Luxuriance in structure and composition is reduced. Brasil, African Highlands, Southeast Asia.
3. *Monsoon Forest*: Tropical and subtropical with a moderate winter dry season. Forest is tall, many-layered, with predominance of deciduous species in the canopy. Central America, India, Southeast Asia.
4. *Temperate Rain Forest* expresses high and constant rainfall in cooler regions. Forest is moderately tall, dense, few-layered, leaves are evergreen, small, or coriaceous. Much moss and lichen. A variant, montane rain forest or cloud forest, is found on tropical mountains. Tasmania, New Zealand, Chile.
5. *Summer-green Deciduous Forest* occupies regions with a pronounced seasonal change of temperature, a cold winter with snow and a mild to warm wet summer. Trees are tall, structure simple, leaves broad, fine and deciduous. Eastern North America, Europe, China.
6. *Needle-leaf Forest* is characteristic of cold regions with long winters and high rainfall. Trees are coniferous, needle or scale-leaved, and may be very large in size. Western North America, Northern Europa, Siberia.
7. *Evergreen Hardwood Forest* characterises regions of 'mediterranean' climate with a dry summer and wet, mild winter. Trees are small (except in Australia) and leaves sclerophyllous. Australia, California, Mediterrranean.
8. *Savanna Woodland* appears under a summer rainfall with a long dry season, i.e., more extreme than monsoon forest. Trees are small, evergreen, in open formation, with a ground layer of tropical bunch-grasses. Brazilian and African plateaux, North Australia.
9. *Thorn Forest and Scrub*. Tropical, dry climates. Trees are small, often thorny and deciduous. The ground layer includes many succulents, annuals and grasses. Brazil, Africa, India.
10. *Savanna* is a moist tropical grassland, with or without trees, and may owe its origin to fire or to adverse soil conditions or both. Pantropical.
11. *Steppe and Semidesert* occur in dry climates with winter-rainfall, i.e., more extreme than evergreen hardwood forest. Open shrublands with annual herbs and grasses, or dry grasslands. (Some tropical dry grasslands have been classed as steppe by other authors). North America, Australia, Russia, Argentina.
12. *Heath*. Like the tropical savanna the heath in temperate regions is governed by fire or adverse soil conditions or both. It is a formation of ericoid shrubs with scattered larger shrubs and small trees. Worldwide, locally.
13. *Dry Desert*. Warm regions of very low rainfall with open vegetation and special plant forms evolved in different parts of the world, e.g. succulent Cactaceae in North America, succulent Liliaceae, Aizoaceae, Euphorbia and Welwitschia in Southern Africa, hummock grasses in Australia.
14. *Tundra and Cold Woodland*. This is the semi-desert of cold regions where there is a short summer growing season. Lichens are especially abundant under sedges and grasses (tundra) or under stunted trees. On rocky areas mosses may be dominant. Northern hemisphere in high latitudes.
15. *Cold desert*. Edge of icecaps, glaciers and permanent snowfields. Vegetation sparse, mainly herbaceous.

Table III

Rübel's formation-types, with Latin, German, and English equivalents

Lignosa - Woody Formations

Pluviilignosa
1. Pluviisilvae - Regenwälder - rain forest
2. Pluviifruticeta - Regengebüsche - rain scrub

Laurilignosa
3. Laurisilvae - Lorbeerwälder - laurel-leaved forest
4. Laurifruticeta - Lorbeergebüsche - laurel-leaved scrub

Durilignosa
5. Durisilvae - Hartlaubwälder - sclerophyll forest
6. Durifruticeta - Hartlaubgebüsche - sclerophyll scrub

Ericilignosa
7. Ericifruticeta - Echte Heiden - heaths

Aestilignosa
8. Aestisilvae - Sommerwälder - summergreen deciduous forest
9. Aestifruticeta - Sommergebüsche - summergreen scrub

Hiemilignosa
10. Hiemisilvae - Regengrüne Wälder - raingreen (monsoon) forest
11. Hiemifruticeta - Regengrüne Gebüsche - raingreen scrub

Aciculilignosa
12. Aciculisilvae - Nadelwälder - needle-leaved or coniferous forest
13. Aciculifruticeta - Nadelgebüsche - needle-leaved scrub

Herbosa - Herbaceous Formations

Terriherbosa
14. Duriherbosa - Hartwiesen, Steppenwiesen - hardgrass prairie and steppe
15. Sempervirentiherbosa - Immergrüne Wiesen - evergreen grassland or meadow
16. Altherbosa - Hochstaudenwiesen - tall herbage or forbland

Aquiherbosa
17. Emersiherbosa - Sumpfwiesen - marsh
18. Submersiherbosa - Submerse Wasserwiesen - submerged aquatics
19. Sphagniherbosa - Hochmoor - peat bog

Deserta - Deserts

20. Siccideserta - Trockeneinöden - dry desert
21. Frigorideserta - Kälteneinöden - cold desert
22. Litorideserta - Strandsteppen - strand (saltspray) steppe
23. Mobilideserta - Wandereinöden - vegetation of dunes, and unstable soils
24. Rupideserta - Felsfluren - vegetation of screes, talus, etc.
25. Saxideserta - Stein- und Holzfluren - vegetation of rocks, tree trunks, etc.

Special Groups

26. Aquerrantia - Phytoplankton - free aquatic organisms, mostly microscopic
27. Solerrantia - Phytedaphon - terrestrial micro-organisms, soil bacteria and fungi
28. Aërerrantia - Phytaëron - aerial microorganisms

phytogeography are those of RÜBEL (1930, 1936) and SCHIMPER (SCHIMPER & FABER 1935). The two systems are summarized as Tables II and III. More recent systems (too detailed to reproduce here) have been proposed by FOSBERG (1961), ELLENBERG & MUELLER-DOMBOIS (1967), and SCHMITHÜSEN (1968). A major treatment of world vegetation by WALTER (1962, 1968) uses no formal classification. The SCHIMPER system seems in many respects the best. RÜBEL's system is more cumbersome and somewhat more artificial. His schematic pairing of forest and shrubland types results in overemphasis of shrublands in the classification, and underemphasis of woodlands (communities of small trees, usually in open spacing and with well-developed undergrowth). RÜBEL refers to his primary types as formation-classes, rather than using the more generally accepted term formation-type. Vegetation maps on a continental or subcontinental scale are almost inescapably based on physiognomy. For applications of formations in vegetation mapping, see KÜCHLER (1967).

Table IV is an effort to coordinate with one another these two classic systems and two current systems (SCHMITHÜSEN 1968, WHITTAKER 1970). The 'subclasses' of SCHMITHÜSEN's classification most nearly correspond to the formation-types of others, rather than his numerous and narrowly defined formation-types. It is possible to find common features as well as divergences among the classifications of different authors. Coordination of different systems is not, however, simple, and some of the types entered in Table IV do not correspond as neatly as their entry in the table might make it appear. Even types given the same name by three or four of the authors (e.g. heath, savanna, tundra) represent three or four different, though overlapping, concepts. Such difficulties reemphasize the subjective choices of the limits and numbers of formations recognized. There is thus a degree of artistry in the classification, or typification, of vegetation physiognomy, though the product of this artistry is valuable to our understanding of world vegetation.

13.3.2 CLIMATIC CORRELATION

Working with the SCHIMPER system, DANSEREAU (1957) sought to relate its units to climate. The formation-types are first rearranged into four major *biochores* – forest, savanna, grassland, and desert. Not everyone will concur with this, especially as DANSEREAU uses 'savanna' in a sense peculiar to himself, but the viewpoint is an interesting one. Within these biochores the formation-types may now be arranged into a triangle representing the range of major

Table IV

A comparison of four physiognomic classifications

Schimper & Faber 1935	Rübel 1930	Schmithüsen 1968	Whittaker 1970
1. Tropical rainforest 2. Subtropical rainforest	1. Rainforest	IA. Evergreen forest	1. Tropical rainforest
4. Temperate rainforest	3. Laurel-leaved forest		3. Temperate giant rainforest 4. Montane rainforest 8. Elfin woodland
6. Needleleaf forest 7. Evergreen hardwood forest	12. Needle-leaved forest 5. Sclerophyll forest		7. Taiga 6. Temp. evergreen forest-- needleleaf sclerophyll
3. Monsoon forest 5. Summergreen deciduous forest	10. Raingreen forest 8. Summergreen forest	IB. Deciduous forest	2. Tropical seasonal forest 5. Temperate deciduous forest
8. Savanna woodland		IIA. Evergreen woodland IIB. Deciduous woodland	11. Temperate woodland
9. Thorn forest and scrub	11. Raingreen scrub	IIC. Xeromorphic woodland IIIC. Xeromorphic shrubland	9. Thorn woodland 10. Thorn scrub
  12. Heath	9. Summergreen scrub 7. Heath 6. Sclerophyllous scrub 13. Needle-leaved scrub 4. Laurel-leaved scrub	IIIB. Deciduous shrubland VIC. Heath (in part) IIIA. Evergreen shrubland	12. Temperate shrublands-- deciduous heath sclerophyll subalpine-needleleaf -broadleaf
10. Savanna 11. Steppe and semidesert	14. Hardgrass prairie	IVA. Savanna IVB. Steppe (in part)	13. Savanna 14. Temperate grassland 19. Cool-temperate desert scrub
13. Dry desert	20. Dry desert	VIIA. Desert	17. Tropical desert 18. Warm-temperate desert
14. Tundra 15. Cold desert	15. Meadows 21. Cold desert	IVC. Meadows VID. Tundra	15. Alpine grasslands 16. Tundra 20. Arctic-alpine desert
(edaphic)	19. Moss moor 2. Rain scrub 17. Marsh	VIE. Moss moor	21. Bog 24. Mangrove swamp 25. Saltmarsh

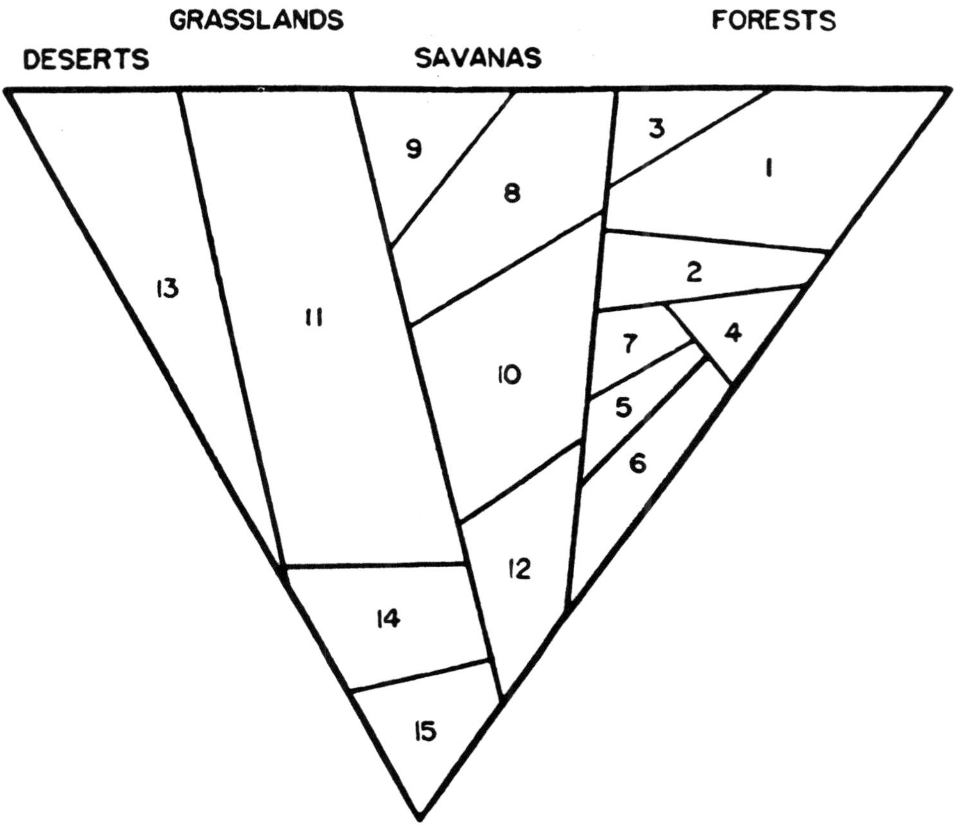

Fig. 1. A bioclimatic diagram arranged according to increasing temperatures (bottom up) and increasing rainfall (left to right) (DANSEREAU 1951). Formation-types are represented by numbers corresponding to those of SCHIMPER & VON FABER (1935), Table II. The upper-left angle is for extreme of heat and drought, the upper-right for the extreme of heat and humidity, the lower angle for extreme cold with low precipitation.

variations in the world's climates (Fig. 1). Fig. 2 represents a somewhat comparable diagram by WHITTAKER (1970). There are various ways in which these treatments, that start with a system of formation-types and then seek to relate these to climate, are imperfect. The formation-types overlap in relation to climate in a complex way. Effects of maritime vs. continental climates and of soil-and-fire influences are, in particular, difficult to deal with adequately. The alternative approach is to start from the other side of the relationship, to seek to define not formation-types as such but major climatic regions that may be characterized both by climate and by physiognomy adapted to that climate. The most widely used systems

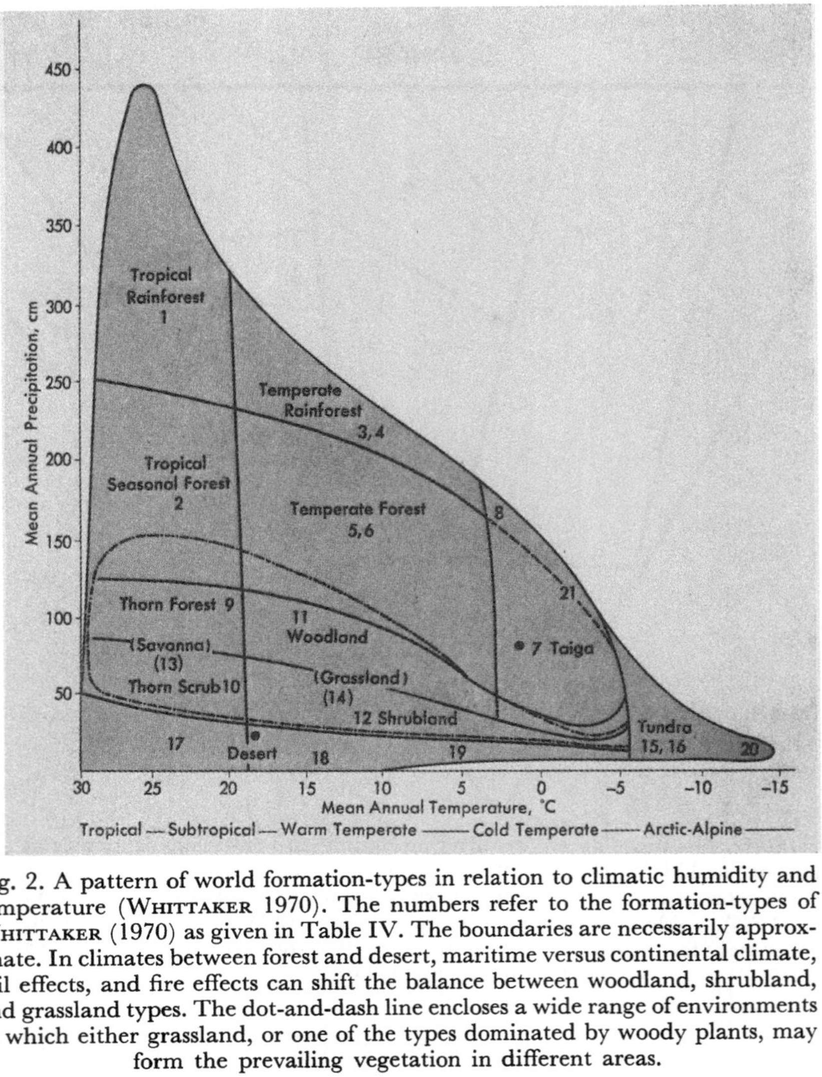

Fig. 2. A pattern of world formation-types in relation to climatic humidity and temperature (WHITTAKER 1970). The numbers refer to the formation-types of WHITTAKER (1970) as given in Table IV. The boundaries are necessarily approximate. In climates between forest and desert, maritime versus continental climate, soil effects, and fire effects can shift the balance between woodland, shrubland, and grassland types. The dot-and-dash line encloses a wide range of environments in which either grassland, or one of the types dominated by woody plants, may form the prevailing vegetation in different areas.

of this sort are those of KÖPPEN (1900, 1923), and THORNTHWAITE (1931, 1948) (see also STEFANOFF 1930, RUMNEY 1968). TROLL (1948, 1961, 1968) and STOCKER (1963) have dealt with the asymmetry of the Northern and Southern Hemispheres, and effects of altitude versus latitude, in the relations of climate and vegetation.

A recent system relating physiognomy to climate is that of HOLDRIDGE (1947, 1967), classifying vegetation of the world into a system of compartments defined by temparature and moisture (Fig. 3). The

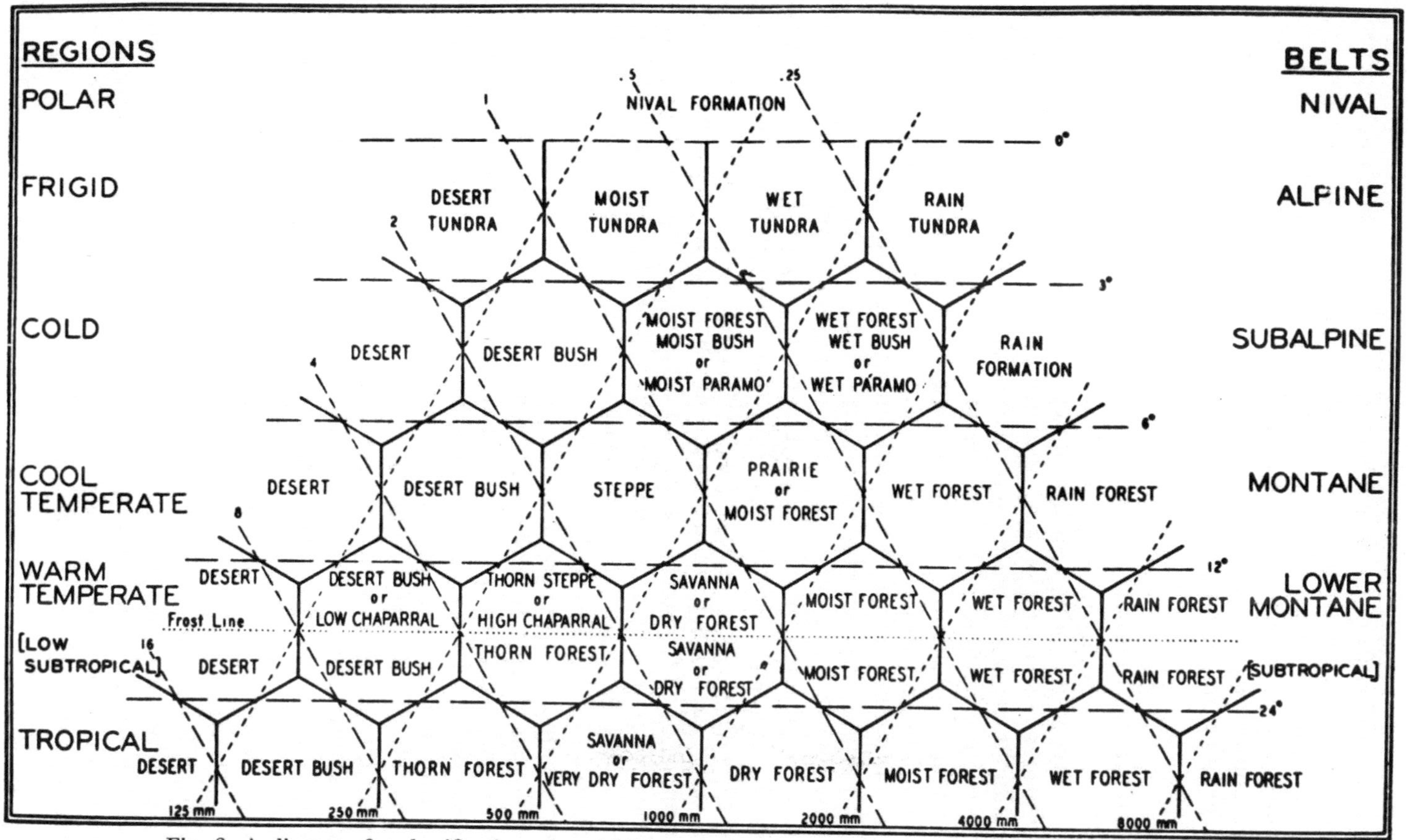

Fig. 3. A diagram for classification of world life zones or plant formations by climate (HOLDRIDGE 1947).

system suffers from its over-schematic character, which juxtaposes in compartments some quite disparate types of vegetation, while also requiring separation into different compartments of other, more similar, types. But all approaches have their difficulties, and HOLDRIDGE's has been gaining favor as an interpretation of tropical vegetation in particular.

13.3.3 DESCRIPTIVE SYSTEMS

An entirely different line of development in the physiognomic treatment of vegetation was opened by KÜCHLER (1947). Classification in its usual sense (here involving grouping entities around abstract type-concepts) was rejected. KÜCHLER proposed instead a system of descriptive notation using four series of symbols — predominant growth-form, height, density and special features — which can be combined so as to show the important features of plant cover. This represented essentially a geographer's approach to the characterisation of plant cover for mapping purposes and is most useful for over-all description of regional vegetation rather than for detailed work on plant communities. KÜCHLER has applied this system, amplified in 1949, in vegetation mapping (1948, 1950). As with RAUNKIAER's biological spectra the simplicity of the system has commended itself to later workers, and despite criticisms which can be levelled at it a recent paper has shown that it could be adapted to use in Australia (ROSS COCHRANE, 1963). In point of fact, as DANSEREAU has shown (1951), KÜCHLER's system for all its practical value to a geographer, is open to considerable criticism by ecologists. Major divisions in the system are of unequal significance and tend to ignore features not found in the North Temperate Zone. DANSEREAU set out to reform the system, making it more logical and comprehensive and appropriate to the needs of descriptive ecologists. He did so successfully but in the process transformed it from its original role as a mapping notation into an elegant ecological characterization which is too elaborate for use in mapping and may very well prove too complicated for widespread use. It is difficult to make any physiognomic classification fully inclusive without also making it cumbersome.

DANSEREAU (1951, 1957) proposed six categories of criteria in place of KÜCHLER's four, viz. growth-form (which he terms life-form), size, function, leaf size and shape, leaf texture, and coverage (Table V). The characters to be distinguished, varying from three to six in each category, are provided with two sets of symbols, the one alphabetical and the other pictorial, the former to be compound-

Table V

Six categories of criteria to be applied to a structural description of vegetation (Dansereau 1957).

1. Form

T	trees
F	shrubs
H	herbs
M	bryoids
E	epiphytes
L	lianas

2. Size

t	tall (T: minimum 25m)
	(F: 2—8m)
	(H: minimum 2m)
m	medium (T: 10—25m)
	(F, H: 0.5—2m)
	(M: minimum 10cm)
l	low (T: 8—10m)
	(F,H: maximum 50cm)
	(M: maximum 10cm)

3. Function

d	deciduous
s	semideciduous
e	evergreen
j	evergreen-succulent; or evergreen-leafless

4. Leaf Shape and Size

n	needle or spine
g	graminoid
a	medium or small
h	broad
v	compound
q	thalloid

5. Leaf Texture

f	filmy
z	membranous
x	sclerophyll
k	succulent or fungoid

6. Coverage

b	barren or very sparse
i	discontinuous
p	in tufts or groups
c	continuous

ed into formulae and the latter for the construction of diagrams. Note that this is no longer a system of major structural types of plants in the sense of Table I; it is a set of intersecting criteria by which plants (and communities) can be characterized without formal classification. To characterize the community a descriptive formula consisting of six letters (one from each of the prime categories) is made up for each synusia or stratum. Structurally very complex communities will qualify for a whole run of six-letter formulae, written with full-stops between them. This does in fact achieve a quite complete description of the community, but the result can be elaborate. The pictorial symbols are used to construct diagrams something like a stylised version of the Richards profile diagram (see below), in which the symbols themselves suggest the shape and size of trees, shrubs, and other plants, and are spaced according to percentage cover (Fig. 4). These diagrams, which should become known as Dansereau Diagrams (Danserograms), have real value for comparative purposes. Because measured profile diagrams involve a certain amount of subjective interpretation and reliance

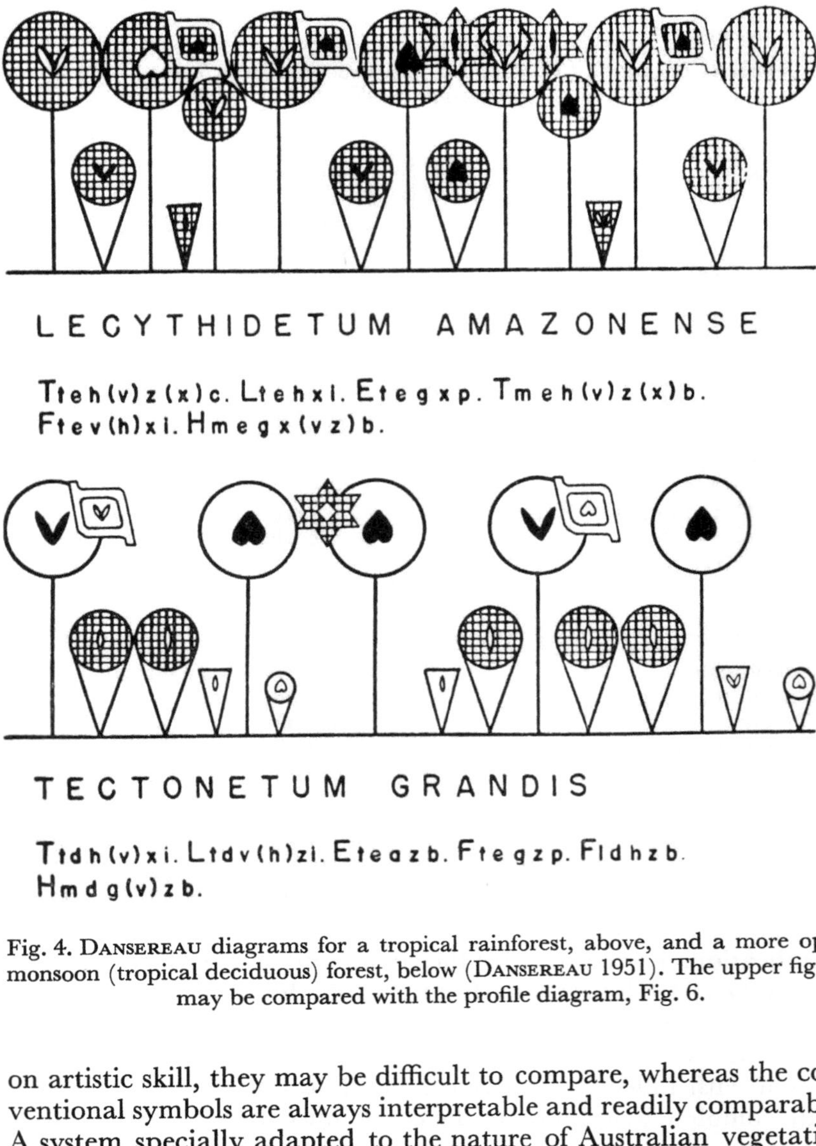

Fig. 4. DANSEREAU diagrams for a tropical rainforest, above, and a more open monsoon (tropical deciduous) forest, below (DANSEREAU 1951). The upper figure may be compared with the profile diagram, Fig. 6.

on artistic skill, they may be difficult to compare, whereas the conventional symbols are always interpretable and readily comparable. A system specially adapted to the nature of Australian vegetation and based on the best features of KÜCHLER and DANSEREAU is now being applied in that country (BEARD 1969, BEARD & WEBB 1972).

13.3.4 FORMATION-SERIES

The fourth approach relates vegetation physiognomy to environmental gradients. BEARD (1944) working in the British tradition to be described below, developed a system of formations for tropical America. Unlike the systems of SCHIMPER and WARMING, which begin with a pre-conceived hierarchy of formations into which examples under study must be fitted, BEARD's system begins with typification of a number of actual communities and then proceeds to consider how they are related. BEARD's formations are more narrowly defined units than those of Tables II and III. The formations are related to one another and environment in terms of sequences along environmental gradients; such a sequence is a *formation-series*. Five formation-series were recognized, two considered to be climatically controlled and three as edaphic. The former are the seasonal series (lowland communities controlled by decreasing rainfall from rainforest through deciduous forests and thorn woodlands to desert) and the montane series (controlled by increasing altitude from lowland to montane rainforest to elfin woodland and the alpine paramo and puna). The edaphic formations were arranged into swamp, seasonal swamp, and dry evergreen series, the last being essentially on shallow soil or rock pavements. These series may be thought of as five directions of divergence from the most highly developed and focal community of most favorable environments, the tropical rainforest. In each case the component formations of the series show progressive reduction of structure and modification of growth-forms outwards toward a theoretical pessimum devoid of vegetation (Fig. 5).

Eleven years later additional information was available for the American tropics, the principal further contribution being that of FANSHAWE (1952). This information enabled BEARD (1955) to revise and expand his formation series into a framework capable in principle of embracing all types of vegetation likely to occur in tropical America. BEARD (1955) pointed out also that vegetation is in fact a continuum, and that the defined formations are only conveniently recognized stages in a continuous gradient. Many communities must be expected to show characteristics intermediate between two formations or even, less commonly, between two formation-series. In a study of temperate mountains WHITTAKER & NIERING (1965) showed the continuity of formations by plotting growth-form coverages in relation to elevation and topography. A formation-series with decreasing elevation, from subalpine forest (taiga) through other forest and woodland types to grassland and desert, was suggested (WHITTAKER & NIERING 1968). WHITTAKER (1970) has also extended BEARD's approach by relating more broadly defined formations to four major climatic gradients for the world. Three formation-series, here termed ecoclines again radiate (upward in mountains, northward, and toward drought) from the tropical rainforest; the fourth is from deciduous forest to desert in

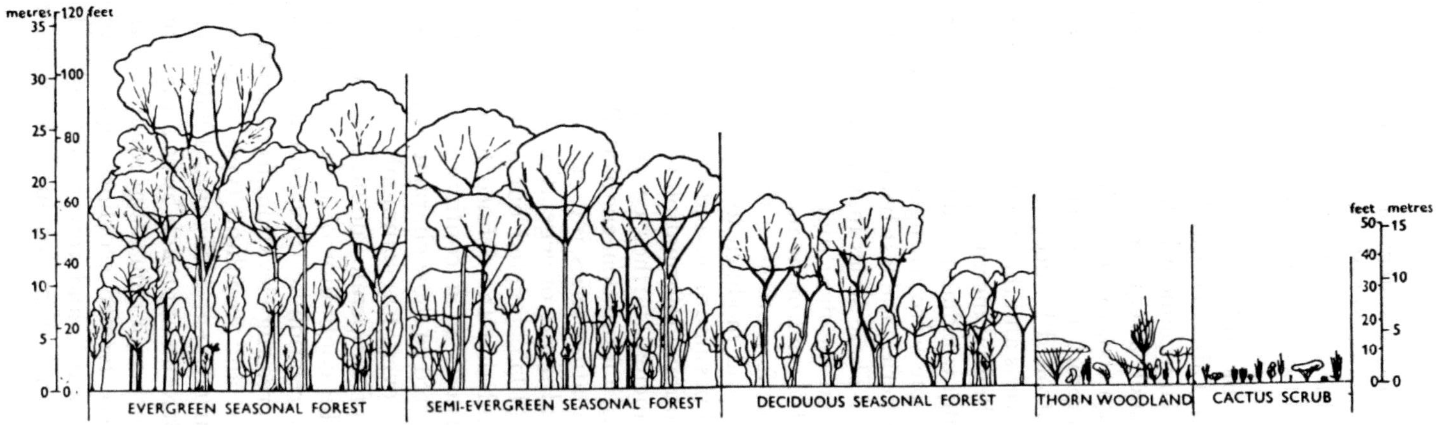

Fig. 5. A formation-series for communities on Trinidad affected by increasing drought from left to right (BEARD 1944). Community structure is represented by profile diagrams in the manner of DAVIS & RICHARDS (Fig. 6). This, the seasonal formation-series, diverges from the tropical rainforest of Fig. 6 along a climatic gradient of increasingly restricted and seasonal precipitation at low elevations in tropical America.

the North Temperate Zone. BEARD's (1944, 1955) most important advance may have been in showing that such higher-level ordination of vegetation is possible, relating formations as clearly characterized physiognomic types to five or more major axes of environmental variation. The system thus provides means of both relating physiognomy to environment, and comparing this relationship for different areas and continents.

13.4 **Physiognomy in British Commonwealth Ecology**

13.4.1 FOREST TYPES AND ASSOCIATIONS

Physiognomic approaches to local vegetation have been most actively employed in British ecology. In the first half of the present century administrators of British Colonial territories, which lay mainly in the tropics, were busy developing resources. Forest services were set up, forest officers proceeded to investigate the forest estates in their charge and to carry out inventories. Taxonomic investigations were made of species present, and ecological investigations were made of forest communities. The word 'forest-type' came into general use, and with the complexity of forest composition in the tropics it was natural that forest types should be based on physiognomy rather than species composition.

In 1924 a British Empire Vegetation Committee was appointed by an Imperial Botanical Conference in London with the object of promoting the study of the plant cover in British countries. While in the end only one of the ambitiously planned regional monographs ever came to fruition (ADAMSON 1938) there was a speedy production of a handbook for ecological workers, '*Aims and Methods in the Study of Vegetation*,' by TANSLEY & CHIPP (1926). The utility of this work, in the event, was diminished by its strongly Clementsian outlook. It represented in fact a Clementsian deviation in British ecology instead of a continuation and expansion of the physiognomic approach which would have been of most value to the forestry workers who were the people most concerned. Classification, to TANSLEY and CHIPP, was centered upon the association as 'the largest unit which consists of a definite assemblage of species (usually with definite dominants) and a definite habitat.' Smaller components of the association could be recognised, under the Clementsian terminology (faciation, location, society and clan, see 14.2). There was, however, no higher classification. The association was considered to possess a characteristic growth-form and structure as well as floristic composition, but the word 'formation' was not mentioned. TANSLEY and CHIPP differed from CLEMENTS only in insisting (following SCHIMPER) upon a distinction between climatic and edaphic climax communities.

The principal weaknesses of this approach were that it proposed to define the association partly by habitat, which in practice was very difficult to assess accurately, and that it lacked physiognomic standards by which different associations could be compared. Field workers dealing with forest types were left to propose their own terminology and to speculate on similarities and differences between their own and other workers' communities. CHAMPION'S (1936) classic work on the forest types of India and Burma is an example of the individualistic approach which became general. TANSLEY and CHIPP did not found a school of ecological thought.

13.4.2 PROFILE DIAGRAMS

The necessary steps to rationalise classification on a physiognomic basis were not taken until DAVIS & RICHARDS (1933-4) adopted the device of the *profile diagram*, a drawing of an accurately measured sample strip of forest (Fig. 6). A sample was to be selected by the field worker as typical of the formation which it is desired to represent and of mature structure, and the sample was to be measured by a definite procedure (quoted from RICHARDS 1952): —'A narrow rectangular strip of forest is marked out with cords, the right angles being obtained with the help of a prismatic compass. In Rain Forest the length of the strip should not usually be less than 200 ft. (61 m,); 25 ft. (7.6 m.) has proved a satisfactory width. All small undergrowth and trees less than an arbitrarily chosen lower limit of height are cleared away. The positions of the remaining trees are then mapped and their diameters noted. The total height, height to first (large) branch, lower limit of crown and width of crown of each tree are then measured. Often it is only possible to obtain these measurements by felling all the trees on the strip, and the trees must be felled in a carefully selected order, so that the heavier trees do not crush the smaller in falling, but it is sometimes possible to make measurements of sufficient accuracy from the ground by means of an Abney level. Felling has the advantage that herbarium material can be collected for the identification of species.'

The field measurements are plotted to produce an accurately drawn diagram which will show the structure of the community in a manner which is impossible by photography in dense forest. For this reason the profile diagram became popular in connexion with

Fig. 6. A profile diagram for primary mixed tropical rainforest, Moriballi Creek, British Guiana, the first application of the profile technique as developed by DAVIS & RICHARDS (1933-4). Only trees over 15 ft (4.6 m) are shown.

studies in tropical forests. After the initial publications by DAVIS & RICHARDS (1933-4) of the profile diagram in relation to studies of rain forest in British Guiana, the same methods were used in Sawarak (RICHARDS 1936) and Nigeria (1939), affording for the first time material for effective quantitative and qualitative comparisons of structure between examples from different parts of the world of tropical forests thought to belong to the same formation-type. A close correspondence was found between the examples of British Guiana and Sarawak, but the Nigerian forest showed differences. The practical utility of the profile diagram for comparative studies was amply demonstrated.

13.4.3 FORMATIONS AND FORMATION-TYPES

A paper by BURTT-DAVY (1938), who had been a member of the original British Empire Vegetation Committee, was read to the British Ecological Society drawing attention to the lack of progress and the general confusion in classification and terminology. After a review of the works of SCHIMPER (1903), WARMING (1909), CHIPP (1926) and CHAMPION (1936), he outlined a set of criteria for comprehensive description of vegetation and proposed a hierarchic physiognomic classification for tropical vegetation with appropriate standardised terminology. The British Ecological Society appointed a committee of three to examine the matter, and their report appeared as RICHARDS, TANSLEY & WATT (1939). These authors decided to 'call attention to certain basic principles' which are quoted as follows:

'There is an urgent demand for the stabilisation of the nomenclature of tropical forest vegetation, in order to avoid the confusion and misunderstanding that result from different (forest) officers calling the same thing by different names and possibly different things by the same name.... The authors hold that such a scheme must be inductive, based on the structure, life-form and flora of the communities themselves, not on mixed criteria which include necessarily vague characterisations of habitat....

'An actual sample of forest to be studied in the field is best regarded in the first place as a particular example (sometimes called a 'stand') of an *association* (dominated by two, several, or more species) or of a *consociation* (dominated by a single species), and characterised by the total *flora*, i.e., the whole collection of species present, though some (sometimes called *characteristic*) have a greater diagnostic or 'indicator' value than others.

'Certain associations and consociations are found to agree in the life-forms (i.e. growth-forms) of their component species, especially of their dominants, though the species may be totally different. This agreement in life-form indicates similarity of *essential habitat,* especially similarity of regional climate, and the associations showing such agreement may be put together in a larger unit – the formation.

'Formations of very well-marked life-form type are so strikingly and characteristically similar and are so obviously determined by similar climates, in whatever part of the world they may occur, that they have long been recognised as together forming distinct entities, which we may call *formation types.*'

RICHARDS, TANSLEY and WATT went on to recommend practical methods of recording structure and growth-form in order to typify formations. The RICHARDS profile diagram was brought in as an integral feature for the expression of structure and at the same time the growth-forms of component species were to be recorded in a quantitative manner. Percentages of the stand, or numbers of individuals per unit area, would be used to express occurrence of special characters such as plank buttresses, stilt roots, pneumatophores, thorny trunks or branches, cauliflory, succulence, peculiarities of bark; or of any special growth-forms present such as palms or cycads, bamboos, rattans and pandanus. Leaf characters are most important, and percentages of species and individuals in each stratum would express occurrence of leaves that are evergreen or deciduous, simple or compound, or of certain size-classes as proposed by RAUNKIAER (1934).

It has been shown that the Clementsian approach to vegetation led to certain difficulties in British practice. CLEMENTS (1916) drew attention to the phenomenon of plant succession which leads to the development of a mature and integrated community. Physiognomic studies, to be of comparative value, must in most cases deal with mature vegetation. A community subject to fire may have a structure directly related to the period since the last burning, and an immature example may be valueless except as a means of showing differences from a known mature example. It is possible that the failure of RICHARD's (1939) example of Nigerian rain forest to agree with the examples from British Guiana and Sarawak was due to disturbance of the Nigerian forest. In general, physiognomic treatments are scarcely relevant to questions of succession, but the British led by TANSLEY (1920) found it necessary to take a stand upon recognition of more than one kind of climax. To CLEMENTS there was but one climax in an area, determined by its climate, all other communities being regarded as in a state of succession toward that climax. TANSLEY objected that young, rapidly evolving communities which were demonstrably in a state of succession should be distinguished from others (e.g. fen, swamp, or saltmarsh) effectively controlled by local soil or water conditions. The latter would be considered by CLEMENTS as theoretically in succession, but in practice they are stable: they are self-maintaining or climax in their own particular habitats. The soils of such communities could not possibly evolve into soils which would support the climatic climax, and thus they should be regarded as edaphic climaxes. Later BEARD (1953) went further in suggesting that some landforms and soils could become senile and over-mature, developing adverse features which led to a devolution from the climatic climax. This was directly opposed to CLEMENT's dictum that the climax can be destroyed but cannot retrogress. Whatever the difficulties which surround the theory of the climax, and they are many (WHITTAKER 1953, and this *Handbook*, Part VIII), a nuclear idea remains important that there is a regionally characteristic vegetation, and it is this which forms the basis for the biogeographic treatment of formations in relation to climate.

13.4.4 FORMATION-SERIES

The methods of RICHARDS, TANSLEY and WATT for physiognomic typification were further developed by BEARD (1944) for comparison and classification of plant formations found in tropical America, only a few of which could be considered rain forest in a narrow sense, thus extending RICHARDS' work on the rain-forest to tropical vegetation in general. As described, BEARD's work led to recognition of a larger number of formations than in some other systems and to the concept of the rain forest as a kind of central or highest type of vegetation possible in the tropical environments, from which five 'formation-series' radiate in response to environmental gradients and interrelate the other formations to one another and the rain forest. The members of each series were united in sharing an essential habitat feature in some degree, either more or less favourable, which gives the formations some impress of common physiognomic characteristics. Thus deciduousness of leaves is a characteristic of the seasonal series, whereas a progressive simplification of structure can be traced through all the members of any series.

The basic principles of his work were expressed by BEARD (1955) in this way: 'The properties of the community express the habitat: the consistent dominance which defines the association expresses a constant local habitat within the regional flora and the consistent physiognomy which defines the formation expresses a constant essential habitat within the tropical zone. Types of habitat which are in general similar permit us further to group formations together into formation-series. There is thus a grouping at three levels: a floristic grouping — the association; a physiognomic grouping — the formation; and a habitat-grouping — the formation-series.'

To conclude this historical review, it has been shown that RICHARDS, TANSLEY and WATT, followed by BEARD, had aimed at the elimination of confusion in classification and terminology in the English-speaking world of tropical ecologists by introducing quantitative methods to define their physiognomic units. Confusion however still reigned in the international sphere, and an attempt to cope with this was produced in 1956 by the Scientific Council for Africa South of the Sahara in a curiously anonymous note appearing in the Publications of the Council under the title '*Phytogeography*.' This publication was the outcome of a conference designed to obtain some unanimity in classification and terminology of African vegetation, as the work done up to that time in different colonial territories by ecologists of different nationalities had produced a kaleidoscopic confusion of approach, method, concepts and nomenclature. The result was a framework similar to BEARD's for tropical America, which could be expected to embrace any plant community of tropical Africa. Twenty-three formations were typified, which is close to BEARD's number of twenty-eight, but there is a considerable difference in emphasis for climatic reasons. Most of tropical America receives rainfall in excess of 1000 mm per annum, whereas most of tropical Africa receives less than this. Thus most of BEARD's formations were moist forests and woodlands and most of the African formations are dry woodlands, savanna and steppe.

In the African publication, each formation is typified with a concise description and a standard name in both English and French, and a standard example taken from previously published work. Profile diagrams are also provided; though these appear to be diagrammatic rather than taken from actual measured samples, they are graphic and effective. The formations are organised into series similar to those of BEARD but the treatment stops short of a fully logical system because of a prior division into two main classes — Closed Forest Formations, and Mixed Forest-Grassland and Grassland Formations — an unnecessary reversion to the days of SCHIMPER and WARMING. Fully logical lowland-climate, montane-climate and edaphic series would each include a full range of types from forest to desert.

This African classification and terminology was followed by KEAY (1959) in the publication of the A.E.T.F.A.T. vegetation map of Africa south of the tropic of Cancer. In a recent paper BEARD

(1967) has discussed its application to tropical Australia. Physiognomic study and classification along the above lines have not yet been much applied outside the tropics. Notable exceptions have been the work of WEBB (1959) in Queensland and ROBBINS (1962) in New Zealand, both working in rain forest. ROBBINS made particular use of profile diagrams whereas WEBB laid the emphasis on growth-forms.

13.5 Floristic and Physiognomic Units

In the past, approaches to vegetation have tended to be more or less sharply divided between the floristic, which may ignore the structure and growth-form of vegetation, and the physiognomic, which need not take account of floristic composition. The former is undoubtedly the more suited to detailed studies of local areas and the latter to broadly-based studies on an international scale. It would seem that a desirable goal for the future should be the marrying of the two approaches into a single system of classification wherein floristic units are combined by their structure into physiognomic units.

Most students of physiognomy have been willing to recognise that their formations are divisible into floristically distinct units. All the British authors so far quoted accepted without question the existence of recognisable community-types characterised by consistent dominants and general composition, which they termed *associations*. At the same time it is true that the very existence of sociality (in the sense of species composition that is consistent between different stands) in tropical forest has been denied by other workers, mainly non-British, with experience in Africa and Malaysia. Thus VAN STEENIS (1958) describes the procedure for belt transects by forest surveyors in Indonesia from which 'it appears, however, that the plots are bewilderingly divergent and ... the picture does not suggest a constant specific composition.' On the other hand BEARD (1949) found from such enumerations in the British West Indian islands that consistent dominance and sociality were regular features of the vegetation, so much so that examples of the same forest type in different islands showed strikingly similar composition. It is evident therefore that sociality in tropical forest is different in different parts of the tropics, and that in areas with very rich floras the pattern tends to be obscured by sheer numbers of species. It has taken the recent work of ASHTON (1964) in North Borneo to demonstrate that modern statistical methods applied to this problem are capable of showing that sociality in a certain sense none the less exists.

The term *association* has been used throughout by British workers for their floristic units, but WHITTAKER (1962) has criticised this on the ground that the usage is not the same as that of the BRAUN-BLANQUET school which claims an exclusive right to the term. WHITTAKER (1962, article 14) suggested that the 'associations' of RICHARDS, TANSLEY and WATT, and BEARD, should be termed *dominance-types*. It is certainly true that the characterisation of the basic floristic unit of vegetation in a certain manner by diagnostic species has become identified with followers of BRAUN-BLANQUET. One may take the view, however, that the word 'association' means simply the basic floristic unit, and there is no reason to discard it if it is being defined in a different way. RICHARDS, TANSLEY and WATT in their full definition of the association quoted above make it clear that the association is known by its dominants but is characterised by its total flora, in which 'some species (sometimes called *characteristic*) have a greater diagnostic or 'indicator' value than others.' In selecting dominants to typify the association we are taking the easiest and most direct way of finding diagnostic species. It seems to the present writer that there is no real conflict. Can we not continue to accept WARMING's definition: 'An association is a community of definite floristic composition within a formation.'?

Different considerations apply to higher-level units. Floristic composition, the basis of the phytosociological approach, is only one of several important attributes of vegetation which merit attention and can be employed as a useful basis of classification. At the present time phytosociologists of the BRAUN-BLANQUET school use floristic groupings which become of less and less practical value toward the higher levels of the hierarchy. The association is the basic unit, alliances have been found useful, but orders and classes are another matter. Higher floristic groupings are applicable only to the geographical area of a particular flora, and even in the widest terms could do no more than extend across a single hemisphere. What valid comparison in floristic terms is there between vegetation units in the northern and southern hemispheres, or even between different parts of the equatorial zone? It does not seem that one can formulate meaningful world systems of vegetation classification except on a physiognomic basis.

13.6 Conclusion

From this review I would draw these conclusions: (i) The earliest approach to vegetation—through community structure or physiognomy—is still a vital and valuable one. It remains the principal basis of treating vegetation on a broad scale in relation to climate, and an important basis of vegetation research in more limited areas (particularly in the early stages of such work). (ii) Classification by physiognomy is subject to the difficulties (subjectivity in choice of community-types and their boundaries, and the conflict between reasonable classification that is incomplete and complete classification that is unmanageable) that afflict all classifications of natural communities. Nevertheless, physiognomy has proved to be the best means of gaining agreement on classification, and of comparing communities, among investigators on different continents. (iii) Physiognomy is adaptable to such different purposes as world-wide generalization about formation-series, and description of community structure without formal classification. The physiognomic approach most effectively relates salient structural characteristics of communities, which must be underlain by functional characteristics, to the qualities of environment to which these characteristics are adapted. (iv) There is advantage, for expressing these relationships, in both typification (in terms of formations and formation-types) and ordination (formation-series). Thus the major perspectives of classification and gradient analysis may well be combined not only in intensive local research, but also in study of the broader relationships of vegetation structure to environment.

13.7.1 Summary

The physiognomic approach to vegetation is based on the study of community structure, as determined by the occurrence of major types of plants (growth-forms). Early sources of the approach led to the central concepts of formation (a major community-type as defined by physiognomy, with consideration also of broad environmental relations, on a given continent), and formation-type (a grouping of formations of convergent structure in response to similar climates on more than one continent). Four major physiognomic approaches are briefly reviewed: (i) Systems of formation-types for the land vegetation of the world, (ii) Treatments seeking joint classification of climate and community physiognomy expressing climate, (iii) Systems for the physiognomic description and characterization of vegetation without formal classification, and (iv) Relation of formations to one another in formation-series along environmental gradients. Emphasis of physiognomy in the study of local vegetation has been productive in British Commonwealth ecology, and has led to effective systems of formations and formation-series for tropical areas particularly. Formations can be reasonably subdivided into floristic units (dominance-types or associations). Floristic treatment of vegetation on a local level can thus be combined with physiognomic treatment, which is almost indispensible for broader biogeographic interpretation of plant communities.

13.7.2 Zusammenfassung

Die physiognomische Betrachtung der Vegetation beruht auf der Untersuchung der Gesellschaftsstruktur, die durch das Auftreten der Hauptformen von Pflanzen (Wuchsformen) bedingt wird. Die Ursprünge dieser Betrachtungsweise führten zu den zentralen Begriffen der Formation (ein Haupt-Gesellschafts-Typus auf einem bestimmten Kontinent, bestimmt durch seine Physiognomie, unter Berücksichtigung der allgemeinen Umweltverhältnisse) und des Formations-Typus (einer Gruppe von Formationen mit ähnlicher (konvergenter) Struktur in Anpassung an ähnliche Klimate auf verschiedenen Kontinenten). Vier Haupt-Betrachtungsweisen auf physiognomischer Grundlage werden kurz besprochen: 1. Systeme von Formationstypen der terrestrichsen Vegetation der Welt, 2. Untersuchungen zur gemeinsamen Klassifikation des Klimas und der dieses ausdrückenden Gesellschaftsphysiognomie, 3. Systeme für die physiognomische Beschreibung und Kennzeichnung der Vegetation ohne formale Klassifikation, 4. Beziehungen zwischen Formationen in Formationsreihen entlang Standorts-Gradienten.

Die Betonung der Physiognomie beim Studium der örtlichen Vegetation ist fruchtbar gewesen in den Commonwealth Ländern und hat zu brauchbaren Systemen von Formationen bzw. Formations-Reihen, besonders in tropischen Gebieten, geführt.

Formationen können zweckmäßig in floristischen Einheiten (Dominanz-Typen oder Assoziationen) unterteilt werden. Die lokale floristische Betrachtung der Vegetation kann so mit einer physiognomischen Betrachtung kombiniert werden, was für eine breitere biogeographische Deutung der Pflanzengesellschaften fast unerlaßlich erscheint.

REFERENCES

Adamson, R. S., – 1938 – The Vegetation of South Africa. Brit. Emp. Veget. Comm., London. 235 pp.

Ashton, P., – 1964 – Ecological studies in the mixed dipterocarp forests of Brunei State. *Oxf. For. Mem.* 25: 1—75.

Beard, J. S., – 1944 – Climax vegetation in tropical America. *Ecology* 25: 127—158.

Beard, J. S., – 1949 – The natural vegetation of the Windward and Leeward Islands. *Oxf. For. Mem.* 21: 1—192.

BEARD, J. S., – 1953 – The savanna vegetation of northern tropical America. *Ecol. Monogr.* 23: 149—215.

BEARD, J. S., – 1955 – The classification of tropical American vegetation-types. *Ecology* 36: 89—100.

BEARD, J. S., – 1967 – Some vegetation types of tropical Australia in relation to those of Africa and America. *J. Ecol.* 55: 271—290.

BEARD, J. S., – 1969 – The vegetation of the Boorabbin and Lake Johnston areas, Western Australia. *Proc. Linn. Soc. N. S. W.* 93: 239—269.

BEARD, J. S. & M. J. WEBB, – 1972 – The vegetation survey of Western Australia, its aims, objects and methods. Part I of Explan. Memoir to vegetation map 'Great Sandy Desert,' 1: 1,000,000 Series. Univ. West. Aust. Press, Perth. (in press.).

BRAUN-BLANQUET, J., – 1951 – Pflanzensoziologie: Grundzüge der Vegetationskunde. 2nd ed. Springer, Wien. 631 pp. 3rd ed. 1964, 865 pp.

BURTT DAVY, J., – 1938 – The classification of tropical woody vegetation-types. *Inst. Pap. Commonw. (Imp.) For. Inst.*, Oxford, 13: 1—85.

CAIN, S. A., – 1950 – Life-forms and phytoclimate. *Bot. Rev.* 16: 1—32.

CAIN, S. A. & G. M. DE OLIVEIRA CASTRO, – 1959 – Manual of Vegetation Analysis. Harper, New York. 325 pp.

CANDOLLE, A. DE, – 1855 – Géographie botanique raisonée. Paris.

CHANPION, H. G., – 1936 – A preliminary survey of the forest types of India and Burma. *Indian Forest Rec.*, New Ser. Silvic. 1 (1): 1—286.

CHIPP, T. F., – 1926 – The Gold Coast forest: A study in synecology. *Oxf. For. Mem.* 7: 1—94.

CLEMENTS, F. E., – 1916 – Plant succession: an analysis of the development of vegetation. *Publs. Carnegie Instn*, Washington 242: 1—512.

DANSEREAU, P., – 1951 – Description and recording of vegetation upon a structural basis. *Ecology* 32: 172—229.

DANSEREAU, P., – 1957 – Biogeography: An Ecological Perspective. Ronald, New York. 394 pp.

DAVIS, T. A. W. & P. W. RICHARDS, – 1933-4 – The vegetation of Moraballi Creek, British Guiana; an ecological study of a limited area of Tropical Rain Forest. *J. Ecol.* 21: 350—384, 22: 106—155.

DRUDE, O., – 1890 – Handbuch der Pflanzengeographie. Englehorn, Stuttgart. 582 pp.

DU RIETZ, G. E., – 1931 – Life forms of terrestrial flowering plants. *Acta phytogeogr. suec.* 3: 1—95.

ELLENBERG, H., – 1956 – Aufgaben und Methoden der Vegetationskunde. In 'Einführung in die Phytologie,' by H. WALTER Vol. 4, pt. 1. Ulmer, Stuttgart. 136 pp.

ELLENBERG, H. & D. MUELLER-DOMBOIS, – 1967 – Tentative physiognomic-ecological classification of plant formations of the earth. *Ber. geobot. Inst. ETH, Stiftg. Rübel*, Zürich 1965-6, 37: 21—55.

FANSHAWE, D. B., – 1952 – The vegetation of British Guiana: a preliminary review. *Inst. Pap. Commonw. (Imp.) For. Inst.*, Oxford, 29: 3—96.

FOSBERG, F. R., – 1961 – A classification of vegetation for general purposes. *Trop. Ecol.* 2: 1—28. Also in 'Guide to the check sheet for ecological IBP areas,' by G. F. PETERKIN. IBP *Handbook* 4: 73—120 Blackwell, Oxford & Edinburgh.

GRISEBACH, A., – 1838 – Ueber den Einfluss des Climas auf die Begränzung der natürlichen Floren. *Linnaea* 12: 159—200.

GRISEBACH, A., – 1872 – Die Vegetation der Erde nach ihrer klimatischen Anordnung. Englemann, Leipzig. 2 vols., 603 & 709 pp.

HOLDRIDGE, L. R., – 1947 – Determination of world plant formations from simple climatic data. *Science, N. Y.* 105: 367—368.
HOLDRIDGE, L. R., – 1967 – Life Zone Ecology. Rev. ed. Tropical Science Center, San José, Costa Rica. 206 pp.
KEAY, R. W., – 1959 – Vegetation Map of Africa South of the Tropic of Cancer. (Engl. & French) Oxford Univ. Press. 24 pp.
KÖPPEN, W., – 1900 – Versuch einer Klassifikation der Klimate, vorzugsweise nach ihren Beziehungen zur Pflanzenwelt. *Geogr. Z.* 6: 593—611.
KÖPPEN, W., – 1923 – Die Klimate der Erde: Grundriss der Klimakunde. de Gruyter, Berlin & Leipzig. 369 pp.
KÜCHLER, A. W., – 1947 – A geographic system of vegetation. *Geogrl. Rev., N. Y.* 37: 233—240.
KÜCHLER, A. W., – 1948 – A new vegetation map of Manchuria. *Ecology* 29: 513—516.
KUCHLER, A. W., – 1949 – A physiognomic classification of vegetation. *Ann. Ass. Am. Geogr.* 39: 201—210.
KÜCHLER, A. W., – 1950 – Die physiognomische Kartierung der Vegetation. *Petermanns geogr. Mitt.* 94: 1—6.
KÜCHLER, A. W., – 1967 – Vegetation Mapping. Ronald, New York. 472 pp.
RAUNKIAER, C., – 1934 – The Life Forms of Plants and Statistical Plant Geography. Clarendon, Oxford. 632 pp.
RICHARDS, P. W., – 1936 – Ecological observations on the rain forest of Mount Dulit, Sarawak. *J. Ecol.* 24: 1—37, 340—360.
RICHARDS, P. W., – 1939 – Ecological studies on the rain forest of southern Nigeria. I. The structure and floristic composition of the primary forest. *J. Ecol.* 27: 1—61.
RICHARDS, P. W., – 1952 – The Tropical Rain Forest: An Ecological Study. Cambridge Univ. Press. 450 pp.
RICHARDS, P. W., A. G. TANSLEY & A. S. WATT, 1939 – The recording of structure, lifeform and flora of tropical forest communities as a basis for their classification. *Inst. Pap. Commonw. (Imp.) For. Inst.*, Oxford. 19: 1—19; also *J. Ecol.* 28: 224—239, 1940.
ROBBINS, R. G., – 1962 – The podocarp-broadleaf forests of New Zealand. *Trans. R. Soc. N. Z. Bot.* 1: 33—75.
ROSS COCHRANE, G., – 1963 – A physiognomic vegetation map of Australia. *J. Ecol.* 51: 639—656.
RÜBEL, E., – 1930 – Pflanzengesellschaften der Erde. Huber, Bern & Berlin. 464 pp.
RÜBEL, E., – 1936 – Plant communities of the world. In 'Essays in Geobotany in Honor of William Albert Setchel,' ed. T. H. GOODSPEED, pp. 263—290. Univ. Calif. Press, Berkeley.
RUMNEY, G. R., – 1968 – Climatology and the World's Climates. Macmillan, New York & London. 656 pp.
SCHIMPER, A. F. W., – 1898 – Pflanzen-Geographie auf physiologischer Grundlage. Fisher, Jena. 876 pp.
SCHIMPER, A. F. W., – 1903 – Plant-geography upon a Physiological Basis. Clarendon, Oxford. 839 pp.
SCHIMPER, A. F. W. & F. C. VON FABER, – 1935 – Pflanzengeographie auf physiologischer Grundlage. 3rd ed. Fisher, Jena. 2 vols., 588 & 1612 pp.
SCHMITHÜSEN, J., – 1968 – Allgemeine Vegetationsgeographie. 3rd ed. de Gruyter, Berlin. 463 pp.
Scientific Council for Africa South of the Sahara, – 1956 – Phytogeography. (French & Engl.) *Publ. Cons. scient. Afr. S. Sahara* 53: 1—33.
STEENIS, C. G. G. J. VAN, – 1958 – Basic principles of rain forest sociology. (French

summm.) In 'Study of Tropical Vegetation', pp. 159—165. *Proc. Kandy Symp.* 1956, Unesco, Paris.

STEFANOFF, B., – 1930 – Versuch zur Darstellung einer parallel Klassifikation der Klimate und der Vegetationstypen. (Bulgarian & German) *Sb. bŭlg. Akad. Nauk.* 26 (3): 1—122. (Engl. summm. by W. B. Turrill, *J. Ecol.* 20: 211—213, 1932).

STOCKER, O., – 1963 – Das dreidimensionale Schema der Vegetationsverteilung auf der Erde. *Ber. dt. bot. Ges.* 76: 168—178.

TANSLEY, A. G., – 1920 – The classification of vegetation and the concept of development. *J. Ecol.* 8: 118—149.

TANSLEY, A. G. & T. F. CHIPP, eds., – 1926 – Aims and Methods in the Study of Vegetation. Brit. Emp. Veget. Comm & Crown Agents for Colonies, London. 383 pp.

THORNTHWAITE, C. W., – 1931 – The climates of North America according to a new classification. *Geogrl Rev., N. Y.* 21: 633—655.

THORNTHWAITE, C. W., – 1948 – An approach toward a rational classification of climate. *Geogrl Rev., N. Y.* 38: 55—94.

TROLL, C., – 1948 – Der asymmetrische Aufbau der Vegetationszonen und Vegetationsstufen auf der Nord- und Südhalbkugel. *Ber. geobot. Inst. Rübel*, Zürich, 1947: 46—83.

TROLL, C., – 1961 – Klima und Pflanzenkleid der Erde in dreidimensionaler Sicht. *Naturwissenschaften* 48: 332—348.

TROLL, C. – 1968 – The Cordilleras of the Tropical Americas. Aspects of climatic, phytogeographical and agrarian ecology. (Span. summm.) *Colloquium geogr.*, Bonn 9: 13—56.

WALTER, H., – 1962, 1968 – Die Vegetation der Erde in ökophysiologischer Betrachtung. Fisher, Jena. 2 vols., 538 & 1001 pp.

WARMING, E., – 1909 – Oecology of Plants: An Introduction to the Study of Plant-Communities. Oxford Univ. Press. 422 pp.

WEBB, L. J., – 1959 – A physiognomic classification of Australian rain forests. *J. Ecol.* 47: 551—570.

WHITTAKER, R. H., – 1953 – A consideration of climax theory: the climax as a population and pattern. *Ecol. Monogr.* 23: 41—78.

WHITTAKER, R. H., – 1962 – Classification of natural communities. *Bot. Rev.* 28: 1—239.

WHITTAKER, R. H., – 1970 – Communities and Ecosystems. Macmillan, New York. 162 pp.

WHITTAKER, R. H. & W. A. NIERING, – 1965 – Vegetation of the Santa Catalina Mountains, Arizona. (II.) A gradient analysis of the south slope. *Ecology* 46: 429—452.

WHITTAKER, R. H. & W. A. NIERING, – 1968 – Vegetation of the Santa Catalina Mountains, Arizona. IV. Limestone and acid soils. *J. Ecol.* 56: 523—544.

14 DOMINANCE-TYPES

Robert H. Whittaker

Contents

14 DOMINANCE-TYPES

14.1 Introduction

As BEARD observes in the preceding article, formations are often divided into subordinate community-types characterized by their dominant species. Many authors have termed these units 'associations,' and BEARD considers this term suitable for them. Community-types recognized by dominant species differ from the associations of the school of BRAUN-BLANQUET, however, in manner of definition and in scale or inclusiveness of the resulting units; moreover the approach through dominance represents a different perspective on what deserves emphasis in the study of vegetation from the floristic system of BRAUN-BLANQUET (1951, article 20 in this volume). The author (WHITTAKER 1962) has suggested that clarity is served by assigning the formal term 'association' to units of the school of BRAUN-BLANQUET, and designating units defined by dominant species as dominance-types or, simply, types.

Such types are among the most widely used of vegetation units. Dominance-types have been used by some authors in all the regional traditions, but they are particularly characteristic of American and British ecology. They have an important place (under the term 'formation') in Russian classifications (article 17). They have been the preferred units of many practical ecologists, for forestry, range and wildlife ecology, and land management particularly. Dominance-types will be discussed both as part of the formal system of CLEMENTS and as more informally used by plant and animal ecologists not of that school.

14.2 The School of Clements

CLEMENTS (1905, 1916, 1928, 1936) sought to fashion a system for treating the vegetation of a continent then rich in plant communities little disturbed by man. The concept of succession of COWLES (1899, 1901) and others was linked with the formation as a unit and with the analogy of the community as an organism. The formation was seen as a complex organism which arises, grows, matures, and dies; the climax formation is the adult organism for which successional communities are developmental stages (CLEMENTS 1928: 125-6).

The formation was also a regional vegetation unit, covering a definite geographic area for which it was the climax, the climate of which it expressed in the growth-form of its dominant species. CLEMENTS' formations corresponded in general to those of other plant geographers; but the idea of succession was used to elaborate a distinctive system of classification.

Formations were divided into associations usually defined by dominant species; associations differed from one another in response to climatic differences within the region of the formation. Associations were divided into consociations characterized by a single dominant species, and fasciations by combinations of dominant species (WEAVER & CLEMENTS 1938). Societies were subordinate units characterized by subdominants, species of growth-form different from the dominants. These terms applied only to communities regarded as climax in CLEMENTS' (monoclimax) sense, and a set of parallel units (formies, associes, consocies, fascies, socies) were applied to successional communities. CLEMENTS' interpretation distinguished from the climatic climax various stabilized communities (proclimaxes) that were not considered true climaxes. Since there was only one climax community in a given region, the successional units were applied to proclimaxes as well as to clearly unstable communities. The lociation, lamiation, sociation, and other units appeared in the system later (CLEMENTS 1936), to produce perhaps the most complex terminology developed by any school. All units below the formation were recognized by dominant (or subdominant) species.

The Clementsian system was for some time the dominant approach to plant communities in the United States, though it was never accepted by all ecologists. It was also widely influential in British ecology (WHITTAKER 1962). As some of CLEMENTS' ideas spread others, notably the organismic analogy and the basing of classification on often hypothetical successional relations, were increasingly criticized and discarded (WHITTAKER 1957, 1962). The work of BRAUN (1938, 1947, 1950, 1956) on the eastern deciduous forests of the United States represents a development from CLEMENTS' approach, without many of the features for which it has been criticized. BRAUN's approach was less formal and deductive than CLEMENTS'. BRAUN used the 'association' concept of CLEMENTS (though recognizing additional eastern forest associations), but placed much less emphasis on classification as such. Though the associations were treated as regional climaxes, these were seen in a perspective quite different from CLEMENTS'—a perspective accepting the existence of a variety of stable communities in a given area and emphasizing the complexity of species distribution and vegetation pattern

(BRAUN 1956). The eastern forests were furthermore seen in a long-term historic view of evolution from Tertiary time to the present. One highly mixed present community-type, the mixed mesophytic forest of the southern Appalachian Mountains, was regarded as nuclear to the eastern forests and most like the ancestral Arcto-Tertiary forests. Other major community-types might be interpreted as derived from the mixed mesophytic by simplification, as 'association-segregates' (BRAUN 1950, 1956, cf. RICHARDS 1952).

CLEMENTS' system has otherwise been of declining influence in the United States in the last two decades. Neither in American nor in British ecology has it been replaced, for most authors, by any other formal system. A majority of ecologists have used, for classification below the formation, dominance-types as units of convenience, part of no formal hierarchy but serving the purposes of research.

14.3 The Meaning of Dominance-Types

14.3.1 THE CONCEPT

A dominance-type is a class of communities defined by the dominance of one or more species . These species are usually the most important ones of the uppermost stratum of the community, but sometimes of a lower stratum of higher coverage. The concept of 'dominant' itself has no clear boundary. In a forest it may be true that the canopy trees are important above all other species for their effects, through microclimate and soil chemistry, on the environments to which other species must be adapted. In an open community such as a desert scrub such effects of major species on the environments of minor ones are less evident. We generally know little of the real importances of species in the sense of effects on one another's populations. 'Dominant' in most cases means only that the species is conspicuous in the community and is high in one of the importance values (coverage, biomass, density, frequency, etc.) by which we seek to express the contribution of species to community structure. In practice a dominance-type unites communities or samples in which some one species or combination of species has high importance values.

14.3.2 IMPLICATIONS OF SPECIES DISTRIBUTIONS

It is then important for understanding of these units to know how dominant species relate to one another. As indicated above (article

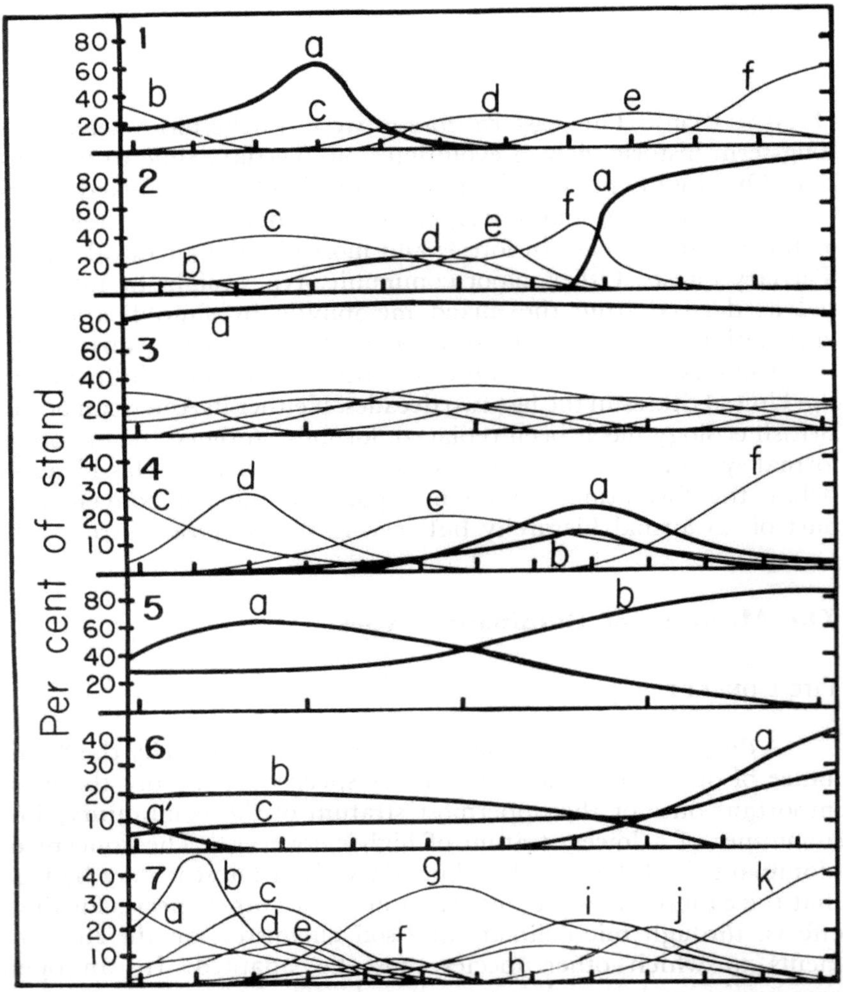

Fig. 1. Distributional relations of species characterizing dominance-types in the Great Smoky Mountains (WHITTAKER 1956, 1962). Population distributions of tree species are plotted by relative densities (per cents of stems 2 cm and over in stands).

1) Dominance of a single species (*Tsuga canadensis*) with a bell-shaped distribution, along the moisture gradient at 1070—1380 m elevation.

2) Dominance of a single species (*Fagus grandifolia*, 'gray' ecotype) with a 'plateau' distribution, along the elevation gradient in mesic sites.

3) Dominance of a single species (*Abies fraseri*) with a broad distribution along the moisture gradient in subalpine forests (1900—2000 m); subordinate species have narrower distributions.

4) Dominance of two species (*Quercus prinus*, *a* and *Castanea dentata*, *b*) with similar distributions in a moisture-gradient transect at 760—1070 m.

5) Dominance of two species (*Abies fraseri*, *a*, and *Picea rubens*, *b*) with dissimilar but broadly overlapping distributions, in a moisture-gradient transect at 1380—1670 m.

6) Dominance in various combinations of three species (*Quercus alba*, *aa'*, *Q. prinus*, *b*, and *Castanea dentata*, *c*) with dissimilar but broadly overlapping distributions, in an elevation transect in subxeric sites.

7) Dominance in various combinations of several species, in a moisture-gradient transect at 760—1070 m. Major species of the cove forest or mixed mesophytic grouping: *Batula alleghaniensis a*, *Tsuga canadensis b*, *Halesia monticola c*, *Acer saccharum d*, *Tilia heterophylla e*, and *Liriodendron tulipifera f*.

3), gradient analysis has provided information on species distributions along environmental complex-gradients. Fig. 1 illustrates for 7 community-types the manners in which their dominant species are distributed along environmental gradients:

1) In a pattern of continuously gradating vegetation, one species (*a*) has a typical, bell-shaped distributional curve and is clearly dominant in a part of the environmental gradient. Species *a* defines a dominance-type, but the limits of this type (whether at 50, 40, 30 per cent of canopy stems, etc.) are necessarily arbitrary.

2) A species (*a*) with a 'plateau' distribution is strongly dominant over a span of the gradient, and its range of dominance is more abruptly separated from that of other species (article 3). The species defines a dominance-type that is relatively discontinuous with other types along one environmental gradient (but may be continuous with other types along other gradients).

3) A dominant species (*a*) has a very wide amplitude along the gradient, but more narrowly distributed undergrowth species permit the recognition of site-types in the Finnish sense (article 15).

4) Two species (*a* and *b*) with similar distributions along the gradient characterize a type in much the same manner as one. The boundaries of the type are again arbitrary, and these two dominants with similar distributions along the moisture gradient are differently distributed in relation to elevation.

5) Two species (*a* and *b*) have dissimilar, but broadly overlapping distributions, dominating a wide range of communities separately or together. A classifying ecologist must choose whether to recognize: a single type dominated by either of the species or both; or two types, one dominated by *a* and one by *b*; or three types as dominated by species, *a*, species *b*, or the combination of *a* and *b*.

6) Three species (*aa'*, *b*, and *c*) have different distributions but overlap broadly. Either a single type including any combination of the dominants, or some number of types characterized by particular dominants or combinations, may be recognized. (The case illustrated is the Quercus-Castanea association of CLEMENTS, within which CONARD (1935) distinguished two associations on Long Island and WHITTAKER (1956) four major types in the Great Smoky Mountains.)

7) Several species (*a* to *f*) have overlapping distributions and often occur together, though in proportions varying from stand to stand. These species are a commodal group in the sense of WHITTAKER (1956); but the classifying ecologist must decide how many species to include in the group, and hence the inclusiveness of the unit the group defines. He may, or may not, choose to recognize as 'segregate' types in the sense of (BRAUN (1950) stands strongly dominated by particular species or combinations.

14.3.3 Limitations

Dominance-types are thus based on species of most varied distributional relationships. A dominant species may be of very wide range along environmental gradients (case 3); the type it defines then comprises stands that are diverse in habitat characteristics and in composition of species other than the dominant. Some other dominant species are narrowly distributed and define fairly homogeneous, localized types. Community-types defined by dominance consequently can be units of widely different scale and inclusiveness. Dominant species provide no secure basis for objective decisions about identity, scale, and number of units on which ecologists can agree. Decisions must be made, from the preferences of individuals, on what species are important enough to contribute to the definition of dominance-types, on what combinations of species sharing dominance are to be accepted as dominance-types, and on where boundaries, if any, are to be placed in their continuous intergradations.

A dominance-type thus is not, and cannot be, a particular, standardized kind of community-unit (Whittaker 1962). Dominance-types are poorly suited to the construction of a formal, hierarchial classification of vegetation. There is convenience in the use of dominance-types because the types can be recognized without intensive study of the full floras of a range of communities. Use of dominance-types often seems appropriate also because they are defined by quantitative relationships and by the species that may be of most interest to ecologists. In comparison with the units of the school of Braun-Blanquet these are not unmixed advantages, for dominance-types are often based on a more limited range of information than are floristic units. There is some justice in the view of members of that school that classification based on dominant species alone is a cruder and more superficial approach to community units than theirs. For systematic classification and for detailed knowledge of community characteristics and relations to environment, the approach through dominance-types has significant disadvantages compared with the system of Braun-Blanquet (article 20).

The choice of dominance-types as the basis of classification should involve acceptance of these limitations. There are reasons many ecologists will continue to accept them and to classify by dominance and physiognomy. For some kinds of study formal classification is not a purpose. Dominance-types may serve well enough to describe some major features of a vegetation pattern, to designate certain communities whose characteristics are being compared, or to summarize some of the results of gradient analysis. Granting the usefulness of dominance-types, some comments on terminology and use of these units follow.

14.4 Application

14.4.1 Definition of Dominance-Types

The preceding implies that dominance-types must be defined by the ecologist's choices of what species and combinations are to characterize types. The lability of the choices provides the means of making the classification work. The options of what combinations or segregations (as in cases 5, 6, and 7) are to be accepted as types permits some adjustment of the number and inclusiveness of the types to suit the classifier's purposes, and to produce units that are reasonably consistent with one another in the ranges of environmental and community variation that each comprises. The process is one of typification, as referred to also by Beard in the preceding article. In the field of variation that is a vegetation pattern, a sufficient number of types are recognized to represent major, recognized differences in community composition (cf. Schmid 1952, Poore 1956, Whittaker 1962). The characteristics of these types may then be incorporated into the definitions of community-types as classes (article 12). These more formal definitions of classes may in turn be so modified and accommodated to one another as to form a workable classification. This classification should provide a reasonable number of units, reasonably comparable in their inclusiveness, and so defined that most or all samples taken from the vegetation pattern can be assigned to one of the units.

14.4.2 Grouping of Dominance-Types

There are a number of ways these types may be grouped into broader units. Among them:

1) By physiognomy into formations. Formations (or biomes) may be used as higher units, within which dominance-types are lower units (Warming 1909, see also preceding article). Formations, as community-types defined by dominant growth-forms and broad environmental relations on particular continents, are in turn grouped into formation-types, comprising similar formations of similar environments on more than one continent. Various difficulties affect the definitions of formations themselves. There are cases in which formations and dominance-types do not very naturally accommodate themselves to a hierarchial, 'genus and species' relationship (Whittaker 1962). In general, however, the flexibility inherent in the definitions of both will permit the creation of suitable units. It is in some cases appropriate to recognize physiognomically defined subformations as a level of classification between formation

and dominance-type. Thus CLEMENTS (WEAVER & CLEMENTS 1929, CLEMENTS & SHELFORD 1939) includes among the North American temperate grasslands tall-grass prairie, mid-grass, and short-grass plains, and California grassland now dominated by annuals. These are termed 'associations' by CLEMENTS; but their definitions are primarily physiognomic (even though CLEMENTS specified major dominants for them), and each includes a number of dominance-types.

2) By species relationships into collective types. Dominance-types may be variously grouped into larger units that are less inclusive than the formation. Dominance-types that share one or more of their major species, or share major species in various directions although none is common to all, or are floristically similar in their subordinate species, or have taxonomically related dominants, may be grouped together. For such units the term collective-type is suggested (WHITTAKER 1962). Comparable units are called associations by British authors (TANSLEY & CHIPP 1926, TANSLEY 1939, CROCKER & WOOD 1947, BEADLE 1948, BEARD 1955), federations by DU RIETZ (1932), alliances by BEADLE & COSTIN (1952). Some of CLEMENTS' associations are collective types in this sense. For example, the Quercus-Carya association of the eastern deciduous forests and the Pinus-Juniperus association of the woodland formation each includes a wide range of combinations of dominant species, linked together by the genera of the dominants.

3) By successional status and prevalence. The Clementsian system distinguished climax associations and successional associes; but many of the latter were stabilized, proclimax communities. The system of BRAUN-BLANQUET (1951) more realistically distinguishes (i) terminal communities ('Schlussgesellschaften,' climatic, regional, or prevailing climaxes) that are stabilized in relation to the climate of a region or elevation belt, (ii) stable communities ('Dauergesellschaften', edaphic or local climaxes, some of the proclimaxes of CLEMENTS) that are also self-maintaining but are adapted to special biotopes in an area, and (iii) unstable, developmental or successional communities. Many difficulties for the Clementsian system resulted from the effort to force on vegetation a dichotomy between communities that were regional and climax, and communities that were local and therefore presumptively successional. Resolution of the difficulty lies in the distinction of three levels of conditions – regional or prevailing climax, local or edaphic climax, and local successional communities. (The fourth combination of regional prevalence with instability is clearly possible.) Many authors and schools have developed concepts of regional (and climax) community-types (WHITTAKER 1953, 1962). Regional types can be defined as units

on different levels – as formations, subformations, collective-types, and dominance-types. The associations of CLEMENTS ranged across the latter three possibilities from dominance-types (e.g. *Fagus grandifolia-Acer saccharum*) to collective types (*Quercus-Carya*) to subformations (some of the grassland associations).

4) By developmental relations into successional series and climax-complexes. A sequence of developmental community-types leading to a stable or climax type in a given kind of habitat is a successional series or sere. A grouping of related successional and climax community-types is a climax-complex (BRAUN-BLANQUET 1951, cf. the formation in the sense of MOSS 1910, TANSLEY 1911, and CLEMENTS 1916, 1936). In a narrower sense, the climax-complex (or successional complex) comprises only a particular climax type plus the various communities of primary successions and disturbance developing toward it. In the broader and more frequently used sense, the climax-complex groups the varied climax and successional communities of a region or elevation zone around a prevailing climax community-type; it is then a complex of successional complexes.

5) By environmental relations. Apart from succession, a chain (or continuum) of dominance-types or other units along a complex-gradient of environment is an ecological series (see also articles 5, 15). Communities that are related to one another in response to an environmental pattern (e. g. a landscape pattern) form a community-complex or -mosaic (cf. KRAUSE 1952). (If successional, as well as spatial, relations are in question, the community-complex is also a climax-complex.) In some applications of ecological series, each section or zone of the series forms a pattern of community-types, one of which prevails and characterizes the zone. Thus along such major complex-gradients as elevation in mountains and tide-levels in the littoral, the ecological series is a chain (or continuum) of community-complexes.

14.4.3 SUBORDINATE UNITS

Division of dominance-types into subtypes may be by different combinations of dominants and subdominants: (i) by marked change in importance value relations (or presence and absence) of the two or several dominant species of the type, (ii) by segregations of particular dominant species or combinations from a larger group of species by which the type is defined, (iii) by subdominants as major species of a stratum other than that of the dominant. The latter subtypes may be, but will not necessarily be, sociations. When forest types are divided by undergrowth composition in relation to site properties, the resulting subtypes may be site-types. Dominance-types may also be divided by groups of character-species or differential-species, and the subtypes may then correspond to lower units of the system of BRAUN-BLANQUET. There are thus possibilities for relating a classification employing dominance-types, which may be rather broad units, to the units of other schools.

14.5 Conclusion

The complexities of communities offer many choices of criteria by which ecologists may seek indication of environment or classification of communities. Choice of which community characteristics are to be used as criteria involves balancing together desired properties of these (WHITTAKER 1954, 1962). The criteria should be: (i) accessible – characteristics of the community that are evident in the field or, at least, can be observed or measured without excessive labor, (ii) significant – important as a characteristic of one community distinguishing it from others, strongly correlated with other characteristics of the community and its environment, and (iii) effective – successful in expressing environmental difference or producing units of classification at the level of intensiveness of a given study. (An accessible and significant criterion may be ineffective in this sense; physiognomy, for example, may be a poor criterion to distinguish among local forests and small site differences.) Differences among schools of ecology represent in part different choices of community characteristics to be emphasized, on the basis of different balances among these desired properties, in relation to different kinds of vegetation and perspectives on vegetation research.

The school of BRAUN-BLANQUET's primary criterion is floristic composition, particularly representation of diagnostic species (character-species and differential-species). As a community characteristic for classification and indication, floristic composition is (i) reasonably accessible, though requiring thorough knowledge of and attention to whole-community flora, (ii) significant, in that the utilization of information on a number of species, particularly those that are narrowly distributed or are near the borders of their distributions, produces information strongly correlated with other community and habitat characteristics, and (iii) effective, since choice of species is flexible, different groups of diagnostic species can serve for different levels of hierarchial classification, or different intensiveness of environmental expression. Dominance, as a primary criterion, is (i) more easily accessible, (ii) significant, but often at a lower or uncertain level because a given dominant may have a wide distribution, or may through disturbance be replaced by another dominant with only limited change in the rest of the community, (iii) often less effective for indication and classification than community composition.

Choice of dominance as a criterion is based on a different balance among these properties, emphasizing accessibility and quantitative importance to some sacrifice of significance in the sense just given and effectiveness. The preceding section suggests a range of units on different levels for classification by physiognomy and dominance — formation-type, formation, subformation and collective type when appropriate, dominance-type, and subtype. By no means is this intended as a system in competition with that of Braun-Blanquet. If a formal, hierarchial classification is the objective, dominance-types are not the way. Yet many ecologists, affected by different interests than phytosociologists, are likely to continue to classify their communities by dominance. For those making such a choice, this discussion is intended to offer clarification of the basis, limitations, and possibilities for such classification, and suggestions on units and concepts.

14.6.1 Summary

In British and American ecology, classification (at levels below the formation) has most often been based on the dominant species of communities. The American author Clements elaborated a formal classification based on successional relations and dominant growth-forms and species. The major units of this system included the formation, the 'association' as a broadly defined regional community-type within the formation, and successional types (associes, etc.). While Clements' work was widely influential, his classification has little current use.

Much current work, however, classifies plant communities into units which are defined by dominants (most conspicuous or important species), and which may be termed 'dominance-types.' The distributions of dominant species overlap with one another in most varied ways. Dominant species are consequently combined in varied, and often continuously intergrading, ways into the communities that must be classified. The ecologist must choose what dominant species, at what levels of importance, are to define dominance-types. Dominance offers no simple and objective criterion for classifying communities, and is poorly suited to the construction of a formal, hierarchial classification.

Dominance-types can be used, however, to construct a workable classification for a given area. Dominance-types may be grouped into formations or subformations, and in some cases into collective dominance-types (groups of dominance-types linked together by shared major species or other characteristics). Dominance-types may be divided either into subtypes by different combinations of their major species, or into some of the units of other schools of ecology (sociations, site-types, or associations and other units of Braun-Blanquet).

These units can be used to form a hierarchy – formation-type, formation, subformation and collective type when appropriate, dominance-type, and subtype. Such a hierarchy should not be regarded as a formal classification comparable to that of Braun-Blanquet. As a basis of classification dominance is more convenient than full floristic composition, but may less effectively express community and environmental relations.

14.6.2 Zusammenfassung

In der britischen und amerikanischen Ökologie wurde die Klassifikation (auf Stufen unterhalb der Formation) meistens auf die dominanten Arten oder Gesellschaften, begründet. Der Amerikanische Autor Clements entwickelte eine formale Klassifikation die auf Sukzessions-Beziehungen und dominanten Wuchsformen und Arten beruht. Die Haupteinheiten dieses Systems umfaßten die Formation, die 'Assoziation' als einen breit gefassten regionalen Gesellschaftstyp innerhalb der Formation, und Entwicklungs-Typen (Associes usw.). Während Clements' Arbeiten weithin von Einfluß waren, wird seine Klassifikation wenig verwendet.

In vielen rezenten Arbeiten werden die Pflanzengesellschaften in Einheiten klassifiziert die durch dominierenden (oder auffallendster oder wichtigen) Arten definiert werden und 'Dominanz-Typen' genannt werden können.

Die Verteilungen von dominanten Arten überlappen einander in sehr verschiedener Weise. Dominante Arten sind infolgedessen in verschiedenen und oft kontinuierlich verlaufenden Weisen in den zu klassifizierenden Gesellschaften vereinigt. Der Ökologe muß wählen mit welchen dominierenden Arten auf welcher Wichtigkeitsstufe, Dominanztypen definiert werden sollen. Die Dominanz bietet kein einfaches und objektives Kriterium für die Klassifikation von Gesellschaften und eignet sich schlecht für die Aufbau einer formalen hierarschisen Klassifikation.

Dominanz-Typen können jedoch dazu dienen, eine brauchbare Klassifikation für ein bestimmtes Gebiet aufzustellen.

Dominanz-Typen können in Formationen oder Subformationen gruppiert und in einigen Fällen durch gemeinsame Haupt-Arten oder andere Merkmale zu kollektiven Dominanz-Typen-Gruppen verbunden werden.

Dominanz-Typen können entweder durch verschiedene Kombinationen ihrer Haupt-Arten in Untertypen oder in einigen Einheiten anderer ökologischen Schulen (Soziationen, Standortstypen, oder Assoziationen und anderen Einheiten von BRAUN-BLANQUET) geteilt werden.

Mit diesen Einheiten kann eine Hierarchie aufgebaut werden: Formations-Typ, Formation, Subformation und, wenn zweckmäßig, Kollektiv-Typ, Dominanz-Typ und Subtyp. Eine solche Hierarchie sollte nicht als eine formale mit derjenigen von BRAUN-BLANQUET vergleichbaren Klassifikation betrachtet werden. Als Klassifikations-Grundlage ist die Dominanz leichter verwendbar als die gesamte floristische Zusammensetzung, aber sie kann Gesellschafts- und Standortsbeziehungen weniger deutlich ausdrucken.

REFERENCES

BEADLE, N. C. W., – 1948 – The vegetation and pastures of western New South Wales, with special reference to soil erosion. Tennant, Govt. Printer, Sidney. 281 pp.

BEADLE, N. C. W. & A. B. COSTIN, – 1952 – Ecological classification and nomenclature. With a note on pasture classification by C. W. E. MOORE. *Proc. Linn. Soc. N. S. W.* 77: 61—82.

BEARD, J. S., – 1955 – The classification of tropical American vegetation-types. *Ecology* 36: 89—100.

BRAUN, E. LUCY, – 1938 – Deciduous forest climaxes. *Ecology* 19: 515—522.

BRAUN, E. LUCY, – 1947 – Development of the deciduous forests of eastern North America. *Ecol. Monogr.* 17: 211—219.

BRAUN, E. LUCY, – 1950 – Deciduous Forests of Eastern North America. Blakiston, Philadelphia. 596 pp.

BRAUN, E. LUCY, – 1956 – The development of association and climax concepts: their use in interpretation of the deciduous forest. *Am. J. Bot.* 43: 906—911.

BRAUN-BLANQUET, J., – 1951 – Pflanzensoziologie: Grundzüge der Vegetationskunde. 2nd ed. Springer, Wien. 631 pp. 3rd ed., 1964, 865 pp.

CLEMENTS, F. E., – 1905 – Research Methods in Ecology. Univ. Publ. Co., Lincoln, Nebr. 334 pp.

CLEMENTS, F. E., – 1916 – Plant succession: an analysis of the development of vegetation. *Publs Carnegie Instn* Washington 242: 1—512.

CLEMENTS, F. E., – 1928 – Plant Succession and Indicators: A Definitive Edition of Plant Succession and Plant Indicators. Wilson, New York. 453 pp.

CLEMENTS, F. E., – 1936 – Nature and structure of the climax. *J. Ecol.* 24: 252—284.

CLEMENTS, F. E. & V. E. SHELFORD, – 1939 – Bio-ecology. Wiley, New York. 425 pp.

CONARD, H. S., – 1935 – The plant associations of central Long Island: a study in descriptive plant sociology. *Am. Midl. Nat.* 16: 433—516.

Cowles, H. C., – 1899 – The ecological relations of the vegetation on the sand dunes of Lake Michigan, I. Geographical relations of the dune floras. *Bot. Gaz.* 27: 95—117, 167—202, 281—308, 361—391.
Cowles, H. C., – 1901 – The physiographic ecology of Chicago and vicinity; a study of the origin, development, and classification of plant societies. *Bot. Gaz.* 31: 73—108, 145—182.
Crocker, R. L. & J. G. Wood, – 1947 – Some historical influences on the development of the South Australian vegetation communities and their bearing on concepts and classification in ecology. *Trans. R. Soc. So. Aust.* 71: 91—136.
Du Rietz, G. E., – 1932 (1930) – Vegetationsforschung auf soziationsanalytischer Grundlage. *Handb. biol. Arb Meth.* 11, 5: 293—480.
Krause, W., – 1952 – Das Mosaik der Pflanzengesellschaften und seine Bedeutung für die Vegetationskunde. *Planta* 41: 240—289.
Moss, C. E., – 1910 – The fundamental units of vegetation: historical development of the concepts of the plant association and the plant formation. *New Phytol.* 9: 18—53.
Poore, M. E. D., – 1956 – The use of phytosociological methods in ecological investigations. IV. General discussion of phytosociological problems. *J. Ecol.* 44: 28—50.
Richards, P. W., – 1952 – The Tropical Rain Forest: An Ecological Study. Cambridge Univ. Press. 450 pp.
Schmid, E.,– 1952 – Natürliche Vegetationsgliederung am Beispiel des Spanischen Rif. *Ber. geobot. Inst. Rübel*, Zürich 1951: 55—79.
Tansley, A. G., ed., – 1911 – Types of British Vegetation, by Members of the Central Committee for the Survey and Study of British vegetation. Cambridge Univ. Press. 416 pp.
Tansley, A. G., – 1939 – The British Islands and Their Vegetation. Cambridge Univ. Press. 930 pp.
Tansley, A. G. & T. F. Chipp, eds., – 1926 – Aims and Methods in the Study of Vegetation. Brit. Emp. Veget. Comm. & Crown Agents for Colonies, London. 383 pp.
Warming, E., – 1909 – Oecology of Plants: An Introduction to the Study of Plant-Communities. Oxford Univ. Press. 422 pp.
Weaver, J. E. & F. E. Clements, – 1929 – Plant Ecology. Mc Graw-Hill, New York. 520 pp. 2nd ed., 1938, 601 pp.
Whittaker, R. H., – 1953 – A consideration of climax theory: the climax as a population and pattern. *Ecol. Monogr.* 23: 41—78.
Whittaker, R. H., – 1954 – Plant populations and the basis of plant indication. (Germ. summ.) *Angew. PflSoziol.*, Wien, Festschr. Aichinger 1: 183—206.
Whittaker, R. H., – 1956 – Vegetation of the Great Smoky Mountains. *Ecol. Monogr.* 26: 1—80.
Whittaker, R. H., – 1957 – Recent evolution of ecological concepts in relation to the eastern forests of North America. *Am. J. Bot.* 44: 197—206.
Whittaker, R. H., – 1962 – Classification of natural communities. *Bot. Rev.* 28: 1—239.

15 THE FINNISH SCHOOL AND FOREST SITE-TYPES

Toomas E. A. Frey

Contents

15 THE FINNISH SCHOOL AND FOREST SITE-TYPES

15.1 Introduction

The classification of vegetation seems an eternal problem, without final answer. It is so because of the existence of regions differing widely not only in climate, soils, and flora, but also in the dynamics of the vegetation as affected by natural disturbance and increasing civilized violence as well. It is hardly possible that a single classificatory procedure should provide for all the problems and patterns with which students of vegetation in different areas must be concerned. The complexity of the task has been met with an equally complex and varied methodology worked out in various countries and phytosociological schools.

It is to be emphasized that ecological methods tend to reflect the properties of the vegetation being studied, and one has to agree with WHITTAKER (1962) that there exists an ecology of ecologists. Hence, any attempt at a universally acceptable procedure for the establishment of classificatory units of world vegetation must fail. At best, perhaps a common platform could be found for different approaches (see HUSTICH 1960). Two major approaches – the physiognomic and the floristic (of BRAUN-BLANQUET) – have gained widest, but far from universal, acceptance. The approach to forest site-types pioneered by CAJANDER has had in vegetation classification a more limited, but still significant role that is to be discussed here.

15.2 Cajander and the Site-Type Concept

15.2.1. DEVELOPMENT OF THE CONCEPT

The Finnish biocoenotic school was founded by A. K. CAJANDER (1879-1943), from wide field experience in Fennoscandia and Siberia with northern vegetation that is comparatively monotonous and fairly poor in species. Early in the century (1901-1906) these extensive areas, unaffected or only slightly disturbed by human activity, were studied by CAJANDER (1903, 1904, 1909a, 1909b 1913) with results that deserved wide attention.

From these studies CAJANDER acquired an excellent perception not only of geographic and habitat relations, but also of vegetation as a dynamic entity. The changes caused by momentary factors (clear cuttings, forest fires, and burnt–over clearings in woods under cultivation) may be striking, and hence the temporary communities in such stands show successional patterns from early pioneer to more stable stages. As a rule, climatic extremes (severe drought and exceptional cold) repeat themselves more frequently and are of modest influence, so that the changes in communities they produce are mostly of small degree and last only for a few years. The more severe disturbances initiate successions that lead back to communities adapted to the average climatic (and edaphic) conditions of their biotopes or habitats, and these communities can be regarded as almost permanent in their composition and structure.

CAJANDER was not the first to make a clear distinction between temporary and permanent communities. However, he was the first to take the stable communities as the basis for the classification of vegetation. He thus anticipated such concepts as the formation of MOSS (1910), TANSLEY (1911), and CLEMENTS (1916), and the climax-complex of BRAUN-BLANQUET (1951). For example, in his early works on the vegetation of the Lena basin (1903, 1904) he recognized the following successional stages for lowland vegetation: (1) sandy barrens, (2) Salicetum viminalis, (3) mixed bushes rich in species, mainly *Salix*, (4) Betuletum odoratae, (5) Piceetum obovatae, (6) Piceeto-Lariceetum dahuricae, the last-named being the most stable community on these sites. Within the limits of the given series, the ecological situation does not differ so much as one might suspect from the contrasting composition and structure of the communities; i.e. the real capacity of the biotope or site remains almost unaltered during successional changes in the communities. Therefore, in so far as the forest type is a unit for practical forestry, the forest type should be described in such a manner that one can identify the type regardless of the facies or the seral stage, and this identification will indicate the site's potential productivity. Hills holds a similar viewpoint when he states: 'Since combinations of macroclimate and landform not only constitute the basic potential productivity but exercise a functional control on the integrated development of vegetation, soil and ecoclimate, combined macroclimate landform units provide the basic classification of ecosystems' (HILLS 1960: 39). It is difficult, however, to use the many, continuously varying qualities of climate, landform, and soil to produce biologically appropriate units of classification. The CAJANDER system uses the expression of these qualities in the composition of stable communities as a means of appropriate and useful classification of the communities, and thereby of the sites which they (and related unstable communities) occupy.

15.2.2. Community Dynamics and Undergrowth

Where forests are concerned, the difficulty is increased by the longevity and competitive capacity of dominant tree species which may call forth long-lasting differences in communities. Nevertheless, as Kalela (1960: 46) writes: 'We should try to find climax communities and to make first-hand investigations in them. The successional phases in which development has progressed farthest follow in the second place, and only after them the short-lived initial stages of the successions. In limiting stand types – especially among the unstable successional communities – we are forced to comparatively broad and collective types, omitting the most temporary and rarely occurring communities. Otherwise the number of stand types will be so great that it would be quite impossible to deal with them or to gain any comprehensive idea of the system as a whole.'

In definition of site-types the undergrowth must be emphasized. The Scandinavian and Baltic area has relatively few tree species and each of these may occur over a wide range of sites with different undergrowth associates. The Finnish system makes use of the *relative* autonomy of different layers, though it does not divide vegetation into separate unistratal associations or unions as does Lippmaa (1933, 1935, 1939; see also Cajander 1909a, Gams 1918, Bolleter 1921, Kujala 1929, Du Rietz 1930, Cain 1936, and article 16 in this volume). Granted that the independence of canopy and undergrowth is only partial, it is still true that the undergrowth provides much more information, more effectively expressing biotope conditions, than the canopy tree species. In some cases one canopy species may be replaced by another in a given kind of site, with little change in undergrowth composition. In other cases the change in undergrowth with canopy species (or age and density) is great, yet the investigator versed in the relations of undergrowth species to site may recognize in this altered undergrowth the characteristics of the stable community that indicates the site-type. It is by knowledge of successional process and of undergrowth species relationships that it is possible, though not always easy, to group the variety of disturbed stands in an area in relation to the stable or normative communities of the site-types.

To put it in other words, the main idea of the Finnish theory of site-types lies in the typification, from the vegetation of a given region, of a limited number of ecosystem-types recognized through the expression of site conditions in stable vegetation, and especially undergrowth composition. Actually the composition and structure of the individual stands representing an ecosystem-type are of a fairly uneven nature and overlap with those of other ecosystems. The site-type as a whole is thus an abstraction based on a series of individual stands which show some divergence caused by climatic, edaphic, cultural and other factors. It should be realized, however, that the variation in disturbed temporary communities is connected with the information about the instability of the ecological situation, whereas the permanent communities carry information on the ecosystem itself.

Variation among stands in a given type of site may be of any degree up to change in the leading dominant species and suppression of undergrowth so that no apparent similarity to a virgin stand remains. But regardless of the degree of disturbance, the given site maintains its basic properties, including its potential productivity and ability to reach the virgin or climax state. Hence, when knowledge of undergrowth and succession have permitted us to assign a disturbed stand to a site-type, we have achieved knowledge–of biotope properties, successional direction, productive potential, suitability for different tree species, and appropriate management, that are by no means explicit in the characteristics of the disturbed stand by itself.

We may quote finally another definition given by CAJANDER himself (1926, 1949): 'All those stands are referred to the same forest type the vegetation of which at or near the time of maturity of the stands and provided the stands are normally stocked, is characterized by a more or less identical floristic composition and by an identical ecologico-biological nature, as well as all those stands the vegetation of which differs from that defined above only in those respects which – being expressions of differences due to age, fellings, etc. – have to be regarded as merely accidental and ephemeral or at any rate as only temporary. Permanent differences call forth a new forest type in cases where they are sufficiently well-marked, or a sub-type in cases where they are less essential, but nevertheless, noticeable.'

15.3 Application in Classification

How to ascertain that a difference in the ecological-biological nature of stands is sufficiently well-marked to indicate a difference in site-type? This question naturally arises and refers to an often misinterpreted point in CAJANDER's theory since, at least at the first glance, he did not give clear enough directions to do so.

15.3.1. Site-Type Classes

The problem of distinguishing forest site-types of the Finnish school is usually settled in three stages. The first stage or level of classification involves *site-type classes*. CAJANDER himself divided the Finnish forests into five such classes: (1) dry-and -poor (heathlands), (2) fresh-mesic (mossy forests), (3) fresh-and-rich (broad leaved forests), (4) wet-and-rich (inundated forests), and (5) wet-and-poor (bog forests) site-types. This division is clearly based on more or less well-known correlations with environmental features. For practical purposes these site-type classes are very useful and readily recognizable, and of evident importance in making a decision when choosing a forest regeneration and management strategy.

Similar or more detailed treatments of classes of site-types have been applied in, e.g., Sweden (ARNBORG 1960), Canada (HEIMBURGER 1941, KUJALA 1945), the United States (HEIMBURGER 1934, CURTIS 1959), the USSR (POGREBNJAK 1929, SUKACHEV 1930, 1931, 1951, VOROB'EV 1953, etc.), and particularly in Estonia (ILVES 1953, KARU & MUISTE 1958, MASING 1969, LÕHMUS 1971). Here it might be of interest to note that already in the preceding centuries colloquial Estonian included as many as ten special words for site-type classes as we now recognize them. Today they are used in their right as terms (Fig. 1).

15.3.2. SITE-TYPES

The second stage is connected with the definition of *site-types* within site-type classes. Due to the circumstances quoted above, the definition should bear on permanent stands and their ground flora, the latter being the most distinctive attribute of Finnish forest typology. Site-types are best recognized by the whole undergrowth flora, for it is the whole composition of species present (and absent) that best indicates the properties of the biotope or habitat. In this emphasis of 'characteristic species composition' the site-type approach may in principle converge with that of the school of BRAUN-BLANQUET (though abundance and frequency relations of species may be given more emphasis). Undergrowth dominants alone are not, however, a sufficient basis for the definition and identification of site-types. A generally important dominant may on occasion be absent (e.g. *Vaccinium myrtillus* from the Myrtillus-type) from a stand that, as judged from over-all floristic composition, belongs to the site-type it characterizes. Yet in practice, in the species-poor vegetation of the North, undergrowth dominants (including bryophytes and lichens) are emphasized in the characterization and naming of site-types. Also in practice, the vitality (growth rate and height) of trees is much used for recognition of site-types, especially by foresters. In areas such as western Canada with more numerous tree and shrub

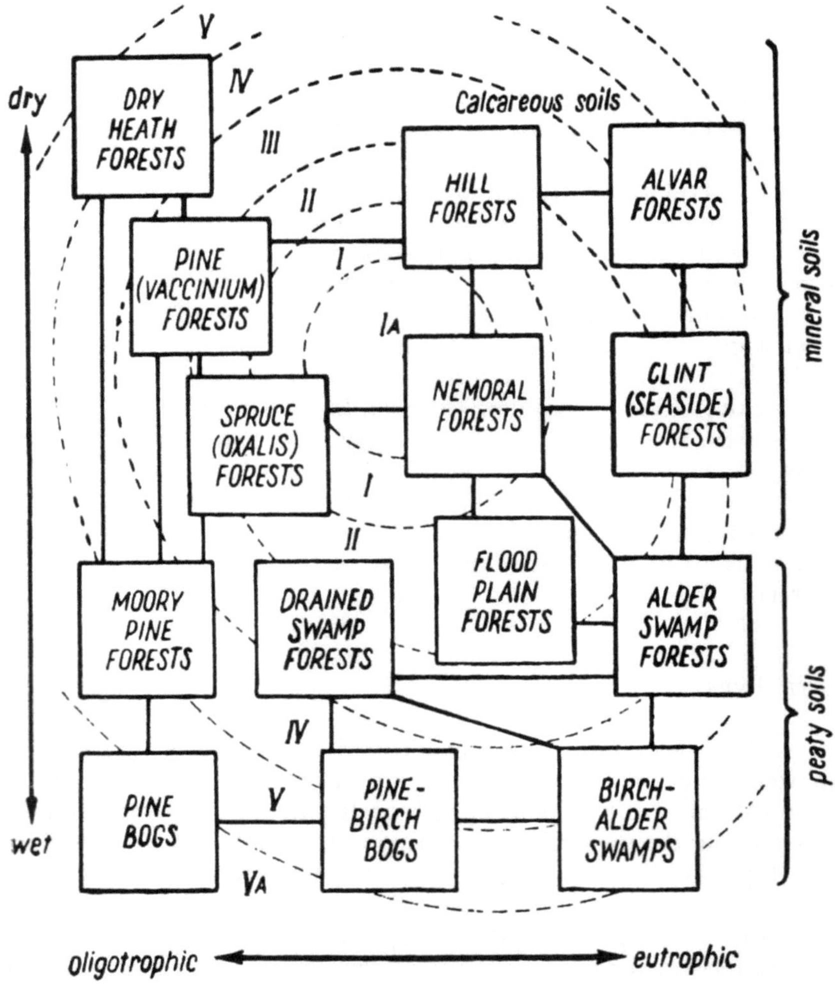

Fig. 1 Estonian forest site-type classes according to MASING (1969). Solid lines indicate the existence in nature of transitional stages between the classes, and interrupted lines the forest quality classes (isobonitets).

species, the indicator value of these has been considered along with that of the lower strata.

CAJANDER in particular was much concerned with dominance and the phenomenon of competition, which he was one of the first to consider. Already in his early papers CAJANDER paid considerable attention to the problem of interspecific competition as expressing the effect of the intermingled ecological factors upon vegetation. He recognized that in the comparatively extreme conditions of boreal forests there are only a few species able to dominate. The other species occurring there must adapt themselves not only to the biotope characteristics but also to the dominant species. From the limited number of dominant species there may result discontinuity in vegetation. Site-types can in any case be characterized, particularly in stable communities whose composition results from competition through a long time, by means of (i) dominant, (ii) constant, (iii) characteristic and (iv) differential species (see CAJANDER 1926: 27, 1930: 27–28). Once established for permanent stands, these species can be readily used in judging the membership of any of the stands representing temporary communities. In secondary forests the tree layer is usually much degraded, and centuries may be needed before the balance is reached again. The ground vegetation, however, generally establishes itself within decades; and in spite of the strong effects of dominant trees upon the other layers, the latter always consist of enough species that this assemblage, especially on better and mesic sites, is a more sensitive indicator than the tree layer itself.

Using dominant undergrowth species or groups of prevailing species (e.g the extraordinary role of lichens in the North), site-types are easily recognizable in poor site-type classes. In his early work CAJANDER (1909a) recognized five site-types for the first class:

(1) Cladina-type (ClT),
(2) Myrtillus-Cladina-type (MClT),
(3) Calluna-type (CT),
(4) Empetrum-Myrtillus-type (EMT),
(5) Vaccinium-type (VT).

Usually there is no problem in identifying types (1), (3) and (5), but the other two are of somewhat mixed nature. The distinctions between these types do not produce real difficulties, however, because in extreme conditions the boundaries are usually well-marked. CAJANDER attributes this phenomenon to the strong competitive relations between the dominant species. The types for dry-poor sites described in areas other than Finland fit in fairly well with CAJANDER's units, regardless of the degree of cultural disturbance in a particular country.

Reasonable agreement between students of vegetation in different areas is easily obtained at the other extreme as well, considering the site-types of mires and bogs. But, the greater the number of species the smaller the role of dominant species; and as soon as one considers mesic conditions, the task turns out to be rather more complicated. The third site-type class was divided by CAJANDER into the following categories:

(1) Geranium-type (GT),
(2) Dryopteris-type (DT),
(3) Oxalis-Maianthemum-type (OMaT),
(4) Sanicula-type (ST),
(5) Aconitum-type (AT),
(6) Vaccinium-Rubus-type (VRT),
(7) Lychnis diurna-type (LT).

All these site-types are rich, of high species diversity and variability within and between stands; and as a result it is difficult to distinguish them. Therefore, the site-types themselves are to some extent speculative in nature. Their boundaries are often gradual, fine-grained and topography-dependent with wide transition zones, so that it is hard to reduce the role of subjective decisions. As a rule, not only dominant and constant but also characteristic and differential species or/and groups of species must be used to define and recognize these site types. It is not very surprising that the fresh-and-rich site types proposed by CAJANDER for the boreal forests of Finland do not coincide with the types suggested for use in the hemiboreal forests of Estonia (Fig. 2). Differences result from both the vegetation itself and the criteria applied. The former point may be illustrated by the fact that LINKOLA (1929), who first applied the Finnish theory to Estonian forests, was forced to introduce a sixth class of site-types into CAJANDER's typology, viz. – (6) dry-and-rich (alvar forests), lacking in Finland but frequent in Estonia as well as in southern Sweden.

Additional difficulties concern the temporary communities of the fresh-and-rich site-type class. It appears that in all types there is a general tendency to form communities poor in species at the early and middle successional stages. The high productivity results in closed and overlapping canopies under which in almost all cases only *Oxalis acetosella* can survive. Other species, excluding mosses, occur in a sparsely scattered manner; and sometimes it is impossible to deduce the basic site-type from the dominance of *Oxalis* and the absence of other herbs. Therefore in Estonia a special site-type, the Oxalis-type, is used. However, it should be borne in mind that an Oxalis-type site does not usually denote more than a site which with age will change into a site of the Oxalis-Maianthemum-type, or passing through the Vaccinium-Rubus-type may acquire the properties of the Dryopteris or even the Sanicula-type (in CAJANDER's terminology).

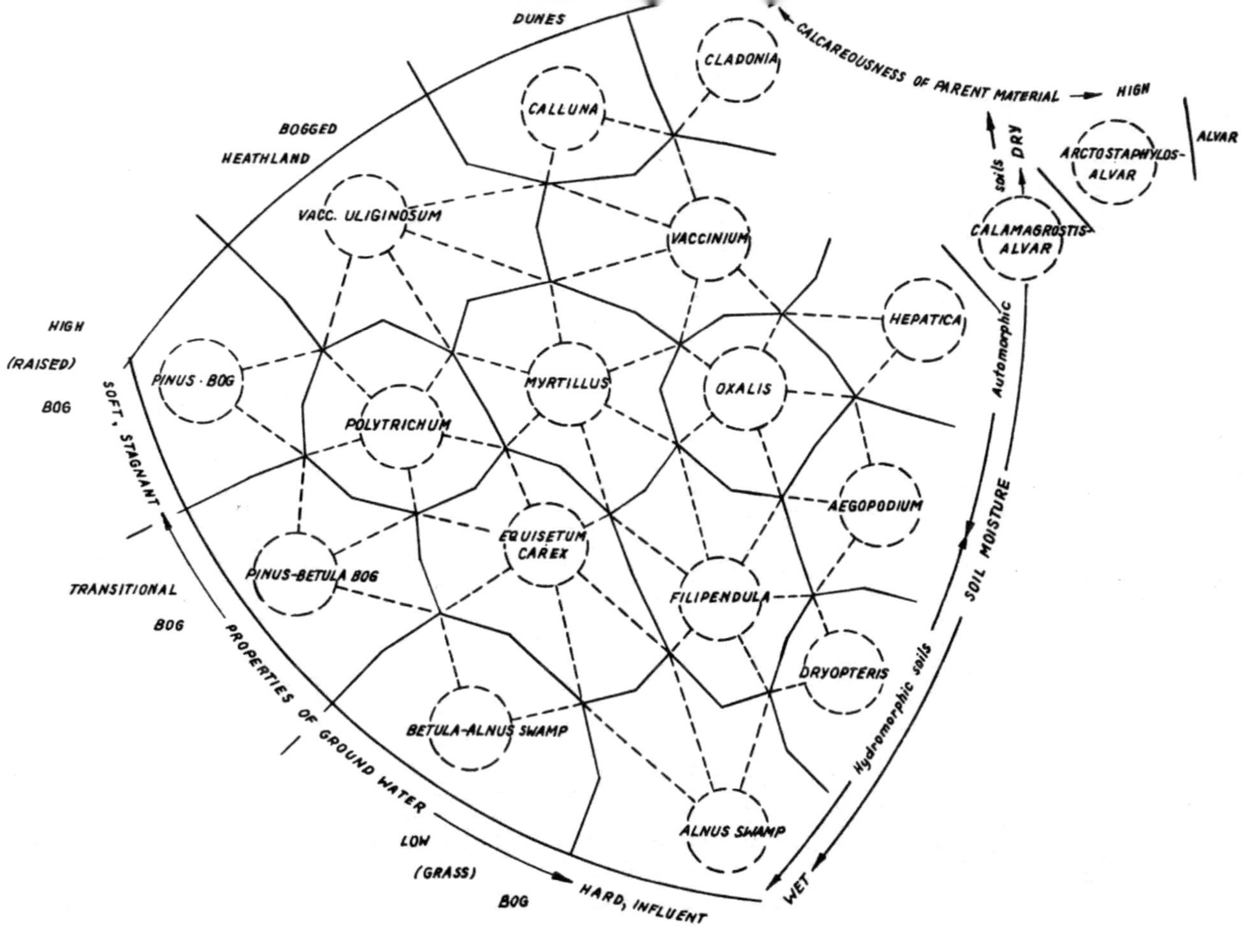

Fig. 2. A chart of forest site-types in relation to one another and leading soil factors in Estonia (Lõhmus 1971). The circles denote the site-type gravity centers defined both by ground vegetation (using the presence of prevalent species weighted by their average cover) and by soil properties. Solid lines indicate the distribution along environmental gradients of the site types, and dotted lines of subtypes, respectively.

15.3.3. Subordinate Units

The third stage of classification involves change in the dominant tree species within the limits of a site-type. In extreme site-types there may be no replacement of the normally dominant tree species by other species. Mesic and rich sites, however, show a variety of dominants or, to put it otherwise, one can find there a great number of cover types or *forest types*. Forest types are thus (unlike site-types) to be distinguished by tree species. Cajander (and the Russians) termed forest types distinguished by both canopy species and undergrowth 'associations'; but the international term for vegetation units defined by the dominant species of the different strata is now *sociation* (articles 17 and 18). For example, in sites having undergrowth of the Myrtillus site-type, four forest types or sociations are distinguished by dominant tree species, the more stable Pinetum myrtillosum and Piceetum myrtillosum, and the temporary Betuletum myrtillosum and Populetum myrtillosum. The Mercurialis-Aegopodium site-type is characterized by as many as six dominant tree species, the corresponding sociations being mainly temporary. For example, *Betula verrucosa* may be replaced by *Populus tremula* before *Picea abies* replaces *Populus* to take its rightful dominance approximately one hundred years after felling.
On the other hand, within the dry-and-rich site-type class (alvar forests) there are four site-types (Karu & Muiste 1958):

(1) Arctostaphylus-Cladonia-type with only *Pinus silvestris*,
(2) Vaccinium-type with *Pinus* or/and *Picea abies*,
(3) Oxalis-Corylus-type with *Pinus*, *Picea*, or *Quercus robur*,
(4) Sesleria-type with the previous three plus *Betula verrucosa* as dominant tree species.

Thus alvar forests include altogether ten forest types or sociations in four site-types.

To sum up the leading principles of the Finnish method of forest site-types, it should be emphasized that the classification of forests into site-types is based on all the similar and distinguishing characters of the stands under consideration (e.g. Tuomikoski 1942, Kalela 1939, 1954, 1960). Usually the question of how widely or narrowly the site-types are to be delimited has been regarded as having only a secondary importance. However, where they are broadly defined Cajander as well as his followers have taken advantage of differentiating them into subtypes (cf. the definition of the site-type above) and variants and/or facies. These units too cannot escape some subjectivity in definition and decision about limits. Yet it should be emphasized that such divisions are based on objective data – the tabular summaries of the stand samples first proposed by CAJANDER (1903) independently of the Central-European investigators.

15.4 Application in Management

It is a fair test of the system that its site-types, using undergrowth as an indicator of the biotope, should predict the forest growth the biotope will support. The relation of site-type to forest growth was investigated by ILVESSALO (1920, CAJANDER & ILVESSALO 1921) and many others. ILVESSALO observed especially the effects of relative favorableness of site on stand density and stand growth. For example, in South Finland major site-types (excluding the hydric) may be arranged in a series of decreasing favorableness – Oxalis-type (OT), Oxalis-Myrtillus-type (OMT), Myrtillus-type (MT), Vaccinium-type (VT), Calluna-type(CT), and Cladonia-type (ClT). When densities of trees by diameter size classes are plotted, for stands of the same age in each of these site-types, a striking relationship appears (Fig. 3). The more favorable the site, the smaller

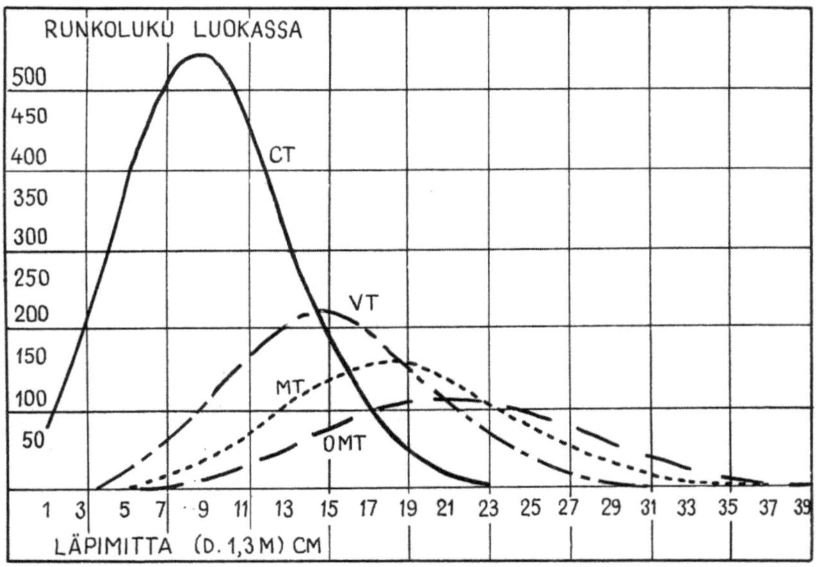

Fig. 3. Mean frequency distributions of tree stems by diameter classes, for 70-year-old *Pinus silvestris* in four forest site-types (CAJANDER & ILVESSALO 1921, Fig. 6). The vertical scale is numbers of stems in 4-cm classes per hectare, the horizontal scale is stem diameter in cm at breast height.

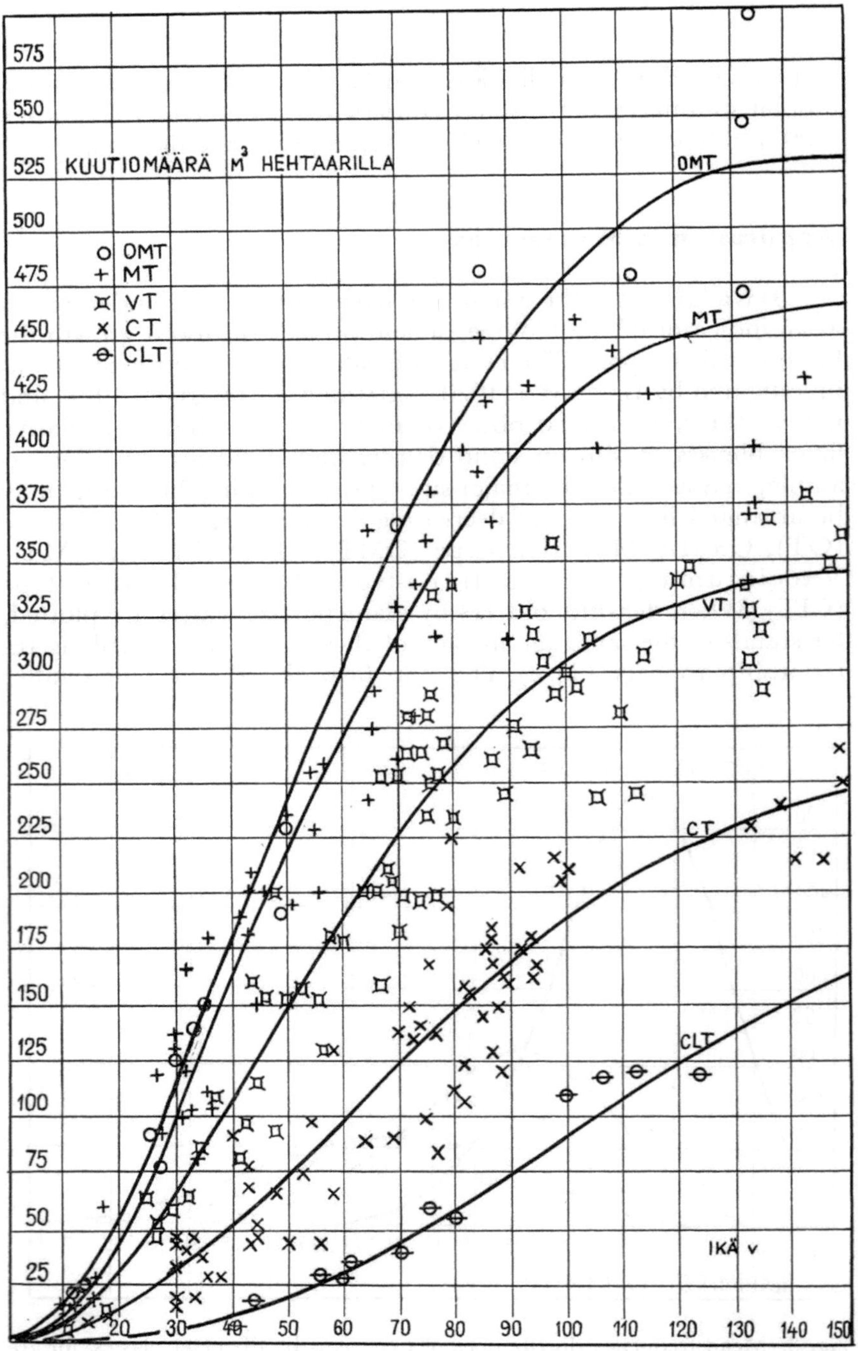

Fig. 4. The relationship of tree growth in *Pinus silvestris* to age in five forest site-types (Cajander & Ilvessalo 1921, Fig. 9, see also Cajander 1949, Figs. 2-4). The vertical scale is accumulated stem wood volume in cubic meters per hectare, the horizontal scale is stand age in years.

the number of trees, the larger the mean diameter, and the broader the dispersion of diameters from the mean. The inverse relation of favorableness to density reflects a widely significant principle of plant competition (YODA et al. 1963, HARPER 1967). The difference in tree sizes must result from difference in growth rate, as represented in Fig. 4. There is dispersion in the growth of different stands of a given site-type at a given age; such dispersions are almost inescapable in ecological phenomena. Yet the essential requirement is met: The site-type designation gives a prediction (within probability limits) of productive potential of the habitat.

CAJANDER thought much less effective a site indication based only on rate of height growth (site quality or 'Bonität' classes). The quality class can only be determined for a given species growing in the biotope at an age and condition suitable for measurement. The site-type, in contrast, can be determined independently of the present tree stand in the biotope. A quality class for one species in a site may offer no prediction of the growth rate of another species there, for there is no fixed proportion between the growth rates of different species in different sites. A site-type designation, however, offers a prediction of growth potential in that site for any species that might be planted there, if the growth rate of that species in the site-type has been measured. A quality class measurement for one species in a site may give little indication of what other species would be suitable to the site. The site-type classification provides an indication of which tree species may reasonably be planted in the site and which of these may best realize the site's potential for forest production at what stocking density. The site-types have much to offer for the skills of forest management.

The objectives of the system for management are those stated outside the Finnish school, by KRAJINA (1960: 55): 'In order that an ecological classification may be used by the forester, silvicultural characteristics might be included, having respect to: productivity and quality of trees; capacity for conservation of soil, water, and other habitat factors; methods of cutting and logging; methods of slash disposal; methods of reforestation; stocking, and methods and time of thinning and pruning; relation to pests; possibility for ranch management; wildlife management. Only the accurate synecological knowledge of forests, concretely registered and mapped for every forest region, will form the sound foundation for any forest management, silvicultural or genetical'. This summary fairly characterizes the nature and practical application of the Finnish theory of site-types in the Scandinavian and Baltic countries, especially in Finland and Estonia.

15.5 Series and Complexes

ILVESSALO's (CAJANDER & ILVESSALO 1921) work implies that in many areas major site-types can be arranged in a sequence of decreasing quality from the fresh-and-rich to the dry-and-poor. Such a sequence is an *ecological series* in the sense of CAJANDER (1903) and Russian authors (Fig. 5). A further result of interest in ILVESSALO's (1922) work is the relation of species diversity to the series from the Oxalis-type to the Cladonia-type in South Finland (Table I). In

TABLE I. Species diversities in forest samples with stands of intermediate ages (ILVESSALO 1922). Samples are mostly 0.2 or 0.25 hectare, but some down to 0.1 ha. Species diversities decrease and proportions of non-vascular plants increase along the moisture gradient from the Oxalis-Type to the Cladonia-Type.

Types:	Oxalis	Oxalis-Myrtillus	Myrtillus	Vaccinium	Calluna	Cladonia
Number of quadrats	21	58	53	38	37	5
Total number of higher plant species	107	105	86	58	28	9
Mean number of higher plant species	37.5	32.0	28.7	18.4	7.6	4.2
Standard deviation, species per quadrat	9.0	8.7	7.3	5.8	3.1	1.7
Per cents of all species,						
higher plants	86.3	84.7	81.9	75.3	63.6	45.0
mosses	10.5	9.7	11.4	14.3	15.9	25.0
lichens	3.2	5.6	6.7	10.4	20.5	30.0

this area species diversity declines with decreasing site quality and productivity (see also CAJANDER 1909a, PALMGREN 1915-17 LINKOLA 1916, 1917, LAKARI 1920), though in some other areas maximum diversity is in the middle of a moisture gradient, or productivity and diversity are inversely related (WHITTAKER 1969). ILVESSALO (1922) gives data also on effects of age of stand and species of canopy dominant on species diversity. PALMGREN (1928) studied the relation of bird communities to a site-type series.

The use of ecological series of site-types as a basis of vegetation interpretation was developed further by KUJALA (1938, 1945). KUJALA considered that, with two major site-types in each of CAJANDER's three classes excluding the hydric, six site-types would in many areas form a series representing the main axis of community response to habitat. For south Finland such a series (from dry to mesic) would be: (1) Calluna-Cladina-type, (2) Vaccinium-type, (3) Myrtillus-type, (4) Oxalis-Myrtillus-type, (5) Oxalis-Majan-

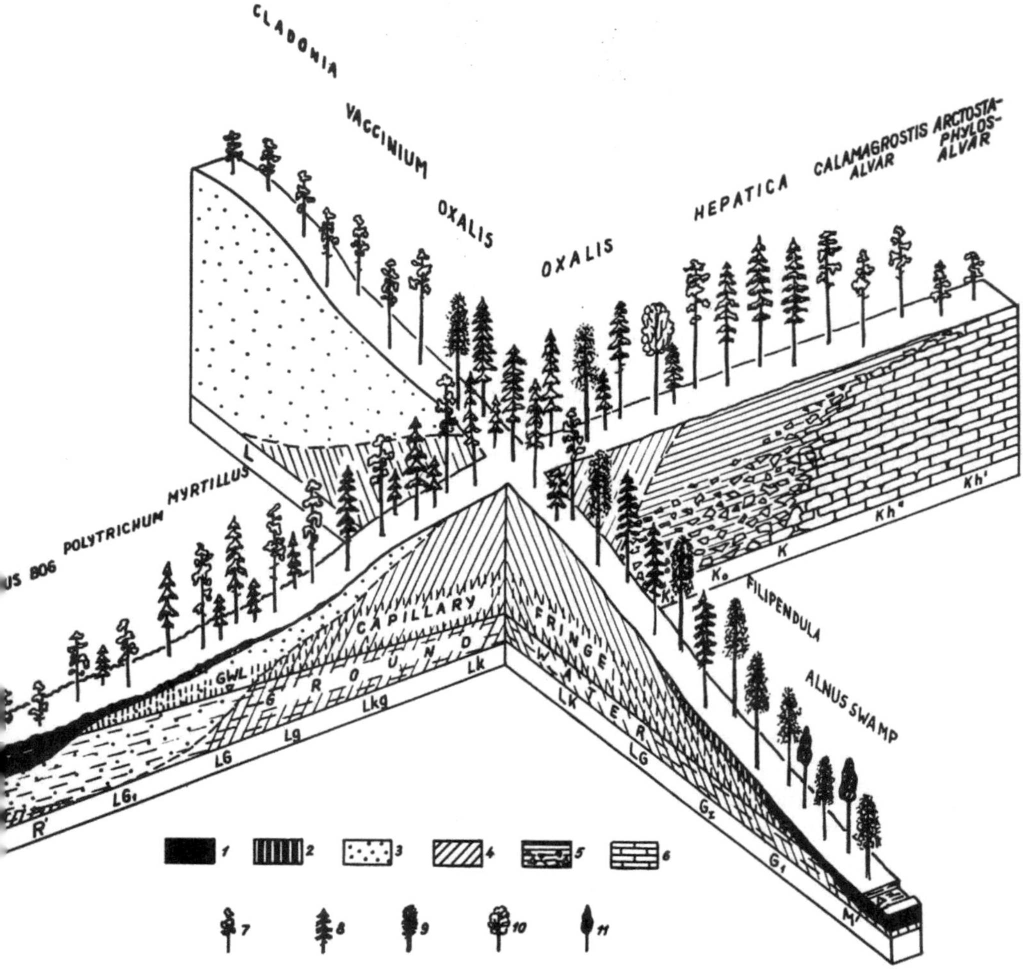

Two ecoclines showing the interrelations of Estonian forest site types and their ecological position
network of complex habitat factors (Lõhmus 1971):1) peat, 2) organic-mineral (peaty) horizon
3) sand, 4) noncalcareous moraine, 5) calcareous moraine, 6) limestone; 7) *Pinus silvestris,* 8) *Picea*
Betula verrucosa, B. pubescens, 10) *Populus tremula,* 11) *Alnus glutinosa*: L – podzol, Lg – gleyed podzol,
ddy podzolic, Lkg – gleyed soddy podzolic, LG – glay podzol, LG_1 – peaty podzol, R' – shallow
g, G_I – podzolic glay, G_1 – peaty glay, M' – shallow low moor, K_1 – brown pseudopodzolic, K_0 – brown
K_1 – brown lessive, Kh'' – shallow limestone rendzina, K' – very shallow limestone rendzina soil.

themum-type, and (6) Oxalis-Filices-type. When one proceeds to a different geographic area, each of these site-types may be replaced by another, related site-type; thus from south to north Finland the Oxalis-Myrtillus-type is replaced by the Dryopteris-Myrtillus-type as the fourth member of the series. These two ecologically equivalent communities are *vicariant* site-types. In north

Finland the types given for south Finland are replaced by vicariants to form the series: (1) Calluna-Myrtillus-Cladina-type, (2) Empetrum-Vaccinium-type, (3) Hylocomium-Myrtillus-type, (4) Dryopteris-Myrtillus-type, (5) Geranium-Dryopteris-type and (6) Filces-type. These two series, comprising corresponding or vicariant types, are *parallel type-series*. KALELA (1961) distinguished six or seven type series each characterizing a forest region in Finland.

A type-series permits one to characterize some main features of the vegetation pattern of an area of northern forest. (Such series may not, however, include hydric or bog communities, and in some areas the vegetation pattern is further complicated by fertility and other soil differences.) Parallel type-series permit comparison of the vegetation patterns of different areas. Thus in his treatment of the Canadian boreal forests KUJALA (1945) compares parallel type-series across the continent. An eastern (Gaspé Peninsula)series resembles the Finnish – (1) Kalmia-Cladonia-type, (3) Vaccinium canadense-type, (4) Cornus-Aralia-type, (5) Oxalis-Cornus-type, (6) Oxalis-Dryopteris-type, (7) Impatiens-Circaea-Athyrium-type. A western (Vancouver Island, maritime-climate) series comprises only – (3) Gaultheria-type, (4) Achlys-Gaultheria-type, (5) Achlys-Tiarella-Aspidium munitum-type. The latter is geographically and floristically distant from the Finnish series and shows an apparently smaller degree of type differentiation along the moisture gradient. KUJALA (1945) further observes that the middle members of the type-series are normally best developed and most extensive in an area and express its climate. These middle members, generally (3) and (4), are consequently *leading-types* (*Leittypen*) for the characterization of regional vegetation; the concept of leading-type converges with that of climatic climax (CLEMENTS 1916) and, perhaps more closely, prevailing or regional climax (WHITTAKER 1953, BRAUN 1956). The characterization of vegetation patterns through type-series may be compared also with that of mountain vegetation through topographic coenoclines (WHITTAKER 1956, 1960, WHITTAKER & NIERING 1965, article 3).

KALELA (1939, 1954, 1958, 1960, 1961, 1962) also developed CAJANDER's system further for the treatment of vegetation patterns, emphasizing both the dynamical and geographical aspects of typology in a regional or zonal approach that is now of major importance in the Finnish school. Each forest site-type, regional or local, comprises a climax community-type and various successional communities developing towards it. The stands belonging to a larger topographic unit form a *stand complex*, i.e. 'the parts of plant cover which are more or less clearly distinguished from the surroundings and composed of numerous stands representing diverse stand types of various sizes and alternating in different ways–being mosaic, belt-like or reticular in shape' (KALELA 1960: 42). Just as stands may be grouped into stand types, he suggests the grouping of *stand complexes* into their *types*. Within each *vegetation region* (KALELA 1939, 1954) the macroclimate is homogenous in broad features and the region is characterized by the prevalence of the same complex-types in ecologically similar localities (see also AHTI et al. 1968). As a rule, the stand types of a vegetation region can be divided into regional (climatic, zonal) and local (edaphic, azonal) plant communities. Phytosociologists have not always paid sufficient attention to this division. In a given region prevailing climate influences the development of certain kinds of soils. The communities that are most successful on the normal and non-extreme soils in a region form together its *regional plant communities*. KALELA thus develops, in the Finnish school, ideas comparable to the community-complex or pattern, landscape-type, and regional or prevailing climax community of other schools.

15.6 The Finnish and Other Approaches

The principles of forest typology proposed by CAJANDER (1909a, 1926, 1930, 1949, CAJANDER & ILVESSALO 1921) were applied, in some cases with modifications, by AALTONEN (1919), ILVESSALO (1920, 1922, 1929, 1967), LAKARI (1920), KUJALA (1921, 1929a, 1929b, 1936a, b, 1945, 1960), LINKOLA (1924, 1929, 1930), KALELA (1941, 1952, 1954, 1958, 1960, 1961), KALLIOLA (1943), KELTIKANGAS (1945, 1959), PERTTULA (1950), JALAS (1950), KALLIO (1957), HÄMET-AHTI (1963), HÖGNÄS (1966), MÄKIRINTA (1968) and others in Finland and elsewhere. Site-types for hydric or bog communities were considered by CAJANDER (1913), LUKKALA (1929), AARIO (1932), PAASIO (1937, 1940, 1941), RUUHIJÄRVI (1960), and EUROLA (1962). The system has been influential outside Finland also (see KALELA 1960, KRAJINA 1960, WHITTAKER 1962, TRASS 1964, ALEKSANDROVA 1969). Closely related approaches in the Scandinavian and Baltic area include the work of MALMSTRÖM (1926, 1936) and NIHLGÅRD (1970) in Sweden, BORNEBUSCH (1923-5, 1929) in Denmark, KIRSTEIN (1929) and MALLNER (1944) in Latvia, and LINKOLA (1929, 1930) in Estonia. Forest site-types were studied in Central Europe by CAJANDER (1909a, 1949, LINKOLA (1924), and KUJALA (1936). North American applications have included ILVESSALO (1929), HEIMBURGER (1934, 1941), KUJALA (1945), SPILSBURY & SMITH (1947), HUSTICH (1949, 1950), and CRANDALL (1958),

AHTI (1959), KALELA (1962), and HÄMET-AHTI (1965). Patagonian forest site-types were studied by KALELA (1941).

Russian forest phytocoenology developed in partial independence of the Finnish school, but was influenced by it. Russian work, from the early writings of MOROSOV (1928) to the recent of SUKACHEV (1944, 1954, 1960) emphasizes the forest stand as a biogeocoenose, a complex whole of organisms and environment and their interrelations. The forest types of SUKACHEV (1928, 1932) and others give equal weight to the canopy dominant and the undergrowth; these are not site-types in the sense of CAJANDER but sociations (termed by the Russians 'associations'). They are, however, treated in terms of ecological series as in Finnish work, though the emphasis has been on series in relation to a number of gradients (see article 17, Fig. 1) rather than on a single series of principal site-types (KUJALA 1945). Russian authors have used ideas of CAJANDER on biologically equivalent sites (SUKACHEV 1934) and twin associations (sociations differing in one of their stratal dominants or unions) (KATZ 1929, 1930). ALEKSANDROVA (1969) has pointed out how CAJANDER's ideas have influenced, or coincided with, the ecological series of SUKACHEV (1925, 1931, 1934, 1938), SOKOLOWA (1935, 1936), and SOKOLOV (1937, 1962), the groups of SMAGIN (1950), the circles of KATS (1936), the 'ingregations' of BYKOV (1962), the sections of MIRKIN (1965), and other Russian interpretations.

A Swedish site-type system was developed by ENEROTH (1936), ARNBORG (1940, 1953) and LINDQUIST (1954). The forest types of this system are sociations, and these are grouped in relation to forest succession in a manner resembling the Finnish. A forest site-type comprises a whole series of plant communities of which only one stage – present under certain conditions, particularly in normal density of the tree layer – is normative for the type. A number of different undergrowth communities are thus grouped together in relation to a kind of site and a normative community which occurs in such sites under certain conditions (ARNBORG 1940). The concept is close to CAJANDER's but as observed by WHITTAKER (1962) 'The site-type approach is thus carried to its logical conclusion; it is a classification not simply of communities, but of sites as characterized by groups of related communities (cf. KALELA 1954).' The Swedish system converges also with the Russian in treatment of stratal communities (unions) as these are variously combined into forest types (sociations), with the latter arranged in ecological series (Fig. 6).

The influence of CAJANDER spread also into Middle Europe (WHITTAKER 1962), but the system has little current use there. The Austrian system of AICHINGER (1951, 1952, 1967, WENDELBERGER 1951) employs a different unit, the *vegetation-development-type*. Devel-

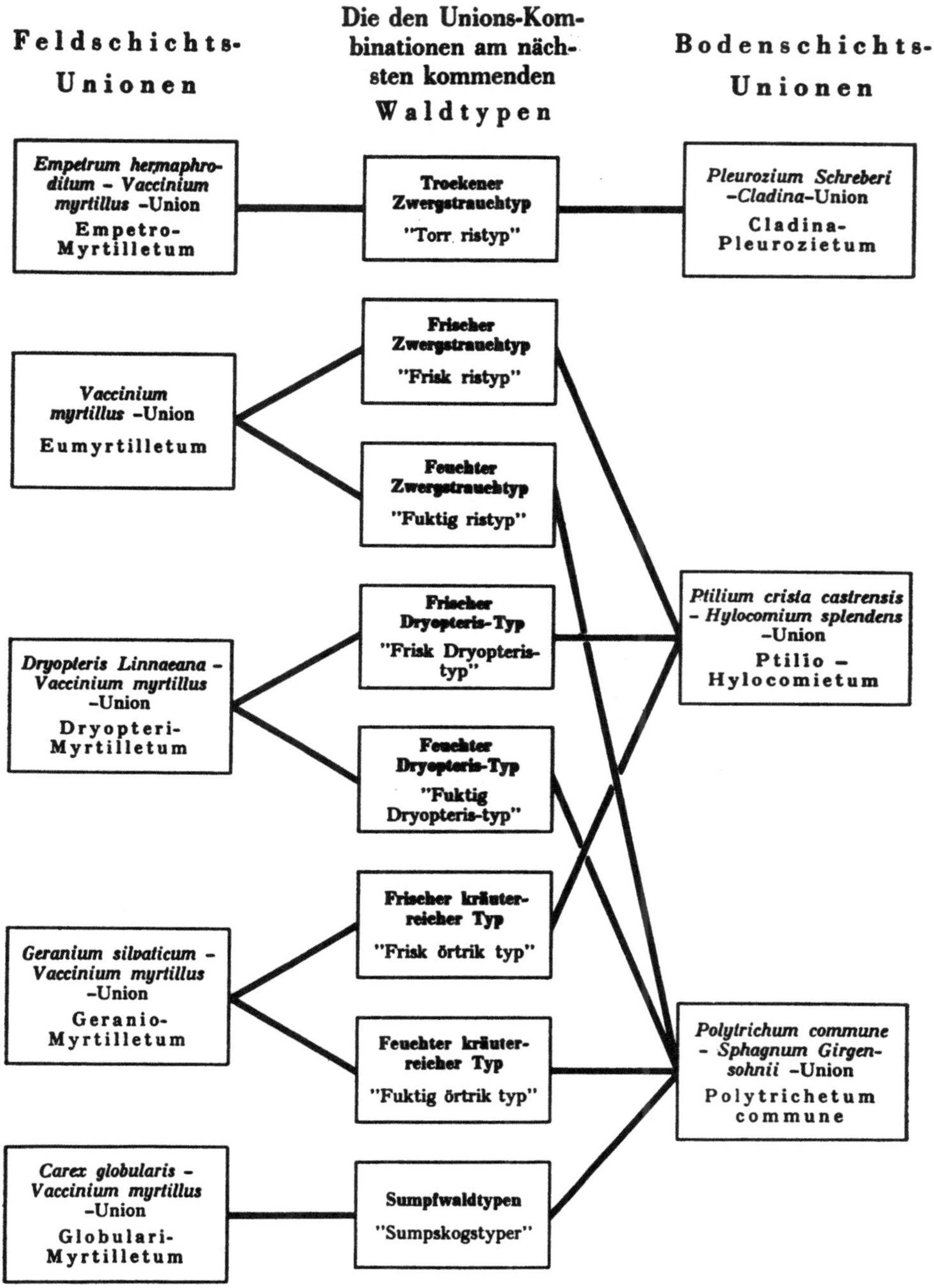

Fig. 6. A system of forest types for a north-Swedish forest (ARNBORG 1940). The middle column gives forest types as formed by different combinations of undergrowth unions – those of the field (herb and shrub) layer on the left and the ground (moss and lichen) layer on the right. Forest types and unions are arranged in relation to the moisture gradient from the dry dwarf-shrub type (*Trockener Zwergstrauchtyp*) through fresh (*Frischer*) and moist (*Feuchter*) types to the wet swamp forest types (*Sumpfwaldtypen*).

opment-types are described by stratal dominance in the manner of sociations, but are defined also by their place in successions. The system, with its strong emphasis of successional relations, is adapted to vegetation heavily and variously influenced by man. Extensive work in applied forestry in middle and southern Europe has applied the BRAUN–BLANQUET system; such work is to be discussed in a succeeding part of the *Handbook*. In general the CAJANDER system has had continued influence only in application to boreal and subalpine forests, the taiga formation-type.

15.7 Conclusion

In historic perspective, CAJANDER himself was influenced by several Scandinavian research workers – e.g. by VAINIO's (1878) deep concern with habitat factors, NORRLIN's (1870) idea of the interdependence of the vegetation and its habitat, NILSSON's (1901) principle of vegetation series based primarily on the character of undergrowth, BLOMQVIST's (1872, 1879) forest growth classes, etc. In developing these and other ideas into a system for the treatment of northern vegetation, CAJANDER influenced and is representative of the whole Northern Tradition (WHITTAKER 1962, TRASS 1964). Many of the ideas of the Finnish school correspond to those of other schools; but the school is distinctive, characterized by its own nuclear concept: the site-type as a grouping of successionally related communities around a mature or normal community-type which indicates site quality and which is to be recognized and related to the successional communities by undergrowth composition. From this central concept have radiated the means of classifying communities, practices of forest management, treatment in terms of ecological series, applications to animal communities, and interpretation of vegetation dynamics and landscape patterns.

It seems broadly true that the emphasis in forest structure shifts downward with increasing latitude, from tropical forests with their floras strongly concentrated in the canopy trees and their epiphytes, through warm-temperate forests with both canopy and undergrowth well developed and mixed in composition, to boreal forests with one or few tree species and with the flora (and in some cases maximum coverage) concentrated in the lower strata. The Finnish and related approaches are suited to the latter part of this spectrum, and the emphasis of stratal relations in the northern schools contrasts with that of floristic composition (or canopy dominance) in warm-temperate and of physiognomy in tropical ecology. We may wonder (see KALELA 1941) whether the Finnish system is not appropriate also for cold-temperate climates of the Southern Hemisphere – for, say, the *Nothofagus* forests of New Zealand. For the Scandinavian and Baltic area at least, CAJANDER's ideas stimulated the development of an approach to vegetation adapted to the ecology of northern forests, and were productive as a means for research on and management of them.

15.8.1. SUMMARY

The Finnish School developed under the leadership of CAJANDER, with the forest *site-type* as a central concept. A site-type groups together mature forest stands of similar undergrowth composition expressing similar site qualities, and unstable stands of the same kinds of sites. In northern forests the canopy tree species less effectively expresses site conditions than does the undergrowth. Site-types are consequently defined largely by the undergrowth of mature or stable stands, and knowledge of undergrowth communities is used to relate unstable stands to stable ones.

With the site-type as the basic unit, vegetation of an area may be classified into (a) three to six site-type-classes expressing broad relationships to soil moisture and fertility, (b) within each class one to several site-types and subtypes, each expressing narrower site differences in distinctive undergrowth communities, and (c) within each site-type or subtype a number of forest types or sociations, defined by dominant species of both canopy and undergrowth and including both stable and unstable communities.

The site-type provides a biological expression of biotope or site characteristics. Recognition of the site-type for a forest stand (which may be disturbed) thus provides indication of the quality of the site, the mature community and the direction of succession toward it, the potential productivity in that site of different tree species, and consequently the appropriate forest management.

Site-types maybe arranged in ecological series along environmental gradients. Several site-types (in many cases about six in one region, excluding hydric sites) thus arranged along the moisture gradient may express major features of vegetational variation in an area. Site-types may also be arranged into patterns including ecological series in relation to more than one environmental gradient. The site-types most widespread in a landscape (those in the middle of the moisture-gradient series) may be considered the leading-types or regional plant communities.

The Finnish approach has been extensively used in the Scandinavian and Baltic area and has influenced Russian forest phytocoenology, but has not been much used in warm-temperate areas. The approach seems best suited to boreal and subalpine forests; for these it is an effective means of ecological investigation and practical management.

15.8.2 Zusammenfassung

Die finnische Schule hat sich unter Leitung von Cajander mit dem Wald-Standorts-Typen (Wald-Typen) als zentralem Begriff entwickelt. Ein Wald-Standorts-Typ umfaßt reife Waldbestände mit ähnlicher Zusammensetzung des Unterwuchses, die ähnliche Standortseigenschaften ausdrückt, und instabile Bestände derselben Standortstypen. In den nördlichen Wäldern drückt die Baumschicht die Standortsbedingungen weniger klar aus als der Unterwuchs. Demzufolge werden die Wald-Standorts-Typen durch den Unterwuchs von reifen oder stabilen Beständen bestimmt und die Kenntnis der Unterwuchs-Gesellschaften wird gebraucht um die instabilen Bestände mit den Stabilen zu vergleichen.

Mit dem Walt-Standorts-Typ als Grundeinheit kann die Vegetation eines Gebietes klassifiziert werden in a. drei bis sechs Wald-Standorts-Typen-Klassen, die allgemeine Beziehungen zur Bodenfeuchtigkeit und Fruchtbarkeit ausdrücken, b. innerhalb jeder Klasse ein bis mehrere Wald-Standorts-Typen und -Untertypen die jede engere Standorts-Unterschiede in bestimmten Unterwuchs-Gesellschaften ausdrücken und c. in jeden Wald-Standorts-Typ oder -Untertyp eine Reihe von Waldtypen oder Soziationen die von dominanten Arten der Baumschicht und des Unterwuchses bestimmt werden und die sowohl stabile wie instabile Gesellschaften unfassen.

Der Wald-Standorts-Typus liefert einen biologischen Ausdruck des Biotopes oder der Standortsmerkmale. Die Erkennung des Wald-Standorts-Typus eines Waldbestandes (der zerstört sein kann) erlaubt die Angabe des Wertes des Standorts, der reifen Gesellschaft und der Entwicklungsrichtung dazu, die potentielle Produktivität von verschiedenen Baumarten auf diesem Standort und dementsprechend die geeignete Wald-Behandlung.

Die Standortstypen können in ökologischen Reihen längs Standorts-Gradienten angeordnet werden. Mehrere Standortstypen (in manchen Fällen etwa sechs, ausschließlich der naßen Standorte) die in solcher Weise längs des Feuchtigkeits-Gradienten angeordnet sind, können die wichtigsten Züge des Vegetationswechsels in einem Gebiet ausdrücken. Wald-Standorts-Typen können auch zu Strukturmustern geordnet werden, die ökologische Reihen in Beziehung zu mehr als einem Standorts-Gradienten umfassen. Die in einer Landschaft am weitesten verbreiteten Standortstypen (diejenigen in der Mitte einer Feuchtigkeits-Gradienten-Reihe) können als die Leittypen oder regionale Pflanzengesellschaften aufgefaßt werden.

Die Finnische Methode hat eine weite Anwendung im Skandinavischen und Baltischen Bereich gefunden und hat die Russische Wald-Phytozönologie beeinflußt, aber sie ist in warm-gemäßigten Gebieten nicht viel gebraucht werden. Die Methode scheint am besten für boreale und subalpine Wäldern geeinget; dort ist sie eine brauchbare Methode für die ökologische Forschung und die praktische Waldpflege.

Acknowledgment

The author thanks A. Kalela for his comments on the article and R. H. Whittaker for his contributions to it during revision.

References

Aaltonen, V. T., – 1919 – Kangasmetsien luonnollisesta uudistumisesta Suomen Lapissa. I. (Germ summ.: Über die natürliche Verjüngerung der Heidewälder im Finnischen Lappland. I). *Comm Inst. forest. fenn.* (*Metätiet. Tutkimuslait. Julk.*) 1: 1–319+1–56.

Aario, L., – 1932 – Pflanzentopographische und paläogeographische Mooruntersuchungen in N-Satakunta. *Fennia* 55(1): 1–179.

Ahti, T., – 1959 – Studies on the caribou lichen stands of Newfoundland. *Ann. Bot., Soc. Zool. -Bot. Fenn. 'Vanamo'* (*Suomal. eläin-ja kasvit. Seur. van. kasvit. Julk.*) 30(4): 1–44.

Ahti, T., L. Hämet-Ahti, & J. Jalas, – 1968 – Vegetation zones and their sections in northwestern Europe. *Ann. bot. fenn.* 5: 169–221.

Aichinger, E., – 1951 – Vegetationsentwicklungstypen als Grundlage unserer land- und forstwirtschaftlichen Arbeit. *Angew. Pfl Soziol.*, Wien 1: 17–20.

Aichinger, E., – 1952 – Fichtenwälder und Fichtenforste als Waldentwicklungstypen. *Angew. Pfl Soziol.*, Wien 7: 1-179.

Aichinger, E., – 1967 – Die Waldentwicklungstypen im Raume von Kirchleerau. (Engl. & French summs.) *Veröff. geobot. Inst. ETH, Stiftg Rübel*, Zürich 39: 187-270, 283, 293.

Aleksandrova, Vera D., – 1969 – Classification of Vegetation. Principles of Classification and Classification Systems of Various Phytocoenological Schools (in Russian) Publ. House 'Nauka,' Leningrad. 274 pp.

Arnborg, T., – 1940 – Der Vallsjö-Wald, ein nordschwedischer Urwald. *Acta phytogeogr. suec.* 13: 128-154.

Arnborg, T., – 1953 – Det nordsvenska skogstypsschemat. 3rd ed. Svenska Skogsvardsför. Förlag, Stockholm. 20 pp.

Arnborg, T., – 1960 – Can we find a common platform for the different schools of forest type classification? *Silva fenn.* 105: 13-15.

BLOMQVIST, A. G., – 1872 – Tabeller framställande utvecklingen af jemnåriga och slutna skogsbestånd af tall, gran och björk. Helsingfors.
BLOMQVIST, A. G., – 1879 – Eine neue Methode, den Holzwuchs und die Standortsvegetation bildlich darzustellen. *Bidr. Känn. Finl. Nat. Folk*, 31: 145-153.
BOLLETER, R., – 1921 – Vegetationsstudien aus dem Weisstannental *Jb. St. Gall. naturw. Ges.* 57(2): 1-140.
BORNEBUSCH, C. H., – 1923-5 – Skovbundsstudier I-III, IV-IX. (Engl. summ.) *Forstl. Fors Vaes. Danm.* 8: 1-148, 181-288.
BORNEBUSCH, C. H., – 1929 – Danmarks skovtyper. *Acta for. fenn.* 34(11): 1-18.
BRAUN, E. LUCY., – 1956 – The development of association and climax concepts: their use in interpretation of the deciduous forest. *Am. J. Bot.* 43: 906-911.
BRAUN-BLANQUET, J., – 1951 – Pflanzensoziologie: Grundzüge der Vegetationskunde. 2nd ed. Springer, Wien. 631 pp. 3rd ed. 1964, 864 pp.
BYKOV, B. A., – 1962 – On the composition of some formations of the ingregations (in Russian). *Trans. Bot. Inst. Kazakh S. S. R.* (*Trudȳ alma-atin. bot. Sada*) 13: 3-27.
CAIN, S. A., – 1936 – Synusiae as a basis for plant sociological field work. *Am. Midl. Nat.* 17: 665-672.
CAJANDER, A. K., – 1903 – Beiträge zur Kenntnis der Vegetation der Alluvionen des nördlichen Eurasiens. I. Die Alluvionen der unteren Lena-Thales. *Acta Soc. Sci. fenn.* 32(1): 1-182.
CAJANDER, A. K., – 1904 – Studien über die Vegetation des Urwaldes am Lena-Fluss. *Acta Soc. Sci. fenn.* 32(3): 1-40.
CAJANDER, A. K., – 1909a – Ueber Waldtypen. *Acta for. fenn.* 1(1): 1-175.
CAJANDER, A. K., – 1909b – Beiträge zur Kenntnis der Alluvionen des nördlichen Eurasiens III. Die Alluvionen der Tornio- und Kemi-Thäler. *Acta Soc. Sci. fenn.* 37(5): 1-224.
CAJANDER, A. K., – 1913 – Studien über die Moore Finnlands. *Acta for. fenn.* 2(3): 1-208.
CAJANDER, A. K., – 1926 – The theory of forest types. *Acta for. fenn.* 29(3): 1-108.
CAJANDER, A. K., – 1930 – Wesen und Bedeutung der Waldtypen. *Silva fenn.* 15(1): 1-66.
CAJANDER, A. K., – 1949 – Metsätyypit ja niiden merkitys. Forest types and their significance. (Finn. & Engl.) *Acta for. fenn.* 56(4): 1-69, 1-71.
CAJANDER, A. K., & Y. ILVESSALO, – 1921 – Über Waldtypen II. *Acta for. fenn.* 20(1): 1-77.
CLEMENTS, F. E., – 1916 – Plant succession: an analysis of the development of vegetation. *Publs Carnegie Instn*, Washington 242: 1-512.
CRANDALL, DOROTHY L., – 1958 – Ground vegetation patterns of the spruce-fir area of the Great Smoky Mountains National Park. *Ecol. Monogr.* 28: 337-360.
CURTIS, J. T., – 1959 – The Vegetation of Wisconsin: An Ordination of Plant Communities. Univ. Wisconsin, Madison. 657 pp.
DU RIETZ, G. E., – 1930 – Vegetationsforschung auf soziationsanalytischer Grundlage. *Handb. biol. ArbMeth.* 11, 5: 293-480.
ENEROTH, O., – 1936 – Om skogstyperna och deras praktiska betydelse. *Handl. LandbrVeck*, Stockholm 1936: 821-837.
EUROLA, S., – 1962 – Über die regionale Einteilung der südfinnischen Moore. *Ann. Bot., Soc. Zool.-Bot. Fenn. 'Vanamo'* (*Suomal. eläin-ja kasvit. Seur. van. kasvit. Julk.*) 33: 1-243.
GAMS, H., – 1918 – Prinzipienfragen der Vegetationsforschung. Ein Beitrag zur Begriffsklärung und Methodik der Biocoenologie. *Vjschr. naturf. Ges. Zürich* 63: 293-493.
HÄMET-AHTI, L., – 1963 – Zonation of the mountain birch forests in northern-

most Fennoscandia. *Ann. Bot. Soc. Zool., Bot. Fenn. 'Vanamo'* (*Suomal. eläin-ja kasvit. Seur. van kasvit. Julk.*) 34(4): 1-127.

HÄMET-AHTI, L., – 1965 – Notes on the vegetation zones of western Canada, with special reference to the forests of Wells Gray Park, British Columbia. *Ann. bot. fenn.* 2: 274-300.

HARPER, J. L., – 1967 – A Darwinian approach to plant ecology. *J. Ecol.* 55: 247-270.

HEIMBURGER, C. C., – 1934 – Forest-type studies in the Adirondack region. *Mem. Cornell Univ.* (*N.Y.*) *agric. Exp. Stn.* 165: 1-122.

HEIMBURGER, C. C., – 1941 – Forest-type classification and soil investigation on Lake Edward Forest Experimental Area. *Canada Dept. Mines Resources, Dominion Forest Serv., Silvic. Res. Note* 66: 1-60.

HILLS, G. A., – 1960 – Comparison of forest ecosystems (vegetation and soil) in different climatic zones. *Silva fenn.* 105: 33-39.

HÖGNÄS, B., – 1966 – Undersökning av skogstyper och beståndsutvickling på Åland. (Engl. summ.) *Acta for. fenn.* 78(2): 1-127.

HUSTICH, I., – 1949 – On the forest geography of the Labrador peninsula. A preliminary synthesis. *Acta Geogr., Helsingf.* 10(2): 1-63.

HUSTICH, I., – 1950 – Notes on the forests on the East coast of Hudson Bay and James Bay. *Acta Geogr., Helsingf.* 11(1): 1-83.

HUSTICH, I., ed., – 1960 – Symposium on forest types and forest ecosystems. *Silva fenn.* 105: 1-142.

ILVES, A., – 1953 – Eesti NSV aıumetsatüübid (Russ. summ.: Estonian upland forest types). In, 'Loodusuurijate Seltsi juubelikoguteos,' (Jubilee Symposium of the Estonian Naturalists Society), pp. 11-49. Tallinn.

ILVESSALO, Y., – 1920 – Tutkimuksia metsätyyppien taksatoorisesta merkityksesta nojautuen etupäässä kotimaiseen kasvutaulujen laatimistyöhön. (Germ. summ.: Untersuchungen über die taxatorische Bedeutung der Waldtypen). *Acta for. fenn.* 15(3): 1-157, 1-51, 1-32, 1-26.

ILVESSALO, Y., – 1922 – Vegetationsstatistische Untersuchungen über die Waldtypen. *Acta for. fenn.* 20(3): 1-73.

ILVESSALO, Y., – 1929 – Notes on some forest (site) types in North America. *Acta for. fenn.* 34(39): 1-111.

ILVESSALO, Y., – 1967 – Luonnonnormaalien metsiköiden kehityksestä Kainuussa ja sen lähiympäristössä. (Engl. summ.: On the development of natural normal forest stands in south-eastern North-Finland). *Acta for. fenn.* 81(5): 1-85.

JALAS, J., – 1950 – Zur Kausalanalyse der Verbreitung einiger nordischen Os- und Sandpflanzen. *Ann. Bot., Soc. Zool.-Bot. Fenn. 'Vanamo'* (*Suomal. eläin-ja kasvit. Seur. van. kasvit. Julk.*) 24(1): 1-362.

KALELA, A., – 1939 – Über Wiesen und wiesenartige Pflanzengesellschaften auf der Fischerhalbinsel in Petsamo Lappland. *Acta for. fenn.* 48(2): 1-523.

KALELA, A., – 1941 – Über die Entwicklung der herrschenden Bäume in den Beständen verschiedener Waldtypen Ostpatagoniens. *Ann. Acad. scient. fenn.* (*Suomal. Tiedeakat. Toim.*)Ser. A, 4 Biol., 3: 1-65.

KALELA, A., – 1952 – Kainuun alueen metsätyypeistä. (Germ. summ.: Über die Waldtypen des Kainuugebietes zwischen Mittel- und Nordfinnland.) *Comm. Inst. forest. fenn.* (*Metsätiet. Tutkimuslait. Julk.*) 40(26): 1-17.

KALELA, A., – 1954 – Zur Stellung der Waldtypen im System der Pflanzengesellschaften. *Vegetatio* 5/6: 50-62.

KALELA, A., – 1958 – Über die Waldvegetationszonen Finnlands. *Bot. Notiser* 111: 353-368.

KALELA, A., – 1960 – Classification of the vegetation, especially of the forests, with particular reference to regional problems. *Silva fenn.* 105: 40-49.
KALELA, A., – 1961 – Waldvegetationszonen Finnlands und ihre klimatischen Paralleltypen. *Arch. Soc. zool.-bot. fenn. 'Vanamo'* (*Suomal. eläin-ja kasvit. Seur. van. Tiedon.*) 16 (Suppl): 65-83.
KALELA, A., – 1962 – Notes on the forest and peatland vegetation in the Canadian Clay Belt Region and adjacent areas. *Comm. Inst. forest. fenn.* (*Metsätiet. Tutkimuslait. Julk.*) 53(33): 1-14.
KALLIO, K., – 1957 – Käenkaali-mustikkatyypin kuusikoiden kehityksestä Suomen lounaisosassa. Taksatoris-liiketaloudellinen tutkimus. (Engl. summ.: On the development of spruce forests of the Oxalis-Myrtillus site type in the south-west of Finland. Forest mensuration and management research). *Acta for. fenn.* 66(3): 1-155.
KALLIOLA, R., – 1943 – Porajärven seudun metsatyypeistä. (Germ. summ.: Über die Waldtypen der Gegend von Porajärvi in Ostkarelien). *Comm. Inst. forest fenn.* (*Metsätiet. Tutkimuslait. Julk.*) 31(2): 1-15.
KARU, A. & L. MUISTE, – 1958 – Eesti metsakasvukohatüübid. (Estonian forest site-types). Tallinn. 44pp.
KATZ, N. J., – 1929 – Die Zwillingsassoziationen und die homologen Reihen in der Phytosoziologie. *Ber. dt. bot. Ges.* 47: 154-164.
KATZ, N. J., 1930 – Die grundlegenden Gesetzmässigkeiten der Vegetation und der Begriff der Assoziation. *Beitr. Biol. Pfl.*, 18: 305-333.
KATS, N. YA., – 1936 – On plant communities of the Caucasian State Reserve in light of the species affinities (in Russian). *Zemlevêdênie* 38(3): 247-280.
KELTIKANGAS, V., – 1945 – Ojitettujen soitten viljavuus eli puuntuottokyky metsätyyppiteorian valossa. (Swed. & Engl. summs.: The fertility of drained bogs as shown by their tree producing capacity, considered in relation to the theory of forest types). *Acta for. fenn.* 53(1): 1-237.
KELTIKANGAS, V., – 1959 – Suomalaisista seinäsammaltyypeistä ja niiden asemasta Cajanderin luokitusjärjestelmässä. (Engl. summ.: Finnish feather-moss types and their position in Cajander's forest site classification). *Acta for. fenn.* 69(2): 1-266.
KIRSTEIN, K., – 1929 – Lettlands Waldtypen. *Acta for. fenn.* 34(33): 1-20.
KRAJINA, V. J., – 1960 – Can we find a common platform for the different schools of forest type classification. *Silva fenn.* 105: 50-55.
KUJALA, V., – 1921 – Havaintoja Kuusamon ja sen eteläpuoliston kuusimetsäalueiden metsä- ja suotyypeistä. (Germ. summ.: Beobachtungen über die Wald- und Moortypen von Kuusamo und der südlich von dort gelegenen Fichtenwaldgebiete). *Acta for. fenn.* 18(5): 1-65.
KUJALA, V., – 1929a – Die Bestände und die ökologischen Horizontalschichten der Vegetation. *Acta for. fenn.* 34(17): 1-26.
KUJALA, V., – 1929b – Untersuchungen über Waldtypen in Petsamo und in angrenzenden Teilen von Inari Lappland. *Comm. Inst. forest. fenn.* (*Metsätiet. Tutkimuslait. Julk.*) 13(9): 1-125.
KUJALA, V., –1936a – Tutkimuksia Keski- ja Pohjois-Suomen välisestä kasvillisuus rajasta (Germ. summ.: Über die Vegetationsgrenze von Mittel- und Nord-Finnland). *Comm. Inst. forest. fenn.* (*Metsätiet. Tutkimuslait. Julk.*) 22(4): 1-95.
KUJALA, V., – 1936b – Waldvegetationsstudien im östlichen Mitteleuropa (Finn. summ.) *Comm. Inst. forest. fenn.* (*Metsätiet. Tutkimuslait. Julk.*) 22(6): 1-115.
KUJALA, V., – 1938 – Metsätyyppien parallelisuudesta. (Germ. summ.: Über Parallelität der Waldtypen). *Comm. Inst. forest. fenn* (*Metsätiet. Tutkimuslait. Julk.*) 27(1): 1-17.
KUJALA, V., – 1945 – Waldvegetationsuntersuchungen in Kanada mit besonderer

Berücksichtigung der Anbaumöglichkeiten kanadischer Holzarten auf natürlichen Waldböden in Finnland. *Ann. Acad. scient. fenn.* (*Suomal. Tiedeakat. Toim.*) Ser. A, 4 Biol., 7: 1-434.
KUJALA, V., – 1960 – Can we find a common platform for the different schools of forest type classification? *Silva fenn.* 105: 56-59.
LAKARI, O. J., – 1920 – Tutkimuksia Pohjois-Suomen metsätyypeistä (Germ. summ.: Untersuchungen über die Waldtypen in Nordfinnland). *Acta for. fenn.* 14(4): 1-85.
LINDQUIST, B., – 1954 – Ein Waldtypenschema für die skandinavischen Buchenwälder. *Angew. PflSoziol.*. [Wien], Festschr. Aichinger 2: 965-970.
LINKOLA, K., – 1916 – Studien über den Einfluss der Kultur auf die Flora in den Gegenden nördlich vom Ladogasee. I. Allgemeiner Teil. *Acta Soc. Fauna Flora fenn.* 45(1): 1-432.
LINKOLA, K., – 1917 – Itä-Karjalan metsätyyppejä koskevia havaintoja. *Acta for. fenn.* 7: 224-245..
LINKOLA, K., – 1924 – Waldtypenstudien in den Schweizer Alpen. *Veröff. geobot. Inst. Rübel*, Zürich 1: 139-224.
LINKOLA, K., – 1929 – Zur Kenntnis der Waldtypen Eestis. *Acta for. fenn.* 34(40): 1-73.
LINKOLA, K., – 1930 – Über die Halbhainwälder in Eesti. *Acta for. fenn.* 36(3): 1-30.
LIPPMAA, T., – 1933 – Taimeühingute uurimise metoodika ja Eesti taimeühingute klassfikatsiooni põhijooni. (Germ. summ.: Grundzüge der pflanzensoziologischen Methodik nebst einer Klassifikation der Pflanzenassoziationen Estlands). *Acta Inst. Hort. bot. tartu.* 3 (4): 1-169, and *Loodusuur. Seltsi Aruanded* 40(1-2): 1-169.
LIPPMAA, T., – 1935 – La méthode des associations unistrates et le système écologique des associations. *Acta Inst. Hort. bot. tartu.* 4 (no. 1-2, art. 3): 1-7, and *Loodusuur. Seltsi Aruanded* (1934), 41(3-4): 205-211.
LIPPMAA, T., – 1939 – The unistratal concept of plant communities (the unions). *Am. Midl. Nat.* 21: 111-145.
LÕHMUS, E., – 1971 – Uusi andmeid Eesti metsatüüpidest (Some new data on Estonian forest types). In: '*Eesti Metsad.*' (in press).
LUKKALA, O. J., – 1929 – Tutkimuksia soiden metsätaloudellisesta ojituskelpoisuudesta. (Germ. summ.: Untersuchungen über die waldwirtschaftliche Entwässerungsfähigkeit der Moore). *Comm. Inst. (quaest.) forest. fenn.* (*Metsatiet. Tuttkimuslait. Julk.*) 15(1): 1-301.
MÄKIRINTA, U., – 1968 – Haintypenuntersuchungen im mittleren Süd-Häme, Südfinnland. *Ann. bot. fenn.* 5(1): 34-64.
MALLNER, F., – 1944 – Vorgeschichte und Werdegang des lettländischen Waldtypensystems. *Z. Forst. u. Jagdw.* 76: 105-136.
MALMSTRÖM, C., – 1926 – The experimental forests of Kulbäcksliden and Svartberget in North Sweden. 2. Vegetation. *Statens Skogsförsöksanst.* [Sweden] *Exkurs.-Ledare* 11: 27-87.
MALMSTRÖM, C., – 1936 – Norrlands viktigaste skogstyper. *Sver. Nat.* 27: 21-41.
MASING, V., – 1969 – Structural analysis of plant cover and classification problems. In 'Plant Taxonomy, Geography and Ecology in the Estonian S.S.R,' pp. 49-59. Tallinn.
MIRKIN, B. M., – 1965 – On the ecological classifications of meadow vegetation in floodplains (Russ. with Engl. summ.). *Bot. Zhur. SSSR.* 50: 324-334.
MOROSOW, G. F., – 1928 – Die Lehre vom Walde. Neuman, Neudamm. 375 pp.
MOSS, C. E., – 1910 – The fundamental units of vegetation. Historical development of the concepts of the plant association and the plant formation. *New Phytol.* 9: 18-53.

NIHLGÅRD, B., – 1970 – Vegetation types of planted spruce forests in Scania, Southern Sweden. *Bot. Notiser* 123: 310-337.

NILSSON, A., – 1901 – Sydsvenska ljunghedar. Stockholm.

NORRLIN, J. P., – 1870 – Bidrag till sydöstra Tavastlands flora. *Notiser ur Förhandl., Sällsk. Fauna Flora fenn.* 11 (N.S.8, 1871): 73-196. (Transl. as: Beiträge zur Flora des südöstlichen Tavastlands. *Acta for. fenn.* 23(4): 15-52, 1923.)

PAASIO, I., – 1937 – Suomen nevasoiden tyyppijärjestelmää koskevia tutkimuksia. (Germ. summ.: Untersuchungen über das Typensystem der Weissmoore Finnlands). *Acta for. fenn.* 44(3): 1-129.

PAASIO, I., – 1940 – Zur Kenntnis der Waldhochmoore Mittelfinnlands. (Finn. summ.) *Acta for. fenn.* 49(2): 1-41.

PAASIO, I., – 1941 – Zur pflanzensoziologischen Grundlage der Weissmoortypen. (Finn. summ.) *Acta for. fenn.* 49(3): 1-84.

PALMGREN, A., – 1915-17 – Studier över lövängsområdena på Åland. I-III. *Acta Soc. Fauna Flora fenn.* 42(1): 1-634.

PALMGREN, P., – 1928 – Zur Synthese pflanzen- und tierökologischer Untersuchungen. *Acta zool. fenn.* 6: 1-51.

PERTTULA, U., – 1950 – Kasvillisuudesta ylisellä Syvärillä sekä siihen etelässä rajoittuvalla Juksovon seudulla. (Germ. summ.: Über die Vegetation am oberen Lauf des Flusses Swir nebst der im Süden anschliessenden Gegend von Juksowo). *Ann. Bot. ,Soc. Zool.-Bot. Fenn. 'Vanamo'* (*Suomal. eläin-ja kasvit. Seur. van. kasvit. Julk.*) 23(6): 1-204.

POGREBNJAK, P. S., – 1929 – Über die Methodik der Standortsuntersuchungen in Verbindung mit den Waldtypen. *Verh. Int. Kongr. forstl. Versuchsanstalten* 2: 455-471.

RUUHIJÄRVI, R., – 1960 – Über die regionale Einteilung der nordfinnischen Moore. *Ann. Bot., Soc. Zool.-Bot. Fenn. 'Vanamo'* (*Suomal. eläin-ja kasvit. Seur. van. kasvit. Julk.*) 31(1): 1-360.

SMAGIN, V. N., – 1950 – An attempt towards a classification of forest types of the southern taiga subzone (in Russian). *Byull. mosk. Obshch. Ispȳt. Prir.*, Ser. Biol. 55(3): 86-89.

SOKOLOV, S. Ya., – 1937 – Advances of forest geobotany (in Russian). *Sov. Bot.* 6: 3-23.

SOKOLOV, S. Ya., – 1962 – Taxonomy of forest associations (in Russian). *Problemȳ Bot.* 6: 110-123.

SOKOLOWA, L. A., – 1935 – Die geobotanischen Bezirke der Onega-Dwina Wasserscheide und der Onega Halbinsel. (Russ. with Germ. summ.) *Trudȳ bot. Inst. Akad. Nauk SSSR*, Ser. 3 (Geobotanica) 2: 9-80.

SOKOLOWA, L. A., – 1936 – Die Vegetation des Rayons der Louchi-Kestenga Chaussee (Karelien). (Russ. with Germ. summ.) *Trudȳ bot. Inst. Akad. Nauk SSSR*, Ser. 3 (Geobotanica) 3: 241-306.

SPILSBURY, R. H. & D. S. SMITH., – 1947 – Forest site types of the Pacific Northwest: a preliminary report. *Tech. Publ. Forest Serv. Br. Columb.* (Dept. Lands & Forests), T. 30: 1-46.

SUKATSCHEW, W. N., – 1925 – Pflanzenassoziationen und Forstbestandestypus (Russian with Germ. summ.). *Trudȳ leningr. lesotekh. Akad.* 32: 39-56.

SUKACHEV, V. N., – 1927 – A Manual for Study of Forest Types (in Russian). 2nd ed., 1930, Moscow-Leningrad. 318 pp.

SUKACHEV, V. N., – 1928 – Principles of classification of the spruce communities of European Russia. *J. Ecol.* 16: 1-18.

SUKACHEV, V. N., – 1931 – The fundamental and leading ideas in studying forest types (in Russian). *Mitt. Leningrad Inst. wissenschl. Forsch. Gebiet Holzindust.* 18(3): 51-71.

Sukatschew, W. N., – 1932 – Die Untersuchung der Waldtypen des osteuropäischen Flachlandes. *Handb. biol. ArbMeth.* 11, 6: 191-250.
Sukachev, V. N., – 1934 – The phytocoenosis – what is it? (in Russian). *Sov. Bot.* 5: 4-18, 47-50.
Sukachev, V. N., – 1938 – The main concepts of the science of vegetation cover (in Russian). In 'The Vegetation of the USSR,' Moscow-Leningrad 1: 15-37.
Sukachev, V. N., – 1944 – On principles of genetic classification in biocenology. (Russ. with Engl. summ.) *Zh. obschch. Biol.* 5(4): 213-227.
Sukachev, V. N., – 1951 – The basic principles of forest typology. Proceedings of the Symposium on Forest Typology, Moscow, pp. 7-19.
Sukatschew, W. N., – 1954 – Die Grundlagen der Waldtypen. *Angew. PflSoziol.* [Wien], Festschr. Aichinger 2: 956-964.
Sukachev, V. N., – 1960 – The correlation between the concept 'forest ecosystem' and 'forest biogeocoenose' and their importance for the classification of forests. *Silva fenn.* 103: 94-97.
Tansley, A. G. ed., – 1911 – Types of British vegetation, by members of the central committee for the survey and study of British vegetation. Cambridge Univ. Press. 416 pp.
Trass, H., – 1964 – Geobotaanika. Ajalugu ja kaasaegsed arengusuunad. (Geobotany: History and Contemporary Trends of Development). MS, Tartu State Univ. 400 pp.
Tuomikoski, R., – 1942 – Untersuchungen über die Vegetation der Bruchmoore in Ostfinnland. 1. Zur Methodik der pflanzensoziologischen Systematik. (Finn. summ.) *Ann. bot., Soc. Zool.-Bot. Fenn. 'Vanamo'* (*Suomal. eläin-ja kasvit. Seur. van. kasvit. Julk.*) 17(1): 1-203.
Vainio, E. A., – 1878 – Kasvistonsuhteista Pohjais-Suomen ja Venäjän-Karjalan rajaseuduilla. (On the flora of the borderland districts between northern Finland and Russian Karelia). *Meddn. Soc. Fauna Flora fenn.* 4(1): 1-161.
Vorob'ev, D. V., – 1953 – Forest Types of the European Part of the U.S.S.R. (in Russian) Kiev. 450 pp.
Wendelberger, G., – 1951 – Das vegetationskundliche System Erwin Aichingers und seine Stellung im pflanzensoziologischen Lehrgebäude Braun-Blanquets. *Angew. PflSoziol.* [Wien] 1: 69-92.
Whittaker, R. H., – 1953 – A consideration of climax theory: the climax as a population and pattern. *Ecol. Monogr.* 23:41-78.
Whittaker, R. H., – 1956 – Vegetation of the Great Smoky Mountains. *Ecol. Monogr.* 26: 1-80.
Whittaker, R. H., – 1960 – Vegetation of the Siskiyou Mountains, Oregon and California. *Ecol. Monogr.* 30: 279-338.
Whittaker, R. H., – 1962 – Classification of natural communities. *Bot. Rev.* 28(1): 1-239.
Whittaker, R. H., – 1969 – Evolution of diversity in plant communities. *Brookhaven Symp. Biol.* 22: 178-196.
Whittaker, R. H. & W. A. Niering, – 1965 – Vegetation of the Santa Catalina Mountains, Arizona. (II) A gradient analysis of the south slope. *Ecology* 46: 429-452.
Yoda, K., T. Kira, H. Ogawa, & K. Hozumi, – 1963 – Self-thinning in overcrowded pure stands under cultivated and natural conditions. *J. Biol. Osaka Cy Univ.* 14: 107-129.

16 SYNUSIAL APPROACHES TO CLASSIFICATION

JAN J. BARKMAN

Contents

16 SYNUSIAL APPROACHES TO CLASSIFICATION

16.1 Introduction

Most of the classifications discussed in this volume apply in principle to communities or biocoenoses — i.e. the whole assemblages of organisms that live together in a particular site or biotope. Since the plants make up by far the largest part of the biomass, as well as the essential structure of land communities, it is reasonable to base our classifications upon the plants. Yet the plant community itself is often a complexly organized system, comprising groups of plants on different levels, differently adapted to light and substrate, differing in seasonal timing, and manner of resource use and function. It is a general observation from the study of plant communities that these groups of plants, adapted to different environmental factors within the community, also may differ in their distributions in relation to habitat factors and climatic gradients. These groups of plants show 'individuality' in their distribution and response to environment, both within and between communities, that justifies study of their composition, ecology, and dynamics, and may even justify a separate classification.

A particular plant community, in its particular local or topographic environment or *biotope*, is a *phytocoenose*. A grouping or class of phytocoenoses that are similar in species composition, or other significant characteristics, is a *community-type* or *coenon*; when the standard units of the school of Braun-Blanquet are in question, community-types are *syntaxa* (associations, etc.). A subdivision of a particular phytocoenose that is based primarily upon life-form, secondarily on position or manner of function of its species in the community, is sometimes termed a *society*.

A phytocoenose may be subdivided in two ways (Barkman 1970). The first, abstract approach groups the species of the phytocoenose according to (i) their structure into growth-form types, (ii) their reaction upon limiting factors of the environment into life-form groups, (iii) their ecological range into ecological groups, (iv) their sociological affinity (association index) into sociological groups. It is also possible to distinguish species groups according to (v) dissemination types, (vi) taxonomic groups, (vii) areal types, etc. Each grouping may be expressed in a spectrum, indicating the percentage share of all the specific groups present in the particular phytocoenose. This spectrum may then be generalized for the syntaxon in question.

A second, concrete approach divides phytocoenoses according to (i) the ecological aspect (microhabitats) into microcommunities, (ii) the spatial aspect into layer communities, (iii) the temporal phases into aspection (seasonal) communities, (iv) on the base of i, ii (and iii) into *societies*. The former approach is synthetic (the species groups must first be established on the basis of comparison with other species and other phytocoenoses), the latter is direct and analytical. It can be carried out within a single phytocoenose. The resulting units are not necessarily spatially distinct in the former case, as the species of one group are often scattered throughout the phytocoenose; but they are usually spatially distinct in the latter case, each society occupying a restricted space within the phytocoenose. Examples of such societies are: the group of tree species populations that make up a forest canopy, a group of herbs in a forest that grow from bulbs or rhizomes below the soil surface to bloom in the spring, a group of lichens that occur together on the bark of a tree trunk between the moss-clad base and the crown. The abstract units into which similar societies (in a number of phytocoenoses) are classified are *synusiae*, and these are the principal concern of this paper.

The study of societies and synusiae is an important alternative and complement to the study of phytocoenoses and syntaxa. I shall consider here the history of the synusial approach, the concepts and terms of greatest usefulness in it, its application to different kinds of communities, and the evaluation of this as an approach to research on communities.

16.2 Historical Development

16.2.1 Early Sources

The recognition and analysis of whole plant communities or phytocoenoses started before that of partial communities. LORENZ (1858) and KERNER (1863) were the first to see the importance of epiphytic and layer communities as fundamental structural elements of plant communities. LORENZ called them 'Komplexe,' and KERNER wrote, 'eine Pflanzenformation ist eine Verkettung von Beständen.' Later he called these 'Bestände' 'Genossenschaften.'

HULT (1881) was the first author to stress the relative independence of layer societies. In a forest with a uniform tree layer for instance two different moss societies may occur side by side, forming a mosaic. The uniform layer he called uniting society ('förenings-bestånd'), the societies of the heterogeneous layer alternating societies ('alternatbestånd'); the whole community is then a twin formation ('tvillings-formation'). He emphasized that any layer may function as uniting society. These ideas were followed by Russian geobotanists. Thus Korjinsky (1888) spoke of 'horizons,' and Katz (1929) classified twin formations into parallel and homologous series. Among other early plant geographers Warming (1909) used the term 'subordinate communities' and Drude (1919) 'Elementarassoziationen.' There is difference of opinion on the meaning of the latter; Braun-Blanquet (1928) considered them to correspond with his subassociations or facies, Du Rietz (1932) with synusiae.

These terms all referred to spatially distinct units within the plant community. Clements (1905), on the other hand, divided plant communities into seasonal aspects and called these 'societies.' In 1907 he narrowed this conception down to spatially distinct units by introducing the condition that societies should have 'a principal subdominant species.' This necessarily restricted his notion to layer communities, but only to subordinate layers, for a dominant species according to his definition occurs only in the uppermost layer. 'In a forest societies are found only beneath the primary layer of trees.' According to his climax theory Clements (1916) divided these units into societies proper (i.e. parts of a climax association) and socies (i.e. parts of a seral unit or associes). In 1928 these notions were further elaborated and a distinction was now made between (i) aspect or seasonal, (ii) layer and (iii) cryptogam societies. As to (iii) Clements remarked that 'distinctions into ground societies, parasitic societies (i.e. those mostly on leaves and herbaceous stems...) and bark societies (...) are convenient, but of minor importance. A distinction based upon life-form, i.e. moss, liverwort, lichen, and fungus is probably of greater value.' (One notes that this is a subdivision based more upon taxonomy (taxocoenoses) than upon life-form.) Apparently spatial isolation, so important a criterion for his layer communities, was not considered by him of much significance for the cryptogam synusiae. Few, if any, other authors have accepted this viewpoint.

16.2.2 The Influence of Gams

The term synusia (German: 'Verein') was first proposed in 1917 in a lecture by Rübel (cf. Rübel 1925) and defined as an *edaphic* unit within a plant association (= 'ökologischer Verein' in the sense of

WARMING), 'for instance a layer community or an epiphytic community.' Modern development of the synusial approach really stems, however, from a most thorough and original treatment of the vegetation science of that time by GAMS (1918). Gams very rightly remarked that we should distinguish between concrete and abstract communities (community-types) and he used the term synusia only in the latter sense. Stands of synusiae were called 'Bestände.' Unfortunately later authors used 'synusia' in both concrete and abstract senses[1]).

GAMS distinguished three types of synusiae, which RÜBEL (1925) considered 'eher hemmend als fördernd für die allgemeine Einführung.' Those of the lowest level ('synusiae of the first grade') were defined as communities of plants or animals, the independent components of which belong to the same life-form and within one (biogeographical) district to the same species. Those of the second grade (level) were defined as communities, the independent components of which belong to different species of the same life-form class and essentially the same temporal aspect. Later authors restricted the term 'synusia' to those of the second grade. GAMS' synusiae of the third grade, finally, are not partial communities at all, but are biocoenoses.

The word 'independent' in GAMS' definitions implies that a synusia of any grade may include dependent organisms of quite different life-forms and vegetation layers, as GAMS pointed out himself. In a pine wood for instance the pine trees form a synusia of the first grade together with their parasitic higher plants (*Viscum album*) and their parasitic and mycorrhizal fungi, as well as phytophagous and gall-forming insects. The synusia would exclude, however, some epiphytic mosses, not because they are wholly autotrophic organisms but because they also occur on soil and rocks (!), thus being independent of the tree bark as a substratum. One notes the difficulty here: the moss *species* is independent of the pine synusia but the epiphytic moss *individual* is not; furthermore, species show varying degrees of dependence on others for support, and/or nutrition.

The inclusion of so many different dependent organisms makes GAMS' synusiae rather impractical. GAMS' life-form classes also were rather too broadly conceived. The Chamaephyta for instance in-

1) Favourable exceptions were DE VRIES and SCHEYGROND (SCHEYGROND 1931) who used the terms 'enkelvoudig verband' (concrete) and 'associatie' (abstract). These terms also, however, are very misleading. The Dutch word 'Verband' also exists in German, where it means 'alliance' (a higher abstract vegetation unit of entire phytocoenoses, Dutch: 'Verbond'); whereas 'association' was, at that time, being used in many different meanings (CLEMENTS, BRAUN-BLANQUET, DU RIETZ), but always for entire phytocoenoses (concrete or abstract).

clude both perennial bryophytes and dwarf shrubs. The clearly distinct moss layer and dwarf shrub layer in a heath or coniferous wood, therefore, belong to the same synusia (second level). This procedure has not been followed by later authors.

It is most curious that few of the later authors (for instance Du Rietz 1932 and Poore 1955) have read Gams' definitions well. When citing Gams, they omit the word 'independent' and therefore exclude the dependent organisms of other life-forms and strata from their synusiae. This misinterpretation may have been caused by Gams' shorter, less clear definition in his table on p. 421 and in his summary on p. 475 (and by the fact that Gams himself quite wrongly suggested, p. 429, that Clements' aspect and layer societies are nothing else than his own synusiae of the first and second grade). But the oversight may also have been influenced by a choice of practicality. The later authors chose the more useful definition of the synusia as a group of species of the same or closely related life-forms within a community. A synusia can, in principle, be a group of 'independent' plants (e.g. the canopy of a forest), or a group of epiphytic species (e.g. mosses or lichens on the bark of the trees), or a group of species that are partly or wholly dependent in nutrition (e.g. the mycorrhizal fungi of the forest floor).

16.2.3 More Recent Conceptions

Braun-Blanquet (1928) defined synusiae as 'communities of species of one life-form group and with uniform habitat requirements'. In this the explicit reference to the habitat of the community is new. Temporal differences are omitted from the definition; yet seasonal aspects do belong to different synusiae for Braun-Blanquet, partly on the basis of difference in life-form. Within a Mediterranean association, for instance, he distinguished a hiemal synusia (hepatics, winter therophytes), a vernal synusia (geophytes and spring therophytes) and an aestival synusia (chamaephytes and hemicryptophytes). In the second (1951) and third (1964) edition of his '*Pflanzensoziologie*' Braun-Blanquet gave almost the same definitions as in 1928. Even in the 1964 edition the term 'life-form group' is not explained, and the chapter on life-forms uses the term in two different meanings, (i) as a subdivision of a life-form class (p. 145) and (ii) as identical with a life-form class (p. 146). His examples on p. 167 show, however, that, in so far as synusiae are concerned, the term 'life-form group' is applied in the latter sense (i.e. identical with the large life-form classes of Raunkjaer). However, the author is not consistent, for he considers moss layer and dwarf shrub layer to belong to different synusiae.

Contrary to most authors BRAUN-BLANQUET's synusiae are based upon life-form and seasonal aspect only, irrespective of floristic composition. DU RIETZ (1930), in contrast, incorporated floristic composition and 'sociological affinity' between the component species into his definition, but neglected life-form and seasonal aspect. His synusiae are identical with layer communities (stratocoenoses in the sense of BALOGH 1938), but the epiphytes are considered to form separate synusiae, the epilithic species not. In 1936 DU RIETZ defined synusiae simply as 'one-layer communities', but in fact his concept had become narrower, for soil-inhabiting bacteria and fungi were considered to constitute two separate synusiae, although living in the same layer (level). In 1957 his definition was more, although not fully, adjusted to his own use, 'a plant community .... either forming one layer or growing epiphytically on other members of the biocoenose.' DU RIETZ (1965) also distinguished various zoosynusiae, some of which inhabit the same vegetation layer and differ only in life-form and in their way of exploiting the habitat. Thus the concept of 'niche' introduced by zoologists was tacitly incorporated in the synusial concept. BARKMAN (1968) used the term 'synusia' in a sense very close to DU RIETZ (1965). However, not only the epiphytes, but also epilithic communities as well as all terrestrial communities on special substrata were considered to be distinct synusiae. In 1970 he also distinguished seasonal synusiae.

ELLENBERG (1956), CAIN & CASTRO (1959) and DAUBENMIRE (1968) used definitions closely similar to BRAUN-BLANQUET's, but ELLENBERG took a narrower view (mosses and lichens separated), whereas CAIN & CASTRO held a slightly broader view, defining a synusia as 'a social aggregate consisting of one *or a few closely related* life-forms occurring together and having a similar ecology' (italics by the present writer). SCAMONI (1963) cherished a quite different opinion, excluding, as he did, the layer communities and restricting synusiae to parts of a plant community inhabiting special microhabitats (including all layers?). His notion seems to be identical with KORCHAGIN's (1964) microcoenosis. Within the phytocoenose KORCHAGIN distinguished not one, but three different levels of structural elements: coenopopulations, synusiae, and microcoenoses. According to him phytocoenoses may be divided vertically into (i) layer synusiae and these may be subdivided into (ii) intralayer synusiae. Among the former he distinguished (ia) layers characterized by the component biomorphs (dominant life-form?), (ib) sublayers differing by the average height of plants within one layer (tall and low herb layer for instance?), (ic) 'canopies', represented by groups of the same species, but of various heights (due to different age classes). Intralayer synusiae may be (iia) societies within a layer, differing in floristic composition, density, etc., (iib) seasonal aspects or aspect societies, (iic) epiphytes. Inhomogeneous phytocoenoses (mosaic complexes etc.) may be divided horizontally into microcoenoses or coeno-elements, differing in composition, etc. Within each microcoenose the layer synusiae are represented by special societies, differing from those of adjacent microcoenoses. Clearly these microcoenoses can be identical with the narrowly defined sociations of DU RIETZ et al. if the component societies have one dominant species each, but this need not be the case. BARKMAN (1968) used the terms 'Mikrogesellschaft' and (1969, 1970) 'microcoenosis' in exactly the same sense as KORCHAGIN without being familiar with his publication. BARKMAN (1970) defined synusiae as parts of a microcoenose (or phytocoenose, if no microcoenoses can be distinguished), formed by plants that belong to the same stratum and do not differ fundamentally in either periodicity or way of habitat exploitation. As a general term covering both microcoenoses and synusiae the word 'merocoenosis' was introduced by MÖRZER BRUINS (1947). Although being applied to (both plant and animal) synusiae only, his definition is wider and includes both horizontal and vertical partitions of biocoenoses, for he speaks only of 'spatially distinct parts of biocoenoses.' Stratocoenoses (the 'horizontaal complex-verband' of DE VRIES in SCHEYGROND 1931) are the vertically separated counterparts of the microcoenoses (the 'verticaal complex-verband' of DE VRIES, l.c.). The former are often identical with synusiae but may also consist of more than one synusia. A survey of the elements of biocoenoses, their interrelations, and current synonyms is given in table I.

16.2.4 THE ATTITUDE OF THE BRAUN-BLANQUET SCHOOL

In the school of BRAUN-BLANQUET a growing tendency can be observed towards giving both stratal and seasonal synusiae more independent syntaxonomical status, although BRAUN-BLANQUET (1964:168) himself has still ignored the problem of taxonomic independence: 'Es besteht natürlich ohne weiteres die Möglichkeit, die Synusien den floristisch gefassten Vegetationseinheiten einzugliedern.'

Originally all synusiae occurring in one particular site were considered parts of one association, the only exception being the epilithes and epiphytes. These were considered to form dependent communities ('abhängige Gesellschaften'), but they were not included in the relevés nor in the association tables of forests, and

Table I. Elements of the Biogeocoenose and their Relations

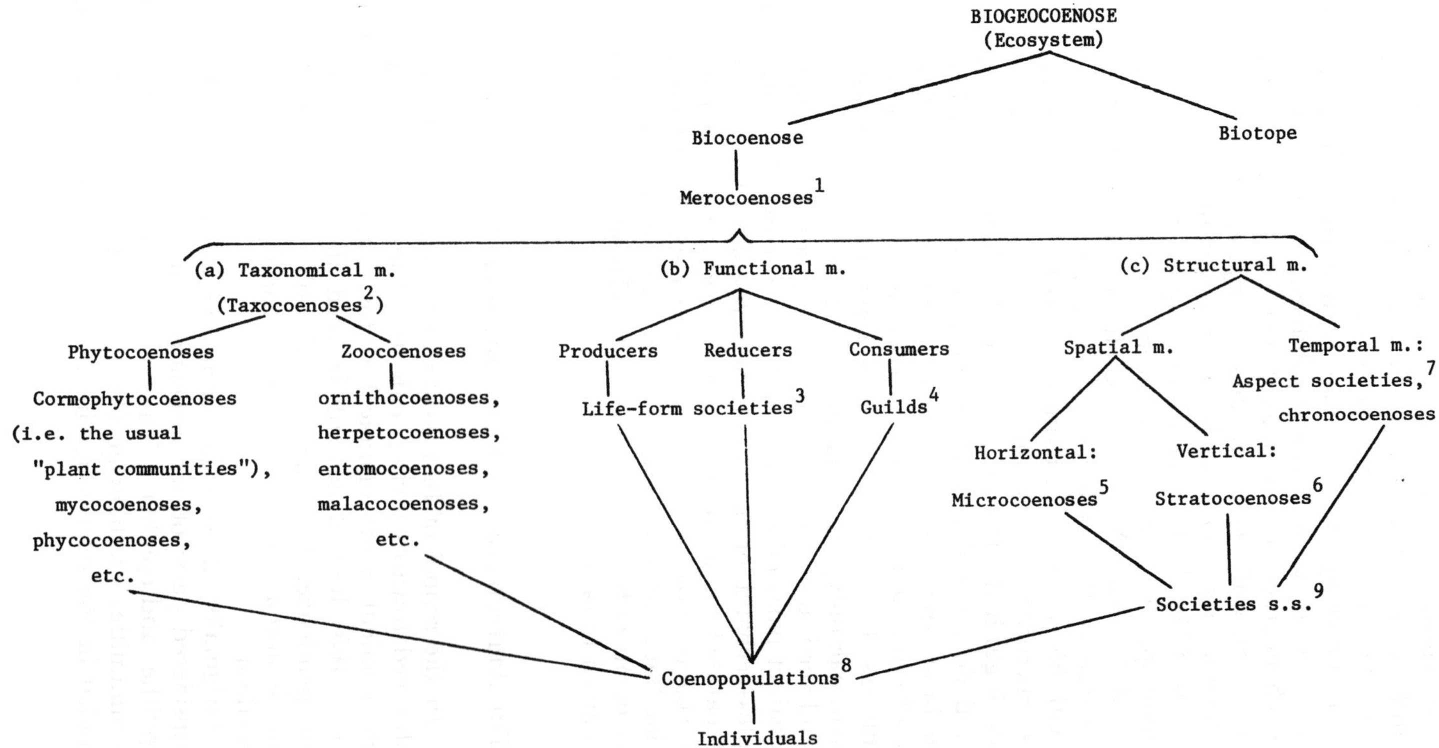

they were described as separate associations. ALLORGE (1922) was the first to publish epiphytic associations, both of woods and wayside trees, followed by OCHSNER (1928) and many others. Later epilithic communities were treated in the same way, starting with MOTYKA (1925) and KLEMENT (1931, 1955). SCHUMACHER (1944), followed by PHILIPPI (1956, 1963), described bryophyte communities on mineral soil in woods as separate associations, whereas the moss communities on humus were still considered part of the forest association. KOPPE (1955) rightly remarked that this is inconsistent; in his opinion none of these bryophyte communities can be considered associations, but all are socions or societates in the sense of DU RIETZ. PIRK & TÜXEN (1949) described a fungus association occurring on cow dung in meadows.

In the BRAUN-BLANQUET school the description of synusial associations was not confined to cryptogam communities, however. TÜXEN (1950) described a number of 'Schleiergesellschaften', which overgrow other plants like a veil, called them associations, and even united then into a separate order (Convolvuletalia sepium). However, these communities root in the same soil as the plants which they overgrow or use as a support, and they cannot be separated from the latter otherwise than as synusiae (BARKMAN 1968,

---

Table 1. (opposite)

Elements of the biogeocoenose are arranged with increasing narrowness of content downward on the vertical, manners of division on the horizontal. Three primary kinds of relationships—taxonomic, functional, and structural (including temporal)—may be used to divide the biocoenose into merocoenoses or community-fractions, which may in turn be subdivided into narrower merocoenoses until one reaches the coenopopulations (of individual species in the biocoenose), each occupying its niche and, finally, the individual organisms of which coenopopulations are composed. To every merocoenose an abstract type or class corresponds; these are in general designated by substituting the suffix -coenon for -coenose (e.g. merocoenon, see Table II). Abstract classes of *societies* in the strict sense, however, are *synusiae* in the sense of this article. References indicated: 1) MÖRZER BRUINS (1947); 2) VAN DER MAAREL (1965); 3) = Synusiae in the sense of BRAUN-BLANQUET, ELLENBERG, DAUBENMIRE, CAIN & CASTRO, not in the sense of GAMS; 4) ROOT (1967); 5) KORCHAGIN (1964), BARKMAN (1969, 1970) = Synusiae sensu SCAMONI (1963); 6) BALOGH (1938) = Synusiae sensu DU RIETZ (1930) = Layer synusiae KORCHAGIN (1964); 7) = Society sensu CLEMENTS (1905); 8) KORCHAGIN (1964); 9) Synusiae sensu DU RIETZ (1965), WILMANNS (1962), BARKMAN (1970) = Intra-layer synusiae KORCHAGIN (1964).

Westhoff 1968, Westhoff & den Held 1969). Yet, the two last-named authors maintain the order Convolvuletalia, 'only because there is no better solution'. Koch & Tüxen (1954, in Oberdorfer 1957) distinguished the floating duckweed communities from the rooting water plants with which they are associated, as a separate class, Lemnetea. Aquatic and epiphytic societies offer special problems which we shall discuss separately below.

A new development in the Braun-Blanquet school was the introduction of seasonal associations. Although Pirk & Tüxen (1949) observed seasonal aspects in a fungus association on cow dung which were totally different as to species composition, they rejected the idea of splitting this association. Oberdorfer (1954), however, distinguished several phanerogamic weed associations in the same arable field, succeeding each other during the year. Hübschmann (1960) distinguished, within certain types of arable land, a summer association of vascular weed plants, and two bryophytic associations, developing in autumn and winter respectively. If this trend is to be continued, we might eventually find ourselves in a situation where one cannot see the wood for the trees, as is clearly suggested by recent developments in taxonomy of aquatic communities (den Hartog & Segal 1964). It is therefore not surprising that this led to reactions both from opponents of the Braun-Blanquet school (Schmid 1941) and from within this school (Höfler 1959 and Zolyomi, Wilmanns, Barkman).

16.2.5 Ranks of Synusial Units, Du Rietz and Lippmaa

Gams was of the opinion that every society (synusia) has its own ecology and its own type of succession. He therefore considered his synusiae (1st and 2nd grade) as the fundamental vegetation units and wrote a monograph based on these units (Gams 1927). Regel (1923) called them 'associations,' a term around which a chaos of confusion existed at that time, and termed types of phytocoenoses 'association complexes.' Until 1930 Du Rietz, too, worked upon the basis of synusial (one-layer) associations, but he then (1930, 1932) proposed to describe both layer communities and phytocoenoses and to unite them into two independent hierarchical systems, although he still considered the former to be the fundamental units. The ranks of (abstract) synusiae proposed were: (1) socion, (2) consocion, (3) associon, (4) federion, (5) subformion, (6) formion, and (7) panformion. A consocion is a layer community with one constant dominant species or two codominant species, and may be combined with any dominant species in the other layers; a socion differs by being restricted to one particular sociation, i.e. a fixed set of dominants in all layers. Therefore a socion is a layer community associated with one particular socion in each other layer, whereas a consocion may be associated with different socia in the other layers. In single-layer phytocoenoses socion, consociation, and sociation are identical. An associon has various dominant species or no dominants at all, but a specific floristic composition. The same applies to the federion. Subformions and formions have a wide geographical range and hence a great floristic variation. They consist of a number of federions connected by gradual transitions and by having a number of species in common. Panformions are characterized by genera or families only. The seven levels of synusial units therefore correspond more or less with (1) sociations, (2) consociations, (3) associations (in the sense of BRAUN-BLANQUET) (4) alliances, (5) orders, (6) classes and (7) class groups (‚Klassengruppen'). It is interesting that DU RIETZ confined all these units to synusiae with a stable character, a viewpoint not shared by other authors.

The development of LIPPMAA's ideas shows exactly the opposite trend. Until 1933 he used the term 'association' both for unistratal and pluristratal community-types, but in 1935 he strongly advocated the use of unistratal associations only and applied this principle to the forest vegetation of Estonia. When a community has more than one layer, not the community as a whole, but each layer should be described and named separately. His arguments were (i) the relative autonomy of the layers, being independent topologically and synsystematically (not ecologically, as he pointed out himself), (ii) the great differences in minimum area of the various layers. LIPPMAA also offered a full hierarchy of synusial units. Within a union (associon) LIPPMAA (1938, 1939) recognized geographic 'facies.' For classification above the union LIPPMAA (1933b, 1939) proposed grouping the unions of a given area that are dominated by the same life-form (or two similar life-forms) and that are ecologically related (i.e. in habitats), into an 'Assoziationsgattung' or genus of unions. Higher units were suggested (association-family, -series, and -class) up to the 'division' comprising all the unions of the earth in which the same life-form or two closely related life-forms are dominant.

At the International Botanical Congress of Amsterdam, 1935, LIPPMAA's ideas were 'rejected' and DU RIETZ' ideas (two independent systems) adopted. Later even LIPPMAA (1938, 1939) was converted. At this congress DU RIETZ (1936) also proposed some changes in terminology, calling both his socions and consocions 'societies' (societas), the associon 'union' and the federion 'federation.' These new terms found wide acceptance among students of moss and lichen communities in Sweden, for instance WALDHEIM (1944, 1947), VON KRUSENSTJERNA (1945) and ALMBORN (1948). On two later occasions (1954, 1965) DU RIETZ again stressed the need for two independent hierarchical systems, one for the phytocoenoses, one for the synusiae. In the discussion following DU RIETZ 's 1965 lecture SCHMITHÜSEN and BARKMAN advocated the analysis of all synusiae, including the epiphytic, when describing a phytocoenose. The present writer has now applied this method to his study of juniper scrub (see below). In 1969 (in WESTHOFF & DEN HELD) he proposed to use, within the synusial system, the units union and federation of DU RIETZ as equivalents of association and alliance. The terms 'subformion, formion and panformion' (which were introduced in 1930 by DU RIETZ only as provisional terms and not maintained in his 1936 publication) were rejected by the present author, both because they suggest a relationship with formations, i.e. physiognomic-structural units, which do not fit into a vegetation system based on floristics, and because they have never been used by other authors. However, the terms order and class proposed by WILMANNS (1962), are already in use for phytocoenose units (syntaxa), so the terms 'ordulus' and 'classicula' were proposed instead (cf. BARKMAN 1970).

16.3 Discussion

The treatment of terms and interpretations of synusial units in the preceding may or may not have exhausted the reader, but it by no means exhausts the literature (see also 16.4.1 and 16.4.2). The proliferation of terms is a source of difficulty for any review attempting a synthetic perspective on synusial approaches. It is as if the students of synusiae had sought to emphasize the importance of their approach by creating a literature of terms exceeding that for phytocoenoses. We shall attempt some selection from this literature as regards (i) the criteria by which synusiae are to be recognized as such and (ii) terms and concepts that might be accepted as having the widest usefulness.

16.3.1 Criteria of Synusiae

Among the many and confusing interpretations of synusiae we may recognize the following criteria: (i) life-form, (ii) spatial separation, (iii) temporal separation, (iv) microenvironment, (v) habitat exploitation and function, (vi) dependence, (vii) composition and dominance, and (viii) taxonomy. The appropriateness of these may be discussed in this sequence.

(i) Life-form was introduced as a primary criterion by Gams (1918) and has been widely used since. Life-forms can be clearly distinguished, and group together species that are related in their major adaptations and functions in the community. Life-forms are more appropriate for separating synusiae than the growth-form concept of physiognomic ecology (article 13). Synusiae defined by life-form have the disadvantage of changing with change in life-form concepts (Wilmanns 1962). Also, we have to choose between different life-form systems (Raunkiaer 1934, Iversen 1936, etc.). So far nearly all synusiae have been based on Raunkiaer's classification. This system groups terrestrial plants primarily by vertical position of the vegetative buds in the unfavourable season (if there is any) in relation to the ground (substratum) surface. The system has disadvantages: (a) It is partly a growth-form system, (b) it is not very useful in ever-humid and warm (equatorial) climates, (c) it is not well adapted to aquatic plants, bryophytes, and lichens, (d) it does not cover the whole plant kingdom, (e) some of its life-form classes are too broadly conceived for our purpose (e.g. the Chamaephyta which comprise a.o. both bryophytes and dwarf shrubs), whereas the subdivisions of these classes are often too narrowly conceived (Chamaephyta reptantia and velantia for instance). These points (a–e) can be amended and partly have been amended by several authors. (f) If the life-form of the plant species is unknown, no synusiae can be distinguished in the growing season, whereas the unfavourable season (winter in the cold temperate region, summer in the Mediterranean etc.) is hardly suitable for vegetation analysis.

The use of Iversen's biological plant types has the disadvantages (a) that experimental work must be carried out before the species can be classified (b) that it is based on a single factor, important though it is, viz. the adaptation of plants to water (moisture, hydrature).

Both life-form systems have the disadvantage that in a forest all trees from seedlings to full-grown specimens would belong to the same synusia, despite their occurrence in layers above the ground with widely different microclimates and also in different root horizons often with very dissimilar physical and chemical soil properties. Besides seedlings and adult plants may have different demands upon the environment. In juniper scrub various shrub species always die as seedlings, whereas *Juniperus communis* itself is often present as a tall shrub only, being unable to germinate under its own canopy. It would seem hardly appropriate to unite these species into one 'shrub synusia.'

(ii) Spatial separation, particularly in a vertical sense (layering), has often been applied as a criterion. It sometimes meets, however, with difficulties, as layers cannot always be clearly distinguished (tropical rain forests (RICHARDS 1952), wet alder woods in the temperate zone, etc.). Distinct stratification is found especially in planted and cultivated woods and one might ask whether it is logical to base a system of layer synusiae upon such communities rather than using undisturbed phytocoenoses as a standard. Clear layering can, however, also be observed in natural subalpine and boreal (subarctic) forests (the latter being often less disturbed than the so-called 'primeval' tropical rain forests!) and in many seminatural phytocoenoses. Another disadvantage is that two layers that are distinct in one phytocoenose, may form one layer elsewhere (see 16.5.1). Usually, however, this phenomenon is restricted to sublayers, e.g. tall and low herb layer.

(iii) Temporal separation is a more controversial basis of distinction since, if the whole life-cycle is considered to represent a species, different seasonal aspects have the same floristic composition. Yet such groups as early-flowering bulb plants and summer grasses, or summer and winter therophytes, are so distinct as to periodicity that separation on at least a synusial level seems appropriate. All the more this holds true of such autonomous partial communities as moss societies developing on ploughed arable fields in winter and societies of therophytes developing only once in a number of years in open spaces of perennial plant communities (e.g. in Australian semideserts, DOING, verbal comm.).

(iv) Microenvironment is an obvious criterion, but needs some comment. The organisms of special substrata like tree trunks, rotten logs, leaves and herbaceous stems, decaying fungi, excrements, game tracks, rabbit holes, molehills, and outcropping rocks are distinctive and deserve recognition in any system. Yet each of these microbiotopes or microhabitats may support an assemblage of organisms of different life-forms and strata. They cannot be considered synusiae by any definitions except SCAMONI's (1963), but must be regarded as higher structural units as pointed out by KORCHAGIN (1964) and BARKMAN (1968), i.e. as microcommunities or microcoenoses each of which may include a number of synusiae. Microbiotopes thus do not define, but may contribute to the delimitation of, synusiae.

(v) Habitat exploitation and function is a more questionable criterion. It is included in the criterion of life-form (i) and partly in those of spatial separation (ii) and periodicity (iii). If life-forms are not used to distinguish synusiae, yet such groups as autotrophic plants, symbiotic (mycorrhiza), saprophytic, and parasitic fungi should be considered to belong to separate synusiae.

(vi) Independence. Syntaxonomic independence (one synusia occurring in various syntaxa) was one of the reasons for Du Rietz and Lippmaa to distinguish synusiae (even as associations). It is a practical reason indeed, but it should not imply that dependent synusiae are not to be accepted, if otherwise they form a clear-cut element in a phytocoenose. The criterion is therefore superfluous.

Ecological dependence was a reason for Gams for not giving the organisms in question the status of synusia, whereas for Clements ecologically 'independent' societies (uppermost layer) were not recognized as such. Neither of these opposing viewpoints seems reasonable, since (a) none of the societies in a phytocoenose is completely independent of the others, (b) there are all possible intergradations from wholly dependent (parasites) to almost independent (tree society), (c) one and the same synusia may occur in both a 'dependent' and an 'independent' position, for example moss societies on both living tree bark and on dead stumps, rocks and soil within a forest and even outside the forest.

(vii) Composition and dominance are alternative (but for some simple communities and for many societies nearly synonymous) criteria of units in any community classification (articles 14 and 20). Once the synusia concept has been defined on whatever basis, composition and dominance come in as important criteria for their delimitation and classification. However, we are here concerned only with the definition of the synusia as a general concept, not of particcular unions. Obviously the above criteria (i), (ii), (iii) and (v) happen to imply a difference in floristic composition as well, although in some cases different strata may consist of the same species. Floristic differences are often also involved in criterion (iv). If, however, in one stratum of a phytocoenose, in an apparently homogeneous microbiotope societies can yet be observed that are consistently and significantly different in composition and/or dominance, these will be distinguished as separate societies whether or not they also differ in life-form, periodicity, or function. Composition therefore is a primary criterion, regardless of the others.

(viii) A few authors distinguished synusiae on the basis of taxonomic groups. Klement (1955) and Ellenberg (1956) for instance separated bryophyte and lichen synusiae. This procedure is not generally accepted and it is not recommended. Higher taxonomic groups are seldom, if ever, homogeneous as to life-form, periodicity, or habitat requirements, nor do they as a rule belong to a single stratum or a single ecological or sociological group. Even moss and herb layer do not coincide with taxocoenoses, for vascular plants like *Tillaea muscosa*, *Sagina procumbens*, and *Cassiope hypnoides* distinctly belong to the moss layer, wherever they occur. Synusiae based on taxonomic groups only, should rather be called taxocoenoses, a term introduced by VAN DER MAAREL (1965) for animal groupings. For purely practical reasons it may be necessary to study taxocoenoses first (we must not forget that all 'plant communities' are in fact taxocoenoses), but our final aim must be to study biocoenoses and synusiae.

16.3.2 PROPOSALS ON CONCEPTS AND TERMS

It will be evident from the discussion in 16.3.1 that some criteria have to be rejected and others meet with serious objections or difficulties in application, with the exception of microhabitat and floristic composition. None of them is in itself a satisfactory basis for the definition of synusiae. Even microhabitat and floristic composition, which need to be incorporated in any definition, only define microcommunities if applied as the sole criteria. Synusiae should be based on these criteria either in combination with life-form or in combination with spatial and temporal separation and habitat exploitation. To the writer's mind the latter solution is the better. A synusia is thus defined as a structural part of a phytocoenose inhabiting (a) a special microhabitat, with (b) a specific floristic composition and consisting of species that (c) belong to the same stratum and that do not differ fundamentally in either (d) periodicity or (e) way of exploitation of their environment. The preceding has also indicated the need for distinction between microcommunities and synusiae. If we omit c, d and e from the above definition, it covers the notion of microcommunity. A microcommunity or microcoenose is thus defined by a special microhabitat and a specific floristic composition (a–b). Horizontal combinations of synusiae of the same stratum (b–c) are called 'stratocoenoses.' If a consistent nomenclature is desired, aspect societies (b–d) could be called 'chronocoenoses.' For taxonomically limited fractions of phytocoenoses the term 'taxocoenose' is already in use. A general term for any of the above partial communities is 'merocoenose,' introduced by MÖRZER BRUINS (1947). All these terms (except synusia) refer to individual stands. Any of these kinds of partial communities can be classified into abstract types. Much confusion has been aroused by using the same word in a concrete and an abstract sense. In analogy with the term 'coenon' (for an abstract whole-community unit or community-type) VAN DER MAAREL (1965) proposed the terms 'merocoenon,' 'stratocoenon,' and 'taxocoenon,' BARKMAN (1970) the term 'microcoenon.' If need be, we could also speak of 'chronocoena.' Unfortunately the term 'synusia' is still being used in both the concrete and abstract sense. The word 'Bestand' used by GAMS for concrete stands of synusiae cannot be recommended because it is not international, and because it is frequently used for entire phytocoenoses (also as its English equivalent, 'stand'). The author and the editor of this chapter suggest that 'synusia' should be applied to abstract units, as GAMS intended, and that 'society' should be revived as the English term for a concrete stand of a synusia (in spite of the different meaning given to 'society' by DU RIETZ (1936), cf. 16.2.5). The German equivalent, 'Verein,' has been more widely used in both abstract and concrete senses, but there would be real advantage in limiting it to the concrete, while 'synusia' and other more formal terms apply to the abstract units. For the principal synusial category in English the term 'union' has been established, especially by DU RIETZ (1936), LIPPMAA (1939) and CAIN & SHARP (1938); it may well be accepted in other languages as an international term for the basic synusial unit, hence paralleling 'association' as the basic unit for phytocoenoses. 'Union' may best apply to broader synusial groupings defined by total species composition, in parallel with the manner in which associations are defined. A subdivision of a union, characterized by differential species, can be called 'subunion' (cf. subassociation). It is proposed here to continue the use of the terms variant, subvariant, vicariant, and subvicariant also for synusial groupings and not to create new terms. A subdivision of a union defined by dominance of (or limited to) a single species, can be called 'socion' (paralleling 'sociation'). If consisting of the dominant species only, the 'socion' is identical with KORCHAGIN's 'coenopopulation.' For higher categories grouping unions I suggest 'federation,' 'ordulus,' and 'classicula' (these being in parallel with alliance or 'Verband,' order, and class for syntaxa) (BARKMAN 1970).

In the effort to bring some order out of confusion, with a number of terms that is sufficient but not excessive, Table II has been prepared after discussion of the problems between author and editor. Table II is void of hierarchial terms for taxo-, micro-, strato-, and chronocoena. The temptation to fill these spaces with new sets of terms has been firmly resisted. It would, for one thing, complicate our system of vegetation units too much. And it is quite possible and sufficient to use simple designations or to describe such merocoena in terms of combinations of synusiae. An additional advantage of the latter method is that it provides us with a flexible system to cope with all problems (cf. 16.4.5).

TABLE II.

Some major concepts and terms for communities and fractions thereof.

	Level of treatment	Concrete objects of study	Abstract units, on any level	Abstract units, hierarchial categories
A.	Whole communities	Community, biocoenose	Community-type, coenon	Various units (see article 12
B.	Plant communities	Plant community, phytocoenose, stand	Syntaxon (for school of BRAUN-BLANQUET)	*association* higher units–alliance, order, class, class group lower units–subassociation, variant, facies; sociation
C.	Any subdivision of the whole community or of the plant community	Subcommunity, merocoenose	Merocoenon	
C. 1.	Taxonomic groupings of species	Taxocoenose	Taxocoenon	
C. 2.	Microbiotope groupings	Microcommunity, microcoenose	Microcoenon	
C. 3.	Layer divisions	Layer community, stratocoenose	Stratocoenon	
C. 4.	Seasonal groupings	Aspect society, chronocoenose	Chronocoenon	
C. 5.	Combinations of C2-4	Society	Synusia	*union* higher units–federation, ordulus, classicula lower units–subunion, variant, vicariant; socion

16.4 **Applications**

16.4.1 VASCULAR PLANTS ON LAND

The synusial approach to whole plant communities on land leads from the proposals of GAMS (1918) and application in the Alps by BOLLETER (1921) and GAMS (1927), to Russian work (see also KORCHAGIN 1964, ALEKSANDROVA 1969, and article 17), and the distinctive Estonian school of LIPPMAA (1931, 1933a, 1933b, 1935a, 1935b, 1935c, 1939), PASTAK 1935, SIRGO 1936, TOMSON 1937 and VAGA 1940.

Gams, as discussed above, recognized three levels or degrees of synusiae, and distinguished these as 'ecological' units from 'topographic' units, such as the association, formed usually from combinations of synusiae. The approach was applied by Bolleter (1921, cf. Gams 1927, 1936), who treated communities both in terms of ecological units or synusiae ('Vereine') and in terms of spatial or topographic units ('Gemeinde') often consisting of several of these. The 'Gemeinde' were some of the more important, among the many possible, combinations of unions and were characterized by the dominant layer and component unions. In the Soviet Union the synusial approach was applied to desert and steppe communities by Keller (1932) and Grossheim (1930). Keller recognized synusial units as 'Genossenschaften'. Grossheim (1930, Grossheim & Prilipko 1929) further elaborated Gams' grades of synusiae with assumptions on the evolution of communities (from loosely organized and simple 'aggregations' (synusiae of the first degree) to more stable, simple 'agglomerations' (of second degree), to pluristratal 'semiassociations' in which the strata retain independence, and pluristratal 'associations' in which the strata have lost independence (the last two are synusiae of the third degree)).

Lippmaa (1933b, 1939) considered that the elemental units of vegetation were 'associations unistrates,' later called 'unions,' confined to a single stratum, with one or two closely related life-forms dominant. A union might consist of many species related by life-form and ecology, or of only one or two species. These single-layer associations were thought to show evident independence of one another, and to combine themselves in a quite variable fashion into forest stands, in which environmental relations were better revealed by the lower strata than by the canopy (Lippmaa 1935a, 1935b, 1935c, cf. article 15). Such complex vegetation as that of a forest was regarded by Lippmaa (1935b) as a complicated assemblage of superimposed 'associations,' with each stratum composed of one or more elemental associations or unions. Since species are differently distributed, floristic composition alone was considered inadequate for the definition of unions. Lippmaa felt that classification of communities must be ecological, and that the union must be characterized by habitat and life-forms as well as floristic composition (Lippmaa 1931, 1933a, 1939). Unions could be grouped and divided to form a full hierarchy of synusial units (see 16.2.5).

Lippmaa (1938, 1939) discussed as an example the Galeobdolon-Asperula-Asarum union of European forest herbs in its distribution, relations to forest communities, division into geographic facies and probable history. Lippmaa's unions were considered by him different and narrower units than the synusiae of Gams.

The union differed also from the stratal components of the sociation (Du Rietz 1930, 1932), the 'socion' and 'consocion'; for the unions were broader units that might be of varied floristic composition (cf. the 'associon' of Du Rietz 1932). The approaches of the Uppsala and Estonian schools were otherwise closely related, since both treated communities in terms of stratal or synusial units and also of 'topographic' units built up of these; the schools differed in which of the units received primary emphasis (see also article 18). Approaches based on stratal units, and complexes of these, were applied also by Regel (1923, 1927) in Lithuania, and de Vries (1926, 1932) in The Netherlands.

The synusial approach has not, however, been widely applied to the vascular plants of land communities. Gams, Bolleter, and Lippmaa et al. appear to have demonstrated an interesting possibility which, however, the great majority of phytosociologists have been content to leave as such while concentrating their primary emphasis on the composition (or stratal structure) of whole phytocoenoses. Thus the synusial approach has faded from active use for the vascular plants of land communities, while at the same time having growing influence on the study of epiphytes, water plants, and soil organisms.

16.4.2 Epiphytic Societies

Until recently epiphytic societies have either been ignored in vegetation analysis or described as autonomous syntaxa (associations etc.) even by those who otherwise support the idea of pluristratal associations. The first has been done by most students of terrestrial vegetation of all schools, the second procedure was followed by moss and lichen specialists with vegetational interests. Until recently the latter usually made only brief mention of the forest types in question, without even indicating their syntaxonomic status. Even when both epiphytic and terrestrial associations were fully described, there was sometimes no mention of their connections (e.g. Allorge 1922, Jovet 1949). Distributions of lichen species were studied in various forest associations by Sulma (1935), and along a continuum of forest types by Hale (1955); but neither of them described either the lichen communities or the forest types. Koskinen (1955) published full vegetation records of lichens on trees, grouped according to forest types, host species and microhabitats on the trees; but he did not name the lichen communities nor did he describe the forest types, which were based on structure and site-type.

A distinct improvement was the work by CAIN & SHARP (1938) who, within each forest association, studied one stand and fully described both qualitatively and quantitatively its tree layer and the terrestrial (soil-), epilithic (rock-), epixylic (wood-), and corticolous (bark-surface) bryophyte societies. However, they did not indicate whether there was any ecological or topographical connection between a particular epilithic society and a particular epixylic or corticolous society. The herb enumeration was incomplete and without quantitative data. A similar study was done by SZCZAWINSKY (1953), and COOKE (1955) who described lichen, moss, and fungus societies for forest associations. BARKMAN (1958b) was primarily concerned with epiphytic community-types. He only briefly indicated the forest alliances and associations in which they were found. WILMANNS (1962) gave detailed enumerations of the terrestrial associations, including grasslands, weed communities, bogs etc., in which the epiphytic communities occurred. Neither of them published full joint descriptions of epiphyte and forest communities nor did they fully analyse the epiphytes of one forest stand or association. The latter has been done by RYDZAK (1961) who compared even different subassociations of forests, but he refrained from describing the epiphytic community-types. His vegetation records are often heterogeneous, comprising as they do the whole vegetation on the tree trunks from 0 to 3 m, including all sides and both fissures and bark plateaux. WILMANNS & BIBINGER (1966), too, analysed all epiphytes of each forest association, but they also described the epiphytic unions as such and determined their distribution on the tree boles (height, aspect) and their frequency percentages on the trees per forest stand. IWATSUKI & HATTORI (1957) in Japan were the first to publish full synoptical tables of records including both terrestrial plants and epiphytes; the same was done by SCOTT & ARMSTRONG (1966) in New Zealand. In the latter study all trees of the sample plot were taken into account, even if they were devoid of epiphytes, and the whole epiphytic vegetation up to 2 m was recorded. If a forest stand is to be analysed as a whole and characterised by its epiphytic vegetation, this is probably the most correct method; but if we want to describe the epiphytic societies as such, smaller and more homogeneous sample plots should be chosen, and the method of WILMANNS & BIBINGER (1966) is to be preferred.

All authors cited above, as well as a number of others agree that certain epiphyte synusiae (as community-types) are restricted to a particular tree species, irrespective of the forest association, whereas others are restricted to a special forest association or alliance, irrespective of the tree species. Only very few are more narrowly restricted — to one tree species in one association. Many, however, occur on both a number of tree species and in a variety of forest associations. Whenever a special linkage is observed, either to trees or forest associations (and above all if the latter are concerned), this linkage is usually of a local nature. Complete linkage of an epiphytic community-type throughout its area is extremely rare. It is only found in endemic unions with very small areas (cf. BARKMAN 1958b). WILMANNS (1962) studied the linkage between epiphytic unions and forest syntaxa, restricting her analysis to southwest Germany. She found that some unions are restricted to (are 'Charakter-Union' of) one association or even subassociation, others to one alliance or order, whereas some are not specific at all. Following GAMS (1918) she called these unions stenocoenotic and eurycoenotic respectively. These terms may coincide with stenoecious or stenotopic and euryoecious or eurytopic respectively, but not necessarily so: a particular forest association may be euryoecious itself and so is in that case the stenocoenotic epiphytic union strongly linked with that association. As a rule forest alliances and suballiances are well characterised by their epiphytic unions, forest associations are usually not. In addition to the 'character-unions,' some epiphytic unions and species can serve well as 'differential-unions' and differential-species for forest alliances, thus contributing to better delimitation of these vegetation types (WILMANNS 1962).

As stated above, it has been a tradition in the BRAUN-BLANQUET school from the very beginning to analyse, name and classify epiphytic societies as separate from forest and other terrestrial phytocoenoses. They were generally considered 'associations,' 'alliances,' etc. CAIN & SHARP (1938), however, called them 'unions' divided into 'facies' and opposed the use of the suffix -etum for these units, since they are not equivalent to the pluristratal associations of which they are part. It is therefore illogical that they had no objection to uniting these unions into unistratal alliances with the ending -ion. WALDHEIM (1944, 1947) and ALMBORN (1948) were more consistent, using the suffixes -etum and -ion although they considered these epiphytic synusiae unions and federations respectively. RICHARDS (1939) on the other hand used the term 'associule' and did not use the suffix -etum. HOSOKAWA (1954) rejected this term except for special studies on epiphyte succession, since it involves a dynamic element in the sense of CLEMENTS (1936). ŞTEFUREAC (1941), PHILLIPS (1951) and BARKMAN (1958b) considered epiphyte societies to be classified into synusiae, yet called the latter 'associations.' FELFÖLDY (1941) who based his units on characteristic species combinations held similar views. Yet he speaks of 'sociations' or 'microsociations' which are in fact pluristratal phytocoenoses, based on dominant species. WILMANNS (1958, 1959) still considered epiphytic communities as associations etc.; but in 1962 she changed her ideas and from then on called them synusiae, although still using the suffixes -etum, -ion, -etalia. In 1966 she extended this interpretation to epilithic and terrestrial cryptogamic synusiae.

Quite different opinions were offered by some Japanese authors, but scarcely accepted by other phytosociologists. HOSOKAWA (1951) found that in tropical forests even synusiae of vascular epiphytes are not correlated with those of vascular terrestrial plants. For this reason he kept them apart and created a series of new terms by adding the prefix aero- to the synusial units of DU RIETZ (1930, 1932), e.g. aerosocion, aeroassocion, etc. OMURA (1952, 1953) distinguished forest associations and within these 'epiphyte-associations' divided into 'epiphyte-sociations' without using the suffix -etum. Both were considered to be one-layer communities. HOSOKAWA (1954) argued that these terms are misleading since they suggest reference to complete phytocoenoses. Curiously enough, he also rejected his own terms ('aeroassocion' etc.) as well as DU RIETZ's terms from which they were derived, because (i) one aeroassocion may occur in different forest associations, (ii) one forest association or even sociation often has several epiphyte associons (unions), even on a single tree, i.e. units of a higher hierarchical level than the corresponding forest syntaxon itself. The concrete epiphyte societies are in his opinion part of a concrete forest phytocoenosis (ecological dependence), whereas the abstract epiphytic units are not part of the abstract forest syntaxa (taxonomic independence). This, of course, was what had been pointed out by DU RIETZ, LIPPMAA and others and the reason why DU RIETZ proposed two independent hierarchies and coined his synusial terms. Most authors, including the writer, therefore see no necessity for a set of new terms such as 'epies,' 'epilia,' and 'epido' for societies, unions, and federations of epiphyte societies, as was proposed by HOSOKAWA (1954, see, however NAKANISHI 1970). The terms introduced by HOSOKAWA (1951) and OMURA (1952) are equally superfluous because the epiphytic synusiae are not fundamentally different from other synusiae, except that they comprise adnate societies on (the mostly dead bark of) other living plants. But the same unions may be found on dead trees and even on rocks. Epilithic moss and lichen synusiae are adnate too, and the epiphytic ones described by these Japanese authors also contain flowering plants.

There are two reasons other than those of HOSOKAWA for questioning whether epiphytic communities are really synusiae. (i) They often consist of different life-forms as conceived by BARKMAN (1958b; cf. his table XVIII, p. 194). But if life-forms are conceived in a broader sense, for instance RAUNKIAER's, this objection is invalidated. Besides, we have seen that for good reasons some authors have based synusiae upon spatial location, especially layering, and not upon life-forms. (ii) Epiphytic synusiae may themselves be several-layered, having for instance a fruticose and a foliaceous or crustaceous lichen layer. These layers may in turn show distributional (and hence, potentially, syntaxonomic) independence (BARKMAN 1958b, p. 331). This implies that, according to our definition, such epiphytic community types are in fact microcoena rather than synusiae (BARKMAN 1969). I have, however, refrained from treating them as such since most epiphytic communities are one-layered, and in others the separation of the layers is obscure, or the layers if distinct are seldom superposed, but form a 'horizontal' mosaic. It is even questionable whether such mosaics are not rather to be treated as successional stages or mixtures of different units.

Summarizing we may state that to-day's trend in phytosociology is to classify epiphytic societies (even if they may be conceived as microcoenoses) into synusiae, and to give them names – also for practical reasons – according to the procedures followed in the classification of phytocoenoses. There is near-unanimous opinion on the last point because: (i) the use of other suffixes than -etum, -ion, etc. would necessitate hundreds of alterations in names of synusial units firmly established in the literature. (ii) it would be necessary to do the same for terrestrial synusiae some of which are unistratal (e.g. Salicornietum strictae) and therefore identical with syntaxa; it would be a quite arbitrary decision which suffix should then be used. (iii) the name of the name-giving species will usually make it clear whether one has to do with a synusia or a syntaxon.

16.4.3 AQUATIC COMMUNITIES

16.4.3.1 *Characteristics*

Aquatic communities offer special problems because they usually have a lower level of organization, and their synusiae are less tightly connected and more often developed as local facies or as pure stands of single species. This is due in part to the fact that differences in specific environments are less stable than in terrestrial communities. The latter show marked vertical gradients both above the soil (microclimate) and in the soil (horizons), whereas water has no vertical gradient except for light and temperature. Water currents tend to obliterate both horizontal and vertical differences in water chemistry, whereas soils may have distinct and permanent horizontal patterns of physical and chemical factors[1]). Finally, floating communities may be driven by wind and currents from one rooting plant community to another. For these reasons the synusial approach has been more widely used here than in terrestrial communities, although as DEN HARTOG & SEGAL (1964) pointed out, all plant and animal communities of a water body belong to the same nutritive cycle. ALMQUIST (1929), who otherwise dealt with biocoenoses only, almost confined himself to the description of synusiae when dealing with fresh-water vegetation. DU RIETZ (1965) followed the same procedure.

Before going into detail it is useful to distinguish with LUTHER (1949) between haptophytes (adnate or non-rooting hydrophytes), rhizophytes (radicant or rooted hydrophytes), planktophytes, and pleustophytes (i.e. microscopic and macroscopic errant or mobile hydrophytes respectively). The distinction of plants into adnate, radicant, and errant life-forms by GAMS is most fundamental and useful in the study of both terrestrial and aquatic vegetation.

16.4.3.2 *Errant Hydrophytes*

The planktophytes and smaller pleustophytes (i.e. the Lemnids and Ricciellids in the sense of DEN HARTOG & SEGAL 1964) differ from all other water plants by forming mobile synusiae, able to shift from one phytocoenose to another. They are also easily and frequently infested with diaspores of other species supplied by wind and birds. For this reason these synusiae are generally considered separate associations, even in the BRAUN-BLANQUET school. Curiously enough, this has been done for planktophytes well before the pleustophyte associations were recognised as such. TÜXEN (1937) for instance still incorporated societies of Lemnaceae in the order Potametalia and even considered *Lemna gibba* a faithful species of the Myriophyllo-Nupharetum.

On the other hand, ALLORGE (1922) distinguished already associations of planktontic algae within aquatic vascular plant communities. He was followed by many others. The reasons for this difference in treatment might be, (i) the fact that planktontic organisms often show vertical movements in addition to horizontal

[1]) Yet VILLERET (1965) claims that the aquatic environment has a mosaic structure and is far from homogeneous. This is likely to hold in particular for shallow stagnant waters, for VILLERET (1954) found an extreme short-distance variation of habitat factors in water at the edge of *Sphagnum* carpets in oligotrophic pools of Brittanny. I have had the same experience in shallow Dutch bog pools.

ones (but *Stratiotes* does the same), (ii) the frequent occurrence of seasonal aspects of these short-lived organisms which may be completely different floristically, (iii) the fact that they often form communities of their own (in large, deep-water basins), (iv) the difference in life-form, (v) the difference in reaction to ecological factors, resulting in syntaxonomic independence, (vi) the completely different methods of investigation, and (vii) the fact that they are studied by other specialists. The last point, although seemingly irrelevant, plays an important role. Many hydrobiologists are not familiar with vascular plant communities, and hardly any phytosociologist is familiar with (microscopic) algae. It is evident from points (vi) and (vii) that in the first stage of investigation planktonic algal communities have to be analysed and synthesized separately, i.e. as taxocoenoses. The same applies to adnate microphyte communities. In a later stage it may be useful to see in how far these communities are linked with and can be incorporated in other water plant communities (SCHMITZ 1965).

Although none of the above points is in itself decisive in requiring distinction of separate plankton phytocoenoses and syntaxa, all contribute to the fact that it may be practical to do so. BRAUN-BLANQUET (1964) is most apodictic in this respect: 'es handelt sich um Biozönosen.' MARGALEF (1947, 1949, 1955), however, considers the whole living community of a water basin as one biocoenosis, but he divides these biocoenoses into 'associations,' some of which are planktontic and comprise microscopic forms, both plants and animals. For him, points (iv) and (v) are obviously not relevant. SCHMITZ (1965), dealing with (mostly adnate) microphytic water communities (microscopic algae), considers them to be synusiae or rather 'microphytocoenoses,' and comments that 'it is hardly recommendable to distinguish several associations in the same locality, yet often impossible to attribute a microphytic community to a particular macrophytic association.' The names of his synusiae are formed in the same way as those of associations (-etum). VILLERET (1965), who speaks of algal associations, doubts their validity as they have been delimited: It is extremely rare, he says, to make a record that fits into one of these abstract units, with the exception of associations of extreme conditions (thermal waters, snow, rapids, salt water sources, etc.). One of the great difficulties in the study of microphytic vegetation is of course that one cannot see the plants. Consequently it is impossible to judge whether the microvegetation is homogeneous or not, nor can we determine the minimum area in the customary way. These difficulties are comparable with those of the study of large algae communities of deep water and the societies of fungi, soil inhabiting or small animals, and animals with nocturnal activity.

As to the communities of smaller pleustophytes, their autonomy has been recognized since KOCH & TÜXEN (in OBERDORFER 1957) created the class of Lemnetea, with one alliance, Lemnion minoris. DEN HARTOG & SEGAL (1964), however, divided this alliance into a Lemnion minoris s.s. of floating pleustophytes (lemnids) and a Lemnion trisulcae of swimming, submerged pleustophytes (ricciellids). Naturally an enormous difference exists between the water surface and the purely aquatic (submerged) environment as habitats. But the same degree of contrast applies to the microclimate of the canopy of a dense forest and its terrestrial moss layer, and likewise to the often acid humus of the latter and the purely mineral, often alkaline subsoil in which the trees root. Such layers should be distinguished, but not as associations (phytocoenotic units).

16.4.3.3 *Adnate Hydrophytes*

Another special group of water plants is formed by the haptophytes. In the terrestrial environment the haptophytic or adnate communities are widely different according to their substratum (living herbs and tree leaves, dead bark, decaying wood, stones and rocks). In the aquatic environment the floristic difference in response to substratum quality may be much less important, probably because the adnate plants absorb water and nutrients over their entire surface from the surrounding water. For practical reasons a distinction is usually made between the haptophytic communities of microscopic algae (see above) and those of larger algae, bryophytes and lichens. Both have, for a long time, escaped the attention of phytosociologists, in particular in the school of BRAUN-BLANQUET. They do not even figure in the survey of water plant communities by LOHMEYER c.s. (1962), nor in that by DEN HARTOG & SEGAL (1964), Yet RÜBEL (1933) already distinguished a separate order of haptophytes, the Encyonematetalia, which, however, only comprised (three) algal communities (as nomina nuda). Macroscopic adnate algal communities have mainly been studied in the littoral zone of sea coasts and brackish waters, often as biocoenoses including sessile animals (barnacles etc.) and semi-sessile animals (molluscs) (for literature see for instance DEN HARTOG (1959) and BRAUN-BLANQUET (1964)).

The first detailed phytosociological description of such communities was published by KORNAŚ & MEDWECKA-KORNAŚ (1950). They distinguished a Fuceto-Furcellarietum which was later split up into a mud-inhabiting rhizophytic Zostero-Furcellarietum and a stone-inhabiting haptophytic Fuco-Furcellarietum (KORNAŚ 1959). Whenever mud and stones or shells are mixed, even in fine-grained patterns, the distinction of their communities on the phytocoenose level seems justified. VAN DEN HOEK (1960) followed the same procedure. DEN HARTOG (1959) is of the opinion, however, that on pebbly bottoms such a distinction is impossible. His associations of rocky sea-coasts, although primarily based upon the dominant alga species of the upper layer, do comprise all layers from large Fucaceae down to crustose red algae and lichens. He stressed the ecological and syntaxonomical independence of some layers and illustrated this by a very instructive scheme (l.c., fig. 14). As DEN HARTOG also pointed out, it is quite possible to classify the separate layers or socions instead of (or in addition to) the often three- or four-layered algal phytocoenoses.

KLEMENT (1955) was of the opinion that lichen communities should be kept strictly apart. When making vegetation records, he even omitted the bryophytes and algae. Thus his marine Verrucarietum maurae contains only lichens, whereas the association table published by DEN HARTOG (1963) also contains four algae. KLEMENT stands rather isolated in this opinion. Fresh-water lichen associations were first described by MOTYKA (1925, cf. KLEMENT 1955). They are all of an amphibious nature. As to the adnate bryophyte communities of fresh water, OCHSNER (1952) still treated them as part of the rhizophytic Potametea, but KRAJINA (1933) and KOCH (1925) considered them autonomous associations and alliances and VON HÜBSCHMANN (1957) even created three classes of fresh-water bryophyte communities (including lichens and mixed communities) and one class of marine associations. Algae were not excluded from the records.

In swift mountain streams bryophytes may actually form pure societies, but in low-land streams of slow currents they are often mixed with phanerogams. *Fontinalis antipyretica*, which VON HÜBSCHMANN considered faithful to the low-land alliance Fontinalion, is also frequent in stagnant water and may then be freely floating or attached to the submerged part of reed culms; the same applies to *Leptodictyum riparium*. In the latter case one could better speak of synusiae within phanerogamic associations. Besides, there is a gradual transition from permanently submerged moss associations to those submerged in winter only, which were also included in the order Fontinaletalia. Now the latter associations, especially VON HÜBSCHMANN's (1957) association of *Leskea paludosa* and *Leptodictyum riparium*, is closely allied to the Chiloscypho-Mnietum which even contains *Fontinalis antipyretica* (BARKMAN 1958b). This, however, is an epiphytic synusia occurring on tree bases that are inundated daily in willow marshes of fresh-water tidal areas. It seems illogical to consider one unistratal community an association and a closely allied unistratal community a synusia, which would mean that they form part of different hierarchical systems. In the writer's opinion this shows that all fresh-water bryophyte 'associations' (phytocoenoses) are best treated as 'unions' (synusiae). These synusiae may or may not be combined with other synusiae into phytocoenoses.

16.4.3.4 *Larger Hydrophytes*

Finally we come to perhaps the most difficult problem, viz. the communities formed by rhizophytes and large pleustophytes. DEN HARTOG & SEGAL (1964) made a strict separation between associations formed by these life-forms. They distinguished 9 classes, 3 of which were new. Apart from the Lemnetea, which form a rather isolated group, three classes of large pleustophytic plants and five classes of rhizophytes were distinguished. This is, however, subject to criticism (cf. BARKMAN 1968, and the reaction by SEGAL in the same issue). Rhizophytes and large pleustophytes have in common that they are not or are hardly mobile. They share the same water mass, the only difference being that the former root in a sand, mud or peat bottom (often a very loose substratum and hardly different from the water itself!), whereas the latter do not. This difference is smaller than it seems, for both groups of plants absorb most of their nutrients from the water, not from the bottom, as was emphasized by DEN HARTOG & SEGAL (1964: 378). Moreover, there are transitions and species that can behave both ways, for instance *Elodea canadensis* and *Myriophyllum spicatum*. In a system of plant communities based primarily on floristic composition, a sharp syntaxonomic separation of these groups as syntaxa would only be justified if it coincided with a clear-cut floristic demarcation. Mixtures do, however, occur: 'it can be a uniform mixture that can hardly be disentangled,' 'vegetations in which rhizophytes and pleustophytes coexist, *have to be* regarded also as mixtures of two vegetation units, even when the vegetation at first glance seems to be homogeneous.' The sharp floristic difference ('the communities of rhizophytes... and pleustophytes ... have not a single species in common') is then a product of the method they use, separating the groups *a priori*. They suggest also as a new criterion success of sexual reproduction, with unwelcome consequences for classification. WESTHOFF & DEN HELD (1969) rejected the classes Ceratophylletea, Utricularietea and Stratiotea. They maintained the Charetea, already proposed by FUKAREK (1961), because in deep lakes the bottom synusiae of Characeae may be the only synusiae present, thus forming phytocoenoses, and because they form a distinct layer, when participating in other phytocoenoses. This too can be criticized by the arguments given above.

16.4.4 Soil-Inhabiting Communities

Although plant sociologists have generally neglected the study of soil-inhabiting organisms, there does exist a respectable number of publications on this topic, mostly, it is true, by non-sociologists. Yet this field of study is by far less advanced than that of the autotrophic macrovegetation, and still less is known about the connections between particular soil communities and the corresponding higher vegetation. There are several reasons for this state of affairs. (i) The study requires special knowledge of these plant groups, which only a few specialists have. None of them is familiar with all groups (bacteria, soil-inhabiting algae, slime moulds, micromycetes, larger fungi). (ii) The number of known species is tremendous. Apinis (1970) mentioned the occurrence of more than 400 species of soil-inhabiting microfungi in a total of five littoral communities, poor in higher plants. Meisel-Jahn & Pirk (1955) recorded 33 species of flowering plants, 21 bryophytes and 121 macrofungi (mushrooms) in spruce forests of the Sauerland (Germany). (iii) Many species have yet to be described. (iv) Determinations of those described are often difficult. Identification often needs the preparation of special cultures in the laboratory (bacteria, algae) or is impossible when the plants are sterile (fungus mycelia), and efforts to induce the latter to form fruit-bodies generally fail. Many of these plants (bacteria, algae) cannot well be preserved in herbaria, others (higher fungi) have often been badly preserved (not dried, no macroscopical description added) or not preserved at all. The literature is scattered, often hard to obtain and expensive, and often also confusing. (v) The plants are invisible in the field (or appear only as fruiting bodies that may be ephemeral), so special sampling methods are needed.

Yet it is important to include them in community analyses. Our knowledge of communities should be based on full species composition, and not only do the fungi alone greatly outnumber the vascular plants and bryophytes in many communities, but they also form an essential part of the biocoenose as humus decomposers. Their role in the biocoenose is highly important in a quantitative sense; in a deciduous wood in England biomass and energy turnover of the fungi are estimated to exceed those of all other lower plants and all animals, taken together, by factors 10 and 4 respectively (SATCHELL, unpublished report).

With regard to higher fungi (macromycetes) it may be remarked that the irregular and often sporadic appearance of fruiting bodies necessitates the sampling of one quadrat at least 6–8 times a year during at least three successive years (LANGE 1948), or every ten days during two years (KALAMEES 1968). However, 5 to 10 years is likely to be a better approximation, according to the present author's experience. The data obtained so far by mycosociologists show that the minimum area for higher fungi is likely to be much greater than for higher plants, i.e. in the order of 1000–5000 sq. m. in a wood, for instance. This makes it difficult to compare the respective synusiae, as it is often hard to find a homogeneous area of woodland of that size. On the other hand, even a large forest area that is homogeneous in its phanerogamic and bryophytic vegetation may be quite heterogeneous for fungi. Their patchy occurrence, and the existence of 'mushroom deserts' and 'mushroom oases' in an otherwise homogeneous forest is well known to mycologists. The effort required to study fungus communities effectively is clearly of a different order of magnitude from that for the usual phytosociological relevés.

Moreover, the limits between fungus and autotrophic communities often do not coincide (cf. for instance LANGE 1948 and TOMILIN 1964), probably because the former are independent of light and strongly dependent on the composition of the humus or the presence of special host species, whereas the latter are more dependent on light and minerals. The relative independence of different synusiae that is one of the themes of this account applies also to fungi vs. green plants (and, probably, to different groups of fungi). GRAHAM (1927) found the same fungus community in two different though neighbouring associations. Both were advanced stages in the succession and grew on soil rich in humus. It is likely that in pioneer communities on more mineral soil fungal synusiae and phanerogamic associations are more closely correlated (ANDERSSON 1950). It is therefore not surprising that HUECK (1953) advocated studying fungus communities separately, at least for the present, for both practical and theoretical reasons (different specialists and methods, different community limits). But he also wrote, 'A synthesis of fungal and autotrophic communities in a later phase of the investigation appears to be desirable, if not necessary.'

In the light of what has been said above and in view of the frequent treatment of moss and lichen synusiae as 'associations,' it is rather surprising that this point of view has not been adopted for fungus communities, except for those on special substrata. Many authors have investigated soil-inhabiting microfungi (for literature see TÜXEN 1964, 1966), but none of them described these as separate associations. As to macrofungi, HÖFLER (1937) regarded them as members of independent communities, because of the lack of topographical correlation with community-types of higher plants, but UBRISZY (1940, 1956) did not recognize the existence of independent fungus associations. FAVRE (1948) used the term 'associations fongiques,' but he explained that it is preferable to consider them synusiae within the larger phytocoenoses, since fungi are heterotrophic and therefore ecologically dependent on higher plants. At least, he says, this should be done until the fungus associations are better known! WESTERDIJK (1949), dealing with lower fungi, however, proposed the term 'association' to designate characteristic combinations of species occurring on certain definable substrata. HUECK (1953) dealing with fungus communities generally, shared this view, stating that 'as the fundamental definition of the term' (i.e. association) 'is unaltered, there appears to be no need to create a new term.' One might remark that synusia and union were not exactly new terms in 1953, and if unions are based on the same (floristic) criteria as associations, this does not imply that the two are identical! SCHWEERS (1949) and ANDREAS (1950) described the mycoflora of certain types of moist to wet, acid, poor grasslands that are rich in bryophytes. In their opinion the special, very characteristic floristic composition of this fungus community justifies the status of association ('*Hygrophorus*-meadow'). This point of view is highly exceptional, however. KALAMEES (1965) is of the opinion that fungal groupings in a forest do not form separate communities. They are part of the biocoenose, but they differ from synusiae by having an entirely different rhythm of development. According to him a new term has to be created, but for the time being they are best regarded as 'consortia' (a term introduced by RAMENSKY).

In the Braun-Blanquet school (cf. TÜXEN et al. 1957) fungus associations are only described as such if they occur on special substrata without other plants, for instance the Fometum igniarii Pirk 1952 on willow trunks, the Trametetum gibbosae Pirk et Tüxen 1957 on rotten beech stumps and the Coprinetum ephemeroidis Pirk et Tüxen 1949 on fresh cow dung. According to this principle PIRK (1950), who published a table of vegetation records of fungi on burnt soil, considered these merely part of the Funarion hygrometricae. Some of the fungi were even considered faithful species of that moss alliance. The practice of naming these fungal societies without higher plants or mosses as 'associations' seems questionable. After all, the beech stumps form part of the forest biocoenose, and the cows form an essential part of the meadow biocoenose. If each fungus society on its special substratum in the forest is to be named an 'association,' the number of these can become overwhelming. It would be more logical and practical to treat fungi through synusiae and to distinguish the latter within a single phytocoenose, according to substratum, position, and function.

Mycorrhizal fungi and parasitic fungi often form synusiae quite independent from other fungi. They are much more linked with certain host species, whereas the saprophytic fungus synusiae seem to be more linked with community-types. However, some mycorrhizal fungi are not at all host-specific (*Lactarius rufus*, *Amanita rubescens*, *A. muscaria* etc.), others are also dependent on other factors, for instance acid soil or limestone, or association variants that occur on wet soil or are poor in grasses (TOMILIN 1964, KALAMEES 1968). Some mycorrhizal-fungi of *Quercus dahurica* are even absent where this tree only develops a shrubby form (TOMILIN 1964). The same applies to parasites. In Eastern Siberia *Uromyces heimerlianus* occurs on various species of *Vicia*. On *Vicia multicaulis* it is only found where it grows in meadows with *Carex schmidtii* and *C. pallida* (TOMILIN 1964). Parasites that are strictly dependent on one host species, may yet occur in various plant associations when the host has a wide habitat range. According to TOMILIN (1964) most saprophytic micromycetes have a wide ecological and sociological range. But saprophytic higher fungi may be very specific, for instance those occurring only on fallen pine cones or beach leaves.

The difference between parasitic and saprophytic synusiae is not always sharp, as parasites may continue their life as saprophytes after the death of their host plant. The soil fungi may form an intergrading spectrum from the purely parasitic, through various kinds of mutualistic and commensal relations, to the saprophytic wholly independent of any living plant; and in the field one cannot see to which group a terrestrial mushroom belongs. It is thus not easy to draw boundaries between the different synusiae for fungi. As a first approach to these, however, it may be reasonable to recognize six broad synusial groupings of fungi in a given forest or scrub phytocoenose: (i) permanently subterranean synusiae of micromycetes, (ii) terrestrial synusiae of saprophytic and mycorrhizal macrofungi with overground fruit-bodies, occurring on humus, (iii) saprophytic fungi on animal excrements and corpses, (iv) parasitic and saprophytic macrofungi on tree bases, decaying stumps and buried wood, (v) epiphytic synusiae of mostly resupinate fungi on the bark of living and dead trunks and branches, (vi) parasitic synusiae on living herbaceous stems, twigs and leaves. Fungi occurring on alien substrata like paper, shoes and other rubbish should be listed separately. In phytocoenoses without woody plants only the groupings (i), (ii) ,(iii) and a fraction of (vi) may be present. In one phytocoenose each category may contain several unions, for instance unions representing different succession stages on decaying stumps. Clearly, further division of these groupings (which seem excessively broad compared with those used for epiphytes) is possible with further study and further knowledge as will be shown below.

16.4.5 An Example: the Dicrano-Juniperetum

16.4.5.1 *Sampling methods*

Vegetation synthesis is dependent on vegetation analysis. Elements that have not been distinguished during the analysis, naturally cannot be described or classified separately. In the Braun-Blanquet school it is customary to use relatively large sample plots (although there is a growing tendency towards smaller plots), and to list the species of each layer separately. On the basis of such relevés it is therefore possible to ordinate or classify the separate layers as well as the whole phytocoenoses, but not the microcoenoses or those societies that form a mosaic within one layer. Hence it is somewhat surprising that in this school layers as a whole have hardly ever been described as separate community-types, whereas microcoena and synusiae within one layer often have. (The reasons why this is so, will be discussed later.)

In order to distinguish such partial communities we must use a different sample procedure with smaller plots. If we want to investigate all elements of a horizontal vegetation mosaic, we can either make vegetation charts, or use line transects, or divide the sample plot into smaller subquadrats of appropriate size (that can be determined by a refined minimal area method, Barkman 1968). The technique has been applied by the present author and his collaborators to investigate herbaceous and bryophyte synusiae within the juniper scrub association on poor sandy soils in the Netherlands and northwest Germany, the Dicrano-Juniperetum (Barkman 1970). On the basis of 1100 small quadrats (30 × 30 cm) 12 microcoena, comprising 18 unions of herbs, bryophytes, lichens, and algae could be distinguished. These were also studied pedologically and microclimatologically. In addition, the synusiae (unions) of fungi, myxomycetes and epiphytes were investigated. The area of each bryophyte or herb society was much too small to have a representative mycoflora. The synusiae of the latter therefore had to be investigated differently, i.e. by noting all fungus species present in the biocoenoses and indicating, for each species. the associated bryophytes and herbs as well as the substratum of the individual fruiting bodies. This led to the distinction of one union of myxomycetes and 15 terrestrial fungus unions.

In total more than 600 plant species were found in this association alone. Some of the main results will be dealt with here, because: (i) They have been published in Dutch; (ii) this is the only study so far of all synusiae (except bacteria and microfungi) within a plant association following the BRAUN-BLANQUET method; (iii) scrub vegetation is highly complex, often more so than even forest vegetation, and therefore most suitable to demonstrate the complex relations between a phytocoenose and its microcommunities and societies.

16.4.5.2 *Synusial Description*

Juniper scrub generally consists of a mosaic of dense shrubs and small open spaces. There are five microbiotopes – Nm, Sm, Sn, dm, and dn. The open spaces differ according to aspect (north and south side of the shrubs). The north side has a moss vegetation (Nm), the south side a narrow strip without vegetation (only fallen needles, Sn) close to the shrub and a broader strip with bryophytes (Sm) at a greater distance from the shrub margin. In the dense scrub moss patches (dm) alternate with patches of bare needles (dn). The latter may have flowering plants and fungi, however. The moss vegetation is different in Sm, dm and Nm. Field measurements show that the floristic differences between these five biotopes are probably due to differences in (i) light intensity, (ii) temperature maxima, (iii) precipitation (throughfall), (iv) potential evaporation, (v) rate of litter fall.

The 18 unions of terrestrial bryophytes, lichens, and herbs are partly controlled by the above microclimatic factors, partly (different synusiae within one of the five biotopes) by pedological factors (pure sand, humus, decaying wood buried in the soil, admixture of oak litter, differences in mineral content of the soil). Each union contains a number of species and is characterised by one species or by a combination of species. Each of the six herb unions is more or less linked with one or more of the 12 moss unions. There are 12 combinations forming micrococena. These micrococena are all linked together by a common shrub union and a common union of lianas. The unions of fungi are considered distinct from the moss synusiae, (i) because of their heterotrophy and quite different periodicity,

(ii) because they either inhabit special microhabitats or form a distinct stratum (The mycelia live either in the rhizosphere of the shrubs and trees, or in the uppermost 2–5 cm of the humus, whereas the herbs root deeper and the bryophytes and lichens live on the very soil surface, as adnate plants.)

There are two mycorrhiza unions, both alien to the juniper scrub proper and occurring only when other trees are present (which is often the case): the mycorrhiza union of broad-leaved trees (*Quercus*, *Betula*) and that of conifers (*Pinus*, *Picea*). The remaining 13 unions are saprophytic and live either in the humus (7 unions) or on special substrata, viz. (i) animal excrements, (ii) rotten fungi, (iii) dead herbaceous plants, (iv) fallen juniper needles, (v) fallen dead twigs and branches, and (vi) buried pine cones. The humus unions are generally linked with special moss (and herb) unions, the mycorrhiza unions and those of special substrata are found scattered throughout the phytocoenose. The same applies to moss unions of special substrata (buried wood, rabbit holes). The alga union of *Prasiola crispa* and *Hormidium flaccidum* (sandy places in Sn) never occurs in combination with other unions, except the shrub union. Four epiphytic unions (bryophytes, algae and lichens) of the junipers are arranged vertically from base to crown; the fifth occurs on bare hard wood. The 4 saprophytic fungus unions on the junipers (dead bark and wood) are mainly arranged horizontally (upper and lower side of slanting trunks respectively). There are also special synusiae on rotten stumps of deciduous trees and of (other) conifers occurring in the juniper scrub. Finally, there are five unions of parasitic fungi, viz. (i) on the base of juniper boles, (ii) on living twigs, (iii) on living needles, (iv) on living herbs and dwarf-shrubs, and (v) on pupae of arthropods in the soil. None of these epiphytic, wood-inhabiting saprophytic, and parasitic synusiae are in any way linked topographically with the terrestrial unions.

Summarizing, we may state that the more than 600 plant species of the Dicrano-Juniperetum can be grouped into 52 unions, 36 of which are terrestrial or soil-occupying. These can be grouped into 12 microcommunities, which belong to 5 microbiotopes. The epiphytic unions are floristically very distinct from the terrestrial, except for the moss and saprophytic fungus synusiae on the extreme bases of the juniper stems. The moss union in question consists of terrestrial species that climb the trees from the ground; the fungus union develops on the bark and then colonises the adjacent soil.

16.4.5.3 *Relations of Synusiae to Associations*

Some synusiae are restricted to one microbiotope or microcommunity, or form a microcommunity of their own, others occur throughout the phytocoenose, either as a continuous separate layer or scattered on special substrata within a certain layer ('azonal' synusiae). The synusiae (unions) are separated either vertically or horizontally. There may also be a (partial) separation in time (herbs in summer, fungi in autumn). Some unions are constantly present, others may be absent from certain stands. The former might be called obligate unions of the association, the latter facultative. The example of the Dicrano-Juniperetum can serve to clarify some notions, for instance the difference between obligate and specific and between facultative and non-specific unions. The moss union of *Hypnum* and *Lophozia* (dm) is constantly present (obligate), but it is not specific (occurs also in other plant associations). The parasitic union of *Gymnosporangium juniperinum* and *G. clavariaeforme* on juniper twigs is specific of the association, but not always present. All unions mentioned here are designated by their name-giving species only, but they contain many more species.

There is also a difference between topographical and ecological dependence. The moss union of *Pseudoscleropodium purum* etc., the herb union of *Sieglingia* and *Galium hercynicum* etc., and the fungus union of *Galerina heterocystis* are topographically linked, because they have similar habitat requirements (north side of shrubs on not too poor sand), but they are not ecologically interdependent. Ecologically dependent unions, for instance the parasitic synusiae of *Juniperus* may be either specific (e.g. the *Lophodermium juniperinum* union on juniper needles) or non-specific (e.g. the *Fomes annosus-Armillaria mellea* - union on the base of the trunks). The latter occurs also in pinewoods and is therefore not topographically dependent on the Dicrano-Juniperetum although it is ecologically dependent. Ecologically dependent unions may even have their optimum elsewhere, e.g. the mycorrhizal synusiae linked with *Pinus* and *Quercus*. A distinction must also be made between topographical linkage of societies (synusiae) to one another within a phytocoenose (syntaxon), and linkage of a particular synusia to the syntaxon. The three moss, herb and fungus unions mentioned above always occur together in the stands of the association, but they also occur outside the Dicrano-Juniperetum, for instance in Violion caninae associations. On the other hand, the *Rhodobryum* moss union and the saprophytic fungus union of *Peniophora laevigata* a.o. are both specific of the association, but they are not necessarily found together within its stands.

From the point of view of classification, the synusiae (unions) that are independent of certain associations, offer the greatest problems. Their widespread occurrence was one of the main reasons to treat them (or all synusiae!) as separate associations. I have called them 'syntaxonomically independent' (BARKMAN 1968) as distinct from ecologically independent synusiae. It is, however, better to speak of syntaxonomic non-specificity (BARKMAN 1970). Besides I made a distinction between syntaxonomic non-specificity and lack of affinity: a certain union may be specific for a certain association, but its nearest floristic affinities may be with unions of quite different associations. It is fundamentally the same problem on a higher hierarchial level, for it means that higher synusial units (federion, ordulus or classicula) into which unions are grouped by floristic affinities may be non-specific in relation to (phytocoenotic) syntaxa even if the unions themselves may be syntaxonomically specific. Both phenomena demonstrate that synusiae are being classified 'unnaturally' when only phytocoenoses are recognized and classified.

16.5 Evalution

Divergent views have been held among phytosociologists and ecologists on the justification and value of the synusial approach (see also WHITTAKER 1962: 34). A number of the arguments have been met with in the foregoing sections, but it seems useful to give here a more coherent discussion. The following is the author's synthesis of comments from the lectures and discussions of the symposia on 'Biosoziologie' and 'Pflanzensoziologische Systematik' (TÜXEN 1965, 1968), special literature on the problem, arguments used implicity in published vegetation systems, and inquiry among and discussions with Dutch vegetationists. The arguments apply to choice among five positions different phytosociologists have taken on synusial treatment of vegetation: (i) opposition to the recognition of synusial groupings, (ii) acceptance of study of societies as an approach strictly subordinate to study and classification of phytocoenoses–i.e. societies might be described within a given phytocoenose, but they would not receive any classification separate from that of the phytocoenoses into syntaxa, (iii) recognition of some synusiae of special substrates as 'associations' while all other societies are left out of consideration, (iv) acceptance of the synusial approach as a complement to the study of whole phytocoenoses, with both societies and phytocoenoses to be classified, separately, and (v) the view that the synusiae should be the principal basis of vegetation study, to which the study of phytocoenoses is secondary or subordinate.

16.5.1 Arguments Against the Synusial Approach

(i) Critics of the synusial approach are often concerned over the multiplication of community-units in addition to an already extensive system of syntaxa for whole-communities. The relative independence is supposed to imply such a multitude of combinations and recombinations of synusiae as to render their treatment as significant units pointless (Cain, 1936). To these arguments it may be responded (a) that if a particular research does not call for the description of synusiae, it is not necessary to do so when dealing with phytocoenoses, (b) the number of synusiae can be reduced by using a rather broadly conceived 'union' concept, (c) the independence makes it probable that the number of unions in a given area might be even less than that of all their possible combinations (the associations). Hence this argument rather works against its advocates.

(ii) A seemingly more fundamental argument, formulated by Gleason (1936) and others, would be the existence of many and manysided ecological interactions between synusiae of apparently disparate habitats: trees, by their shade, moisture, and chemistry of leaf litter, affect the growth of herbs, mosses and fungi, and the latter in turn may affect the germination and growth of tree seedlings, etc. The phytocoenose is thus regarded as a closed functional system, the synusiae are not. If this were an argument, the study of the ecology of individuals and populations of one species would not be justified either. Besides, it is not true. Phytocoenoses are only a part of ecosystems and even ecosystems are open systems, interacting with each other.

(iii) More important is the observation that synusiae that are distinct in one situation, seem in others to merge together or to have a different composition. Floristic and ecological differences between unions may gradually disappear from one region to another, as occurs with the difference between hollows and hummocks in oligotrophic bogs from north Finland to west Ireland (Palsa bog - raised bog - flat bog - blanket bog ecocline). Even in a restricted area herb unions that are clearly separable by either a mosaic relation (intracommunity pattern) or by height, in one association, may form a single union in another association or even another stand of the same association. It has also been observed that species may interchange their position from one layer synusia to another, for instance in bog communities. Besides, consistently distinct layer unions above the soil need not correspond with root layers, so that one species may belong to different unions above and in the soil, e.g. in the Rosmarino-Lithospermetum (Barkman, 1958a).

To the critic this may mean that synusiae should not be distinguished even where the distinction is clear. To the student of synusiae the occasional blurring of boundaries no more invalidates classification on the synusial level than parallel phenomena do on the level of syntaxa.

(iv) Finally, where layers are distinct, all species of higher layers participate in all lower layers above the soil and in all shallower root layers in the soil during a part of their life-cycle, in the case of natural, self regenerating phytocoenoses. This is the main reason for the ZM school not to regard layer synusiae as distinct. Incidentally, this fact would plead for the distinction of life-form synusiae instead of layer synusiae (cf. 16.3.1).

16.5.2 Arguments for the Synusial Approach

These are partly of a practical, partly of a theoretical nature. The practical arguments can be summarized as follows:

(i) The necessity of taxonomic specialization by investigators nearly always compels separate study of vascular plants and fungi; demands on investigators' time may require that really intensive analysis of lichen and bryophyte synusiae be carried out by someone other than the student of vascular plants.

(ii) Different research methods, including different kinds of samples accommodated to different sizes, positions and seasonal timings of organisms, must be applied to phytocoenoses and to different synusiae. In a forest the minumum area of the moss layer is much smaller than that of the tree layer, the minimum area of the terrestrial macrofungi considerably larger. A small plot therefore is by no means representative for the tree layer and the fungi, a plot large enough for a complete species' list of the latter is not suitable for the moss layer, as it will often comprise different moss societies or — to put it another way — it is not suitable because the moss layer is too heterogeneous.

The theoretical arguments are partly typological (iii-iv), partly ecological in nature (v-viii).

Typological arguments. (iii) Ignorance of synusiae or subordination to syntaxa seems illogical as the two may be equivalent regarding (a) size of the concrete stands, (b) sharpness of the boundaries between the stands, (c) floristic differences between adjacent stands (societies, phytocoenoses) or between abstract types (unions versus associations), (d) structural differences between stands or types. In fact, the boundaries between societies are often sharper than between phytocoenoses, for the simpler and more strongly dominated by one species is a community or fraction thereof, the more likely is its boundary to appear definite. In the ZM school, however, boundaries between clones or aggregates (families) of different species are not considered of any phytosociological importance, unless correlated with other floristic or site differences (cf. point vii). Difference in structure between adjacent societies may be as great as that between phytocoenoses accepted as different 'formations'. It is not surprising to find that in the ZM school such societies (as opposed for instance to alternating herb societies) have been treated as syntaxa of the rank of alliances or higher, for instance the Calluno-Genistion pilosae and the Violion caninae (Nardo-Galion) which may form a mosaic in dry heath and which belong to different formations (IV and V) of SCHMITHÜSEN (1961) respectively (cf. however, BARKMAN, 1968, p. 32!).

(iv) Many synusiae are distributed in ways that do not conform to those of the syntaxa in which they occur. To recognize in phytosociology only syntaxa is thus to fail to recognize types and floristic affinities of existing vegetation units on a lower organisational level. This syntaxonomic non-specificity of synusiae has been frequently referred to above. Even specific unions, limited to certain associations, cause difficulties if we try to classify the latter into alliances etc., for the unions in question may be more closely allied with unions of quite different alliances. This syntaxonomic non-affinity (BARKMAN, 1968) is in fact only a matter of non-specificity on a higher level (non-specific federation). An extreme case is that of mobile synusiae (cf. point Vb). These problems can only be solved by classifying synusiae and syntaxa separately.

Ecological arguments.

(v) Some synusiae do not belong to the ecosystem proper of the phytocoenose in which they are found, either (a) because they depend on energy, nutrient or diaspore supply from outside the biocoenose or (b) because the plants themselves may freely enter and leave the biocoenose.

Typical examples of (a) are moss societies near forest edges with inblowing dust and in forest brooks. Both need a constant supply with nutrients from outside the forest. They are usually not considered part of the forest biocoenose (association) and are often considered to represent separate associations. This is open to criticism, for all biocoenoses receive and many are dependent on nutrient supply from outside, either partially (for instance sloping mire associations) or wholly (cave and deep sea biocoenoses). More often than not it is difficult to say whether this supply is indispensable or not. It seems practical, however, not to consider those groupings of species as part of the phytocoenose that live only near the edge, do not perform their life-cycle and are wholly dependent on diaspore supply from other phytocoenoses (vicinism). This is certainly true also of errant (floating) societies (b), like air and water plankton communities and floating water plants (Lemnion), that can be driven in and out of a given phytocoenose by wind and water currents. Such 'dysaptic complexes' (WOLAK, 1970) can be no more considered part of the biocoenose (not even as synusiae) than birds or insects that happen to fly over a vegetation stand or rest there temporarily. They are ecologically independent.

(vi) Many synusiae appear both as phytocoenoses in their own rights and as components of another phytocoenose. In the latter case they do belong to the phytocoenose, but the relation is not obligate, for instance the Rosmarino-Lithospermetum ericetosum occurs in the same region (southern France) in almost the same composition with ('R.-L. pinetosum') and without a canopy of *Pinus halepensis*. Another example are the treeless and tree savannahs in Surinam where the trees are the only difference (VAN DONSELAAR 1965). More often there is an obligate dependence in one region, independence in another region. This holds true for herb and dwarf-shrub synusiae that need the shade of a tree layer in a dry climate, but can be self-supporting in a moist climate or on a north facing slope. Thus the *Rhododendron ferrugineum* union forms an understory in subalpine *Larix-Pinus cembra* woods and a community of its own above the timber line. The same applies to the Swedish *Salix glauca-S. lanata-S. lapponum* thickets occurring both above the timber line ('forest without trees' of the Swedish botanists) and in the subalpine *Betula tortuosa*-woods, and to the treeless Querco-Blechnetum ('Q.-B. extrasilvaticum') in Ireland.

In all examples mentioned it seems illogical to consider such unions as independent associations in one case and to ignore them as vegetation units of whatever kind, when being part of other associations, as is often done.

(vii) Even wholly dependent synusiae may differ in microhabitat. Vertical habitat differences (between layers) are obvious, but the individuality of the layers may be obscured by discrepancy between corresponding layers above and in the soil, and by different life stages of the plants (16.5.1, (iii) and (iv)). Horizontal variation in habitat may be either due to primary or to secondary differences in site factors, the latter as a result of biotic heterogeneity (BARKMAN 1968). Societies based on primary differences (e.g. rocky outcrops in the forest floor) are nearly always regarded as distinct associations, those based on secondary site differences (for instance variation in light climate on the soil as a result of variation in canopy density;

special conditions around tree bases) are often not regarded as such. Yet epiphytic societies and hummocks and hollows in bogs, which belong to the second category, are always separated and rightly so. If we realize that the biocoenose also includes animals, even game tracks, rabbit holes, mole hills and ant heaps belong to the secondary environmental differences. Besides, it is often difficult to know whether a case of biotic heterogenity is not after all due to minor initial (primary) variations in abiotic site factors. It seems therefore more practical to consider all microhabitats as separate societies of the same phytocoenose. However, there is one restriction to be made: the microhabitats should be sufficiently distinct in floristic respect. Otherwise one might consider every gregarious plant species to have its own society (union), for every species has its specific influence on the environment.

(viii) Even in the same microhabitat plants may differ considerably as to function and performance. It seems reasonable to separate such groups as autotrophic, parasitic and saprophytic synusiae, but not as syntaxa. The differences are not sharp. Parasitic fungi may continue their life as saprophytes after the death of their host tree. Among flowering plants the existence of hemiparasites is well-known and many so-called autotrophic plants also absorb organic substances. Difference in habitat exploitation is often expressed in different periodicity. It seems practical to separate the well marked cases, like therophytes (annuals) and perennials, especially if the former appear only once in a number of years (semi-deserts).

The phenomena discussed in points (vii) and (viii) are all part of the broader phenomenon of the evolution of species toward difference in niche (position and manner of function in relation to other species within the community, see article 3). Synusiae comprise groups of species that have evolved toward similar positions, and uses of resources and time, in their communities. Synusiae can thus like the guilds of zoologists (Root 1967) be regarded as niche-groupings. Synusiae are of interest not only for descriptive analysis of communities but also for interpretation of community organization.

16.5.3 Conclusion

The author finds these arguments in favor of the synusial approach convincing. It is not the case that synusiae are so bound to one another and particular syntaxa, that study of the latter gives sufficient knowledge of the former. It would be convenient to phytosociologists if such were the case; but all the respects in which we have observed that synusiae differ from phytocoenoses and one another–distributional and temporal relations, structure and composition, environmental relations and function, and scale and suitable research methods–imply the appropriateness of accepting synusiae as research objects in their own rights.

The fact that synusiae are both parts of phytocoenoses, interacting with and in many cases dependent on other synusiae, and yet in distribution are relatively independent of particular phytocoenoses and other synusiae, is paradoxical. The paradox is the same as that which applies to species, as these are both interrelated in communities, and individualistic in distribution (WHITTAKER 1962, article 3). As species, despite their interactions, are distributed differently, each according to its own characteristics and ways of relating to environment, so are synusiae, as groups of species responding to different environmental factors, differently distributed. Both sides of the paradox are true, but there is no real contradiction; for the kinds of interactions that occur among species (or synusiae) do not (with some exceptions) imply that they should have the same distributions (WHITTAKER 1962). The significance of both interactions and relative distributional independence implies that both phytocoenoses and societies are appropriate objects of study. We thus decide for viewpoint (iv) among those above (16.5)–acceptance of both approaches to communities as useful and productive, with societies and phytocoenoses to be separately classified into independent hierarchies of synusiae and syntaxa (DU RIETZ 1930, 1932; BARKMAN 1958b, 1970; WILMANNS 1962, 1970).

Parts of communities that might be subject to synusial treatment form a continuous size spectrum from the canopy trees of the forest, to the bacteria of decay in dead microorganisms. They form a second spectrum of relative independence from the fully autotrophic and dominant trees, to the wholly dependent fungal parasites and pathogens. There is no evident way to draw a boundary limiting synusiae to larger fractions of communities or to species to some degree independent of other species. Only judgment and practicality can determine what synusiae are to be accepted as useful for research. Such judgment must exert some limit on the number of synusiae given formal recognition as taxonomic units, if overwhelming numbers of such units are not to be described. Apart from formal designation of units, however, there is no reason why a student should not divide the societies of a community as finely as his particular research makes appropriate. The problems of synusial taxonomy–of distinguishing among multidirectionally related and often intergrading objects and avoiding an excessive number of classes–are not different in character from the problems of the taxonomy of phytocoenoses.

Though the taxonomic problems of societies and communities are similar, their ecological relations are different. Much may be learned from study of synusiae, as phytosociology seeks means for more penetrating investigation of some aspects of communities than that based on syntaxa. As a spectrum permits us to analyse much about composition of light and wave lengths of absorption that is not evident in the color of the light, so synusiology permits us to analyse much about the composition of a community and the character of its species relationships that is not evident in the phytosociology of the community as a whole.

16.6.1 Summary

An alternative to the study and classification of plant communities as wholes is the approach through societies and synusiae. A society is defined here as a part of a plant community with a specific floristic composition, consisting of plants that belong to the same layer (if layers can be distinguished), occupy the same microhabitat (if these are present) and have essentially the same periodicity and way of habitat exploitation. A synusia is an abstract unit grouping together similar societies that occur in a number of plant communities (phytocoenoses). Different societies of the same microhabitat are grouped together as microcommunities or microcoenoses (abstract type: microcoenon).

The study of synusiae was originated by Gams (1918), and the influence of the idea spread through different schools of phytosociology with a proliferation of terms. The problems of criteria for defining synusiae and related units and of the appropriate concepts and terms, are discussed; and terms are summarized in Tables I and II. The principal unit for hierarchial classification of societies is the 'union,' defined by floristic composition and analogous to the association. Subordinate units may be termed 'subunion,' 'variant,' 'vicariant' and 'socion,' higher units 'federation,' 'ordulus,' and 'classicula.' The system of synusiae is thus broadly parallel to, but independent of, the system of units (syntaxa) for classification of phytocoenoses. It is proposed here to form the scientific names of the synusiae according to the rules generally accepted for syntaxa, i.e. -etosum for subunions, -etum for unions, -ion for federations, -etalia for orduli, -etea for classiculae.

Applications of synusial approaches are reviewed for (i) vascular plants on land, (ii) societies of epiphytes, (iii) aquatic communities and (iv) soil-inhabiting communities, and (v) as an example, a particular community, a juniper scrub association of the Netherlands and Northwest Germany that the author has studied in detail (unpublished). Though synusiae have had little application to vascular plants in land communities since the work of LIPPMAA, they have been of increasing importance in the study of other kinds of organisms and communities.

Synusiae that do not correspond in distribution to the syntaxa in which they occur may be termed 'syntaxonomically non-specific.' Such independence of distribution of synusiae, compared with syntaxa and with one another, is a general phenomenon of communities and a justification of the synusial approach. Some synusiae, however, are syntaxonomically specific; thus a particular synusia may be a character-union for a given association. Other justifications for the synusial approach include the relative ecological independence of many synusiae and the wide differences among synusiae of a given community in composition and structure, in ecological and functional relationships, and in the appropriate scales for samples to study them.

It is true both that synusiae may be related to one another by interactions among their species when they occur together in the same phytocoenoses, and that they are different or 'individualistic' in their responses to environment within the community and in their distributions. The study of synusiae is consequently a w ell-justified complement to the study of phytocoenoses and syntaxa. The analysis of a community through its synusiae permits study of relationships among species and kinds of communities that cannot generally be obtained through study limited to observation and classification of phytocoenoses as wholes.

16.6.2 ZUSAMMENFASSUNG

Eine andere Möglichkeit für das Studium und die Klassifikation von Pflanzengesellschaften als Ganzheiten ist die Untersuchung von Sozietäten und Synusien. Eine Sozietät oder Verein wird hier umschrieben als ein Teil einer Pflanzengesellschaft mit einer besonderen floristischen Zusammensetzung, die aus Pflanzen derselben Schicht besteht, (wenn Schichten unterschieden werden können), welche den selben Kleinstandort (Mikrobiotop) einnehmen, (wenn solche vorhanden sind), und die im wesentlichen dieselbe Periodizität und dieselbe Art der Ausnützung des Standorts haben. Ein Synusie ist eine abstrakte Einheit, die ähnliche Sozietäten zusammenfaßt, die in einer Anzahl von Pflanzengesellschaften (Phytozönosen) vorkommen, unabhängig von der Frage ob diese Phytozönosen zu den selben oder zu verschiedenen Gesellschaftstypen (Syntaxa) gehören. Verschiedene Sozietäten des selben Klein-Standorts werden als Mikro-Gesellschaften oder Mikrozönosen (abstrakter Typus: Mikrozönon) zusammengefaßt. Die Beziehungen zwischen Gesamt-Gesellschaften und Teil-Gesellschaften bis zur Stufe von Pflanzen-Individuen werden in Tabellen I und II veranschaulicht.

Das Studium von Synusien wurde von Gams (1918) begonnen, und der Einfluß dieses Begriffes breitete sich mit einer Aufspaltung der Termini durch verschiedenen Schulen der Pflanzensoziologie aus. Das Problem der Kriterien für die Definition von Synusien und verwandten Einheiten und der zugehörigen Begriffe und Termini, wird diskutiert; die Termini werden zu Tabelle II zusammengefaßt. Die Grund-Einheit für die hierarchische Klassifikation der Sozietäten ist die 'Union', die analog der Assoziation durch die floristische Zusammensetzung definiert wird. Untergeordnete Einheiten können 'Subunion', 'Variante,' 'Vikariante' und 'Sozion', höhere Einheiten 'Federation,' 'Ordulus' und 'Classicula' genannt werden. Das System der Synusiae ist also parallel mit, aber unabhängig, von dem System von Einheiten (Syntaxa) für die Klassifikation von Phytozönosen. Es wird hier vorgeschlagen die wissenschaftliche Namen der Synusien in Übereinstimmung mit den allgemein für Syntaxa angenommenen Regeln zu bilden, d.h. -etosum für Subunionen, -etum fur Unionen, -ion für Federationen, -etalia für Orduli, -etea für Classiculae.

Anwendungen von Synusial-Untersuchungen sind möglich bei 1) Land-Gefäßpflanzen, 2) Epiphyten-Sozietäten, 3) Wasser-Gesellschaften und 4) bodenbewohnende Gesellschaften. Zur Erläuterung der Methoden und der verwickelten Beziehungen zwischen Synusien wird ein Beispiel einer besonderen Gesellschaft, einer Juniperus-Gestrupp-Assoziation der Niederlande und Nordwest-Deutschlands gegeben, die der Autor eingehend untersucht hat (unpubliziert). Obwohl Synusien seit der Arbeiten Lippmaa's wenig Anwendung auf Gefäßpflanzen in terrestrischen Gesellschaften gefunden haben, sind sie für das Studium von anderen Typen von Organismen und Gesellschaften immer wichtiger geworden.

Synusien, die in ihrer Verbreitung nicht mit den Syntaxa, in denen sie auftreten, übereinstimmen, können 'syntaxonomisch nicht-spezifisch' genannt werden. Eine solche Unabhängigkeit in der Verbreitung der Synusien, im Vergleich mit den Syntaxa und miteinander, ist eine allgemeine Erscheinung von Gesellschaften und eine Rechtfertigung für die Synusialbetrachtung. Einige Synusiae sind aber syntaxonomisch spezifisch; so kann eine besondere Synusia eine Charakter-Union für eine bestimmte Assoziation sein. Andere

Rechtfertigungen für die synusiale Betrachtung sind u.a. die relative ökologische Unabhängigkeit vieler Synusiae, die großen Unterschiede zwischen Synusien einer bestimmten Gesellschaft in Zusammensetzung und Struktur, in ökologischen und funktionellen Beziehungen und in den geeigneten Maßstäben für Aufnahmen um sie zu untersuchen.

Es ist wahr, daß Synusien aufeinander bezogen sein können durch Wechselwirkungen zwischen ihren Arten, wenn sie zusammen in denselben Phytozönosen vorkommen und daß sie verschieden oder 'individualistisch' sind in ihrer Antwort auf den Standort innerhalb der Gesellschaft und in ihrer Verbreitung. Das Studium der Synusien ist also eine wohl-berechtigte Ergänzung zum Studium von Phytozönosen und Syntaxa. Die Analyse einer Gesellschaft durch ihre Synusien erlaubt die Beziehungen zwischen Arten und Gesellschaftstypen zu studieren, was im allgemeinen nicht durch ein auf Beobachtung und Klassifikation der Phytozönosen als Ganzes beschränktes Studium erreicht werden kann.

REFERENCES

Aleksandrova, V. D., – 1969 – Classification of Vegetation: Principles of Classification and Classification Systems of Various Phytocoenological Schools. (In Russian). Naukov, Leningrad. 275 pp.

Allorge, P., – 1922 – Les associations végétales du Vexin francais. *Revue gén. Bot.* 33: 481—544 etc.; 34: 71—79 etc.

Almborn, O. – 1948 – Distribution and ecology of some South Scandinavian lichens. *Bot. Notiser*, Suppl. 1 (2): 1—354.

Almquist, E., – 1929 – Upplands vegetation och flora. *Acta phytogeogr. suec.* 1: 1—622.

Andersson, O. – 1950 – Larger fungi on sandy grass heath and sand dunes in Scandinavia. *Bot. Notiser*, Suppl. 2 (2): 1—89.

Andreas, C. H., – 1950 – De Hygrophorusweide een associatie. *Fungus* 20. 66—68.

Apinis, A. E., – 1970 – Das Verhalten der Pilze in bestimmten Graslandgesellschaften. (Engl. summ.) In 'Gesellschaftsmorphologie (Strukturforschung),' ed. R. Tüxen, *Ber. Symp. Int. Ver. Vegetationskunde*, Rinteln 1966: 172—186.

Balogh, J., – 1938 – Biosoziologische Studien über die Spinnenfauna des Sashegy (Adler-Berg) bei Budapest. *Festschr. Strand* 4: 464—499.

Barkman, J. J., – 1958a – La structure du Rosmarineto-Lithospermetum helianthemetosum en Bas-Languedoc. *Blumea*, Suppl. 6: 113—136.

Barkman, J. J., – 1958b – Phytosociology and Ecology of Cryptogamic Epiphytes. Van Gorcum, Assen. 628 pp.

Barkman, J. J., – 1968 – Das synsystematische Problem der Mikrogesellschaften innerhalb der Biozönosen. (Engl. summ.) In 'Pflanzensoziologische Systematik,' ed. R. Tüxen, *Ber. Int. Synp. Vegetationskunde*, Stolzenau/Weser 1964: 21—53.

Barkman, J. J., – 1969 – Epifytengemeenschappen. Ch. XIV in 'Plantengemeenschappen in Nederland,' by V. Westhoff & A. J. den Held, pp. 272—286. Thieme, Zutphen.

Barkman, J. J., – 1970 – Enige nieuwe aspecten inzake het probleem van synusiae en microgezelschappen. (Germ. summ.) *Meded. Landbouwhogeschool Wageningen* 5: 85—116.

BOLLETER, R. – 1921 – Vegetationsstudien aus dem Weisstannental. *Jb. St. Gall. naturw. Ges.* 57 (2): 1—140.

BRAUN-BLANQUET, J., – 1928 – Pflanzensoziologie: Grundzüge der Vegetationskunde. 1st ed. Springer, Berlin. 330 pp. 2nd ed., Springer, Wien, 1951, 631 pp. 3rd ed., 1964, 865 pp.

CAIN, S. A., – 1936 – Synusiae as a basis for plant sociological field work. *Am. Midl. Nat.* 17: 665—672.

CAIN, S. A., & G. M. DE OLIVEIRA CASTRO – 1959 – Manual of Vegetation Analysis. Harper, New York. 325 pp.

CAIN, S. A. & A. J. SHARP – 1938 – Bryophytic unions of certain forest types of the Great Smoky Mountains. *Am. Midl. Nat.* 20: 249—301.

CLEMENTS, F. E., – 1905 – Research Methods in Ecology. Univ. Publ. Co., Lincoln. 334 pp.

CLEMENTS, F. E., – 1907 – Plant Physiology and Ecology. Holt, New York. 315 pp.

CLEMENTS, F. E., – 1916 – Plant succession: an analysis of the development of vegetation. *Publs Carnegie Inst.* Washington 242: 1—512.

CLEMENTS, F. E., – 1928 – Plant Succession and Indicators. Wilson, New York. 453 pp.

CLEMENTS, F. E., – 1936 – Nature and structure of the climax. *J. Ecol.* 24: 252—284.

COOKE, W. B., – 1955 – Fungi, lichens, and mosses in relation to vascular plant communities in eastern Washington. *Ecol. Monogr.* 25: 119—180.

DAUBENMIRE, R., – 1968 – Plant Communities. Harper & Row, New York. 300 pp.

DONSELAAR, J. VAN, – 1965 – An ecological and phytogeographic study of northern Surinam savannas. *Wentia* 14: 1—163.

DRUDE, O., – 1919 – Die Elementar-Assoziation im Formationsbilde. *Bot. Jb.* 55 (Beibl. 122): 45—82.

DU RIETZ, G. E. – 1930 – Classification and nomenclature of vegetation. *Svensk Bot. Tidskr.* 24: 489—503.

DU RIETZ, G. E., – 1932 – Vegetationsforschung auf soziationsanalytischer Grundlage. *Handb. biol. ArbMeth.* 11, 5: 293—480.

DU RIETZ, G. E., – 1936 – Classification and nomenclature of vegetation units 1930—1935. *Svensk Bot. Tidskr.* 30: 580—589.

DU RIETZ, G. E., – 1954 – Die Mineralbodenwasserzeigergrenze als Grundlage einer natürlichen Zweigliederung der nord- und mitteleuropäischen Moore. *Vegetatio* 5/6: 571—585.

DU RIETZ, G. E., – 1957 – Vegetation analysis in relation to homogeneousness and size of sample areas. 8*me Congr. int. bot.*, Paris 1954, *C.-R. Séances* Sect. 7 et 8 (Suppl.): 24—35.

DU RIETZ, G. E., – 1965 – Biozönosen und Synusien in der Pflanzensoziologie. (Engl. summ.) In 'Biosoziologie,' ed. R. TÜXEN, *Ber. Int. Symp. Vegetationskunde*, Stolzenau/Weser 1960: 23—42.

ELLENBERG, H., – 1956 – Aufgaben und Methoden der Vegetationskunde. In 'Einführung in die Phytologie' by H. WALTER, Vol. 4, Pt. 1. Ulmer, Stuttgart, 136 pp.

FAVRE, J., – 1948 – Les associations fongiques des hauts-marais jurassiens et de quelques régions voisines. *Beitr. Kryptog. Flora Schweiz* 10 (3): 1—228.

FELFÖLDY, L. – 1941 – A debreceni Nagyerdö epiphyta vegetációja. (Germ. summ.: Die Epiphytenvegetation des Waldes 'Nagyerdö' bei Debrecen). *Acta geobot. hung.* 4 (1): 35—73.

Fukarek, F., – 1961 – Die Vegetation des Darss und ihre Geschichte. *Pflanzensoziologie* 12: 1—321. Fischer, Jena.
Gams, H. – 1918 – Prinzipienfragen der Vegetationsforschung: Ein Beitrag zur Begriffsklärung und Methodik der Biocoenologie. *Vjschr. naturf. Ges. Zürich* 63: 293—493.
Gams, H., – 1927 – Von den Follatères zur Dent de Morcles: Vegetationsmonographie aus dem Wallis. *Beitr. geobot. Landesaufn. Schweiz* 15: 1—760.
Gams, H., – 1936 – Beiträge zur pflanzengeographische Karte Österreichs. 1. Die Vegetation des Grossglocknergebietes. *Abh. zool.- bot. Ges. Wien* 16 (2): 1—79.
Gleason, H. A. – 1936 – Is the synusia an association? *Ecology* 17: 444—451.
Graham, V. O., – 1927 – Ecology of fungi in the Chicago region. *Bot. Gaz.* 83: 267—287.
Grossheim, A. A. – 1930 – Zur Frage nach dem Zustandekommen der Pflanzendecke. *Beitr. Biol. Pfl.* 18: 225—268.
Grossheim, A. A. & L. I. Prilipko, – 1929 – Geobotanical survey of the Karabakh Steppe, Azerbaijan (In Russian with Engl. summ.). *Trudy geobot. Obsled. Pastb. SSR Azerbaidzh.* 4: 3—130.
Hale, M. E. Jr., – 1955 – Phytosociology of corticolous cryptogams in the upland forests of southern Wisconsin. *Ecology* 36: 45—63.
Hartog, C. den, – 1959 – The epilithic algal communities occurring along the coast of the Netherlands. *Wentia* 1: 1—241.
Hartog, C. den, – 1963 – Enige waterplantengemeenschappen in Zeeland. (Engl. summ.) *Gorteria* 1: 155—164.
Hartog, C. den, & S. Segal, – 1964 – A new classification of the water-plant communities. *Acta bot. neerl.* 13: 367—393.
Hoek, C. van den, – 1960 – Groupements d'algues des étangs saumâtres méditerranéens de la côte française. *Vie et Milieu* 11 (3): 390—412.
Höfler, K., – 1937 – Pilzsoziologie. *Ber. dt. bot. Ges.* 55: 606—622.
Höfler, K., – 1959 – Ueber die Gollinger Kalkmoosvereine. *Sber. öst. Akad. Wiss.*, Math.-nat. Kl. 168: 541—582.
Hosokawa, T., – 1951 – On the nomenclature of Aerosynusia. *Bot. Mag.*, Tokyo 64: 107—111.
Hosokawa, T., – 1954 – On the *Campnosperma* forests of Kusaie in Micronesia, with special reference to the community units of epiphytes. *Vegetatio* 5/6:351—360.
Hübschmann, A. von, – 1957 – Zur Systematik der Wassermoosgesellschaften. *Mitt. flor.-soz. ArbGemein*, Stolzenau, N.F. 6/7: 147—151.
Hübschmann, A. von, – 1960 – Einige Ackermoos-Gesellschaften des nordwestdeutschen Gebietes und angrenzender Landesteile und ihre Stellung im pflanzensoziologischen System. *Mitt. flor.-soz. ArbGemein*, Stolzenau, N.F. 8: 118—123.
Hueck, H. J. – 1953 – Myco-sociological methods of investigation, *Vegetatio* 4 (2): 84—101.
Hult, R., – 1881 – Försök till analytisk behandling af växtformationerna. *Meddn Soc. Fauna Flora fenn.* 8: 1—155.
Iversen, J., – 1936 – Biologische Pflanzentypen als Hilfsmittel in der Vegetationsforschung — Ein Beitrag zur ökologischen Charakterisierung und Anordnung der Pflanzengesellschaften. *Medd. Skalling-Lab.*, København 4: 1—224.
Iwatsuki, Z. & S. Hattori, – 1957 – Studies of the epiphytic moss flora of Japan. 8. The bryophytic communities in the *Pinus pumila* association of central Japan. *J. Hattori bot. Lab.* 18: 70—77.
Jovet, P., – 1949 – Le Valois, phytosociologie et phytogéographie. Soc. d'Ed. d'Enseign. Sup., Paris. 389 pp.

KALAMEES, K. – 1965 – On problems and methods in mycosociology (In Russian with Engl. summ.). *Akad. Nauk, Eston. SSR*, Tartu: 14—22.

KALAMEES, K., – 1968 – Mycocoenological methods based on investigations in the Estonian forests. *Acta Mycol.* 4 (2): 327—335.

KATZ, N., – 1929 – Die Zwillingsassoziationen und die homologen Reihen in der Phytosoziologie. *Ber. dt. bot. Ges.* 47: 154— 164.

KELLER, B., – 1932 – Die Methoden zur Erforschung der Ökologie der Steppen- und Wüstenpflanzen. *Handb. biol. ArbMeth.* 11, 6: 1—128.

KERNER VON MARILAUN, A., – 1863 – Das Pflanzenleben der Donauländer. Wagner, Innsbruck. 348 pp.

KLEMENT, O., – 1931 – Zur Flechtenflora des Erzgebirges. Die Umgebung von Komotau. *Beih. bot. Zbl.* 48 (2): 52—96.

KLEMENT, O., – 1955 – Prodromus der mitteleuropäischen Flechtengesellschaften *Beih. Repert. Spec. nov. Regni veg.* 135: 5—194.

KOCH, W., – 1925 – Die Vegetationseinheiten der Linthebene unter Berücksichtigung der Verhältnisse in der Nordostschweiz. *Jb. St. Gall. naturw. Ges.* 61 (2): 1—146.

KOPPE, F., – 1955 – Moosvegetation und Moosgesellschaften von Altötting in Oberbayern. *Reprium nov. Spec. Regni veg.* 58 (1/3): 92—144.

KORCHAGIN, A. A., – 1964 – Synusial structure of forest communities. *Proc. X Int. Bot. Congr.*, Edinburgh 1964, Abstracts, p. 403.

KORJINSKY, S. J., – 1888 – La limite nord de la région du tschernosjom dans la partie orientale de la Russie Européenne (Russian). *Trudy Obshch. Estest. imp. kazan. Univ.* 18 (5): 1—253, 22 (6): 1—201.

KORNAŚ, J., – 1959 – Sea bottom vegetation of the Bay of Gdańsk off Rewa. *Bull. Acad. pol. Sci.*, Cl. II, Sér. Sci. biol. 7: 5—10.

KORNAŚ, J. & A. MEDWECKA-KORNAŚ, – 1950 – Associations végétales sousmarines dans le Golfe de Gdańsk. *Vegetatio* 2: 120—127.

KOSKINEN, A. – 1955 – Ueber die Kryptogamen der Bäume, besonders die Flechten, im Gewässergebiet des Päijänne sowie an den Flüssen Kalajoki, Lestijoki und Pyhäjoki. Floristische, soziologische und ökologische Studie I. Helsinki. 176 pp.

KRAJINA, V., – 1933 – Die Pflanzengesellschaften des Mlynica-Tales in den Vysoke Tatry (Hohe Tatra). *Beih. bot. Zbl.*, Abt. 2, 50: 774—957, 51: 1—224.

KRUSENSTJERNA, E. VON, – 1945 – Bladmossvegetation och bladmossflora i Uppsalatrakten. (Engl. summ.) *Acta phytogeogr. suec.* 19: 1—250.

LANGE, M., – 1948 – The agarics of Maglemoose: study in the ecology of the agarics. *Dansk bot. Ark.* 13 (1): 1—141.

LIPPMAA, T., – 1931 – Pflanzensoziologische Betrachtungen. *Acta Inst. Horti bot. tartu.* 2 (3/4): 1—32, and *Loodusuur. Seltsi Aruanded* 38 (1/2): 1—32.

LIPPMAA, T., – 1933a – Aperçu général sur la végétation autochtone du Lautaret (Hautes-Alpes) avec des remarques critiques sur quelques notions phytosociologiques. (Eston. summ.) *Acta Inst. Horti bot. tartu.* 3 (3): 1—108.

LIPPMAA, T., – 1933b – Taimeühingute uurimise metoodika ja eesti taimeühingute klassifikatsiooni põhijooni (Eston. with Germ. summ.: Grundzüge der pflanzensoziologischen Methodik nebst einer Klassifikation der Pflanzenassoziationen Estlands). *Acta Inst. Horti bot. tartu.* 3 (4): 1—169.

LIPPMAA, T., – 1935a – La méthode des associations unistrates et le système écologique des associations. *Acta Inst. Horti bot. tartu.* 4 (1/2, art. 3): 1—7.

LIPPMAA, T., – 1935b – Une analyse des forêts de l'île estonienne d'Abruka (Abro) sur la base des associations unistrates. *Acta Inst. Horti bot. tartu.* 4 (1/2, art. 5): 1—97.

LIPPMAA, T., – 1935c – La méthode des associations unistrates et le système écologique des associations. *Proc. Zesde Int. Bot. Congr.*, Amsterdam 1935, 2: 109—110.

LIPPMAA, T., – 1938 – Areal und Altersbestimmung einer Union (*Galeobdolon-Asperula-Asarum-U.*) sowie das Problem der Charakterarten und der Konstanten. *Acta Inst. Horti bot. tartu.* 6 (2): 1—152.

LIPPMAA, T., – 1939 – The unistratal concept of plant communities (the unions). *Am. Midl. Nat.* 21: 111—145.

LOHMEYER, W. c.s., – 1962 – Contribution à l'unification du système phytosociologique pour l'Europe moyenne et nord-occidentale. *Melhoramento* 15: 137—151.

LORENZ, J. R., – 1858 – Allgemeine Resultate aus der pflanzengeographischen und genetischen Untersuchung der Moore im präalpinen Hügellande Salzburgs. *Flora* 41: 209—221, 225—237, 241—253, 273—286, 289—302, 344—355, 360—376.

LUTHER, H., – 1949 – Vorschlag zu einer ökologischen Grundeinteilung der Hydrophyten. *Acta bot. fenn.* 44: 1—15.

MAAREL, E. VAN DER, – 1965 – Beziehungen zwischen Pflanzengesellschaften und Molluskenfauna. (Engl. summ.) In 'Biosoziologie,' ed. R. TÜXEN, *Ber. Symp. Int. Ver. Vegetationskunde*, Stolzenau/Weser 1960: 184—198.

MARGALEF, R., – 1947 – Los methodos para la investigacion de las comunidades acuaticas adnadas y especialmente las formadas por organismos microscopicos (perifiton, pecton). *Collnea bot.*, Barcinone 1 (3): 247—259.

MARGALEF, R. – 1949 – Las asociaciones de algas en las aguas dulces de pequeño volumen del Noreste de España. (Engl. summ.) *Vegetatio* 1 (4/5): 258—284.

MARGALEF, R., – 1955 – Comunidades bióticas de las aguas dulces del noroeste de España. (Engl. summ.) *Publnes. Inst. Biol. apl.*, Barcelona 21: 5—86.

MEISEL-JAHN, S. & W. PIRK, – 1955 – Über das soziologische Verhalten von Pilzen in Fichten-Forstgesellschaften. *Mitt. flor.-soz. ArbGemein.*, Stolzenau, N.F. 5: 59—63.

MÖRZER BRUINS, M. F., – 1947 – Over levensgemeenschappen. (Engl. summ.) Diss. Kluwer, Deventer. 195 pp.

MOTYKA, J., – 1925 – Die Pflanzenassoziationen des Tatra-Gebirges. II. Die epilithischen Assoziationen der nitrophilen Flechten im polnischen Teile der Westtatra. *Bull. int. Acad. Pol. Sci. Lett.*, Cl. Sci. Math. Nat., Sér. B, 1924: 835—850.

NAKANISHI, S., – 1970 – The bryophyte-lichen vegetation of Mt. Hakusan area. *Scient. Stud. Hakusan Nation. Park* Japan, pp. 174—200. Ishikawa-ken, Japan.

OBERDORFER, E., – 1954 – Ueber Unkrautgesellschaften der Balkanhalbinsel. *Vegetatio* 4 (6): 379—411.

OBERDORFER, E., – 1957 – Süddeutsche Pflanzengesellschaften. *Pflanzensoziologie* 10: 1—564. Fischer, Jena.

OCHSNER, F., – 1928 – Studien über die Epiphytenvegetation der Schweiz. *Jb. St. Gall. naturw. Ges.* 63 (2): 1—106.

OCHSNER, F., – 1952 – Moose in den Pflanzengesellschaften des Languedoc.*Ber. schweiz. bot. Ges.* 62: 106—122.

OMURA, M., – 1952 – A synecological study on the epiphyte communities of lichens in the subalpine forests of Titibu mountain region in Japan. *Bull. Soc. Pl. Ecol.*, Tokyo 1 (4): 176—181.

OMURA, M., – 1953 – Synecological studies on the epiphyte communities of lichens in the temperate forests of the Iide and Titibu mountain regions in Japan. *Mem. Fac. Sci., Kyushu Univ.*, Ser. E, 1 (3): 147—157.

PASTAK, E., – 1935 – Harilaiu Taimkate (Engl. summ.: The vegetation of the peninsula of Harilaid, Estonia). *Acta Inst. Horti bot. tartu.* 5 (1/2, art. 3): 1—44.

PHILIPPI, G., – 1956 – Einige Moosgesellschaften des Südschwarzwaldes und der angrenzenden Rheinebene. *Beitr. naturk. Forsch. Südw. Dtl.* 15 (2): 91—124.

PHILIPPI, G., – 1963 – Zur Kenntnis der Moosgesellschaften saurer Erdraine des Weserberglandes, des Harzes und der Rhön. *Mitt. flor.-soz. ArbGemein.* Stolzenau, N.F. 10: 92—108.

PHILLIPS, E. A., – 1951 – The associations of bark-inhabiting bryophytes in Michigan. *Ecol. Monogr.* 21: 301—316.

PIRK, W., – 1950 – Pilze in Moosgesellschaften auf Brandflächen. *Mitt. flor.-soz. ArbGemein.*, Stolzenau, N.F. 2: 3—5.

PIRK, W., – 1952 – Die Pilzgesellschaft der Baumweiden im mittleren Wesertal. *Mitt. flor.-soz. ArbGemein.*, Stolzenau, N.F. 3: 93—96.

PIRK, W. & R. TÜXEN, – 1949 – Das Coprinetum ephemeroides, eine Pilzgesellschaft auf frischem Mist der Weiden im mittleren Wesertal. *Mitt. flor.-soz. ArbGemein.*, Stolzenau, N.F. 1: 1—7.

PIRK, W. & R. TÜXEN, – 1957 – Das Trametetum gibbosae, eine Pilzgesellschaft modernder Buchenstümpfe. *Mitt. flor.-soz. ArbGemein.*, Stolzenau, N.F. 6/7: 120—126.

POORE, M. E. D., – 1955 – The use of phytosociological methods in ecological investigations. I. The Braun-Blanquet system. *J. Ecol.* 43: 226—244.

RAUNKIAER, C., – 1934 – The Life Forms of Plants and Statistical Plant Geography. Clarendon, Oxford. 632 pp.

REGEL, C., – 1923 – Assoziationen und Assoziationskomplexe der Kola Lappmark. *Bot. Jb.* 58: 607—635.

REGEL, C., – 1927 – Zur Klassifikation der Assoziationen der Sandböden. *Bot. Jb.* 61: 263—284.

RICHARDS, P. W., – 1939 – The bryophyte communities of a Killarney oakwood. *Annls. bryol.* 9: 108—130.

RICHARDS, P. W. – 1952 – The Tropical Rain Forest: An Ecological Study. Cambridge Univ., New York, 423 pp.

ROOT, R. B., – 1967 – The niche exploitation pattern of the blue-gray gnatcatcher. *Ecol. Monogr.* 37: 317—350.

RÜBEL, E., – 1925 – Vorschläge zur Untersuchung von Buchenwälder. *Veröff. geobot. Inst. Rübel*, Zürich, Beibl. 3: 1—35.

RÜBEL, E., – 1933 – Versuch einer Uebersicht über die Pflanzengesellschaften der Schweiz. *Ber. geobot. Inst. Rübel*, Zürich 1932: 19—30.

RYDZAK, J., – 1961 – Tree lichens in the forest communities of the Białowieża National Park. (Polish & Russ. summ.s) *Annls Univ. Mariae Curie-Skłodowska*, Lublin, C 16 Sect. C. Biol., 16 (2): 17—47.

SCAMONI, A., – 1963 – Einführung in die praktische Vegetationskunde. 2nd ed. Fischer, Jena. 236 pp.

SCHEYGROND, A., – 1931 – Het plantendek van de Krimpenerwaard. IV. Sociographie van het hoofd-associatie-complex Arundinetum-Sphagnetum. *Ned. kruidk. Archf.* 1932 (1): 1—184.

SCHMID, E., – 1941 – Vegetationsgürtel und Biocoenose. *Ber. schweiz. bot. Ges.* 51: 461—474.

SCHMITHÜSEN, J., – 1959 – Allgemeine Vegetationsgeographie. In 'Lehrbuch der allgemeinen Geographie,' E. Obst, vol. 4: 1—261. de Gruyter, Berlin.

SCHMITZ, W. – 1965 – Die Soziologie aquatischer Mikrophyten. (Engl. summ.) In 'Biosoziologie,' ed. R. TÜXEN, *Ber. Symp. Int. Ver. Vegetationskunde*, Stolzenau /Weser 1960: 120—139.

SCHUMACHER, A. – 1944 – Ueber Calypogeia arguta im Bergischen Lande. Beil. z. 14. *Rundbr. Zentralstelle für Vegetationskartierung*, Stolzenau.

SCHWEERS, A. C. A. – 1949 – De Hygrophorusweide, een associatie. *Fungus* 19: 17—18.

SCOTT, G. A. M. & J. M. ARMSTRONG, – 1966 – The altitudinal sequence of

climax vegetation on Mt. Anglem, Stewart Island. *N.Z. J. Bot.* 4 (3): 283—299.
SIRGO, V. – 1936 – Emajõe alamjooksul Peipsiäärsel madalikul asuvaist taimeühinguist (Engl. summ.: Plant unions of the swamps at the mouth of the River Emajogi). *Acta Inst. Horti bot. tartu.* 5 (1/2, art. 4): 1—64.
ŞTEFUREAC, T. I., – 1941 – Cercetări sinecologice şi sociologice asupra Bryophytelor din codrul secular Slatioara (Bucovina). (French summ.) *Anal. Acad. rom.*, Bucureşti, ser. III, 16 (Mem. 27): 1—197.
SULMA, T., – 1935 – Beiträge zur Ökologie und Verbreitung der Flechten auf dem Lubliner Hügelland. *Bull. Int. Acad. pol. Sci. Lett.*, Cl. Sc. Math. Nat., Sér. B (1), 1935 (1—3): 77—100.
SZCZAWINSKY, A., – 1953 – Corticolous and lignicolous plant communities in the forest associations of the Douglas Fir forest on Vancouver Island. Thesis, Univ. of Brit. Columbia, Vancouver. 283 pp.
TOMILIN, B. A. – 1964 – Environmental factors affecting the distribution of fungi among plant communities (In Russian with Engl. summ.) *Bot. Zh. SSSR* 49 (2): 230—239.
TOMSON, A., – 1937 – Sõrve taimkate (French summ.: La végétation de la presqu' île de Sõrve, Estonie). *Acta Inst. Horti bot. tartu.* 6 (1, art. 1): 1—87.
TÜXEN, R., – 1937 – Die Pflanzengesellschaften Nordwestdeutschlands. *Mitt. flor.-soz. ArbGemein. Niedersachsen* 3: 1—170. *Jber. naturh. Ges. Hannover*, 1929/30-1935/36, 81—87: 1—170.
TÜXEN, R., – 1950 – Grundriss einer Systematik der nitrophilen Unkrautgesellschaften in der Eurosiberischen Region Europas. *Mitt. flor.-soz. ArbGemein.*, Stolzenau, N.F. 2: 94—175.
TÜXEN, R., – 1964-6 – Bibliographia Phytosociologica Cryptogamica. Pars II: Bibliographia mycosociologica I, II. *Excerpta bot.* B, 6 (2): 135—160; 6 (3): 161—178, 7 (3): 220—224.
TÜXEN, R. (ed.), – 1965 – Biosoziologie. *Ber. Symp. Int. Ver. Vegetationskunde*, Stolzenau/Weser, 1960, 350 pp. Junk, The Hague.
TÜXEN, R. (ed.), – 1968 – Pflanzensoziologische Systematik. *Ber. Symp. Int. Ver. Vegetationskunde*, Stolzenau/Weser, 1964, 346 pp. Junk, The Hague.
TÜXEN, R., A. VON HÜBSCHMANN, & W. PIRK, – 1957 – Kryptogamen- und Phanerogamen-Gesellschaften. *Mitt. flor.– Soz. Arb. Gemein.*, Stolzenau/Weser, N.F. 6/7: 114—118.
UBRISZY, G. DE, – 1940 – Adatok a Nyírség gombavegetációjának ismeretehéz. (French summ.: Contribution à la connaissance de la végétation mycologique de Nijirseg). *Acta geobot. hung.* 3: 65—78.
UBRISZY, G. DE, – 1956 – Ujabb vizsgálatok az erdötipusok talajlakó nagygombáinak társulasi viszonyairól. (Germ. summ.: Neuere Untersuchungen über die Cönologie bodenbewohnender Grosspilze der Waldtypen). *Acta bot. hung.* 2: 391—424.
VAGA, A., – 1940 – Fütotsönoloogia põhiküsimusi (Engl. summ.: On some fundamental problems in phytocoenology). *Acta Comment Univ. tartu*, Ser. A, 35 (6): 1—152.
VILLERET, S., – 1954 – Contribution à la biologie des Algues des tourbières a Sphaignes. *Bull. Soc. scient. Bretagne* 29: 1—246.
VILLERET, S., – 1965 – Quelques aspects biologiques de la répartition des Algues d'eau douce. (Engl. summ.) In 'Biosoziologie,' ed. R. TÜXEN, *Symp. Int. Ver. Vegetationskunde* Stolzenau/Weser 1960: 140—149.
VRIES, D. M. DE, – 1926 – Het plantendek van de Krimpenerwaard I. Phytosociologische beschouwingen: Begrippen, wetten, bouwbeschrijvende methodiek. *Ned. Kruidk. Archf.* 1925: 215—275.
VRIES, D. M. DE, – 1932 – Grondslag van een Nederlandse plantensociographische naamgeving. *Ned. kruidk. Archf.* 1931: 517—527.

WALDHEIM, S., – 1944 – Mossvegetationen i Dalby-Söderskogs Nationalpark. *K. Svenska VetenskAkad. Avh. Naturskydd.* 4: 1—142.
WALDHEIM, S., – 1947 – Kleinmoosgesellschaften und Bodenverhältnisse in Schonen. *Bot. Notiser,* Suppl. 1 (1): 1—203.
WARMING, E., – 1909 – Oecology of Plants: An Introduction to the Study of Plant Communities. Oxford Univ., Oxford. 422 pp.
WESTERDIJK, J., – 1949 – The concept 'association' in mycologie. *Antonie van Leeuwenhoek* (J. Microbiol. Serol.) 15: 187—189.
WESTHOFF, V., – 1968 – Einige Bemerkungen zur syntaxonomischen Terminologie und Methodik, insbesondere zu der Struktur als diagnostisches Merkmal. (Engl. summ.) In 'Pflanzensoziologische Systematik,' ed. R. TÜXEN, *Ber. Symp. Int. Ver. Vegetationskunde,* Stolzenau/Weser 1964: 54—70.
WESTHOFF, V. & A. J. DEN HELD, – 1969 – Plantengemeenschappen in Nederland. Thieme, Zutphen. 324 pp.
WHITTAKER, R. H., – 1962 – Classification of natural communities. *Bot. Rev.* 28: 1—239.
WILMANNS, O., – 1958 – Zur standörtlichen Parallelisierung von Epiphyten- und Waldgesellschaften. *Beitr. naturk. Forsch. SüdwDtl.* 17 (1): 11—19.
WILMANNS, O., – 1959 – Epiphytengesellschaften Nordgriechenlands im Vergleich mit denen Mitteleuropas. *Phyton* 8 (1/2): 175—182.
WILMANNS, O., – 1962 – Rindenbewohnende Epiphytengemeinschaften in Südwestdeutschland. *Beitr. naturk. Forsch. SüdwDtl.* 21 (2): 87—164.
WILMANNS, O., – 1970 – Kryptogamen-Gesellschaften oder Kryptogamen-Synusien? In 'Gesellschaftsmorphologie (Strukturforschung),' ed. R. TÜXEN, *Ber. Symp. Int. Ver. Vegetationskunde,* Rinteln 1966: 1—7.
WILMANNS, O. & H. BIBINGER, – 1966 – Methoden der Kartierung kleinflächiger Kryptogamengemeinschaften. *Bot. Jb.* 85 (3): 509—521.
WOLAK, J., – 1970 – Complexe horizontal et complexe vertical. (Germ. & Engl. summs.) In 'Gesellschaftsmorphologie (Strukturforschung),' ed. R. TÜXEN, *Ber. Symp. Int. Ver. Vegetationskunde,* Rinteln 1966: 136—141.

17 RUSSIAN APPROACHES TO CLASSIFICATION OF VEGETATION

VERA D. ALEKSANDROVA

Contents

17 RUSSIAN APPROACHES TO CLASSIFICATION OF VEGETATION

17.1 Introduction: Brief Historical survey

17.1.1 The Early Period

The first suggestions of classifying vegetation in Russia can be found in works of early naturalists (Tatishchev 1740, see Tatishchev 1950, Krasheninnikov 1755, Engelmann 1810, and others) and in papers by travelers–geographers and floristic botanists of the middle of the XIX century (Ruprecht 1845, Trautvetter 1849-1851, Borshchev 1865, Middendorf 1867, and others). Typological characteristics of vegetation were considered at that time in only a fragmentary manner, and usually in connection with floristic and general description of particular regions. Appearance of investigations which focused special attention upon geobotanical classification dates back to the 1880's when these problems were recognized primarily in two areas–forest typology and studies of steppe vegetation.

The peculiarities of forest typology in Russia were due to the demands of forestry activities (Genko 1889, see Genko 1902 Gutorovich 1897, Morozov 1912, Serebrennikov 1913, Krüdener 1916, and others). The result was an essentially ecological classification: a 'forest type' comprised forest stands grouped on the basis of similarities in the envoronment and resemblance in stand quality. A certain forest type consequently might include forest stands with different dominant tree species. At the same time Russian classification of forests was being influenced by foreign scientists, thus Gordyagin (1900) employed a method similar to that of the Uppsala school in his investigations of the West Siberia forests. Along with Russian terms he used Latin terminology in defining forest types and proposed for every association 'a principal group of species' on the basis of species constancy.

Classification of steppe vegetation (Krasnov 1886 et al. Korzhinski 1888 et al. Tanfilyev 1897 et al., Kostychev 1900, Gordyagin 1900-1, and others) was developing at this time principally in connection with the zonal division of the vast steppe areas of Russia. Terms like 'formation', 'association', and 'vegetation type' were used; but no real classification was expressed thereby, for these terms were used to designate any plant communities.

In 1902 FLEROV & FEDCHENKO published their classification, comprising all vegetation types of the European part of Russia. The major unit of classification was 'a group of plant communities'; and several 'groups of communities' were suggested: 1) water group, 2) mire group, 3) forest group, 4) steppe group, 5) meadow group, 6) bare slopes group, and 7) cultured (anthropogenic) group. Each 'group' was further subdivided into 'communities' or 'formations' (these terms were employed as synonyms). Classification was based upon environmental relations of communities and on the dominant species; dynamics of vegetation was also considered. However, no theoretical justifications for use of taxonomic categories and definition of units of vegetation were proposed.

17.1.2 DEVELOPMENT OF MODERN VIEWPOINTS

The fundamentals for developing a theoretical background for Russian classification of vegetation were established by SUKACHEV and RAMENSKI. They both expressed their views (at the same time) in December 1909 in reports in Moscow (SUKACHEV 1910, RAMENSKI 1910); SUKACHEV proceeded from the idea of discrete units of vegetation, whereas RAMENSKI conceived vegetation as continuous. RAMENSKI (1915-1953) developed further his theory of vegetative cover taken as a continuum and delineated a variety of quantitative approaches to studies of vegetation, including an ingenious method of ordination (more detailed description thereof can be found in article 5 in this volume).

SUKACHEV repeated in his report (1910) the definitions of community and formation offered in his earlier paper 'Forest formations and their relationships in the Bryansk woods' (1908): 'The plant community ('Einzelbestand' of SCHRÖTER, 'association locale' of FLAHAULT, forest stand in forestry) is taken to mean an association of organisms in which mutual influence is exerted by plants on one another and between plants and the environment. The plant formation ('Bestandtypus' of SCHRÖTER, forest stand type in forestry) comprises those plant communities of a given geographic region which are characterized by essentially the same ecology, i.e. displaying essentially similar interrelations between plants as well as between the latter and the environment' (SUKACHEV 1908: 5-6). In his report in 1909 SUKACHEV (1910: 150) repeated almost literally this definition, already referred to as characteristic of the 'Leningrad school', and implying simultaneously in the single term 'community' (in the middle 1930's and later, 'phytocoenose') both the chorological unit of the plant cover (stand, 'Einzelbestand') and the phytosociological concept of associated growth of plants (community, 'Pflanzengemeinschaft'). After the appearance of the proceedings of the Brussels Congress (FLAHAULT & SCHRÖTER 1910), SUKACHEV began to use in the latter sense the term 'association' instead of 'formation.'

In 1915 SUKACHEV wrote a paper containing a first suggestion of a coordinate scheme of plant associations in relation to environmental gradients, which was further developed in 1927 and in other works (Fig. 1). This method gained much support from Russian scientists and is widely employed in classification of forests and other vegetation types (cf. Fig. 2). SUKACHEV (1917, 1928, et al.) proposed a hierarchic system of taxonomic categories for communities the lowest unit of which was the association, the main unit of middle rank the formation, while the broadest unit was the vegetation type. The terminology proposed by SUKACHEV, and the meanings he implied thereby, have become most widely accepted in Soviet geobotany.

The classification of vegetation has been further enriched with the introduction of the notion of 'placor' by VYSOTSKI (1909, 1930, et al.). The word is derived from Greek (πλάξ or πλάκα and ορος) – upland habitats which are adequately drained under given macro-

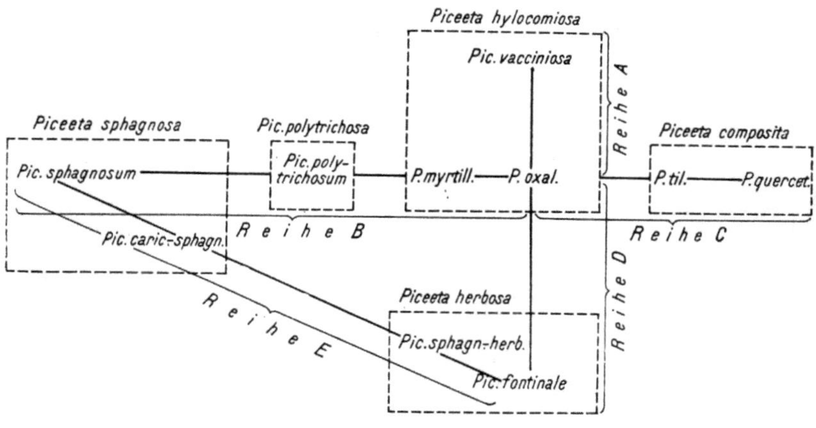

Fig. 1. A compound ecological series for spruce forest (Piceeta) according to SUKACHEV (1928: 218, fig. 48, cf. 1932).P.oxal – Piceetum oxalidosum, P.til. – Piceetum tiliosum, P. quercet. – Piceetum quercetosum, P. sphagn.- herb.– Piceetum sphagnoso-herbosum, P. myrtill. – Piceetum myrtillosum, Pic. caric.-sphagn. – Piceetum caricoso-spagnosum. Spruce forest with oxalis (Piceetum oxalidosum) is the central or climax type, from this radiate series ('Reihe'): A – toward drier, nutrient-poor soils; B – toward wetter soils with stagnant water; C – toward soils with more favorable nutrient content; D – toward wet soils with moving water; E – connects the latter with the bog forests.

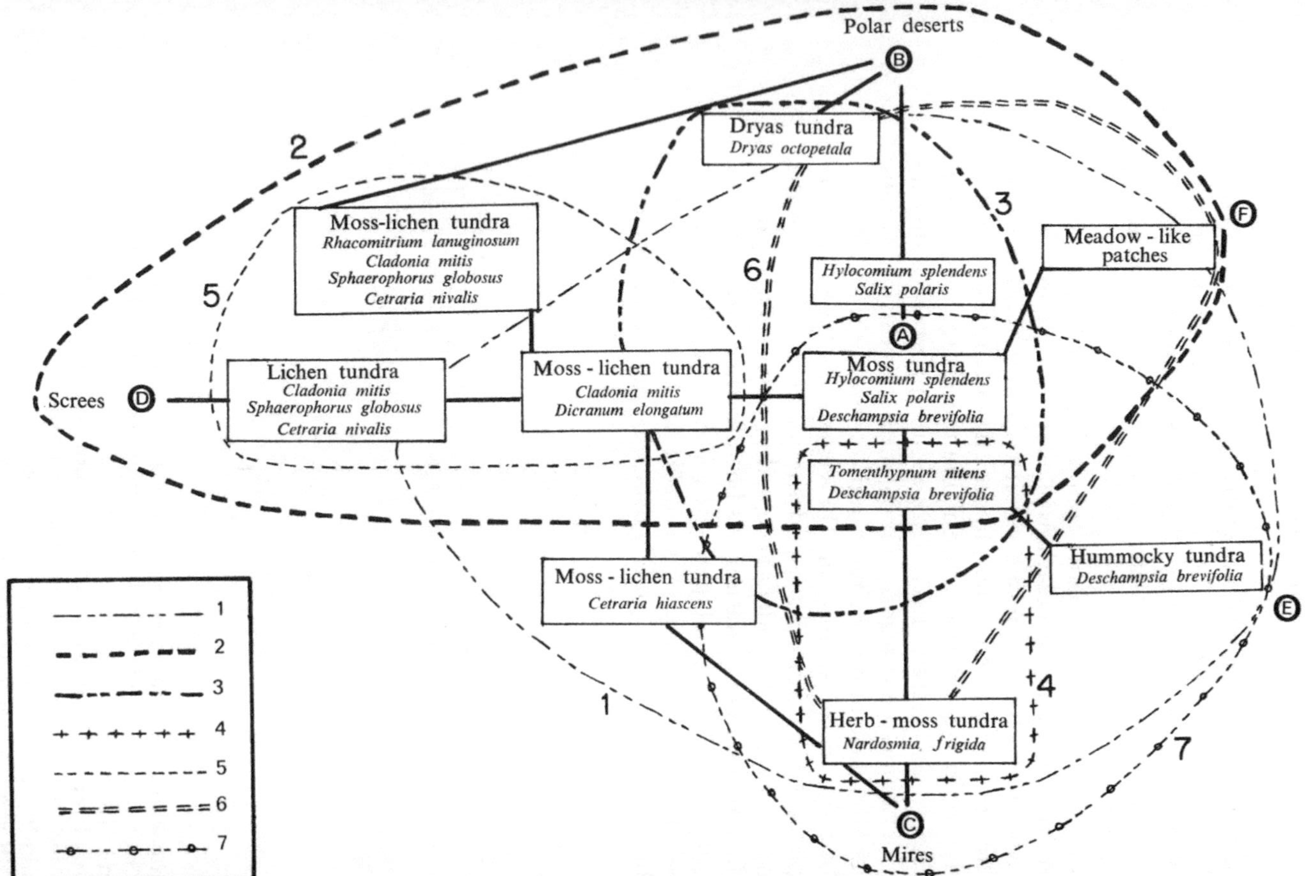

Fig. 2. A compound ecological series for tundra communities, and the distributional areas occupied by some plant species on the Southern Islands of Novaya Zemlya (ALEKSANDROVA 1956: 231, Fig. 10). The moss tundra community at *A* is central to the whole pattern, from this radiate ecological series: *AB* – toward drier sites with a thinner snow cover; *AC* – toward wet, poorly drained sites with heavier snow cover; *AD* – toward screes and rockslopes, showing the effect of increase in rock content of the soil; *AE* – toward increasing of solifluction; and *AF* – toward enrichment of the soil with nitrogen by animals. The distributions (ecologo-phytocoenological areas) of some major plant species in the community pattern are outlined: 1 – *Deschampsia brevifolia* R. Br., 2 – *Papaver polare* (A. Tolm.) Perf., 3 – *Dryas octopetala* L., 4 – *Nardosmia frigida* Hook.,

climatic conditions. Making use of this definition and referring to ideas expressed by SIBIRTSEV (1900) on 'zonal' soils, PACZOSKI (1915) introduced notions of zonal, azonal and extrazonal vegetation; ALEKHIN (1915: 409) independently of the latter proposed the terms zonal and intrazonal. 'Zonal' vegetation, occupying placors, is analogous to the climax concept in the sense of CLEMENTS. 'Azonal' vegetation (which can be seen in nonplacoric habitats of any zone, e.g. flood plain meadows) and 'intrazonal' vegetation (which occurs only within certain zones, e.g. sphagnum bogs) are together loosely analogous to the subclimax concept of CLEMENTS. 'Extrazonal' vegetation meets its analogue in the terms pre- and postclimax (CLEMENTS 1936); this is vegetation 'spreading outside its zone' (ALEKHIN 1936: 366), e.g. forest stands on north-facing slopes in the steppe zone. However, in contrast to CLEMENTS, Russian authors used these concepts without requiring successional relationships between placoric and non-placoric formations.

ALEKHIN, who was in controversy with SUKACHEV on a number of points, formed the 'Moscow school' in Russian geobotany including DOKHMAN, KATS, URANOV and others. Disagreements with SUKACHEV involved both concepts and terminology; ALEKHIN argued in particular against the double meaning of the term 'community' as defined by SUKACHEV. Scientists of the Moscow school extensively studied the methodology of investigation and treatment of data. The significance of the means of detecting analytical characteristics of plant communities was emphasized (ALEKHIN 1932, et al.); and a critical evaluation in practice was applied to the methodological recommendations of the Uppsala school concerning the constancy concept, minimal area and others. KATS (1933, et al.) contributed to the studies of the association (= sociation) as a small, homogeneous unit defined on the basis of the dominants of layers. DOKHMAN (1930, 1936, et al.) developed principles for detecting and classifying of community complexes.

In the 1920's in the Ukraine POGREBNYAK (1929, et al.) investigated classification in his own way, basing his classification on ordinating communities along gradients of two factors: moisture and nutrient condition of soil. The forest type, as conceived by representatives of this school (cf. the Finnish school), comprises forest sites with the same climax-type and all successional stands, including areas without any trees (ploughed fields and the like).

The system of concepts and ideas for classification was developing in Russia under the primary influence of Russian ecological and genetical traditions, derived especially from ideas of the great Russian naturalist DOKUCHAEV which influenced MOROZOV, SUKACHEV, and their contemporaries. On the other hand, there was marked influence of the phytocoenologists of northern Europe, particularly those of the Uppsala school (HULT, SERNANDER, DU RIETZ, and others), and WARMING, as well as CAJANDER, SCHRÖTER, DRUDE, BROCKMANN-JEROSCH, and RÜBEL. The American trend of classifying of vegetation by dynamics as developed by COWLES and CLEMENTS, and the approaches to classification offered by BRAUN-BLANQUET in the southern Europe, were not popular among Russian geobotanists at that time. The development of principles of classification of vegetation in Russia was influenced also by the vast territories of the country and the great variety of vegetation found there.

In 1934 a discussion took place in Leningrad on fundamental problems in phytocoenology. Leading Soviet phytocoenologists participated in this: SUKACHEV, ALEKHIN, SHENNIKOV, OVCHINNIKOV, and others. Reports of this discussion were published in '*Soviet Botany*', 5, 1935. The principal attention was given to discussion of concepts and terms pertaining to the main purposes of phytocoenology, and the meaning of the lower units of classification of vegetation; the different approaches of the 'Leningrad school' (SUKACHEV) and 'Moscow school' (ALEKHIN) to these problems were defined and expressed. In these years a series of works appeared concerning major units of vegetation classification (KOROVIN 1934, VASILYEV 1935, GORODKOV 1935, ILYNSKI 1935, PROZOROVSKI 1935, SHENNIKOV 1935, ALEKHIN 1936, 1938, and others). In these same years (the 1930's) a system of ideas emerged as characteristic of vegetation clasification in Soviet geobotany. Despite some still existing controversy, there are common traditional ideas shared by all Soviet geobotanists to this day that differ from those of foreign schools.

17.1.3 MORE RECENT TRENDS

In the 1940's [1]) LAVRENKO used principles of classification based upon dominant ecological variants of the growth-forms (ecobiomorphs) for classification of steppes (1940), compilation of small-

[1]) In these years SUKACHEV began to write on the biogeocoenose (SUKACHEV 1942, 1945, et al.). This concept cannot be directly employed in classification of vegetation, since the biogeocoenose is not the subject of phytocoenology but of a different discipline – biogeocoenology – that concerns the biogeocoenotic geosphere (SUKACHEV 1964), i.e. the ecosphere in the sense of COLE (1958). A biogeocoenose comprises along with the phytocoenose the animals thereof, and the soil, atmospheric gases and water in a state of mutual exchange with the living organisms. 'Biogeocoenose' can be considered either as a synonym of the holocoen in the sense of FRIEDERICHS (1930), or as an ecosystem within the boundaries of one phytocoenose (of one stand) (LAVRENKO & DYLIS 1968).

scale maps of vegetation (1941, et al.), and regional geobotanic characterization of the USSR (1947). These years saw also the development of genetical classification (LESKOV 1943, SOCHAVA 1944, 1945, OVCHINNIKOV 1947, et al.). Of particular importance was the introduction of the phratria concept (SOCHAVA 1944, et al.), extensively used by SOCHAVA and his school in classification of vegetation and vegetation mapping.

Towards the end of the 1950's publications appeared resuming discussion of the vegetation continuum (PONYATOVSKAYA 1959, 1961, VASILEVICH 1960, 1966a, 1968a, 1969, RABOTNOV 1963, ALEKSANDROVA 1965, 1969, TRASS 1966, FREY 1967, DOLAKHANOV 1970, and others). In the 1960's there has been a tendency toward revision of traditional ideas on classification and the concepts of association and formation, and toward exploring new methods of classification (VASILEVICH 1963b, 1966b, 1968c, 1969, 1971, NITSENKO 1963, 1966, MIRKIN 1965, TRASS 1965, 1971, IPATOV 1966, 1969, ISACHENKO 1967, 1969, NORIN 1966, 1970, FREY 1966b, 1967, 1971, KARAMYSHEVA 1967, SOCHAVA 1968, 1970, KARAMYSHEVA & RACHKOVSKAYA 1968, RYSIN & KOVALENKO 1968, APALIA 1968, 1971, LUKICHEVA & SABUROV 1971, and others). Further publications deal with quantitative geobotany and application of mathematical methods to classification of vegetation (VASILEVICH 1960, 1963a, 1967a, 1967b, 1969, 1971, FREY 1963, 1967, 1968, 1971, NESHATAEV 1968, 1971a, VOLKOVA 1969, SAMOYLOV 1970, NORIN 1971, MIRKIN 1970a, and others).

In order to describe Russian approaches to classification of vegetation further, the discussion will divided as follows: the objects of classification, the association, the formation, and the higher units of vegetation.

17.2 Objects of Classification

17.2.1 PHYTOCOENOSES

The main object of classification of vegetation is considered to be a phytocoenose, as understood by SUKACHEV (1957: 15) 'The 'phytocoenose, or plant community, is taken to mean any plot of vegetation on the given area uniform in its composition, synusial structure, and the pattern of interactions among plants and between these and the environment.' This definition, repeatedly expressed by SUKACHEV with but minor modifications (1954: 316, et al.), and similar definitions given by SHENNIKOV (1964: 12, et al.) are most widely accepted in the USSR. As already been mentioned, this definition of 'phytocoenose', or 'community' implies double meaning: the chorological unit (stand, 'Bestand') and phytosociological concept (plant community, 'Pflanzengemeinschaft') are called by the same term. In a discussion with SUKACHEV lasting over 20 years, ALEKHIN proposed to divide these two concepts – the second to be referred as a 'plant community' (or 'phytocoenose'), the first to be termed a 'parcel of association' (other terms have been suggested, e.g. 'phytochora' – NORIN, 1970). ALEKSANDROVA (1969: 22-36) considers it reasonable to associate with the term 'phytocoenose' its original meaning of a chorological ('topographical') unit, or stand, for which this term had been suggested by GAMS (1918: 421), [1]) with the word 'community' to serve as a general, rankless term for definition of any vegetation unit both chorological and synecological. The above given definition by SUKACHEV is, however, the most widely accepted in Russian geobotany.

Since the phytocoenose as a chorological unit is defined on the basis of similarity or homogeneity of vegetation within the limits of a given parcel of vegetation cover, the meaning of phytocoenose depends upon the way the concept of 'similarity' is defined. YAROSHENKO (1961: 50) writes: 'As introduced by SUKACHEV into the phytocoenose definition, the character of similarity of composition and interaction of structural parts is not so readily discernible as might appear at first glance. The trouble is that the so-called mosaic phytocoenoses are too often met in nature; these display a pattern of patches differing from each other in species composition and structure. In such a case similarity of the phytocoenose as a whole must be understood to mean only a similar mosaic'. The different elements of a mosaic in the phytocoenose were called 'microgroups' or 'microaggregations' (YAROSHENKO 1961, et al.) or 'microphytocoenoses' (LAVRENKO 1959). 'Each microgroup includes only a small part of the whole community, but it includes all the community's layers at a given place, and in this respect the microgroup or microphytocoenose differs from the 'synusia' (YAROSHENKO 1961: 57). A question then arises should any 'microcoenoses' be considered as separate phytocoenoses, or as fragments from different phytocoenoses installed within the given phytocoenose? LAVRENKO (1959:31) uses the following criterion: If a plant cover contains a patch which differs from the surrounding vegetation then, if the roots of plants surrounding this patch join beneath it, we have a microgroup

[1]) The first who used the term 'phytocoenose' was PACZOSKI (1915, 1921), however he implied another meaning by this term, namely, he proposed to call 'phytocoenose' a stand with one species (e.g. stand of *Phragmites communis*), while the 'community' was a stand comprising different species. Note also that PACZOSKI (1896, 1921) was the first to use the term 'phytosociology.'

(microphytocoenose) that is part of a mosaic phytocoenose. If, on the other hand, the roots of plants surrounding the patch in question contact it only along its boundary, the patch is to be regarded as a different phytocoenose or a fragment from another phytocoenose; and in such a case we see not a mosaic phytocoenose, but a phytocoenose-complex or its fragments. (A fragment of a phytocoenose is a plot smaller than minimal or representative area.) This criterion is considered crucial in discriminating mosaic-structure and complex-structure: The term 'mosaic structure' denotes non-uniformity of composition within a phytocoenose, 'complex-structure' non-uniformity of a higher rank, the elements of which represent different phytocoenoses or their fragments. Searches for objective criteria for detecting phytocoenoses and their boundaries in the field are being continued with the aid of statistical methods (Nitsenko 1948, 1971, Bykov 1957: 50, Vasilevich 1967a, 1971, Lobanova 1971, Trotsenko 1971, and others).

17.2.2 Synusiae

Within a phytocoenose, as its distinctive parts, synusiae are recognized. The term 'synusia' was defined so broadly by Gams (1918) that confusion in use of the concept resulted (see also article 16). In the USSR the term is used with two different meanings.

Alekhin (1936: 325-327, et al.) applied 'synusia' to an assemblage of species of similar environmental relations and life-forms within a given phytocoenose. This concept is close to the 'ecological groups' of Ellenberg, (1950, 1952) the 'bioecogroups' of Apalia (1968), and the like. But the term 'synusia' has been more widely accepted in the USSR in the sense of Lippmaa (1933, 1939, et al.), for the concept initially termed by him the 'unistratal association.' Synusiae in the sense of Lippmaa are communities of plants growing together which, in the course of long-lasting natural selection, have become adapted to co-habitation under the conditions of certain environmental relations. This gives deep phytocoenological meaning to the synusia and justifies use of synusiae as natural units of classification. The relative independence of synusiae is readily illustrated by such phenomenon as Sochava (1930: 28) described in the Ural mountains and called the 'incumbation' (overlaying) and 'decumbation' (segregation) of layers and formation of 'incumbational (overlay) series'. Fruitful investigations in the classification of synusiae in the sense of Lippmaa are being carried out in the USSR by Trass (1964, et al.) and his disciples (Martin, 1967, et al.). Trass offered a system of units for classification of synusiae (Table I) employing the units of Lippmaa and, in part, the terminology proposed by DU RIETZ (1930: 497). These taxonomic units were used by MIRKIN (1970b, et al.) in classification of forests with regard to their ground layer.

Table I

System of taxonomic units for classification of synusiae (monosynusial units) and phytocoenoses (polysynusial units) by TRASS (1964:104). (See also article 16, Table I.)

Monosynusial units	Polysynusial units
Society	Sociation
Union	Association
Federation	Association-group
Formion	Formation
Panformion	Formation-group
Type of monosynusial vegetation	Type of polysynusial vegetation

The smallest part of the plant cover that can be distinguished as a phytocoenological unit is the 'coenocell' that IPATOV (1966) described as 'an elementary unit of social life of plants.' The coenocell comprises all plants found in the area of immediate influence by a single plant; it can also be defined as an assemblage of all plants growing within the 'phytogenic field' of biotic interactions with a single individual, as described by URANOV (1965). Recently VASILEVICH (1971) provided statistical evidence for an elementary pattern phenomenon in plant cover called by him a 'coenoquant.'

17.2.3 Complexes and Continua

The importance of units of the plant cover broader than the phytocoenose for classification is acknowledged by Soviet geobotanists. Such units are complexes (Dimo & Keller 1907, Du Rietz 1921, 1930, Dokhman 1930, 1936, et al.) and compound complexes (Dokhman l.c., et al.). All these units are termed by Isachenko (1967, 1969, et al.) 'combinations': microcombinations, mesocombinations, macrocombinations. Macrocombinations are similar to the concept of 'Fliesen' of Schmithüsen (1968: 126, 236). Recently significant contributions have been made in the USSR to classification based on units of this kind (Isachenko 1967, 1969, Guricheva et al. 1967, Karamysheva & Rachkovskaya 1968, 1969 et al.), especially for large-scale mapping of vegetation.

In Soviet geobotany renewed attention has been given to the problem of the degree of objectivity of these units, as affected by the vegetation continuum conception (Vasilevich 1960, 1966a, 1968a et al., Ipatov 1966, Trass 1966, Frey 1967, Aleksandrova 1965, 1969, Dolukhanov 1970, et al.). It has been concluded that, although continuity is an inherent and crucial feature of vegetation, the continuum is not structureless, not amorphous, not homogeneous. Vegetation forms a structured, patterned continuum, heterogeneous and unequal in its various parts, with regular repetition under similar conditions of similar patterns of plant populations. There are, moreover, cases both of continuous transitions from one population pattern to another, and of more or less 'steep' transition from one 'isotropic' phytocoenose to another (Vasilevich 1967a: 92, see also Nitsenko 1948, 1971). Relative discontinuity in a vegetation continuum is related to the scale of investigation (Masing & Trass 1963, Sochava 1965a: 4, Vasilevich 1968a, et al.). There is no conflict between the conception that phytocoenoses exist as parts of partly or wholly continuous patterns of vegetation, and treatment of those phytocoenoses as objects of classification (cf. Ramenski 1924, 1938, Whittaker 1956, 1967, and articles 3 and 5 in this volume).

Thus, for classification of vegetation Russian geobotanists use a number of units, each of which has its significance and its own area of application. These units are listed above. The principal object of classification is considered to be the phytocoenose or plant community; the abstract classes or community-types include the association, formation, and other units of the hierarchic system into which phytocoenoses are classified.

17.3 Units of Classification

17.3.1 Associations

Russian geobotanists in the 19th century used both the terms 'association' and 'formation' without any real taxonomic meaning. The definition of the association concept and the theoretical foundation for distinguishing associations were originally presented by SUKACHEV. In his initial papers SUKACHEV called the small typological unit of vegetation, distinguished on the base of layer dominants, a 'formation' (SUKACHEV 1908: 56, 1910: 150, cf. HULT 1881). Following the appearance of the proceedings of the Brussels Congress (FLAHAULT & SCHRÖTER 1910) he changed this term in favour of 'association'. At the later Amsterdam Congress the term 'sociation' was accepted for this unit (DU RIETZ 1936). Throughout the following it should be kept in mind that the Russian 'association' corresponds in general to the 'sociation' of Western authors.

In the 1930's much attention was given to the theoretical and methodological foundation of the association concept as a small, homogeneous unit, and to the technique for distinguishing of such associations, in the works of KATS (1929, 1933, et al.); the same attitude was shared by LESKOV (1943) and more recently by NITSENKO (1963, et al.). The majority of the Soviet geobotanists, however, are inclined to give the term 'association' not to the smallest homogeneous unit, but to a broader one, to a group of phytocoenoses 'with a given system of variation' (SHENNIKOV, 1964: 395). The association as a such a unit was used by SUKACHEV (1928, 1931, et al.), ALEKHIN (1935, 1936, 1938); BYKOV (1957), SOCHAVA (1961, 1968), MASING & TRASS (1963), MIRKIN (1968), APALIA (1968, 1971), SAMOYLOV (1970), SHELYAG-SOSONKO (1971), and others. It is considered possible to distinguish subdivisions of this broader association: subassociations (SUKACHEV 1928: 119, SHENNIKOV 1964: 392, et al.), association variants (VASILEVICH 1968b, APALIA 1968, 1971, MATVEEVA 1968, et al.), narrowly defined sociations (ALEKHIN 1935, 1938; MASING & TRASS 1963, TRASS 1964, SHENNIKOV 1964, et al.), plant combinations (VASILEVICH 1963b, 1968b), dominant modifications (MIRKIN 1968), etc.

An effort toward substantiating the meaning of the association as a natural unit was offered by VASILEVICH (1968b: 872). VASILEVICH defines the plant association as the first (the smallest) natural group of communities, the taxonomic distinctiveness of which is shown statistically: a group, forming a more or less isolated cluster of stand-samples in an abstract multi-dimensional model (Fig. 3). Units lower than the association (association variants, etc.) are under-

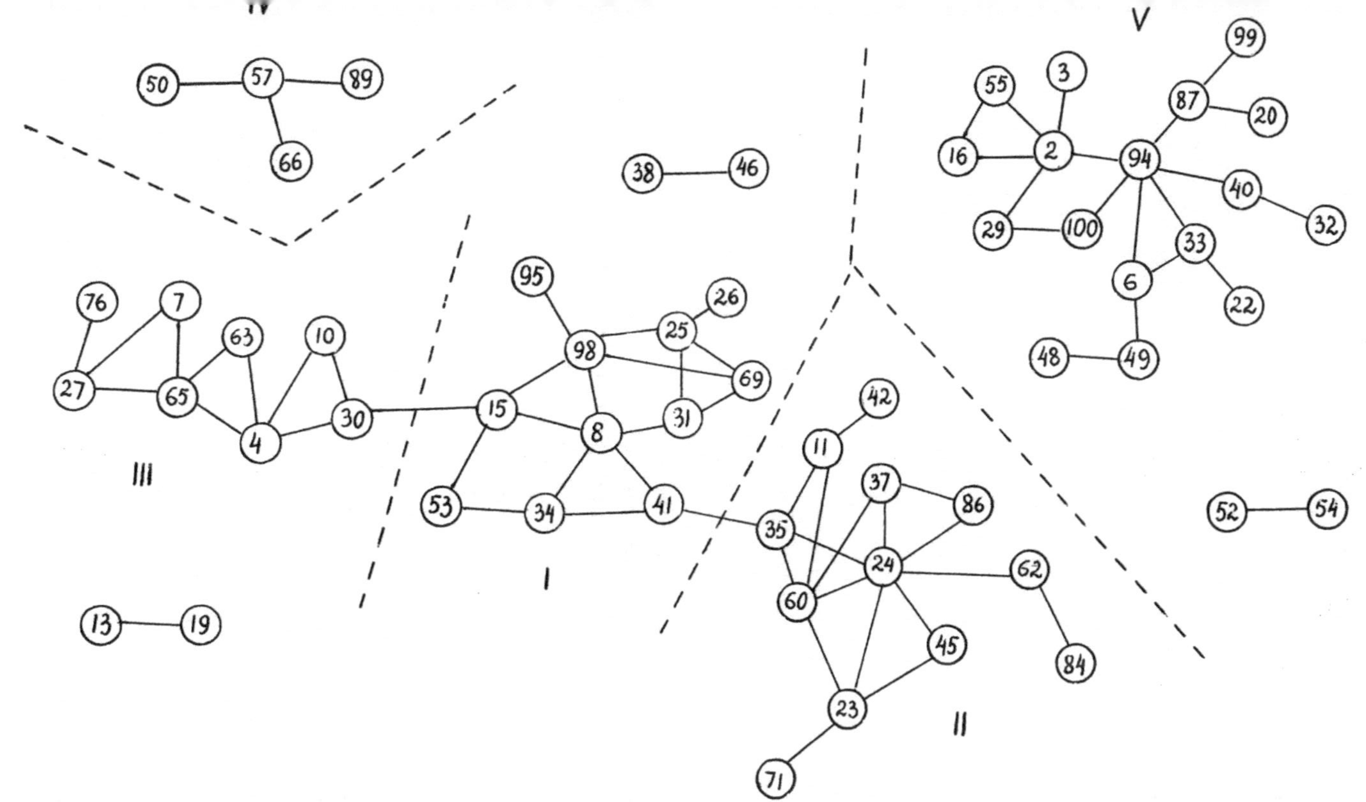

Fig. 3. The clusters of samples of coniferous and parvifoliate forests in a multi-dimensional geometrical model (VASILEVICH, 1967b: 48, fig. 1). Samples (phytocoenoses) differing less than by 25 R^2 value are joined by a line. I – V – group numbers. Dashed lines separate different groups. The clusters might be considered as typological units or noda of the forests studied. $R^2 = (\sqrt{a_1}-\sqrt{a_2})^2 + (\sqrt{b_1}-\sqrt{b_2})^2 + (\sqrt{c_1}-\sqrt{c_2})^2$.......... where a_1, b_1, c_1, ... are coverages (per cents) of species in the first sample, a_2, b_2, c_2,,..... coverages of the same species in the second sample.

stood as parts (to be considered as artificial) of this natural group. Evidently a similar but intuitive attitude was shared by SUKACHEV, who called the association 'a lower taxonomic unit' but described subassociations within it.

The manner of distinguishing associations according to the layer dominants is traditional for Russian classification of vegetation, and is most widely accepted. A number of Soviet geobotanists, however, do not share this view. RAMENSKI (1938: 333, 1950, 1952, et al., article 5) most energetically argued against classification by dominants, emphasizing use of the complete list of species and basing the systematization of phytocoenoses on ordination. DOKHMAN (1954, 1960) also used for distinguishing associations not the dominants (at least not only the dominants), but a 'character-group of associated species'; a similar approach was used by DOLUKHANOV (1957) and ISACHENKO & RACHKOVSKAYA (1961). The inadequacy of using dominants for vegetation classification was pointed out also by TRASS (1965). A 'dominant-determinant' approach to classification was elaborated by MIRKIN (1965, et al.), and a similar principle was successfully employed by SAMOYLOV (1970) in classification of meadow communities. Distinction of associations on the basis of bioecogroups, rather than dominants, was developed by APALIA (1968, 1971), etc. For the classification of meadows and other herbaceous phytocoenoses, where the relative abundance of species may vary considerably with annual fluctuations, the concepts of 'ecological replacement groups' and 'change dominant-complex' are proposed and used for classification of phytocoenoses (RABOTNOV 1963, 1965, NITSENKO 1963, MIRKIN 1968), In his recent papers SOCHAVA (1968, et al.) gives the term 'epiassociation' to a unit still broader including the climax community with successional stages (SOCHAVA 1968: 17, 21).

The technique for distinguishing plant associations, characteristic of the Russian approach to classification of vegetation, is described by NESHATAEV (1971b). A collection of stand-samples is taken through geobotanical analysis of either sample plots (quadrats) or within natural boundaries of phytocoenoses as estimated visually. The choice of stand-samples is the same as that suggested by BRAUN-BLANQUET (1964: 24): samples are chosen at typical stands, taking the 'typical' to mean the observed repeated correspondence of similar patterns of species with similar biotopes. In this way the very choice of stand-samples predestines the distinction of certain associations. The second step is tabulation of the samples, which have already been referred to one or several related associations. Examination of such a table (no formal procedure is implied) makes possible clarifying the associations and their distinctions, and permits their definitions to be stated. The third step is the description of association characteristics according to dominants and subdominants, position in ecological and ecologo-phytocoenological series, etc; also analysis of the composition of life-forms and ecological and geographical species groups is performed, and the analytic and synthetic characters of the association are determined (PONYATOVSKAYA 1952, SHENNIKOV 1964: 395-404, etc.).

The names of associations are formed in Soviet geobotany in two ways. The first implies a binary name: the noun, first, is the genus of the dominant with the suffix '-etum'; while the adjective following the genus is the dominant in a subordinate layer with the suffix '-osum'. In case two dominants exist in one layer both are included in the name of the association, so that the name of one (in the dative) is united with the other through a dash, for example: Tilieto-Quercetum aegopodiosum, Piceetum hylocomioso-myrtillosum. When further specification of the dominant is needed, species names may be incorporated as follows: Spahgneto fusci-Sphagnetum magellanicum eriophorosum vaginati, and the like. For subassociations the name is expanded thus: Nardetum sphagnosum subass. agrostosum caninae, and so on. Sometimes the adjective can be derived from common features of the association in question, e.g.: Quercetum robori inundatum, Nardetum alpinum, Nardetum purum, Nardetum pastoralis, etc. (SHENNIKOV 1964).

The second method, rather widely employed in the USSR, was advocated by ALEKHIN. The name of an association is composed of the dominants and subdominants of each layer, with the dash separating the dominants of different layers while the plus sign joins codominants of one layer,e.g.: ass. Pinus silvestris+Picea abies–Vaccinium myrtillus– Pleurozium schreberi+Hylocomium splendens. This system permits quite effective characterization of community structure by the major species of all the vegetation layers; but (a) it is applicable only to the smallest homogeneous units (sociations), and (b) it becomes quite artificial when the plant cover has no clearly defined layers. MIRKIN and SAMOYLOV, to distinguish associations according to dominants and ecological determinants, employ a system of underlining. Thus SAMOYLOV (1970) in the names of associations underlines dominants with a solid line, ecological determinants (indicator species) with a dashed line; species belonging to one associated group (statistically determined) are united with a dash, those belonging to different groups with a plus.

The traditional method as outlined above is used most widely in the USSR, but an alternative that is developing uses random sample choice and statistical analysis of the set of stand-samples collected (VASILEVICH 1960, 1963a, b, 1967a, b, 1969, 1971, FREY, 1963, 1966, 1967, 1971, KRAUKLIS & MEDVEDEV 1966, NESHATAEV 1967, 1971a, MARTIN 1967, 1969, VOLKOVA 1969, IPATOV 1969, MEDVEDEV 1969, MIRKIN 1970a, SAMOYLOV 1970, BUŠS 1971, LOBANOVA 1971, NORIN 1971, TROTSENKO 1971, et al.).

17.3.2 FORMATIONS

The formation is an important concept in the system of units characteristic of Russian approaches to classification of vegetation. This term was used at the end of the 19th and beginning of the 20th centuries for different kinds of categories. Thus, for example, GORDYAGIN (1901) like KORZHINSKI distinguished 'four steppe formations, namely the meadow-steppe, the shrub-steppe, the Stipa-steppe and the stony-steppe,' understanding the formation as a large physiognomic division of vegetation; SUKACHEV, on the other hand, in his papers of 1908 and 1910 considered a formation to be the smallest typological unit. After the latter unit had been given the name 'association,' the term 'formation' found use in Russian geobotanical literature for the principal unit of intermediate rank, comprising all communities with the same species dominant. This definition of the formation concept has been firmly established in Russian geobotany: 'Associations with the major layer composed of the same dominant species are united into a formation. Thus, to the formation Pineta silvestris belong all pine forest associations with *Pinus silvestris* dominant, however different the lower layers may be. Various meadows with *Alopecurus pratensis* dominant constitute the formation Alopecureta pratensis. Existence of two dominants in the major layer distinguishes certain formations, e.g. Pineto-Betuleta, and so on' (SHENNIKOV 1964: 410-411). Similar definitions of 'formation' were given by ALEKHIN (1936: 350, 1944: 165), LAVRENKO (1959: 61, 1968: 56), VORONOV (1963: 299), and others. Thus the Russian use of 'formation', like that of 'association,' differs from usage standard in the West, where the formation is a physiognomic unit (see article 13). The Russian formations are dominance-types in the sense of WHITTAKER (1962, article 14), and correspond to the concepts of association and consociation of many American and British authors. Despite the rather artificial means of definition, the formation as a dominance-type is convenient because of the simplicity of its definition and its apparent objectivity (any researcher may classify stand-samples into the same units). These practical advantages explain the fact that this unit contributed so much to generalizing the whole body of evidence gained by Russian geobotanists over the period of 1930's to 1950's. Particularly extensive geobotanical investigations were made in the 1930's, after the war toward the end of the 1940's and in the 1950's when aeromethods were used. As a result an inventory or registration of natural meadow and pasture lands of all zones was accomplished (LARIN 1965, et al.); registration of the forests of the country has been done (TSEPLYAEV 1965, et al.); and a registration of bogs comprising detailed data of their areas and distribution was possible (NIKONOV 1955, et al.). A great many vegetation maps of middle and small scale have been compiled (LAVRENKO 1941, 1950, LAVRENKO & SOCHAVA 1956, et al.); numerous regional monographs and reviews have been published (KOROVIN 1934, GORODKOV 1935, SHENNIKOV 1938, ZINSERLING 1938, LAVRENKO 1940, PROZOROVSKI 1940, SOCHAVA 1948, LAVRENKO & SOCHAVA 1956, et al.); and geobotanical characterization of regions of the USSR has been accomplished (LAVRENKO 1947). Primarily because of the system of classification based on dominants it has been possible to synthesize in a short period of time the results of registration of vegetation over the whole vast and most variable territory of the USSR.

In the 1960's however, along with the development of more detailed investigations the problem of the formation concept in the USSR was exposed to a critical reconsideration, particularly in connection with large-scale mapping of vegetation (SOCHAVA 1962, 1965a, ISACHENKO 1962, 1969, et al.). Some geobotanists in Russia refuse to accept the definition of formations by dominants and find more useful ecological and floristic criteria, structural characteristics of communities, and their successional relationships (DOKHMAN 1954, 1960, KOLESNIKOV 1956, DOLUKHANOV 1957, SOCHAVA 1961, YAROSHENKO 1961, et al.). Other scientists suggest discarding the formation as a category in the hierarchy of vegetation units (MIRKIN 1965, 1968, SAMOYLOV 1970). Instead of the traditional hierarchy–vegetation type, formation, association group, association–, MIRKIN employs the following scheme–vegetation type, vegetation subtype, section of the subtype (association class), association group, association, dominant modification. Along with this he considers it possible to retain the concept of 'formation' as a useful rankless unit.

17.3.3 HIGHER UNITS

The term 'vegetation type' designates the principal unit of major rank in Russian geobotany. Acceptance of this term in the sense of BROCKMANN-JEROSH & RÜBEL (1912) is due essentially to the works

Table II

Classification scheme for pine forests as made by URANOV

(ALEKHIN 1938)

Taxonomic categories	Phytocoena, community-types
Vegetation type	Lignosa
Formation class	Aciculilignosa
Formation group	Aciculisilvae
Formation	Pineta silvestris
Association group	Pineta silvestris hylocomiosa
Association	Pinus silvestris - Vaccinium vitis - idaea - Hylocomia
Sociation	Pinus silvestris - Vaccinium vitis - idaea - Pleurozium schreberi + Hylocomium splendens

of ALEKHIN (1936, 1938, 1944, et al.). The hierarchy of subordinate units in the classification scheme suggested by ALEKHIN is exemplified in Table II. Such an approach to major units was justified in the initial stages of geobotany, as a means for preliminary systematization of the rather incomplete data on plant cover of great regions of the Earth. Considering this type of classification SOCHAVA (1944: 4) very reasonably wrote: 'Doubtless this approach employs very superficial physiognomical features. . . There is no doubt of the artificial nature of the major physiognomic units of such a classification.'

Most widely used in the USSR as major units of classification are the vegetation types in the sense of LAVRENKO (1940, 1947, 1959, et al.). According to LAVRENKO (1947: 10), 'a vegetation type is determined essentially on the basis of ecologic-biological characteristics of the dominant species in the coenosis, or édificateurs.' This approach, called by LAVRENKO (1959: 62) 'physiognomic-ecological,' allows distinction of types of communities by structure and environmental relations and as units of landscape significance, such as steppe, meadow, savanna etc, instead of such abstract concepts as Lignosa, Herbosa, and the like. The steppe vegetation type, for instance, is defined by LAVRENKO (1954: 174) as follows: 'Les steppes, en tant que type de végétation, comprennent les associations d'Herbacées vivaces microthermiques et xérophiles (resistantes au froid et à la sécheresse), parmi lesquelles les Graminées cespiteuses tiennent la première place.' Consequently, the types of vegetation of LAVRENKO are similar to the formations of American and West Europaean authors, as defined by physiognomic-ecological criteria. The weak point of this approach is that rather seldom do we observe in nature communities with marked dominance of a single growth-form or ecobiomorph, frequently two or more ecobiomorphs share dominance. That is especially true for the vegetation of boreal bogs, tundras, savannas etc. ALEKSANDROVA (1971) proposed as a complementary criterion the 'character synusia,' i.e. a synusia that is not always a dominant one, but the presence of which is particularly characteristic of a given vegetation type. Thus, for the tundra vegetation type observed on automorphic soils of tundra-zone placors the character synusiae are those of hekistotherm (sensu CANDOLLE 1874), hypoarctic (sensu TOLMACHEV 1932, YURTSEV 1966) and arctoalpine dwarf shrubs and half-prostrate (sensu SEREBRYAKOV 1964) shrubs, with more or less abundant mosses.

Along with the physiognomic-ecological approach most widely accepted in the USSR a historic-genetical approach has gained popularity; in this the principal criterion is the historical (florogenetic and phylocoenogenetic) relationship between communities. The idea of classification of vegetation based on floristic or florogenetic relationships originated in Russia very long ago (KRASNOV 1888, PACZOSKI 1891, 1915, KORZHINSKI 1888, SUKACHEV 1915, 1928, SINSKAYA 1933, LESKOV 1943, et al.). The most fruitful developments of this genetic approach to classification have been those of OVCHINNIKOV (1947, 1957, et al.) and SOCHAVA (1944, 1945, 1961, 1965b, 1967, et al.). OVCHINNIKOV proposed the concept of florocoenotype; in his definition a 'florocoenotype is an assembly of plant formations, dominants of which have developed through the common adaptive evolution under certain long-lasting physiographic conditions' (OVICHINNIKOV 1947: 18-19). In case of the Central Asia vegetation 24 florocoenotypes have been distinguished by this author.

SOCHAVA has developed a system of classification the central concept of which is the 'phratria' of formations (Greek φρατρια), defined as follows: 'Formation classes are united into phylogenetic series, or complexes, which can be given the name of phratriae of plant forma-

Table III

The system of vegetation classification proposed

by SOCHAVA (1961:15)

Taxonomic categories	The Olenek-Vilyui plateau	The Amur-Zeya region	The Kazakhstan melkosopochnik
Association	Moss (Tomenthypnum nitens, Aulacomnium turgidum) dryas (Dryas crenulata) larch light forests	Herbs (Carex sutschanensis, Scabiosa fischeri, Tanacetum sibiricum) lespedeza oak forests	Grass-forbs (Stipa rubens, Stipa lessingiana) steppe with mesophytic element in flora
Association group	Moss-lichen-dwarf shrub larch light forests	Herb-lespedeza oak forests	Forbs-Stipa rubens steppe
Subformation	Central Siberia subformation	Upper Amur subformation	Northern Kazakhstan subformation
Formation	Northern taiga larch light forests (Larix dahurica)	Quercus mongolica forests	Stipa lessingiana steppe
Formation group	Northern taiga larch forests and light forests	-	Stipa-steppe
Formation class	Forests and light forests of Larix dahurica	Manchurian broad-leaved forests	Transvolga-Kazakhstan caespitose grass steppe
Phratria of formations	Angara phratria	Manchurian phratria	Transvolga-Kasakhstan phratria
Vegetation type	Boreal	Nemoral	Steppe type
Suite of vegetation types	Arcto-herb-forest suite		
System of vegetation types	Northern extratropical		
Phytosphere			

tions. Phratriae include formation classes that are related phylocoenogenetically, but frequently are diverse physiognomically. One phratria comprises, as a rule, forest, shrub, and herbaceous formation classes related phylogenetically' (SOCHAVA 1944: 11). The hierarchy of categories employed by SOCHAVA is summarized in Table III. Under the term 'vegetation type' SOCHAVA implies meaning different from that understood in the physiognomic-ecological approach: 'vegetation types' in SOCHAVA's sense are formed as a result of grouping not phytocoenoses but the vegetation complexes (combinations) of landscapes and regions; it is thus related to the 'Pflanzendivisionen' of SCHMITHÜSEN (1968: 315). Thus, for example, the 'boreal vegetation type' according to SOCHAVA is formed through grouping vegetation combinations composed of taiga forests, bogs, meadows and their anthropogenic successional communities.

The system offered by SOCHAVA reveals in the course of its classification by phylocoenogenetical considerations, that different types of communities display different regional relationships. For example, moss-lichen-dwarf shrub light forests of *Larix dahurica* are found only within the boundaries of the Angara phratria. SOCHAVA (1964a, b, 1967) has developed a system of classification of the world vegetation including 98 phratriae. The broadest unit of vegetation defined by BRAUN-BLANQUET, the 'Gesellschaftskreis' (community-circle), in many respects reminds one of the phratria of SOCHAVA, although the Kreis is broader than the phratria. Thus, for example, the vegetation of Europe (tundra not included) is described by BRAUN-BLANQUET as three Kreisen, while SOCHAVA describes in the same area 17 phratriae; LEBRUN (1947) distinguished 6 Kreisen in Africa, SOCHAVA (1964a: 13) 22 phratriae. Physiognomical convergence revealed in ecologically related conditions (even in the case of complete floristic non-identity) permits a different classification according to physiognomic-ecological principles, as described above in LAVRENKO's system and discussed in article 13 in this volume.

17.4.1 SUMMARY-CONCLUSION

1) The physiognomic-ecological approach to classification of vegetation is the most widely used in Russian geobotany. The greatest support is given to the system used by LAVRENKO, which is based upon the dominants of the major stratum: on the ecobiomorph (growth-form) of the dominant for distinguishing higher units and on dominant species for distinguishing middle and lower units of vegetation.

2) In classifying vegetation for regional description and mapping, the concepts of zonal, intrazonal and extrazonal vegetation are widely accepted. The concept of zonal vegetation relates to that of placor habitats and is analogous to CLEMENTS' climax-formation (cf. WHITTAKER 1953: 42, WALTER 1954 et al.); however no successional relationship of zonal with non-placoric intrazonal (subclimax) and extrazonal (preclimax and postclimax) formations need be implied.

3) The principal lower unit of classification in Soviet geobotany is the 'association,' distinguished according to stratal dominants. It is closely related to the sociation of Scandinavian phytocoenologists, although it is, as a rule, a somewhat broader unit.

4) Usage of the term 'formation,' significantly differs from that by the West European and American scientists. A majority of Soviet geobotanists employ 'formation' for a major category of middle rank, distinguished by the dominant species of the upper layer, and consequently close to the association and consociation of Anglo-American ecologists (CLEMENTS, TANSLEY).

5) It is characteristic of the Russian approach to use the term 'vegetation type' to designate higher units of vegetation. A majority of Soviet geobotanists use it for broad ecologic-physiognomical units such as meadows, steppe, savannas, and the like, distinguished according to the dominant growth-form. The vegetation type in this sense is similar to the formation of British and American authors. SOCHAVA (1944) has developed a distinctive system including as higher units one termed 'vegetation type' (which is, however, understood not in the physiognomic-ecological but in a regional-historical sense and comprises communities with different dominant ecobiomorphs), and as major divisions of vegetation types the 'formation phratriae.'

6) Methods for collecting vegetation samples in the field and classifying them have, in the traditional Russian approach, been based on the idea of discontinuity of both the objects of classification (phytocoenoses, as stands) and taxonomic units (associations, etc.), as well as upon choice in the field of sample plots considered 'typical' of taxonomic units. In the 1960's a critical revision of traditional concepts has led to increasing acceptance of two alternative approaches: (i) Vegetation has been treated as a continuum (as originally proposed by RAMENSKI many years ago), and conceived as a heterogeneous, structured pattern of community intergradation often including relative discontinuity. (ii) Random choice of the stand samples in the field is combined in many works with statistical treatment of the data obtained.

17.4.2 ZUSAMMENFASSUNG

1) In der Russischen Geobotanik wird am meisten die physiognomisch-ökologische Betrachtung der Vegetationsklassifikation verwendet. Das von LAVRENKO begründete System wird am häufigsten angewandt. Dieses System basiert auf den Dominanten der Hauptschicht, auf dem Ökobiomorph (Wuchsform) der Dominan-

ten für die Unterscheidung der höheren Einheiten und auf den dominanten Arten für die Unterscheidung der mittleren und niederen Vegetationseinheiten.

2) In Vegetationsklassifikationen für regionale Beschreibungen und Kartierungen werden die Begriffe der zonalen, intrazonalen und extrazonalen Vegetation weitgehend akzeptiert. Der Begriff der zonalen Vegetation hängt mit dem 'Plakor-Habitat' zusammen und ist der Clementschen Klimax-Formation analog (cf. WHITTAKER 1953: 42, WALTER 1954 et al.); Sukzessions-Beziehungen brauchen aber nicht zwischen zonalen und non-plakorischen intrazonalen (Subclimax) bw. extrazonalen (Präklimax und Postklimax) Formationen zu bestehen.

3) Die wichtigste niedere Klassifikations-Einheit in der Sovietischen Geobotanik ist die 'Assoziation', die nach den Schicht-Dominanten unterschieden wird. Sie ist der Soziation der Skandinavischen Phytozönologen sehr ähnlich, obwohl sie im allgemeinen eine etwas breitere Einheit darstellt.

4) Die Anwendung des Terminus 'Formation' ist von derjenigen der west-Europäischen und amerikanische Wissenschaftler sehr verschieden. Die Mehrheit der Soviet-Geobotaniker verwenden 'Formation' für eine Hauptkategorie mittleren Ranges, die auf Grund der dominanten Art der oberen Schicht unterschieden werden und dementsprechend der Assoziation und Konsoziation der anglo-amerikanischen Ökologen (CLEMENTS, TANSLEY) ähnlich ist.

5) Es ist charakterisch für die russische Betrachtung den Terminus Vegetationstyp für die Kennzeichnung der höheren Vegetationseinheiten zu verwenden. Die Mehrheit der Soviet-Geobotaniker verwendet ihm für breite ökologisch-physiognomische Einheiten wie Wiesen, Steppen, Savannen, die nach der dominanten Wuchsform unterschieden werden. Der Vegetationstyp in diesem Sinne ist der Formation der britischen und amerikanischen Autoren ähnlich. SOCHAVA (1944) hat ein 'abweichendes' System entwickelt, das als höhere Einheit die 'Vegetationstypen' (die aber nicht im physiognomisch -ökologischen sondern in einem regionalhistorischen Sinne gemeint sind und Gemeinschaften mit verschiedenen dominanten Ökobiomorphen umfassen) und als Haupt-Divisionen der Vegetationstypen die 'Formations-Phratriae' enthält.

6) Methoden für die Sammlung von Vegetationsanalysen im Gelände und deren Klassifikation waren in der traditionellen russischen Betrachtung auf die Idee der Diskontinuität sowohl der Klassifikationsobjekte (Phytozönosen und Bestände) und der taxonomischen Einheiten (Assoziationen usw.) wie auf die Wahl

der im Gelände für taxonomische Einheiten typisch erachteten Probeflächen gegründet. In den 60er Jahren hat eine kritische Revision der traditionellen Begriffe zu einer wachsenden Annahme zweier verschiedenen Betrachtungen geführt: (1) Die Vegetation wird als ein Kontinuum (wie ursprünglich von RAMENSKI vor vielen Jahren vorgeschlagen) betrachtet und als ein heterogenes Strukturmuster von Gemeinschaftsüberlappung aufgefasst, das allerdings relative Diskontinuität enthalten kann. (2) Zufällige Wahl der Probeflächen im Gelände wird in manchen Arbeiten mit einer statischen Behandlung der erhaltenen Daten kombiniert.

REFERENCES

ALEKHIN, V. V., – 1915 – Types of Russian steppe (In Russian). *Izv. Bot. sada Petra Velikogo* 15(3-4): 405-432.

ALECHIN W. W., – 1932 – Die vegetationsanalytischen Methoden der Moskauer Steppenforscher. *Handb. Biol. ArbMeth.* 11, 6: 335-373.

ALEKHIN, V. V., – 1935 – Main concepts and main units in phytocoenology (In Russian). *Sov. Bot.* 1935(5): 21-34.

ALEKHIN, V. V., – 1936 – Vegetation of the USSR in its main zones (In Russian). In: 'Fundamentals of Botanical Geography,' H. WALTER & V. ALEKHIN, pp. 306-715. Biomedgiz, Moscow & Leningrad.

ALEKHIN, V. V., – 1938 – Schemas de classification de la végétation (In Russian). *Sov. Bot.* 1938(3): 96-98.

ALEKHIN, V. V., – 1944 – Geography of plants (In Russian). Sovetskaya Nauka, Moscow. 455 pp.

ALEKSANDROVA, V. D. (ALEXANDROVA, V. D.), – 1956 – Vegetation of the southern Island of Novaya Zemlya between 70° 56' and 72° 12' n.l. (In Russian). *Rastit. kraĭn. Sev. SSSR* i ee Osv. 2: 187-306.

ALEKSANDROVA, V. D., – 1965 – On the problem of distinguishing phytocoenoses in a vegetative continuum. (In Russian with Engl. summ.) *Bot. Zh. SSSR* 50(9): 1248-1259.

ALEKSANDROVA, V. D., – 1969 – Classification of Vegetation: Principles of Classification and Classification Systems of Various Phytocoenological Schools (In Russian). Nauka, Leningrad 275 pp.

ALEKSANDROVA, V. D., – 1971 – Principles of zonal subdivision of Arctic vegetation (In Russian with Engl. summ). *Bot. Zh. SSSR* 56(1): 3-21.

APALIA, D. K., – 1968 – Distinguishing of plant associations on the base of bioecogroups (In Russian). Abstract of a report. See: NORIN, 1968.

APALIA, D. K., – 1971 – Distinguishing of plant associations on the base of the indicator bioecogroups. (In Russian). In, 'Metody vydeleniya rastit. assotsiatsi,' pp. 38-51. Nauka, Leningrad.

BORSHCHOV, I. G., – 1865 – Materials for botanical geography of the Aralo-Caspian region (In Russian). *Zap. imp. Akad. Nauk* 7: 1-90.

BRAUN-BLANQUET, J., – 1964 – Pflanzensoziologie: Grundzüge der Vegetationskunde. 3 Aufl. Springer, Wien. 865 pp.

BROCKMANN-JEROSCH, H. & E. RÜBEL., – 1912 – Die Einteilung der Pflanzengesellschaften nach ökologisch-physiognomischen Geschichtspunkten. Engelmann, Leipzig. 72 pp.

BUŠS, K. K., – 1971 – Some problems of biometric analysis of anthropogenic successions on drained forest areas (In Russian with Engl summ..) *Bot. Zh. SSSR* 56(1): 22-30.

BYKOV, B. A., – 1957 – Geobotany (In Russian). 2nd ed. Izd. Akad. Nauk Kazak SSR, Alma-Ata. 381 pp.

CANDOLLE, ALPH DE, – 1874 – Constitution le règne végétal de groupes physiologiques. *Archives Sci. phys. nat.*, Geneva 50: 1-38.

CLEMENTS, F. E., – 1936 – Nature and structure of the climax. *J. Ecol.* 24: 252-284.

COLE, L., – 1958 – The ecosphere. *Scient. Amer.* 198(4): 83-96.

DIMO, N. A. & B. A. KELLER, 1907 – In the semi-desert region (In Russian). Izd. Saratovsk. gubern. Zemstva, Saratov. 215 pp.

DOKHMAN, G. I., – 1930 – Phytosociological analysis of vegetation of the Starobyelskie steppe (In Russian). *Isv. ass. Nauchno-issled. inst. physico-math. fak. I Moskow Univ.* 3(2-A): 212-232.

DOKHMAN, G. I., – 1936 – On some classification units of the vegetation complexes (In Russian with Germ. summ.) *Zemlevêdênie* 38(3): 294-326.

DOKHMAN, G. I., – 1954 – Vegetation of Mugodzhary (In Russian). Geografgiz, Moscow. 235 pp.

DOKHMAN, G. I., – 1960 – On the system of diagnostic characters of vegetation units (In Russian with Engl. summ.) *Bot. Zh. SSSR* 45(5):637-648.

DOLUKHANOV, A. G., – 1957 – Problems of the coenological forest classification in connection with convergence features in vegetation (In Russian). *Tez. dokl. deleg. s'ezda VBO*, Leningrad, 4(2): 18-23.

DOLUKHANOV, A. G., – 1970 – The questions of typology of mountain forests in connection with the nature of relative continuity of their plant cover (In Russian with Engl. summ.) *Trudý mosk. Obshch. ispýt Prir*, old. biol. 38: 24-33.

DU RIETZ, G. E., – 1921 – Zur methodologischen Grundlage der modernen Pflanzensoziologie. Holzhausen, Wien 267 pp.

DU RIETZ, G. E., – 1930 (1932) – Vegetationsforschung auf soziationsanalytischer Grundlage. *Handb. biol. ArbMeth.*, 11, 5: 293-480.

DU RIETZ, G. E., – 1936 – Classification and nomenclature of vegetation units 1930-1935. *Svensk Bot. Tidskr.* 30: 580-589.

ENGELMANN, G. I., – 1810 – The Theoretical and Practical Manual for Drainage of Lands (In Russian). St. Petersburg. 291 pp.

FLEROV, A. F. & B. A. FEDCHENKO, – 1902 – Handbook for Studying Plant Communities of Middle Russia (In Russian). Izd. Sabashnikovykh, Moscow. 184 pp.

FREY, T., – 1966 – On the significance of Czekanowski's index of similarity. *Zastosow. Mat.* 9(1): 1-7

FREY, T. E. A., – 1963 – On the use of the correlation coefficient of interspecific relations (In Russian). *Bot. Zh. SSSR* 48(2): 235-239.

FREY T. & L. VÕHANDU, – 1966 – Uus meetod klassifikatioomühikute püstitaniseks (In Estonian with Engl. summ.) *Eesti NSV Tead. Akad. Toim.* biol seer. 4: 565-576.

FREY, T. E.-A., – 1967 – On the mathematical-phytocoenological methods of classification of vegetation (In Russian). Tartu State University, Tartu. 18 pp.

FREY, T. E., – 1968 – Cluster analysis of the sample plots descriptions with 'K' criterion (In Russian). Abstract of a report. See: NORIN, 1968.

FREY, T. E., – 1971 – Cluster analysis of the sample plots descriptions with 'K' criterion (In Russian). In, 'Metody vydeleniya rastit. assotsiatsi,' pp. 226-241. Nauka, Leningrad.

FRIEDERICHS, K., – 1930 – Die Grundfragen und Gesetzmässigkeiten der land- und forstwirtschaftlichen Zoologie insbesondere der Entomologie. Parey, Berlin. 417 pp.

GAMS, H., – 1918 – Prinzipienfragen der Vegetationsforschung: Ein Beitrag zur Begriffsklärung und Methodik der Biocoenologie. *Vjschr. naturf. Ges. Zürich* 63: 293-493.

GENKO, N. K., – 1902 – Description of the Białowieza (In Russian). *Lĕsnoĭ Zh.* S-Peterb. 5: 1012-1056, 6: 1269-1302.

GORDYAGIN, A. YA., – 1900-1901 – Materials for studying soils and vegetation of the Western Siberia (In Russian). *Trudy Obshch. Estest. imp. kazansk. Univ.*, Kazan 34(3):3-222, 35(2): 223-528 (+36).

GORODKOV, B. N., – 1935 – Vegetation of the Tundra Zone of the USSR (In Russian). Izd. Akad. Nauk SSSR, Moscow & Leningrad. 141 pp.

GURICHEVA, N. P., Z. V. KARAMYSHEVA & E. I. RACHKOVSKAIA, – 1967 – On the compilation of a legend to a large-scale vegetation map in the desert-steppe subzone of the Kazakhstan (In Russian). *Geobot. Kartogr.*, Leningrad 1967: 57-67.

GUTOROVICH, I. I., – 1897 – Notes of a northern forester (In Russian). *Lĕsnoĭ Zh.* S-Peterb. 2: 216-228, 5: 789-799.

HULT, R., – 1881 – Försök till analytisk behandling af växtformationerna. *Meddn Soc. Fauna Flora fenn.* 8: 1-155.

ILYNSKI, A. P., – 1935 – The higher taxonomic units in geobotany (In Russian). *Sov. Bot.* 1935(5): 49-65

IPATOV, V. S., – 1966 – On the concept of phytocoenosis and an elementary cell of the social life of plants (In Russian). *Vest. leningr gos. Univ.* 15, ser, biol., 3: 56-62

IPATOV, V. S., – 1969 – On the choice of typical objects in geobotanical investigations (In Russian). *Problemij Bot.* 11: 13-21.

ISACHENKO, T. I., – 1962 – Principles and methods of generalisation in large, middle and small-scale geobotanical mapping (In Russian). In: 'Printsipy i metody geob. kartogr', pp. 28-46. Moscow Leningrad.

ISACHENKO, T. I., – 1967 – On the cartography of serial rows and microbelts in valleys and lake-basins (In Russian). *Geobot. kartogr.*, Leningrad 1967:42-56.

ISACHENKO, T. I., – 1969 – The structure of the vegetation cover and mapping (In Russian). *Geobot. kartogr.*, Leningrad 1969:20-33.

ISACHENKO, T. I. & E. I. RACHKOVSKAYA, – 1961 – Principal zonal types of the Northern Kazakhstan steppe Vegetation (In Russian). *Trudij bot. Inst. Akad. Nauk SSSR*, ser. 3 (Geobotanika) 13:133-397

KARAMYSHEVA, Z. V., – 1967 – The experience in the treatment of the descriptions of sample areas of steppe communities by the method of Braun-Blanquet (In Russian). *Bot. Zh. SSSR*, 52(8): 1132-1144.

KARAMYSHEVA, Z. V. & E. I. RACHKOVSKAJA, – 1968 – On the complication of a small-scale map of a steppe territory of Kazakhstan (In Russian). *Geobot. kartogr.* Leningrad, 1968: 5-21.

KARAMYSHEVA, Z. V. & E. I. RACHKOVSKAYA, – 1969 – A vegetation map of the middle part of the Tersakkan river basin (In Russian). In: 'Biologicheskie issledovaniya v Kazakhstane,' I: 481-485. Nauka, Leningrad.

KATZ, N. J. (KATS), – 1929 – Die Zwillingsassoziationen und die homologen Reihen in der Phytosoziologie. *Ber. dt. bot. Ges.* 47: 154-164.

KATZ, N. J., – 1933 – Die Grundprobleme und die neue Richtung der Phytosoziologie. *Beitr. Biol. Pfl.* 21:133-166. (In Russian, *Byull. mosk. Obshch. Ispýt. Prir.*, otd biol. 43(2): 188-221, 1934).

KOLESNIKOV, B. P., – 1956 – A review of the forest formation of Primorye and Priamurye (In Russian). In, 'Akad. V. N. Sukachevu k 75-let. so dnya rozhdeniya,' pp. 286-305. Izd. Akad Nauk SSSR, Moscow & Leningrad.

KORZHINSKI, S. I., – 1888-91 – Northern boundary of the chernozem region of the East part of the European Russia in botanic-geographical and pedological points of view (In Russian). *Trudij Obshch. Estest. kazan. Univ.* Kazan, 18(5): 1-253, 22(6): 1-201.

KOROVIN, E. P., – 1934 – Vegetation of the Central Asia and Southern Kazakhstan (In Russian). Moscow & Tashkent. 480 pp. 2nd ed., 1961-2, Tashkent, 2 parts, 452 & 547 pp.

KOSTYCHEV, P. A., – 1900 – Relations between soils and some plant formations (In Russian). *Dnev. S'ězda russk. Estestvoisp. Vrach.* 8(9): 6-7.

KRASHENINNIKOV, S. P., – 1755 – Description of the Kamchatka Land (In Russian). St. Petersburg. 2 vols., 438 & 319 pp.

KRASNOV, A. N., – 1886 – Geobotanical investigations in the Kalmyk steppe (In Russian). *Izv. Russk. Geogr. Obshch.* 22(1): 1-22.

KRASNOV, A. N., – 1888 – An outline of development of flora of the southern part of the Eastern Tian Shan (In Russian). *Zap. Russk. Geogr. Obshch.* 19: 1-413.

KRAUKLIS, A. A. & YU. O. MEDVEDEV., – 1966 – The principles of showing vegetation in series of large-scale maps in connection with mapping dynamics of natural environment (In Russian). *Geobot. kartogr.* Leningrad 1966: 26-35.

KRÜDENER, A. A., – 1916 – Fundamentals of the forest types classification. I and II (In Russian). In, 'Materily po izuch. russk. lesa,' pp. 1-318. Petrograd.

LAVRENKO, E. M., – 1940 – Steppe of the USSR (In Russian). In, *Rastitelnost SSSR* 2: 1-206 Moscow & Leningrad.

LAVRENKO, E. M., – 1941 – An Explanatory Text to the Vegetation Map of the USSR, scale 1: 5 000 000 (In Russian). Moscow & Leningrad. 132 pp.

LAVRENKO, E. M. (ed.), – 1947 – Geobotanical Regionalization of the USSR (In Russian). Izd Akad. Nauk SSSR, Moscow & Leningrad. 152 pp.

LAVRENKO, E. M. (ed.), – 1950 – Vegetation map of the European part of the USSR, scale 1: 2 500 000. Explanatory text (In Russian). Izd. Akad. Nauk SSSR, Moscow & Leningrad 288 pp.

LAVRENKO, E. M., – 1954 – Les steppes de la région stéppique Eurasienne (Géographie, dynamisme, histoire). *Essais de Bot.* 1: 157-194. Éd. l'Acad. Sci. URSS, Moscou & Léningrad.

LAVRENKO, E. M., – 1959 – Main regularities of plant communities and the ways of their investigation (In Russian). In, *Polevaya geobotanika* 1: 13-78. Izd. Akad. Nauk SSSR, Moscow & Leningrad

LAVRENKO, E. M., – 1968 – On the next task of the exploration of the plant cover in connection with botanico-geographical regionalization of the USSR (In Russian). In, 'Osnovn. Probl. Sovrem. Geobot,' pp. 45-68. Nauka, Moscow & Leningrad.

LAVRENKO, E. M. & N. V. DYLIS, – 1968 – Achievements and current tasks in the investigation of terrestrial biogeocoenoses in the USSR. (In Russian). *Bot. Zh. SSSR*, 53(2): 155-167.

LAVRENKO, E. M. & V. B. SOCHAVA, – 1956 – Plant cover of the USSR. Explanatory text to 'Geobotanical Map of the USSR' scale 1: 4 000 000. (In Russian). Izd. Akad. Nauk SSSR, Moscow & Leningrad. 2 parts, 460 & 971 pp.

LARIN, I. V., – 1965 – Natural meadows and pastures and how became they reestablished. (In Russian). In, *Probl. Sovrem. Bot.* 1: 172-205. Izd. Akad. Nauk SSSR, Moscow & Leningrad.

LEBRUN, J., – 1947 – La végétation de la plaine alluviale au sud du lac Édouard. *Inst. Parcs Natl. Congo Belge, Explor. Parc Natl. Albert,* 4 Mission Lebrun 1: 1-800, Bruxelles.

LESKOV, A. I., – 1943 – Principles of the natural system of plant associations (In Russian). *Bot. Zh. SSSR* 28(2): 37-52.

Lippmaa, T., – 1933 – Taimeühingute uurimise metoodika ja eesti taimeühingute klassifikatsiooni põhijooni (Eston. with Germ. summ.: Grundzüge der pflanzensoziologischen Methodik nebst einer Klassification der Pflanzenassoziationen Estland). *Acta Inst. Horti bot. tartu.* 3, 4: 1-169.

Lippmaa, T., – 1939 – The unistratal concept of plant communities (the unions). *Am. Midl. Nat.* 21: 111-145.

Lobanova, V. F., – 1971 – The quantative analysis of the boundaries of some meadow phytocoenoses of Karelia. (In Russian with Engl. summ.). In, *Kolichestv. metody analiza rastit.*, 2: 178-184. Riga.

Lukicheva, A. N. & D. N. Saburov, – 1971 – Methods of grouping samples into plant associations in connection with the landscape structure (In Russian). In, 'Metody vydel. rastit. assotsiatsi'. Nauka, Leningrad.

Martin, Yu. L., – 1967 – Development of lichen synusia on the glacier moraines of the Polar Ural (In Russian). Izd. Akad. Nauk SSSR, Sverdlovsk. 22 pp.

Martin, Yu. L., – 1969 – Ordination method in lichenocoenology (Russian with Engl. summ.). In, 'Kolichestv. metody analiza rastit.', pp. 83-86. Tartu.

Masing, V. V. & H. H. Trass, – 1963 – Elaboration of some theoretical problems in the works of Estonian geobotanists. (In Russian with Engl. summ.). *Bot. Zh. SSSR* 48(4): 473-485.

Matveyeva, N. V., – 1968 – The structure of the vegetation of the main types of tundras in the middle course of the river Piasina (the Western part of the Taimyr Peninsula). (In Russian with Engl. summ.) *Bot. Zh. SSSR* 53(11): 1588-1603.

Medvedev, Yu. O., – 1969 – Some phytocoenological and dynamic peculiarities of the dark-coniferous forest (In Russian with French summ.). In, 'Yuzhnaya tayga Priangarya,' pp. 219-262. Nauka, Leningrad.

Middendorf, A. E., – 1867 – Siberische Reise. Bd 4, pp. 525-783, Die Gewächse Sibiriens. Akad. Wissensch., St. Petersburg.

Mirkin, B. M., – 1965 – On the ecological classification of the meadow vegetation of flood-plains (In Russian with Eng. summ.). *Bot. Zh. SSSR* 50(3): 324-334.

Mirkin, B. M., – 1968 – The criteria of dominants and determinants in the classification of phytocoenoses (In Russian with Engl. summ.). *Bot. Zh.SSSR* 53(6): 767-778.

Mirkin, B. M., – 1970a – Introduction to Quantitative Methods of Vegetation Analysis (In Russian). Izd. Bashkirsk, gos. Univ., Ufa. 87 pp.

Mirkin, B. M., – 1970b – Main Problems of Geobotany (In Russian). Izd. Bashkirsk, gos. Univ., Ufa. 88 pp.

Morozov, G. F., – 1912 – Forest Science. I. Introduction to Forest Biology (In Russian). St. Petersburg. 83 pp.

Neshataev, Yu. N., – 1967 – An attempt of the large-scale geobotanical mapping of the forest-steppe oak forest (In Russian). *Uchenije Zap. leningr. gos. Univ.* 331: 87-117.

Neshataev, Yu. N., – 1968 – A selective-statistical method for distinguishing plant associations (In Russian). Abstract of a report. See: Norin, 1968.

Neshataev, Yu. N., – 1971a – A selective-statistical method for distinguishing plant associations (In Russian). In, 'Metody vydel. rastit. assotsiatsiy,' pp. 181-205. Nauka, Leningrad.

Neshataev, Yu. N., – 1971b – Method for grouping of vegetation samples in the curriculum of the Geobotany Department of the Leningrad State University (In Russian). In, 'Metody vydel. rastit. assotsiatsiy,' pp. 23-37. Nauka, Leningrad.

NIKONOV, M. N., – 1955 – Regionalisation of the peatbogs in conncetion with its utilization in national economy (In Russian). *Trudij Inst. Lesa, Akad, Nauk SSSR Mosk.* 31: 49-63.

NITSENKO, A. A., – 1948 – The boundaries of plant associations in nature (In Russian). *Bot. Zh. SSSR* 33(5): 478-495.

NITSENKO, A. A., – 1963 – On some controversial theoretical problems in geobotany (In Russian with Engl. summ.). *Bot. Zh. SSSR* 48(4): 486-501.

NITSENKO, A. A., – 1966 – On the criteria for distinguishing plant associations (In Russian with Engl. summ.). *Bot. Zh. SSSR* 51(8): 1085-1098.

NITSENKO, A. A., – 1971 – An attempt to apply quantiative analysis to the problem of causality of border-lines in vegetation cover (In Russian with Engl. summ.). In, 'Kolichstvennye metody analiza rastitelnosti,' pp. 216-220. Riga.

NORIN B. N., – 1966 – On the zonal types of vegetational cover in the Arctic and Subarctic zones (In Russian with Engl. summ.). *Bot. Zh. SSSR.* 51(11): 1547-1563.

NORIN, B. N., – 1968 – A conference on the methods of distinguishing of plant associations held in Leningrad on May 15-17, 1967 (In Russian). *Bot. Zh. SSSR* 53(6): 872-878.

NORIN, B. N., – 1970 – The functional structure of plant aggregations in the forest-tundra zone (In Russian with Engl. summ). *Bot. Zh. SSSR* 55(2): 170-183.

NORIN, B. N., – 1971 – Application of the similarity coefficient to classification of the forest-tundra microaggregations (In Russian) In, 'Metody vydeleniya rastit, assotsiatsi,' pp. 206-225. Nauka, Leningrad.

OVCHINNIKOV, P. N., – 1947 – About principles of classification of vegetation (In Russian). *Soobshch. Tadzhiksk. fil. Akad. Nauk SSSR*, Stalinabad 2: 18-23.

OVCHINNIKOV, P. N., – 1957 – The florocoenotypes and their value for classification of vegetation in the Middle Asia. (In Russian). In: *Tezisy dokl. S'ezda VBO*, Leningrad, 7: 28-36.

PACZOSKI, J. K., – 1891 – Stages of floral development (In Russian). *Vestnik estetstvozn.*, Kiev, 8: 261-270.

PACZOSKI, J. K., – 1896 – Zycie gromadne roślin. *Wszechświat* 15. Warszawa. (Reprinted with Engl. transl., 'Social life of plants,' *Bibljot. Bot.*, Krakow, 2: 1-40, 1930).

PACZOSKI, J. K., – 1915 – Vegetation of the Kherson district. I. Forests (In Russian). Izd. Olkhovnikova, Kherson. 202 pp.

PACZOSKI, J. K., – 1921 – Fundamentals of Phytosociology (In Russian). Izd. Stud. komit. S-Kh. technikuma, Kherson. 346 pp.

POGREBNYAK, P. S. (POGREBNJAK P. S.), – 1929 – Über die Methodik der Standortsuntersuchungen in Verbreitung mit Waldtypen. *Verh.* II *Int. Kongr. forstl. Versuchanstalten*

PONYATOVSKAYA, V. M., – 1952 – Of several years' investigation of the herb communities of the Southern Kirghiz forest belt (In Russian). *Trudy bot. Inst. Akad. Nauk SSSR*, ser. 3 (Geobotanika) 8: 156-240.

PONYATOVSKAYA, V. M., – 1959 – On two trends in phytosociology (In Russian). *Bot. Zh. SSSR* 44(3): 402-407. Engl. transl. in *Vegetatio*, 10:373-385, 1961).

PROZOROVSKI, A. V., – 1935 – Zonal types of the Soviet Central Asia deserts (In Russian). *Isv. Russk. Geograf. Obshch.* 67(3): 318-331.

PROZOROVSKI, A. V., – 1940 – Semideserts and deserts of the USSR (In Russian). In, *Rastitelnost SSSR* 2:207-480. Izd. Akad Nauk SSSR, Moscow & Leningrad.

RABOTNOV, T. A., – 1963 – An attempt to use of the idea of vegetation continuum in study of the plant cover of Wisconsin (In Russian). *Byull. mosk. Obshch. Ispýt. Prir.*, otd biol. 68(4): 147-151.

RABOTNOV, T. A., – 1965 – On the character of the structure of polydominant meadow coenoses (In Russian with Engl. summ.). *Bot. Zh. SSSR* 50(10): 1396-1408.

RAMENSKI, L. G., – 1910 – The comparative method for ecological study of plant communities (In Russian). *Dnev. S' êzda russk. Estestvoisp. Vrach.*, Moscow 12(4): 389-390.

RAMENSKI, L. G., – 1915 – About the quantative study of plant cover (in Russian). In, *Materialy po organiz. i kulture kormov. ploshchadi* 12: 105-140. St. Petersburg.

RAMENSKI, L. G., – 1924 – Main regularities of the plant cover (In Russian). *Věstnik opijtnogo dêla Stredne-Chernoz. Obl.*, Voronezh pp. 37-73.

RAMENSKI, L. G. (RAMENSKY, L. G.), – 1929 – On the method of comparative treatment and systematization of plant lists and other objects determined by several factors with unlike actions (In Russian). *Trudij Soveshch. geobot.-lugov. sozvan.*, (*Gos.*) *lugovoi Inst.* (Transl. in *Beitr. Biol. Pfl.* 18: 269-304, 1930).

RAMENSKI, L. G., – 1932 – Die Projektionsaufnahme und Beschreibung der Pflanzendecke. *Handb. Biol. ArbMeth.* 11, 6: 137-190.

RAMENSKI, L. G., – 1938 – Introduction to the Complex Soil-Geobotanical Investigation of Lands (In Russian). Selkhozgiz, Moscow. 620 pp.

RAMENSKI, L. G., – 1950 – Classification of lands according to their plant cover (In Russian). *Problemy̆ Bot.* 1: 484-512.

RAMENSKI, L. G., – 1952 – About some principles of the current geobotany (In Russian). *Bot. Zh. SSSR* 37(2): 202-211.

RAMENSKI, L. G., – 1953 – On ecological study and systematization of plant communities. (In Russian). *Byull. mosk. Obshch. lspy̆t. Prir.*, otd. biol. 58(1): 35-54.

RUPRECHT, F., – 1845 – Flora Samoedorum cisuralensis. St .Petersburg. 67 pp.

RYSIN, L. P. & Z. M. KOVALENKO, 1968 – On the possible use of the methods of the BRAUN-BLANQUET school in our geobotanical investigations. (In Russian with Engl. summ.). *Byull. mosk. Obshch. lspy̆t. Prir.*, otd. biol. 73(1) 93-114.:

SAMOYLOV, YU. I., – 1970 – A tentative classification of meadows in the floodplains of the river Msta. (In Russian with Engl. summ.) *Bot. Zh. SSSR* 55(10): 1419-1431.

SCHMITHÜSEN, J., – 1968 – Allgemeine Vegetations Geographie. 3 Aufl. de Gruyter, Berlin. 463 pp.

SEREBRYAKOV, I. G., – 1964 – Life-forms of the higher plants and their study (In Russian) In, *Polevaya Geobot.* (*Field geobotany*) 3: 146-208. Izd. Akad. Nauk SSSR, Moscow & Leningrad.

SEREBRENNIKOV, P. P., – 1913 – The forest-types and their value in forestry (In Russian). *Lêsnoi Zh.* S-Peterb. 1-2: 39-72.

SHELYAG-SOSONKO, YU. R., – 1971 – The oak-woods of Podolia (In Russian). *Bot. Zh. SSSR* 56(4): 512-516.

SHENNIKOV, A. P., – 1935 – Principles of the botanical classification of meadows (In Russian). *Sov. Bot.* 1935(5): 35-49

SHENNIKOV, A. P., – 1938 – Meadows vegetation of the USSR (In Russian). In, *Rastitelnost SSSR* 1:429-647. Izd. Akad Nauk SSSR, Moscow & Leningrad.

SHENNIKOV, A. P., – 1964 – Introduction to Geobotany (In Russian). Izd. Leningr. gos. Univ., Leningrad. 447 pp.

SIBIRTSEV, N. M., – 1900 – Pedology. I. (In Russian). Izd. Skorokhodova, St. Petersburg. 212 pp.

SINSKAYA, E. N., – 1933 – The principal features of the forest evolution of the Caucasus, in connection with the history of species (In Russian with Engl. summ.). *Bot. Zh. SSSR* 18(5): 370-406, (6): 487-515.

SOCHAVA, V. B., – 1930 – A timber line in the Lyapin Ural mointains (In Russian). *Trav. Mus. bot. Acad. Sci. Russ.* 22: 1-47.
SOCHAVA, V. B., – 1944 – An attempt of the phylogenetical systematics of plant associations. (In Russian). *Sov. Bot.* 1944(1): 3-18.
SOCHAVA, V. B., – 1945 – Phratriae of plant formations of the USSR and their phylocoenogenesis (In Russian). *Dokl. Akad. Nauk SSSR.* 47: 60-64.
SOCHAVA, V. B., – 1948 – Geographical connections of the plant cover on the territory of the USSR (In Russian). *Uchenȳe Zap. Leningr. pedagog. Inst.* 73: 3-89.
SOCHAVA, V. B., – 1961 – Problems of classification of vegetation, typology of physico-geographical facies and biogeocoenoses (In Russian). *Trudȳ Inst. Biol.* Sverdlovsk (Uralsk. fil. Akad. Nauk SSSR) 27: 5-22.
SOCHAVA, V. B., – 1962 – Mapping problems in geobotany (In Russian). In, 'Printsipy i metody geobot. kartograf,' pp. 5-27. Izd. Akad. Nauk SSSR, Moscow & Leningrad.
SOCHAVA, V. B., – 1964a – A new world vegetation map (In Russian). *Geobot. kartogr.*, Leningrad 1964: 3-15.
SOCHAVA, V. B., – 1964b – Vegetation (of world and continents). Explanatory text (In Russian). In, 'Fiziko-geogr. atlas mira,' pp. 280-283. Izd. Akad. Nauk SSSR, Moscow.
SOCHAVA, V. B., – 1965a – Current problems of large-scale vegetation cartography (In Russian). *Geobot. kartogr.*, Leningrad 1965: 3-10.
SOCHAVA, V. B., – 1965b – Vegetation. *Soviet Geogr.* (Rev. Transl.) 6(5-6): 348-359.
SOCHAVA, V. B. (SOTCHAVA, VICTOR), – 1967 – Carte universelle de la végétation du monde. *Rev. roumaine biol., Sér. bot.* 12(2-3): 239-242.
SOCHAVA, V. B., – 1968 – Plant communities and dynamics of natural systems. (In Russian). *Doklady Instit. geogr. Sibiri i Dalnego Vostoka, Irkutsk* 20: 12-22.
SOCHAVA, V. B., – 1970 – Theoretical principles of steppe geosystems topology. (In Russian). In, 'Topologiya stepnykh geosistem,' pp. 3-12. Nauka, Leningrad.
SOUKATCHEV (see SUKACHEV).
SUKACHEV, V. N., – 1908 – Forest formations and their interrelations in the Bryansk woods (In Russian). *Trudȳ po lesn. opȳtn. Delu Ross.* 9: 1-61.
SUKACHEV, V. N., – 1910 – On the plant formation (In Russian). *Dnev. S'êzda russk. Estestvoisp. Vrach.*, Moscow 12(2): 150, *Pochvovêdênīe* 12(4): 393, 1910.
SUKACHEV, V. N., – 1915 – Introduction to Plant Community Science (In Russian). Izd. Panafidinoi, Petrograd. 127 pp.
SUKACHEV, V. N., – 1917 – About terminology in the plant community science (In Russian). *Zh. russk. bot.Obshch.* 2(1-2): 1-19.
SUKACHEV, V. N., – 1927 – Manual for Study of Forest Types (In Russian). Novaya derevnya, Moscow. 166 pp. 2nd ed., 1930, Moscow & Leningrad, 318 pp. 3rd ed., 1931, Selkhozgiz, Moscow & Leningrad, 328 pp.
SUKACHEV, V. N., – 1928 – Plant Communities (Introduction to Phytosociology) (In Russian) 4 ed. Kniga, Leningrad & Moscow. 232 pp.
SUKATSCHEW, W. N., – 1932 – Die Untersuchung der Waldtypen der osteuropäischen Flachlandes. *Handb. biol. ArbMeth.* 11, 6: 191-250.
SUKACHEV, V. N., – 1942 – Idea of development in phytocoenology (In Russian). *Sov. Bot.* 1942(1-2): 5-17.
SUKACHEV, V. N., – 1945 – Biogeocoenology and phytocoenology. *Dokl. Akad. Nauk SSSR* 47(6): 447-449.
SOUKATCHEV, V. N., – 1954 – Quelques problèmes théoriques de la phytocénologie. In, *Essais de botanique*, 1: 310-330. Ed. l'Acad. Sci, URSS, Moscou & Léningrad
SUKACHEV, V. N., – 1957 – General principles and a program for study of forest

types (In Russian). In,'Method. ukazan. k izuchen. tipov lesa,'V. N. SUKACHEV, S. V. ZONN & G. P. MOTOVILOV, pp. 9-63. Izd. Akad. Nauk SSSR, Moscow.

SUKACHEV, V. N., – 1964 – Main concepts of forest biogeocoenology (In Russian). In, 'Osnovy lesn. biogeotsenologii,' pp. 5-59. Moscow.

TANFILIEV, G. I., – 1897 – Physico-geographical regions of the European Russia (In Russian). *Trudȳ Voln. econom. obshch.* 1, 33 pp. Izd. Demakova.

TATISHCHEV, V. N., – 1950 – Selected Works on the Russian Geography (In Russian). Izd. Akad. Nauk SSSR, Moscow & Leningrad. 248 pp.

TOLMACHEV, A. I., – 1932 – Flora of the central part of the Eastern Taimyr. I. (In Russian). *Trudȳ polyarn. Kom.* 8: 1-126.

TRASS, H. H., – 1964 – Application and validity of the synusial method in phytocoenology (In Russian with Engl. summ.) In, 'Izuchenie rastit, ostrova Saaremaa,' pp. 82-111, Izd. Akad. Nauk Est. SSR, Tartu.

TRASS, H. H., – 1965 – Importance of the plant communities dominants to classification of vegetation (In Russian). *Problemȳ sovremennoi botaniki* 1: 247-250. Nauka, Moscow & Leningard.

Trass. H. H., – 1966 – On vegetation discontinuity and continuity (In Russian with Engl. summ.). *Trudȳ mosk. Obshch. ispȳt. Prir.*, otd. biol. 27: 167-182.

TRASS, H. H., – 1971 – The development of a quantitative approach to the investigation of vegetation (In Russian with Engl. summ.). *Bot. Zh. SSSR* 56(4): 457-464.

TRAUTVETTER, E. R., – 1849-51 – Die Pflanzengeographischen Verhältnisse des Europäischen Russlands. Häcker, Riga. 3 parts ,51, 64, & 82 pp.

TROTSENKO, G. V., – 1971 – The application of the ordination method for analysis of the boundaries among some tundra associations (In Russian with Engl. summ.). In, *Kolichestvennye metody analiza rastit.* 2: 288-292. Riga.

TSEPLYAEV, V. P., – 1965 – Forestry in the USSR (In Russian). Lesnaya promyshlennost, Moscow. 408 pp.

URANOV, A. A., – 1965 – The phytogenic field (In Russian). *Problemȳ sovremennoi botaniki*, 1: 251-254. Nauka, Moscow & Leningrad.

VASILEVICH, V. I., – 1960 – On application of statistical methods to study of plant associations (In Russian). *Vest. leningr. gos. Univ.* 9, ser. biol. 2: 64-70.

VASILEVICH, V. I., – 1963a – A tentative morphological analysis of the meadow continuum (In Russian). *Bot. Zh. SSSR* 48(11): 1653-1659.

VASILEVICH, V. I., – 1963b – Statistical approach to plant association (In Russian). *Trudȳ bot. Inst. Akad. Nauk SSSR*, ser. 3 (Geobotanika), 15: 94-105.

VASILEVICH, V. I., – 1966a – A study of vegetational continua (In Russian with Engl. summ.). *Trudy mosk. Obshch, ispȳt, Prir.*, otd. biol. 27: 59-69.

VASILEVICH, V. I., – 1966b – What is the natural classification (In Russian). In, 'Filosofsk. problemy sovrem. biol,' pp. 177-190, Moscow & Leningrad.

VASILEVICH, V. I., – 1967a – About the methods of analysis of the boundaries of phytocoenoses (In Russian with Engl. summ.). *Byull. mosk. Obshch. Ispȳt. Prir.*, otd. biol. 72(3): 85-93.

VASILEVICH, V. I., – 1967b – A continuum in the coniferous and parvifoliate forest of the Karelian isthmus (In Russian). *Bot. Zh. SSSR* 52(1): 45-53.

VASILEVICH, V. I., – 1968a – Commentary on the continuum concept of vegeation. *Bot. Rev.* 34(3): 312-314.

VASILEVICH, V. I., – 1968b – Principles for distinguishing of plant associations. Abstract of a report. See: NORIN, 1968.

VASILEVICH, V. I., – 1968c – Natural classification in phytocoenology (In Russian with Engl. summ.). *Trans Tartu State Univ.* (*Tartu Ülik. Toim.*) 211: 136-147.

VASILEVICH, V. I., – 1969 – Statistical Methods in Geobotany (In Russian). Nauka, Leningrad. 231 pp.

VASILEVICH, V. I., – 1971 – The representative area of the local abundance and coenoquant concept (In Russian with Engl. summ.). In, *Kolichestv. metody analiza rastitelnosti* 2: 33-38. Riga.
VASILYEV, YA. YA., – 1935 – A concept of 'forest type' and a scheme of the forest types classification (In Russian). *Sov. Bot.* 1935(1): 36-63.
VOLKOVA, V. G., – 1969 – Detailed plans of vegetation and the method of complex ordination (In Russian). *Geobot. kartogr.* Leningrad 1969: 42-51.
VORONOV, A. G., – 1963 – Geobotany (In Russian). Vysshaya shkola, Moscow. 273 pp.
VYSOTSKI, G. N. (VYSSOTZKY, G.), – 1909 – On phyto-topological maps, methods of making them and their practical value (In Russian). *Pochvovêdênie* 11(2): 97-124.
VYSOTSKI, G. N. (WYSSOTZKY, G. N.), – 1930 – Skizzen über die hydrologischen Grundlagen der Bodenkunde. *Pochvovêdênie* 1930(4): 5-31.
WALTER, H., – 1954 – Klimax und zonal Vegetation. *Angew. Pfl Soziol.*, Wien, Festschrift Aichinger 1: 144-150.
WHITTAKER, R. H., – 1953 – A consideration of climax theory: the climax as a population and pattern. *Ecol. Monogr.* 23: 41-78.
WHITTAKER, R. H., – 1956 – Vegetation of the Great Smoky Mountains. *Ecol. Monogr.* 26: 1-80.
WHITTAKER, R. H., – 1962 – Classification of natural communities. *Bot. Rev.* 28(1): 1-239.
WHITTAKER, R. H., – 1967 – Gradient analysis of vegetation. *Biol. Rev.* 42: 207-264.
YAROSHENKO, P. D., – 1961 – Geobotany (In Russian). Izd. Akad. Nauk SSSR, Moscow & Leningrad. 474 pp.
YURTSEV, B. A., – 1966 – Hypoarctic botanico-geographical belt and the origin of its flora (In Russian). *Komarov. Chten.* 19, 93 pp. Nauka, Moscow & Leningrad.
ZINSERLING, Y. D., – 1938 – Vegetation of mires (In Russian). In, *Rastitolnost SSSR* 1: 355-428. Izd. Akad. Nauk SSSR, Moscow & Leningrad.

18 NORTH EUROPEAN APPROACHES TO CLASSIFICATION

Hans Trass and Nils Malmer

Contents

18 NORTH EUROPEAN APPROACHES TO CLASSIFICATION

18.1 Introduction

In the four Scandinavian countries – Sweden, Norway, Denmark and Iceland – and the four Baltic countries – Finland, Estonia, Latvia, and Lithuania – vegetation study and the development of classifications have proceeded along a number of paths. There are, however, linkages among the several schools of vegetation science in the Scandinavian and Baltic area, because of which WHITTAKER (1962) grouped them together in a 'Northern Tradition.' The leading plant geographers and phytocoenologists of these countries, e.g. WARMING and RAUNKIAER in Denmark, POST, SERNANDER, FRIES, and DU RIETZ in Sweden, NORDHAGEN and DAHL in Norway, CAJANDER in Finland, and LIPPMAA in Estonia have to a considerable extent influenced vegetation study not only in Northern Europe but, in numerous cases, on a broader scale. On the role of these North-European naturalists in vegetation studies WHITTAKER (1962: 38) comments, 'The magnitude of the contribution of the smaller nations at the Scandinavian and Baltic area (and Switzerland) is one of the striking features of the history of ecology as a whole.'

Other articles in this volume review the Finnish school (article 15) and the synusial approach of which Lippmaa's work in Estonia is a part (article 16). In Scandinavia itself the tradition of intensive local study of vegetation is longest, and problems of classification have been subject to most extensive investigation, in Sweden. We shall therefore in this review concern ourselves primarily with the 'school of Uppsala' and other Swedish approaches, secondarily with other Scandinavian and Baltic work.

18.2 Historical Background

18.2.1 PERIODS

In Swedish vegetation study four periods can be distinguished: (i) from the first half of the 18th century to the middle of the 19th century – from LINNÉ to POST – i.e., the time when vegetation groupings first made their appearance in Swedish natural science, (ii) from the middle of the 19th century to the beginning of the 20th century – from Post until the foundation of the Uppsala Institution of Plant Biology (1914) i.e. the formation of the inductive approach in Swedish vegetation science, (iii) from the main work of Fries (1913) and foundation of the Institution to World War II – i.e. the golden age of the Uppsala school, and (iv) the post-war period – i.e. the time when the Uppsala school merged in other trends and schools and its influence weakened.

18.2.2 Carl von Linné

As a number of Du Rietz' studies (1942c, 1957a, 1957b) have indicated, several elements of vegetation study can be found already in Linné's (1707–1778) works. Linné realized the great diversity of plant cover and described a number of its types – fens and bogs, eutrophic and oligotrophic lakes, coniferous forest belts in mountains. He also presented a typology of the habitats of plants ('stationes plantarum'), characterized indicator plants, etc.

18.2.3 Hampus von Post

According to Du Rietz (1921) Hampus von Post (1822–1911) was the founder of phytocoenology in Sweden. 'He was the first to distinguish communities of different ranks and to apply an exact method to bounded quadrats for the analysis of these, to introduce a growth-form system useful (in contrast to Humboldt's) for European vegetation, and to oppose consciously the deductive approach to plant sociological research in the field.' (Du Rietz 1921: 49). Post (1951) was of the opinion that in the study of vegetation one should proceed primarily from its composition. For that purpose he described sample plots of definite size, estimating relative importance of each species according to a 5-degree abundance scale. He applied the term 'locale' ('vegetationslokaler') to the main unit of vegetation, with relative homogeneity as the main feature of this unit. He divided locales into subunits and united them into his highest category, the 'group' ('vegetationsgrupper'). In Central Sweden he distinguished six vegetation groups–coniferous forests, deciduous forests, meadows, cultural vegetation, fresh water vegetation, and salt water vegetation.

18.2.4 Ragnar Hult

Ragnar Hult (1857–1899), a Swede by nationality who lived in Finland, a lecturer at the University of Helsinki who wrote most of his studies in Swedish, is considered by Du Rietz (1921: 57) one of the founders and most outstanding students of phytocoenology. His major work was an 'Essay toward analytical treatment of plant formations' (in Swedish, Hult 1881).

Hult was an antagonist of the so-called 'deductive' approach to research, believing that vegetation units distinguished according to habitat conditions are artificial and even pseudo-scientific. Hult writes (1881: 8–9): 'We can thus in work that has the investigation of the physiognomy of vegetation as its goal, distinguish two directions: the deductive and the inductive ... The former assumes as known the external conditions on which the occurrence of each species is dependent. It thus builds from the general principle that, so long as the distributional relations of species are equivalent, to the same complex of local relationships must correspond the same formation... The inductive school proceeds from the observation of the manner of grouping of plant communities as this appears in nature. Through analysis of the composition of formations it seeks the laws for these, and through the analysis of external relationships under which formations occur it seeks the laws for their distribution on the earth's surface. It comes from actualities to principles, from the particular to the general.' When studing plant communities Hult proceeded from rather detailed structural divisions – he distinguished between layers, main-forms and vegetation-forms. He describes his vegetation units as 'formations' (e.g., Pineta cladineta, Cladineta pura, etc.) by strictly observing their growth-form structure, and for illustrative purposes representing their structure in graphs. In his later writings Hult paid more attention to environmental conditions along with his physiognomic methods. In his analysis of the alpine plant communitives of Northern Finland (Hult 1885) he accepted 29 'formations.' Though these are physiognomic units, defined by stratal structure, they are narrow units on the level of current sociations or associations, not formations. Among these 'formations' he recognized two large groups – multistratal and unistratal formations. This was a classification surpassing in detail and effectiveness all preceding treatments of the plant cover.

18.2.5 Rutger Sernander

In 1886 Hult organized a plant geographic field seminar in

Finland where he demonstrated his methods. Among the participants was a 20-year-old Swedish botanist, Rutger SERNANDER (1866–1944), who after his return home began to propagate HULT's views in Sweden. SERNANDER (1894, 1898) emphasized the importance of small sample plots of definite size in studying plant communities, and introduced a percentage scale for estimating the importance values of plant species. His many-sided investigations inspired a lively interest in vegetation studies and classification among botanists in Sweden. During the first decade of the present century dozens of papers were published in Sweden describing vegetation types, proceeding primarily from POST's, HULT's and SERNANDER's methods. The most important of these papers came from the University of Uppsala where, under the influence of SERNANDER, the 'Uppsala school' of vegetation study developed.

18.3 The Uppsala School

18.3.1 DEVELOPMENT: GUSTAV EINAR DU RIETZ

Characteristics of the Uppsala school emerged during a period of years as a result of intensive research and discussions among Swedish botanists. The formation of school's ideas was greatly influenced by the Plant Biological Seminar organized by SERNANDER in 1907 and by the Uppsala Institution of Plant Biology – Växtbiologiska Institution – that won world-wide fame after its foundation by SERNANDER in 1914 (R. E. FRIES 1950). In the years 1913–1921 a number of papers contributing to the development of phytocoenology were published, among them both monographs of field research and analyses of community theory.

In 1913 Thore C. E. FRIES (1892–1936) published his classic monograph on alpine and subalpine vegetation in Lapland (FRIES 1913). Here FRIES took into regular use the term 'association.' The association was regarded as a vegetation type uniform, on the whole, in physiognomy and floristic composition; whereas the formation was characterized by physiognomic correspondence without reference to floristic composition. These small and numerous 'associations' represented the emergence, in the line of evolution from the units of POST and HULT, in essentially its modern form of the unit that was to be known as the 'sociation.' Associations were grouped into 'series,' broader units that had been proposed by NILSSON (1902, see 18.4.4).

FRIES' paper was by 1920 followed by a series of monographs, many of them on mountain vegetation – SAMUELSSON (1917a, 1917b),

MELIN (1917), TENGWALL (1920), and SMITH (1920) – and the following decade produced monographs dealing with diverse communities – Osvald 1923, 1925, MALMSTRÖM 1923, BLOMGREN & NAUMANN 1925, DU RIETZ 1925a, b, ALMQUIST 1929, BOOBERG 1930, and the Norwegian work of NORDHAGEN (1923, 1928). Some of these authors agreed with FRIES in considering floristic-physiognomic classifications as correct and natural (SMITH, TENGWALL), while others prefered ecological (environmental) classification. Thus for example, SAMUELSSON (1917a: 31) writes that the natural distribution of plant communities of a definite area can be obtained only if their ecology has been studied as thoroughly as possible. Otherwise, these aims will be inaccessible to us and we confine ourselves to groupings, to a smaller or greater extent artificial. Ideas of this kind were criticized by DU RIETZ, TENGWALL, OSVALD, and other representatives of the emerging Uppsala school.

From field experience and discussion a *theory* of phytocoenology – a coherent set of principles and concepts linked with one another and field technique – took shape. The leader in the formulation of this theory, the major figure in advocating the distinctive approach to vegetation of the school of Uppsala, was G. E. DU RIETZ (1885–1967), who stated the viewpoint in a series of papers: DU RIETZ (1917) discussing the 'association-complex' and other concepts; DU RIETZ, FRIES, & TENGWALL (1918) criticizing the deductive approach and presenting principles for the determination of minimum-area, with a survey of Scandinavian plant formations; DU RIETZ, FRIES, OSVALD, & TENGWALL (1920) outlining the theory of constant species and the principles for their application in the delimitation of vegetation units; and DU RIETZ (1921), the major work summarizing and generalizing the views of the Uppsala school.

The last two are most important in stating the theory, which we may summarize. Each association is considered to possess a number of species (or at least one species) that occur in all the plots of the association, provided these plots exceed the minimum area. The association may be defined by these species, the 'constants,' which occur in 90 percent or more of samples, either locally or throughout the range of the association. The number of constants normally exceeds the number of species in the intermediate constancy classes; the association is thus made up of a nucleus of constant species, forming the dominant part of the vegetational structure of the community, and a variable group of incidental and accompanying species. The association was thought also to be characterized by a definite minimum area for sampling, related to its richness in species and the size of individual plants. When sample size was increased below the minimum area, the number of constants increased rapidly; but the number of constants remained the same with further increase in sample size beyond the minimum area. Associations were thought to contact one another at sharp boundaries, where the group of species characterizing one abruptly replaced the group characterizing the other; the transition between associations was normally extremely small in relation to the area of the associations themselves. From these principles a full definition of the association could be formulated (Du Rietz et al. 1920, Du Rietz 1923): An association is a complex of species-combinations which recur with especial frequency in nature and possess a common nucleus of species (constants) almost never lacking and present in more or less definite quantitative relations; this complex is as a rule sharply bounded in relation to other comparable species-combinations – i.e. through the lack or relative rarity of intermediate species-combinations. We may consider how the component ideas of the theory were treated by further study. We shall discuss only some aspects of the polemic battles of the northern and southern schools, for which the interested reader may see also Du Rietz (1924, 1928), Du Rietz & Gams (1924), Kylin (1926), Nordhagen (1928), Faegri (1937), Braun-Blanquet (1925), Rübel (1925, 1927), Wangerin (1925), Pavillard (1927, 1935), Lüdi (1928), and the summaries of Whittaker (1962), Aleksandrova (1969) and Shimwell (1971).

18.3.2 Boundaries between Communities

One of the most important theoretical assertions of the Uppsala school, and of Du Rietz (1921, 1922, 1924) particularly, was the normal occurrence of sharp boundaries between plant communities. Du Rietz (1921: 195–6) concluded that, '...boundaries are, as a rule, astonishingly sharp and distinct...diffuse boundaries and slowly intergrading, continuous transitions – such as might well be characteristic of associations of sites with continuous change in environmental factors – have not yet been found in any exactly analyzed case.' In the course of time only the wording and the imperative character of this assertion changed (1923, 1924). Having studied the vegetation of other areas (Swiss Alps), Du Rietz admits that narrow transitional belts can exist between stands in the areas greatly disturbed by man, but considered that phytocoenoses are integral in nature and have sharp boundaries. Du Rietz (1923) states that as among plant species there are also among associations 'good' and 'weak' ones, but he regards the former as typical and prevalent.

Since in general Du Rietz preferred compromises and took

into consideration the results obtained by others, his steadfastness to this tenet should be considered exceptional. It may be pointed out that by the time he started his researches a number of ecologists and plant sociologists had already published data showing the obscurity of boundaries between communities and the frequently continuous character of plant cover (RAMENSKY 1915, 1918, 1924, see article 5, NEGRI 1914, GLEASON 1926). DU RIETZ's concept of sharp boundaries was criticized by a number of his contemporaries (FRÖDIN 1922, KYLIN 1923, NORDHAGEN 1928, LÜDI 1928), and was later opposed by the continuum theory as this was worked out and penetrated into several ecological schools (see summaries: ALEKSANDROVA 1965, VASSILEVITCH 1966, GOODALL 1963, MCINTOSH 1967, NITZENKO 1969, TRASS 1966, WHITTAKER 1962, 1967). The attachment to the idea of sharp boundaries may be explained psychologically by the fact that the denial of this tenet would have led to the necessity of renouncing other basic principles of classification closely associated with the problem of boundaries. In this connection it should be pointed out that several of DU RIETZ' colleagues, leaders of great schools, are or were antagonists of the continuum theory (BRAUN-BLANQUET 1928b, PAVILLARD 1928, NICHOLS 1929, see also WHITTAKER 1962: 82, SUKATSCHEV 1964).

18.3.3 MINIMUM AREA CONCEPT

Each association was considered by DU RIETZ to have a characteristic minimum area. Using his own and other Uppsala phytocoenologists' data, DU RIETZ (1921: 174) drew the conclusion that in Scandinavia minimum areas for associations are as follows: in floristically poor forests 1—4 m², in floristically rich forests slightly larger, in shrubby formations, heaths, bogs, pure lichen formations and presumably on floristically poor meadows they are, as a rule, under 1 m², in floristically rich shrubby formations and on most of the meadows 1—4 m², and on floristically richest meadows over 4 m². One notes that the ideas of minimum area and constant species are coupled: The minimum area is that in which all the constants occur, and the constants are the species that occur in the minimum area.

The minimum area concept, as based on constants, was criticized by a number of phytocoenologists and ecologists (NORDHAGEN 1922, PEARSALL 1924, LIPPMAA 1931, 1933, CAIN 1934, 1938, 1943, ASHBY 1936, SCHENNIKOV 1937, RAMENSKY 1937, VESTAL 1949, GOODALL 1952 – a good bibliography, 1961, GREIG-SMITH 1964, URANOV 1966, et al.). Several authors reached the conclusion that the sizes of minimum areas presented by Uppsala researchers were unreal, being frequently considerably larger in nature (Konovalov & Povarnitzyn 1927, Noskova 1928, Konovalov 1935). Schalyt (1935) showed that in the steppe the minimum area is not determinable; a more or less steady increase in the constants takes place as the analysis area is enlarged. Instead of the minimum area based on the constants, the species-area curve ('Artenzahl-Arealkurve') method has been widely applied (Braun-Flanquet 1928a, 1951, Cain 1938, Tüxen 1970). Du Rietz (1957a: 11) points out that '...minimal area should not be confused with the minimal area (Minimalraum) of Braun-Blanquet, which rests on the species-area-curve, not on the constant-area-curve as our minimal area.' Du Rietz (1957c: 12) goes on to say that '...the importance of the minimal area concept was obviously somewhat overrated in the vigorous discussions of the twenties. Nevertheless, the investigations mentioned had the important practical result that Scandinavian phytosociologists went over from the use of large and not very homogeneous sample areas to smaller, purer and more homogeneous ones...'

On the other hand, the minimal area concept based not on constants but on the whole floristic composition of the community has also been criticized as not mathematically demonstrable (see Greig-Smith 1964). Goodall (1961: 191) writes that '...the hypothesis of continuous increase in variance is inconsistent with the concept of a homogeneous plant community, but fully in accord with that of vegetational continua.' There have been several attempts, not evidently successful, to express minimal area more objectively (Cain 1943, Goodall 1954, Vestal 1949, Archibald 1949, et al.), but we shall not dwell upon them here. The term 'representative area' has been widely applied in Soviet phytocoenology (Ramensky 1924, 1937, 1938). Thus it is necessary to distinguish among three notions – minimum area (constants), minimal area (floristic composition) and representative area (complex of features in Soviet practice). The representative area is the minimal sample size that embraces all basic features of a given community, including floristic composition, quantitative interrelations between species, and structure (Uranov 1966); but no satisfactory statistical method for the determination of this complex notion has been worked out. Perhaps, for reasons involving fundamental characteristics of communities, it is not possible to do so.

18.3.4 Rule of Constancy

The notion of constants was introduced by the Swiss naturalist

BROCKMANN-JEROSCH (1907); DU RIETZ made this concept central to his theory of the association. He stressed repeatedly that the constants characteristic of each association are already noticeable in case of relatively small quadrats, while in further enlarging these the constants do not increase in number (DU RIETZ 1921). Illustrating the constant, accessory and accidental species in graphs (op. cit., Fig. 5, 6, 7, 8, 9) DU RIETZ asserted that there is an abrupt division or cleft between the constants and other species. The property stated for constants – that they gain constancy already in small quadrats and are consequently sharply distinguished from the rest of the species – is considered by DU RIETZ as a most essential feature of the association. DU RIETZ (1921: 200) held the view that constancy relations come into being as a result of the competition among plant species. Constants serve as the nucleus from which it is necessary to proceed in the study and grouping of plant communities.

The Uppsala school's belief in constants and their essential role in classifying communities has been much criticized (ARRHENIUS 1921, NORDHAGEN 1924, KYLIN 1926, LIPPMAA 1931, 1933, 1938, ASHBY 1936, VAGA 1940, KONOVALOV & POVARNITZYN 1927, SCHENNIKOV 1937, NOSKOVA 1928, RAMENSKY 1937, URANOV 1925, 1966, SCHALYT 1935, GREIG-SMITH 1964, GOODALL 1952, 1961, VESTAL 1949, et al.). Vegetation *lacks* the clear division between constants and other species groups that the Uppsala phytosociologists thought they had found. NORDHAGEN (1922: 43) wrote: 'The assertion of DU RIETZ, that there is a fundamental distinction between constants and accessory species, is quite untenable.' KYLIN (1926: 158) asserted: 'The constancy concept of the Swiss school is in principle correct, which one cannot say of that of the Uppsala school. If, however, we accept the strong requirements of the constants in the Uppsala school and add these as criteria to the constancy concept of the Swiss, then I think we retain something useful for phytosociology.' In plant sociological studies calculation of the constancy (or presence) degree of each species is one basis of estimating the degree of homogeneity of a set of samples for a community-type. In this use the role of constancy is reduced from a theoretical concept and essential characteristic of vegetation units, to just one of several diagnostic features of such units.

LIPPMAA (1938) subjected the nature of constants to thorough analysis. Having compiled the distribution maps of the Galeobdolon-Asperula-Asarum union on the basis of character species and constants, he found that '*In the areal definition of a Union it is not equivalent whether one proceeds from character-species or from the constants.* Actually, only the character-species permit an areal definition.

The constant species are often widely distributed species, that only because of their abundance are always recurring in stand samples and are in this way constant, though this constancy of course in no way precludes their occurrence as constants in one or more other unions as well.' (LIPPMAA 1938: 83–84).

18.3.5 DEFINITION BY DOMINANTS

The abandonment of the cleft between constants and other species (with the implications of this for the concepts of the minimum area and the association)was a decisive setback for the Uppsala theory. Already, however, associations had been recognized by the dominant species of strata; and concept was now accommodated to practice – the association was to be defined by dominants (DU RIETZ 1930a, b, 1932, 1936). The artillery of criticism pivoted to follow the school into its new position. Meanwhile the Uppsala 'association' was renamed the 'sociation' (DU RIETZ 1930a, b, 1936).

A sociation was defined as a particular combination of stratal units, with each stratal unit in turn being characterized by the dominance of one or more species. (The stratal units in question have been at different times termed 'consocions,' 'socions,' and 'unions,' see DU RIETZ 1930a, b, 1936 and article 16.2.5). In practice the sociation was characterized by the dominant species of the different strata; a different and broader unit, the consociation, was characterized by the dominant species of a single layer (usually the uppermost) while the other layers might be heterogeneous to any degree. Definition of the sociation by stratal dominants did not imply that every possible combination of dominants became a different sociation; for in a given layer a sociation might be dominated by a number of species occurring in more or less variable quantities. (DU RIETZ 1930b: 376). Nevertheless, it is evident that the number of stratal combinations may be very large, particularly in multistratal vegetation. Numerous examples can be presented to illustrate the abundance of sociations: OSVALD (1923) distinguished 164 in a bog five by eight miles in area, KATZ (1928) found 81 on six small fens in the Moscow Region, BOOBERG (1930) 79 on a fen in Northern Sweden, see also MATVEYEVA (1967), etc. We shall give only a modest example of the combinations of dominants into sociations from Estonian work (Table I, MASING 1969). Five 'associations' are each characterized by *Pinus sylvestris* and one or two other socions or stratal dominants as indicated in the heading of the table; and each association groups together sociations differing in stratal structure in other respects. Thus the nine roman nu-

TABLE I

Sociations described in Estonian bogs (MASING 1969)

Sociations presented in the columns of the table belong to the following associations: I Pinus sylvestris – Calluna vulgaris – Sphagnum Ass. II Pinus sylvestris – Ledum palustre – Sphagnum Ass. – III Pinus sylvestris – Ledum palustre – Pleurozium Schreberi Ass. IV Pinus sylvestris – Calluna vulgaris Ass. V Pinus sylvestris – Cladonia Ass.

Dominant of tree layer, density	Dominants field layer	Dominants of moss layer			
		Sphagna (Sphagnum angusti-folium etc. (excl. S. fuscum)	Forest mosses (Pleurozium Shreberi etc.)	Lichens (Cladonia species, subgenus Cladina)	Absent (Very thin layer)
dense					
	Ledum palustre	II	III	—	III
	Chamaedaphne calyculata	II	III	—	III
	Vaccinium uliginosum	II	III	—	III
Pinus sylvestris	Vaccinium myrtillus	—	III	—	III
(as a raised	Vaccinium vitis-idaea	—	III	V	III
bog form)	Calluna vulgaris	I	IV	V	—
	Empetrum nigrum	I	IV	—	IV
	Rubus chamaemorus	I	—	—	—
	A very sparse layer	II	—	V	—
sparse					

meral III's in the table are nine sociations grouped together into the Pinus sylvestris-Ledum palustre-Pleurozium schreberi 'association,' that differ either in dominance of *Pleurozium schreberi* or occurrence of only a sparse moss layer on the one hand, and in dominance of *Ledum palustre* or one of four other shrub species in the field layer on the other hand. The roman numerals in the body of the table thus represent twenty-two different sociations, and these do not exhaust the combinations that could be recognized as such.

From such observations the basing of classification on dominant species has been widely criticized (see also BECKING 1957: 424, WHITTAKER 1962, and article 14). PAVILLARD (1935: 212) wrote 'The practical result of the initiative of the Upsala group was ... to pulverize the vegetation of their respectively countries into an innumerable multitude of minute rudimentary groups ... generally covering a very restricted area ...'. GOODALL (1953: 41) – 'Communities defined on the dominant species of the highest stratum only are apt to be too broad for general use; if they are defined on a combination of dominants in each stratum, the vegetation is often divided into a large number of inconveniently small units.' VAGA (1940: 150) – 'Sociation sensu DU RIETZ is a piece of multistratal vegetation theoretically cut out in such a manner that each of its strata (socion) consists of one or more dominant species. The number of such theoretical units may be very great and therefore 'sociation' can not be considered as the fundamental multistratal unit.' In the Soviet Union as well, where formerly classification by dominants almost completely prevailed (excluding RAMENSKY and Estonian authors), this practice has been opposed and the basing of associations on a complex of features advocated (RAMENSKY 1952, NITZENKO 1961, MASING & TRASS 1963, TRASS 1963a, 1963b, 1965b, MIRKIN 1965, 1968, et al.).

It by no means follows from these criticisms that study of sociations with the Uppsala methods is useless. Sociations express the structural variation of plant communities, in some vegetation they are effective means for the detailed study of those communities. But as observed by VAGA above, these units in their multiplicity are not 'natural,' fundamental,' units in terms of which all vegetation or even northern vegetation must be conceived. Neither in constancy nor in dominance did the Uppsala school find a redoubt in which its units, as expressions of a theory of the structure of vegetation in general, could be defended.

18.3.6 CLASSIFICATION UNITS

There has thus been a progression in the criteria to be used in defining vegetation units from constants through dominants towards diagnostic species resembling those of the School of BRAUN-BLANQUET. Representatives of the Uppsala school in their first papers regarded *constant species* and *physiognomy* as primary features for distinguishing the main unit of vegetation classification (DU RIETZ et al., 1920: 18 – 'An association is a plant community with certain constants and a certain physiognomy'). DU RIETZ (1930b: 307) later referred to *constant dominants* 'A sociation is a stable phytocoenose of essentially homogeneous species composition, that is at least with constant dominants in each layer.' – and to *dominants* as the criteria for distinguishing the basic units. DU RIETZ (1932: 63–64) – 'A consocion is a relatively homogeneous plant population of species that all belong to the same layer, dominated by a certain species or a few species together. Such a consocion can, either alone or with one or more other consocions of other layers, form a sociation – that is, the plant community forming the whole vegetation of its site and formed by either a certain consocion or a combination of certain consocions of different layers.' DU RIETZ (1957c: 25) – 'The lowest unit in the series of biocoenoses, the sociation, is a biocoenose with a definite and homogeneous composition in each layer represented, i.e. each layer has the same dominant or dominants in every part of the sociation concerned, and the general list of species is at least not more variable than in the association of which the sociation forms a part.' While emphasizing dominants and sociations, Uppsala phytosociologists were already devoting attention to diagnostic species and associations. Associations may currently be distinguished on the basis of two types of diagnostic species (see 18.4.3 and article 20) – character-species ('Leitarten,' 'Kennarten') and differential-species ('Scheidearten,' 'Trennarten') (ALBERTSON 1946 GJAEREVOLL 1956, SJÖRS 1956, GILLNER 1960, et al.).

The most essential clarification was made possibly by the understanding that there might be different basic vegetation units as defined by dominants or by diagnostic species. At the Sixth Botanical Congress in Amsterdam in 1935 (DU RIETZ 1936, CAIN 1936) three resolutions were accepted: (i) To use the term *sociation* for vegetation units characterized mainly by dominance in the different layers, in the sense of Scandinavian phytosociologists; (ii) To use the term *association* for vegetation units characterized mainly by characteristic- and differential-species in the sense of Zürich-Montpellier (i.e. of BRAUN-BLANQUET), or at least for units of the same order of sociological value; *subassociation* and *facies* can, where appropriate, be used for their subordinate units; and (iii) To unite sociations and associations into *alliances* in the sense of Zürich-Montpellier, and the alliances into higher units. It was thus accepted that the association in the sense of BRAUN-BLANQUET was suitable for the conditions of southern Europe and as the general unit, whereas the appropriateness of the sociation was more limited, applying especially to the relatively poor flora of Scandinavia and other areas where community-types are best defined by dominance.

During the earlier period of the Uppsala school only a few units, mainly 'associations,' 'formations,' and NILSSON's – 'series,' were distinguished. By the end of the nineteen-twenties a complicated system of classification units was worked out (DU RIETZ 1929, 1930a, 1930b, 1932). A certain rapprochement to BRAUN-BLANQUET's school was thus taking place; BRAUN-BLANQUET's fundamental unit, the association, was accepted, but his hierarchy (alliance, order, class) was not yet applied. The units for the classification of *phytocoenoses* were as follows – sociation, consociation, association, federation, subformation, formation, panformation. Units for *synusiae* were – socion, consocion, associon, federion, subformion, formion, panformion. Among the *complexes* of phytocoenoses were distinguished – mosaic complexes, belt-complexes, vegetation regions, vegetation zones, and horizons. In 1932, after presenting in detail his views on vegetation classification (Du Rietz 1930b), Du Rietz (1932: 79—80) wrote in an optimistic manner: 'Whereas sociations and consociations are characterized through their dominants, and often also through other constants, associations and federations are given unity through the sociological affinity of the different dominants of their component consociations and through character-species in the sense of Braun-Blanquet (1928a: 52). I give below a first effort toward a synopsis of lichen communities on rocks in the wet-halophyte zone of eastern Sweden, which should above all show in a concrete example, *that it is by no means impossible to combine study of the sociations of Northern phytosociology with that of the associations and alliances of the school of Braun-Blanquet.*'

But Du Rietz had to go further in his pursuit of compromise. In 1935 at the Sixth International Botanical Congress in Amsterdam (Du Rietz 1935, 1936) he introduced Braun-Blanquet's units (alliances in the first place). By the time of the Eight International Botanical Congress in Paris, 1954, he had worked out the following system of units: panformation, formation, sub-formation (order in the sense of Braun-Blanquet), alliance, suballiance, association, subassociation, consociation and sociation (Du Rietz 1957c). Meanwhile Nordhagen (1937, 1943) and others had also found it possible to combine use of sociations as lower units with the higher units of the school of Braun-Blanquet, or to adapt the full hierarchy of that school to northern vegetation (see also Dahl 1957, Gillner 1960, Ivarsson 1962, 1971, Kielland-Lund 1967, and Hallberg 1971).

13.3.7 The Outcome for the Uppsala School

Thus were many of the distinctive ideas of the Uppsala school abandoned. The theory of the community-unit formulated by Du Rietz 1921, Du Rietz et al. 1920) was phytosociology's clearest and most promising statement of what Whittaker (1956, 1962) has called the 'association-unit theory' – the assertion that there exist among natural communities clearly-defined, sharply-bounded 'natural' units, inherent in the structure of vegetation and therefore necessarily the basis of phytocoenological method. The 'association' of the school of Braun-Blanquet is a working assumption, more justified by its usefulness than by any systematic, theoretical interpretation (Braun-Blanquet 1932, 1951). If asserted as a theory about a 'natural' unit, as distinguished from a useful convention, the Braun-Blanquet association is as vulnerable to criticism on the basis of the principles of species individuality and community continuity as was the Uppsala sociation (*vide* discussions in, BRAUN-BLANQUET 1925, 1928b, DU RIETZ & GAMS 1924, LENOBLE 1926, 1928, ELLENBERG 1954, POORE 1956, WHITTAKER 1962). It was a signal endeavor of DU RIETZ as a scientist to offer more than a working assumption: a theory, an interpretation of the *meaning* of associations as phenomena. Under scrutiny, however, the theory suffered only misfortune. The assumption of sharp boudaries, the rule of constancy, the concept of minimum area, the interpretation of the association as a 'concrete' unit, and the underlying belief that the interactions among species would organize them into discrete units – all these fell before research that revealed aspects of the complexity and continuity of species and community relationships (see articles 2, 3).

As the theory was abandoned the practice of basing vegetation study on the constants lost its justification, and the school's practice developed from the doctrine of constants through the emphasis of dominants to acceptance of diagnostic species as bases of classification at higher levels, at least. In the process the school of Uppsala has largely become part of the expanding sphere of the school of BRAUN-BLANQUET. There is no longer a school of Uppsala as such; Swedish phytosociology now represents more a tradition than a well-defined method. A persistent feature of this tradition is the stress on small-scale variation in vegetation and the interest in vegetation units of low rank, for example the prominent position of sociations and unions in vegetation description. This emphasis may be ascribed not only to the tradition and its methods, but also to a flora rather poor in number of species and therefore including many vegetation types with few and poorly specialized species. It is often difficult to find species fulfilling the requirements for character-species, at least on the level of associations.

Thus there is no longer antagonism in principle, only difference in emphasis regarding small-scale units, quantitative sampling, stratal structure, recognition of dominance, and application of diagnostic species. We shall consider next the Swedish approach that has thus developed from the school of Uppsala, before considering further the Scandinavian and Baltic context of Swedish work.

18.4 **Swedish Approaches**

18.4.1 SAMPLING TECHNIQUE

Phytosociological description of vegetation has always been founded on field sampling alternating with studies of the relevés at home. The detailed sample procedures of Du Rietz (1921: 215–216, 1930b: 370–371, 1942a) have been more or less closely followed by most authors. As in other schools the sampled areas of vegetation have been rather arbitrarily selected. For example Sjörs (1948: 281) states that the sampling areas 'were selected in such a way that the vegetation within a square should always be as homogeneous as possible and that the different facies of the association should be represented as fully as possible in the squares.' Du Rietz (1957c: 31) makes the same recommendation for sociations.

There has always been a strong demand for homogeneity not only of individual samples but also of the set of samples representing a sociation. For the individual sample Du Rietz (1957c: 31) recommends that 'each small quadrat must be more homogeneous than the sociation as a whole.' Relative homogeneity or consistency with one another of the samples in a set representing a community-type is 'homotoneity' in the sense of Dahl (1957). Vegetational analyses presented (e.g., Osvald 1923, Figs. 21, 26, 31, 43, 44 and corresponding tables) show that not only did analyzed samples always have to be referred to the same community-type, but also that the same dominating species had to be found in each of the layers represented in those samples, i.e., a sort of quantitative homotoneity (Ivarsson 1962). Such a high demand for sample consistency may bring about an important selection in the choice of sampling areas. During the last twenty years, however, the demand must have been considerably less and the representativeness of the material probably correspondingly better. Compare, for example, the samples of Osvald (1923) and Sjörs (1948). More stress has been laid on a qualitative consistency based on species presence (Ivarsson 1962: 26, Malmer 1962: 48, Tyler 1969a: 28).

Scandinavian phytosociologists have sometimes worked in a simple way, noting only the species occurring within an area of variable size with homogeneous vegetation, in some cases completing the species list with information about dominants and estimates of quantity and frequency (e.g. Melin 1917, Malmström 1937, Waldheim 1944a, Pettersson 1958, Lindgren 1970). Such species lists have been recommended as a first step or means of study when time does not permit further analysis (Du Rietz 1930b: 383–385, 1957c). More generally, however, small square analysis (Du Rietz 1942a, 1957c) has always been characteristic of Scandinavian phytosociology. This is a detailed analysis of exactly delimited small sample areas of a definite size and usually (but not necessarily) in the form of a square. Their size should be larger than the minimum area (see 18.3.3), which in this case means the smallest area of a plant community in which all its constant species occur. This definition thus deviates from that given by the BRAUN-BLANQUET school (e.g. BRAUN-BLANQUET 1951: 88, ELLENBERG 1956: 18). The size of sampling areas most commonly used has been 1 m^2. However, larger sampling areas have sometimes been used, for example, 4 m^2 in species-rich meadows and 16 m^2 in forest phytocoenoses, but hardly ever larger ones than this. Smaller sampling areas are frequently used, for example 0.25 m^2 in epilithic lichen vegetation and in bog vegetation since the 1940's. The epiphytic vegetation on trunks has been investigated with still smaller sampling areas varying from 0.2 to 0.04 m^2.

In earlier work the sampling areas were arbitrarily placed wherever the community under investigation was found. They could be placed either near each other on the same site or on several separate localities (DU RIETZ 1930b: 431). Such a procedure is still common, but today often the small square analysis is carried out within larger areas (usually 50—1000 m^2) representing a homogeneous stand or 'segment' of the investigated community or, for example, on bogs often a mosaic complex of different communities (DU RIETZ 1957c). Within such a stand a number of small squares are placed in such a way that will they give a picture of the variation of the vegetation within the stand. Only a few authors have used a random or systematic distribution of the quadrats. If a sufficient number of quadrats are laid out in a given stand, species occurrence in the stand is effectively measured as *frequency*, the percentage of samples from the stand in which a species is present. In early work of the Uppsala school both frequency in this sense and the percentage of samples each representing a different stand in which a species occurs (which percentage is a *constancy*) were termed 'constancy,' and some of the statements about the 'rule of constancy' are actually statements about frequencies.

TABLE II
Coverage Scales

	HULT-SERNANDER-DU RIETZ		BRAUN-BLANQUET
Degree of cover	Part of the area covered	Middle of cover class	Part of the area covered
1	at most 1/16	1/32	less than 1/20
2	1/16–1/8	3/32	1/20–1/4
3	1/8–1/4	6/32	1/4–1/2
4	1/4–1/2	12/32	1/2–3/4
5	more than 1/2	24/32	more than 3/4

In small square analysis the coverage is estimated for each species or group of species present. Most authors have used the HULT-SERNANDER-DU RIETZ five-degree scale for this purpose (DU RIETZ 1921: 225), as given together with the BRAUN-BLANQUET (1951) scale in Table II. Earlier this scale was less distinctly defined (cf. MELIN 1917: 5, OSVALD 1923: 43), but since about 1920 this scale has been used to estimate just coverage. One notes that this scale does not involve number of shoots, or amount of plant biomass, or sociability, and in that way deviates from many other scales used in phytosociology.

Several authors have made minor additions to this HULT-SERNANDER-DU RIETZ scale. OSVALD (1923 etc.), for example, used + or — after the figure for the degree of cover in order to indicate a higher or lower cover than the middle of the cover class. Some authors have extended the scale by dividing either or both of the classes 1 and 5. Thus, for example, GILLNER (1960), IVARSSON (1962) and TYLER (1969a, 1969b) indicate when there are only one or two shoots of a species and the cover much less than 1/16 of the sampling area. Class 5 has by several authors (e.g. STEEN 1954, IVARSSON 1962, SJÖGREN 1961, 1964, TYLER 1969a) been divided in two in somewhat different ways. GILLNER (1960) and ANDERSSON (1970) have used the BRAUN-BLANQUET cover scale (BRAUN-BLANQUET 1951: 59). Several Finnish authors (e.g. KALELA 1939, TUOMIKOSKI 1942, HAVAS 1961) have directly estimated the coverage in per cent.

Also other methods for a quantitative estimation of the plant species in the quadrat analysis have been used. (See also Danish work of RAUNKIAER, 1909-10, 1912, 1918, and BÖCHER, 1936). LAGERBERG (1916) and later MALMSTRÖM (1923) used another method with a systematic distribution of 25 or 50 small (0.1 m^2) subquadrats within a larger sample quadrat of 25 m^2. The point quadrat method published by LEVY & MADDEN (1933) and used by LINDQUIST (1931) and JULIN (1948) is rather similar. In this case the cover of a species is estimated from its percentage occurrence at 100 systematically arranged points within an exactly delimited area of 1 m^2. Even quite exact measurements of the cover for phytosociological purposes have been performed (THUNMARK 1931).

Most authors have treated the whole phytocoenoses even if DU RIETZ (1936, 1957c) following GAMS (1918) and LIPPMAA (1935, 1939) has strongly emphasized the different layers as independent units. NORDHAGEN (1943: 32 ff.), however, takes an opposite position. A synusial approach has been adopted mainly for lake-vegetation (e.g. DU RIETZ et al. 1939 and KAARET 1953) and for bryophyte, lichen and algal communities (see article 16). Irrespective of the treatment of the stratification in the vegetation there has always been a strong demand for the completeness in the species lists viz., at least all macrophytes (vascular plants, bryophytes, lichens, and some large algae) occurring within the sampling area should be listed.

18.4.2 Tabular Arrangement

From the beginning tables worked out by Scandinavian phytosociologists have had a very characteristic form from which the individual authors usually have deviated only slightly. The species occurring in each layer of the vegetation studied are grouped together. In terrestrial phytocoenoses rarely more than four layers are distinguished, viz., the tree layer, the shrub layer, the field (herb and dwarf-shrub) layer, and the bottom (ground surface) layer (Du Rietz 1921: 133—134). Within each layer the species are arranged according to their growth-forms in alphabetic order. In terrestrial vegetation the most commonly distinguished growth-forms are trees, shrubs, dwarf-shrubs, herbs, graminoids, bryophytes, and lichens. This means that within a vegetation table of this type a species has a fixed position in relation to the other species depending on its growth-form and layer. In such a table no consideration is given as to whether species characterize the community, nor to constancy, frequency, coverage, or any other quantitative aspect. Separate systems have been worked out for the stratification and growth-forms of aquatic vegetation (Du Rietz 1921: 134, 1930b: 390—392, 1932, 1940, Thunmark 1931, Wassén 1966: 24—27; cf. also Luther 1949 and article 16).

In a vegetation table the samples from the same locality or site usually are brought together. This is especially important when a number of small square analyses have been performed within the same stand. In that way it is possible to compare with one another at the same time the quadrats from a stand and the different stands. During the last decades many authors have presented survey tables of plant communities in which only the frequencies of the species in the community (Waldheim 1947), or the frequencies together with some indication of coverages (Sjörs 1954, Gjaerevoll 1956, Persson 1961, Malmer 1962 etc.) are given. Such survey tables usually exclude species which occur only with low frequency in all communities. Often the species and the plant community-types are arranged in such a way that the table illustrates vegetational gradients distinguished during the investigation.

18.4.3 Community-Types of Low and Middle Rank

In Scandinavian phytosociology the greatest interest has always been concentrated on plant communities of low rank, perhaps due to the field technique adopted. As indicated above, the sociation (until 1930 designated 'association,' cf. Du Rietz 1930b: 304 and 307, or 'soziotypus' according to Nordhagen 1922 and Malmström 1923) was regarded as the fundamental unit, at that time characterized by fixed constant species and physiognomy (Du Rietz et al. 1920, Du Rietz 1921, Osvald 1923). More than one constant species could dominate an association (Du Rietz 1921: 186), and variants were sometimes distinguished by differences among the constant species. Later authors have given stronger importance to dominance when distinguishing sociations (18.3.6); and sociations were designated by naming the dominants in each layer, for example, the 'Calluna vulgaris-Cladonia rangiferina-silvatica-Soziation' (Du Rietz 1930b: 434).

Groups of sociations ('Assoziationen') with close floristic and physiognomic relationship have been designated 'Assoziationsgruppen' (Du Rietz 1921, Osvald 1923). Sociations with a common dominant in one layer but not in the others may be brought together in a consociation (Du Rietz 1930b: 311, 1932, Thunmark 1931, Albertsson 1946; cf. Warén 1926), but this unit has not been much used.

From about 1935 onwards (Du Rietz 1936, 1942a, 1942b) classifications have given less emphasis to just the dominant and constant species. More stress is put upon the whole community composition, manifested as a regularly repeated combination of species. As especially important however, have been regarded the two kinds of indicator species, viz., species confined to only one plant community-type (character-species) and species confined to one of two community-types compared with each other (differential-species) (cf. 'ledart' and 'skiljeart,' Du Rietz 1942a, 1942b). Within each of these types it is possible to distinguish species which are exclusive, i.e. entirely confined to one community only or to one of two communities being compared with each other, and species that are only preferential, i.e. occurring as a dominant or with high frequency in one community-type but not confined to that community-type (Sjörs 1948, 1954). This system of diagnostic species is not the same as that of Braun-Blanquet, and generally differential-species have had greater importance than character-species, at least on lower levels. Community-types delimited in this way are, however, more or less equivalent to associations in the sense of the Braun-Blanquet school. Most authors have used this term for designating them, but some others have used the neutral term 'plant community,' e.g. MÖRNSJÖ (1969).

As units of lower rank within an association several authors have used 'variants,' which have been treated separately because of minor differences in species composition (PERSSON 1961, MALMER 1962, SONESSON 1970). Less attention has been paid by them to phytosociological units corresponding to sociations, even if the latter have been frequently used as subordinate units. DU RIETZ (1942a: 125—126) and ALBERTSSON (1946) distinguish sociations within associations, whereas NORDHAGEN (1943) and GJAEREVOLL (1949, 1950, 1956) try to find character- and differential-species making it possible to group sociations into associations in the sense of Middle-European plant sociologists. The associations are designated by using the names of one or a few of the species characteristic or of importance for the association in any other way. Generally the species names are used directly, for example the Pinus-Carex globularis-Spaghnum parvifolium-association (SJÖRS 1948: 140). It is less common to use a suffix to species names such as the common -etum and -etosum.

The authors using a synusial approach have followed many of the same ideas. Thus, for example, LINDQUIST (1931, 1938, 1954) dealing with the field and bottom layers separately in South Scandinavian broad-leaved forests distinguishes societies (in 1931 called 'socions'), and unions (only 1938) with reference to dominant species, constants and physiognomy. The 'isozionen' (DAHLBECK 1945) from seashore meadows are similar in character. JULIN (1948), ARNBORG (1940, 1943) and IVARSSON (1962, 1971) have instead used indicator species for separating the unions of the different layers in forest vegetation or areas where forest vegetation is colonizing a formerly open landscape. WALDHEIM (1944b, 1947), KRUSENSTJERNA (1945), and SJÖGREN (1961, 1964) have treated bryophyte and lichen communities of different phytocoenoses according to these general principles, SJÖGREN (1964: 16—23) also the field layer of broad-leaved forests. Article 16 further reviews synusial approaches.

18.4.4 HIGHER GROUPINGS

There has been weaker interest in the systematic arrangement of the basic vegetational units into higher ones in Scandinavian phytosociology. In the earlier Scandinavian phytosociological works the lower units were grouped by their physiognomy, stratification and dominating growth-forms. DU RIETZ offered a more elaborate system of formations (Du Rietz 1921: 136–140, 1925b, Osvald 1923, Malmström 1923) and formal hierarchies based on dominance (Du Rietz 1930b, 1936, see also 18.3.6), but these were not successful. Several modern authors have grouped associations (e.g. Du Rietz 1942b, Albertsson 1946, 1950, Gjaerevoll 1956, Persson 1961, Sjögren 1961, 1964, Malmer 1962, Mörnsjö 1969, Sonesson 1970) or sociations (Nordhagen 1937, Kalliola 1939) with reference to their species composition, especially the diagnostic species, into 'alliances' or 'federations.' Often neutral terms like 'vegetation,' 'series,' or common names as 'carr,' 'grassland,' 'heath,' etc. are used instead of a technical designation, thus avoiding definite ranking of the community-type. Nordhagen (1937, 1943, cf. also 1928:90 and 465–467) is one of the few Scandinavians who has worked out a complete hierarchic system including not only alliances but also orders and classes.

The reverse principle, viz., a successive division of higher vegetational units into lower ones, has also been used. 'Any plant community and its natural delimitation is established through direct field studies of the unit in question, . . . not . . . through arranging them' (i.e., the lower units) 'in groups with reference to their greater or smaller similarity,' (Du Rietz 1942a: 7, see also 1949: 299). This approach has been most highly developed for bog vegetation, a panformation within which, for example, the main formation of the boreal bogs is divided into two formations, viz., Ombrosphagnetea and Sphagno-Drepanocladetea (Du Rietz 1949, 1954). These units correspond to classes in the sense of the Braun-Blanquet school and are subdivided into subformations (or orders) alliances, suballiances, etc. Sjögren (1961, 1964) in a similar manner has distinguished first the federations of bryophyte communities and then within them the unions.

Plant communities (mainly representing sociations and associations) occurring close together within an area, for example on a bog, have often been brought together in higher units called complexes ('Assoziationskomplexen' in Du Rietz et al. 1918, Du Rietz 1921, Osvald 1923; 'Phytocoenosenkomplexen' in Du Rietz 1930b; 'zonations' in Malmer 1962). The floristic and physiognomic relationships between the communities within a complex may be rather weak in spite of their close ecological relationship. Several of the higher units in the bog vegetation system of Du Rietz referred to above have more the character of complexes than floristically distinguished units (Malmer 1968). Du Rietz (1936, 1957c) even argues for regarding a phytocoenose as a vegetational complex of two or more synusiae.

Most authors have regarded all community-types they describe, as well as their indicator species, as of only local significance, viz., representing only the investigated district. Knowledge of related vegetation in other areas is usually not considered when defining the vegetation units (e.g. Sjörs 1948). The studies on the regional differentiation of the plant communities over larger areas therefore usually are carried out through comparisons with other authors and the units distinguished by them when treating similar vegetation. This is in accordance with the opinion put forward by Du Rietz (1942a, 1949) that vegetation regions should be regarded as phytocoenose complexes of very high rank, and that the subdivision of their phytocoenoses should be independent, using different units for each region. Many times there may exist close floristic relationships between units widely separated in this way. The approach is in contrast with that of the school of Braun-Blanquet, in which not regional vegetation complexes or patterns, but a formal hierarchy that should be consistent in application to different regions, is given primary emphasis.

Several authors, especially those dealing with bog vegetation, have instead of putting the lower units of vegetation into a hierarchial system followed Tuomikoski (1942) and arranged them according to directions of variation in the vegetation (Sjörs 1948, Persson 1961, Malmer 1962) or vegetational gradients (Malmer 1965, 1968, Mörnsjö 1969, Andersson 1970, Sonesson 1970; see also Havas 1961). This gives a multidimensional, coordinate framework in relation to which it is possible to place each vegetational unit. Especially in mosaic complexes, such a system is very informative. It may reflect very well the characteristic features and variation of the vegetation and offer important advantages for studies on the relationship between vegetation and environment. Such a coordinate system can be changed into a conventional hierarchy by using the different gradients, one by one, as bases of dividing the vegetation into units (Tuomikoski 1942).

Nilsson (1902) divided terrestrial vegetation into higher units called 'series,' which have been of great importance for several later Swedish authors. From the beginning they were characterized as ecosystems, but today they are treated as purely vegetational units each including both wooded and non-wooded vegetation types. Sjörs (1956) recognized four units of this type. The heath series comprises plant communities dominated by dwarf-shrubs and grasses with dry and narrow leaves (e.g. *Deschampsia flexuosa*, *Nardus stricta*) together with mosses and lichens. In the meadow series dwarf-shrubs and lichens are few or lacking, herbs usually common and the grass species broad-leaved. The steppe series (cf. also Sterner 1922 and Böcher 1945: 123) comprises the few units of vegetation found in Sweden related to the south-east European steppe and limestone cliff vegetation. All the mire vegetation in the sense of Du Rietz (1954) is brought together in the mire series.

A terminology related to Nilsson's system, is the common separation of 'poor' and 'rich' vegetation types. These terms are used only to characterize vegetation types according to their species composition and do not state anything about the habitat conditions or the number of species (cf. Sjörs 1948: 281, Malmer 1960: 111–112, 1962: 33, Persson 1961: 106 ff). Following this system Du Rietz (1945a) divided lake vegetation into poor lake vegetation (corresponding to the *Lobelia*-lakes or the oligotrophic lakes) and rich lake vegetation (*Potamogeton*-lakes or the eutrophic lakes) and also divided corticolous lichen vegetation (Du Rietz 1945b, Almborn 1948) into poor bark vegetation (e.g. Physodion) and rich bark vegetation (e.g. Xanthorion). According to this terminology the heath series includes poor vegetation types, the meadow and steppe series rich vegetation types. The mire series includes both poor vegetation types (bogs and poor fen vegetation) and rich vegetation types (rich fen vegetation, cf. Du Rietz 1949, 1954, Malmer 1965, Sonesson 1970).

One of the most striking features in the Scandinavian vegetation is the strong difference between areas with soils poor in lime (mainly Archean areas) where the number of species found is much less than in areas with soils rich in lime. This is consistently the case from the south Scandinavian lowlands to the mountains in the north. The division of vegetation into a heath series and a meadow series or into poor and rich vegetation types well reflects this important vegetational difference. This poor – rich gradient is, together with the vegetational gradient in response to the water regime of the site, often used as a coordinate system for a first arrangement of phytocoenoses (cf. e.g. Gjaerevoll 1956) – see Fig. 1.

The lack of clearly defined higher units in Scandinavian vegetation science results partly from incomplete knowledge of the vegetation. One of the drawbacks of the Scandinavian school is that the time-consuming methods, especially in vegetation types rich in species, limit the extent of investigations. For this reason it takes a long time before a general survey is obtained. For several vegetation types and regions in Sweden there are, however, up-to-date articles summarizing results of the investigations of the different vegetation types with references and tables. As examples we may mention studies on forest vegetation (Sjörs 1965), mire vegetation (Malmer 1965, Sjörs et al. 1965, Persson 1965), mountain vegetation (Gjaerevoll & Bringer 1965, Persson 1965), steppe vegetation (Albertsson 1950), salt marshes (Gillner 1965, Tyler

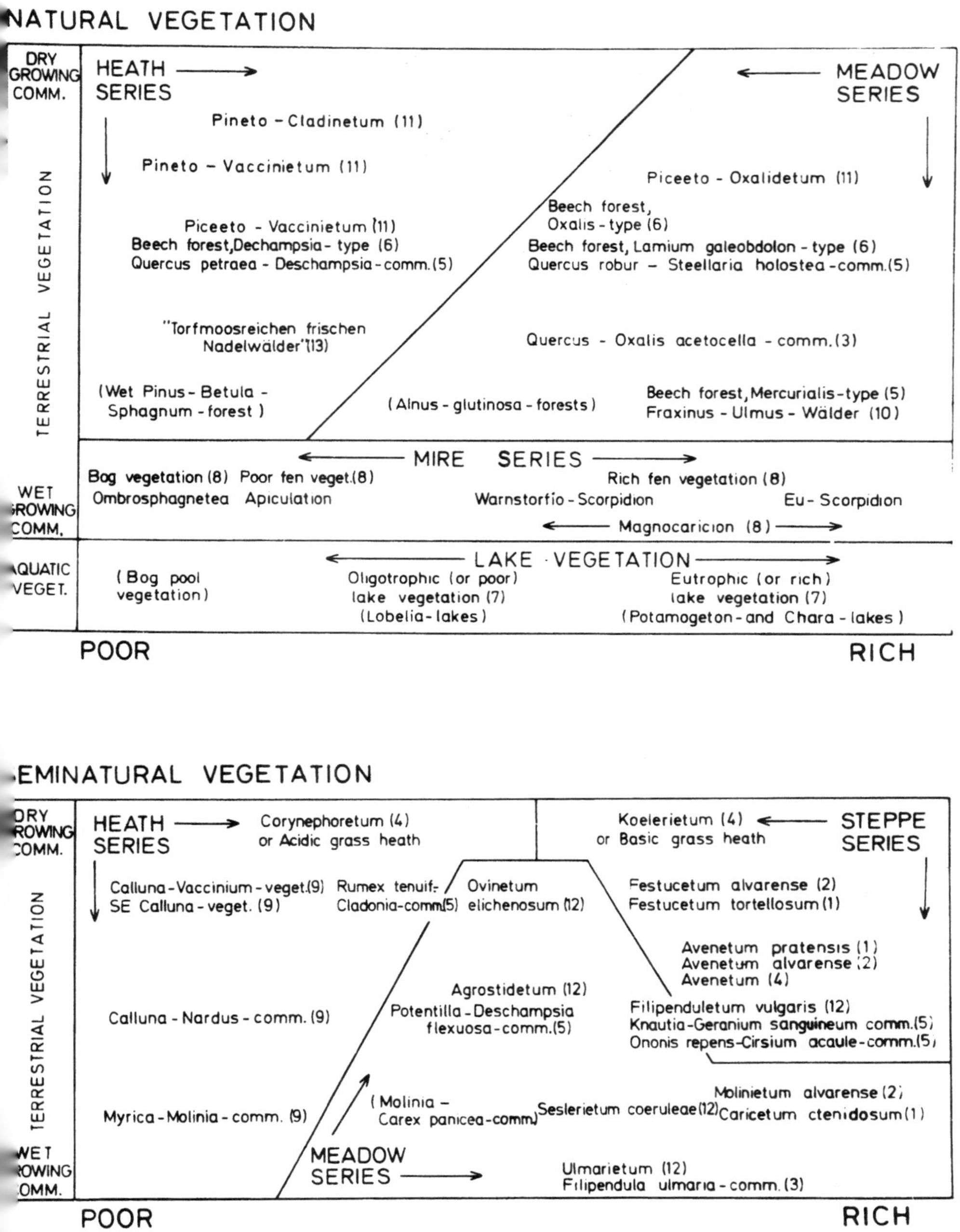

g. 1. Tentative, schematic figure illustrating some vegetational units of low and middle nk from the Nemoral and Boreo-Nemoral regions of South Sweden in relation to vegetation ies along the moisture gradient and the poor-rich nutrient gradient of vegetation. See ther discussion in the text! Explanations: The figures after the names of the vegetation units er to the following literature (either original articles with tables or survey articles with ther references), viz., 1 and 2 to Albertsson 1946 and 1950; 3 to F. Andersson 1970; to 0. Andersson 1950, 5 to Berglund 1963, 6 to Lindgren 1970, 7 to Lohammar 1965, 8 to almer 1965, 9 to Malmer et al. 1965, 10 to Sjögren 1964, 11 to Sjörs 1965, 12 to Steen 54 and 13 to Waldheim 1944a. For communities within parentheses no modern literature references are available.

1969b), epilithic vegetation (KRUSENTSJERNA 1965), dwarf shrubs (MALMER et al. 1965), and the different vegetation types of coastal landscapes (BERGLUND 1963, HALLBERG & IVARSSON 1965).

18.5 Other Scandinavian and Baltic Approaches

18.5.1 THE FINNISH SCHOOL

A. K. CAJANDER of Finland ranks with POST, HULT, and SERNANDER as one of the pioneers of Scandinavian and Baltic phytocoenology. CAJANDER (1903a, 1903b, 1903c) used an early form of the compiled tables of vegetation samples which appear in most later studies, and he established the manners of naming associations (1903a) and sociations (1903c). CAJANDER recognized 'associations' by dominant species or combinations of dominants; and on the basis of characteristics of other layers he distinguished 'facies' within these, corresponding more nearly to the sociations of Uppsala (CAJANDER 1903a, 1922, CAJANDER & ILVESSALO 1921). The concept of 'ecological series' or chains of communities along environmental gradients, characteristic of much later Finnish and Russian work, (articles 5 and 15.5) appeared in his study of the relations of alluvial vegetation to water level (CAJANDER 1903a). The contribution most closely associated with his name, however, is the system of forest site-types based on the relations of undergrowth vegetation to biotope conditions (CAJANDER 1909, 1949). The approach is discussed further in article 15.

18.5.2 THE ESTONIAN SCHOOL

DU RIETZ reached the conclusion that the layers of phytocoenoses are relatively independent units and should be classified as such. Together with GAMS (1918, 1927, 1933) DU RIETZ became an advocate of the synusial approach to communities through units comprising species of a single stratum that are relatively homogeneous in their life-forms and ecological relations. Whereas DU RIETZ advocated that both whole phytocoenoses and stratal societies should be classified, separately, the Estonian school of LIPPMAA (1933, 1935, 1938, 1939, VAGA 1940, article 16) fully shifted emphasis to the stratal units. The resulting synusial classification has not been extensively applied to vascular plant communities, but has been influential in Sweden (18.4.3) and elsewhere, particularly for the classification of thallophyte societies and aquatic plant communities (article 16). More recent Estonian applications of the synusial approach are discussed by TRASS (1964); for further review of contributions to phytocoenology in the Estonian S.S.R. see TRASS (1965a, 1970), MASING & TRASS (1963), and MASING (1969).

18.5.3 DANISH SCHOOLS

The early influence of Denmark was exerted through two of the great plant geographers, WARMING and RAUNKIAER. Along with his contribution to the physiognomic study of vegetation (WARMING 1895, 1896), WARMING (1909) had a major influence in fixing the concept of the 'formation' as a physiognomic unit, defined not by species but by plant growth-forms. RAUNKIAER (1909-10, 1910, 1934) was one of the pioneers of quantitative approaches to vegetation, and shared with the Uppsala school the emphasis on vegetational statistics and the use of small vegetation units. In contrast to WARMING, RAUNKIAER (1909-10, 1918) applied the term 'formation' to narrowly defined units, in general characterized and named by their dominant species; the formation was later (1928) defined as comprising stands agreeing in the occurrence of 'frequency-dominants,' species present in more than 80 percent of the quadrats representing a formation. Formations could be grouped into 'series' of formations by their dominant life-forms (1909-10); and a hierarchy of 'groups,' 'classes,' and 'branches' of formations was also suggested (1918). The 'formations' of RAUNKIAER and Danish phytosociologists following him correspond more nearly to the sociation than to the formations and associations of other schools. Convergence with other schools was sought by IVERSEN (1936), who identified the formation of RAUNKIAER with the sociation of DU RIETZ and grouped sociations into 'Soziation-Verbände' corresponding more nearly with the associations of BRAUN-BLANQUET. SØRENSEN (1948) sought to relate the various units through quantitative comparison of vegetation samples, using an index (article 6, formula 40) that has been widely applied.

In early work of BÖCHER (1933) in Greenland, community-types were defined by frequency dominants and termed associations, and these were grouped into formation-types in the sense of RÜBEL (1930). In later work BÖCHER (1943, BÖCHER et al. 1946) has used the sociation as a unit, together with 'types' and 'series' characterized by dominating species, physiognomy, and ecology. BÖCHER (1933, 1938) has been concerned also with the geographic relations of species and developed this approach in his monograph (1954) on vegetation complexes of Southwest Greenland. In place of the character-species of BRAUN-BLANQUET, four other types of diagnostic species are recognized: (i) Area species, or area-geographical differential species, that occur in comparable communities and situations in different areas, and by their presence and absence distinguish the communities of these different areas. (ii) Climatic species or climatic indicators that indicate, by their broader geographic distribution and climatic relations, something of the affinities of the community in which they occur. (iii) Habitat species or ecological differential species that in a given area are associated with particular edaphic and micro-climatic conditions, and hence may be character-species for the communities of these. (iv) Ecogeographical guiding species, that are of special significance for their indication of both local site or biotope, and broader geographic or climatic, affinities of a community in which they occur. These concepts were used for definition of higher units in a hierarchy from sociations through sociation groups, vegetational types, and vegetational complexes, to vegetational regions.

18.5.4 ICELANDIC VEGETATION

The Arctic tundra is usually thought a single formation or formation-type, but within this treeless vegetation there is diversity of physiognomy with changing dominance of growth-forms – dwarf-shrub, lichen, moss, and graminoid (grass and sedge) – in response to differences in biotope. The criteria of 'formations,' as major units characterized by similarity of physiognomy and broad environmental relations (article 13), can be applied to define within the tundra smaller-scale units that we might perhaps call 'microformations.' Thus in the predominantly tundra vegetation of Iceland OSTENFELD (1889, 1905) distinguished his vegetation types or 'formations' by combinations of biotope characteristics and physiognomy. Within these types lower-level 'facies' and 'associations' were characterized by their major species. The ideas of OSTENFELD and his successors were combined with those of the school of RAUNKIAER in the work of the Danish author MØLHOLM HANSEN (HANSEN 1930). Quantitative samples were taken by the methods of RAUNKIAER and classified into 'formations' as recognized by Icelandic authors. Species were classified into seven geographic groups or areal types, and 'formations' were characterized by life-form and areal-type percentages, or spectra. The distinctive Icelandic classification of arctic vegetation was applied in a series of studies by STEINDÓRSSON (1936, 1945, 1946, 1954) recognizing 'formations' and within these narrowly defined dominance-types, defined by one or more major species, as 'associations' or sociations. The work of HANSEN was, as observed by WHITTAKER (1967) a notable early venture in direct gradient analysis. 'Formations' were arranged in ecological series – by moisture conditions, snow-cover persistence, and elevation – in extensive tables that showed the changes in species populations, species diversities, and representation of RAUNKIAER life-forms and areal types along these gradients. The Danish school thus possessed, in the quantitative methods of RAUNKIAER, use of ecological series by HANSEN (1930, 1932) and similarity measurements by SØRENSEN (1948), the beginnings of what might have become a development of gradient analysis.

18.5.5 NORWEGIAN MOUNTAIN VEGETATION

In the nineteen-twenties the outstanding Norwegian phytocoenologist Rolf NORDHAGEN was one of DU RIETZ' opponents who criticized the rule of constancy, the concept of sharp boundaries, and other views of the latter (NORDHAGEN 1923, 1924, 1928; cf. the responses of DU RIETZ 1925a, 1928). NORDHAGEN (1954: 81) has said about the development of his views that up to 1928 he distinguished vegetation units on the basis of life-forms of the dominating physiognomic species, while afterwards he came to the conclusion 'that such grouping in many ways was unnatural, especially with regard to the ecological or causal side of the problems.' Having been on joint excursions to North Africa, the Alps, and the Carpathian mountains with BRAUN-BLANQUET, SZAFER, and other Middle European phytosociologists, NORDHAGEN got acquainted with basic principles of the Zürich-Montpellier school and associated himself with them. He introduced character- and differential-species as well as the fidelity concept into his classification of vegetation units – associations and alliances. At the same time NORDHAGEN (1954: 82), however, keeps to his standpoint that '... the lower or basic vegetation units must be founded upon *dominance* alone, not only of phanerogams, but of lichens and mosses as well.' In his works (NORDHAGEN 1923, 1937, 1943, 1954) he describes in great detail numerous plant communities, particularly those of the mountain areas. It is noteworthy that Nordhagen attaches much importance to *alliances* as relatively broad groupings regarded as natural units.

We have seen that the views of DU RIETZ and NORDHAGEN – as well as the latter's colleagues FAEGRI (1934, 1937), GJAEREVOLL (1949, 1956), and DAHL (1957) – on the problems of vegetation classification draw nearer and largely coincide from the nineteen-thirties on. This circumstance provided Du Rietz with sufficient grounds to speak of a Swedish-Norwegian school. At the same time differences in minor issues survive, e.g., the application of synusial method, as well as in classification problems. Dahl's (1957) work on mountain vegetation, 'Rondane,' used the Braun-Blanquet classification, but employed both dominants and differential-species for the characterization of units on the level of the association. Dahl also discussed the occurrence of discontinuity in some, and intergradation in other parts of his alpine vegetation, applied the Sørensen (1948) similarity index as an aid to classification, and treated problems of species diversity and homogeneity in communities (see also Dahl & Hadač 1941, 1949, Dahl 1960).

18.6 Conclusion

What is finally to be said on interpretation of Scandinavian and Baltic phytocoenology and the extent of its contribution? There is, first, the effect that Whittaker (1962) terms the ecology of ecological schools – the effect of the kind of vegetation studied and the kinds of problems encountered on a school's choice of ideas and techniques. Dealing with northern vegetation that is on the whole relatively poor in species, with marked species dominance and often well-defined stratal differentiation, the northern authors naturally took directions different from those pursued in southern Europe. The emphases of small samples, careful quadrat technique, species dominance, and stratal structure common as themes to several of the northern schools may be thought responses to these vegetational circumstances. Yet northern vegetation is itself diverse enough; and effects of this diversity are evident (along with those of the ideas of leaders of schools) in the development of quite different approaches to the mountain vegetation of Sweden and Norway, the lowland forests of Finland and Estonia, the tundra of Iceland and Greenland, and the heaths and commons of Denmark. Looking superficially at the history of the Northern and Southern Traditions in European phytosociology, there is a marked contrast: in the North, a scientific effort branching into numerous schools; in the South a number of schools among which that of Braun-Blanquet increasingly prevailed, unifying the efforts of phytosociologists in a great achievement.

Two approaches to an object of study may be distinguished: that of composition, which may be represented in the listing of components, and that of structure, emphasizing the quantitative and spatial relations of those components. It cannot be said that the northern schools neglected composition or that the southern schools were uninterested in structure; yet the difference in emphasis is significant. The school of BRAUN-BLANQUET concerned itself with composition and the use of diagnostic species; from the fact that among many experiments in classification this was, over all, the most successful has resulted the ascendancy of that school. The northern schools emphasized various different aspects of structure; and their approaches to structure proved less adaptable to other vegetation than the approach through composition. The approach of BRAUN-BLANQUET could be accommodated, with change of emphasis, to northern vegetation more reasonably than approaches adapted to the structure of northern vegetation could be used in other areas.

The influence of the northern schools has been exerted in other directions. From the North primarily have come the development of physiognomic ecology, quantitative techniques and quadrat studies, the study of life-forms, synusial approaches to communities, detailed forest typology, and beginnings of gradient analysis. The South has prevailed in its construction a of formal system of classification; but the North has produced a wide range of approaches including classification but extending beyond this to many ways of studying and describing communities that have affected the study of vegetation throughout the world. They have influenced Russian phytocoenology profoundly (ALEKSANDROVA 1969, article 18) and English-language ecology substantially, and have not been without influence on the Southern Tradition. A northern phytocoenologist may thus both grant the achievement of the school of BRAUN-BLANQUET, and value contributions from his own tradition that have served well for the study of northern vegetation and much influenced the study of plant communities elsewhere.

18.7.1 SUMMARY

One of the main lines of development in the study of northern vegetation led from HULT and POST through SERNANDER to DU RIETZ and the school of Uppsala. The Uppsala school used small sampling areas of defined size for detailed study of vegetation, and classified these samples into relatively narrow units defined (in later practice) by the dominant species of the different strata. While these units have been called 'associations' by some earlier authors, the internationally accepted term for them is now 'sociation.'

The Uppsala school developed, in connection with these units, a theory of the structure of vegetation. Sociations were considered

(i) to be characterized by a nucleus of species that were 'constant' (occurring in 90% or more of the samples of a sociation) and were sharply distinguished from other, accompanying species, (ii) to possess a definite minimum area as a sample size sufficient to include the constant species, and (iii) to be separated by sharp boundaries from other sociations. Research on these concepts forced abandonment of the theory.

Swedish phytosociology is now more a tradition in the approach to northern vegetation than a distinct school. Vegetation study in this tradition is rather more concerned with analysis of small sample plots, with stratal structure, and with quantitative relations and species dominance. The sociation was earlier used as a unit for detailed study, but today more interest is focused on associations comparable with those of BRAUN-BLANQUET. Swedish phytosociology has been on the whole little concerned with formal classification, and a number of different approaches have been used for grouping communities on higher levels.

Other Scandinavian and Baltic approaches include the use of forest site-types in Finland (CAJANDER), stratal or synusial units in Estonia (LIPPMAA), classification of Norwegian mountain vegetation (NORDHAGEN, DAHL), development in Denmark of quantitative approaches (RAUNKIAER) and study of species distributions (BÖCHER, classification by small-scale tundra 'formations' in Iceland (STEINDÓRSSON), and work anticipating gradient analysis (CAJANDER & ILVESSALO, MØLHOLM HANSEN, SØRENSEN, TUOMIKOSKI). The northern schools have produced no general system of classification as widely successful as that of BRAUN-BLANQUET, but have made major contributions through development of classifications appropriate to northern vegetation and approaches to the structure of communities that have been influential throughout the world.

18.7.2 ZUSAMMENFASSUNG

Eine der Hauptlinien in der Entwicklung des Studiums der nordischen Vegetation führte von HULT und POST über SERNANDER zu DU RIETZ und der Uppsala-Schule. Die Uppsala-Schule verwendete kleine Aufnahmeflächen bestimmter Grösse für die detaillierte Untersuchung der Vegetation und klassifizierte diese Aufnahmen in ziemlich eng gefaßten Einheiten (in der späteren Praxis) durch dominante Arten der verschiedenen Schichten. Nachdem diese Einheiten von einigen früheren Autoren 'Assoziationen' genannt wurden, werden sie jetzt nach internationalem Gebrauch als 'Soziationen' bezeichnet.

Die Uppsala-Schule entwickelte in Zusammenhang mit diesen Einheiten eine Theorie der Vegetationsstruktur. Von den Soziationen wurde behauptet: (i) sie seien gekennzeichnet durch einen Kern von Arten, die 'konstant' (d.h. in 90 % oder mehr der Aufnahmen einer Soziation vertreten) sind und sich scharf von anderen, begleitenden Arten unterscheiden, (ii) sie haben ein bestimmtes Minimum-Areal im Sinne einer Aufnahmefläche, die groß genug ist um die konstanten Arten zu enthalten, und (iii) sie seien durch scharfe Grenzen von anderen Soziationen getrennt. Die Nachprüfung dieser Thesen zwang zur Aufgabe dieser Theorie.

Die schwedische Pflanzensoziologie ist jetzt eher eine Tradition in der Betrachtung der nordischen Vegetation als eine eigene Schule. Die vegetationskundliche Forschung in dieser Tradition beschäftigt sich mit der Analyse kleiner Aufnahmeflächen, mit der Schichtung, mit quantitativen Beziehungen und der Art-Dominanz. Während früher die Soziation als Einheit für Einzel-Untersuchungen verwendet wurde, wird dagegen heute das Interesse auf die mit der BRAUN-BLANQUET-Auffassung vergleichbaren Assoziationen konzentriert. Die schwedische Pflanzensoziologie hat sich im Allgemeinen wenig mit formaler Klassifikation beschäftigt, und verschiedene Möglichkeiten zur Gruppierung von höheren Einheiten wurden verwendet.

Andere skandinavische und baltische Arbeiten beschäftigen sich mit den Wald-Standorts-Typen in Finland (CAJANDER), Schichten- oder Synusium-Einheiten in Estland (LIPPMAA), der Klassifikation norwegischer Gebirgsvegetation (NORDHAGEN, DAHL), in Dänemark mit quantitativen Betrachtungen (RAUNKIAER) und dem Studium der Arten-Verteilung (BÖCHER), der Klassification durch klein-masstäbliche Tundra- 'Formationen' in Island (STEINDÓRSSON), und der Gradienten-Analyse vorgreifenden Arbeiten (CAJANDER & ILVESSALO, MØLHOLM HANSEN, SØRENSEN, TUOMIKOSKI). Die nordischen Schulen haben zwar kein allgemeines Klassifikations-System geschaffen, das ebenso erfolgreich gewesen wäre wie das von BRAUN-BLANQUET, aber sie haben durch die Entwicklung der Klassifikationen und das für die nordische Vegetation geeignete Studium der Gesellschaftsstruktur bedeutende Beiträge geleistet die in der ganzen Welt Einfluss ausgeübt haben.

Editor's note

To represent northern phytosociology in this book separate articles by Professors TRASS and MALMER were requested by the editor, and the general editor and publishers. The editor has shortened both articles and combined them into this version. Section 18.4 was contributed by MALMER and the rest of the article primarily by TRASS, while the editor also has contributed to some sections. The two authors deserve any credit, and the editor will accept any blame, for the result of this collaboration at a distance between Sweden and the Estonian S.S.R., Germany and The Netherlands, and Ithaca, New York.

REFERENCES

ALBERTSSON, N., – 1946 – Österplana hed, ett alvarområde på Kinnekulle. (Germ. summ.). *Acta phytogeogr. suec.* 20: 1—267.

ALBERTSSON, N., – 1950 – Das grosse südliche Alvar der Insel Öland: eine pflanzensoziologische Übersicht. *Svensk bot. Tidskr.* 44: 269—331.

ALEKSANDROVA, V. D., – 1965 – On the problem of distinguishing phytccoenoses in a vegetative continuum (In Russian with Engl. summ.) *Bot. Zh. SSSR* 50 (9):1248—1259.

ALEKSANDROVA, V. D., – 1969 – Classification of vegetation: Principles of classification and classification systems of various phytocoenological schools (In Russian). 'Nauka', Leningrad. 276 pp.

ALMBORN, O., – 1948 – Distribution and ecology of some South Scandinavian lichens. *Bot. Notiser Suppl.* 1 (2): 1—252.

ALMQUIST, E., – 1929 – Upplands vegetation och flora. *Acta phytogeogr. suec.* 1: 1—622.

ANDERSSON, F., – 1970 – Ecological studies in a Scanian woodland and meadow area, southern Sweden. I. Vegetational and environmental structure. *Op. bot. Soc. bot. Lund* 27: 1—190.

ANDERSSON, O., – 1950 – The Scanian sand vegetation—a survey. *Bot. Notiser* 1950: 145—172.

ARCHIBALD, E. E. A., – 1949 – The specific character of plant communities. II. A quantitative approach. *J. Ecol.* 37: 274—288.

ARNBORG, T., – 1940 – Der Vallsjö-Wald, ein nordschwedischer Urwald. *Acta phytogeogr. suec.* 13: 128—154.

ARNBORG, T., – 1943 – Granberget: En växtbiologisk undersökning av ett sydlappländskt granskogsområde med särskild hänsyn till skogstyper och föryngring. *Norrl. Handbibl.* 14: 1—282.

ARRHENIUS, O., – 1921 – Species and area. *J. Ecol.* 9: 95—99.

ASHBY, E., – 1936 – Statistical ecology. *Bot. Rev.* 2: 221—235.

BECKING, R. W. – 1957 – The Zürich-Montpellier school of phytosociology. *Bot. Rev.* 23: 411—488.

BERGLUND, B., – 1963 – Vegetation på ön Senoren. II. Landvegetationen. *Bot. Notiser* 116: 31—80.

BLOMGREN, N. & E. NAUMANN – 1925 – Untersuchungen über die höhere Vegetation des Sees Straken bei Aneboda. *Handl. K. Fysiogr. Sällsk. Lund*, N.F. 36 (6): 1—51, *Acta Univ. Lund.* Avd. 2, 21 (6): 1—51.

BÖCHER, T. W., – 1933 – Studies on the vegetation of the east coast of Greenland between Scoresby Sound and Angmagssalik (Christian IX. s Land). *Meddr Grønland* 104 (4): 1—132.

BÖCHER, T. W., – 1936 – Om en metode til undersøgelse af konstans, skudtaethed og homogenitet. *Bot. Tidsskr.* 43: 278—304.

BÖCHER, T. W. – 1938 – Biological distributional types in the flora of Greenland: a study on the flora and plant-geography of South Greenland and East Green-

land between Cape Farewell and Scoresby Sound. (Danish summ.) *Meddr. Grønland* 106 (2): 1—339.
Böcher, T. W., – 1943 – Studies on the plant geography of the North-Atlantic heath formation. II. Danish dwarf shrub communities in relation to those of northern Europe. *Biol. Skr.*, K. danske Vidensk. Selsk. 2 (7): 1—130.
Böcher, T. W., – 1945 – Beiträge zur Pflanzengeographie und Ökologie dänischer Vegetation. II. Über die Waldsaum- und Graskrautgesellschaften trockener und halbtrockener Böden der Insel Seeland mit besonderer Berücksichtigung der Strandabhänge und Strandebenen. (Engl. summ.) *Biol. Skr.*, K. danske Vidensk. Selsk. 4 (1): 1—163.
Böcher, T. W., – 1954 – Oceanic and continental vegetational complexes in Southwest Greenland. *Meddr Grønland* 148 (1): 1—336.
Böcher, T. W., T. Christensen & M. S. Christiansen, – 1946 – Slope and dune vegetation of North Jutland. I. Himmerland. *Biol. Skr.*, K. danske Vidensk. Selsk. 4 (3): 1—78.
Booberg, G., – 1930 – Gisselåsmyren. En växtsociologisk och utvecklingshistorisk monografi över en jämtländsk kalkmyr. *Norrl. Handbibl.* 12: 1—329.
Braun-Blanquet, J., – 1925 – Zur Wertung der Gesellschaftstreue in der Pflanzensoziologie. *Vjschr. naturf. Ges. Zürich* 70: 122—149.
Braun-Blanquet, J., – 1928a – Pflanzensoziologie: Grundzüge der Vegetationskunde. Springer, Berlin. 330 pp.
Braun-Blanquet, J., – 1928b – À propos d'associations végétales. *Archs Bot., Bull. mens.*, Caen 2 (4): 67—68.
Braun-Blanquet, J., – 1932 – Die Pflanzensoziologie in Forschung und Lehre. I. Pflanzensoziologische Forschungsprobleme. *Biologe* 1: 175—180, and *Communs Stn. int. Géobot. médit. alp.*, Montpellier 14.
Braun-Blanquet, J., – 1951 – Pflanzensoziologie: Grundzüge der Vegetationskunde. 2nd ed. Springer, Wien. 631 pp. 3rd ed., 1964, 865 pp.
Brockmann-Jerosch, H., – 1907 – Die Pflanzengesellschaften der Schweizeralpen. I. Die Flora des Puschlav (Bezirk Bernina, Kanton Graubünden) und ihre Pflanzengesellschaften. Leipzig, Englemann. 438 pp.
Cain, S. A., – 1934 – Studies on virgin hardwood forest: II. A comparison of quadrat sizes in a quantitative phytosociological study of Nash's Woods, Posey County, Indiana. *Am. Midl. Nat.* 15: 529—566.
Cain, S. A., – 1936 – Synusiae as a basis for plant sociological field work. *Am. Midl. Nat.* 17: 665—672.
Cain, S. A., – 1938 – The species-area curve. *Am. Midl. Nat.* 19: 573—581.
Cain, S. A., – 1943 – Sample-plot technique applied to alpine vegetation in Wyoming. *Am. J. Bot.* 30: 240—247.
Cajander, A. K., – 1903a (1906) – Beiträge zur Kenntnis der Vegetation der Alluvionen des nördlichen Eurasiens. I. Die Alluvionen des unteren Lena-Thales. *Acta Soc. Sci. fenn.* 32 (1): 1—182.
Cajander, A. K., – 1903b (1906) – Studien über die Vegetation des Urwaldes am Lena-Fluss. *Acta Soc. Sci. fenn.* 32 (3): 1—40.
Cajander, A. K., – 1903c – Beiträge zur Kenntnis der Vegetation der Hochgebirge zwischen Kittilä und Muonio. *Fennia* 20 (9): 1—37.
Cajander, A. K., – 1909 – Ueber Waldtypen. *Acta for. fenn.* 1 (1): 1—175.
Cajander, A. K., – 1922 – Zur Begriffsbestimmung im Gebiet der Pflanzentopographie. *Acta for. fenn.* 20 (2): 1—8.
Cajander, A. K., – 1949 – Forest types and their significance. *Acta for. fenn.* 56 (4): 1—71.
Cajander, A. K. & Y. Ilvessalo, – 1921 – Über Waldtypen II. *Acta for. fenn.* 20 (1): 1—77.
Dahl, E., – 1957 – Rondane: mountain vegetation in South Norway and its

relation to the environment. *Skr. norske Vidensk-Akad.*, Mat.-naturv. Kl., 1956 (3): 1—374.
DAHL, E., – 1960 – Some measures of uniformity in vegetation analysis. *Ecology* 41: 805—808.
DAHL, E. & E. HADAČ, – 1941 – Strandgesellschaften der Insel Ostøy im Oslofjord: eine pflanzensoziologische Studie. *Nytt Mag. Naturvid.* 82: 251—312.
DAHL, E. & E. HADAČ, – 1949 – Homogeneity of plant communities. *Studia bot. čechoslov.* 10: 159—176.
DAHLBECK, N. – 1945 – Strandwiesen am südöstlichen Öresund. *Acta phytogeogr. suec.* 18: 1—168.
DU RIETZ, G. E., – 1917 Några synpunkter på den synekologiska vegetationsbeskrifoningens terminologi och metodik. (Germ. summ.) *Svensk bot. Tidskr.* 11: 51—71.
DU RIETZ, G. E., – 1921 – Zur methodologischen Grundlage der modernen Pflanzensoziologie. *Akad. Abhandl., Uppsala.* Holzhausen, Wien. 267 pp.
DU RIETZ, G. E., – 1922 – Die Grenzen der Assoziationen: Eine Replik an John Frödin. *Bot. Notiser* 1922: 90—96.
DU RIETZ, G. E., – 1923 – Der Kern der Art- und Assoziationsprobleme. *Bot. Notiser* 1923: 235—256.
DU RIETZ, G. E., – 1924 – Studien über die Vegetation der Alpen, mit derjenigen Skandinaviens vergleichen. *Veröff. geobot. Inst. Rübel, Zürich* 1: 31—138.
DU RIETZ, G. E., – 1925a – Zur Kenntnis der flechtenreichen Zwergstrauchleiden im kontinentalen Südnorwegen. *Handl. Svenska Växtsociol. Sällsk.* 4: 1—80.
DU RIETZ, G. E., – 1925b – Die regionale Gliederung der skandinavischen Vegetation. *Handl. Svenska Växtsociol. Sällsk.* 8: 1—60.
DU RIETZ, G. E., – 1928 – Kritik an pflanzensoziologischen Kritikern. *Bot. Notiser* 1928: 1—30.
DU RIETZ, G. E., – 1929 – The fundamental units of vegetation. *Proc. 4th Int. bot. (Plant Sci.) Congr.*, Ithaca 1926, 1: 623—627.
DU RIETZ, G. E., – 1930a – Classification and nomenclature of vegetation. *Svensk bot. Tidskr.* 24: 489—503.
DU RIETZ, G. E., – 1930b (1932) – Vegetationsforschung auf soziationsanalytischer Grundlage. *Handb. biol. ArbMeth.* 11,5 (2): 293—480.
DU RIETZ, G. E., – 1932 – Zur Vegetationsökologie der ostschwedischen Küstenfelsen. *Beih. bot. Zbl.* 49 (Erg.-Bd.): 61—112.
DU RIETZ, G. E., – 1935 – Classification and nomenclature of vegetation units 1930—1935. *Proc. Zesde Int. bot. Congr.*, Amsterdam 1935, 2: 104—105.
DU RIETZ, G. E., – 1936 – Classification and nomenclature of vegetation units 1930-1935. *Svensk bot. Tidskr.* 30: 580—589.
DU RIETZ, G. E., – 1940 – Das limnologisch-thalassologische Vegetationsstufensystem. *Verh. int. Verein. theor. angew. Limnol.* 9: 102—110.
DU RIETZ, G. E., – 1942a – Växtgeografins grunder (Mimeographed). (This paper has been rewritten several times but is only partly printed as Du Rietz 1957c. It has been available for most authors dealing with Swedish vegetation and has influenced their work. For this paper the last edition from 1961 has been used.)
DU RIETZ, G. E., – 1942b – Rishedsförband i Torneträskområdets lågfjällsbälte. (Germ. summ.) *Svensk bot. Tidskr.* 36: 124—146.
DU RIETZ, G. E., – 1942c – Linné som fjällväxtgeograf. *Årsskr. Svenska Linné-Sällsk.* 25:
DU RIETZ, G. E., – 1945a – Nitella Nordstedtiana i två uppländska sjöar. *Svensk bot. Tidskr.* 39: 83—94.
DU RIETZ, G. E., – 1945b – Om fattigbark- och rikbarksamhällen. *Svensk bot. Tidskr.* 39: 147—150.

Du Rietz, G. E., – 1949 – Huvudenheter och huvudgränser i svensk myrvegetation. (Engl. summ.) *Svensk bot. Tidskr.* 43: 274—309.
Du Rietz, G. E., – 1954 – Die Mineralbodenwasserzeigergrenze als Grundlage einer natürlichen Zweigliederung der nord- und mitteleuropäischen Moore. *Vegetatio* 5/6: 571—585.
Du Rietz, G. E., – 1957a – Linnaeus as a phytogeographer. *Vegetatio* 7: 161—168.
Du Rietz, G. E., – 1957b – Linné som myrforskare. *Uppsala Univ. Årsskr.* 5: 1—80.
Du Rietz, G. E., – 1957c – Vegetation analysis in relation to homogeneousness and size of sample areas. *Compt. Rend. Rapp. Commun., Int. bot. Congr.*, Paris 1954, sect. 7—8: 24—40. (see also Du Rietz 1942a).
Du Rietz, G. E., T. C. E. Fries & T. Å. Tengwall, – 1918 – Vorschlag zur Nomenklatur der soziologischen Pflanzengeographie. *Svensk bot. Tidskr.* 12: 145—170.
Du Rietz, G. E., T. C. E. Fries, H. Osvald & T. Å. Tengwall, – 1920 – Gesetze der Konstitution natürlicher Pflanzengesellschaften. *Vetensk. prakt. Unders. Lappl.* (Luossavaara-Kiirunavaara Aktiebolag), Flora och Fauna 7: 1—47.
Du Rietz, G. E. & H. Gams, – 1924 – Zur Bewertung der Bestandestreue bei der Behandlung der Pflanzengesellschaften. *Vjschr. naturf. Ges. Zürich* 69: 269—280.
Du Rietz, G. E., A. G. Hannerz, G. Lohammar, R. Santesson, and M. Waern 1939 – Zur Kenntnis der Vegetation des Sees Tåkern. *Acta phytogeogr. suec.* 12: 1—65.
Ellenberg, H., – 1954 – Zur Entwicklung der Vegetationssystematik in Mitteleuropa. *Angew. PflSoziol.*, Wien, Festschr. Aichinger 1: 134—143.
Ellenberg, H., – 1956 – Grundlagen der Vegetationsgliederung I. Aufgaben und Methoden der Vegetationskunde. In, 'Einführung in die Phytologie' ed. H. Walter 4(1): 1—136. Ulmer, Stuttgart.
Faegri, K., – 1934 – Über die Längenvariationen einiger Gletscher des Jostedalsbre und die dadurch bedingten Pflanzensukzessionen. *Bergens Mus. Årbok*, Naturv. rekke 1933 (7): 1—255.
Faegri, K., – 1937 – Some recent publications on phytogeography in Scandinavia. *Bot. Rev.* 3: 425—456.
Fries, R. E., – 1950 – A Short History of Botany in Sweden. Almqvist & Wiksell, Uppsala. 162 pp.
Fries, T. C. E., – 1913 – Botanische Untersuchungen im nördlichsten Schweden: Ein Beitrag zur Kenntnis der alpinen und subalpinen Vegetation in Torne Lappmark. *Vetensk. prakt. Unders. Lappl.* (Luossavaara-Kiirunavaara Aktiebolag), Flora och Fauna 2: 1—361.
Frödin, J., – 1922 – Les limites des associations: une réponse à Einar Du Rietz. *Bot. Notiser* 1922: 149—154.
Gams, H., – 1918 – Prinzipienfragen der Vegetationsforschung: Ein Beitrag zur Begriffsklärung und Methodik der Biocoenologie. *Vjschr. naturf. Ges. Zürich* 63: 293—493.
Gams, H., – 1927 – Von den Follatères zur Dent de Morcles: Vegetationsmonographie aus dem Wallis. *Beitr. geobot. Landesaufn. Schweiz* 15: 1—760.
Gams, H., – 1933 – Die Stellung der Waldtypen im Vegetationssystem. *Forstarchiv* 9: 53—59.
Gillner, V., – 1960 – Vegetations- und Standortsuntersuchungen in den Strandwiesen der schwedischen Westküste. *Acta phytogeogr. suec.* 43: 1—198.
Gillner, V., – 1965 – Salt marsh vegetation in Southern Sweden. *Acta phytogeogr. suec.* 50: 97—104.
Gjaerevoll, O., – 1949 – Snøleievegetasjonen i Oviksfjellene. (Engl. summ.) *Acta phytogeogr. suec.* 25: 1—106.

GJAEREVOLL, O., – 1950 – The snow-bed vegetation in the surroundings of Lake Torneträsk, Swedish Lappland. *Svensk bot. Tidskr.* 44: 387—440.

GJAEREVOLL, O., – 1956 – The plant communities of the Scandinavian alpine snow-beds. *Skr. norske Vidensk-Akad.* 1956 (1): 1—405.

GJAEREVOLL, O. & K. G. BRINGER, – 1965 – Plant cover of the alpine regions. *Acta phytogeogr. suec.* 50: 257—268.

GLEASON, H. A., – 1926 – The individualistic concept of the plant association. *Bull. Torrey bot. Club* 53: 7—26.

GOODALL, D. W., – 1952 – Quantitative aspects of plant distribution. *Biol. Rev.* 27: 194—245.

GOODALL, D. W., – 1953 – Objective methods for the classification of vegetation. I. The use of positive interspecific correlation. *Aust. J. Bot.* 1: 39—63.

GOODALL, D. W., – 1954 – Minimal area: a new approach. *Compt. Rend. Rapp. Commun., Huit. Int. bot. Congr.*, Paris 1954, 7—8: 19—21.

GOODALL, D. W., – 1961 – Objective methods for the classification of vegetation. IV. Pattern and minimal area. *Aust. J. Bot.* 9: 162—196.

GOODALL, D. W., – 1963 – The continuum and the individualistic association. (French summ.) *Vegetatio* 11: 297—316.

GREIG-SMITH, P., – 1964 – Quantitative Plant Ecology. 2nd ed. Butterworths, London. 256 pp.

HALLBERG, H. P., – 1971 – Vegetation auf den Schalenablagerungen in Bohuslän, Schweden. *Acta phytogeogr. suec.* 56: 1—149.

HALLBERG, H. P. & R. IVARSSON, – 1965 – Vegetation of coastal Bohuslän. *Acta phytogeogr. suec.* 50: 111—122.

HANSEN, H. MØLHOLM, – 1930 – Studies on the vegetation of Iceland. In, 'The Botany of Iceland,' ed. J. L. A. KOLDERUP ROSENVINGE & E. WARMING 3 (pt. I, no. 10): 1—186. Frimodt, Copenhagen.

HANSEN, H. MØLHOLM, – 1932 – Nørholm Hede, en formationsstatistisk Vegetationsmonografi. (Engl. summ.) *K. danske Vidensk. Selsk. Skr.*, Nat. Math. Afd., Ser. 9, 3 (3): 99—196.

HAVAS, P. – 1961 – Vegetation und Ökologie der ostfinnischen Hangmoore. *Ann. bot. Soc. Zool.-Bot. Fenn. 'Vanamo' (Suomal. eläin-ja kasvit. Seur. van. kasvit. Julk.)* 31 (2): 1—188.

HULT, R., – 1881 – Försök till analytisk behandling av växtformationerna. *Meddn Soc. Fauna Flora fenn.* 8: 1—155.

HULT, R., – 1885 – Blekinges vegetation: ett bidrag till växtformationernas utvecklingshistoria. *Meddn Soc. Fauna Flora fenn.* 12: 163—252.

IVARSSON, R. – 1962 – Lövvegetationen i Mollösunds socken. (Germ. summ.) *Acta phytogeogr. suec.* 46: 1—197.

IVARSSON, R., – 1971 – Lövvegetationen på Lindö och Kalvö i norra Bohuslän. I. Allmän översikt. *Svensk bot. Tidskr.* 65: 1—38.

IVERSEN, J., – 1936 – Biologische Pflanzentypen als Hilfsmittel in der Vegetationsforschung: ein Beitrag zur ökologische Charakterisierung and Anordnung der Pflanzengesellschaften. *Meddr Skalling-Lab.*, København 4: 1—224.

JULIN, E., – 1948 – Vessers udde, mark och vegetation i en igenväxande löväng vid Bjärka-Säby. (Germ. summ.) *Acta phytogeogr. suec.* 23: 1—186.

KAARET, P. – 1953 – Wasservegetation der Seen Orlången und Trehörningen. *Acta phytogeogr. suec.* 32: 1—64.

KALELA, A., – 1939 – Über Wiesen und wiesenartige Pflanzengesellschaften auf der Fischerhalbinsel in Petsamo Lappland. *Acta for. fenn.* 48 (2): 1—523.

KALLIOLA, R., – 1939 – Pflanzensoziologische Untersuchungen in der Alpinen Stufe Finnisch-Lapplands. *Ann. bot., Soc. Zool.-Bot. Fenn. 'Vanamo' (Suomal. eläin-ja kasvit. Seur. van. kasvit. Julk.)* 13 (2): 1—328.

KATZ, N. J., – 1928 – Zur Kenntnis der Niedermoore im Norden des Moskauer Gouvernments. *Beih. Repert. Spec. nov. Regni veg.* 56: 1—79.

KIELLAND-LUND, J., – 1967 – Zur Systematik der Kiefernwälder Fennoscandiens. *Mitt. flor.-soz. ArbGemein.* N.F., Stolzenau 11/12: 127—141.

KONOVALOV, N. A., – 1935 – On representative area of some oak associations (In Russian), *Trudy leningr. Obshch. Estest.* 64 (2): 40—55.

KONOVALOV, N. A. & V. A. POVARNITZYN, – 1927 – On the methods of statistical phytosociological study of forest associations (In Russian). *Izv. Lesn. Inst.* 35: 1—20.

KRUSENTJERNA, E. VON, – 1945 – Bladmossvegetation och bladmossflora i Uppsalatrakten. (Engl. summ.) *Acta phytogeogr. suec.* 19: 1—250.

KRUSENSTJERNA, E. VON, – 1965 – The growth on rock. *Acta phytogeogr. suec.* 50: 144—148.

KYLIN, H., – 1923 – Växtsociologiska randanmärkningar. *Bot. Notiser* 1923: 161—234.

KYLIN, H., – 1926 – Über Begriffsbildung und Statistik in der Pflanzensoziologie. *Bot. Notiser* 1926: 81—180.

LAGERBERG, T., – 1916 – Markflorans analys på objektiv grund. *Meddn St. SkogsförsksAnst.*, Stockholm 11: 129—200 & XV—XXIV.

LENOBLE, F., – 1926 (1927) – À propos des associations végétales. *Bull. Soc. bot. Fr.* 73: 873—893.

LENOBLE, F. – 1928 – Associations végétales et espèces. *Archs Bot., Bull. mens.*, Caen 2 (1): 1—14.

LEVY, E. B. & E. A. MADDEN, – 1933 – The point method of pasture analysis. *N.Z. Jl. Agric.* 46: 267—279.

LINDGREN, L., – 1970 – Beech forest vegetation in Sweden—a survey. *Bot. Notiser* 123: 401—424.

LINDQUIST, B., – 1931 – Den skandinaviska bokskogens biologi. (Engl. summ.) *SkogsvFör. Tidskr.*, Stockholm 29: 179—532.

LINDQUIST, B., – 1938 – Dalby Söderskog: en skånsk lövskog i forntid och nutid. (Germ. summ.) *Acta phytogeogr. suec.* 10: 1—273.

LINDQUIST, B., – 1954 – Ein Waldtypenschema für die skandinavischen Buchenwälder. *Angew. PflSoziol.*, Wien, Festschr. Aichinger 2: 965—970.

LIPPMAA, T., – 1931 – Pflanzensoziologische Betrachtungen. *Acta Inst. Horti bot. Univ. tartu.* 2 (3/4): 1—32, and *Loodusuur. Seltsi Aruanded* 38 (1/2): 1—32.

LIPPMAA, T., – 1933 – Taimeühingute uurimise metoodika ja Eesti taimeühingute klassifikatsiooni põhijooni (Eston. with Germ. summ.: Grundzüge der pflanzensoziologischen Methodik nebst einer Klassifikation der Pflanzenassoziationen Estlands) *Acta Inst. Horti bot. tartu* 3 (4): 1—169, and *Loodusuur. Seltsi Aruanded* 40 (1/2): 1—169.

LIPPMAA, T., – 1935 – Une analyse des forêts de l'île estonienne d'Abruka (Abro) sur la base des associations unistrates. *Acta Inst. Horti bot. tartu.* 4 (1/2, art. 5): 1—97, and *Acta Commen'. Univ. tartu.* A 28 (1): 1—97.

LIPPMAA, T., – 1938 – Areal und Altersbestimmung einer Union (Galeobdolon-Asperula-Asarum-U.) sowie das Problem der Charakterarten und Konstanten. *Acta Inst. Horti bot. tartu.* 6 (2/3): 1—152.

LIPPMAA, T., – 1939 – The unistratal concept of plant communities (the unions). *Am. Midl. Nat.* 21: 111—145.

LOHAMMAR, G., – 1965 – The vegetation of Swedish lakes. *Acta phytogeogr. suec.* 50: 23—47.

LÜDI, W., – 1928 – Der Assoziationsbegriff in der Pflanzensoziologie, erläutert am Beispiel der Pflanzengesellschaften des Tansbodengebietes im Lauterbrunnental. *Biblthca bot.* 96: 1—93.

Luther, H., – 1949 – Vorschlag zu einer ökologischen Grundeinteilung der Hydrophyten. *Acta bot. fenn.* 44: 1—15.

Malmer, N., – 1960 – Some ecological studies on lakes and brooks in the South Swedish uplands. *Bot. Notiser* 113: 87—116.

Malmer, N., – 1962 – Studies on mire vegetation in the archean area of southwestern Götaland (South Sweden). I. Vegetation and habitat conditions on the Åkhult mire. *Op. bot. Soc. bot. Lund* 7 (1): 1—322.

Malmer, N., – 1965 – The southern mires. *Acta phytogeogr. suec.* 50: 149—158.

Malmer, N., – 1968 – Über die Gliederung der Oxycocco-Sphagnetea und Scheuchzerio-Caricetea fuscae: Einige Vorschläge mit besonderer Berücksichtigung der Verhältnisse in S-Schweden. (Engl. summ.) In, 'Pflanzensoziologische Systematik,' ed. R. Tüxen, *Ber. Symp. int. Vereinig. Vegetationskunde*, Stolzenau/ Weser 1964, 8: 293—305.

Malmer, N., B. E. Berglund, J. Ericson, L. Påhlsson, & G. Rasmusson, – 1965 – The south-western dwarf shrub heaths. *Acta phytogeogr. suec.* 50: 123—130.

Malmström, C., – 1923 – Degerö stormyr. En botanisk, hydrologisk och utvecklingshistorisk undersökning över ett nordsvenskt myrkomplex. *Meddn St. SkogsförsAnst.*, Stockholm 20: 1—206.

Malmström, C., – 1937 – Tönnersjöhedens försökspark i Hallandt. Et bidrag till kännedomen om sydvästra Sveriges skogar, ljunghedar och torvmarker. *Meddn St. SkogsförsAnst.*, Stockholm 30: 323—528.

Masing, V., – 1969 – Structural analysis of plant cover and classification problems. In, 'Plant Taxonomy, Geography and Ecology in the Estonian S.S.R.' pp. 49—59. 'Valgus,' Tallinn.

Masing, V. & H. Trass, – 1963 – Elaboration of some theoretical problems in the works of Estonian geobotanists (In Russian with Engl. summ.). *Bot. Zh. SSSR* 48 (4): 473—485.

Matveyeva, E. A., – 1967 – The Meadows of Soviet Pribaltic (Comparative Analysis). 'Nauka,' Leningrad. 336 pp.

Mc Intosh, R. P., – 1967 – The continuum concept of vegetation. *Bot. Rev.* 33: 130—187.

Melin, E., – 1917 – Studier över de norrländska myrmarkernas vegetation med särskild hänsyn till deras skogsvegetation efter torrläggning. *Norrl. Handbibl.* 7: 1—426. Almqvist & Wicksells, Uppsala.

Mirkin, B. M., – 1965 – On the ecological classifications of meadow vegetation in foodplains (In Russian with Engl. summ.). *Bot. Zh. SSSR* 50 (3): 324—334.

Mirkin, B. M., – 1968 – The criteria of dominants and determinants in the classification of phytocoenoses (In Russian with Engl. summ.). *Bot. Zh. SSSR* 53 (6): 767—778.

Mörnsjö, T., – 1969 – Studies on vegetation and development of a peatland in Scania, south Sweden. *Op. bot. Soc. bot. Lund* 24: 1—187.

Negri, G., – 1914 – Le unità ecologiche fondamentali in fitogeografia. *Atti (R.) Accad. Sci., Torino* 49: 1089—1105, 1174—1198.

Nichols, G. E., – 1929 – Plant associations and their classification. *Proc. 4th Int. bot. (Plant Sci.) Congr.*, Ithaca 1926, 1: 629—641.

Nilsson, A., – 1902 – Svenska växtsamhällen. *Tidskr. Skogshushållning* 30: 127—147.

Nitzenko, A. A., – 1961 – On the phytotopological classification of the plant cover (In Russian). *Trudy Inst. Biol., Ural. fil., Sverdlovsk* 27: 29—37.

Nitzenko, A. A., – 1969 – The problem of continuity and intermittence of vegetation cover (In Russian with Engl. summ.). *Zh. obshch. Biol.* 30 (4): 387—397.

NORDHAGEN, R., – 1923 – Vegetationsstudien auf der Insel Utsire im westlichen Norwegen. *Bergens Mus. Årb.*, Naturv. rekke 1920—21 (1): 1—149.

NORDHAGEN, R., – 1924 – Om homogenitet, konstans och minimiareal. *Nytt Mag. Naturvid.* 61: 1—51.

NORDHAGEN, R., – 1928 – Die Vegetation und Flora des Sylenegebietes I. Die Vegetation. *Skr. norske Vidensk-Akad.*, Mat.-naturv. Kl. 1927 (1): 1—612.

NORDHAGEN, R., – 1937 – Versuch einer neuer Einteilung der subalpinen-alpinen Vegetation Norwegens. *Bergens Mus. Årb.*, Naturv. rekke 1936 (7): 1—88.

NORDHAGEN, R., – 1943 – Sikilsdalen og Norges Fjellbeiter: en plantesosiologisk monografi. *Bergens Mus. Skr.* 22: 1—607.

NORDHAGEN, R., – 1954 – Vegetation units in the mountain areas of Scandinavia. *Veröff. geobot. Inst. Rübel, Zürich* 29: 81—95.

NOSKOVA, T. A., – 1928 – Minimum area in the forest associations (In Russian). *Dnevnik Vses. s'esda bot.* : 253—254.

OSTENFELD, C. H., – 1889 – Skildringer af Vegetationen i Island. I-II. *Bot. Tidsskr.* 22: 227—253.

OSTENFELD, C. H. – 1905 – Skildringer af Vegetationen i Island. III-IV. *Bot. Tidsskr.* 27: 111—122.

OSVALD, H., – 1923 – Die Vegetation des Hochmoores Komosse. *Handl. Svenska Växtsociol. Sällsk.* 1: 1—436.

OSVALD, H., – 1925 – Zur Vegetation der ozeanischen Hochmoore in Norwegen. *Handl. Svenska Växtsociol. Sällsk.* 7: 1—106.

PAVILLARD, J., – 1927 – Les tendences actuelles de la phytosociologie. *Archs Bot., Bull. mens.*, Caen 1 (6): 89—112.

PAVILLARD, J., – 1928 – Espèces et associations. *Archs Bot., Bull. mens.*, Caen 2 (4): 68—72.

PAVILLARD, J., – 1935 – The present status of the plant association. *Bot. Rev.* 1: 210—232.

PEARSALL, W. H., – 1924 – The statistical analysis of vegetation: a criticism of the concepts and methods of the Upsala school. *J. Ecol.* 12: 135—139.

PERSSON, Å., – 1961 – Mire and spring vegetation in an area north of Lake Torneträsk, Torne Lappmark, Sweden. I. Description of the vegetation. *Op. bot. Soc. Bot. Lund* 6 (1): 1—187.

PERSSON, Å., – 1965 – Mountain mires. *Acta phytogeogr. suec.* 50: 249—256.

PETTERSSON, B., – 1958 – Dynamik och konstans i Gotlands flora och vegetation. *Acta phytogeogr. suec.* 40: 1—288.

POORE, M. E. D., – 1956 – The use of phytosociological methods in ecological investigations. IV. General discussion of phytosociological problems. *J. Ecol.* 44: 28—50.

POST, H. VON, – 1851 – Om vextgeografiska skildringer. *Bot. Notiser* 1851: 110—127, 161—187.

RAMENSKY, L. G., – 1915 – The problem of quantitative study of herb cover (In Russian). *Materialy po organiz. i kulture kormov. ploshchadi* 12: 105—140.

RAMENSKY, L. G., – 1918 – Study of meadows in Voronesh gouvernements (In Russian). *Mat. po jest-ist. issl. Voron. gub.* : 63—93.

RAMENSKY, L. G., – 1924 – Main regularities of the plant cover and their study (In Russian). *Vêstnik opytnogo dêla*, Voronezh, pp. 37—73.

RAMENSKY, L. G., – 1937 – Calculation and Description of the Vegetation (on the Basis of Projection Method) (In Russian). Moscow. 100 pp.

RAMENSKY, L. G., – 1938 – Introduction to the Complex Pedologic-Geobotanic Investigation of Landscapes (In Russian). Selkhozgiz, Moscow. 620 pp.

RAMENSKY, L. G., – 1952 – On some principal positions in contemporary geobotany (In Russian). *Bot. Zh. SSSR* 37 (2): 181—201.

RAUNKIAER, C., – 1909-10 – Formationsundersøgelse og Formationsstatistik. *Bot. Tidsskr.* 30: 20—132.
RAUNKIAER, C., – 1910 – Statistik der Lebensformen als Grundlage für die biologischen Pflanzengeographie. *Beih. bot. Zbl.*, Abt. 2, 27: 171—206d.
RAUNKIAER, C., – 1912 – Measuring apparatus for statistical investigations of plant-formations. *Bot. Tidsskr.* 33: 45—48.
RAUNKIAER, C., – 1918 – Recherches statistiques sur les formations végétales. *Biol. Meddr*, K. danske Vidensk. Selsk. 1 (3): 1—80.
RAUNKIAER, C., – 1928 – Dominansareal, Artstaethed og Formationsdominanter. *Biol. Meddr*, K. danske Vidensk. Selsk. 7 (1): 1—47.
RAUNKIAER, C., – 1934 – The Life Forms of Plants and Statistical Plant Geography. Clarendon, Oxford. 632 pp.
RÜBEL, E., – 1925 – Betrachtung über einige pflanzensoziologische Auffassungsdifferenzen: Verständigungsbeitrag Schweden-Schweiz. *Veröff. geobot. Inst. Rübel, Zürich.* Beibl. 2: 1—12.
RÜBEL, E., – 1927 – Einige skandinavische Vegetationsprobleme. *Veröff. geobot. Inst. Rübel, Zürich* 4: 19—41.
RÜBEL, E., – 1930 – Pflanzengesellschaften der Erde. Huber, Bern-Berlin. 464 pp.
SAMUELSSON, G., – 1917a – Studien über die Vegetation der Hochgebirgsgegenden von Dalarne. *Nova Acta. R. Soc. Scient. upsal.*, Ser. 4, 4 (8): 1—252.
SAMUELSSON, G., – 1917b – Studien über die Vegetation bei Finse im inneren Hardanger. *Nytt Mag. Naturvid.* 55: 1—108.
SCHALYT, M. S., – 1935 – Rule of constancy and minimum area in steppes of the U.S.S.R. (In Russian). *Sov. Bot.* 1: 8—36.
SCHENNIKOV, A. P., – 1937 – Theoretical geobotany during last 20 years (In Russian). *Sov. Bot.* 5: 58—94.
SERNANDER, R., – 1894 – Studier öfver den gotländska vegetationens utvecklingshistoria. *Diss., Uppsala.* 112 pp.
SERNANDER, R., – 1898 – Studier öfver vegetationen i mellersta Skandinaviens fjälltrakter. 1. Om tundraformationer i svenska fjälltrakter. *Öfvers. Förhandl., K. (Svenska) Vetensk-Akad.* 6: 325—367.
SHIMWELL, D. W., – 1971 – Description and Classification of Vegetation. Sigdwick & Jackson, London. 322 pp.
SJÖGREN, E., – 1961 – Epiphytische Moosvegetation in Laubwäldern der Insel Öland (Schweden). *Acta phytogeogr. suec.* 44: 1—149.
SJÖGREN, E., – 1964 – Epilithische und epigäische Moosvegetation in Laubwäldern der Insel Öland (Schweden). *Acta phytogeogr. suec.* 48: 1—184.
SJÖRS, H., – 1948 – Myrvegetation i Bergslagen. (Engl. summ.) *Acta phytogeogr. suec.* 21: 1—299.
SJÖRS, H., – 1954 – Slåtterängar i Grangärde finnmark. (Engl. summ.) *Acta phytogeogr. suec.* 34: 1—135.
SJÖRS, H., – 1956 – Nordisk växtgeografi. Scandinav. Univ. Books, Oslo, København, Stockholm, Helsingfors. 229 pp. 2nd ed. 1967, 240 pp.
SJÖRS, H., – 1965 – Forest regions. *Acta phytogeogr. suec.* 50: 48—63.
SJÖRS, H., F. BJÖRKBÄCK & NORDQVIST, – 1965 – Regional ecology of mire sites and vegetation. *Acta phytogeogr. suec.* 50: 180—188.
SMITH, H., – 1920 – Vegetationen och dess utvecklings-historia det centralsvenska högfjällsområdet. *Norrl. Handbibl.* 9: 1—238. Almqvist & Wiksells, Uppsala.
SONESSON, M. – 1970 – Studies on mire vegetation in the Torneträsk area, Northern Sweden. III. Communities of the poor mires. *Op. bot. Soc. bot. Lund* 26: 1—120.
SØRENSEN, T. A., – 1948 – A method of establishing groups of equal amplitude in plant sociology based on similarity of species content and its application to

analyses of the vegetation on Danish commons. *Biol. Skr.*, K. danske Vidensk. Selsk. 5 (4): 1—34.

STEEN, E., – 1954 – Vegetation och mark i en uppländsk beteshage med särskild hänsyn till betesgångens inverkan. *Meddn St. JordbrFörs.*, Stockholm 49: 1—146

STEINDÓRSSON, S., – 1936 – Om Vegetationen paa Melrakkasljetta i det nordøstlige Island. *Bot. Tidsskr.* 43: 436—483.

STEINDÓRSSON, S., – 1945 – Studies on the vegetation of the central highland of Iceland. In, 'The Botany of Iceland,' ed. J. L. A. KOLDERUP ROSENVINGE & E. WARMING 3 (pt. 4, no. 14): 345—547. Munksgaard, Copenhagen.

STEINDÓRSSON, S., – 1946 – Contributions to the plant-geography and flora of Iceland. IV. The vegetation of Ísafjarðardjúp, northwest Iceland. *Acta nat. islandica* 1 (3): 1—32.

STEINDÓRSSON, S., – 1954 – The coastline vegetation at Gásar in Eyjafjörður in the North of Iceland. *Nytt Mag. Bot.* 3: 203—212.

STERNER, R., – 1922 – The continental element in the flora of South Sweden. *Geogr. Annlr* 1922: 221—444.

SUKATSCHEV, V. N., – 1964 – Fundamental concepts in forest biogeocoenology (In Russian). In, 'Osnovy lesn. biogeotsenologii' (Fundamentals of Forest Biogeocoenology), pp. 5—49. 'Nauka,' Moscow.

TENGWALL, T. Å., – 1920 – Die Vegetation des Sarakgebietes. I. In, 'Naturwiss. Untersuch. des Sarakgebirges in Schwed.-Lappland, geleitet von Dr. Axel Hamberg' 3 (4): 269—436. Fritzes, Stockholm.

THUNMARK, S., – 1931 – Der See Fiolen und seine Vegetation. *Acta phytogeogr. suec.* 2: 1—198.

TRASS, H., – 1963a – Probleme des Vegetationsklassifikationen der waldlosen Nieder- und Übergangsmoore Estlands (In Russian with Germ.summ.). *Tartu Riikl. Ülik. Toim.* 145, Bot.-al. Tööd 7: 60—73.

TRASS, H., – 1963b – On the typology of the dominants of plant communities (In Russian with Engl. summ.). *Byull. mosk. Obshch. Ispyt. Prir.*, otd. biol. 68 (5): 29—36.

TRASS, H., – 1964 – Application and validity of the synusial method in phytocoenology (In Russian with Engl. summ.). In, 'Izuchenie rastit. ostrova Saaremaa,' pp. 82—111. Tartu.

TRASS, H., – 1965a – Trends of development in the Estonian ecology and some theoretical problems (In Finnish). *Luonnon Tutk.* 69 (3): 110—120.

TRASS, H., – 1965b – On significance of dominants of plant communities in classification of vegetation (In Russian). *Problemy sovrem. Bot.* 1: 247—250.

TRASS, H., – 1966 – On vegetation discontinuity and continuity (In Russian with Engl. summ.). *Trudy mosk. Obshch. ispyt. Prir.*, Otd. biol. 27: 167—182.

TRASS, H., – 1970 – Coenoelements in plant communities (In Russian with Engl. summ.). *Trudy mosk. Obshch. ispyt. Prir.*, Otd. biol. 38: 184—193.

TUOMIKOSKI, R., – 1942 – Untersuchungen über die Vegetation der Bruchmoore in Ostfinnland. I. Zur Methodik der pflanzensoziologischen Systematik. (Finnish summ.). *Ann. bot. Soc. Zool.-Bot. Fenn. 'Vanamo'* (*Suomal. eläin-ja kasvit. Seur. van. kasvit. Julk.*) 17 (1): 1—203.

TÜXEN, R., – 1970 – Einige Bestandes- und Typenmerkmale in der Struktur der Pflanzengesellschaften. (Engl. summ.). In, 'Gesellschaftsmorphologie (Strukturforschung),' ed. R. Tüxen, *Ber. Symp. int. Vereinig. Vegetationskunde*, Rinteln 1966, 10: 76—107.

TYLER, G., – 1969a – Studies in the ecology of Baltic sea-shore meadows. II. Flora and vegetation. *Op. bot. Soc. bot. Lund* 25: 1—101.

TYLER, G., – 1969b – Regional aspects of Baltic shore-meadow vegetation. *Vegetatio* 19: 60—86.

URANOV, A. A., – 1925 – Materialien zu einer phytosoziologischen Beschreibung der Hegesteppe in Gouvernement Pensa im Lichte des Gesetzes der Konstant (In Russian with Germ. summ.). Otd. ohr. prir. glavnauki NKP 7, Tr. no izutsch. zapov. : 1—40.

URANOV, A. A., – 1966 – Number of species and area (In Russian with Engl. summ.). *Trudy mosk. Obshch. ispyt. Prir.*, Otd. biol. 27: 183—204.

VAGA, A., – 1940 – Fütotsönoloogia põhiküsimusi (Eston. with Engl. summ.: On some fundamental problems in phytocoenology). *Acta Inst. Horti bot. tartu.* 7 (1/2): 1—152, and *Acta Comment Univ. tartu.*, Ser. A, 35 (6): 1—152.

VASSILEVITSCH, V. I., – 1966 – A study of vegetational continua (In Russian with Engl. summ.). *Trudy mosk. Obshch. ispyt. Prir.*, Otd. biol. 27: 59—69.

VESTAL, A. G., – 1949 – Minimum areas for different vegetations: their determination from species-area curves. *Illinois biol. Monogr.* 20 (3): 1—129.

WALDHEIM, S., – 1944a – Die Torfmoosvegetation der Provinz Närke. *Handl. K. fysiogr. Sällsk. Lund*, N.F. 55 (6): 1—91, *Acta Univ. Lund*, N.F., Avd. 2, 40 (6): 1—91.

WALDHEIM, S., – 1944b – Mossvegetationen i Dalby-Söderskogs nationalpark: ett bidrag till kännedomen om Skånes bryofytvegetation. *K. svenska Vetensk. Akad. Avh. Naturskydd.* 4: 1—142.

WALDHEIM, S., – 1947 – Kleinmoosgesellschaften und Bodenverhältnisse in Schonen. *Bot. Notiser Suppl.* 1 (1): 1—203.

WANGERIN, W., – 1925 – Beiträge zur pflanzensoziologischen Begriffsbildung und Terminologie. I. Die Assoziation. *Beih. Repert. Spec. nov. Regni veg.* 36: 3—59

WARÉN, H., – 1926 – Untersuchungen über sphagnumreiche Pflanzengesellschaften der Moore Finlands unter Berücksichtigung der soziologischen Bedeutung der einzelnen Arten. *Acta Soc. Fauna Flora fenn.* 55 (8): 1—133.

WARMING, E., – 1895 – Plantesamfund: Grundtraek af den økologiske Plantegeografi. Philipsens, Kjøbenhavn. 335 pp.

WARMING, E., – 1896 – Lehrbuch der ökologischen Pflanzengeographie: Eine Einführung in die Kenntnis der Pflanzenvereine. Borntraeger, Berlin. 412 pp.

WARMING, E., – 1909 – Oecology of Plants: An Introduction to the Study of Plant-Communities. Oxford Univ. Press, Oxford. 422 pp.

WASSÉN, G., – 1966 – Gardiken: Vegetation und Flora eines lappländischen Seeufers. *K. svenska Vetensk. Akad. Avh. Naturskydd.* 22: 1—142.

WHITTAKER, R. H., – 1956 – Vegetation of the Great Smoky Mountains. *Ecol. Monogr.* 26: 1—80.

WHITTAKER, R. H., – 1962 – Classification of communities. *Bot. Rev.* 28 (1): 1—239.

WHITTAKER, R. H., – 1967 – Gradient analysis of vegetation. *Biol. Rev.* 42: 207—264.

19 NUMERICAL CLASSIFICATION

David W. Goodall

Contents

19. NUMERICAL METHODS OF CLASSIFICATION

19.1 Introduction

Though most classification of vegetation has been based on more or less subjective approaches (see, for instance, the review by Whittaker, 1962), there has been an increasing tendency in the past two or three decades to use numerical methods. This has largely been due to the greater objectivity that could be claimed for them. Let it be understood that this objectivity does not imply that an element of choice does not enter into the selection of methods (Lambert & Dale 1964); it means that, given the same stands and the same data from them, a procedure can be defined which can be applied unequivocally by anyone who understands it, with the assurance that the same result will be obtained. In other words, subjectivity is removed from an important part of the whole operation.

The development of numerical methods for classifying vegetation has been greatly encouraged by the development and ready availability of digital computers, which have taken much of the drudgery out of the extensive calculations often involved. The rapid development of numerical taxonomy over the past fifteen years (see Sokal & Sneath 1963) has contributed to the development of parallel methods in plant ecology. Numerical methods for the classification of vegetation are discussed in a number of recent reviews and monographs, including Williams & Dale (1965), Gounot (1969), Pielou (1969), Vasilevich (1969), and Goodall (1970).

19.2 General Concepts

19.2.1 Classification Principles

Classes can be defined by extension or intension — by a listing of the items included, or a definition of their range of variation. Except in the higher echelons (to which I shall return later), extensive classification has no place in the classification of vegetation — or indeed in biological classifications of any sort. The number of members, or potential members, of any class is virtually unlimited, and in consequence it is only through their attributes (in the broadest sense) that the class membership can be defined. In general, the classification aims at defining a set of classes in which members of the same class shall be 'as alike as possible,' members of different classes 'as unlike as possible,' to quote WILLIAMS & DALE (1965). And in vegetation, as in most biological subject matter, similarity is a matter of degree. No class of interest is likely to be homogeneous — it will only be less heterogeneous than the set of objects as a whole.

By dividing an assemblage of objects into classes, one has the possibility of making a wide variety of statements about them with great economy. By saying that an object in the room falls into the class *dog*, one saves oneself the trouble of specifying that it has four legs and a tail, claws, teeth, fur, warm blood and all the other features which distinguish members of that class from the classes *table*, *carpet*, *lamp*, *man*, *canary* and so forth. A special-purpose classification ('external classification' — WILLIAMS & DALE, 1965) is directed to prediction of a limited and specified range of attributes; others are of interest only in so far as they facilitate the identification of the class to which an unknown object belongs. A general-purpose classification ('internal classification' — WILLIAMS & DALE, 1965), at which one is usually aiming in biology, is intended to predict as wide a range of attributes as possible.

In classifying vegetation, the attributes one wishes to predict certainly include the floristic composition (qualitative and quantitative) and many features of the abiotic environment — climate and soil. The vegetational characteristics should be predictable at any season from the class membership. Predictions concerning the zoological and microbiological components of the ecosystem may also be needed. Thus a general-purpose classification is usually the aim.

It should be remembered that a general-purpose classification, like the proverbial jack-of-all-trades, is master of none — it is suboptimal for the prediction of any particular variable. An optimal general-purpose classification is optimal only in relation to a particular weighted function of the various variables to be predicted; if the weights are changed, the optimal classification may well be different.

In the primary classification of vegetation samples or stands, we are not interested in the samples in their own right. We are interested in them only *qua* samples — as representatives of a hypothetical larger population of similar stands or parts of the landscape that might equally well have been sampled — and our conclusions about their classification are of interest only if the conclusions can be applied to the population as a whole. This means that an extensive classification is not wanted except in so far as it can be translated into intensive terms, and hence become applicable to the parent population.

Once a set of primary classes has been established, the situation becomes different. A further (secondary) classification of these classes is concerned with this complete definable and delimited population of classes, and in consequence sampling questions do not arise. Thus, at this level an extensive classification is quite appropriate, and it is not essential that it should be expressed in terms of attribute values.

It is sometimes said that stands or samples of vegetation are 'over-defined' — that there is much 'redundancy' in their description; the observations made on them are more numerous than the minimum theoretically needed to allot them among a limited number of classes. This implies, however, a failure to distinguish the problem of allocation from that of classification. It is true than the number of variables and their alternative values by which a vegetation sample could be defined is far greater than are ever likely to be measured, and far greater than are needed to distinguish it from any others in the set. If, however, the set of samples is regarded as representative of a population, the same principles apply as in sampling practice generally — every additional sample contributes to and improves the estimation of population parameters, and there is no true redundancy. A particular classification procedure may fail to make use of some part of the data; these data may indeed then be regarded as redundant for that procedure—but this implies that the procedure is sub-optimal, not that the stands are over-defined.

19.2.2 The Geometrical Model of Classification

If one considers the set of stands to be classified as a set of points in a multidimensional space, the axes of which represent the different variables specifying the stands, it is clear that any subdivision of this space generates a classification. The stands within each cell thus delimited constitute a class (and, indeed, any classification of the stands (assumed intensive) can be so represented). The value of such a classification will then depend on the precision with which the values of the variables can be predicted for an arbitrary stand of which nothing is known but its class membership. Where classification is based on presence or absence of one or more species, or on the dominance of a particular species (in which case the dividing hypersurfaces are perpendicular to one of the axes), the underlying assumption is that numerous other variables are correlated with these species, so that class allocation by this one (or these few) species will mean that many other variables tend to be defined more or less precisely at the same time. In the same way, in a successful ordination (see articles 10, 11), the main axes are highly correlated with many of the variables, so that a division into classes by hyperplanes at right angles to the main ordination axes would permit high predictability for many variables.

In the preceding paragraph no account has been taken of the distribution in the hyperspace of the points representing stands. They may, with suitable scaling of the variables, be distributed in a hypersphere. This is quite improbable in a practical situation, for it would imply that the variables are uncorrelated. Much more likely is a hyperellipsoid with its axes inclined to those of the coordinate system, or some more complicated hypersolid. Wherever the points are clustered around a single centre, with their density decreasing progressively away from it, there is no 'natural' subdivision, and a choice of hypersurfaces dividing the stands into classes must be arbitrary. It may be, however, that the points representing the stands are distributed in several clusters, each separately hyperspheroidal or hyperellipsoidal, and separated by zones where the density of points is relatively low. In such a case, division by hypersurfaces passing through these zones of low density would give a classification which would have a claim to being 'natural' in some sense, and which would, on average, be expected to give greater predictability than any other arbitrary division into the same number of classes. (This appeals intuitively, but no demonstration of its truth has ever been seen; and indeed an unequivocal demonstration of the existence, in empirical data, of the partial discontinuities[1] postulated is far from straightforward, as will be shown below.)

If the stands to be classified constitute a representative sample of an underlying population, and the population has discontinuities in its distribution, then the sample may be expected to reflect these discontinuities. A classification of the sample set based on discontinuities is then likely to lead to optimal conclusions regarding the classification of the population itself. A selective sampling procedure, on the other hand, may generate discontinuities which do not correspond with discontinuities in the parent population, and the same classification technique may in this case give results which are far from optimal when applied to the population.

Even where the parent population shows discontinuities which are reflected in the sample set, one should interpret them with cir-

1) Williams (see Lambert & Dale 1964) has proposed a definition of discontinuity which would require each class of stands to have no species in common with any other, and Lambert & Dale would use 'heterogeneity' for other cases. This appears an unduly restrictive definition of discontinuity, however, and very few instances listed in the literature of plant sociology would qualify.

cumspection. The discontinuities in the vegetation may be the result of varying representation of the biotope space in the area sampled, so that there are some intermediate habitat types which are under-represented, while those on either side are well represented. The vegetation characteristic of the latter types of habitat will then appear as clusters separated by a discontinuity. For purpose of prediction within the area sampled, the resulting classification will be unexceptionable; but one must not proceed to deductions about the interpretation of vegetational discontinuity, ignoring the habitat discontinuity on which it was based.

19.2.3 How Many Classes?

For the primary classification of a set of entities, there are three types of decision to be made, though these decisions may be linked. They are:

(i) How many classes should be recognized?
(ii) How are they to be delimited?
(iii) To which class is each entity to be allotted?

If the approach is that outlined in the preceding paragraphs, in which the classes are defined intensively, the answer to (iii) automatically follows from (ii); if, as often happens, the class is *extensively* defined within the particular set of samples studied, allocation of new samples may require a new decision procedure.

If the data show no inherent discontinuities, the number of classes to be recognized (question (i)) is an arbitrary decision, though the arbitrariness may be masked by throwing the blame on to the 'stopping rule' of a divisive procedure (see below). In principle, it is clear that, as one increases the number of classes, the predictions that can be based on class membership become more precise, but the allocation of a stand to its appropriate class becomes more difficult. As the size of classes decreases, the variance or information between classes increases at the expense of that within classes, so that in the extreme, when each class consists of a single stand, all the variance is between classes. Where no natural discontinuities exist the choice of how far to subdivide is one of practical convenience rather than of principle. On the other hand, where class delimitation is based on discontinuities in the distribution of points, the number of classes need not be an arbitrary decision, but may be decided by the data. Even here, however, the discontinuities may themselves vary continuously in their distinctness, so that, even though the first one or two are perfectly clear, one comes to a point where decision as to whether or not a discontinuity exists is arbitrary. The arbitrariness may, indeed, be pushed one stage further off by defining a statistical test to determine whether a discontinuity will be recognized or not, and in this case the arbitrary element will be the choice of significance level.

19.2.4 Hypotheses and Classification

Where, as in the primary classification of vegetation samples, we are interested in drawing conclusions as to the structure of a parent population, it is often appropriate to set up a null hypothesis and to test it before proceeding to classify the data. The appropriate null hypothesis (Goodall 1966b, c) is clearly that the samples were drawn from a single population, with variation but without discontinuity. If this null hypothesis is upheld by the test, classification must be regarded as an arbitrary procedure, serving practical ends only, and probably following different rules from a 'natural' classification based on discontinuities in the underlying population.

As will be seen below, many of the classification procedures proposed in the literature have no statistical element and do not lend themselves to the testing of null hypotheses. In such cases, it has been said that the procedures are to be regarded as ways, not of testing hypotheses, but of generating them (e. g. Lambert & Dale 1964). However, too little attention has been given to the type of hypothesis which might be generated, or how it might be tested. A very suitable type of hypothesis would be

i) If a new sample is collected, and is allotted to this or that class by such and such criteria, then the values of such and such other attributes in it will have expected values specified in one vector, with error variance specified in another vector; or

ii) The population from which these samples are drawn is composed of such and such classes, the attributes in each of which are defined by specified multivariate distributions.

The first of these hypotheses is purely operational, and implies nothing about the structure of the parent population, whereas the second explicitly describes the parent population. Either can be tested by taking new samples from the defined population. It should be emphasized that if numerical classification of vegetation or other objects is regarded as a procedure for generation of hypotheses, then the hypotheses generated should be made explicit in a testable form, and their generation should be followed by the performance of a test.

Where one is considering the classification process in terms of hypotheses, and hopes to develop objective tests, it is wise to use separate sets of data for building the hypotheses and for testing them. If this is not done, there is a considerable risk that the test will be invalid. For certain restricted types of hypothesis, statistical procedures exist whereby some part of the formation of a hypothesis (estimation of parameters) can make use of the same data as are used to test it; this applies, for instance, to regression analysis but is rarely likely to be true for procedures in vegetation classification, and the safest plan is to use separate sets of data for the two processes.

19.2.5 Heterogeneity and Discontinuity

As stated above, if the data contain discontinuities they can serve as the basis for a 'natural' classification; if on the other hand they are continuous, a classification is bound to be arbitrary. The approach suited to these two situations is likely to be different. Accordingly, a first step in primary classification should be to determine whether or not the data are homogeneous. As suggested by Brisse & Grandjouan (1971), this indicates the desirability of ordination as a prelude to classification. Ordination will not only enable the continuity or discontinuity of the data to be studied, but will also give information on the number of clusters ('natural' classes) to be recognized, and on their shape.

The recognition of discontinuities in multivariate data is far from simple. With continuous variables, an acceptable approach might be through a null hypothesis of multivariate normal distributions. The estimation of parameters in the latter case would require a provisional allocation of samples among the two classes, and this allocation should be that which, in some sense, maximizes the variance between classes and minimizes that within them.

For binary attributes (such as species presence) it might be best to reduce the dimensionality by an ordination, which at the same time will replace the original binary variables by a smaller number of quasi-continuous variables. These new variables may then be treated in the same way as described in the preceding paragraph. Even for continuous variables, ordination in fewer dimensions may prove a convenient preliminary to tests for discontinuity.

19.3 Variables and Weighting

As in vegetation classification generally, a wide range of variables can be used in numerical classification. Most commonly they are floristic variables, the commonest of all being the presence and absence of each species — a binary variable. Any variable ('importance value') expressing the quantity in which each species occurs in the vegetation can also be used, relative or absolute — cover, basal area, density, biomass, etc. — or any combined measure such as the synthetic importance value of the Wisconsin school (e. g. CURTIS & McINTOSH 1951). For simplicity in computation, continuous variables such as cover may be converted into binary variables, by allotting the alternative values according to whether or not the quantities exceed a specified threshold (e.g. NOY-MEIR et al. 1970). Rankings have rarely been used; they deserve more attention, for subjective rankings of species can be observed quickly and perhaps more reliably than quantitative assessments of the proportional or absolute contribution of a species to the vegetation.

It has often been suggested (LAMBERT & DALE 1964, 1967, ORLOCI 1968, NOY-MEIR et al. 1970, GREIG-SMITH 1971) that species presence is an appropriate variable to use where the set of stands to be classified is highly heterogeneous, and that in such cases quantitative measures add little useful information. In more restricted and less heterogeneous sets of data, on the other hand, variables expressing the quantitative contribution of the different species to the vegetation may be preferable.

Species variables are not the only way in which vegetation can be described. The species can, for instance, be divided into ecological groups, responding similarly to factors of the environment (article 2). Or they can be grouped in terms of growth-form, and the proportion of each growth-form in terms of cover, biomass, etc. used as variables in the classification process; or the division of species could be in terms of geographical origin, or disseminule type; or physiognomic characters, closely linked with the proportions of different life-forms, could be used (WEBB et al. 1970).

Finally, the environmental factors associated with the vegetation could also be used as variables in the classification of vegetation by numerical methods, as they have often been used in non-numerical methods at the formation level (article 13).

The results of classification procedures may depend greatly on the weighting accorded to the different variables. It seems intuitively likely that this dependence may be unimportant if the classification is based on discontinuities in the data. In other cases, however, the effects may be considerable.

Weighting may be implied by the variable used and the method of analysis, without any explicit consideration. If, for instance, the variables used are species presence and absence only, many methods of analysis tacitly allot equal weight to each species. The same is true if the variables are standardized measures of quantity. If they are unstandardized, on the other hand, the tacit weighting may be proportional to the average quantity in which the species occurs.

An extreme case of differential weighting is to use data for some of the species only, and ignore the rest. Not infrequently, for computational convenience, species occurring in few samples have been ignored (e.g. GOODALL 1954, WEBB et al. 1967b) — though this may make the recognition of distinct vegetation types represented by few samples more difficult (GOODALL 1969). In species-rich vegetation, where each vegetation type is well represented in the sample set, however, sampling of the species present by life form, for instance, or at random, is unlikely greatly to affect the resulting classification (WEBB et al. 1967b, 1970, NOY-MEIR et al. 1970).

In numerical taxonomy, emphasis has often been placed on the 'Adansonian principle' of equal weight for all attributes. This goal is illusory when the attributes to be recorded are selected subjectively — which is usual in taxonomy proper. In vegetation, at least the different species present constitute an objectively definable list of variables; and if we adopt a common measure for them — biomass, cover, etc. — their respective weighting is decided in a biologically meaningful way.

WILLIAMS & DALE (1965) have claimed that the scale of the axes for continuous variables may affect the results of classification. It is clear that a proportional change in scale cannot generate discontinuities where there were none, though this could easily happen with some types of non-linear transformation. If, on the other hand, the change in scale affects the weighting used for maximization, the optimum division into arbitrary classes would be affected. And the results of the various classification procedures without significance tests described below may also depend on scale.

It has been claimed with some justice (e.g. LAMBERT & DALE 1964) that absence of a species from a sample should not be considered simply as the zero limit of a quantitative variable, but is qualitatively different from any positive value. Absence may arise in a sample in two ways. Either one may be sampling a population from which the species is absent; or the species may be present in the population but happen to be absent from the sample. The problem is akin to that of the 'contamination' of a Poisson distribution by additional zeros. The presence of a species is positive evidence that environmental conditions fall within the range of tolerance of that species; its absence is not positive evidence to the contrary.[1])

It has been suggested that the situation be handled by using two

1) Similarly, HALL (1970) suggests that species present in high quantity in the sample are those more relevant to its classification.

variables for each species — one a binary variable expressing presence or absence only, the other a real variable taking positive values only, and indeterminate when the first variable is zero. This is done, for instance, in the matrix partition method of WILLIAMS & DALE (1962). The probabilistic methods of GOODALL (1966a, d, 1968, 1971), which are distribution-free and do not depend on 'distance' measures, can also deal satisfactorily with this difficulty.

19.4 Sample Size Effects

Since vegetation is structured, the results of classification may depend on the size of the samples classified. If samples are not less than the *minimal area* for the vegetation (defined as that of a square with the side equal to the distance at which variance between samples ceases to be a function of their spatial separation — see GOODALL 1961), then the results of classification should be independent of their size (NOY-MEIR et al. 1970). If they are smaller, then the classification will in part reflect the mosaic structure of the vegetation, in addition to differences among vegetation types. The assumption that a minimal area exists — that the vegetation is homogeneous at a certain scale — cannot however be taken for granted (GOODALL 1961), and where it is untrue one must accept the dependence of classification on sample size at all levels.

19.5 Clustering Procedures

Methods used to identify the clusters which should be distinguished (together often with the interrelations among them) are commonly called 'clustering procedures'.

One of the earliest of these procedures was that of CZEKANOWSKI (1909), originally applied to the classification of ethnological material, but used for vegetation classification by Polish ecologists in particular (e. g. MOTYKA 1947, MATUSZKIEWICZ & MATUSZKIEWICZ 1956).

The first step is to calculate a matrix of similarities between pairs of stands. Any index of similarity (see article 6) may be used, but MOTYKA's choice was that of KULCZYŃSKI (1928). The order of the stands in the matrix is then rearranged by inspection so that as many as possible of the high similarities are close to the main diagonal (article 7). When this is done, it will often be found that the stands fall into groups, similarities within which are high, while those between different groups are mainly low. These groups then constitute the classes of the classification.

Re-ordering of a matrix is also an important step in the Zürich-Montpellier method of vegetation analysis (see article 20), but here the matrix is one of species × stands (relevés). Re-ordering is normally carried out by inspection, but numerical methods for the purpose have been devised (Moore et al. 1967, Lieth & Moore 1971), and lead to the arrangement of stands in a sequence where breaks (separating classes of stands) can easily be distinguished.

Both these approaches have the drawback of assuming that the classes can be arranged in a one-dimensional sequence. Otherwise it will not be possible to re-order the matrix so that similarities decrease regularly away from the main diagonal. If the classes are highly discrete, this may not matter — all similarities between classes will be much lower than those within classes, and the classes will be readily recognizable as separate groups even though their order is uncertain.

An alternative numerical basis for subjective classification of a set of vegetation samples can be provided by an ordination. Any of the standard methods of ordination (see article 11) can be used in this way, the purpose being to reduce the dimensionality of the system, and hence make its characteristics more readily open to inspection.

There are rather few examples in the literature where an ordination has been divided arbitrarily to give classes. Brown & Curtis (1955) divided a forest gradient into five ranges of their weighted-average index, and Looman (1963) divided a one-dimensional ordination of grassland samples into seven classes.

The preparation of an ordination in two or three dimensions and its graphical representation often bring clustering and discontinuities to light, and thus form the basis for a 'natural' classification. An example is the distinction between swale and dune communities in Goodall's (1954) study of the Victorian Mallee.

In contrast to the mixed numerical and subjective procedures outlined above, purely numerical procedures are usually organized in step-wise fashion. They lead to an arrangement of the samples in a hierarchical pattern which can be expressed as a tree diagram or *dendrogram*, reflecting both the successive steps and the presumptive relationships among the classes distinguished (Figs. 1, 2).

It is clear that a hierarchical system or dendrogram is not logically equivalent to a distribution of samples into classes. It establishes a pattern of closeness of relationship among the samples, without specifying dividing lines which distinguish classes. To distinguish a set of classes on the basis of a dendrogram requires the establishment of a 'stopping rule', which will define a point beyond which distal branches of the dendrogram will be regarded as mem-

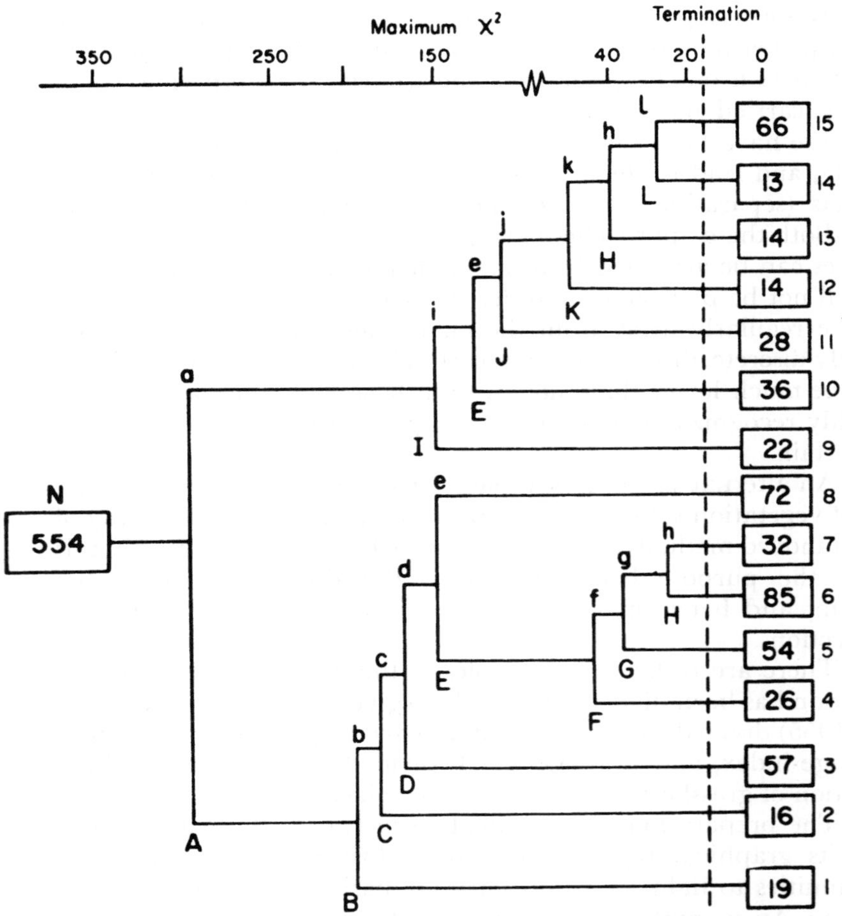

Fig. 1. A dendrogram, representing classification of 554 quadrats comprising 98 species from wet-land vegetation on the Isle of Arran, Scotland, by normal association analysis (CRAWFORD & WISHART 1967, Fig. 1). The numbers in the boxes are numbers of quadrats in the terminating classes; the numbers outside the boxes are reference numbers for these classes. The letters at the nodes represent the species on which a given division of the quadrats was based, with presence of a species indicated by a capital letter and its absence by a small letter: A, *Molinia caerulea*; B, *Anthoxanthum odoratum*; C, *Cirsium palustre*; D, *Rhacomitrium lanuginosum*; E, *Sphagnum* spp.; F, *Juncus acutiflorus*; G, *Myrica gale*; H, *Calluna vulgaris*; I, *Iris pseudacorus*; J, *Glaux maritima*; K, *Sagina procumbens*; L, *Alnus glutinosa*.

bers of the same class. For classifying a set of samples representing one or more underlying populations, the stopping rule should be probabilistic (WEBB et al. 1967a). It may, for instance, depend on a measure of the homogeneity of the set of samples considered as

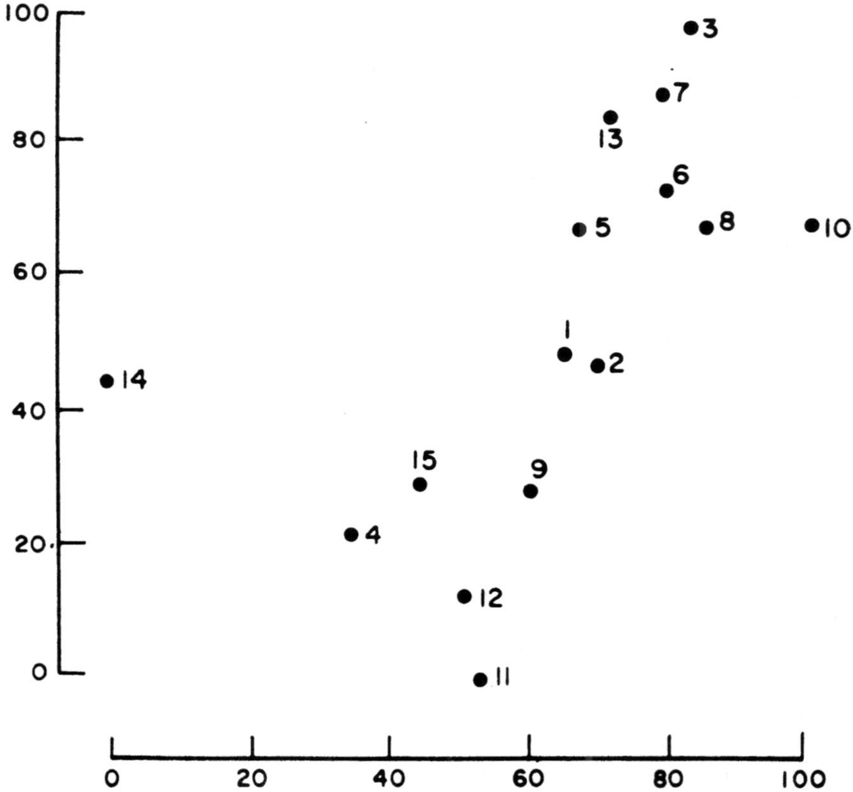

Fig. 2. A Wisconsin comparative ordination of the numbered classes of quadrats derived by association analysis as shown in Fig. 1. (Crawford & Wishart 1967, Fig. 3). The *x* axis is based on classes 14 and 10, the *y* axis on classes 11 and 3. In this case the association analysis, as a monothetic technique, does not effectively reproduce the relationships indicated in the ordination, as a polythetic technique (cf. Greig-Smith et al. 1967, Flenley 1969, Ward 1971).

belonging to a single class. Often, however, the stopping rule is much more arbitrary, and may depend on the value of some statistic calculated from the samples without regard to probabilistic considerations. Czekanowski (1909), for instance, used the minimum similarity within a set as a stopping rule. Williams & Lambert (1960) used a fixed value for max. χ^2 in association analysis (see below) in the same way, whereas in the earlier form of this procedure Goodall (1953) had used χ^2 as a probabilistic stopping rule, and Noy-Meir et al. (1970) seem to have adopted a rather similar practice.

A variety of techniques is available in deriving a hierarchical structure or dendrogram from a primary matrix of attribute values for each of the samples to be classified. The first step is usually to convert this primary matrix into another which is simpler in the sense that either the distinctions between samples, or those between attributes, have been eliminated. That is, one may first convert the primary matrix of samples-by-attributes into one of attributes-by-attributes, and classify the stands on the basis of the interrelations shown among the attributes. These are often known as R-techniques (the 'transposed-structuring' techniques of LAMBERT & DALE (1964), or the 'derived-structuring' methods of WILLIAMS & DALE (1965)). Other methods derive a secondary sample-by-sample matrix in which each cell represents some measure of similarity or 'distance' between the two samples in question (see article 6); these methods are known as Q-techniques ('self-structuring' techniques of LAMBERT & DALE, or 'direct-structuring' methods, according to WILLIAMS & DALE).

In considering numerical procedures for classification a distinction is often made between *monothetic* and *polythetic* methods. In the former, attributes are considered singly, and each dividing hyperplane, in our geometrical model, is consequently perpendicular to one of the axes. In a polythetic procedure, on the other hand, a number of different attributes are taken into account simultaneously, and the dividing hypersurface is consequently likely to be oblique to the axes, and may be curved. In numerical classification, a well-known method using the monothetic approach is that known as 'association analysis' (section 19.5.1.1 below); a descriptive analogue is classification of vegetation by the dominant species (article 14), or strict application of the presence or absence of a single characteristic species. Since the absence of a particular species may be a matter of chance, there is a substantial random element in the allocation of a particular sample by monothetic criteria (NOY-MEIR et al. 1970). If the species is annual, there may also be important seasonal differences in classifications based on this principle (HOPKINS 1968). Polythetic classifications are much less affected by such influences.

Another distinction often made in the literature of numerical taxonomy is that between *divisive* and *agglomerative* approaches. A divisive approach is one which begins with the complete set of entities to be classified, and divides them progressively into smaller and smaller groups. An agglomerative approach proceeds from a single entity, considers which other entity is most 'similar' to it in some defined sense, then which third entity is most 'similar' to these two, and thus proceeds to build up a cluster of entities around that which was first selected. At some point it may be decided that none of the remaining entities is similar 'enough' (a 'stopping rule') to be associated with the cluster being formed, and a new cluster is initiated. It will be noted that both types of procedures develop *extensively* — the classes are defined in the first instance by their membership, though afterwards attempts may be made to define the classes intensively, and thus provide a principle of allocation for any future samples.

In considering the various classification procedures that have been proposed and used for vegetation, we shall take first those that are mainly divisive, and then those which are mainly agglomerative. The distinction is not a firm one – a procedure may include steps of both types, and it is likely that combined procedures of this sort will be favoured as numerical classification becomes more sophisticated.

WILLIAMS et al. (1966), in considering criteria for choice of a classification procedure, suggest that the use of a hierarchical system implies that the user is specifically interested in 'the actual path of fusion'. In the present discussion, however, we are distinguishing the primary recognition of clusters from consideration of their interrelations, so that the hierarchical system is merely regarded as a convenient means to the end of cluster recognition, and the path of fusion is not a matter of interest nor a suitable criterion for choice of method.

A drawback of the hierarchical approach (LAMBERT & WILLIAMS 1966) is that the decisions made are irrevocable. If the rules o the clustering algorithm, at a particular point in the process, lead to a certain division or a certain fusing of groups, this can never be corrected by subsequent actions within the strictly hierarchic procedure. This has led some workers to amend it by allowing for the possibility of fusion among clusters that have been separated and have become associated with different branches of the dendrogram (GOODALL 1953, GREIG-SMITH et al. 1967, CRAWFORD & WISHART 1968).

19.5.1 DIVISIVE METHODS

These include both monothetic and polythetic procedures. Some have – or can have – a statistical basis. At any stage one may set up a null hypothesis that an existing group of samples could have arisen by sampling from a single population with attributes distributed in some specified fashion – perhaps independently of one another. Then this grouping is divided further only if the null hypothesis is rejected at the chosen significance level.

19.5.1.1 *Association Analysis*

This is a monothetic divisive approach which had its origin in the concept that, within a homogeneous community, species are uncorrelated as between different samples. Consequently, criteria for division were sought which would give classes within which interspecific correlations were as few as possible.

As originally described by GOODALL (1953), the interspecific associations were tested for each pair of species by 2×2 tables (see section 6.2), using either the χ^2 test or the exact hypergeometric test. Of those species which had significant associations with others, that which occurred in the largest number of samples was selected, and the samples were divided into two groups according to whether this species was present or absent. The same test was then applied to the set of samples in which the first criterion species was present, and if it still proved heterogeneous it was divided on the presence of a second criterion species; the process was continued until the final group had a matrix of associations such as might have arisen by chance. The remaining samples, lacking one or more of the criterion species for the first group, were then reconsidered in the same way, and the process was continued until the samples had been divided into a number of groups, each of which had no associations significant by the test used. Finally, the groups were re-combined by pairs; if no significant associations re-emerged, the combined group was retained.

This technique was modified by WILLIAMS & LAMBERT (1959, 1960) in four ways. In the first place, the null hypothesis of a homogeneous group of stands was abandoned, and with it the testing of significance of a 2×2 table. Instead, a fixed level of χ^2 was adopted as a stopping rule, further subdivision being abandoned if the maximum value of χ^2 was less than this. Second, at each stage, the species selected as the criterion for division was that which had the maximum sum of χ^2 values with all other species. As WILLIAMS & LAMBERT pointed out, this maximized the effect of the subdivision in reducing other associations within the resulting groups. Thirdly, the procedure was made strictly hierarchical in that the group in which a criterion species was absent, as well as that in which it was present, was divided further without recombination. Finally, WILLIAMS & LAMBERT abandoned tests for possible recombination of homogeneous groups at the end—which fitted in with their abandonment of a significance criterion, and made the procedure purely divisive. Fig. 1 shows a dendrogram from association analysis.

NOY-MEIR et al. (1970) point out that association analysis is likely to be much more satisfactory as a classification procedure in species-rich vegetation than where only a few species are present. Like other monothetic procedures, the risk of misclassification is considerable if the criterion species happens to be absent by chance.

A variant of this method was proposed by CRAWFORD &

WISHART (1967) under the title 'group analysis.' Successive divisions are based on the presence or absence of particular species, but these species are selected according to the frequency with which they occur in the presence of numerous others. If S_i is the proportion of the total species which occurs in the *i*th sample, $\sum_j$ the sum of S_i for all samples in which species *j* occurs, P_j is the proportion of such samples, and $\sum.$ the sum of S_i over all samples, then the statistic that forms the criterion for choice of which species should serve as basis for dividing the set of samples is

$$\mu'^2 = 4\ (\textstyle\sum_j - P_j \sum.)^2$$

Division ceases when $\sum.$ exceeds half the number of samples in the set. CRAWFORD & WISHART tested this method on a random set of data; no clear-cut classes were found, but the data divided into very small groups—which, they claimed, served to indicate its value.

It will be noted that this method does not provide for test of a null hypothesis, and the 'stopping rule' is an arbitrary one. The rationale of the method in giving special weight to species occurring mainly in species-rich samples seems open to question. Some species-poor communities in pioneer habitats are at least as distinctive as those in more favoured situations.

In a subsequent paper (CRAWFORD & WISHART 1968), an adjustment of the original classification was introduced by examining the values of S_i for each sample within the group to which it was allotted. Where S_i was less than an arbitrary value β, the corresponding value for this sample was recalculated on the assumption that it was reclassified with each of the other groups; it was then attached to the group for which S_i was maximal. Another rule limited the minimum size of a group; this resulted in a smaller number of groups which were more compact than a similar number of groups would have been without the reclassification process.

While reclassification to avoid wrong allocation by chance seems to have merit, particularly in a monothetic procedure, there seems no particular reason for limiting the minimum size of a group, and thus not recognizing the rarer (but perhaps very distinctive) communities included in the sampling.

19.5.1.2 *Elimination of Outliers*

Among divisive procedures the simplest is the removal from the collection of a single sample which for some reason is regarded as differing most markedly from the norm represented by the rest. If the null hypothesis is entertained that the collection of samples can be regarded as taken from a single population in which the presence and variations in quantity of the different species are uncorrelated, a method has been described (GOODALL 1966a) for determining the probability that any particular sample will have a specific composition differing as widely from the norm as it does; this probability is termed the *deviant index*. If a sample is found to differ 'significantly' from the rest by this test, it is reasonable to regard it as representing a different type of vegetation, and hence to exclude it from the class formed, presumptively, by the rest. If this test is applied repeatedly, the bulk of the samples can be reduced to a homogeneous core in which no one sample need be considered as an outlier by this test (cf. article 5). Possible relationships among the outliers successively eliminated may then be tested.

It is of course possible, in a small group, that a subset of samples differ from the rest, but that none individually has a significant deviant index because the homogeneous part of the group is 'contaminated' by other members of the subset. A method has been described (GOODALL 1966d) for testing whether any two samples differ together from the norm of the group as a whole; and the same test may be generalized to larger subsets.

A simpler version of the deviant index (GOODALL 1969) depending on presence and absence only was applied to the removal of outliers from a set of vegetation samples in which a large number of species were recorded only in one or two samples.

19.5.1.3 *Method of Edwards and Cavalli-Sforza*

EDWARDS & CAVALLI-SFORZA (1965) described a divisive polythetic method for determining optimal dichotomous subdivisions of a set of samples. All possible dichotomies are considered, and the intragroup sum of squared distances determined:

$$\sum_{i=1}^{M-1} \sum_{j=i+1}^{M} \sum_{k=1}^{S} (x_{ik} - x_{jk})^2 \qquad (2)$$

where x_{ij} is the value of the j th attribute in the i th sample, and there are S attributes with M samples in the group in question. That dichotomy for which this sum is minimal is chosen, and the process is repeated on the group thus divided.

This is a laborious process, because of the number of possible dichotomies to be considered, and the authors do not suggest it be used where the total number of samples exceeds 15. It can be applied to either quantitative or binary data (and is considerably simplified in the latter case), but is sensitive to the scale used for the different attributes.

No probabilistic treatment was proposed to determine whether clusters should be separated, and it is clear that the authors envisaged it as a method of classifying the elements of a defined population rather than a set of samples which might not in fact need classifying at all.

The procedure leads to an analysis of variance. But, since the classification is not defined *a priori* but developed as a result of the operation of the procedure, the usual tests for comparison of variance do not apply. Presumably it would be possible (perhaps by Monte Carlo methods) to develop criteria for the partition of variance, optimal in the sense described, based on the null hypothesis of a single population; but so far this has not been done.

19.5.1.4 *MacNaughton-Smith's Method*

MacNaughton-Smith et al. (1964) proposed a rather similar polythetic divisive method, which seems to have been little applied. A measure of dissimilarity between samples or groups of samples is defined — in fact, the weighted Euclidean distance proposed by the same authors (Williams et al. 1964). That sample which is most dissimilar from the rest is identified, and any other samples which are more similar to this sample than to the rest of the set are added to it, one by one, to form a distinct class. When no more samples are more similar to this segregate group than to the rest, the process is repeated within each of the two classes thus distinguished.

The drawback of this procedure, as in that of Edwards & Cavalli-Sforza, is that no significance tests are incorporated. This is an improvement which could, however, be introduced with little difficulty, by making use of the probabilistic indices described below (section 19.5.2.4) instead of weighted Euclidean distance.

19.5.1.5 *Information Analysis*

If one considers a set of M samples in which S species are recorded, the information content or entropy can be expressed (Williams et al. 1966) as:

$$I = SM \log M - \sum_{j=1}^{S} [a_j \log a_j + (M - a_j) \log (M - a_j)] \quad (3)$$

where a_j is the number of samples in which the j th species is recorded. Within any sub-set of m' samples, the information content will be less, and if the set of M samples is divided into two, the sum of the information content within the two sub-sets will be less than that of the set as a whole. The reduction in information content through subdivision can thus be regarded as a measure of the 'efficiency' of the subdivision. Successive divisions like this will result in a hierarchical system or dendrogram.

Successive divisions, at each stage maximizing the reduction in information content, could thus constitute an algorithm for divisive polythetic classification. This, however, comes against the same difficulty as the procedure of EDWARDS & CAVALLI-SFORZA above — that the number of partitions to be tested is very large. LANCE & WILLIAMS (1968) accordingly proposed a monothetic procedure using the concept of information. A division of samples according to the presence or absence of each species is considered, the reduction of information content calculated in each case, and that division selected which gives the greatest reduction.

If the information content of a set of M samples is I, and division of the set into two portions at random gives the information content of the two sub-sets as I', so that the decrease in information content through their separation (or the increase through their fusion) is:

$$\Delta I = I - I' \tag{4}$$

then the quantity $2\Delta I$ is distributed as χ^2 with n degrees of freedom. LAMBERT & WILLIAMS (1966) suggest that this could be used as a probabilistic stopping rule. Since, however, the hierarchical divisions are chosen so as to maximize ΔI at every step, a significance test of a null hypothesis of homogeneity of the group before subdivision would need to be based on the distribution of the *maximum* of this quantity within a set of the size in question.

Information-theory methods have usually been applied to binary data — the presence and absence of species. As PIELOU (1969) points out, they could equally be applied to quantitative data if the different variables (species quantity, for instance) can all be expressed in terms of a common integral measure — such as dry weight to the nearest gram, cover to the nearest unit per cent, number of individuals (ORLOCI 1969), or frequency in sub-samples of a fixed size (ORLOCI 1970, DALE 1971). On the other hand, the earlier suggestion by ORLOCI (1968) that arbitrarily chosen ranges of a continuous variable could be treated as alternative values of a qualitative variable involves loss of the relevant ordering of these values, and in consequence seems unlikely to be satisfactory.

It should be borne in mind that information-theory methods

contain built-in assumptions regarding weighting, depending on the particular variables used. In ORLOCI's (1970) method, for instance, species occurring only at low frequency will have rather little effect on clustering.

19.5.2 Agglomerative Methods

In contrast with the methods described above, these proceed from the individual samples, and combine them in progressively increasing groups of decreasing internal similarity until the whole set of samples is combined. Again, a dendrogram results.

In applying the agglomerative procedures, the clusters may be developed according to various rules. One of the commonest is that of the 'nearest neighbour' — that is, decisions as to whether a new sample should be added to a cluster, or two clusters should be joined, are based on the similarity of the two most similar samples involved in the fusion, irrespective of the size or extent of the cluster. An alternative often used is a centroid measure — that is, the individual samples in the cluster are ignored and are replaced by their centroid, a hypothetical sample for which each variable has the mean value of all samples in the cluster. Yet another approach is to apply whatever pairwise measure of similarity of distance is selected to every possible pair involved, and to minimize the mean of these measures in the new cluster. Finally, there is the rather rarely used 'furthest-neighbour' criterion, in which the most dissimilar members of the two clusters to be amalgamated are considered, thus minimizing the maximum dissimilarity or distance within the resulting larger cluster.

19.5.2.1 *Simple Similarity Methods*

The agglomerative procedure used by SNEATH (1957) for bacterial classification deserves mention for its historical importance in numerical taxonomy, though it has only rarely been used to classify vegetation (GILBERT & WELLS 1966). It is polythetic, being based on a similarity coefficient between the entities to be classified (the JACCARD (1901) 'coefficient of community', (39) in article 6, was used, but the procedure is not tied to any particular measure of similarity). That pair of entities with the highest similarity are combined, and combination proceeds using successively smaller similarities to associate each as yet unallotted entity, either with another single entity or with a group already recognized. It is a nearest-neighbour procedure, so that two groups are combined if the highest similarity between a member of each reaches the required level.

Other methods based on pair-wise similarities among stands can also be used. West (1966), for instance, used Sokal & Sneath's (1963) 'weighted pair-group method', in which successive amalgamations are based not on the similarity to the nearest neighbour in the cluster, but on the mean similarity to all members of it.

19.5.2.2 *Orloci's Agglomerative Technique*

The principle underlying this technique of Orloci's (1967) is at every stage to minimize the variance within groups, and maximize that between groups. It is equally applicable to presence-and-absence and to quantitative data. A distance function between vegetation samples j and k defined as:

$$D_{jk} = \sqrt{\sum_i \left(\frac{x_{ij}}{v_j} - \frac{x_{ik}}{v_k}\right)^2} \tag{5}$$

where x_{ij} is the value of the i th variable in the j th sample, and v_j is the length of the j th vector, that is $\sqrt{\sum_i x^2_{ij}}$. Then the dispersion within a group containing M samples is

$$Q = \left\{\sum_{j=1}^{M} \sum_{k=j+1}^{M} D^2_{jk}\right\} /M \tag{6}$$

The procedure starts by combining that pair of samples for which Q is minimal. Then either another pair of samples is combined, or a third sample is combined with the existing pair, the decision depending on which operation gives a minimal increase in the sum of Q values within the resulting group(s). The process of forming groups, or combining them, continues step by step until the whole set of samples is combined into a single group, the choice of combination at each stage being that which minimizes the increment in intragroup dispersion, and hence maximizes the residual dispersion between groups.

Orloci suggests that, when a hierarchy has been constructed in this way, one may examine it to determine the significance level of the separation at each point. In effect, the procedure amounts to an analysis of variance, the increment in dispersion at each stage in agglomeration being a variance with one degree of freedom. Since however a fixed variance has been partitioned according to a particular set of rules which, *ex hypothesi*, put the variances in increasing order of magnitude, the question is: what is the probability that the observed, or a greater, proportion of the variance should be accounted for by the first division; that the observed, or a greater, proportion of the *residual* variance should be accounted for by the second division; and so forth. These questions are not on all fours with those normally asked in an analysis of variance, and no formal answer appears yet to have been produced. As an approximate test, however, one may determine the ratio

$$(n-g)Q_1/Q_r \tag{7}$$

where a total of n samples have already been partitioned into g groups with an intra-group dispersion of Q_r, and Q_1 is the increment in Q which results from the proposed amalgamation. If this ratio does not exceed the variance ratio ($n_1=1$, $n_2=n-g$) corresponding with a probability of

$$1-(1-\alpha)^{2/\{g(g-1)\}} \tag{8}$$

where α is the protection level selected, then the amalgamation is acceptable.

19.5.2.3 *Hall's Heterogeneity Method*

A similar method proposed by Hall (1967a, 1967b) is based on a heterogeneity function expressing the differences in attribute values (species quantity, etc.) within a group of samples, the contribution of each species to the heterogeneity function being weighted by its average presence in that group, as a proportion of its average presence in all samples. The clustering procedure is agglomerative, association of samples or groups at each stage being such as will minimize the increase in heterogeneity within the group. Class distinctions are recognized when the increase in heterogeneity at a given link is abruptly greater than usual. It is a drawback of this, like many other clustering methods, that no satisfactory test of a null hypothesis is possible, the distribution of the heterogeneity function being unknown. It is, of course, conceivable that a nonparametric test for the increase in heterogeneity at a given linkage could be developed. So far this has not been done.

In applying this method, Hall comments (1967b) that association of a single stand with an existing cluster is much more likely to occur than a combination of two existing clusters of about equal size, even if the single stand is much more unlike either cluster than they are like one another. Thus there is a tendency to 'chaining' — the association of stands one by one with existing clusters — rather than the progressive expansion of several clusters of comparable size as the process of agglomeration proceeds.

19.5.2.4 *A Probabilistic Approach*

A method for classification based on probabilistic considerations has been described by GOODALL (1966a, 1966d, 1968, 1971), though applications to the classification of vegetation have not yet been published. It is based on the null hypothesis that the samples are derived from a single population in which the attributes are uncorrelated. Methods are derived for calculating the probability that the set of attributes in any individual will deviate from the norm as much as they do ('deviant index'), that the sets in any two individuals will deviate similarly from the norm as much as they do ('similarity index'), and that the set of attributes of an individual will resemble those of a specified group as much as they do ('affinity index'). Each of these indices can handle qualitative, ordered or quantitative variables and combine the evidence from them.

With the aid of these tools, it is possible to examine a set of stands and determine whether they can (within the significance limit selected) be regarded as a sample from a homogeneous population. If not, the most deviant samples may be removed one by one until the residue is homogeneous. Alternatively, if two samples have a signifiant similarity index, they can be treated as the nucleus of a cluster, and the affinity of other samples with this cluster nucleus is tested. Those which show a significant affinity are added to the cluster, and the process continues, again until the residue is homogeneous, with separation of additional clusters if indicated. As mentioned, it is an assumption of this method that the variables are uncorrelated; if this is untrue for the variables as observed, they may be replaced by a new set of variables derived from them by transformation and rotation of axes.

19.5.2.5 *Information Analysis*

A divisive method using the concept of information has been described above (section 19.5.1.5). An agglomerative treatment (WILLIAMS et al. 1966, LAMBERT & WILLIAMS 1966, WEBB et al. 1967a, ORLOCI 1971) may be based on the same principle. One may ask what combination of samples into clusters (or of clusters already recognized into larger clusters) will result in a minimum increase in information within the resulting larger groups — that is, in effect, which combination of samples or groups will unite those which are most similar. This procedure can also be continued, of course, at each stage combining two single samples, one sample with a group, or two groups, choosing whichever fusion gives the minimum increase in intra-group information of all the possible alternatives. Thus again a hierarchy of groups is built up, comparable with that obtained by the divisive approach.

ORLOCI (1970) suggests two alternative algorithms for clustering in this way, using functions respectively taking into account the difference in distribution of each species considered separately

$$\sum_i \sum_j f_{ij} \log_e \frac{c f_{ij}}{f_{i\cdot}} \qquad (9)$$

or the differences if the species are considered in the context of the overall records for each stand

$$\sum_i \sum_j f_{ij} \log_e \frac{f_{\cdot\cdot} f_{ij}}{f_{i\cdot} f_{\cdot j}} \qquad (10)$$

where f_{ij} is the number of sub-samples in which the i th species was recorded in the j th stand (out of a total of c), and the dot subscript indicates summation. If the first expression is used, differences between stands in the sum of frequencies for all species would contribute to the clustering decisions, in the latter case they would depend only on their *relative* frequencies. In the example given by ORLOCI, however, the dendrograms resulting from use of the two functions were very similar.

LAMBERT & WILLIAMS (1966) point out that their information-analysis technique has a tendency to associate with one another samples which differ widely from the larger groups present, even though their characteristics may be very diverse, and though they may share more features with one or other of the larger groups than they have in common. It also tends to combine in one group a number of samples which agree only in containing a small number of species. These features suggest that the results of information analysis should not be accepted uncritically.

19.6 Cluster Shape and Dendrogram Shape

In the geometric model we have been using, the cluster of points representing a class may take a wide variety of shapes. In the simplest case, the density of points will decrease outwards from a centre in a nested series of hyperspherical envelopes. In other cases, they will be hyperellipsoids. This distinction will depend on the scaling and orientation of the axes; by appropriate rotation and scaling any hyperellipsoid can be transformed into a hypersphere. Other figures are more distinctive — hyperellipsoids with one or more axes curved, for instance; or star-shaped figures, with a dense central body from which a number of extensions radiate.

The shape of a cluster representing a vegetational class may have considerable interest to the plant sociologist, and is well worth study. This can only be done effectively if numerous samples are available; it calls for an ordination to be applied separately to the samples allocated to each class (GREIG-SMITH et al. 1967), followed by an inspection of the results of ordination, or formal tests if any particular hypotheses about cluster shape appear to be of interest.

Cluster shape should not be confused with the shape of a dendrogram. Comments have frequently been made on the occurrence of 'chaining' in some dendrograms — in agglomerative procedures, a cluster may tend to grow by accretion of individual samples one by one, rather than by fusing with other clusters. This may indeed arise if the cluster is a very elongated one; but this cluster shape is by no means a necessary condition for dendrogram chaining, and chaining seems to be more a feature of the technique than of the data (LANCE & WILLIAMS 1967). Objections to it seem to be practical (in terms of increased computational effort) rather than theoretical.

19.7 Classification of Groups of Samples

In principle, the first step in classification is the recognition of classes among which the samples are allocated, without regard to any hierarchical or other relationship among the classes thus delimited. In practice, as we have seen, the two operations are often combined. In association analysis or information analysis, for instance, no distinction is made between the procedure for distinguishing ultimate groups (if indeed there is any stopping rule at which ultimate groups are delimited, as often there is not) and that for combining the ultimate groups into broader ones. This lack of distinction between cluster recognition and cluster classification is to be regretted, for the same procedures are not necessarily optimal for the two processes. The considerations involved may be different, and in any case some rescaling of the variables may be appropriate (WILLIAMS & DALE 1965).

If the primary classes or groups of samples represent sub-populations separated from one another by discontinuities, then the question of the relationship among these classes is a substantive scientific one. If they are in any case artificial segregates of a single population, then the question of their interrelationship may have little interest.

One common way in which the relationship between classes is expressed is in terms of a higher-order classification — a hierarchy of classes. A set of classes, A_i, form a hierarchy if

(1) for every A_i, A_j, there is an A_k for which

$$(A_i \subset A_k) \wedge (A_j \subset A_k) \tag{11}$$

and

(2)
$$(A_i \subset A_k) \wedge (A_i \subset A_j) \rightarrow (A_j \subset A_k) \vee (A_k \subset A_j) \tag{12}$$

Thus, in a purely hierarchical arrangement, the relationship between two classes is expressed in terms of class membership, and not quantitatively.

A distinction has sometimes been drawn between hierarchical and multidimensional or reticulate relationships among vegetation classes (e.g. POORE 1955). These views of the relationships between classes are not mutually exclusive. A system in multidimensional space could be expressed as a hierarchy, but in so doing one loses information regarding the distance and directional relationships among entities classified.

A hierarchical structure for a set of classes can be established by any of the step-wise procedures used to identify primary classes and delimit their membership. Since, however, we are concerned with a defined population rather than a set of samples, probabilistic considerations are hardly relevant.

In a hierarchical system the question may arise of the comparability of levels of subdivision in the different branches of the dendrogram. This can be — and often is — decided by arbitrary levels of the similarity or distance index on which the hierarchy is based; if the two groups into which a higher group is being divided have a similarity less than a certain arbitrary value, then they are at a higher hierarchical level ('order,' for instance); if they are more similar the groups are at a lower level ('alliance,' say). For such decisions, however, the arbitrariness of the choice of similarity value for each hierarchical level, its essential incomparability in different contexts, and the further arbitrary choice of a similarity index itself, should be emphasized.

In the spatial model, a higher-order class is simply a group of clusters in spatial proximity. This suggests that, rather than develop a hierarchical arrangement for the classes, it might be more informative and more fruitful to express their spatial relations in quantitative terms. This is done, for instance, in the various plexus models discussed elsewhere (article 7), though an adequate representation in this form would require almost as many dimensions as there are classes.

Spatial relations among classes can often be represented more simply if one first applies ordination to the primary classes. Ordination procedures are discussed elsewhere in this volume (articles 10, 11), and any making use of quantitative variables could be applied to the values specifying the centroids of each class. Often only the first two or three axes of the ordination are statistically significant or meaningful, and in this case the plexus construction can be based on them and the residual variation can be treated as an error term.

Yet another method of representing the relations among classes, which however has been little used, is by cross-classification. This could take the form of two or more quite independent hierarchical classifications of the same set of classes — which might, for instance be based respectively on habitat features and on successional status.

19.8 Classification and Ordination

In the early days of applying numerical methods to vegetational classification, ordination methods were also being developed concurrently, and there was much confusion as to their relative roles. It is now, however, generally recognized that the two approaches can be appropriately applied to the same data (Figs. 1, 2), and that to some extent the choice is a matter of taste, and should be guided by the intended use of the results. For vegetational mapping, for instance, classification is greatly to be preferred. Where the vegetational characteristics of a site are required in terms of a simplified vector of real variables — for further statistical analysis, say — ordination is clearly the treatment of choice.

It is difficult to grasp effectively the results of an ordination where the number of axes is large. It has accordingly been suggested that classification is appropriate in the first instance where very heterogenous data are included in the same analysis, but that ordination should then be applied to each of the classes thus distinguished. On the other hand, arguments have been adduced in favour of an initial ordination as a guide to classification (e.g. BRISSE & GRANDJOUAN 1971). On balance, it seems best to use a rough ordination in few dimensions to indicate discontinuities, cluster shape and number, and perhaps to show where dividing lines between classes should be drawn. Ordination should then be applied afresh to these more manageable groups separately; as GREIG-SMITH (1971) points out, their main axes may not be parallel, so that to treat them in a single analysis may be unfruitful.

19.9 Evaluation and Comparison

Numerical procedures for the classification of vegetation have often been judged by comparison with accepted subjectively-developed classifications (e.g. HALL 1970). WILLIAMS et al. (1966) similarly state that the results of numerical classification should 'suggest groupings which are ecologically meaningful when tested by appeal to other relevant information from outside.' These points of view indicate appropriately that numerical classification, of vegetation as of other objects, is still at an early stage of development, and has not yet reached the point where it can be freed from the leading-strings of common sense and ecological experience.

A formal comparison of the merits of two alternative classifications should be based on the objectives to which they are directed. If they are intended as general-purpose classifications, then the criterion should be the prediction of a randomly-chosen attribute in a randomly-chosen stand; and the basis of their evaluation should be the success with which they do so. Since, however, this success may vary greatly over the domain covered (between classes; between variables; and between different ranges of the same variable), a weighting function to be applied to the differences between true and predicted values would be required for a rigorous treatment.

Closely akin to this criterion is the suggestion often made that the classification should be able to predict, for the samples used in developing the classification, values of variables which have not been involved in it. If the classification was based on only the tree species present, for instance, its evaluation might be based on the precision with which it predicts the composition of the field layer. There is a risk of bias here in that the variables used for the test are likely to be correlated with those used in developing the classification. And, as in the previous case, a weighting function would be needed.

Where comparison of two classifications is not for the purpose of evaluation and choice, but simply to express the extent to which they agree or disagree, an information-theory approach seems promising; the joint information in the two classifications could be compared with that peculiar to the two classifications (cf. ESTABROOK 1967). An alternative approach by WILLIAMS et al. (1969) was to determine the common elements in a set of twelve classifications by enumerating the times each pair of stands (among ten) occurred in the same class out of four. One of these pairs occurred in the same class in all twelve classifications, indicating that this pair at least were consistently associated throughout — for the chance of such an event by random assortment of stands among classes would be quite small. This approach could clearly be developed and extended to cover other comparisons between classifications; but tests based on information theory seem, in principle, likely to be more sensitive.

In the long run, evaluation of classification *procedures* should be based on evaluation of the classifications they generate: by their fruits ye shall know them. One more direct suggested criterion, however, has been based on stability. If the procedure is applied independently to two sets of samples, of which the larger includes the smaller, the class membership of samples in the smaller set should coincide in the two classifications. While one may agree that this property is desirable, there seems no reason to regard it as a sufficient criterion.

Comparative studies of classification procedures have generally been concerned with those producing a hierarchical structure, and have been limited to visual and descriptive comparisons of the shape of the dendrogram, and the arrangement of the samples classified at the ends of the branches. Since few of these procedures provide for significance tests of a suitable null hypothesis, tests of significance of differences between them are also lacking.

Even where clustering procedures are non-probabilistic and so are inappropriate as a basis for conclusions regarding a parent population from which a sample was drawn, they are likely in fact to be used for this purpose. Consequently, it might be appropriate to set up an artificial population with known properties — consisting perhaps of several distinct classes, each with specified means and distributions for each variable; take repeated random samples from this artificial population; apply several classificatory procedures to these samples; and study how well the results agreed with the structure of the known population from which the samples were drawn. By applying this approach to a number of artificial populations differing in structure, it might be possible to arrive at conclusions as to the general value of these procedures when applied to samples from unknown populations.

If classification procedures are regarded primarily as methods of generating hypotheses, then the appropriate test of the value of alternative procedures should be based on tests of the validity of the hypotheses they respectively generate. These tests should clearly be based on the same material, and applied to structurally similar hypotheses generated by the different procedures. So far, this has not been done. Different procedures have been applied to the same data, but the classifications resulting have not been put in the form of alternative hypotheses suitable for comparative testing, and *a fortiori* no such comparative tests have performed.

19.10 **Conclusions**

Classification of vegetation differs from classification of organisms in that, in most groups of organisms, the process of evolution has led to the formation of natural classes (species). The organisms in each species share a common gene pool and hence have a limited range of phenotypic variation; since gene flow between species is limited or absent, there tend to be sharp discontinuities in phenotype. In vegetation, on the other hand, there is no comparable genetic basis for differences in attribute sets between different types of stand, the barriers to movement of individuals between adjacent stands are much less marked than barriers to gene flow between species, and in consequence discontinuities between stand types do not arise as clearly from the logic of vegetation organization as they do from the logic of speciation.

The purpose of numerical classification of vegetation is the same as that of classification in general — to identify sets of stands for which many of the same statements will be true, and thus to reduce redundancy and economize effort in describing the variety of the stands classified. This purpose may be facilitated if there are discontinuities in the distribution of the attributes which characterize the stands; if they exist,these discontinuities are likely to be discovered and used by any efficient classification, numerical or otherwise.

Numerical classification can claim advantages over more traditional methods if

(i) it is quicker;

(ii) it uses less skilled labour;

(iii) it is more objective;

(iv) it more consistently identifies and uses natural discontinuities; or

(v) it uses optimization procedures.

These possible advantages will be considered *seriatim.*

Though at first glance it would seem likely that use of a digital computer could speed up the classification process, this is not necessarily true. The experienced field ecologist is often, half-consciously, developing and refining his conceptions of classification during the actual process of collecting stand data, whereas the computer can only start work after these data have been collected and converted to the form required for input. On the other hand, some non-numerical methods of classification involve steps which could be mechanized with great advantage. In particular, the work involved in rearranging stand × species tables in studies by the Zürich-Montpellier school (article 20) could be performed better and more quickly by computer.

The use of less skilled labour for numerical classification is a real possible advantage. In the simpler types of vegetation, records of stand composition can be made by relatively unskilled labour. If then the process of analyzing these records and sorting them into classes can be given to a suitably programmed computer, skilled labour may be needed only in programme preparation. The resulting classifications might, however, have value only as evaluated and interpreted in the light of field experience and understanding.

It is also true that numerical classification is more objective than most traditional methods. The use of numerical methods requires that the classification process be defined exactly and in great detail and, once this algorithm has been written, it contains no ambiguity and allows no scope for individual judgement or choice. Anyone applying the same algorithm or computer program to the same data must arrive at the same conclusions. Thus all subjectivity is removed from this part of the classification process — though, of course, the choice of the particular algorithm used was, at an earlier stage, the subject of individual judgement.

An explicit search for discontinuities calls for a probabilistic approach, and consequently most of the techniques for numerical classification, being deterministic, do not meet this need in any direct fashion; in fact the intuitive approach, which will often look for discontinuities sub-consciously, may perform better in this respect. Even some of the probabilistic approaches described above, based on a model of uncorrelated attributes, will not distinguish an elongated cluster from two separate clusters. Consequently, though a consistent and objective search for discontinuities could be an advantage of numerical methods, it is not in fact realized in most of the existing techniques.

Optimization of a classification, in the sense that the classes predict attribute values of their members as precisely as a division into a given number of classes could do, may be approached by some intuitive classifications; but a consistent attainment of the optimum can only result from application of a numerical method. Many of the clustering methods described include optimization elements; but the quantity maximized or minimized is often only rather indirectly related to the intra-class variance. And, in the many hierarchical procedures, optimization of a quantity at each step of the hierarchy may not ensure its optimization in the final groupings. These difficulties seem to be avoided by the procedure of ORLOCI (section 19.5.2.2), which however is unsatisfactory when it comes to a 'stopping rule.'

The theoretical advantages of numerical over traditional classification seem thus to be more in potential than in performance. One may expect further development of probabilistic methods which will be based on natural discontinuities in the data, and will perform an optimal allocation of samples adjacent to the discontinuities. Meanwhile, advocacy of the use of numerical methods must be based more on economy in time and skill than on their theoretical advantages.

It has been mentioned that it may be useful to precede a numerical classification by an ordination, which will make clearer the main directions of variation, and the broad structure of the data collected. It has, moreover, been suggested that classes which have been distinguished should subsequently be ordinated as a means of establishing and presenting their mutual relationships. In this event, it may be asked, what is the function or value of a classification process sandwiched between these two ordination operations? The answer lies simply in the convenience, for human thought, of the class concept. It is much easier to think or write about a group of stands considered simply as members of a class, whose variation within the class can conveniently be ignored, than as individuals placed within a continuum. In a continuum concept, all statements about the composition or behavior of a stand would have to be expressed as a function of position in the continuum, whereas in terms of a class concept they can be absolute for each class. The biggest advantage, however, probably comes in relation to mapping. To map the spatial distribution of a number of mutually exclusive classes is easy. To map position in a continuum in relation to the two-dimensional space of the earth's surface would require a series of overlay contour diagrams, one for each axis of the continuum, and would be far more difficult to interpret.

Though preliminary ordination is desirable, it is likely that many investigators will prefer to proceed directly to classification, using one of the existing clustering methods. Choice among them should depend, among other things, on computing costs (LAMBERT & DALE 1964); but the risks of chance misclassification with monothetic methods (see above) argue for one of the polythetic methods, despite their usually greater computing load. The polythetic methods have the advantage of greater flexibility; and, in fact, they include monothetic division as a possible, but improbable, outcome. Where quantitative variables are used, the best choice seems to be either the agglomerative method of ORLOCI (section 19.5.2.2) or the divisive method of MAC NAUGHTON-SMITH et al. (section 19.5.1.4); for binary variables (presence and absence of species) one may use the agglomerative method based on information theory (section 19.5.2.5); and for mixed variables the probabilistic method described in section 19.5.2.4 may be appropriate. There is great need for extensive trials of the power of these different techniques to detect known discontinuities in artificial populations, and the differences in power revealed could then be balanced against differences in cost.

As for the choice of variables, there seems reason to support the view that, where the number of species and the range of variation to be covered are large, species presence may supply all the information needed for classification, and quantitative variables may even cloud the issue. If only a fraction of the species is being used in the analysis, care should be taken in their selection. It cannot be assumed that the less frequent species can safely be ignored (GOODALL 1969); a better rule might be to omit from the analysis those species which show little association with others.

Where the group of stands to be classified cover only a rather limited range of variation, or where the vegetation is poor in species, quantitative variables will generally be needed for the additional information they contain. Thus, it may often be appropriate to perform some inital subdivisions on the basis of species presence and absence, and then to examine these classes in greater detail, using quantitative measures of species participation.

It should be emphasized again, though, that direct application of clustering methods should be regarded as a rather makeshift procedure. As further experience in this field accumulates it is to be expected that techniques will develop involving ordination in the first instance, which will permit discontinuities and clusters to be identified in a single operation, rather than by a hierarchical procedure, and based on tests of successively more complex null hypotheses regarding the cluster structure of the underlying population of stands from which a sample has been taken.

19.11.1 SUMMARY

Considering stands of vegetation to be represented by points in a multidimensional system of which the axes are the variables by which the stands may be described, classification amounts to the division of this space into discrete cells by hypersurfaces. If the distribution of points is interrupted by discontinuities or regions of low density, the dividing hypersurfaces should follow these discontinuities. Even if there are no such discontinuities, division of vegetation space into arbitrary classes may still serve practical purposes.

It is recommended that the distribution of points in vegetation space — simplified by an ordination — should be studied as a preliminary to numerical classification. If the distribution is continuous, an arbitrary number of classes may be chosen, and dividing hypersurfaces selected which will minimize the total intra-class variance. If discontinuities are revealed, they should be identified, and the classes delimited in reference to them.

The various 'clustering' methods in vogue should be regarded as approximations to the ideal procedure described above. They are usually hierarchic, proceeding either by successive divisions of the complete group of stands, or by succesive amalgamation of smaller, then larger, groups. Monothetic divisive procedures, in which each successive division is based on the value of a single variable, are usually simplest and most expeditious, but are less reliable than the polythetic methods (depending on several variables simultaneously) which, for practical reasons, are usually agglomerative. All these clustering methods may be improved by abandoning the strictly hierarchical principle, and at the end of the process re-examining the resulting classes for possible recombination or for transfers of stands from one class to another. Since the stands classified are only a sample of the population of stands to which the classification is intended to apply, probabilistic tests are highly desirable as part of the classification procedure.

Any relevant variables may be used for classification, but the floristic variables have most usually been employed — either some measure of the quantity of each species present, or a binary variable expressing its presence or absence only. For classifications covering a wide range of stands the latter is recommended; but when the focus narrows quantitative variables are more useful.

Once primary classes of stands have been recognized the interrelations among the classes may be investigated, either by ordination or by a further classification procedure, which should now be nonprobabilistic.

9.11.2 Zusammenfassung

Wenn Vegetationsaufnahmen durch Punkte in einem vieldimensionalen System dargestellt werden, dessen Achsen die Variabeln sind, mit welchen die Bestände beschrieben werden können, läuft die Klassifikation auf die Teilung dieses Raumes durch Hyperflächen in diskrete Zellen hinaus. Wenn die Verteilung der Punkte durch Lücken oder durch Regionen von geringer Dichte unterbrochen wird, sollten die abgesonderten Hyperflächen diesen Diskontinuitäten folgen. Auch wenn keine solche Unregelmäßigkeiten vorkommen, kann die Einteilung des Vegetations-Raumes in willkürliche Klassen noch praktischen Zwecken dienen.

Es wird empfohlen, die Verteilung von Punkten im Vegetations-Raum — die durch eine Ordination vereinfacht wird — als Vorarbeit der numerischen Klassifikation zu untersuchen. Wenn die Verteilung stetig ist, können eine willkürliche Zahl von Klassen und trennende Hyperflächen ausgewählt werden, welche die gesamte Zwischen-Klassen-Varianz auf ein Minimum einschränken. Wenn Diskontinuitäten gefunden werden, sollten diese identifiziert und die Klassen in Beziehung zu ihnen begrenzt werden.

Die verschiedenen beliebten Schwarm-('clustering') Methoden sollten als Annäherungen des oben beschriebenen idealen Verfahrens aufgefaßt werden. Sie sind meistens hierarchisch und entspringen entweder fortarbeitender Teilung der gesamten Gruppe von Aufnahmen oder aufeinanderfolgender Zusammenfügung kleiner und dann größeren Gruppen. Einseitige (monothetische) Teilungs-Verfahren bei denen jede folgende Teilung auf den Wert einer einzigen Variabelen begründet ist, sind gewöhnlich am einfachsten und am schnellsten durchführbar, aber sie sind weniger zuverlässig als die vielseitigen (polythetischen, auf mehreren Variabelen beruhenden) Methoden, die gleichzeitig und aus praktischen Gründen meistens zusammenballend (agglomerativ) sind. Alle diese Schwarm-Methoden können verbessert werden, wenn das streng hierarchische Prinzip verlassen wird, und am Ende des Verfahrens die sich ergebenden Klassen auf mögliche Wiedervereinigung oder auf Übertragungen von Aufnahmen von der einen in die andere Klasse erneut untersucht werden. Weil die eingeordneten (klassifizierten) Aufnahmen nur eine Stichprobe der Gesamtheit von Beständen darstellen, auf welche die Klassifikation angewandt werden soll, sind Wahrscheinlichkeits-Prüfungen als Teil des Klassifikationverfahrens hochst erwünscht.

Jede wichtige Variabele kann für eine Klassifikation verwendet werden, aber die floristischen Variablen sind am meisten verwendet worden — entweder als Maß für die Menge jeder vorhandenen Art oder als eine binäre Variable, die nur ihre Anwesenheit oder ihr Fehlen ausdrückt. Für Klassifikationen eines weiten Bereiches von Aufnahmen wird die letzte Methode empfohlen; aber wenn der Brennpunkt sich verengt, sind quantitative Variablen nutzlicher.

Wenn Grund-Klassen von Aufnahmen erkannt worden sind,

können die Beziehungen zwischen den Klassen untersucht werden, sei es durch Ordination oder durch ein weiteres Klassifikations-Verfahren, welches nur non-probabilistisch sein sollte.

REFERENCES

BRISSE, H. & G. GRANDJOUAN, – 1971 – Adaptation d'une méthode de classification multivariable par similitudes à l'écologie végétale en milieu naturel. I. Exposé de la méthode. *Oecol. Pl.* 6: 163—187.

BROWN, R. T. & J. T. CURTIS, – 1952 – An ordination of the upland forest communities of southern Wisconsin. *Ecol. Monogr.* 22: 217—234.

CRAWFORD, R. M. M. & D. WISHART, – 1967 – A rapid multivariate method for the detection and classification of groups of ecologically related species. *J. Ecol.* 55: 505—524.

CRAWFORD, R. M. M. & D. WISHART, – 1968 – A rapid classification and ordination method and its application to vegetation mapping. *J. Ecol.* 56: 385—404.

CURTIS, J. T. & R. P. MCINTOSH, – 1951 – An upland forest continuum in the prairie-forest border region of Wisconsin. *Ecology* 32: 476—496.

CZEKANOWSKI, J., – 1909 – Differentialdiagnose der Neandertalgruppe. *Korresp Bl. dt. Ges. Anthrop.* 40: 44—47.

DALE, M. B., – 1971 – Information analysis of quantitative data. In, 'Statistical Ecology,' ed. G. P. PATIL, E. C. PIELOU & W. E. WATERS. Pa. State Univ., University Park 3: 133—148.

EDWARDS, A. W. F. & L. L. CAVALLI-SFORZA, – 1965 – A method for cluster analysis. *Biometrics* 21: 362—375.

ESTABROOK, G. F., – 1967 – An information theory model for character analysis. *Taxon* 16: 86—97.

FLENLEY, J. R., – 1969 – The vegetation of the Wabag region, New Guinea Highlands: a numerical study. *J. Ecol.* 57: 465—490.

GILBERT, N. & T. C. E. WELLS, – 1966 – Analysis of quadrat data. *J. Ecol.* 54: 675—685.

GOODALL, D. W., – 1953 – Objective methods for the classification of vegetation. I. The use of positive interspecific correlation. *Aust. J. Bot.* 1: 39—63.

GOODALL, D. W., – 1954 – Objective methods for the classification of vegetation. III. An essay in the use of factor analysis. *Aust. J. Bot.* 2: 304—324.

GOODALL, D. W., – 1961 – Objective methods for the classification of vegetation. IV. Pattern and minimal area. *Aust. J. Bot.* 9: 162—196.

GOODALL, D. W., – 1966a – Deviant index - a new tool for numerical taxonomy. *Nature, Lond.* 210: 216.

GOODALL, D. W., – 1966b – Classification, probability and utility. *Nature, Lond.* 211: 53—54.

GOODALL, D. W., – 1966c – Hypothesis testing in classification. *Nature, Lond.* 211: 329—330.

GOODALL, D. W., – 1966d – A new similarity index based on probability. *Biometrics* 22: 882—907.

GOODALL, D. W., – 1968 – Affinity between an individual and a cluster in numerical taxonomy. *Biométr.-Praxim.* 9: 52—55.

GOODALL, D. W., – 1969 – A procedure for the recognition of uncommon species combinations in sets of vegetation samples. *Vegetatio* 18: 19—35.

GOODALL, D. W., – 1970 – Statistical plant ecology. *Ann. Rev. Ecol. Syst.* 1: 99—124.

GOODALL, D. W., – 1971 – Cluster analysis using similarity and dissimilarity. *Biométr.-Praxim.* 11: 34—41.

GOUNOT, M., – 1969 – Méthodes d'étude quantitative de la végétation. Masson, Paris. 314 pp.

GREIG-SMITH, P., – 1971 – Application of numerical methods to tropical forests. In, 'Statistical Ecology,' ed. G. P. PATIL, E. C. PIELOU & W. E. WATERS. Pa. State Univ., University Park 3: 195—204.

GREIG-SMITH, P., M. P. AUSTIN, & T. C. WHITMORE, – 1967 – The application of quantitative methods to vegetation survey. I. Association-analysis and principal component ordination of rain forest. *J. Ecol.* 55: 483—503.

HALL, A. V., – 1967a – Methods for demonstrating resemblance in taxonomy and ecology. *Nature, Lond.* 214: 830—831.

HALL, A. V., – 1967b – Studies in recently developed group-forming procedures in taxonomy and ecology. *J. S. Afr. Bot.* 33: 185—196.

HALL, A. V., – 1970 – A computer-based method for showing continua and communities in ecology. *J. Ecol.* 58: 591—602.

HOPKINS, B., – 1968 – Vegetation of the Olokemji Forest Reserve, Nigeria V. The vegetation on the savanna site with special reference to seasonal changes. *J. Ecol.* 56: 97—115.

JACCARD, P., – 1901 – Distribution de la flore alpine dans le Bassin des Dranses et dans quelques régions voisines. *Bull. Soc. vaud. Sci. nat.* 37: 241—272.

KULCZYŃSKI, S., – 1928 – Die Pflanzenassoziationen der Pieninen. *Bull. int. Acad. pol. Sci. Lett.*, Cl. Sci. Math. Nat., Ser. B., 1927 (Suppl. 2): 57—203.

LAMBERT, J. M. & M. B. DALE, – 1964 – The use of statistics in phytosociology. *Adv. ecol. Res.* 2: 59—99.

LAMBERT, J. M. & W. T. WILLIAMS, – 1966 – Multivariate methods in plant ecology. IV. Comparison of information-analysis and association-analysis. *J. Ecol.* 54: 635—664.

LANCE, G. N. & W. T. WILLIAMS, – 1967 – A general theory of classificatory sorting strategies. I. Hierarchical systems. *Comput. J.* 9: 373—380.

LANCE, G. N. & W. T. WILLIAMS, – 1968 – Note of new information-statistic classificatory program. *Comput. J.* 11: 195.

LIETH, H. & G. W. MOORE, – 1971 – Computerized clustering of species in phytosociological tables and its utilization for field work. In, 'Statistical Ecology,' ed. G. P. PATIL, E. C. PIELOU & W. E. WATERS. Pa. State Univ., University Park 1: 403—422.

LOOMAN, J., – 1963 – Preliminary classification of grasslands in Saskatchewan. *Ecology* 44: 15—29.

MACNAUGHTON-SMITH, P., W. T. WILLIAMS, M. B. DALE, & L. G. MOCKETT, – 1964 – Dissimilarity analysis: a new technique of hierarchical sub-division. *Nature, Lond.* 202: 1034—1035.

MATUSZKIEWICZ, W. & A. MATUSZKIEWICZ, – 1956 – Materiały do fitosocjologicznej systematyki ciepłolubnych dąbrów w Polsce (Germ. summ.: Zur Systematik der Quercetalia pubescentis-Gesellschaften in Polen). *Acta Soc. bot. Pol.* 25: 27—72.

MOORE, G. W., W. S. BENNINGHOFF, & P. S. DWYER, – 1967 – A computer method for the arrangement of phytosociological tables. *Proc. Ass. Comput. Mach.* 297—299.

MOTYKA, J., – 1947 – O zadaniach i metodach badań geobotanicznych (French summ.: Sur les buts et les méthodes des recherches géobotaniques). *Annls. Univ Mariae Curie-Skłodowska*, Lublin, Sect. C. Suppl. 1: 1—168.

NOY-MEIR, E., N. H. TADMOR, & G. ORSHAN,– 1970 – Association analysis of desert vegetation. *Israel J. Bot.* 19: 561—591.

ORLOCI, L., – 1967 – An agglomerative method for classification of plant communities. *J. Ecol.* 55: 193—206.

ORLOCI, L., – 1968 – Definitions of structure in multivariate phytosociological samples. *Vegetatio* 15: 281—291.
ORLOCI, L., – 1969 – Information analysis of structure in biological collections. *Nature, Lond.* 223: 283—284.
ORLOCI, L., – 1970 – Analysis of vegetation samples based on the use of information. *J. theoret. Biol.* 29: 173—189.
ORLOCI, L., – 1971 – Information theory techniques for classifying plant communities. In, 'Statistical Ecology,' ed. G. P. PATIL, E. C. PIELOU & W. E. WATERS. Pa. State Univ., University Park 3: 259—279.
PIELOU, E. C., – 1969 – An Introduction to Mathematical Ecology. Wiley-Interscience, New York. 286 pp.
POORE, M. E. D. ,– 1955 – The use of phytosociological methods in ecological investigations. III. Practical applications. *J. Ecol.* 43: 606—651.
SNEATH, P. H. A., – 1957 – The application of computers to taxonomy. *J. gen. Microbiol.* 17: 201—226.
SOKAL, R. R. & P. H. A. SNEATH, – 1963 – Principles of Numerical Taxonomy. Freeman, San Francisco. 359 pp.
VASILEVICH, V. I., – 1969 – Statistical Methods in Geobotany (In Russian). Nauka, Leningrad. 232 pp.
WARD, S. D., – 1971 – The phytosociology of *Calluna-Arctostaphylos* heaths in Scotland and Scandinavia. I. Dinnet Moor, Aberdeenshire. *J. Ecol.* 58: 847—863.
WEBB, L. J., J. G. TRACEY, W. T. WILLIAMS, & G. N. LANCE, – 1967a – Studies in the numerical analysis of complex rain-forest communities. I. A comparison of methods applicable to site/species data. *J. Ecol.* 55: 171—191.
WEBB, L. J., J. G. TRACEY, W. T. WILLIAMS, & G. N. LANCE – 1967b – Studies in the numerical analysis of complex rain-forest communities. II. The problem of species-sampling. *J. Ecol.* 55: 525—538.
WEBB, L. J., J. G. TRACEY, W. T. WILLIAMS, & G. N. LANCE, – 1970 – Studies in the numerical analysis of complex rain-forest communities V. A comparison of the properties of floristic and physiognomic-structural data. *J. Ecol.* 58: 203—232.
WEST, N. A., – 1966 – Matrix cluster analysis of montane forest vegetation of the Oregon Cascades. *Ecology* 47: 975—980.
WHITTAKER, R. H., – 1962 – Classification of natural communities. *Bot. Rev.* 28: 1—239.
WILLIAMS, W. T. & M. B. DALE, – 1962 – Partitioned correlation matrices for heterogeneous data. *Nature, Lond.* 196: 602—603.
WILLIAMS, W. T. & M. B. DALE, – 1965 – Fundamental problems in numerical taxonomy. *Adv. bot. Res.* 2: 35—68.
WILLIAMS, W. T. & J. M. LAMBERT, – 1959 – Multivariate methods in plant ecology. I. Association-analysis in plant communities. *J. Ecol.* 47: 83—101.
WILLIAMS, W. T. & J. M. LAMBERT, – 1960 – Multivariate methods in plant ecology. II. The use of an electronic digital computer for association-analysis. *J. Ecol.* 48: 689—710.
WILLIAMS, W. T., M. B. DALE, & P. MACNAUGHTON-SMITH, – 1964 – An objective method of weighting in similarity analysis. *Nature, Lond.* 201: 426.
WILLIAMS, W. T., J. M. LAMBERT, & G. N. LANCE, – 1966 – Multivariate methods in plant ecology. V. Similarity analysis and information-analysis. *J. Ecol.* 54: 427—445.
WILLIAMS, W. T., G. N. LANCE, L. WEBB, J. G. TRACEY, & M. B. DALE, – 1969 – Studies in the numerical analysis of complex rain-forest communities. III. The analysis of successional data. *J. Ecol.* 57: 515—535.

20 THE BRAUN-BLANQUET APPROACH

Victor Westhoff and Eddy van der Maarel

Contents

20 THE BRAUN-BLANQUET APPROACH

20.1 Introduction

We shall, in this last chapter, treat the floristic-sociological or BRAUN-BLANQUET approach to classification and interpretation of communities. Before elaborating the details of the approach, we may state its essence in three ideas:

(i) Plant communities are conceived as types of vegetation, recognized by their *floristic composition.* The full species compositions of communities better express their relationships to one another and environment than any other characteristic.

(ii) Amongst the species that make up the floristic composition of a community, some are more sensitive expressions of a given relationship than others. For practical classification (and indication of environment) the approach seeks to use those species whose ecological relationships make them most effective indicators; these are *diagnostic species* (character-species, differential-species, and constant companions).

(iii) Diagnostic species are used to organize communities into a *hierarchical classification* of which the association is the basic unit. The vast information with which phytosociologists deal must, of necessity, be thus organized; and the hierarchy is not merely necessary but invaluable for the understanding and communication of community relationships that it makes possible.

The reader will see how far the elaboration of this three-part theme has led the members of the 'school of BRAUN-BLANQUET.' The results are (like community relationships) complex. We fear lest the reader should lose sight of the heart of the approach — the floristic perspective — for the abundance of technical details that follow. The latter are only the skeleton of the approach. Furthermore, if we do not press the physical comparison too far, its body is the still-growing corpus of basic and applied research and the resulting knowledge of and insight into communities. This review must deal mainly with the skeleton of the approach; we cannot in the space we have discuss adequately the kinds of understanding of communities that have come from it. We ask the reader to recognize wherein this account must stop short, and to seek if he is interested a further feeling for what the approach offers by observing or reading studies applying it. On the other hand, we shall pay some attention (especially in sections 20.8 and 20.9) to recent developments in phytosociology that are not part of the BRAUN-BLANQUET approach as such but may be of value since they introduce concepts and methods that relate the floristic-sociological approach to other approaches.

Textbooks and reviews on the floristic-sociological approach have been published in most European countries, Argentine, India, Japan and the U.S.A. (see the bibliography by MAAREL, TÜXEN & WESTHOFF, 1970, Exc.[1]). Recent accounts include GOUNOT (1969), WESTHOFF & HELD (1969), and KNAPP (1971). Useful accounts in English are BECKING (1957), WHITTAKER (1962), KÜCHLER (1967) and SHIMWELL (1972).

20.2 **History**

The origin and history of the approach have been reviewed elsewhere and can be stated only briefly here (see GAMS 1918, BRAUN-BLANQUET 1921, and DU RIETZ 1921, as well as BECKING 1957, WHITTAKER 1962, and SHIMWELL 1972). The systematic description of plant communities and the idea of community types can be traced from great students of plant geography, HUMBOLDT (1805), SCHOUW (1823), HEER (1835) and GRISEBACH (1838). From their work two main approaches developed, the physiognomic and the floristic. The physiognomic approach developed as its principal unit the formation, a unit characterized by physiognomy or vegetation structure (POST 1862, GRISEBACH 1872, WARMING 1895; see article 13). Through the work of students dealing with smaller-scale units (esp. LECOQ 1844, 1855, THURMANN 1849, LORENZ 1858, KERNER 1863 and DRUDE 1890, in southern and central Europe, POST 1862, HULT 1881 and CAJANDER 1903 in northern Europe) there developed the essential idea of the floristic-sociological approach: plant communities as units of classification based primarily on species composition. From LORENZ (1863), MÖBIUS (1877) and DAHL (1908) a parallel approach to biotic communities can be traced. (See BALOGH 1958 and WHITTAKER 1962 for review and references.)

Much of the further development leading to the BRAUN-BLANQUET approach was centered in Zürich (STEBLER & SCHRÖTER 1893, SCHRÖTER 1894, SCHRÖTER & KIRCHNER 1902, BROCKMANN-

[1]) This and many other bibliographies to be mentioned in this contribution have been published in *Excerpta Botanica Sect. B Sociologica* (R. TÜXEN Ed). They will not be listed in the references, but indicated with 'Exc'.

Jerosch 1907) and Montpellier (Flahault 1893, 1898, 1901, Pavillard 1901, 1912). A main outcome of this period was a full hierarchy for vegetation classification—vegetation-type, formation-group, formation, subformation, stand-type down to local variants (Nebentypen) and geographic variants (Fazies). The stand-type was called the 'association' and considered the basic unit. Flahault & Schröter (1910) agreed on the following definition, presented to the Third International Botanical Congress in Brussels, 'An association is a plant community of definite floristic composition, presenting a uniform physiognomy, and growing in uniform habitat conditions. The association is the fundamental unit of synecology.' (translation following Pavillard 1935b).

From this background Braun-(Blanquet) carried out a monographic study of alpine vegetation (1913). In this, and the essay by Braun-Blanquet & Furrer (1913), attention was focused on 'Charakterpflanzen' or character species — species that possess 'fidelity' (relative restriction) to a given association. Gradmann (1909) had advocated the approach through floristic composition and the use of character species ('Leitpflanzen'). The key ideas in Braun-Blanquet's treatment were: (i) The study of communities should be based on a fundamental unit, comparable to the species. (ii) This unit should be the association, and associations should be defined by their possession of character-species. (iii) Each association consists (like a species) of 'individuals,' and the association (like the species) can be described from samples of its individuals. (iv) Each sample ('Aufnahme,' relevé) should be chosen so as to represent adequately such an individual, and it should include analysis of the complete species assemblage. (v) Associations should be grouped into higher units not by physiognomy, but by floristic composition.

Additions to the approach came in further publications. Braun-Blanquet (1915) added the 'Assoziationsgruppe' (later called the Verband, alliance) as a unit above the association, also defined by character-species. Braun-Blanquet (1918) added the subassociation, as a deviation from the typical association expressed in a constant floristic difference, and the facies as a subordinate unit possessing merely quantitative differences. Braun-Blanquet (1921) outlined essentially the full system, including the analytical scales (20.4.5) and the 'sociological progression' (20.3.4) for arranging communities by their levels of organization. Braun-Blanquet (1925) pressed the claims of fidelity or 'Gesellschaftstreue' as the key to vegetation systematics, against the disagreement of northern schools. In the same essay he introduced the 'characteristic species combination' (charakteristische Artenkombination after SCHMID, 1923) as the ultimate community diagnosis, comprising both character-species and constant companion species. KOCH (1925) and BRAUN-BLANQUET & JENNY (1926) added the concept of differential-species for the characterization of subordinate units, and the order as a unit above the alliance. BRAUN-BLANQUET & PAVILLARD (1922, 1925, 1928) in their *Vocabulaire* codified the analytic and synthetic procedures of the approach. Finally the science of plant sociology and its application were spelled out more fully in the first edition of the textbook of BRAUN-BLANQUET (1928, 1932).[1])

From these beginnings the influence of the approach spread through western and central Europe. We cannot review that spread, but will mention such leaders as PAVILLARD, ALLORGE and MOLINIER in France, TÜXEN and OBERDORFER in Germany, SZAFER and PAWŁOWSKI in Poland, FURRER in Switzerland, SOÓ in Hungary, KLIKA in Czechoslovakia, HORVAT, HORVATIĆ and WRABER in Yugoslavia, BORZA in Rumania, GIACOMINI and TOMASSELLI in Italy, BOLOS in Spain, LEBRUN in Belgium, and DE LEEUW in the Netherlands. References for these authors are found in BRAUN-BLANQUET (1964, 1968). BRAUN-BLANQUET's own work began at Zürich and continued from 1927 onwards at Montpellier, where he still leads the S.I.G.M.A., Station Internationale de Géobotanique Méditerranéenne et Alpine. A second centre was established 1932 at Stolzenau, West Germany under Reinhold TÜXEN, a pre-eminent leader of the approach and former director of the Zentralstelle (later, Bundesanstalt) für Vegetationskartierung, who has done much to give the approach its value in application. From TÜXEN's centre the approach spread to various non-European countries as well, and notably to Japan.

During its spread the approach has been known by several names: French-Swiss (or Swiss-French) school, Zürich-Montpellier school, Middle European-Mediterranean school, and Sigmatism (TANSLEY 1922, DU RIETZ 1936, BRAUN-BLANQUET 1959, 1968, EGLER 1954 ,BECKING 1957, WHITTAKER 1962, TÜXEN 1969a). Most of these terms lack specificity or are unclear. Moreover we feel that the time has passed for the word 'school' with its implications of a fixed system. We prefer in this article the most direct name, viz. BRAUN-BLANQUET approach, whilst we may characterize the essential ideas in the term, the 'floristic-sociological approach.'

---

[1]) One notes through this development of the full range of concepts the continuing contribution of BRAUN-BLANQUET himself, that has given the approach its present character (GAMS 1972 to the contrary).

20.3 General Concepts

20.3.1 Concrete vs. Abstract Units

Much dispute has centered on the nature of the plant community. This will be briefly reviewed as far as the BRAUN-BLANQUET approach is concerned. (See further PAVILLARD 1935, WESTHOFF 1951, 1970, WHITTAKER 1962, BRAUN-BLANQUET 1964.)

Concepts of the plant community include: (i) the organismal concept (CLEMENTS 1936, TANSLEY 1920): the community as a 'superorganism.' (ii) the concept of social structure (PACZOSKI 1930) and many early Russian authors such as SUKATSCHEW (1929). (iii) the individualistic concept (GLEASON 1926): the community as a changeable mixture of 'individualistically' distributed plant species. (iv) the concept of population structure (WHITTAKER 1953, 1962, 1970): the community as a system of interacting species and vegetation as a complex population pattern.

The BRAUN-BLANQUET approach takes a practical, intermediate position that recognizes the heterogeneity of species distributions but emphasizes nonetheless the interactions between plants in the community, which has a certain individuality because of relative discontinuities between communities in the field. Definitions of plant community or phytocoenose range from the more superficial, 'any collection of plants growing together which has as a whole a certain unity' (TANSLEY 1935), to the more profound, 'a plant community (+animal community = biotic community) is a working community; the species composition of which is in a sociologic-dynamical equilibrium in competition for space, minerals, water and energy, in which each component affects all others and which is characterized by harmony between environment and production and phenomena expressing life in form, colour and temporal course' (TÜXEN 1957).

As shown by TÜXEN's definition, the community concept may be broadened from the plant community to the biotic community of plants and animals: producers, consumers and decomposers. In a further broadening of perspective the biotic community (= biocoenose) plus its environment or biotope is treated as a functional unit, the ecosystem. The concept of ecosystem has become most familiar from its expression in English-language ecology by TANSLEY (1935), but the concept is largely identical with FRIEDERICH's (1927, 1958) 'holocoene' and the 'biogeocoenose' of SUKACHEV (e.g. 1954, 1960) and other Russian authors. Though by far the greatest development of the BRAUN-BLANQUET approach has been in application to plant communities, biotic communities are clearly amenable to parallel study (section 20.7.2). Tüxen (1957, 1965b) has emphasized this possibility and stressed the influence of Friederichs and Thieneman (e.g. 1956) on his own views.

The term plant community and its German equivalent 'Pflanzengesellschaft' have been used in both concrete and abstract senses, which has caused dispute and confusion (see Tansley 1920, 1935, Du Rietz 1921, Alechin 1926, Pavillard 1935, Westhoff 1951, 1965, Whittaker 1956, 1962). The floristic-sociological approach has always stressed the distinction between the concrete and the abstract community. The classification units were, naturally, abstract units. Concrete plant communities were often, especially in forestry, referred to as stand (Bestand), whilst concrete representatives of associations were even called association-individuals (Pavillard 1912, Braun-Blanquet & Pavillard 1922). This term has been much disputed and criticized (most thoroughly by Whittaker 1962) and we consider it now as of only historic interest.

Westhoff (1951, 1965) has acknowledged the distinction between concrete and abstract communities in separate definitions, which were linked to a general definition of vegetation. For the concrete plant community the term phytocoenose (Gams 1918) was proposed, for the abstract community the term phytocoenon (Maarel 1965, specifying the general term 'coenon' proposed by Barkman et al. 1958, Westhoff et al. 1959).

The term (phyto)coenon may be proposed as a suitable international term which may replace the terms community-type (Whittaker 1956, 1962) and nodum. The latter term has been suggested by Poore (1956, 1962), but has been used in a more specific meaning by Williams & Lambert (1961) in their 'nodal analysis' (cf Ivimey-Cook & Proctor 1966).

For biotic communities parallel terms: biocoenose and biocoenon may be used (Maarel 1965). For partial communities Mörzer Bruyns' (1950) term merocoenose can be used with merocoenon as the abstract equivalent. Layer communities could be called, stratocoenose-stratocoenon. For specific subcommunities consisting of plants of the same stratum, life-form and seasonal relations the couple, society-synusia may be reserved. (See further article 16.)

In conclusion we present Westhoff's (1951, 1970) definitions in a slightly adapted form. Vegetation is defined as a system of largely spontaneously growing plant populations, growing in coherence with their sites and forming an ecosystem with these sites and all other forms of life occurring in these sites. (Thus are excluded all assemblages of mobile plants and collections of plants growing in arrangements set up by man such as flowerbeds and arboreta.)

A phytocoenose is defined as a part of a vegetation consisting of interacting populations growing in a uniform environment and showing a floristic composition and structure that is relatively uniform and distinct from the surrounding vegetation.

A phytocoenon is defined as a type of phytocoenose derived from the characterization of a group of phytocoenoses corresponding with each other in all characters that are considered typologically relevant.

20.3.2 Floristic-sociological Classification Units

Phytocoena include such various kinds of vegetation units as formations defined by physiognomy, dominance-types defined by major species, forest site-types defined by undergrowth composition, and noda derived by quantitative comparisons or numerical procedures. The approach of Braun-Blanquet has its own, formal hierarchy of units, of which none of those just mentioned is a part. The fundamental unit of the hierarchy is the *association*, a unit that corresponds in function to the species as the fundamental unit of idiotaxonomy, or the classification of individual organisms. The word 'association' has had a long, complex, and argument-afflicted history as different schools sought to determine its meaning to their preference (Whittaker 1962, Shimwell 1972, see also articles 14, 17 and 18). In an earlier definition by Braun-Blanquet (1921) 'the association is a plant community characterized by definite floristic and sociological (organizational) features which shows, by the presence of character-species (exclusive, selective, and preferential) a certain independence.' Meijer Drees (1951) defined the association as ' a plant community identified by its characteristic taxon combination, including one or more (local) character-taxa or differentiating taxa.' A similar conception was agreed on during a Symposium on Plant Sociological Systematics at Stolzenau in 1964 (in Tüxen 1968b, especially Oberdorfer 1968) and a later colloquium at Rinteln (see Dierschke 1971). We shall return to characterization of the association (20.6.2) but emphasize that, in keeping with the floristic-sociological perspective, the association is defined by its *characteristic species combination* including character- and differential-species as well as companions (Begleiter) with high presence values (over 60%). The Sixth Botanical Congress at Amsterdam, 1935, accepted definition of the association by characteristic or differential species in the sense of Braun-Blanquet as one of three resolutions given in article 18.3.6.

As basic units associations, like species, are to be grouped into a hierarchy of higher units. Associations are classed into alliances, alliances into orders, orders into classes, and classes into divisions (20.6.4-5); associations are divided into lower units of the hierarchy (20.6.7). All these units are coena as defined above; but the coena of the formal system of BRAUN-BLANQUET may be termed *syntaxa* (BARKMAN et al. 1958, WESTHOFF et al. 1959). Thus the parallelism in the classification of organisms and phytocoenoses is realized: to the species as a fundamental unit corresponds the association, to the taxon for units on any level corresponds the syntaxon, to (idio)taxonomy for the practice of classification corresponds syntaxonomy, to (idio)systematics as the broader study of relationships among organisms corresponds synsystematics.

20.3.3 DIAGNOSTIC SPECIES

Syntaxa are defined by diagnostic species (character-species, differential species, and constant companions). *Character-species* are species that are relatively restricted to the stands (or samples) of a given phytocoenon, and therefore characterize it and indicate its environment (Fig. 1). In ideal, a group of character-species is used for the characterization of a syntaxon of the BRAUN-BLANQUET classification on any level from the association to the class. To serve as character-species for an association, a species should have a relatively narrow distribution even if it is not simply restricted to the association. (Degrees of restriction to a given syntaxon, which are termed degrees of *fidelity* or Treue, are discussed in 20.5.6.) Note that the concept does not say that the species need be important in phytocoenoses; very minor species may have diagnostic value.

The German term is 'Charakterart'; this has been variously translated into English as characteristic species (BRAUN-BLANQUET 1932), which seems unsatisfactory, faithful species (POORE 1955a, BARKMAN 1958a, BEEFTINK 1962, MOORE 1962, WESTHOFF 1959), and character-species (BECKING 1957, WHITTAKER 1962). The latter, most direct translation, is our preference. HEIMANS (1939) introduced 'kensoort' in Dutch to avoid the germanism 'karaktersoort'; and the Dutch kensoort returned to German as *Kennart* (TÜXEN 1950: 99). Most phytosociologists writing in German now use that term, which is accepted in the third edition of BRAUN-BLANQUET's (1964) text. Syntaxa are most often defined by the fundamental units of idiotaxonomy — species —, but this is not always the case. Sometimes plant subspecies, varieties, or ecotypes may contribute to the definition of lower-level syntaxa or geographically vicariant

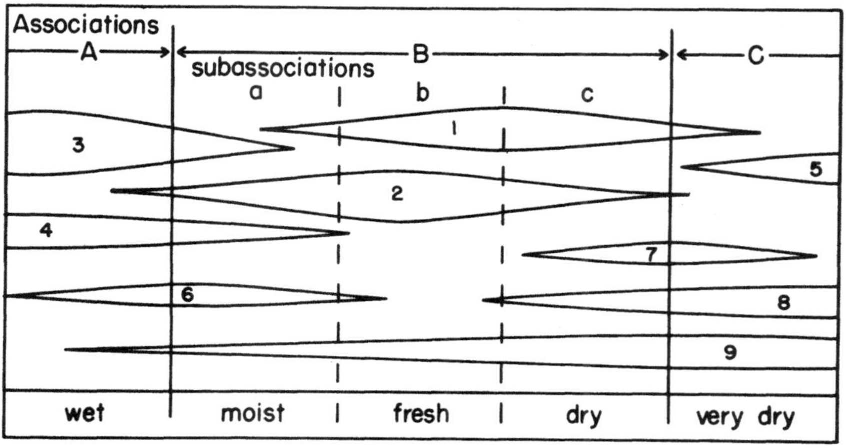

Fig. 1. Diagnostic species along a moisture gradient. Species 1 and 2 are character-species for association *B*, and have their populations centered in (or largely confined to) that association. Species 3 and 4 are character-species for association *A*, and species 5 is a character-species for association *C*. Species 4 and 6 are differential-species for subassociation *a* of association *B*, and species 7 and 8 are differential-species for subassociation *c* of association *B*. In each case presence of the differential-species distinguishes the moist, or dry, subassociation from the "typical" (fresh) subassociation *b*. Species 9 is a more widely distributed species that might help characterize association *B* as a constant companion for that association. Species 9 might also be a character-species for a higher syntaxon, such as an alliance uniting associations *B* and *C* with other associations.

associations. Sometimes a genus or subgenus may be used to define a higher syntaxon, as the alliance Spartinion is defined by the fidelity of any species of the genus *Spartina* (BEEFTINK 1965, 1968, see also PIGNATTI 1968a: 72, WESTHOFF apud PIGNATTI 1968a: 74). It is therefore appropriate to broaden character-species to character-taxon (BARKMAN et al. 1958) as we shall do here following Dutch use of 'kentaxon' (WESTHOFF & HELD 1969). The German equivalent is of course 'Kenntaxon'.

Character-species characterize syntaxa by their normal occurrence in phytocoenoses of that syntaxon, contrasted with their absence less frequent occurrence or smaller total estimate in phytocoenoses of all other syntaxa. It is possible also to distinguish closely related syntaxa by the presence and absence of certain species without concern for the broader distributions of those species. Two subassociations of an association may be characterized by the fact that samples of one normally include species 4 and 6, whereas samples of the other normally lack these species (Fig. 1). These species function as *differential-species* that distinguish the two subassociations. Differential-species, as the second class of diagnostic species, define syntaxa on the basis of the distributional boundaries of species (without regard to fidelity for the syntaxa in question), and are used primarily to define lower syntaxa.

Since being introduced by KOCH (1925) and BRAUN-BLANQUET (1925) as 'Differentialart', the concept has entered English as differential species (BRAUN-BLANQUET 1932, WHITTAKER 1962) or differentiating species (BECKING 1957), whilst in German the term *Trennart* has come into general use (TÜXEN 1950, 1952, BRAUN-BLANQUET 1959, 1964a) in parallel with Kennart. The use of differential-species has been further developed by SCHWICKERATH (1931, 1942, 1954a) and TÜXEN (1937, 1950, 1952). Differential-species and Trennart are synonyms, and these are largely synonymous also with differentiating species; but we shall here use 'differentiating species' for the more informal use of diagnostic species of unassigned syntaxonomic position in the synthetic phase of research (20.5), and speak of differential-species and differential-species groups for the formal characterization of syntaxa (20.6). A further recent development is the concept of 'character-combination' (after the Dutch kencombinatie) introduced by WESTHOFF & HELD (1969) following BEEFTINK (1965, 'differentiating species combination'). The essential concept is the exclusiveness to a syntaxon of the combination of species (or taxa), whilst none of them need be a character-taxon.

A third group of species with diagnostic value are the constant companions (20.5.3) which occur in most relevés of a syntaxon but are not designated as character- or differential-species. Constant companions are added to the character-taxa to form the 'characteristic species (taxon) combination' for associations and higher taxa. Although BRAUN-BLANQUET (1925) introduced this concept rather early and repeatedly (e.g. 1959, see MOORE 1962) stressed its relevance there has been doubt amongst critics (e.g. POORE 1955-1956). Of course, as BRAUN-BLANQUET (e.g. 1932: 68) remarked for associations, these syntaxa are established best which possess a high proportion of both characteristic and constant species.

20.3.4 SOCIOLOGICAL PROGRESSION

BRAUN-BLANQUET (1921) sketched an 'arrangement of the plant communities according to their sociological progression.' The idea of this arrangement was again a parallel with idiobiology, specifically with the arrangement of taxonomic groups by evolutionary level. Levels of structural and organizational development were chosen as criteria for arranging phytocoenoses from moving aero- and hydroplankton communities on the lowest level to the tropical rain forest at the highest level. Other communities are arranged between these according to increasing stratification, complexity, presence of dependent communities, species richness, diversity of growth-forms, and assumed intensity of species interactions. Thus BRAUN-BLANQUET (1921) summed up many of the structural-successional trends much later commented on by ODUM (1969)! The scheme was later modified in detail, and increasing stability was added as a criterion. BRAUN-BLANQUET (1964) arranges higher syntaxa (classes and orders) in the progression.

WAGNER (1968) called the sociological progression a 'purely artificial division principle' (cf. ELLENBERG 1963). Of course, no phylogenetic meaning should be imputed to its sequence. Yet the sociological progression is useful, as a principle at right angle to the levels of the hierarchy, in arranging syntaxa in surveys. Its use is illustrated in vegetation monographs by TÜXEN (1937), OBERDORFER (1957), ELLENBERG (1963), OBERDORFER et al. (1967), WESTHOFF & HELD (1969), and others. It may be of interest that it has appeared also as an axis of broad-scale ordination of syntaxa (MAAREL 1972a, see 20.9.2).

20.3.5 NATURAL CLASSIFICATION

The problem of 'natural' classification has been discussed in another article (12.1.2). It has been accepted by BRAUN-BLANQUET and others that associations are abstract units. A degree of continuity or gradualness in transitions between phytocoena is recognized, but considered not to preclude classification. 'We are convinced that the plant cover of the earth in all its dimensions can be divided into phytosociological groupings of higher and lower rank; their delineation may be either sharp or less sharp and gliding.' (BRAUN-BLANQUET and TÜXEN in comment on GOODALL 1963, translated.)

American research in gradient analysis has led to interpretations that may seem in conflict with those of BRAUN-BLANQUET. The American work of WHITTAKER (1951, 1954, 1956, 1962, 1967, articles 2 and 3) and CURTIS (CURTIS & McINTOSH 1951, CURTIS 1959, McINTOSH 1967, article 7) developed in parallel to Russian work of RAMENSKY (1930, article 17) toward the concept expressed by WHITTAKER (1970) as 'the population structure of vegetation.' In this concept the population structure of the individual phytocoenose may be analysed through dominance-diversity and species-importance relations, and niche differentiation among plant species expressed in stratification and functional differences among them. The phytocoenose is then conceived as a 'system of interacting, niche-differentiated and partially competitive species.' The vegetation of an area, on the other hand, is to be approached through gradient analysis, studying the manner in which species populations are distributed along environmental gradients and combined into phytocoenoses. Species are differently distributed from one another, so that undisturbed phytocoenoses intergrade continuously in most areas. 'In relation to patterns of environmental gradients, communities form complex and largely continuous population patterns.' (WHITTAKER 1970, article 3.9.)

It is further recognized that, 'Because of environmental interruptions and some relative discontinuities inherent in vegetation itself, the pattern may also be considered a complex mixture of continuity and relative discontinuity' (WHITTAKER 1956: 32). Vegetational continuity by no means precludes useful classification (WHITTAKER 1956, 1962, 1970). Gradient analysis implies a vision of vegetation, regarded as a coherent pattern of intergrading phytocoenoses, different from that of phytosociologists who tend to see it in terms of a typology of phytocoena (see, however, 20.7.4 on complexes). Yet we think that the difference between the concept of the gradient approach and BRAUN-BLANQUET's is clearly one of degree — of emphasis of continuity vs. discontinuity where both are present, of species individuality vs. species groupings where both are realistic, and of gradient analysis vs. classification where both are possible. We judge the difference between less extreme students of gradient analysis, and of the BRAUN-BLANQUET approach, to be one of emphasis and perspective, not one of fact or understanding.

Given the individuality of species distributions and some degree of continuity between phytocoenoses, classification is not strictly natural in the sense defined in article 12.1.2. BRAUN-BLANQUET (1951a: 561, 1959: 147) has considered argument on the naturalness of classification pointless. TÜXEN (1955), in response to a criticism of ELLENBERG (1954b), interpreted phytosociological classification through the concept of *types* as ideal concepts, recognized in an empirical way from 'correlation concentrates,' i.e. groups of correlated characters. That which is evident and characteristic of a type is always its nucleus, not its periphery; types are not pigeonholes but foci in a field of variation. GLAHN (1968, see also RAUSCHERT 1969) elaborated the concept of vegetation type, distinguishing as its three aspects: (i) The vegetation type as identity: repetitions of certain observations are approached via intuitive integration, resulting in vegetation-type concepts based on recurring combinations of species. (ii) The vegetation type as maximal correlative concentration: The joint floristic-sociological and ecological approach leads to types as correspondences of recurring species combinations with recurring combinations of environmental factors. (iii) The vegetation type as systematic category: A hierarchic system is feasible as a result of an integrative, inductive process grouping the initial types into higher and higher ranks on the basis of common species combinations. The similarity of this statement to the theory of classification developed at greater length by WHITTAKER (1962) may be noted. Relative naturalness of classification in this perspective is to be judged by success in embodying the maximum number of significant relationships among phytocoena and species in the structure of the hierarchy. We assert that compared with other classifications the BRAUN-BLANQUET approach, with its floristic emphasis and maximum use of species distributional relationships for classification, does not fall short.

20.4 **Analytical Research Phase**

20.4.1 RECONNAISSANCE AND CHOICE OF PLOTS

An analysis of the vegetation is as a rule preceded by a preliminary survey of the area, when this is little known to the investigator. This reconnaissance (see CAIN & CASTRO 1959) includes a study of the general vegetation pattern, and the establishment of the apparent relations of the various vegetation types with geology, topography and soil conditions. The next step, 'primary survey,' including a superficial description of the major communities, is mostly passed over in the BRAUN-BLANQUET approach.

A vegetation analysis starts with the choice of stands (phytocoenoses) on the basis of the reconnaissance. Within the stand one sample plot is laid down, often covering a large part of the stand. The analysis of this plot is called the *relevé*. The BRAUN-BLANQUET approach has often been criticized for the subjectivity of its sampling procedure. However, subjectivity of stand choice must be accepted in the procedures of many empirical sciences. A selection of relevés is desired that will effectively represent the variation in the vegetation under study, the samples being so chosen that they will not represent different phytocoena disproportionately and will not include mixed, incomplete, or unstable stands. For this purpose of equitable representation of different kinds of communities with most useful relevés, a subjective, 'stratified' sample selection is far superior to sample choice by random points on a map.

Relevés may be undertaken without classification as a purpose. They may then serve for the study of vegetation dynamics on a given area (periodicity, fluctuation, or succession) or another ecological purpose, e.g. the study of the ecology of a certain taxon or ecotype by describing its pattern in relation to the pattern of the vegetation. In most cases, however, relevés are intended to be used for some form of classification or ordination. In that situation, the only preconception in this choice is the demand for uniformity of the stand.

'Uniformity' may better express what is sought than 'homogeneity.' This chapter is not the place for a treatment of the mathematical aspects of homogeneity in vegetation; we may refer to POORE (1955, 1956), ELLENBERG (1956), DAHL (1957, 1960), BECKING (1957), LAMBERT & DALE (1964), BRAUN-BLANQUET (1964), MAAREL (1966b) and GOUNOT (1969) for general considerations, to GOODALL (1952, see also his bibliography, 1962 Exc.), DAHL & HADAČ (1949), DAHL (1957), GREIG-SMITH (1964) for aspects of homogeneity in the distribution of plant individuals, to GOODALL (1961) and GREIG-SMITH (1964), for pattern analysis, to RAUNKIAER (e.g. 1934), GUINOCHET (1955), DAHL (1957), CAIN & CASTRO (1959) for the relation between homogeneity and frequency distribution and to CURTIS (1959), GODRON (1966) and GOUNOT (1969) for infrastand similarities and information measures as approaches to homogeneity.

The first condition is that no obvious structural boundaries or variation in stratification are visible within the stand. The second criterion is uniform floristic composition. It is usual to look for joint patterns of dominant and/or abundant species and then to delimit a stand where qualitative changes in patterns occur, i.e. where one or more species drop out and others come in. In many cases an experienced field worker is able to judge this rapidly. In many cases however the changes in pattern are quantitative rather than qualitative. The species composition differs little from one site to an adjacent one, but the relative proportions of abundance and cover do vary. It is usual to make separate relevés in any case where the abiotic habitat factors show a clear discontinuity or at least a marked transition.

Many species cause pattern heterogeneity only by their growth form or their aggregation or shoot clustering. In such a case one may try at first to study this biotically heterogeneous pattern as one single stand, i.e. with one relevé. However, the situation becomes different as soon as the crowding of one major species leads to the establishment of one or more other species which appear to be locally bound to it. A simple example is the establishment of a dwarf shrub patch in a herbaceous vegetation, e.g. a clone of *Salix repens* in open dune grassland. When one or more species are found particularly bound to that dwarf shrub patch we prefer to consider the latter as a separate phytocoenose which has thus to be analysed with a separate relevé. However, there are exceptions to this rule. E.g. OBERDORFER (1970, Canaries) and WERGER (1973, South Africa) report stable savanna mosaics consisting of a dwarf-shrub and grassland and a low tree and shrub phytocoenose. Such mosaics are considered one vegetation type, since one does not find locally more extensive patches of either phytocoenose without adjacent patches of the other one.

In intermediate or doubtful cases, such as a swamp community with tall tussocks of sedges (*Carex hudsonii*) alternating with wetter hollows, it will be advisable, at least when the type of vegetation pattern is unknown, to follow both procedures, viz. analysing the pattern as a whole as well as relevés of tussocks and hollows separately. It will then turn out later which relevé is more useful for the classification purpose.

20.4.2 BOUNDARIES

As was said before, the recognition of distinct stands may presuppose the occurrence of discontinuities in the field. Although according to the opinion of BRAUN-BLANQUET workers such discontinuities are mostly to be observed, it is obvious that boundaries between stands are less sharp in some cases than in others. In such cases the boundary may be detected by means of a belt transect analysis or a 'line taxation.' Such transects consist of a series of small quadrats laid down at right angles to the extension of the boundary zone. By comparison of quantitative data on species occurrences in the quadrats the boundary zone can be discerned from the uniform phytocoenoses. MAAREL (e.g., MAAREL & LEERTOUWER 1967) and FRESCO (1972) refined boundary analysis by constructing 'differential profiles,' in which can be shown degrees of change between adjacent quadrats along a gradient. For similar techniques see article 5.4.4.

In general we may distinguish between two types of boundary zones which are, according to the relation theory of LEEUWEN (1965–1970, WESTHOFF 1971a, b, WESTHOFF & LEEUWEN 1966, SHIMWELL 1972), the limes convergens and the limes divergens. The *limes convergens* zone, or convergent limit, is characterized by sharp

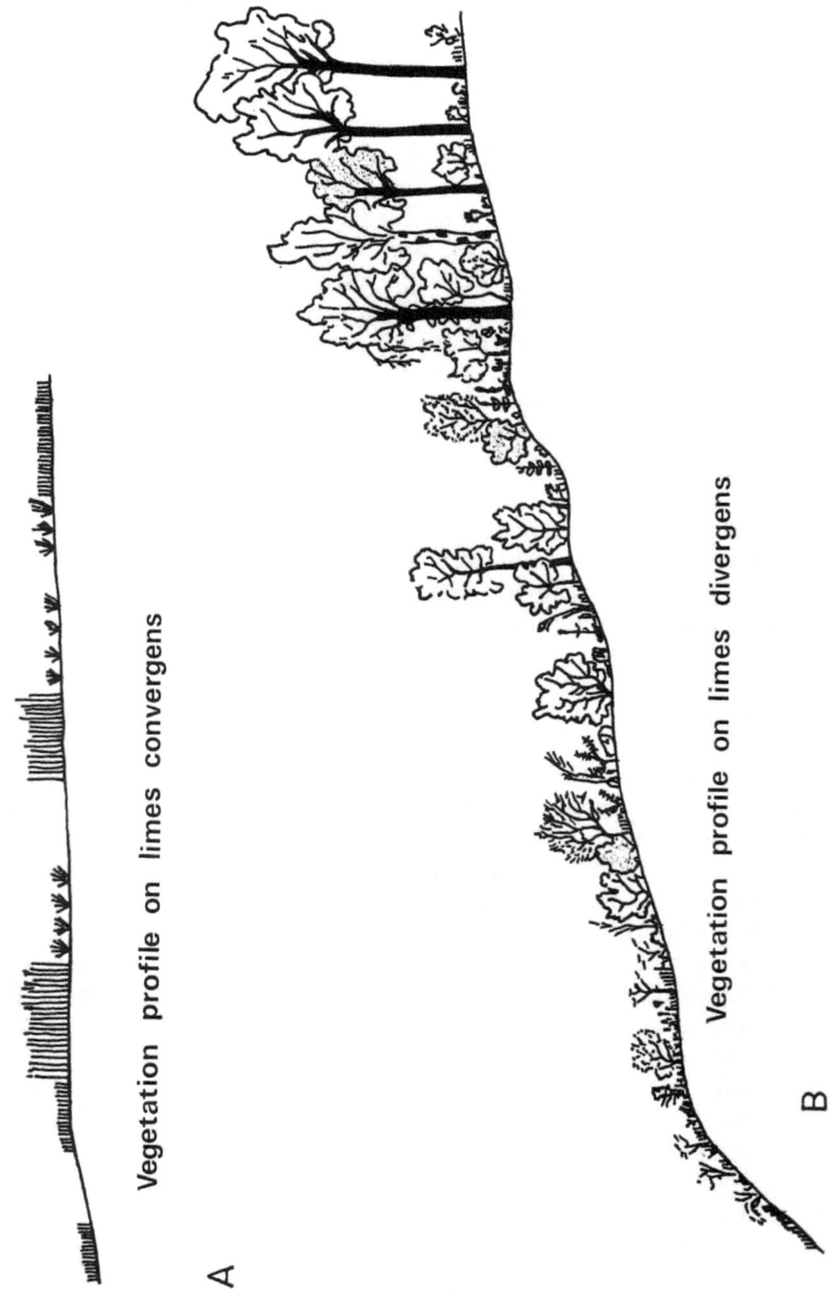

Fig. 2. Vegetation profiles corresponding to the limes convergens and limes divergens zones (LEEUWEN 1965).

vegetational boundaries on either side of which uniform phytocoenoses occur, with few species represented, some by many individuals, in coarse-grained patterns. The corresponding environment is unstable, i.e. with sharp fluctuations in vertical direction (e.g. water-table or phreatic level) or in horizontal direction (e.g. deposition of flood marks). A typical synsystematical unit of this type is the Agropyro-Rumicion crispi.

The *limes divergens* zone, or divergent limit, is characterized by numerous small-scale boundaries in phytocoenoses merging continuously into each other, with many species generally represented by few individuals in fine-grained patterns. The corresponding environment is stable and determined by a gradient. A typical series of divergent synsystematical units is Mesobromion-Trifolion medii-Carpinion betuli. Fig. 2 shows a vegetation profile belonging to these types of boundary. According to MAAREL (e.g. 1966a), LEEUWEN (1966) and WESTHOFF (1971a) the ecotone concept of LIVINGSTONE and CLEMENTS (e.g. WEAVER & CLEMENTS 1938, CURTIS 1959) could be applied to the limes convergens zone, whilst the ecocline concept as derived from HUXLEY's cline concept (WESTHOFF 1947, MEIJER DREES 1951) may be considered an equivalent of the limes divergens zone.

The delimitation of stands within ecotones generally gives no difficulties, except for pattern problems already discussed. Gradients of vegetation-and-biotopes (ecoclines) can be approached by belt transects following the direction of variation, which transects have to be divided into sample plots (not necessarily quadrats) small enough to be uniform. Recently JAKUCS (1972) gave examples of this approach in 'grassland-shrub-woodland' transitions within the range of the Quercetea pubescenti-petraeae in Hungary.

20.4.3 MINIMAL AREA AND PLOT SIZE

The size of the sample plot is largely dependent on the structure of the vegetation under study, but may be affected also by the size of the stand. In many cases the stand is sufficiently small to be analysed largely in one relevé. In many other cases, however, this is not possible, and then the plot should be large enough that all species of regular occurrence in the stand should be present in the sample plot. In extremely uniform phytocoenoses which are very poor in species, e.g. salt marshes, where differences in abundance and cover are of major concern, these differences should be considered in establishing the size of the sample plot.

These conditions refer to the minimal area concept (see also articles 5.2.1, and 18.3.3). Without treating this concept in detail we may give some comments relevant to the phytocoenose description. First the minimal area as an analytical concept (for the phytocoenose) has to be distinguished from the synthetic minimal area (for the phytocoenon). The analytic minimal area should be defined for a stand under study as a representative area, e.g. as an adequate sample of species of regular occurrence in the stand; the size decision may, but will not necessarily, be based on the total number of species in the stand or the usual minimal area curve (BRAUN-BLANQUET 1964, CAIN & CASTRO 1959, DU RIETZ 1921, 1930, GOODALL 1952, 1961, GREIG-SMITH 1964, HOPKINS 1957, MAAREL 1966b, TÜXEN 1970c and 1970 Exc., VESTAL 1949). A synthetic minimal area, in contrast, is subject to many current definitions (e.g. BRAUN-BLANQUET 1964, CAIN & CASTRO 1959, and WESTHOFF 1951 — who explicitly defined it as 'the minimal surface which has as a rule to be occupied by a sample of a plant community if the normal specific assemblage will be able to develop'). The size decision in this case involves not merely the stand itself but the 'normal' composition of stands representing a phytocoenon. It is clear that before judging this in the field one should know from synthesis the characteristic species combination. Nevertheless many authors, including BRAUN-BLANQUET (1964), describe the normal species-area determination, applied analytically to the stand, as the way to find the minimal area.

TÜXEN (1970c) presented numerous examples of species-area curves ending in a horizontal part, implying that at the sample size where the horizontal part begins, full species representation or saturation has been reached. Tüxen consequently spoke of 'saturated communities,' which he found among homogeneous communities of various organization levels. He cited various authors who found curves of the same type, including DU RIETZ (1921).

Most authors, e.g. BRAUN-BLANQUET (1964), CAIN & CASTRO (1959), HOPKINS (1955), and MAAREL (1966b) have found curves that never reach an asymptote. It is true that many of their curves refer to either relatively small or relatively large areas. It remains unclear in which situation which type of species relation can be expected — the ARRHENIUS-type of a log-log relation (cf. PRESTON 1962), the ROMELL-type of a log-linear relation (cf. HOPKINS 1955, DAHL 1957, WILLIAMS 1964, MAAREL 1966b), or the KYLIN-type of saturation relation. BALOGH (1958), and MAAREL (1966b), quoting FREY (1928), who presented the three types, interpreted them as referring to very uniform and species-poor environments (limes convergens type!), highly variegated and

TABLE I

Minimal area values in square meters for various communities

Epiphytic communities	0.1–0.4
Terrestrial moss communities	1–4
Hygrophylous pioneer communities (Isoeto-Nanojuncetea)	1–4
Dune grasslands (Koelerio-Corynephoretea)	1–10
Salt marshes (Asteretea tripolii)	2–10
Pastures (Lolio-Cynosuretum)	5–10
Mobile coastal dune communities (Ammophiletea)	10–20
Hay meadows (Arrhenatheretalia)	10–25
Heathlands (Nardo-Callunetea)	10–50
Alpine meadow and dwarfshrubs (Elyno-Seslerietea)	10–50
Calcareous grasslands (Festuco-Brometea)	10–50
Chaparral, temperate sclerophyll shrubland	10–100
Weed communities (Secalietea)	25–100
Scrub communities (Rhamno-Prunetea)	25–100
Steppe communities	50–100
Temperate deciduous forest (Querco-Fagetea)	100–500
Mixed deciduous forest (North America)	200–800
Tropical secondary rainforest	200–1000
Tropical swamp forest	2000–4000

species-rich situations (limes divergens type), and intermediate types or (BALOGH) 'normal homogeneous stands', respectively.

Table I presents some minimal area values for community types, taken from various sources. (See TÜXEN, 1970 Exc. for a bibliography). They range from a few dm² for certain epiphytic communities to one hectare or more for climax tropical rain forests. In general the plot size is taken somewhat larger than the minimal area. It will be clear that plot size should not vary too much within one vegetation type. The shape of the plot is not standardized and may depend on the situation. If possible, a quadrat of rectangular shape is to be preferred.

20.4.4 DESCRIPTION OF STRUCTURE AND STRATA

Vegetation layering is an important structural character. Mostly only four principal layers are distinguished as the tree, shrub, herb, and moss layers. The latter is also named field layer or thallophyte layer (which is less adequate, because many of the larger fungi do not belong to it). These principal layers may be further subdivided. In the relevé as many layers are distinguished as is considered appropriate or necessary. For each layer height and coverage degree in per cent, mostly with 5–10% intervals, are estimated. For tree and shrub layers the age is estimated and additional data on stem diameter, number of dead or fallen trees, and occurrence of epiphytic communities are gathered.

Stratification can be more accurately described with the help of diagrams in which the total or combined cover of each layer is indicated (Hult 1881, Braun-Blanquet 1928, cf. Cain & Castro 1959, Knapp 1971). These diagrams are related to vertical profiles, as have frequently been presented for tropical rain forest (see article 13.4.2) and by Zonneveld (1960) for the freshwater tidal delta of the Rhine. Both however have their own special uses. The profile can show exactly the structure of a given phytocoenose; in the coverage-stratification diagram a pattern typical for a phytocoenon can be generalized (Cain & Castro 1959). A more formal method is that of Dansereau (1957, Dansereau & Arros 1959) for the description and recording of vegetation on a structural basis (see article 13.3.3).

It is possible to classify different strata into separate synusial units. This approach is treated in detail by Barkman (article 16). The Braun-Blanquet approach however considers the main strata of a given stand as a single phytocoenose which has to be analysed as a whole, at least as far as terrestrial communities are concerned. The main arguments are: (i) The different strata are rooting in a common substratum; (ii) The layers are ecologically closely interrelated; (iii) The plants of all other layers have originally been part of the field layer and they must pass through one or more layers as they develop from seedlings to their mature life-form stature. The layered community is, then, a dynamic totality (see Webb et al. 1967b on the vertical integration of the rain forest).

The situation in aquatic communities of higher plants is less clear however. Here the coherence of layers is much looser, or absent. Du Rietz (1930a) has proposed a stratification scheme, which was modified by Hartog & Segal (1964) (see also article 16.4.3). Layers are distinguished according to rooting or non-rooting of plants and the positions of leaves in relation to the water surface. The layers are named by growth-forms after representative genera, e.g. the isoetid layer.

An additional analysis can be made of subterranean layers. For a treatment of root stratification see Braun-Blanquet (1964: 59–62). Bibliographies of root studies were given by Wilmanns (1959 Exc., 1966 Exc.) and Tüxen & Wilmanns (1973 Exc).

20.4.5 Floristic Analysis

20.4.5.1 *Cover and Abundance*

The description of structure is followed by an inventory of taxa, at least of phanerogams, pteridophytes, bryophytes and lichens.

Table II

The combined estimation (cover-abundance) scale of Braun-Blanquet, compared with the cover-abundance scale of Domin (See Evans & Dahl 1955). The subdivisions 2m, 2a, and 2b are proposed by Barkman et al. (1964). The ordinal transformation is discussed in 20.5.4.

Braun-Blanquet, cover-abundance			ordinal transform	Domin	
				+	one individual, reduced vigor
r	one or few individuals		1	1	rare
+	occasional and less than 5 % of total plot area		2	2	sparse
1	abundant and with very low cover, or less abundant but with higher cover; in any case less than 5 % cover of total plot area		3	3	< 4 %, frequent
2	very abundant and less than 5 % cover, or 5–25 % cover of total plot area				
	2m	very abundant	4		
	2a	5–12.5 % cover, irrespective of number of individuals	5	4	5–10 %
	2b	12.5–25 % cover, irrespective of number of individuals	6	5	11–25 %
3	25–50 % cover of total plot area, irrespective of number of individuals		7	6 7	26–33 % 34–50 %
4	50–75 % cover of total plot area, irrespective of number of individuals		8	8	51–75 %
5	75–100 % cover of total plot area, irrespective of number of individuals		9	9 10	76–90 % 91–100 %

The taxa are listed according to the layer in which they grow. Plants which appear to be structurally transgressive, i.e. occurring in more than one layer, have to be recorded in each of these layers separately.

Taxa occurring only outside the sample plot (but within the stand) are noted in parentheses. Next, the quantitative occurrence of each taxon is estimated. In the Braun-Blanquet approach two criteria are considered most useful: abundance and coverage degree. Abundance relates to the density of the individuals of a given species in a plot. Cover degree is measured as the vertical projection of all aerial parts of plants of a given species as a percentage of the total plot area. The term 'dominance,' often used as a synonym for coverage, is less appropriate.

Abundance and cover degree are usually estimated together in a single 'combined estimation' or 'cover-abundance' scale. The five-point scale in Table II, from BRAUN-BLANQUET (1928, see also article 18.4.1) is in general use. Several authors (e.g. TUOMIKOSKI 1942, DOING 1954, EVANS & DAHL 1955, BARKMAN et al. 1964) used more detailed scales of combined estimation; the BRAUN-BLANQUET symbol 2 especially was refined. The more elaborate scales are useful for special purposes, e.g. for an accurate record of change of abundance and cover by succession on permanent sample plots in the course of years (DOING 1954). However, they often suggest more accuracy than can really be justified. Only values obtained by one investigator should then be compared (cf. CAIN & CASTRO 1959: 142–143, MAAREL 1966b). We have recently introduced as a refinement of scale interval 2 (taken from BARKMAN et al. 1964) the scale subdivisions 2m, 2a, and 2b given in Table II. This elaboration, which brings the total number of scale values to 9, has proved to be useful and reliable.

20.4.5.2 *Sociability*

Sociability or gregariousness is an expression of horizontal pattern of species. It is a measure of the degree of clustering (contagion) of the plant units of a species. A plant unit (WILLIAMS 1964) may be an individual or a shoot or a sprout-forming part of an individual. Measurement of sociability goes back (again) to HEER (1835).

In the floristic-sociological approach sociability is estimated with the following scale (BRAUN-BLANQUET 1928, 1932, 1951, 1964).

1. growing solitary, singly.
2. growing in small groups of a few individuals, or in small tussocks (caespitose), e.g. *Corynephorus canescens* in shifting sands.
3. growing in small patches, cushions or large tussocks, e.g. *Carex hudsonii* as a hummock builder in eutrophic swamps; *Silene acaulis* and *Saxifraga oppositifolia* in alpine swards.
4. growing in extensive patches, in carpets or broken mats, e.g. stands of *Hedera helix*, *Lamium galeobdolon*, *Asperula odorata* etc. in deciduous temperate forest.
5. growing in great crowds or extensive mats completely covering the whole plot area; mostly pure populations, e.g. *Erica tetralix* in *Erica*-heath; *Sphagnum rubellum* or *S. pulchrum* in raised bogs.

A variant of scale value 5 (5̣̇, loose or open 5) was proposed for populations with a cover degree of over 75% consisting, however, of plants which are sufficiently separated as to leave space for other species. Similar variants can be used for loose cover of 51–75% (4) and 26–50% (3) (MELTZER & WESTHOFF 1942, BRAUN-BLANQUET 1964).

In the relevé sociability values are written immediately behind the combined estimation values, e.g. *Scirpus maritimus* 1.1.

During the last decade several investigators (see FUKAREK 1964) expressed the opinion that the diagnostic value of sociability has been overestimated and that a certain sociability degree is a specific character of most taxa. Other authors (BRAUN-BLANQUET 1964, SCAMONI & PASSARGE 1963, WESTHOFF 1965) do not agree with this view. Only a few species have a fixed degree of aggregation based upon their innate manner of growth. The degree of gregariousness of most species is much influenced by habitat conditions and competition, and therefore is of major phytosociological importance. Sociability may also change considerably during the course of a succession; many examples are given by the authors quoted above. On the other hand, the variation in sociability will be correlated with variation in cover degree to some extent.

Sociability is commonly considered as an expression of vitality. However in various situations this is not so. JAKUCS (1970, 1972) remarked that character-taxa of the Trifolio-Geranietea (thermoxerophilous woodland fringe communities or 'Saumgesellschaften') tend to grow with sociability 1 in their optimal habitat, whilst they form polycorms with reduced vitality in suboptimal habitats. Facies of species are often connected with extreme or disturbed habitats (see 20.6.7).

20.4.5.3 *Vitality and Fertility*

Further variables in the performance of a species are its vitality and its fertility, representing vegetative and generative development respectively. BRAUN-BLANQUET (e.g. 1932, 1964) developed a scale of relative 'thriving' (Gedeihen) with four categories indicated by symbols (BRAUN-BLANQUET 1932).

●, 1 Well developed, regularly completing the life cycle (an extraordinary vitality is indicated with 'lux', luxurious).

⊙, 2 With vegetative propagation but not completing the life cycle.

○, 3 Feeble with low vegetative propagation, not completing the life cycle.

○○, 4 Occasionally germinating but not vegetatively propagating.

BRAUN-BLANQUET (1964) slightly altered the symbols and gave them numbers from 1 to 4, as above.

BARKMAN et al. (1964), following VARESCHI (1931) and ZOLLER (1954), stated that for many species vitality and fertility are independent or even negatively correlated parameters. They proposed separate scales for these.

20.4.5.4 *Periodicity*

In addition to vitality the seasonal phase in the life cycle of each plant, its 'phenological state' is recorded. Various scales are in use, the first one being that of GAMS (1918). BRAUN-BLANQUET (1964: 67 and 510) presented two scales (which are very similar). As was stated by GAMS (1918), ELLENBERG (1939, 1954a) and others (see also the bibliography by BALÁTOVÁ-TULAČKOVÁ, 1970 Exc.) one incidental record during the relevé is really not sufficient, a complete phenological diagram should be desired for each community. For a relevé at a given time, however, appropriate abbreviations may be used — v. (vegetative), fl. (flowering), fr. (fruiting), etc.

20.4.6 RELEVÉ PROTOCOLS

The various structural and floristic data discussed so far are written in standardized form either in field note books or on special protocol forms. These notes are preceded by some notes on the following items:

1. Date; running number; topographic locality (as detailed as possible); altitude; exposure and inclination; geology of substrate. If possible the location should be indicated on a detailed map.
2. Size and shape both of the plot and of the entire uniform stand; character of adjacent vegetation; soil profile; phreatic level; depth and differentiation of root system.
3. Character and intensity of human and animal influence, e.g. pasturing, burning, mowing, treading, manuring, irrigation.

Table III presents an example of a relevé, taken from MELTZER & WESTHOFF (1942). For each species are recorded the scale numbers for cover-abundance (before the period) and sociability (after the period), the phenological abbreviation, and the vitality symbol.

TABLE III

Protocol of a relevé (translated from MELTZER & WESTHOFF 1942)

Nr. 39462. 1st August 1939. Terschelling, Bessenplak S of beacon near beach mark 6. Gridnr. G5.61.43 in IVON-system (Institute for Vegetation Research in the Netherlands). Stand very uniform, Empetrum heath on slope of 6 m tall parabolic dune, exposition NNE, inclination 30°.

Habitat:	shadowed, moist soil, by day not strongly heated and rarely desiccating. Slight shifting of sand. Little human and animal influence.	
Profile:	A_0:	2 cm semi-decayed material.
	A_1:	5 cm dark humus containing sand.
	C:	bright, white dune sand.
Sample plot		100 sq m.

Herb layer cover 100 %, 20–40 cm			
Polypodium vulgare	2.3	v.	●
Empetrum nigrum	4.4	fr.	●
Hieracium umbellatum	1.1–2	fl.	●
Festuca rubra subvar. arenaria	+.1	fr.	●
Hypochoeris radicata	+.1	fr.	●
Calamagrostis epigeios	+.1	fl.	●
Jasione montana	+.1	fl.	●
Carex arenaria	1.1	v.	⊙
Ammophila arenaria	2.2	v.	○
Salix repens	+.2	fr.	●
Viola canina var. dunensis	r	v.	⊙
Moss layer cover 100 %, 2–5 cm			
Hypnum cupressiforme var. ericetorum	3.3	v.	⊙
Pleurozium schreberi	3.3	v.	⊙
Dicranum scoparium	2.3	v.	⊙
Mnium hornum	2.3	v.	●
Lophocolea bidentata	2.2	fr.	●
Eurhynchium stokesii	+.3	v.	⊙
Plagiothecium denticulatum	+.2	fr.	●
Polytrichum juniperinum	+.3	v.	○
Peltigera canina	+.2	v.	●
Parmelia physodes	+.1	v.	⊙
Cladonia alcicornis	+.2	v.	⊙

20.5 Synthetical Research Phase

20.5.1 STEPS IN SYNTHESIS

The analysis of stands is only the first step in the description of vegetation units. After relevés have been collected, they must be compared. This is the start of the synthetical phase which leads to the distinction of coena and, if wished for, the final classification of syntaxa. To this purpose, a number of relevés are tabulated in a matrix, which is usually named a relevé table or community table.

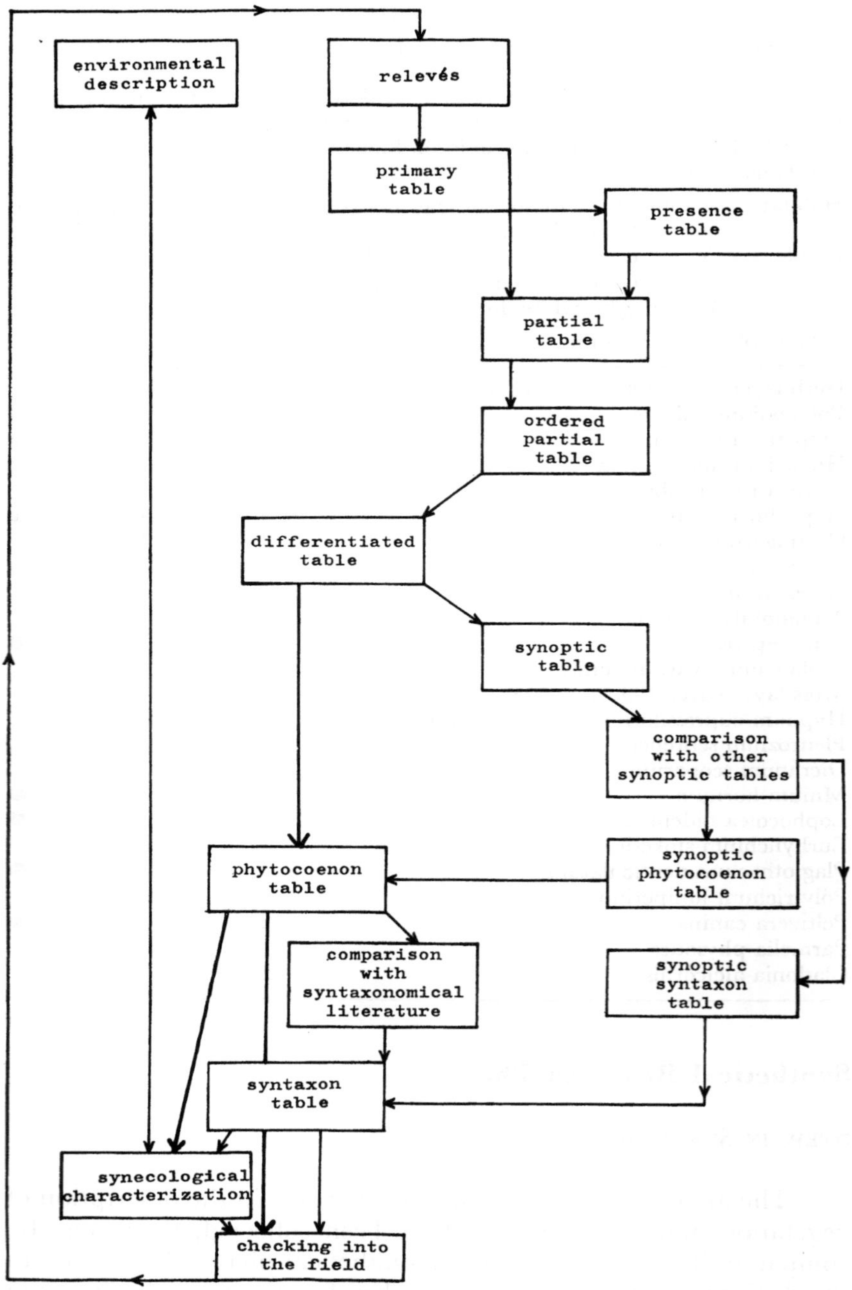

Fig. 3. Scheme of steps in the synthetic procedure.

Such a table is also called a primary or raw table (Rohtabelle). Each primary table is then rearranged into a structured table in which one or more uniform phytocoena are distinguished and characterized. After consultation of the relevant syntaxonomical literature (and possible further revision), this phytocoenon table or community table may then be presented as a syntaxon table, often for an association.

When environmental data such as soil analyses are available the vegetation types arrived at can be characterized synecologically. Finally the coena, or syntaxa, are checked in the field, e.g. whilst mapping the vegetation of the area under investigation or during the reconnaissance of a neighbouring area. This may lead to the collection of new relevés. Thus vegetation classification in the BRAUN-BLANQUET approach is essentially an iterative process, or one of successive approximation in the sense of POORE (1956). Fig. 3 shows the various steps in synthesis. SHIMWELL (1972) also outlines in English (cf. KNAPP 1958 and ELLENBERG 1956) the steps of the procedure.

The checking of units is an essential part of the method; this point seems not to be recognized by all critics of the BRAUN-BLANQUET approach (e.g. POORE 1955–56). MOORE (1962) explained that the misapprehension that the fidelity concept rests on a circular argument (e.g. POORE 1955a) is based on the failure to recognize the checking step: 'In present practice, associations are not distinguished in the field at all, but only when editing the tables of relevés. The first step involves describing uniform tracts of vegetation, *not* representative stands of a presumed association. Only when sufficient relevés have been accumulated and analysed one can discern the associations. Of course, in observing continental phytosociologists at work in their own homeland whose vegetation they know, one may be misled as to their methodology. They will now have reached the second stage of checking the reality of units already distinguished. They will not necessarily make this clear to a visiting inquirer.'

20.5.2 PRIMARY TABLE

In the primary table all taxa are listed at the left hand side of a sheet of squared paper; to each relevé a single vertical column is assigned. Each item in any column should contain either a dot or dash (in the case that a taxon is absent in the corresponding relevé) or else a combined estimation value (or an actual coverage per cent) and preferably a sociability value too. It may be convenient at this preliminary stage to group the species under separate headings according to strata and to separate phanerogams and cryptogams. The addition of new species from the later relevés produces a characteristic tailing off to the right of the table. Each row of the table represents a species (in a given stratum). Comparison of the rows of the table enables us to judge the distribution of any taxon over the relevés; comparison of the columns may lead to preliminary conclusions on relative similarities of the relevés.

The primary table should be rearranged several times, in order to establish groupings of relevés by rearrangement of columns as well as to group taxa with similar distributions in the table by rearrangement of rows. Before we discuss this rearrangement procedure (20.5.5) some concepts and techniques should be discussed.

20.5.3 Presence, Constancy, Homotoneity

Presence is the occurrence of a taxon in a vegetation table. It is usually measured as a degree by the number of relevés in which the taxon occurs (regardless of abundance and cover) expressed as a percentage of the total number of relevés compared. *Presence degree* (Stetigkeit) can thus be calculated from any number of relevés, regardless of the difference in size of the plots.

When plots of equal size are compared the corresponding per cent occurrence for a taxon is named 'constancy' (Konstanz). (Frequency, in contrast with these, is an analytical concept dealing with the distribution of a taxon within subsamples from a given stand.) Presence degree and constancy are determined from stands of different and often widely distributed localities. Constancy determinations on plots of equal size were especially favored by the Uppsala school (article 18), starting with Du Rietz et al. (1920), although the first such measurements, by Brockmann-Jerosch (1907) had used plots of various sizes. In the Braun-Blanquet approach presence degree has been termed 'Stetigkeit' and distinguished from 'Konstanz'; but confusion has resulted from the translation of both concepts as 'constancy' in some statements in English (e.g. Braun-Blanquet 1932, Moravec 1971).

Constancy and presence degree can be given in exact percentages, or in percentage classes. It is usual to distinguish five classes, noted in Roman figures:

Class:	Percentage of plots in which taxon occurs:
I	1–20
II	21–40
III	41–60
IV	61–80
V	81–100

The numbers of taxa falling into these classes are often presented in a 'constancy diagram.' Such diagrams are important characteristics of vegetation units and provide useful tools to check the uniformity or homogeneity of a table. This synthetic homogeneity concept, however, should be distinguished from analytical homogeneity. BARKMAN (1958a: 316–317) designated these concepts as 'intensive homogeneity' and 'extensive homogeneity.' However, as DAHL (1970, apud TÜXEN, 1970c: 101) pointed out, the latter term has been designated as *homotoneity* (NORDHAGEN 1943, DAHL 1957, 1960). This term was long overlooked in the BRAUN-BLANQUET approach, but now receives common use (e.g. TÜXEN 1970c, MORAVEC 1971). Homogeneity is an analytic concept, based on comparing different plots of the same size taken from an individual stand, whereas homotoneity is a synthetic concept, based on comparing similar plots from different stands of the same community-type or phytocoenon.

Homotoneity (see also 20.8.4) has been judged with the help of constancy diagrams; especially in North-European approaches (see further article 18). The classical interpretation, based on RAUNKIAER's (e.g. 1934) 'law of frequency,' of a constancy diagram is that the following relation between the constancy classes exists: $S_{I} > S_{II} > S_{III} \gtreqqless S_{IV} < S_{V}$ (the reversed J-shape). In very homotoneous tables class V may equal class I (U-shape). When classes III and IV include more species than class V, the table is considered heterotoneous (cf. GUINOCHET 1955, CAIN & CASTRO 1959, MORAVEC 1971). According to MAAREL (1972, in MAAREL & TÜXEN 1972: 209), the reasoning of WILLIAMS (1964) and observations of BARKMAN (1958a) showed that constancy class I is more or less dependent on the total number of relevés. Between the other classes the following relation exists in homotoneous tables: $S_{IV}+S_{V}/S_{II}+S_{III}$ is slightly over 1, whilst $S_{III}+S_{IV}+S_{V}/S_{II}$ is mostly about 2.

A second feature with which the homotoneity of a table can be easily checked is the variation in the number of taxa within the relevés of the table. In most cases well-developed stands of a phytocoenon do not differ much in number of species. A high variance of that number is reason to suppose that the table is heterotoneous, which may mean that more than one phytocoenon is represented in it (Hofmann & Passarge 1964). After the relevés have been classed into groups according to their number of taxa, we may obtain the frequency distributions of these classes, which will indicate this variation. A table comprising relevés belonging to two different coena may produce a curve with two or more peaks (Knapp 1971).

Apart from superficial inspection of homotoneity some simple measures have been proposed on the basis of constancy figures and species numbers (see Moravec 1971 for a survey). The following symbols, all corresponding to one vegetation table, are used: M is the number of relevés: S_m is the total number of species and $\bar{S}$ the mean number of species per relevé; S_v, etc. is the number of species falling into constancy class V, etc. C_i refers to the constancy per cents for species of the constancy class(es) indicated by the subscript to the summation sign, or for the prevalent species P — those belonging to the group of species with the highest constancy values, the number of which (S_p) approaches the average number of species ($\bar{S}$).

Amongst the measures based on constancy figures are:

1. The ratio S_V/S_{IV} (Dahl 1957).
2. Basic homotoneity coefficient $\frac{1}{S_m}\sum_{IV+V} C_i$ (Moravec 1971).
3. 'Homogeneity value' $\frac{1}{S_p}\sum_{p} C_i$ (Raabe e.g. 1952).
4. Index of homogeneity (Curtis 1959) $\dfrac{\sum_{p} C_i}{\sum_{Sm} C_i S_m}$
5. Mean constancy for all species $\sum_{S_c} C_i/S_m$

6. Corrected homotoneity coefficient (Moravec 1971). This is coefficient 2 with an 'oscillation factor,' determined by the difference in species number between the richest and the poorest relevé.

Tüxen (1970c) presented curves for various communities ending in a horizontal line (section 20.4.3); one may derive a 'minimal relevé number' from such curves. In fact Tüxen based his saturation quotient on tables containing about this minimal number of relevés. Other observations suggest that the number of species grows continuously with the number of relevés (e.g. Etter 1949, Barkman 1958a, Dahl 1957, 1960, Williams 1964). Dahl (1961) based an 'index of uniformity' on this relation (see 20.8.4) and BARKMAN adapted KLEMENT's heterogeneity index by omitting species with a constancy of less than 10 %.

7. 'Saturation quotient' $100\bar{S}/S_m$ (TÜXEN 1970c).

As Tüxen mentioned this measure is the reciprocal of the homogeneity (i.e. heterogeneity) value of DAHL & HADAČ (1941). WHITTAKER (1972) suggested use of $BD = (S_m/\bar{S}) - 1$ as a measure of beta diversity (article 3.6). PFEIFFER's (1957) homogeneity value was reduced to TÜXEN's measure by MORAVEC. KLEMENT (1941) used the same measure under the same name as DAHL & HADAČ did, as was mentioned by BARKMAN (1958a), who concluded that this index is largely dependent on the total number of relevés in the table.

In conclusion we think it would be more realistic to base a saturation quotient explicitly on species with a minimum constancy value, e.g. species from constancy classes II–V, and to measure the rate of increase in total species number separately. For measurement of homotoneity we suggest the mean similarity coefficient for the relevés (see 20.8.5).

20.5.4 SPECIES WEIGHTS

Species may be weighted in synthetical procedures such as fidelity determination, assignment of syntaxa to higher units and calculation of spectra. Weighting is usually with the combined estimation value or some transformation of this. SCHWICKERATH (1931) used 'Artmächtigkeit' — and when species groups were concerned 'Gruppenmächtigkeit' (species and group importance value, MAAREL 1972b) — based on BRAUN-BLANQUET figures and arbitrary numerical values for symbols + and r. TÜXEN & ELLENBERG (1937) and BRAUN-BLANQUET (1946) used the Deckungswert (cover value) and Gruppenwert (group value) by taking average coverage percentages for BRAUN-BLANQUET values 2–5 and arbitrary values for 1, + and r. The latter procedure has received more application than that of SCHWICKERATH, although TÜXEN & ELLENBERG stressed its limited applicability, which was also criticized by WESTHOFF (1947), MEIJER DREES (1949), SISSINGH (1950).

DAGNELIE (1960) suggested the use of an arc-sine transformation as is usual with percentage scores. His rounded figures appear to be identical with the original BRAUN-BLANQUET values. Various other transformations have been applied, including those of ETTER (1949), BARKMAN (1958a), BARKMAN et al. (1964), MAAREL (1966b), SCHMID & KÜHN (1970), MOORE (MOORE & O'SULLIVAN 1970) and LONDO (1971). (See MAAREL, 1972b for a review.)

Without discussing this problem here (see MAAREL 1972b) we are in favour of a simple 'ordinal transformation' of the extended nine-point BRAUN-BLANQUET scale (Table II). This produces also partial accordance with the DOMIN scale in its modification by EVANS & DAHL (1955), which is much used in Scandinavia and Britain.

20.5.5 TABLE REARRANGEMENTS

The development of phytosociological tables has been stressed by the leaders of the floristic-sociological approach as a crucial procedure. E.g. BRAUN-BLANQUET (1951a, b) wrote (transl.): 'Appropriately elaborated association tables are comparable with thorough species diagnoses. By means of the tables the considerable detailed work from the minute floristic analyses of the lower vegetation units becomes accessible and evaluable. From the table it also appears whether one has worked seriously and reliably; the tables are the proper touch-stone of the concerned plant sociologist.'

At the same time the table method is said to be difficult to learn, at least by written instructions. TÜXEN & PREISING (1951) spoke of 'special aptitudes, a view of the interrelations in the tabular picture, and broad experience in the sociology, synecology and syngenetics of plant communities,' that are all needed for mastering the ordering of tables. Although personal instruction in the technique is important, as numerous pupils of BRAUN-BLANQUET and TÜXEN can confirm, it may be admitted that early BRAUN-BLANQUET sociology paid insufficient attention to explicit statements of the table techniques.

KNAPP (1948, 1958) and ELLENBERG (1956) presented detailed accounts of the table technique as it had been developed by TÜXEN in the 1930's at the Zentralstelle für Vegetationskartierung. ELLENBERG's scheme (1956: 46, slightly altered) is as follows (see Fig. 3):

(1) rearrangement of the first matrix to a 'presence table';
(2) tracing of table-differentiating species, mostly with the aid of 'partial tables';
(3) rearrangement of taxa and relevés according to the empirically established occurrence of groups of differentiating species; the result is a 'differentiated table';
(4) compilation of every relevant differentiated vegetation table into a 'synoptic table' ('Übersichtstabelle');
(5) determination of fidelity by comparing an adequate number of

synoptic tables ('zusammenfassende Übersichtstabellen' in the sense of Ellenberg l.c.);

(6) rearrangement of the differentiated table into a final syntaxon table ('charakterisierte Tabelle' in the sense of Ellenberg l.c.).

Similar schemes were presented by Moore (1962), Scamoni & Passarge (1963), Fukarek (1964), Küchler (1967), Gounot (1969), Knapp (1971) and Shimwell (1972). The steps 1, 2 and 3 are the basic ones; they are essential for any local classification of relevés into phytocoena. Step 4 is preliminary to 5. Steps 5 and 6 have to do with the next level of investigation: the establishment of a (perhaps provisional) syntaxonomical classification.

Tüxen (1970b) recommended to start with a preliminary procedure aiming at the exclusion of relevés with deviant numbers of species that suggest fragmentary or heterogeneous stands. Such relevés are thus excluded from successive rearrangements and fidelity calculations and are not considered in an association diagnosis.

The first rearrangement of a primary table aims at listing the species after their degree of presence. In the presence table it may become apparent, whether the relevés vary irregularly, or whether certain combinations of species are found recurrently. In the latter case such groups may be more or less mutually exclusive and serve as groups of differentiating species. Those are not likely to be found in the presence classes V and I; they will mostly belong to the intermediate classes II, III and IV. The supposed differentiating species are underlined or boxed in the presence table. These differentiating species, which are provisional sets of diagnostic species (see 20.3.4), are used to rearrange the table.

The next step is to set up a partial or extracted table in which only the species thought to have diagnostic value are given and used to rearrange both columns and rows. During the rearrangement of a provisional partial table, the differentiating species are written on the left hand side of the paper and the rows arranged in their corresponding groups, and the relevés (columns) are rearranged to show a diagonal order of species groups from the left to the right of the table. To simplify the rearrangement of the relevés dictation or transfer strips are used.

When the partial table has been worked out, the complete table (thus containing again the temporarily omitted species) is rearranged accordingly so that the differentiating groups appear clearly. Now we have the 'differentiated table.' The species with a high presence values indicate the degree of coherence or homotoneity of the table. The species with intermediate and lower presence values used in differentiating groups represent the heterotoneity of the table. After comparing the sizes of the two groups, one may decide how many phytocoena the table includes and whether the table should be divided, or combined with others for further treatment.

Step 4 from Ellenberg's scheme comprises the replacement of each uniform table or partial table by a column in which for each participating species the presence degree is indicated (either as a percentage or a class value). Such a table (which may consist of only one column) is called Übersichtstabelle by Ellenberg but is also known as Sammeltabelle, synthetische Tabelle and Stetigkeitstabelle. Following a recent denomination by Doing (1969a) and Tüxen (1970b) we propose the neutral name 'synoptic table.'

After comparison of the synoptic table with those from other types of vegetation from the same region an idea can be formed about the local diagnostic species groups in the table under study (step 6). The differentiating groups used may represent recognized diagnostic groups, or may represent approximate groupings that should be revised on the basis of broader knowledge of diagnostic groups, or may represent diagnostic groups for newly recognized phytocoena. With those species, as well as the constant taxa, a phytocoenon may then be identified or described. Here we could speak of a synoptic phytocoenon table, from which the corresponding relevé-table or 'phytocoenon table' (Gesellschaftstabelle) can be rewritten. When moreover the synoptic table is compared with the literature and with tables from similar vegetation types from other regions, character-taxa may be found and a syntaxonomic interpretation may follow. The synoptic table now serves as a synoptic-syntaxon-table and the corresponding (rewritten) relevé-table is now a syntaxon-table. The latter will often be an association table. The fidelity research implied in steps 5 and 6 will be discussed in 20.5.7-8.

The rearrangement method might be criticized with the objection that the table pattern may be 'an artefact imposed by the clever shuffling of data and not mirroring any reality in nature,' as it has been expressed by Moore (1962) in his refutation of this objection. In reality, the characterized table is, first, an effective organization of information, an ordering of species and relevés in relation to one another (and biotope properties). Second, it is considered a working hypothesis to be tested by further observation. It should receive confirmation: (i) from further floristic data, critically and impartially collected; (ii) from mapping the units distinguished; and (iii) from the evidence of relationships between the vegetational units and environment. The groups of differentiating taxa may be presumed to indicate ecological differences in one or more specific major biotope factors or correlated sets of factors, such as phreatic level, humus, carbonate content, amount of soil hardening; salt content, or frequency and duration of periodical inundation by sea water; exposure; shade; rate of pasturing or mowing; etc. The resulting hypotheses have to be checked by continued ecological research. If the work has been well done, the differentiating groups should have indicator value in relation to environmental factors (see 20.7.6), and the phytocoena as kinds of communities should bear consistent, reliable relations to kinds of biotopes.

20.5.6 Mechanical Means for Rearrangement

The 'classical' rearrangement technique by manual ordering of taxa and relevés, as it has been rendered above, is a time-consuming procedure and a potential source of errors. Attempts to simplify and mechanize the procedure have been made by several investigators. The most simple means is the use of cardboard strips, usually for the relevés and thus for the columns of the table.

Various systems of blocks (wooden, plastic, or aluminium) have been developed (Wilmanns 1959, Margl 1967, Müller et al. 1972). Trentepohl (1968) showed a magnetic device which seemed to be less vulnerable to manual disordering of the blocks.

For tables with a large number of relevés or a large number of species the use of punched cards is obvious. Three types of cards are in use — the edge-punched card, the visual punched card, and the machine punched card. Cottam & Curtis (1948) initiated use of cards with the third and most advanced type. Examples of the use of edge-punched cards are found with Emberger et al. (1957), Gounot (1957) and Zonneveld (early sixties, internal undated reports Netherlands Soil Survey Institute).

Ellenberg & Cristofolini (1964) intensively used visual punch cards, mainly for sorting relevés. The corresponding cards are laid on a light-table and then compared with a standard card — e.g. one containing the characteristic species combination of a certain syntaxon. Ellenberg (1968) published a version of this technique where combined estimation values of species are taken into consideration, however at the expense of space on the card. According to Ellenberg, this apparatus is only efficient if the number of relevés is large, at least 200 to every table. As Müller et al. (1972) stated, this technique may be useful where the vegetation has already largely been classified and the vegetation units described, and where additional relevés need only to be arranged and included in existing tables. Further use of machine punched cards in phytosociology has been described by DAGNELIE (1960), BECKING (1961), GROENEWOUD (1965) and many others; whilst recently a number of computer-based table rearrangement programmes have been developed; this will be discussed in 20.8.1 and 20.8.6.

20.5.7 DETERMINATION OF FIDELITY

The next step in the synthetic procedure is the determination of fidelity. By comparing a number of tables — if possible all available tables covering the vegetation of a given area — it is possible to discern character-species in the sense discussed in 20.3.3. As indicated there, the criteria are the distribution of the species in different coena; fidelity is the degree to which a species is concentrated in one coenon, vs. dispersed with more even occurrence in several coena. Species distributions in the tables are observed on the basis of measures already discussed — presence degree, combined estimation value, vitality, and sociability. It should be emphasized that these measures are relevant to a species' fidelity only as they differ between different coena. Character-species can as well be minor species as major ones; there is no necessary relation between fidelity and dominance. In the classification of European forests, more significance is attached to the fidelity of shrubs, herbs, and mosses, both because tree species are often widely distributed in different coena, and because disturbance has brought extensive replacement of natural tree species.

Five degrees of fidelity have been distinguished (BRAUN-BLANQUET 1921, 1928, 1932, 1951, 1964) in the relation of a species to the given coenon.

A. Character-taxa.

Fidelity degree 5: *Exclusive taxa* (treue): taxa completely or almost completely confined to one phytocoenon (vegetation unit);

Fidelity degree 4: *Selective taxa* (feste): taxa occurring with clear preference for one phytocoenon but also, though with a considerably lower presence degree, in other coena;

Fidelity degree 3: *Preferential taxa* (holde): taxa present in several coena, perhaps with about equal presence degree, but with a higher combined estimation value and (or) with a higher vitality degree in one particular coenon;

B. Companions.
Fidelity degree 2: *Indifferent taxa* (vage): taxa without pronounced preference for any coenon.

C. Accidentals.
Fidelity degree 1: *Strange taxa* (fremde): taxa having a definite presence degree optimum and mostly also a cover-abundance optimum outside the considered coenon. These are often accidental intruders from neighbouring coena or relics from a coenon that preceded in succession.

TABLE IV

Determination of fidelity according to SZAFER and PAWŁOWSKI (BRAUN-BLANQUET 1932)

F = fidelity degree;
A = cover-abundance combined estimation;
C = presence or constancy class;
V = vitality.

F	in phytocoenon under consideration C	A	in comparable phytocoena C	A
5	IV–V	3–5	I–II	+–2(1)
	IV–V	+–2	I	+–2
	I–III	+–5	absent or very rare	
4	IV–V	3–5	II–III (IV) (relic or pioneer)	+–2(1)
	IV–V	+–2	II–III	+–1(2)
	III–IV	+–2	I–II (III)	+–1(2)
	I–III	+–2	I (rare)	+
3	I–V C, A various	3–5 V normal	I–V C, A lower	+–2 V reduced
2	C, A, V various		similar	
1	I V reduced – outskirts and disturbed parts of the stand(s) –	+–1	higher	

As early as 1927, SZAFER & PAWŁOWSKI gave a quantitative outline for the determination of the fidelity class. It has been taken over, with slight modifications, by BRAUN-BLANQUET (1928, 1932, 1951, 1964) and it is presented here with slight alterations (Table IV). In this table the comparison is between a phytocoenon that is being studied and comparable phytocoena. It seems logically incorrect to specify in such a table 'the given association' versus 'other associations,' as BRAUN-BLANQUET (l.c.) has done in all editions of his manual.

BECKING (1957) has modified the scheme of SZAFER & PAWŁOWSKI (1927) considerably, especially in omitting the possibility of obtaining fidelity degree 5 with a low presence degree. However, it has been crucial in the BRAUN-BLANQUET approach that faithful species need not be constants, and apart from diagnostic implications (see 20.6.1) one cannot simply change this principle.

20.5.8 GEOGRAPHICAL EFFECTS ON FIDELITY

Since the beginning of the BRAUN-BLANQUET approach (BRAUN-BLANQUET 1921) it has been recognized that the diagnostic value of character-taxa is geographically limited. Naturally enough this recognition has become more and more general with the growing geographical extension of phytosociological knowledge. The vegetation of a region with a uniform climate and a uniform geological history usually shows great uniformity in the ecological amplitudes and therefore in the sociological positions of its species. Most species however occur in larger, climatically and geologically heterogeneous areas and present different ecological amplitudes or habitat relations in different parts of their areas. In many cases these differences in local habitat response within the species are correlated with genetic differences: we are dealing then with ecotypic variation. In other cases genetic differences, if present, are not evident. Geographic variation in habitat response may involve such shifts in topographic or other local distribution, as tend to compensate for difference in climate (law of relative habitat constancy, cf. WALTER & STRAKA 1970, article 3.2).

A number of species which behave as xerophytes in Central Europe, e.g. *Bromus erectus*, *Koeleria gracilis*, *Carex humilis*, *Peucedanum cervaria*, are bound to humid soils of cool northern exposures in the Mediterranean. Species which are real woodland plants in a continental climate may present much wider amplitudes in oceanic climates: e.g. *Osmunda regalis*, *Listera ovata*, *Dryopteris filix-mas*. On the contrary a number of thermoxerophilous species are indifferent to soil conditions in Central Europe, whereas in the cooler and more humid climate of NW-Europe they are strictly confined to calcareous soils. *Iris pseudacorus*, a species of wet eutrophic swamps in Central and Western Europe, grows in moderately damp grassland in the perhumid Western Irish climate. *Schoenus nigricans*, in Central Europe faithful to calcareous marshes or fens, occurs in Ireland both in fens and in ombrotrophic blanket bogs. Even within a much more limited geographical range the differences in ecological amplitude may be considerable. Within The Netherlands,

Orchis morio is faithful to the Mesobromion of dry chalk in Southern Limburg, whereas it is bound to moorland (Calluno-Genistetum molinietosum) in the Pleistocene part of the country, to mesohalinic peat soil (community of *Orchis morio* and *Ophioglossum vulgatum*) in the brackish Holocene parts of Holland (sensu stricto) and to a narrow ecotone at the margin of wet heather dune slacks on the West Frisian islands (cf. WESTHOFF & HELD 1969). *Hymenophyllum peltatum* is in the major part of Ireland bound to damp oakwood, but on the extremely oceanic cliff of Slievemore on Achill Island (Co. Mayo) it thrives freely in the open air in between *Calluna vulgaris* and *Empetrum nigrum* (PRAEGER 1934).

Consequently, one and the same species can be a character-taxon for association *A* in one area and for association *B* in another one. BRAUN-BLANQUET (1964: 100) concluded that the validity and diagnostic value of character-taxa for a given association are mostly restricted to a climatically uniform area. It follows that a typology of character-taxa with respect to their geographical extension of their validity may be useful.

BRAUN-BLANQUET (1928, 1932) gave a first distinction between general and regional character-taxa. Later (1951a) he mentioned absolute, territorial and local character-taxa (cf. ELLENBERG 1956). The distinction between territorial and local was unclear, however; apparently it refers to the size of the range of the association in which the taxon shows fidelity. As BARKMAN (1958a) pointed out the replacement of general by absolute suggests a change of criterion, i.e. from range of fidelity to degree of fidelity, which was in fact realized in BRAUN-BLANQUET's definition of 'Absolute Treue.' With BARKMAN we think that these two criteria should be used independently. A third criterion is the relation between the range of the character-taxon and the range of the corresponding syntaxon.

MEIJER DREES (1951) elaborated this criterion and came to four possibilities: 1) ranges of taxon and syntaxon coincide, 2) taxon range forms part of syntaxon range, 3) syntaxon range forms part of taxon range and 4) ranges differ but overlap. MEIJER DREES maintained BRAUN-BLANQUET & MOOR's term 'regional' which he applied to all cases in which the taxon is a character-taxon throughout the common range of taxon and syntaxon, contrasted with the term 'local' for cases in which the taxon is a character-taxon only in part of the common area.

The corresponding scheme was taken over by BECKING (1957, 1961) who incorporated it moreover in a complex scheme in which also the degrees of fidelity were considered. BARKMAN (1958), who wished to maintain the more obvious meaning of the terms local and regional, adopted the eight types of MEIJER DREES (1951), but largely renamed them. Finally KNAPP (1971), who only mentioned MEIJER DREES and BECKING, refined the division into four types of geographical character-taxa in the taxon-bound sense of MELTZER & WESTHOFF (1942): viz. 'local', 'territorial or regional,' 'region-' and 'absolute.' The following scheme may be given as a summary:

i) Local character-species, with fidelity restricted to an area that forms only part of the total area of the syntaxon.

ii) Regional character-species, with fidelity for the total area of the syntaxon, but with the species area exceeding that of the syntaxon. Subcategories could be distinguished in relation to BRAUN-BLANQUET's phytogeographical typology including provinces and sectors.

iii) General character-species, with fidelity for the total area of the syntaxon, and the species area coinciding with the syntaxon area.

20.5.9 PHYTOCOENON CHARACTERIZATION

After a primary table has been rearranged into a phytocoenon table, presence values have been calculated, and the fidelity and differentiating value of the taxa towards neighbouring communities have been established, the corresponding phytocoenon can be characterized. First the general structure can be described. Mostly the presence class of each species is indicated at the right hand side of the table, in some cases the presence per cent. To the right of the presence figure the range of the combined cover-abundance estimation values may be given.

Then, according to procedures described above the faithful, differentiating and constant taxa may be indicated. The diagnostic value may be only provisional, when the relevés come from a limited area or when insufficient other material could be compared. We are dealing here with the first of three phases of systematic phytosociology, viz. the local phase. The taxa that are of lowest presence degree and lack diagnostic value (mostly of presence class I) are often presented in an addendum to the table. In many cases the description of phytocoena does not immediately lead to a formal classification. Without this, the phytocoenon tables may summarize much of what phytosociology seeks to understand of relations amongst species, kinds of communities, and environments. When data are thus organized into phytocoenon tables, however, the stage is set for the syntaxonomic treatment next to be discussed.

e V

ation table of the Artemisietum maritimae in the SW Netherlands by W.G. Beeftink

	Artemisietum maritimae typicum								
	Initial phase with Artemisia maritima					Typical form			
		4)		5)					
of relevé	53136	53134	53133	52390	52371	52338	54081	53022	55058
y[1]	Ka	Kå	Ka	Ka	Ka	He	Ko	Ba	Ka
[2]	cr	cr	cr	dike	dike	cr	cr	ab	cr
y flood-water[3]	p	p	p	p	p	p	e	p	p
	2-9-53	2-9-53	2-9-53	21-8-53	20-8-52	5-8-52	13-8-54	5-8-53	5-8-55
in m^2	2x12	2x12	3x10	2½x20	4x12	2x25	3x20	3x15	3x12
e in %	98	95	95	95	100	100	100	100	100
er-taxon of the association									
ia maritima	5.4	4-5.4	5.4	4.5.5	4.3-4	2.1	2-3.1-2	2.1-2	2.1-2
ntial-taxa of the subassociations									
maritima	-	-	-	-	-	-	-	-	-
s stolonifera var. compacta									
salina	-	-	-	-	-	-	-	-	-
x hastata	-	-	-	2-3.2	3.3	-	-	-	-
er-taxa Armerion maritimae									
rubra f. litoralis	+.2	+.2	1.2-3	-	+.2	5.5	5.5	5.5	5.5
aritima	-	-	-	-	-	+.1	-	+.1-2	-
gerardii	-	-	-	-	-	-	-	-	-
is strigosa	-	-	-	-	-	-	-	-	-
er-taxa Puccinellion maritimae									
lia maritima	+.2	+.2	-	-	+.1-2	-	-	-	+.2
e portulacoides	2.2-3	2-3.3	2.2-3	+.1-2	1.1-2	+.1-2	1.1-2	+.2	1-2.2
ia scorpioides	-	-	+.2	-	-	-	-	-	-
r-taxa Glauco-Puccinellietalia									
ria marginata	-	+.1	-	-	-	-	1.1	1.2	+.1-2
vulgare ssp. vulgare	-	-	+.2	-	+.1-2	+.1-2	1.1-2	+.2	1.1-2
r-taxa Asteretea tripolii									
ipolium	+.1	+.1-2	+.1	-	+.1	+.1	1.1-2	2.1-2	1-2.2
in maritima	-	-	-	-	-	-	+.2	-	+.2
maritima	+.2	+.2	+.2	-	+.1	+.1-2	-	+.2	+.2
r-taxa Thero-Salicornion, Thera-Suaedion									
tinion									
ia europaea	-	-	-	-	-	-	-	-	-
aritima	-	+.1	-	+.1	+.1-2	-	+.1	-	+.1
townsendii	-	+.2	-	-	-	-	-	-	-
xa									
a pungens	-	-	-	+.1-2	1.1-2	1-2.1	-	-	+.1
littoralis	-	-	-	+.1	-	-	-	-	-
erenne	-	-	-	-	-	-	-	-	-

20.6 **Syntaxonomical Research Phase**

20.6.1 Syntaxon Table Characterization

The syntaxonomical research phase starts when a phytocoenon is to be fitted into the hierarchic system of syntaxa. The phytocoenon table is then interpreted by consulting the relevant syntaxonomic literature, especially synoptic syntaxon tables. Questions in this are: (i) Which already described association can be recognized in the characteristic taxon combination of the phytocoenon; (ii) which lower units could be recognized on the basis of the established differentiating taxa; (iii) which taxa can be recognized as character- or differential-taxa from higher units already distinguished?

Careful treatment of these questions may lead to the conclusion that the phytocoenon under study is one already recognized, or that it must be described as a new syntaxon, or that part of the existing hierarchy must be redefined. Here, as Ellenberg (1956: 57) said 'begins the range of the 'tact' of which Braun-Blanquet, 1951: 18 spoke.' The expression 'sociological view' ('soziologischer Blick') is relevant. Though some of the decisions on phytocoenon similarity are objective, the essential role of judgement and experience in a typological discipline should be recognized.

The results of these considerations are expressed in the structuring and labeling of the phytocoenon table, which now becomes a syntaxon table. Usually the highest unit treated in a given syntaxon table is the association. In such an association table the subordinate units distinguished are presented as well. Table V presents an example of an association table of an already recognized association, viz. the Artemisietum maritimae, as it occurs in a region not systematically described before, the southwestern Netherlands.

When a number of related associations have been described, their regional descriptions might be presented together in a synoptic syntaxon table. An example of such a table is presented in Table VI. Both tables, composed by W. G. Beeftink (cf. Beeftink 1962, 1965, 1968), whose cooperation is gratefully acknowledged, will be discussed in the subsequent paragraphs.

20.6.2 Characterization of the Association

As was discussed in 20.3.2 the association concept of the Braun-Blanquet approach was always based on the presence of a characteristic taxon combination, including character-species. However, as the number of described associations grew, the role of character-taxa gradually diminished whilst the diagnostic importance of differential-taxa increased. In some association descriptions character-taxa are not mentioned separately but are given together with the differential-taxa ('Kenn- und Trennarten'). This led to the character-combination concept which in fact recognized the possibility that none of the participating taxa is a character-taxon in a strict sense, but rather the correlated occurrence of taxa is the essential 'character' of the phytocoenon.

Braun-Blanquet (1964: 122) seemed to confirm this development when he stated that 'the essential association features' should be found in a stand in order to assign it to that association and then defined this as follows: 'primarily the normal characteristic species combination should be present, that is a minimum number of character- and differential-species and some of the more important companions.' However, during recent colloquia and symposia of the International Society for Plant Geography and Ecology the presence of at least one character-taxon was demanded by others, especially by Oberdorfer (1968); see also the following discussions, during the 1964 symposium at Stolzenau (Tüxen 1968: 132–141) and Dierschke (1971). Dierschke (1971) even stated that 'the basic unit for the Prodromus is the association, the rank of which is determined by the constant occurrence of at least one character-species and which is defined by its characteristic species combination.' This definition may be considered the standard from which, however, vegetational circumstance may force some departures.

20.6.3 Geographic Problems with Associations

Many of the problems with character-taxa result from complexities in the geographic behaviour of species in relation to coena, as discussed above (20.5.8). The difficulties led Ellenberg (1954b) to speak of 'the crisis of the character species doctrine' and to propose a much more local delimitation of associations without general character-species. As Ellenberg (1956) added, the use of geographically restricted 'local character-species' for associations is then appropriate. The local establishment of associations emphasized by Ellenberg (1954b) had, in fact, been common practice by then for some 20 years, as Tüxen (1955) stated in a reply to Ellenberg, whilst referring to Braun-Blanquet & Moor (1938) and Tüxen (1937). In fact Oberdorfer (1957, but also

1968 and OBERDORFER et al. 1967) also worked with local character species, though always in close connexion with alliance character-species.

This local association approach as such has been criticized from another point-of-view by SCHWICKERATH (1942, 1954a, 1963). This author emphasized that geographically differentiating taxa should be used in the delimitation of syntaxa, but whenever a series of geographically different syntaxa should have one or more character-taxa in common they should be grouped together into one association. SCHWICKERATH especially criticized the splitting up of the Xerobrometum by BRAUN-BLANQUET & MOOR (1938).

To solve the problem use of different terms for more localized, vs. more widely distributed, phytocoena has been suggested. KNAPP (1942, 1948, 1958) introduced the concept of chief-association (Hauptassoziation) for the type of association SCHWICKERATH was referring to. MEIJER DREES (1951) defined it as follows: the smallest unit possessing absolute or regional character-species. Absolute = general or regional is meant here in the sense of BRAUN-BLANQUET & MOOR (1938) and MEIJER DREES, i.e. not occurring outside the association's range. KNAPP (1971) later mentioned the concept rather incidentally as referring to character-species that are valid 'at least within an entire flora-region.' The more local units comprised in a Hauptassoziation were called Gebietsassoziationen, or 'regional associations'. PASSARGE (1968) and PASSARGE & HOFMANN (1968) elaborated the suggestion by KNAPP towards a separate system of historic-geographical units besides the usual edaphic-ecological units. 'Regional' (territorial or provincial) associations could then be distinguished as parallel but subordinated units to the 'elementary association,' with which the chief-association is identical. PASSARGE & HOFMANN (1968) subjected the regional association to a trinary nomenclature as is applied in zoological systematics for geographical races.

This brings us back to OBERDORFER (1957, 1968) who sharply distinguished between 'association,' 'regional-association' and 'geographical race.' (See also MÜLLER 1968). A regional association ('Gebietsassoziation') has a limited distribution-area; it is characterized by a combination of regional (or local) character-taxa, corresponding with SCHWICKERATH's geographically differentiating taxa, and general alliance character-taxa. A geographical race is a regional expression of a larger, regional or general association, without such a typical combination of territorial and alliance character-taxa. As OBERDORFER (1968) remarked, it will often be very difficult to distinguish between the two levels. Geographical races have not yet found a place in the formal hierarchy but have nonetheless frequently been described, particularly in Germany (e.g. OBERDORFER 1957), Czechoslovakia (e.g. NEUHÄUSL & NEUHÄUSLOVÁ-NOVOTNÁ 1972) and Poland (e.g. MATUSZKIEWICZ 1962).

We have thus a surfeit of terms and concepts. To resolve this matter we refer to the remarks on character-taxa (20.5.8) and adopt the distinctions made by OBERDORFER and others under three terms for units of expanding geographic extent: the local association, regional association, and (general) association. Clearly these intergrade, but in principle they should possess local, regional and general character-taxa, respectively. Groups of related regional associations, which are in fact vicariant associations (cf. PASSARGE 1968) could be called 'vicariant association groups'; this term seems more appropriate than either chief or elementary association. Another acceptable name for the vicariant association-group is collective association (cf. MEIJER DREES 1951).

We shall now briefly discuss the Artemisietum maritimae association as presented in Table V. It is characterized by one character-taxon, *Artemisia maritima* which has, according to Table VI, a high fidelity degree (at least in the investigated area) and which is, according to the synchorological data presented by BEEFTINK (1965) as well as chorological data on the species, a territorial character-taxon. It follows from Table VI that most of the salt marsh associations in the southwestern Netherlands are typified by few but generally 'good' character-taxa (fidelity degrees 4 or 5). The normal characteristic taxon combination of the Artemisietum maritimae in the southwestern Netherlands consists of *Artemisia maritima*, *Festuca rubra* f. *litoralis*, *Limonium vulgare* ssp. *vulgare*, *Aster tripolium* and *Plantago maritima*. BEEFTINK has listed most of the taxa under a specific head so that the companions are few and simply called 'other taxa'. Note the four relevés at the right end of the table, that lack the character-taxon and are considered a fragmentary syntaxon.

20.6.4 HIGHER UNITS (ALLIANCE, ORDER, CLASS)

When a new association is recognized, it must be placed in the system of higher units. The assignment of an association to an alliance (and other higher units) is primarily based on comparison of floristic relationships. As in plant taxonomy (cf. DAVIS & HEYWOOD 1963) the terms relation(ship) and affinity refer only to the possession of common characters. They do not refer to syngenetical (successional) links; we may compare with SOKAL & SNEATH's

(1963) emphasis of phenetic relationships. Group importance values of the character- (and differential-) taxa of the relevant alliances are calculated, either on a presence-absence or on a quantitative basis, and compared (cf. WESTHOFF 1947, RAABE 1957). It may occur that the affinity towards two alliances (or orders or classes) is almost equal. An example of close relationship towards two alliances is presented by the Artemisietum maritimae. Generally the Armerion maritimae alliance is the obvious higher unit, as follows from tables V and VI, but the subassociation Artemisietum maritimae typicum is also clearly related to the Puccinellion maritimae. In fact the association was once assigned to the Puccinellio-Salicornion alliance (BRAUN-BLANQUET & LEEUW 1936) from which the Puccinellion maritimae was split off, but later (e.g. TÜXEN 1937, BEEFTINK 1965, WESTHOFF & HELD 1969) it was considered part of the Armerion maritimae. In cases of doubt equal use is made of structural, physiognomical and synecological considerations (cf. OBERDORFER 1957, WESTHOFF & HELD 1969).

It may also be the case that in a little-known area a number of new associations are described; and some of these may bear so little relation to known alliances (etc.) that they should be classified into new higher units. As was introduced in 20.3.2, various higher units above the association level, viz. the alliance, the order, and the class are characterized in the same way as the association. Species equally occurring in a group of related associations but faithful to that group, become character-taxa for higher units to which they are faithful. Contrary to most associations the higher units generally do possess a number of regional or general character-taxa of high fidelity. When an alliance character-taxon shows a preference for one association within an alliance, it may be used both as alliance character-taxon and as differential-taxon for an association within the alliance. The species is then termed a 'transgressive character-taxon.' Species may of course bear similar character- and differential- relations to orders vs. alliances, and classes vs. orders. Interpolated units like the suballiance and suborder are characterized by character-taxa and/or differential-taxa. Their use is to be avoided; but the suballiance has had some use.

Formal definitions for higher syntaxa and instructions for their ranking have not been agreed upon (cf. PIGNATTI 1968b, DIERSCHKE 1971). In practice the number of related associations tends to determine the number of the higher units classifying these. Growth in number of associations expresses its influence upward, in growth in the number of alliances, etc. For example, TÜXEN (1937) described 94 associations and 41 alliances for northwest Germany;

in TÜXEN (1955) the numbers were 189 and 76. In the Netherlands the number of described alliances grew from 39 (WESTHOFF et al. 1946) to 85 (WESTHOFF & HELD 1969). PIGNATTI (1968b) called the attention to the 'inflationary' character of this phenomenon. His suggestive title stimulated much discussion (TÜXEN 1968b: 89—97), which produced few solutions apart from plans for cooperation and standardization now being realized in the European Prodromus of plant communities.

20.6.5 UNITS ABOVE THE CLASS

The growing number of classes, as well as the extension of the BRAUN-BLANQUET approach to various, and floristically different parts of the world, lead to the introduction of units above the class. The grouping above the class originally proposed by BRAUN-BLANQUET, the 'vegetation circle' was a phytogeographical rather than a sociological unit. BRAUN-BLANQUET (1959) and TUXEN (1970a) on a suggestion of SCHMITHÜSEN sought to establish a new unit, the *class group* (cf. TÜXEN 1963: 213—218).

The class group (TÜXEN 1970a) is a set of territorially defined, largely vicariant classes analogous with the 'vicariant association group' (20.6.3). The characterizing taxa are in this case *genera* with vicariant species in different floristic provinces or regions; for example the classes Querco-Fagetea silvaticae (Europe), Querco-Fagetea grandifoliae (North America), and Querco-Fagetea crenatae (East Asia) form a class group: Querco-Fagea, with the suffix -ea probably taken from JAKUCS (e.g 1972). Curiously, BRAUN-BLANQUET (1964) presented the class group under a different name 'community kingdom' (Gesellschaftsreich) and apparently as a somewhat different phytogeographical concept. Although he first (p. 140) stated that this unit 'takes together classes floristically related by vicariant species and identical higher categories', he concluded that community kingdoms should coincide with SCHMITHÜSEN'S (1961) vegetation kingdoms established by means of vicariant classes.

TÜXEN'S class group should not be confused with the geographically determined class-group of SCAMONI et al. (1965). However, according to the elaboration given by PASSARGE (1968) and PASSARGE & HOFMANN (1968), these class groups correspond actually to the usual European classes. It is questionable whether PASSARGE & HOFMANN'S (1968) complicated system of two hierarchies can be integrated in the BRAUN-BLANQUET system. TÜXEN (1971) strongly rejected such an integration.

Three other treatments of higher units deserve mention. (i)

Chapman (1952, 1959) used terms like Coeno-Salicornietalia for groups of higher units dominated by species of one widely distributed genus. For such a unit the name Coeno-Salicornietum should be preferred. (ii) The concept of 'vegetation type' was developed by Hadač (e.g. 1962, 1967) for groups of more or less similar classes. The vegetation type is based on the occurrence of common species (indicating or 'bezeichnende' species). However, the floristic basis of these units seems to be rather loose, and as Hadač admits the vegetation types are 'practically formations'. (iii) Jakucs (1961) also used the class group concept, which he later (e.g. 1972) named 'division' and for which he introduced the suffix -ea. Hadač (1967) concluded that the division is largely identical with his vegetation type, which term should then have priority. The identity seems to be real, but for it the term 'vegetation type' is least descriptive and most confusing; *division* should be preferred. Bolos (1968) presented a system of divisions for classes occurring in Spain. As term and concept the division is consonant with the hierarchy.

In conclusion, a division may be defined as a syntaxon above the class level that unites related classes within a floristic region (or province) on the basis of common division character-taxa. The character-taxa may be species, or genera, or both. Let us call this a 'vertical' unit in the sense that it unites lower units sympatric in a given region. The class-group concept, in contrast, is comparable with the vicariant association group in that it joins units on the same level from different floristic territories. It is in this sense a 'horizontal' unit. Its character-taxa will more usually be genera than species. Therefore the name 'vicariant class group' may be proposed, defined as a group of classes allopatric in geographic occurrence, but linked by the occurrence of vicariant species in one or more genera. Since two units would be distinguished in this way, it may be questioned whether the suffix -ea should be used for both. It seems logical to reserve this suffix for the division, for which it was proposed.

The definition of the division is not physiognomic. We suggest that, the difference in definition notwithstanding, divisions may converge in practice with the formations of Anglo-American ecology (article 13), as broad physiognomic units limited to a given region or continent. Vicariant class groups, in contrast, would link floristically related phytocoena of different continents. Thus the Querco-Fagetea-group would comprise forest phytocoena with genera in common in Europe, eastern North America, and western Asia, whilst a Spartinetea-group would comprise a set of more narrowly defined phytocoena (*Spartina* associations) in the salt marshes of the world.

20.6.6 Horizontal and Vertical Classification

Thus we follow the suggestion of Knapp (1948), to classify independently by two principles, the one on an edaphic-ecological ('vertical') and the other on a historic-geographical ('horizontal') basis. This suggestion has been amply discussed (Westhoff 1950, Meijer Drees 1951, Barkman 1958a, Beeftink 1965, Scamoni et al. 1965, Passarge & Hofmann 1968, Passarge 1968). All authors agree with the incorporation of both viewpoints in syntaxonomy. Meijer Drees emphasized the vertical approach, because syntaxonomy should build upward from locally classified associations. Beeftink suggested that the choice of the criterion would depend on the peculiarities of the vegetation concerned: for some phytocoena with disjunct distributions the stronger floristic relations and preferred grouping would be horizontal whereas for other phytocoena the more natural grouping would be vertical. Passarge & Hofmann use both principles on each level, with the geographical units subordinated to the corresponding edaphic units, which makes the system very complicated (cf. Tüxen 1971, in his critical review of Passarge & Hofmann 1968).

We would modify Beeftink's suggestion in recommending that the main-axis of the classification should be 'vertical' on all levels of the hierarchy. Given the main direction of the classification as vertical, secondary to this, 'horizontal' vicariant groupings may be recognized on any syntaxonomic level. We have mentioned such groupings for associations (20.6.3) and higher syntaxa (20.6.5); we may now trace the problem and the principle into the lower syntaxa.

20.6.7 Lower Units (Subassociation, Variant, Facies)

Associations are on the one hand joined into higher units, and on the other divided into subordinate units. As in idiotaxonomy the former procedure is compulsory, the latter facultative. The lower syntaxa are mostly characterized by differential-taxa. Usually, also, the syntaxa below the association level are described as deviations from average or typical situations. Thus to complete the syntaxonomy on a certain level a 'typical' subassociation (variant, etc.) and one or more 'deviant' syntaxa are described and characterized in parallel.

Variations within an association have been approached in three directions; to the edaphic-ecological ('vertical') and historic-geographical ('horizontal') may be added a syndynamical (successional) point-of-view. It has been proposed repeatedly that different syntaxa should correspond to these approaches each of which could form a hierarchy of subordinate units. Thus the subassociation might be based on local edaphic or micrometeorological differences within the association, the variant on geographical or climatological differences (with the terms vicariant and geographical race as synonyms), and the phase on differences in successional status. This distinction is in accordance with the original approach of Braun-Blanquet (1928, 1932, Braun-Blanquet & Pavillard 1928). Schwickerath (1942) and Duvigneaud (1946, 1949) distinguished five types of differential-taxa on the basis of differences in nutrient status, moisture conditions, successional status, geographical area, and human influences.

The system was further extended by the incorporation of the sociation, a unit defined by the dominant species of its strata (see article 18). This suggestion of Nordhagen (e.g. 1937) was acknowledged by the 1935 Botanical Congress (see Du Rietz, 1936). In the Netherlands it has become common practice to describe species-poor communities without character-species but with one or more dominants as sociations or consociations (Westhoff 1947, 1949, Beeftink 1965, Westhoff & Held 1969). Contrary to Du Rietz (1930b, 1936, 1965) the sociation and consociation are considered of association-rank (Braun-Blanquet 1955, Maarel et al. 1964). They may be assigned to an alliance and higher syntaxa (Westhoff & Held 1969).

From the beginning many communities have been described without receiving a definite syntaxonomic position. They are often called 'community of ......' ('Gesellschaft von ......') (cf. Ellenberg 1956, Neuhäusl 1963, Fukarek et al. 1964). These units can be either provisional local coena, which might later be ranked in the system when more evidence is available, or coena without character-species or dominants. Barkman et al. (1958) and Westhoff et al. (1959), proposed the term 'consortium' for the latter units.

An additional term for subordinate units, 'form,' is used in different ways for syntaxa of lower rank. It may mean a unit of no specified rank (as seems to be the case with Schubert 1960, for example). It is also used in relation to altitude (Oberdorfer 1957, 1968, Braun-Blanquet 1964) or moisture conditions (Duvigneaud 1946). The 'Ausbildungsform' is considered the lowest unit in the hierarchy apart from the facies by Tüxen (1970a, cf. Schubert 1960, Passarge 1968).

The distinction of local-edaphic and geographic units has been maintained by Braun-Blanquet (1951, 1964), Meijer Drees (1951), Oberdorfer (1957, 1968), Barkman (1958b) and others.

Under the influence of Tüxen (1937) the variant became a subunit under the subassociation (Braun-Blanquet 1951a, 1964) rather than a differently defined (geographic) unit. In line with this practice we suggest as the downward extension of the primary hierarchy, for units based on local variations: association, subassociation, variant, subvariant, and facies. The last two may be often unneeded, but facies may be ecologically significant and useful as indicators in some cases (Krause 1954). Geographic races of lower syntaxa would be separately indicated by the territories in which they occur (Oberdorfer 1957, Müller 1968).

The *subassociation* is indicated with the suffix -etosum. Of the various ecological possibilities and corresponding symbols of Duvigneaud (1946, 1949) *d* and *δ* may be used for subassociations differentiated by nutrient and moisture status, respectively. Schwickerath (e.g. 1942), Duvigneaud (1946) and Meijer Drees (1951) showed how subassociations of an association may in fact indicate the transitions of that association towards related associations. The subassociation 'typicum' may then be considered as the 'nucleus' of the association (cf. Fig. 1); in fact this term is commonly used for the subassociation that lacks differential-species. Westhoff (1965, Westhoff & Held 1969) observed that many subassociations of the latter type were relatively poor in species, and particularly poor in character-species of the association. Such units were indicated as 'inops' (= having shortage of); the 'typical' Querco-Betuletum is an example of such an 'inops'-subassociation.

The next lower rank is called the *variant*. According to Braun-Blanquet (1964: 124) a variant would not be characterized by differential-species, but 'either by a strong prominence of certain species, which cannot be considered as differential-taxa, or by a slightly deviating species assemblage.' This criterion is, however, diagnostically too vague. Most phytosociologists share the opinion that a variant also must be characterized by differential-species. The same holds for the next lower rank, the subvariant. The variant-differentiating taxon (taxa) may be differentiating only because of its 'strong prominence,' i.e. its higher combined estimation values in one variant compared with others and not because it is clearly present resp. absent. In most cases the variant is distinguished within the subassociation; but it is also possible to divide an association directly into variants, if the mutual differences are considered too small to justify the rank of subassociation. Below the variant the subvariant may be used if needed.

The lowest unit is the *facies*. It is not even characterized by differential-taxa; it thus does not employ diagnostic-species and is not part of the formal hierarchy. A facies is usually characterized by the dominance, in a high cover degree (scale 4 or 5), of one of the species belonging to the normal floristic assemblage of an association. It is not usual to consider a stand as a facies if this dominance is a normal feature of that association. Since, for instance, dominance of *Fagus sylvatica* is normal for any association assigned to the alliance Fagion sylvaticae, it is not appropriate to construct such a coenon as 'a *Fagus* facies of the Fagetum.' A facies, therefore, is a deviation phenomenon. It may be brought about by special, and sometimes by extreme conditions of abiotic factors, but in many cases it is the result of human disturbance (e.g. MELTZER & WESTHOFF 1942, KNAPP 1971). Examples are a facies of *Rubus* sect. *Sylvatici* or *Rubus* sect. *Heteracanthi* in a stand of Querco-Carpinetum, or a facies of *Acorus calamus* in a stand of Scirpo-Phragmitetum. It is often possible to deduce an effect of human disturbance from the appearance of a certain facies.

20.6.8 NOMENCLATURE

Formal nomenclature of syntaxa started with the naming of associations by BRAUN-BLANQUET (1913), following precedents of CAJANDER (1903), BROCKMANN-JEROSCH (1907) and RÜBEL (1912). The first general proposals were discussed at the Brussels Botanical Congress (FLAHAULT & SCHRÖTER 1910); the nomenclature of associations and other syntaxa developed gradually thereafter. With the growth of the number of described syntaxa the need for standardization and rules increased. A start toward systematic nomenclature was made in 1933 (BRAUN-BLANQUET 1933b), and DAHL & HADAČ (1941) proposed a coherent set of rules based on the code of botanical nomenclature. During the Stockholm Botanical Congress BARKMAN (1953a) proposed a general set of rules and MEIJER DREES (1951) a complete system of nomenclature. These were discussed in a number of papers in *Vegetatio* vol. 4. Contributions by BACH et al. (1962) and RAUSCHERT (1963) appeared later, and BRAUN-BLANQUET (1964) reported on the discussion and presented 7 rules. The next phase started during the Symposium on Plant Sociological Systematics at Stolzenau, 1964, where MORAVEC (1968) introduced a set of 26 articles and urged a general acceptance of these amongst active plant sociologists. A special Working Group is now preparing a code for the new Prodromus project (see TÜXEN 1968b: 152, MORAVEC 1968, 1969, 1971, NEUHÄUSL 1968).

Some instructions for the construction of names and some proposed general rules for nomenclature of syntaxa follow.

According to MORAVEC (1971) five necessary principles are:
1. Each syntaxon (with definite rank, position and delimitation) has only one correct name.
2. Each name can be correctly used for one syntaxon only.
3. The correct name is established according to rules based on the priority principle.
4. The association is the fundamental nomenclatural unit (syntaxon).
5. The validity of nomenclatural rules is retroactive.

20.6.8.1 *Construction of Names*

For the standardization of syntaxon names, the following procedure is in general use.

To the generic part of the names of one or two (not more) characteristic (not necessarily faithful) taxa of a syntaxon a suffix is added. These suffixes are specific for the different syntaxonomic ranks, see Table VII. The suffix '-etum' for the association goes back to classical Latin and has been in use since HUMBOLDT (1805). The suffix '-ion' for the alliance was proposed by MOSS (1910) and taken over by BRAUN-BLANQUET (1921: 347). The suffix '-etalia' for the order was presented in BRAUN-BLANQUET (1928). For the class, up to 1932 (BRAUN-BLANQUET, FULLER & CONARD) no specific suffix was used, since the classes were designated by circumscriptions such as 'communities of maritime dunes' (later the class Ammophiletea). In 1934 MEIER & BRAUN-BLANQUET proposed the suffix '-etales' for the class, but (BRAUN-BLANQUET et al. 1939) this suffix was later changed into '-etea'. The suffix '-etosum' was introduced by BEGER (1922).

If we are dealing with an association or alliance that is sufficiently characterized by one taxon (e.g. by a faithful dominant character-taxon), the genus name of the taxon is used with the appropriate suffix (-etum, -ion), followed by the species name in the genitive. For example the association Ericetum tetralicis is named after *Erica tetralix*, and the alliance Alnion glutinosae after *Alnus glutinosa*.

In some cases the specific epithet has been replaced by a geographical adjective indicating the area characteristic of the syntaxon, e.g. Agropyretum boreoatlanticum for the association of *Agropyron junceum* (syn. *Elytrigia juncea*) of the North Atlantic Coast. In this case this has been done to distinguish it from a different association of *Agropyron junceum* occurring in the Mediterranean. In recent decades, however, there is a tendency to avoid such geographical names as far as possible and to replace them by bigeneric names. In the case of orders and classes and, though less frequently, alliances the specific epithet is omitted when the context leaves no doubt – for example, 'Fagion' instead of 'Fagion sylvaticae' in a European context.

When a syntaxon was to be named after two characteristic (not necessarily faithful) taxa, the name of the second taxon was provided with the appropriate suffix; whilst the name of the first taxon was joined with the second one by the suffix -eto. Examples are the association Querceto-Betuletum, the alliance Alneto-Ulmion, the order Glauceto-Puccinellietalia, and the class Querceto-Fagetea. However in many cases this suffix was abbreviated to a connecting vowel after the stem of the first name, e.g. Alneto-Padion became Alno-Padion. Bach et al. (1962) proposed to do this systematically, mainly on the argument that, for example, 'Querceto-Betuletum' could mean a mixture of, or transition between, a *Quercus* stand and a *Betula* stand, which is of course something quite different. This proposal has been widely accepted. However, the correct choice of the connecting vowel, especially in the case of the third Latin declension and with names derived from Greek words, gives difficulties. Bach et al. (1962) and Rauschert (1963) gave directions for a large number of cases.

There are two reasons for choosing a bigeneric name. The first and more obvious is that a combination of two names should characterize the syntaxon better. The first name may be a dominant, generally the second name represents the character-taxon considered diagnostically most important. The other reason was already mentioned, viz. monogeneric names with a geographical adjective are replaced by bigeneric names, in which the first name comes from a species characteristic for the geographical region originally indicated. The example given above, the Agropyretum boreo-atlanticum, has been renamed in fact the Minuartio-Agropyretum (Tüxen 1955).

In some cases the nomenclature is at variance with the standard procedure. The following deviations should be mentioned:

(1) Contractions. It has become general use to follow the proposal by Koch (1925) to replace clumsy terms like 'Potamogetonion' (and, later on, also 'Potamogetonetalia' and 'Potamogetonetea') by 'Potamion', 'Potametalia' et 'Potametea'. However, Oberdorfer et al. (1967) proposed to return to the complete and cumbersome names.

(2) In special cases descriptive adjectives or substantives are in-

cluded in the name. 'Magnocaricion' indicates an alliance characterized by tall sedges; 'Macrophorbio-Alnetum' is meant to characterize a moist woodland (carr) association in which 'macrophorbiae' (tall forbs, Hochstauden) form a group of differentiating species.

(3) Finally, sometimes specific epithets have been used instead of generic ones. This was common practice in the early days of alpine phytosociology; BROCKMANN-JEROSCH (1907) and RÜBEL (1912), for example, used the names 'Curvuletum' and 'Firmetum' instead of 'Caricetum curvulae' and 'Caricetum firmae.' Later this practice has been applied only very rarely; a rather recent example is 'Alno-Padion' (KNAPP 1942, MATUSZKIEWICZ & BOROWIK 1957), to designate an alliance characterized by *Alnus glutinosa* and *Prunus padus*.

20.6.8.2 *Validity, Changes and Authors' Names*

Various rules have been proposed for the settlement of the validity of syntaxon names. A name should have been published after 1900 in printed form, available at least in generally accessible libraries. Publication of a name should be combined with an adequate diagnosis.

For associations and lower-rank units this would include the presentation of a table with the complete floristic composition of at least three relevés and the assignment of a type-relevé. For higher units the listing of character- and differential-taxa and the assignment of a type-unit of the next lower level would be required. Such types are called nomenclatural types.

When two or more correctly published names appear to refer to one and the same syntaxon, the oldest name has priority. When a syntaxon is divided into two or more units of the same rank, the original name goes to that new unit that contains the nomenclatural type.

When the valid name of a taxon used in a syntaxon name changes, corresponding change in the syntaxon name should be considered. In any case a syntaxon name should be changed when one of the name-giving taxa appears to be a younger homonym of a different legitimate name. For example, the Isoëtetum setacei should become the Isoëtetum delilei, since *Isoetes setacea* (Box ex) DELILE appeared to be a homonym of *Isoetes setacea* LAM. = *I. echinospora* DUM. and was then renamed *I. delilei* ROTHMALER.

No agreement has been yet reached on the – still very frequent – simple nomenclatural changes. E.g. *Carex nigra* (L.) REICH-

ARD appears to be the correct name for *C. fusca* ALL. Should the order Caricetalia fuscae now be renamed the Caricetalia nigrae? MORAVEC (1968) would say not; WESTHOFF would say yes and in fact (WESTHOFF & HELD 1969) changed this name.

Rules for the citation of authors' names have also been proposed, more or less parallel to those of idiobotanical nomenclature. The name of the author(s) who published a valid name is added to that name together with the year of publication. When this name refers to a previously published but invalid name, the earlier author's name is added in the form, Thero-Airion TÜXEN ex OBERDORFER 1957. When a name is changed, e.g. when the syntaxonomic level is changed, the author of the changed name is added between brackets. When a syntaxon is given a new contents or delimitation, both old and new authors are mentioned connected with the abbreviation em. (= emendavit).

These proposals were advanced by MORAVEC (1968). PIGNATTI (1968b: 87) recommended dropping authors' names to avoid the description of new syntaxa by authors who primarily want to have their names immortalized. In the discussion following, p. 89—97, there was agreement on omitting author's names when no confusion could occur, especially with higher units and when a prodromus can be referred to.

20.6.9 SCHEME OF SYNTAXA

All in all phytosociological nomenclature may be considered a necessary evil. Contrary to the opinion of DOING (1962, 1966, 1970) that a rigid system of nomenclature should be abandoned for practical reasons, we would emphasize that the value of the system for ordering information and communicating among phytosociologists outweighs its difficulties. We would further state that descriptions of formal phytosociological syntaxa should either be published in strict accordance with nomenclatural rules, or they should not be published as such.

Table VII presents the current syntaxa and their nomenclature, as discussed so far. Examples together with their denominating taxa are added. The variant (and the facies) are given here without suffixes. The geographically determined syntaxa (20.6.3, 20.6.5), which have not yet been agreed upon, are not included.

TABLE VII
Scheme of Syntaxa

Levels and units of the formal hierarchy from highest (Division) to lowest subvariant, with suffixes and examples of the construction of names after denominating taxa.

Syntaxon	Suffix	Examples	Denominating taxa
Division	-ea	Querco-Fagea	genus Fagus
Class	-etea	Phragmitetea	Phragmites australis
		Querco-Fagetea silvaticae	Quercus robur Fagus sylvatica
Order	-etalia	Littorelletalia	Littorella uniflora
		Festuco-Sedetalia	Festuca and Sedum L. div. spp.
Alliance	-ion	Agropyro-Rumicion crispi	Agropyron repens, (syn.: Elytrigia repens) and Rumex crispus
		Alnion glutinosae	Alnus glutinosa
Suballiance	-ion (-esion)	Ulmion carpinifoliae (Ulmesion)	Ulmus carpinifolia
Association	-etum	Ericetum tetralicis	Erica tetralix
		Elymo-Ammophiletum	Elymus arenarius, Ammophila arenaria
Subassociation	-etosum	Arrhenatheretum elatioris brizetosum	Briza media
Variant		ibid., Salvia variant	Salvia pratensis
Subvariant		ibid., Bromus subvariant	Bromus erectus

20.7 Extension of the Approach

20.7.1 OTHER GEOGRAPHIC AREAS

From its center of origin in the Alps and the western Mediterranean the BRAUN-BLANQUET approach spread into many European countries and Japan. Larger parts of the Mediterranean and Euro-Siberian, and smaller parts of the Sino-Japanese, floral regions have been synsystematically described. Still, a considerable part of the original working area is insufficiently known, particularly the eastern Mediterranean, Russia and Scandinavia (see the map presented by DIERSCHKE 1971). Expansion of the approach's influence continues, and we shall give a short survey of more recent literature applying the approach outside its European homeland. The survey cannot be exhaustive; only regions will be mentioned in which substantial work has been done. For further references the reader is referred to *Excerpta Botanica Sectio Sociologia* (bibliographic contributions to which are indicated below as 'Exc.'). This series can also serve as a survey of phytosociological work in the three regions mentioned above, with bibliographies for almost all European countries.

Japan may deserve special attention. The phytosociological study of Japan was initiated by SUZUKI/TOKIO and developed especially by MIYAWAKI and his pupils (see MIYAWAKI 1960 and 1966a, 1971a Exc.). Of special interest is the manner in which the Japanese communities are compared with Mediterranean and Indo-Malaysian communities. Other comparative studies in Japan include OHBA (1972), in which various vicariant salt marsh associations were described, partly belonging to the 'European' class Asteretea tripolii; TÜXEN, MIYAWAKI & FUJIWARA (1972) on the class Oxycocco-Sphagnetea in which Europe and North America (work of R. KNAPP and A. DAMMAN) are included; and TÜXEN (1966a) on holarctic Honckenyo-Elymetea communities. These studies were all based on extensive relevé tables and close cooperation between European and Japanese phytosociologists. From these studies the concept of vicariant class group developed (see TÜXEN 1970a and 20.6.5). Vegetation mapping on a phytosociological basis has reached a high level in Japan (see MIYAWAKI 1966b, 1971b, Exc. for a survey of maps).

The *arctic-subarctic* region has only occasionally been described. Iceland is becoming relatively well-known; a general survey is in preparation by R. TÜXEN based on phytosociological excursions by HADAČ, BÖTTCHER, the International Society for Plant Geography and Ecology, and others. Newfoundland was partly described by DAMMAN (1964). The research on Oxycocco-Sphagnetea by DAMMAN in Canada was already mentioned under Japan as was the study of TÜXEN (1966a). See also STEINDÓRSSON (1966, Exc.) for Iceland, BÖCHER (1961, Exc.) for Greenland, and HANSON (1959, Exc.) for Alaska.

The *west and central Asiatic* (Irano-Turanian) region has been described in a fragmentary way. An extensive survey of xerophytic and summer-dry hygro- and mesophilous communities of Afghanistan was presented by GILLI (1969, 1971), who could reasonably establish associations, though not as sharply delimited as in Europe because of the large number of species in most communities. HARTMANN (1968, 1972) published accounts on mountain grassland and scrub communities.

For the *Macaronesian* region the Canaries are relatively well-known through work of RIVAS-GODAY & ESTEVE CHUECA (1965), OBERDORFER (1965), LEMS (1968), and SUNDING (1969), as well as

LOHMEYER & TRAUTMANN (1970) who reviewed this literature (see also references in SUNDING 1969-70, Exc.). Studies on the Azores and the Cape Verdes were listed by PINTO DA SILVA & TELES (1962, Exc.). OBERDORFER (1970) reported on some plant communities on the Canaries and described complex, stable, mosaics of succulent scrubs and open sclerophyllous woods, which are amongst the most complicated communities of the world and should be studied as complexes (cf 20.7.4 and the South-African Kingdom below).

Rather numerous, though still scattered phytosociological studies have been carried out in the *North-American* and *Canadian* regions, starting with the description of Long Island by CONARD (1935) including 71 associations. CONARD (1952) also published on the vegetation of Iowa and DANSEREAU (1957) on the middle St. Lawrence Valley. TOMASELLI (1958, quoted in KÜCHLER 1967) described plant communities in eastern Kansas. Monographs are concentrated on forest communities (except for the studies of LOOMAN e.g. 1969 on grasslands). They include those of MEDWECKA-KORNAŚ (1961) in the Montreal area, GRANDTNER (1966) in Quebec, KORNAŚ (1965) in North Carolina, CRISTOFOLINI (1967) in Tennessee, JANSSEN (1967) in Minnesota and KRAJINA (1969) in British Columbia. KNAPP (1957, see also 1965) presented a preliminary survey of higher phytosociological units, viz. 72 classes, some of which were divided into orders and later (KNAPP 1964) extended it to the whole holarctic Kingdom.

The *Indo-Malaysian* and the *Polynesian* subkingdoms are practically phytosociological terra incognita. Some studies have been made in India (BHARUCHA & MEHER-HOMJI 1963, Exc., GUPTA 1966, 1967, Exc., MEHER-HOMJI 1969, Exc., MEHER-HOMJI & GUPTA 1972a,b, Exc.).

The *African* territory has been studied rather intensively, here and there. French studies on the North African desert region culminate in the monograph of QUÉZEL (1965), which presents a coherent survey of over 100 desert communities largely fitted into a syntaxonomic system (see also ROUSSINE & SAUVAGE 1961, Exc.). Relatively intensive studies have been performed in the West African rain forest region by LEBRUN (e.g. 1960), LEBRUN & GILBERT (1954) and SCHNELL (1952). The associations and higher units described are all based on relevés and range from xerophytic and hydrophytic pioneer communities to climax forests. LEBRUN concludes that the BRAUN-BLANQUET approach is very well applicable. Savanna studies in this area include those of DUVIGNEAUD (1949), ADJANOHOUN (1962), and DEVRED (1956). In Moçambique a first phytosociological inventory was carried out by GOMES PEDRO & GRANDVAUX BARBOSA (1955), see also PINTO DA

SILVA & TELES (1962, Exc.). KNAPP (1965, 1966a) presented preliminary surveys of higher units for West, Central and East Africa and reviewed the literature for some countries (KNAPP 1969—70, 1971a,b, Exc.).

Phytosociological research in *South Africa* started recently and is now expanding rapidly (EDWARDS 1967, TAYLOR 1969, ZINDEREN BAKKER 1971, WERGER 1973, see WERGER for references). WERGER et al. (1972) presented the first account of plant communities in the Cape Kingdom, including vegetation types of the famous 'Fijnbos' formation. Their experiences with the BRAUN-BLANQUET approach in this extremely varied mosaic of species-rich communities are comparable with that in European xerotherm woodland-scrub-borderline communities (cf. JAKUCS 1961, 1972).

The *Caribbean* region has also been treated in general phytosociological studies, e.g. KNAPP (1965). Detailed descriptions of savanna vegetation in Surinam have been published by DONSELAAR (1965, 1969), who found the BRAUN-BLANQUET approach applicable. The North Surinam open savannas could be grouped into one class. A special problem was formed by the floristic similarities between community types of deviating structure. DONSELAAR considered the floristic relations as decisive, as did WERGER et al. (1972) for similar situations in Fijnbos communities of the Cape. Earlier Surinam work is mentioned in BOERBOOM (1970, Exc.). *Brazilian* rain forest descriptions are very limited (VELOSO 1962 in BRAUN-BLANQUET 1964, DONSELAAR 1965). The *Andean* region has received more attention; OBERDORFER (1960), SCHMITHÜSEN (1960), ESKUCHE (1968, 1969) described various scrub and forest communities, and KOHLER (1968, 1970) desert and coastal communities. See also ESKUCHE (1967, Exc.) for this, and adjacent regions in South America. KNAPP(1966b) presented a survey of higher vegetation units in *Patagonia*.

The BRAUN-BLANQUET approach has thus been applied in nearly all floristic regions, and to almost all types of vegetation. The approach should in principle be universally applicable; and in fact no difficulties have been met in its expansion out of Europe more fundamental than those encountered in Europe. The principal directions of difficulty are: (i) In species-poor vegetation in the North and the Antarctic, character-taxa may be difficult to find and use, and species dominance (and the sociation as a unit) may be more emphasized (see article 18). (ii) Rich tropical vegetation confronts the approach with such a wealth of species as to make application more difficult; physiognomy has been more emphasized in the tropics as a means of recognizing phytocoena. (iii) Delimitation of uniform stands becomes difficult in mosaic structures of open xerothermic scrubs and woodlands, such as found in the Canaries and South Africa (cf. 20.4.1).

20.7.2 Biotic Communities

Approaches to the classification of animal communities are diverse (Whittaker 1962); applications of the Braun-Blanquet approach have mostly been concentrated on the analysis of one or more animal groups in plant communities already described. General bibliographies were presented by Rabeler (1957b, 1964, Exc.). More or less complete descriptions of biotic communities are restricted to marine benthic environments, in which animals and plants were used together to construct a hierarchy according to Braun-Blanquet principles (Roger Molinier 1960).

The following conclusions may be drawn from work on biotic communities (cf. Braun-Blanquet 1951, 1953, 1964, Rabeler 1937, 1960, 1962, 1965, Westhoff & Westhoff-Joncheere 1942, Mörzer Bruyns 1950, Whittaker 1962, Maarel 1965, Tüxen 1965a and Barkman article 16 in this volume):

1. Species from most animal groups form characteristic combinations, often indicated as 'communities.' Mostly they are merely taxocoenoses (cf. Maarel 1965).
2. The local distribution pattern of such groupings may coincide with that of the plant communities in which the groupings have been established, but mostly it is of a larger, sometimes of a smaller scale.
3. Most animal groupings are primarily bound to specific abiotic factors like soil surface moisture or to vegetation structural features (or, rather, to the micrometeorological features determined by them).
4. The plant community can be characterized by animal species groups, although the number of faithful animal species is low. As long as the synusial structure of animal communities is not sufficiently known, the phytocoenoses are the most realistic framework for biocoenotic studies. This implies that attempts like that of Quézel & Verdier (1953) to create a separate hierarchy of 'communities' of members of one – or a few – animal groups is unrealistic when these communities are in fact taxocoenoses.

20.7.3 Microcommunities and Synusiae

Since these community-types are treated by Barkman (article

16), we refer only to some current literature. The BRAUN-BLANQUET approach has been applied to various kinds of microcommunities and synusiae. Bibliographies have been composed for: epiphytic communities (BARKMAN 1962, Exc., 1966, Exc.), marine algal communities (HARTOG 1967 Exc.), epigeic moss and lichen communities (HÜBSCHMAN & TÜXEN 1964, Exc., TÜXEN 1964, Exc., 1968/1969, Exc.) and soil inhabiting communities (APINIS 1969, Exc.) Classification systems have been developed by HARTOG (1959) and BOUDOURESQUE (1971) for marine algal communities; HARTOG distinguished associations which were grouped into formations, BOUDOURESQUE presented a complete hierarchic system. Broad classification systems of moss and lichen communities have been presented by KLEMENT (1955), KOPPE (1955), BARKMAN (1958), WILMANNS (1962). From these studies one may conclude that the BRAUN-BLANQUET system is generally applicable to microcommunities and synusiae, and that it is advantageous to approach the latter through a classification separate from that for phytocoenoses (see also article 16.5 for references).

20.7.4 COMMUNITY COMPLEXES

A 'complex' is a set of contiguous or continuous communities forming a mosaic or pattern. Complexes can thus be studied on various levels from the microtopography of bogs to whole landscapes. Small-scale topographic complexes may often be studied without direct concern for successional relationships, even though the communities may have developmental relations to one another (as in bogs) or to the potential natural vegetation (SEIBERT 1968). In situations where man has drastically changed the vegetation pattern as well as the substrate or in geologically young systems such as coastal dunes it may be more realistic to approach complexes directly in their local mosaic or zonation structure. A difficult type of complex is the 'superposition complex' (Überlagerungskomplex) within which rapidly changing community patterns occur, e.g. with aquatic or with ephemeral communities (cf. MÜLLER 1970).

For communities related by successional process the concept of community ring (Gesellschaftsring) was mentioned by KRAUSE (1952) and developed by SCHWICKERATH (1954b) and SCHMITHÜSEN (1961). The ring comprises the series of communities that are syndynamically related to a terminal community. SEIBERT (1968) translated the concept as 'circle of communities', a term that unfortunately invites confusion with BRAUN-BLANQUET's (1932, 1964) 'circle of vegetation' or 'Vegetationskreis'. The English-language terms 'climax-complex' and 'successional complex' (see article 14.4.2) may well be preferred.

Study of broader topographic and edaphic complexes in relation to climax theory was initiated by TÜXEN. (For an English account of TÜXEN's views, see KÜCHLER 1967). TÜXEN & DIEMONT (1937, see also TÜXEN, 1933) developed the parallel concepts of 'climax swarm' and 'climax group' for spatially contiguous stable communities (Dauergesellschaften) that occur in the same climate, but differ in consequence of direction of exposure in mountains (swarm), or soil parent material in lowland areas (group). TÜXEN (1933) proposed also the term 'paraclimax' for a widespread stable community the characteristics of which are determined by soil development, and not climate alone. TÜXEN furthermore recognized a series of climax-regions for Europe, each characterized by a number of character-associations bound to it in their distributions, together with companion-associations (Begleitassoziationen) of wider distribution in more than one climax-region. The latter idea is a development beyond the climax-complex of BRAUN-BLANQUET (1928, 1964) and others; it characterizes regions not by a single, somewhat hypothetical climatic climax (Schlussgesellschaft), but by associations used as diagnostic-communities.

TÜXEN (1937) also recognized the coherence between the various substitute communities (Ersatzgesellschaften), both the completely cultural and the semi-natural ones, which may occur in the space that could be (and may actually partly be) occupied by a terminal community. Such locally coherent communities were consequently called 'contact communities.' The recognition of the coherence of substitute communities in their potential development towards one terminal community led TÜXEN (1956) to the concept of *potential natural vegetation.* This concept is defined in relation to a given habitat, as the vegetation that would finally develop (terminal community) if all human influences on the site and its immediate surroundings would stop at once, and if the terminal stage would be reached at once. With the latter restriction abiotical (e.g. macroclimatological) changes during this development are meant to be excluded. Although in many cases the potential natural vegetation may be identical with the original vegetation and interpretable as such, the distinction of this concept is useful for the following reasons: (i) There may have been no 'original' vegetation at all, since man has influenced the ecosystem under consideration at least as long as the present climate (and thus the present climax complex) lasts; (ii) Man may have induced irreversible changes in the ecosystem, so that the presumed 'original' vegetation can never establish again; (iii) reconstruction of the original vegetation may theoretically be possible in some cases, but precarious in practice.

Further discussion of climax interpretation is beyond our purposes here; we refer to WHITTAKER (1953), DANSEREAU (1957), SCHMITHÜSEN (1961) and BRAUN-BLANQUET (1964). We mention as significant, however, the phytosociological study of community complexes over wide geographic areas by KRAUSE (1952). KRAUSE observes the consistency of community composition over extensive areas together with the occurrence of vicariant communities, particularly between eastern and western Europe. The community-complex of a given area is shaped both by its macroclimate, and by local factors producing mosaic-like community relationships. Regions are characterized by different complexes, and different prevailing or dominant communities. Complexes give way to other complexes through geographic distance in three ways: (i) compression of communities into smaller areas and exclaves, (ii) increasing floristic impoverishment until a given community can be traced only in infrequent fragments, and (iii) the replacement of dominant communities by vicariants suited to other climates. For further discussion of vegetation areas and regions see SCHMITHÜSEN (1961), BRAUN-BLANQUET (1964), and SEIBERT (1968).

20.7.5 APPLIED PHYTOSOCIOLOGY

The floristic-sociological approach has been applied in many neighbouring disciplines. Early indications of this application were presented by BRAUN-BLANQUET (1930) and by TÜXEN (in BRAUN-BLANQUET & TÜXEN 1932) who observed several non-botanical sciences with which phytosociology might have mutual relationships: geology, hydrology, geomorphology, climatology, soil science, animal sociology (see 20.7.2), geography, archaeology, palaeobotany (particularly palynology), nature conservation and land use planning. Applications of phytosociology were often based on vegetation maps. Above all, under TÜXEN the Zentralstelle (later Bundesanstalt) für Vegetationskartierung (Federal Institute for Vegetation Mapping) at Stolzenau/Weser, West-Germany, developed vegetation mapping and its application in land interpretation and management and thereby contributed considerably to the acceptance of phytosociology as a significant applied, as well as basic science. Many studies at the Institute, models of such research, can be found in the series *Angewandte Pflanzensoziologie* (Stolzenau/Weser) and *Mitteilungen der Floristisch-Soziologische Arbeits-*

gemeinschaft (Neue Folge, Stolzenau). See also TÜXEN (1956) for a survey of maps, BRAUN-BLANQUET (1969) for a survey of TÜXEN's work, and KÜCHLER (1967) for an English account of floristic-sociological vegetation mapping.

Since applied phytosociology will be the subject of other volumes of the *Handbook*, we present here only a survey of main applied floristic-sociological literature and bibliographies. General publications on agriculture and forestry include textbooks by KNAPP (1949) and ELLENBERG (1950, 1952, 1954a) as well as the series *Angewandte Pflanzensoziologie* (Stolzenau) and the symposia on Vegetation Mapping, Anthropogenic Vegetation, Plant Sociology and Palynology, Landscape Ecology and Experimental Phytosociology of the International Society for Plant Geography and Ecology (TÜXEN 1963, 1966b, 1967c, 1968a, 1969b). Further general references are found in *Excerpta Botanica, Sectio Sociologica.*

An important general conclusion may be drawn from applied research: detailed floristic-sociological description of vegetation combined with adequate analysis of its environment provides a basis for applied phytosociology in many fields of research which can be of great social importance, at least in the cultural landscapes of Europe.

20.7.6 STRUCTURAL CONSIDERATIONS

BRAUN-BLANQUET established the association on a floristic-sociological basis, but since the floristic-sociological characters of an association are supposed to reflect all other characters a floristicsociologically uniform association might be expected to be structurally uniform as well. The original association concept thus implied the physiognomic uniformity of the association. As WESTHOFF (1967) showed, it is not always true that floristically consistent units are also structurally uniform. Many examples are known of considerable structural differences – even on the formation level – within associations or higher syntaxa (e.g. BARKMAN, 1958b). In a number of cases the discrepancy between floristic and structural uniformity has been solved by a refined syntaxonomic treatment.

An example is the Salicornieto-Spartinetum BRAUN-BLANQUET & DE LEEUW (1936) of NW-Atlantic mud flats. This association was considered a mosaic of patches of two different life forms, hence of two different structural types. New relevés taken separately from the two kinds of patches revealed them to be floristically different, which led to separated associations (cf. BEEFTINK 1962, 1965). These associations were in fact assigned to separate classes, viz. Spartinetea and Thero-Salicornetea (e.g. BEEFTINK, 1962). A second example from a very different situation is the classification of scrub and mantle communities separately from the woodland classes to which they used to be assigned. TÜXEN (1952) created for these the order Prunetalia spinosae which he later (1962) placed in a separate class Rhamno-Prunetea (after RIVAS GODAY & BORJA CARBONELL).

Besides attempts to use structural criteria in floristic-sociological classification various suggestions for an integration of the latter system with physiognomic systems have been put forward. A connexion between the syntaxonomical class and the physiognomic formation has been sought by DOING (1962) and PASSARGE (1966, 1968, SCAMONI et al. 1965). PASSARGE (1966) adopted a formation definition that was based on the original concept of Grisebach (See BEARD, article 13), as well as a growth-form definition from SCHMITHÜSEN (1961): a phytosociological formation is a sociological vegetation unit dominated by related growth-forms and hence showing a uniform physiognomy. Informal arrangements of classes within formations have been presented by ELLENBERG (1963) and WESTHOFF (1967), WESTHOFF & HELD (1969). One could also think of connexions between formations and either of the floristic-sociological units above the class, viz. the division and the vicariant class group.

20.7.7 INDICATOR GROUPS

One of the essential bases of applied phytosociology is the use of indicator groups of species. It is consistent with – indeed it expresses – the perspective of the BRAUN-BLANQUET approach that species are used, on the basis of their distributions, as indicators of biotope characteristics and other factors, and that when possible groups of species are thus used. Groups of species may often give much more effective indication of environmental factors than dominant species, or other single species alone, or abundance relations that may be much affected by disturbance. Discussions of indicator use of species have been given by, among others, DUVIGNEAUD (1946), ELLENBERG (1950), WHITTAKER (1954a) and TÜXEN (1970b). We may consider indicator species in four contexts – in connexion with synsystematic, biotope or ecological, successional, and geographic indication. In each of these, but particularly in the biotope-ecological function, it is possible to use indicator species in a perspective of either classification or gradient analysis.

20.7.7.1 *Indicators for Classification*

One direction of development – the synsystematic use of groups of species as indicators for syntaxa – is implicit in the BRAUN-BLANQUET approach. The concept was developed further, however, with the distinction of differentiating taxon groups (SCHWICKERATH 1942, 1954a, DUVIGNEAUD 1946), and the idea of sociological groups (SCAMONI & PASSARGE 1959, 1963, DOING 1962, 1969b). The latter are established as groups of species that are distributionally related, and that consequently occur in and characterize, particular syntaxa; the idea is related to that of 'commodal' species (WHITTAKER 1956). SCAMONI et al. (1965) constructed a sort of 'sociological profile' by arranging relevés for phytocoena and determining the distribution of sociological groups over the units compared. Treatment in terms of syntaxonomical groups has been developed by SEGAL & WESTHOFF (1959) and MAAREL (1969 et seq.).

Clearly, groups of species should indicate characteristics of biotopes or habitats, as well as the particular phytocoena that occur in those biotopes. The concept of ecological groups — sets of species that, because of their similar distributional response to environmental factors, tend to occur together and to indicate properties of biotopes — may be traced from DUVIGNEAUD (1946, 1949). Ecological groups have had most extensive application by ELLENBERG (1950, 1952, 1956), who relates them both to environmental factors and to the syntaxa of the BRAUN-BLANQUET approach. By describing species responses to various important factors — such as pH, lime, moisture, nitrogen — one may arrive at integrated ecological characterizations of species (ELLENBERG 1956, compare article 5.4.1.) Ecological groups have been used also by investigators of the Centre d'Études Phytosociologiques et Écologiques at Montpellier (GOUNOT 1969, GUILLERM 1971, DAGET et al. 1972).

A third application is recognition of geographic indicator groups that express geographic-historic influences and may characterize vicariant phytocoena. SCHWICKERATH (1942) and OBERDORFER (1957) have used geographic groups in dealing with vicariant associations. Given classification of species into geographically defined groups, or areal types, it is possible to characterize communities by geographic spectra, based on representation of these groups. The geographic perspective has been developed by MEUSEL et al. and by MØLHOLM HANSEN and BÖCHER in the Danish school (see article 18.5.3-4), and in the United States WHITTAKER (e.g. 1954b, 1960) has compared areal-type spectra of communities.

20.7.7.2 *Indicators for Gradient Relations*

Community spectra may intergrade, perhaps continuously, along major axes of biotope characteristics, geography and climate, and succession. Spectra (per cent representation of different species groups) may consequently be used to indicate position along these axes. There seems to have been little application of indicator groups to the study of succession, though the recognition of 'decreaser, increaser, and invader' groups in response to grazing disturbance (article 4.2.1) should be noted. Study of gradients of geographic spectra also has been limited, though ELLENBERG (1950) deals with climatic indicator groups, and article 3 (Fig. 8) illustrates a gradient in geographic spectra. A third possibility, use of representation of ecological groups to indicate position along environmental gradients, has had such extensive development as to be beyond review here. Use of ecological groups in direct ordination is discussed in articles 2—5 and 8; use in indirect ordination in article 9. The most extensive application in phytosociology is that of ELLENBERG (1950, 1952, 1956). ELLENBERG classified European species (weeds in agricultural fields, and plants of meadows and pastures) into ecological groups on the basis of their responses to various environmental factors. Weighted averages (see 2.2.3.1) for samples expressed their positions in relation to these factors. At the same time the relation of the species to syntaxa was known, and it was consequently possible to interpret the relations of the syntaxa to one another and environment. The technique thus created a multidimensional direct ordination, and belongs also to section 20.9.1.

20.8 Numerical Techniques

There remain to be discussed two (closely related) recent developments in phytosociology – numerical techniques and ordination. For review of similarity measurements and approaches to numerical classification we refer the reader to GOODALL's articles (6 and 19) in this volume. We discuss briefly, however, work in this direction employing the BRAUN-BLANQUET approach.

20.8.1 STORAGE OF RELEVÉS

Although the numbers of relevés treated in syntaxonomical studies may be very high – often several thousands – the number of storage systems is low. Most individual phytosociologists and their institutes work with an archive of relevé note books and tables. Such is the case at syntaxonomic centres like the Station Internationale de Géobotanique Méditerranéene et Alpine (S.I.G.M.A.) at Montpellier and the former Bundesanstalt für Vegetationskartierung at Stolzenau (data from the latter are now at TÜXEN's Arbeitsstelle für theoretische und angewandte Pflanzensoziologie at Todenmann über Rinteln, West Germany). Some archives have been set up with visual punched cards. The only such archive of some size seems to be that of the Geobotanical Institute Rübel at Zürich including about 5500 forest relevés from Switzerland (ELLENBERG 1968).

Besides incidental use of machine punched cards in the application of computer techniques, two archive systems based on computer hardware are known to us: the system of the C.E.P.E., Centre d'Études Phytosociologiques et Écologiques at Montpellier and the system of the Working Group for Data-Processing of the International Society for Plant Geography and Ecology. The C.E.P.E. system (see EMBERGER 1968) has the following characteristics: (i) The basic unit is the relevé, which is defined as the whole of ecological and phytosociological observations at a particular site. (ii) General vegetation structure, situation of the site, general climate and soil are described by means of measurements or estimations of numerous factors. (iii) For each factor a code is devised with, in most cases, 10 classes. (iv) All coded determinations of general features are punched in three 'parent' relevé cards, 'Cartes-Maîtresse-Relevé' (CMR) of which columns 1—71 (except for CMR 3) are used. (v) Each taxon from the relevé is treated separately for each vegetation layer in which it has been observed. Coded determinations include vegetation layer, abundance, and dominance (in separate 0—9 codes), pattern within the site, phenological state, vitality and life-form. The corresponding punched cards are called 'detailed cards', Cartes Détail (CD) and used only from column 72 onwards. (vi) Species are enumerated from 0001—4779 according to the Quatre flores de France by FOURNIER (1961). Cryptogams are provisionally enumerated from 5000 onwards. (vii) Data per species per layer are combined with CMR data, which results in three combined cards.

In this way a relatively large archive arises, which has the advantage of being rapidly accessible to all kinds of spectrum and correlation calculations.

The 'Working Group' was established in 1969 during the International Symposium at Rinteln after a proposal by S. PIGNATTI, G. CRISTOFOLINI and D. LAUSI (Trieste). One major aim (see MAAREL 1971) was the treatment of relevés in such a way that the data could be readily stored and retrieved. An outline of the system has been presented by CRISTOFOLINI et al. (1969); its main characteristics are: (i) The basic unit is a full relevé table, either published or in manuscript form. (ii) Rows of relevé tables are punched – from column 21 onwards. Cover-abundance values per species are coded either as direct BRAUN-BLANQUET symbols or in a 0—9 transformation code (Table II). (iii) In the first 20 columns coded data on publication, table number in publication, and taxon number are punched. (iv) Taxa are coded with 7 digits, 4 for the genus, 3 for the species – as yet no space is left for infraspecific taxa. The genus enumeration is according to the world survey *Genera Siphonogamarum* by DALLA TORRE & HARMS (1900—1907, reprint 1963). Species enumeration is as yet restricted to European species (Russia not included), according to *Flora Europaea* as far as possible, otherwise according to a provisional enumeration based on six large standard floras of Europe. Additional provisional codes for cryptogams are being developed. (v) A separate punched card is used for bibliographical data on the paper (or set of unpublished tables). (vi) A separate set of punched cards is used for general data on the site and its situation, for each relevé in each table. This part of the system is still in development. The Montpellier system, which is already elaborated in this respect, could be taken as a starting-point.

The Working Group decided to concentrate activities on salt marsh communities. Some 3400 relevés (including 576 Spartinetum relevés) have been stored in punched card decks and on magnetic tape.

20.8.2 SPECIES CORRELATION

The study of species correlation was introduced by VRIES (e.g. VRIES et al. 1954, DAMMAN & VRIES 1954). The plexus of grassland species he presented (article 7, Fig. 3, p. 169) was taken over in the textbooks of ELLENBERG (1956) and BRAUN-BLANQUET (1964), who underlined VRIES' conclusion 'that the resulting species groups broadly coincide with the associations of the Zürich-Montpellier school.'

Measurement of species distributional similarities may be based on either binary or quantitative data. Distributional similarity based on binary (presence and absence) data may be termed 'species association,' that based on quantitative representation in relevés, may be termed 'species correlation' (see article 6). In the BRAUN-BLANQUET approach the quantitative weightings may be either the abundance-cover scale (20.4.5.1) in individual relevés (or a transformation of that scale), or presence degrees (20.5.3) in different phytocoena. Species association seems to be more appropriate for the detection of species groups in relevés representing an extensive range of community difference, whilst correlation studies are more appropriate for intensive study within the range of broadly overlapping species (cf. GREIG-SMITH 1964). The use of species correlation in the establishment of sociological groups is obvious, but phytosociologists using such groups have hardly applied this technique. The studies of HEGG (1965; see article 7, Fig. 7, p. 174) and STOCKINGER & HOLZNER (1972) may, however, be mentioned as examples (see also article 7).

The detection of ecological groups has been particularly developed by phytosociologists using factor analysis and principal component analysis. DAGNELIE (1960, see also article 9) described the use of factor analysis for the establishment of sociological groups, and the use of joint species-environmental factor analysis for the establishment of ecological groups. GODRON (1966) used species correlations for the checking of 'imbricating' ecological groups; GUILLERM (1971) and DAGET et al. (1972) constructed ecological profiles of species against environmental factors. LACOSTE & ROUX (1971) and ROMANE (1972) used 'factor analysis of correspondences' in which floristic and environmental variables were analysed in combination to arrive at ecological-sociological groups.

Another such method was developed by FRESCO (1971) as part of his 'compound analysis.' He constructed overlapping species groups based on similarities between species with respect to their loadings on eigenvalues extracted in subsequent factor analyses. This method could be applied to very large data sets, e.g. the first selection of salt marsh data (1296 relevés, 245 species). From this salt marsh treatment (with an arbitrary similarity level) 65 such groups were derived which could be interpreted roughly as alliance character-species groups.

20.8.3 FIDELITY TESTS

Although it seems obvious to use tests on the significance of the exclusiveness or differentiating value of species (or species groups) for a given vegetation type, phytosociologists have hardly done so. MEIJER DREES (1949) suggested the use of the *t*-test for the significance of the difference between the group amounts of species groups in different relevés of a set. In his example the relevés were assigned to a particular association and the two species-groups examined were syntaxonomic groups for two orders; the problem was the assignment of the association to one or the other order. For such situations Goodall (1953b) used Fisher's discriminant function, which is more powerful but also more laborious.

Goodall (1953b) also proposed an index of fidelity for a species based on the ratio of constancy values for that species in two communities, which could be tested with the χ^2 test or by exact calculation of P. In addition he suggested the use of this index as an 'indicator value' for the comparison of the constancy of a species in one community with that in various other communities — Goodall spoke of 'all other communities in the area.' In usual phytosociological terms Goodall's indicator value is the real fidelity index and the application of this measure to the two-community comparison would be an 'index of differentiation.' Such tests may have limited value, particularly because of the non-random selection of relevés from which they are derived (Goodall 1953b). Still, their use may be experimented with for syntaxon diagnoses from tabular summaries.

20.8.4 Relevé Similarity

Measurements of similarity between relevés have long been accepted as one basis of grouping these into phytocoena (Braun-Blanquet, 1928, 1932). Various coefficients of similarity and dissimilarity are in use in phytosociology, including the formulas of Jaccard and Sørensen (article 6, formulas 39 and 40). Recently an anonymous 'similarity ratio,' which was introduced by Wishart (1969) in Clustan, a set of classification programmes, was used by Kortekaas & Maarel (1972). This formula reads

$$\sum_i x_i y_i / \sum_i x_i^2 + \sum_i y_i^2 - \sum_i x_i y_i ,$$

and is thus a generalisation of Jaccard's formula. Comparatively little use has been made of direct dissimilarity measures (Groenewoud 1965, Maarel 1966b). Similar formulas, most of them derived from Sørensen's formula, are in use for the comparison of sets of relevés or the calculation of average similarity within a set (cf. Sørensen 1948, Raabe 1952, Češka 1966, 1968, Frydman & Whittaker, 1968, Maarel, 1969). Weighting of species may use presence values or a combination of presence with an importance value. Ellenberg (1956) as well as Pignatti & Mengarda, see Pignatti 1964, suggested the use of a similarity coefficient in the assignment of a relevé to a type by comparing it with a 'standard' or average relevé, e.g. one consisting only of the normal characteristic species combination. Hofmann & Passarge (1964) and Scamoni et al. (1965) determined similarity between relevés and sets of relevés with reference to the sociological groups ('group affinity').

20.8.5 Measurement of Homotoneity

Homotoneity we have defined as the homogeneity of a community table, the relative consistency of the relevés the table comprises (20.5.3). In addition to the simple measures mentioned in 20.5.3, the mean similarity coefficient for the relevés, as calculated in a direct way by Češka (1966), is a convenient measure, particularly in classifications based on relevé similarity. Either the average of all interrelevé similarity values within a table, or the average relevé similarity with the average for the table (i.e. the centroid of the relevé cluster) may be used. See also article 6 and 20.8.4 and 20.8.8 on similarity measures.

Alternative techniques based on species relationships include the use of positive interspecific correlation by Goodall (1953a), association analysis (Williams & Lambert 1959), and information analysis (Williams et al. 1966) (see article 19.5.1.5 and 19.5.2.5). Maarel (1966b) derived an information measure that is related to that of Williams et al. and appeared to be a useful heterogeneity coefficient in a study of a local grassland community complex. Godron (1966) developed an information measure, based on Brillouin's information formula, for determination of the heterogeneity within a series of contiguous quadrats in the field. The frequencies of the species occurring in the set determine the value of this measure. It could also be used for heterotoneity in sets of relevés and in that case is rather similar to the formula of Williams et al. (1966).

Dahl (1957, 1960) derived an index of uniformity from the Fisher model (logarithmic series) for species-individual and species-area relations ($S = \alpha \log_e (1+N/\alpha)$, S is number of species, N number of individuals, and α a diversity index, see Williams 1964). Dahl considered the species-relevé relation to be similar and defined the diversity parameter for a table as $(S_m—\bar{S})/\alpha \log M$ (see symbols and formulas in 20.5.3). The index of uniformity is accordingly defined as $\bar{S}/\alpha$. A clear relationship with Curtis' (1959) index was shown by Dahl (1957).

20.8.6 Numerical Classification

Of the various numerical classification techniques which are described in article 19, only a few are in use in quantitative phytosociology. Direct use of similarity matrices was developed by Polish phytosociologists (MATUSZKIEWICZ 1948, FALIŃSKI 1960) and by GUINOCHET (1955, GUINOCHET & CASAL 1957), who spoke of the 'differential analysis of CZEKANOWSKI' (after the Polish originator of the approach). See further article 7.3.1.

Agglomerative clustering techniques (19.5.2) based on similarity coefficients like the JACCARD and SØRENSEN indices were used by SØRENSEN (1948), LOOMAN & CAMPBELL (1960), MAAREL (1966b), MOORE & O'SULLIVAN (1970), MOORE et al. (1970) and KORTEKAAS & MAAREL (1972). Agglomerative techniques can be profitably combined with re-allocation techniques (cf. WISHART, 1969). The relevé-groups arrived at in these classifications could generally be typified with characteristic species combinations and connected with syntaxonomical units.

IVIMEY-COOK & PROCTOR (1966) applied nodal analysis (association analysis followed by an inverse species clustering) to salt marsh, fen and woodland data. They obtained tables with clear blocks of relevés and species and concluded that the results confirmed the arrangement of the data arrived at by traditional phytosociological methods. The technique was considered particularly useful for the detection of differentiating species. Moreover they found minor lines of division which had not been obvious in previous studies.

20.8.7 TABLE REARRANGEMENT

A successful nodal analysis should produce an ordered table. A number of special techniques for table rearrangement have been developed, from the early work of BENNINGHOFF & SOUTHWORTH (1964) to the more recent of MOORE (1972, MOORE et al. 1970, MOORE & O'SULLIVAN 1970). The following computer programmes are known to us:

i) BENNINGHOFF & SOUTHWORTH (1964), G. W. MOORE et al. (1967), LIETH & MOORE (1971). This programme finds species clusters by application of LIN's algorithm for the 'travelling salesman problem.' It is applied to species with intermediate presence in the table. Additionally, relevés are sorted. Final table ordering is by hand.

ii) J. J. MOORE (1971, last version). This programme 'PHYTO' rearranges both species and relevés. It finds pairs of species with intermediate presence values and high co-occurrence values. Opposing pairs are used as differentiators. Further arrangements are dictated by the user, either through suspected mutually exclusive species or through preferred order of species and relevés.

iii) SPATZ (1969, 1972 last version). This programme forms groups of relevés on a similarity basis. Differentiating species are defined as species with presence of > 50 % in one group and < 10 % in all others. Output includes a synoptic table with synthetic figures.

iv) SCHMID & KÜHN (1970). This programme calculates D^2 values and forms groups on lowest D^2 basis. Differentiated tables can be dictated.

v) ČEŠKA & ROEMER (1971). This programme finds species-relevé groups through iteration. Diagnostic species are defined as in programme 3 (values vary from 50 vs. 10 to 66 vs. 33). A relevé is considered a member of a group if it contains at least 50 % (or 66 %) of the diagnostic species of that group. Groups are arranged in order of size and within-group similarity. Final arrangement is usually by hand.

vi) JANSSEN & MAAREL (1972, JANSSEN 1972). This programme is based on programmes HIERAR and RELOC of the CLUSTAN set (WISHART 1969) with a choice of 10 similarity coefficients. It starts with an allocation of relevés to groups, either at random or as determined by previous classifications and phytosociological experience. Group arrangement is based on group properties, particularly the relation between relevé number and species number, or alternatively on position along the first component of a principal components analysis of groups. Species are arranged so as to form a diagonal structure in the table. Significance of positive differentiating value of species in one or more groups is determined by a χ^2 test. Final arrangement is usually through MOORE's PHYTO-programme.

vii) STOCKINGER & HOLZNER (1972). This programme finds species groupings through similarity analysis. Relevés are sorted according to occurrence of species groups.

It follows from this description that all programmes require the personal finishing touch of the user; this is felt to be not an inconsistency but rather a matter of efficiency. Most designers claim a considerable gain in accuracy, since re-writing tables by hand is no longer necessary. The speed of the programmes will depend on type of computer and size of the table. Programmes i, ii, v, and vii are most adapted to tables with comparatively many relevés and few species, programmes iii and iv to tables of the reverse form. Programme vi has a larger capacity, but requires preliminary experience to be fully efficient.

20.8.8 Numerical Syntaxonomy

Numerical syntaxonomy has been mainly restricted to the characterizing of various syntaxon levels by average similarity. Various coefficients of similarity and dissimilarity are in use in phytosociology. 'Coefficients of community' comparing samples by the per cents of their species shared have been used most, with about equal interest in the formulas of Jaccard and Sørensen (article 6, formulas 39 and 40). Formulas using importance values for species (cover-abundance in relevés, especially) include the 'percentage similarities,' notably the Czekanowski index (article 6, formula 50). Similar formulas are in use for comparing sets of relevés (representing different phytocoena), or calculating average similarity within sets (see Sørensen 1948, Raabe 1952, Češka 1966, 1968, Frydman & Whittaker 1968, Maarel 1969, 1972a). Comparison of sets of relevés may use presence per cents, or mean cover-abundance values, for the species in these sets. Numerical classification is further discussed in article 19.

Sørensen (1948) concluded that grassland groupings with similarity levels of 0.40 roughly corresponded with syntaxa on the alliance level. Ellenberg (1956) stated that the average similarity of relevés within one association is between 0.25 and 0.50, whereas subunits may be distinguished by levels above 0.50 (Jaccard values, which are 10% lower than Sørensen values on the average). Looman & Campbell (1960) calculated Sørensen values of > 0.70 within subunits belonging to one grassland association, whereas values between subunits were all < 0.50. Hofmann & Passarge (1964) presented group affinity values between various woodland associations and subassociations. Within associations values were mostly > 0.60, between typical subassociations of related associations values were mostly between 0.30 and 0.50. Raabe (1952) obtained affinity values (Kulczyński coefficient) between associations and alliances of weed, salt marsh and alpine communities. Within-alliance values were between 0.40 and 0.50, between-alliance values were 0.20 to 0.40. Within salt marsh associations values varied more widely, from 0.30 to 0.80, and here the influence of geographical distance between local representatives of associations was evident.

Neuhäusl & Neuhäuslová-Novotná (1972) studied within- and between- group similarities (Sørensen values) of many woodland associations of the alliance Carpinion. Within-group similarities of locally established associations and lower units were generally over 0.55. When larger areas were involved, these values were lower but still over 0.45. Similarities between associations and between lower units were always lower, mostly 0.10 to 0.20. These calculations made possible some improvement in the delimitation of Carpinion associations (which, as the authors stated, must be checked by renewed phytosociological table studies). KORTEKAAS & MAAREL (1972) carried out a numerical analysis of European *Spartina* communities involving an agglomerative classification with relocation of 576 relevés. They arrived at four separate dendrograms each comprising relevés dominated by one of the *Spartina* species. On various levels of the hierarchy groups were inspected for character-species and compared with the existing syntaxonomy (BEEFTINK & GÉHU, in prep). For these communities values of the similarity indices between 0.40 and 0.60 could be connected with the association level, 0.61—0.70 with the subassociation level and 0.71—0.80 with the variant level. Figure 4 presents the dendrogram for the *Spartina maritima* relevés. Each 'subassociation' had exactly one 'good' differential-taxon. The examples suggest that in related communities a reasonable parallel between syntaxonomical level and similarity level can be established. Furthermore, the

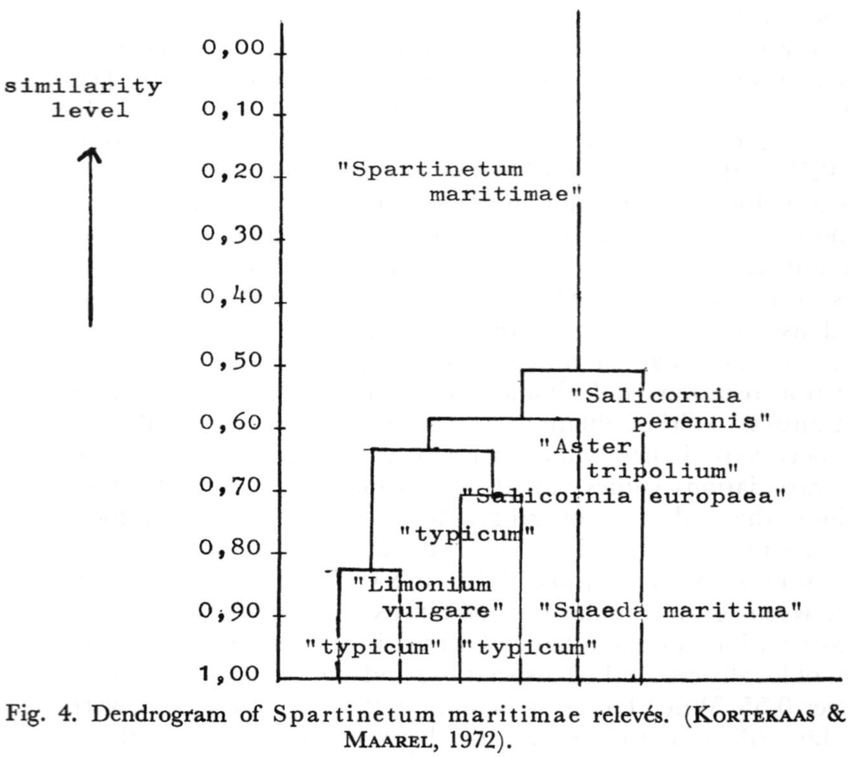

Fig. 4. Dendrogram of Spartinetum maritimae relevés. (KORTEKAAS & MAAREL, 1972).

similarity values for a given syntaxonomic level in different vegetation types are similar, suggesting that various authors have made comparable intuitive judgments of relative similarity for various kinds of communities. Similarity measurements should thus be thought a useful aid to synsystematics.

20.9 **Ordination**

'Ordination' refers to the arrangement of entities (generally samples, or species) in a uni- or a multidimensional order (RAMENSKY 1930, GOODALL 1954, articles 1 and 2.) Ordination is often considered to stem from the concept of vegetation as a continuum, whereas the floristic-sociological classification may seem to assume discontinuity. It may consequently be argued whether ordination is appropriate in phytosociology. Even if we assume for the moment that associations are generally discontinuous with one another, there is no reason ordination cannot be used as an aid to understanding the relations of relevés and lower syntaxa to one another and environment, within the association. On a different level, ordination and the perspective of gradient analysis can well be applied to the relations of associations and higher syntaxa as wholes, to one another and environment. We feel consequently that ordination is a fully acceptable supplementary approach in phytosociology.

There has been some dispute, mainly between BRAUN-BLANQUET (1939, 1951b, 1955) and GAMS (1918, 1941, 1954) on the question 'linear or multidimensional system in plant sociology.' GAMS emphasized the multidimensional relationships of plant communities, without indicating the way of constructing an appropriate system, as BRAUN-BLANQUET remarked. WAGNER (1954, 1968) observed that many syntaxa show multidimensional relationships to various syntaxa of the next higher level. The 'flexibility' ELLENBERG (1954b) demanded at lower syntaxonomic levels could be thought of in this perspective. Surely it is true both that linear arrangements into hierarchies are possible and valuable, and that study of multidirectional relationships may be rewarding.

20.9.1 INFORMAL ORDINATION

A number of approaches that represent ordination but do not employ similarity measurements have been applied in phytosociology. In many ordered phytocoenon tables a suspected underlying environmental gradient is chosen as a basis for ordering that may be completed by inspection of ecological groups (MOORE et al. 1970). Often a moisture gradient is chosen and a 'dry' a 'typical' and a 'moist' subassociation are arranged within an association table. The unidimensional approach known as compositional ordination (article 2) has a phytosociological precursor in the approach of SCHWICKERATH (1931 et seq.), in which relevés of one association were arranged on the basis of representation of diagnostic species groups. The development of weighted-average ordinations by ELLENBERG (1950, 1952) has been referred to (20.7.7.2).

More abstract two- or many-dimensional schemes including Russian examples (discussed in article 5) were recommended by GAMS (1941, 1961). The approach of DUVIGNEAUD (1946) was an early floristic-sociological ordination without numerical basis. A number of phytosociologists have presented patterns ('mosaic charts', 2.3.3) of phytocoena in relation to environmental gradients (see DUVIGNEAUD 1946, WAGNER 1954, 1968, ELLENBERG 1952, Fig. 5, 1963, ZONNEVELD 1960, HEGG 1965, and article 18, Fig. 1, p. 555).

20.9.2 FORMAL ORDINATION

The first ordination of relevés on the basis of their similarities is found in the Polish approach of MATUSZKIEWICZ & TRACZYK (1958), and FALIŃSKI (1960). They constructed 'dendrites' as arrangements that can be ecologically interpreted (cf. GAMS 1961, 1967, and article 7).

Ordination in the sense of axis construction was first applied in phytosociology by DAGNELIE (1960) on beechwoods (see article 9). A comparable factor analysis on heathlands was presented by FRESCO (1969). GROENEWOUD (1965), MOORE et al. (1970) and MAAREL (1972a) applied principal component analysis to similarity matrices. The Wisconsin polar ordination of BRAY & CURTIS (1957), or slightly modified forms, was applied in Dutch work (unpublished MSc theses, Universities of Utrecht and Nijmegen, MAAREL 1966b, 1969, MAAREL & LEERTOUWER 1967, LONDO 1971) and further by ROGERS (1970). LACOSTE & ROUX (1971) applied factor analysis of correspondences (see 20.8.2) to relevés assigned to various subalpine associations and subassociations.

Ordination can be applied also to phytocoena (or syntaxa), with each phytocoenon treated as a composite sample with its species composition summarized as presence per cents or mean importance values. One may then conceive of environmental gradients

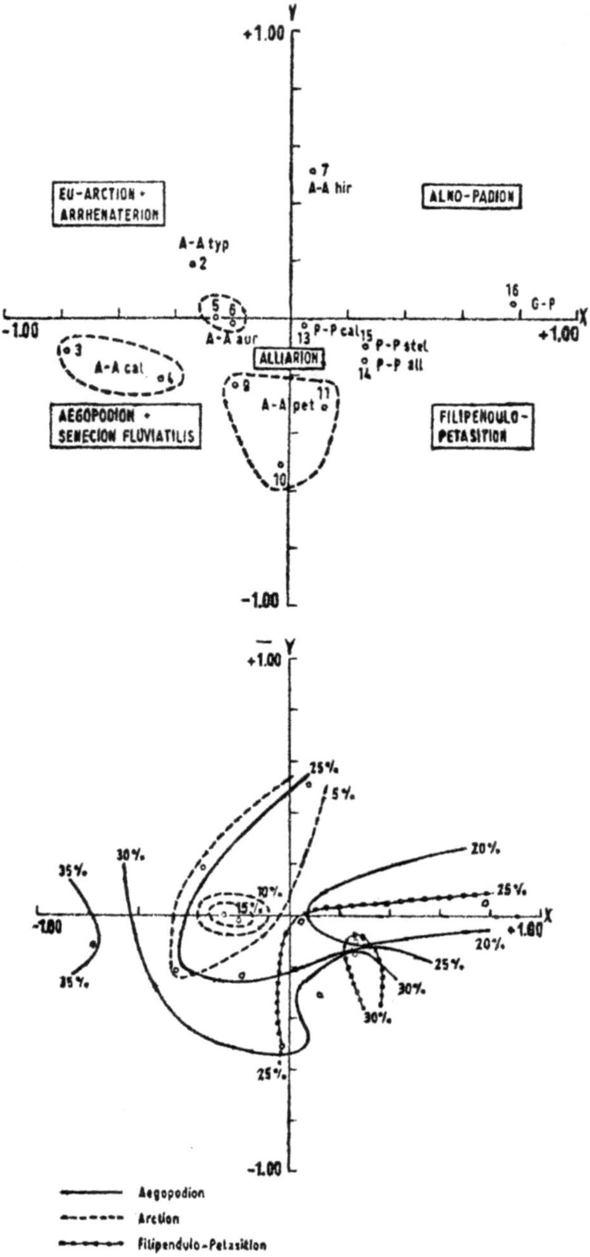

Fig. 5. Ordination of nitrophilous edge communities described by TÜXEN (1967a), from MAAREL (1969).

relating these phytocoena as a multidimensional habitat space (GOODALL 1963, WHITTAKER 1967) in which the centroids of the phytocoena (as clusters of relevés) are ordinated. ORLOCI (1966) carried out such an ordination by principal component analysis; GITTINS (1965), CRAWFORD & WISHART (1966, 1967), FRYDMAN & WHITTAKER (1968), MAAREL (1969), and WHITTAKER (1972) applied polar ordination to community-types. Mostly the similarities between types were measured by presence or constancy values. All the resulting ordinations were effective and easily interpretable. An effective approach to interpretation is the plotting of contour lines for biotope measurements (e.g. FRYDMAN & WHITTAKER 1968), to determine the relationship between axes and environmental factors, and the plotting of community-types and representation of diagnostic groups (FRYDMAN & WHITTAKER 1968, MAAREL 1969, WHITTAKER 1972) to show the relation of these to environment and one another. Fig. 5 illustrates the latter technique (see further articles 2, 8, and 9).

When no appropriate environmental data are available it is possible to interpret the ordination by plotting the distribution of syntaxonomical groups. Figure 5 presents an example. Nitrophilous edge communities, bordering woodlands and scrubs, as described by TÜXEN (1967a) are ordinated and the ordination space is characterized by various alliances. The associations are reasonably separated in the ordination space, except for the Agropyro repentis-Aegopodietum. The subassociation with *Chaerophyllum hirsutum*, in particular, is distinct and has a considerable representation of Alno-Padion species. Patterns of three alliances are presented. Lines with equal percentage amounts of one alliance are called 'isocenes' and outlined. The greatest difference in the set of types is between the relatively dry, open nitrophilous Agropyro repentis-Aegopodietum calystegietosum sepium from anthropogenic edges at low altitudes, and the relatively moist and shaded Geranio phaei-Petasitetum along streamlets in the montane region. WHITTAKER (1972) and FRYDMAN & WHITTAKER (1968) presented a similar approach with similar results. Besides species populations, species groups, and environmental measures, various spectra can be plotted in an ordination field. RAUNKIAER life-forms, for example, showed clear distribution patterns when thus plotted by FRYDMAN & WHITTAKER (1968).

Other uses of ordination include applications to succession (MAAREL 1969, LONDO 1971) and ordinations based not on species composition but on structural characters, life-forms or sociological (syntaxonomic) groups (KNIGHT 1965, GOFF & COTTAM 1967, MAAREL 1969, WHITTAKER 1972). MAAREL (1972a) used

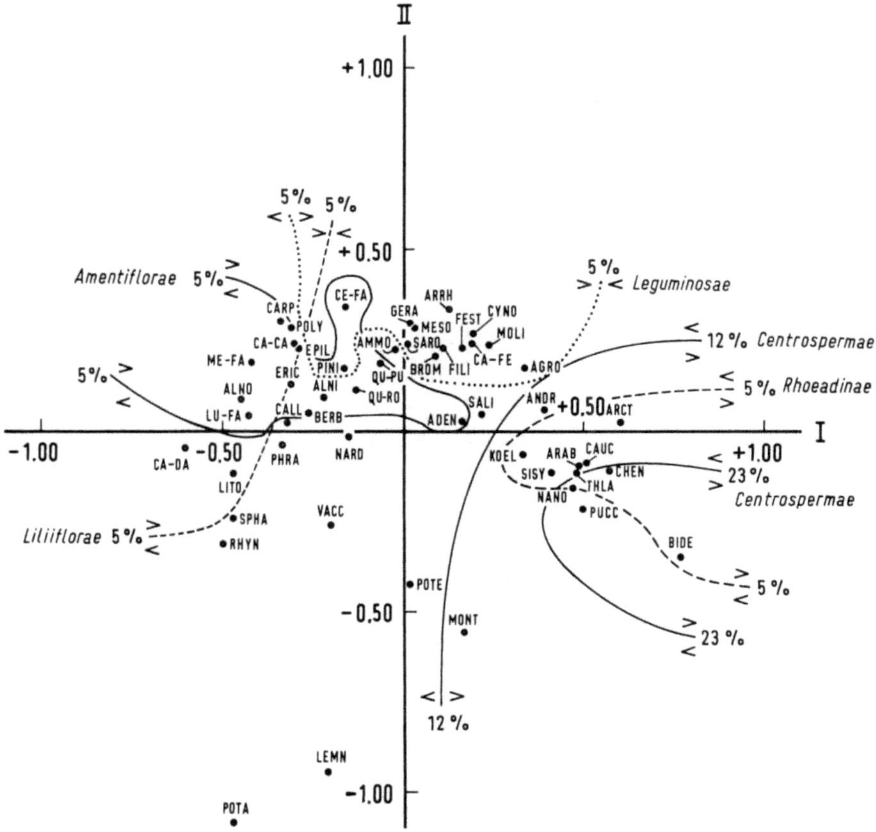

Fig. 6a. Ordination of 51 European associations and lower syntaxa on the basis of their plant order spectra, dimensions 1 and 2. (Maarel, 1972a).

higher idiotaxonomical units as attributes to compare and ordinate vegetation types. Figure 6 presents some results of an ordination of 51 associations and subassociations belonging to 49 alliances, and thus covering almost the entire variation in central and western Europe. The corresponding similarity matrix was based on representations of plant orders in synoptic tables, mainly from Oberdorfer (1957). The first line of variation shows the sociological progression from pioneer comminities, with alliances such as the Bidention (BIDE) and Nanocyperion (NANO) on the right, to mature communities like the Fagion (—FA) and Carpinion (CARP) on the left. Most grasslands, marshes, and heathlands are intermediate. This variation is illustrated by isolines of various idiotaxonomic orders, e.g. Centrospermae vs. Liliiflorae. The second axis involves a moisture gradient from wet (Eu-Potamion,

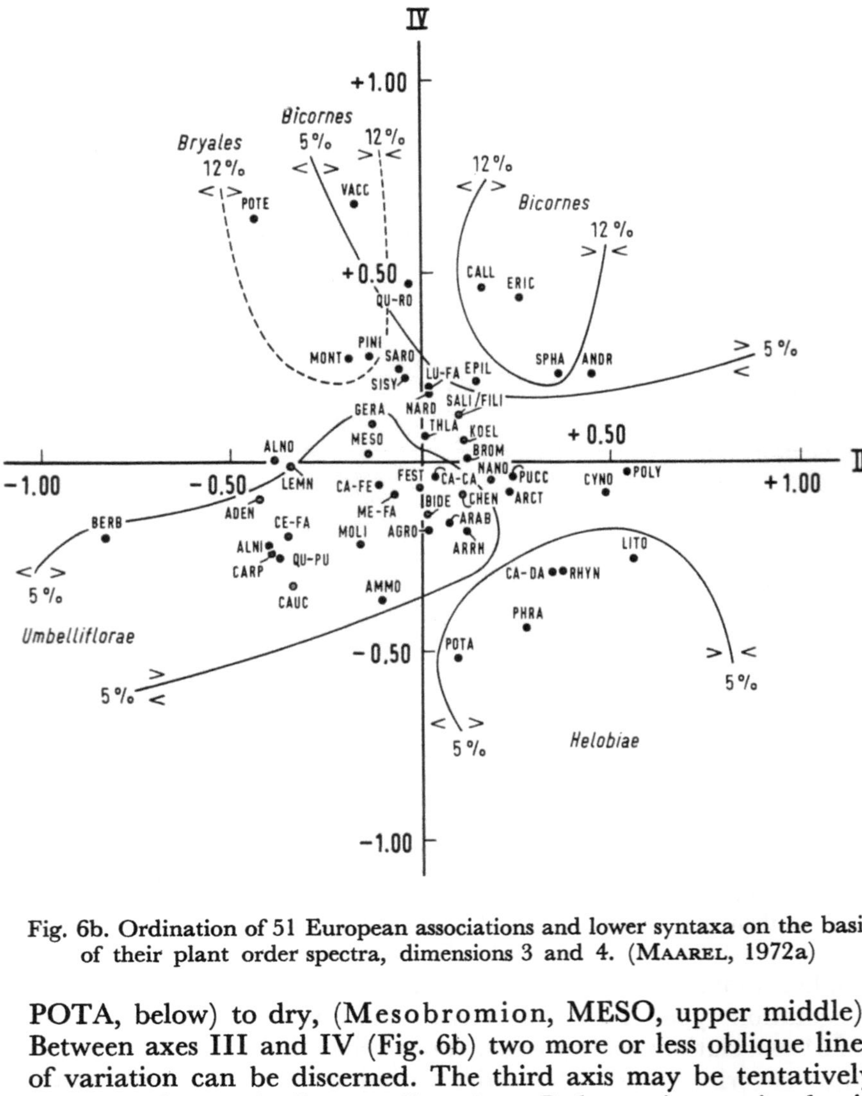

Fig. 6b. Ordination of 51 European associations and lower syntaxa on the basis of their plant order spectra, dimensions 3 and 4. (MAAREL, 1972a)

POTA, below) to dry, (Mesobromion, MESO, upper middle). Between axes III and IV (Fig. 6b) two more or less oblique lines of variation can be discerned. The third axis may be tentatively interpreted as a shading gradient from Polygonion avicularis (POLY) and Phragmition (PHRA) below, to Cardamino-Montion (MONT) and Vaccio-Piceion (VACC) above, from prevalence of Helobiae to that of Bryales. The fourth axis runs from raw humus soils with a relatively high C/N ratio to soils with a low C/N ratio and high microbiological activity (Ericion tetralicis, ERIC, with Bicornes, upper right, to Caucalion, CAUC, with Umbelliflorae, lower left). We regard Fig. 6 as representing only a first venture in broad-scale ordination of phytocoena; its significance may be in its demonstration of possibility.

20.10 Conclusion

In perspective, these numerical techniques and ordinations should be seen as ancillaries to the Braun-Blanquet approach. We re-emphasize the essential ideas with which we began: the floristic approach to understanding, the functional value of diagnostic species for classification and environmental indication, and the utility of the hierarchical classification for ordering knowledge and expressing understanding. Given this as the approach, then the numerical techniques serve to reduce labour, and ordinations to enhance interpretation. Ordination without prior classification has accomplished significant research in American ecology, but ordinations in different areas are not easily coordinated with one another. It is the European experience that the Braun-Blanquet approach provides such coordination of the work of different investigators through its classification, whilst the classification provides a context that increases effectiveness of ordination. On the one hand prior classification makes possible the ordination of phytocoena, rather than an often unmanageable number of relevés; on the other hand the known relationships of species groups and phytocoena to one another and environment give basis for interpreting an ordination. The ordination should, in turn, express and further clarify those relationships.

We return to another theme – the complexity of the Braun-Blanquet approach as a technical system. The complexity has causes; these include the numerousness of species and phytocoena, the multifarious distributional relations of species to one another and phytocoena, and the many directions of environmental, geographic, and developmental relationships amongst species and phytocoena. The complexity of the Braun-Blanquet system is thus a response of science to the complexity of vegetation. It is no real statement of preference that the approaches through physiognomy and species dominance are 'simpler'; it is not necessarily true that they are, and to the extent that they are, they are only because they do not treat and coordinate so wide a range of information. We thus express our judgement (or our bias) that in the study of vegetation one approach has most fully faced the demands for a method that is both detailed and generalizing, both locally intensive and integrative of the results from different local areas – and that that one approach is Braun-Blanquet's.

This judgement we must balance with another. The Braun-Blanquet approach is very demanding of research effort by many investigators, if the vegetation of extensive areas is to be known through it. The effort has been possible in Europe and some other areas but is not yet possible everywhere — there are not phytosociologists enough. The approach may be particularly difficult to initiate in an area lacking prior studies to offer suggestions toward classification and diagnostic species groupings, unless the initiator is experienced in the approach. This difficulty is one reason for the importance of approaches through physiognomy and dominance in many areas; by these approaches knowledge of vegetation may be gained that might not otherwise be obtained at all. It is true also that the physiognomic and other approaches may reveal vegetational relationships that are important, but are outside the main concerns of the BRAUN-BLANQUET approach. The statement that BRAUN-BLANQUET's is the most fully developed approach to vegetation is thus no rejection of the contributions of others amongst the many ways for the ordination and classification of communities.

20.11.1 SUMMARY

(20.1) Amongst the central ideas of the approach are that: (i) classification and interpretation of communities should be based on their full floristic composition, (ii) with emphasis on diagnostic species, whose relative restriction to samples characterizes communities and indicates their environments and (iii) which may be used to organize the communities into a formal, hierarchical classification. (20.2) This floristic-sociological approach to vegetation had its origin in southern European phytosociology centered in the cities of Zürich and Montpellier, but especially in the ideas of BRAUN-BLANQUET.

(20.3) It is important to distinguish particular, concrete plant communities or phytocoenoses, and abstract classes of plant communities; the term phytocoenon is recommended for the latter. The phytocoena of the formal BRAUN-BLANQUET hierarchy are termed syntaxa, in analogy with the taxa into which individual organisms are classified. Syntaxa are characterized by diagnostic species of three types: character-species are centered in or relatively restricted to a particular syntaxon compared with all others, and therefore characterize it and indicate its environment; differential-species distinguish two closely related syntaxa by presence in most samples of the one and absence in most samples of the other; constant companions are not restricted to a given syntaxon but help to characterize it and indicate its relationships to higher units. In addition to the hierarchy, syntaxa may be arranged along a sociological progression from simple and poorly organized to complex and highly differentiated communities. The BRAUN-BLANQUET classification is considered natural in the sense that its syntaxa (though their limits may be arbitrary) are consistent with a large number of relationships between environment, species populations, and biocoenoses.

(20.4) The research procedures include three phases — analytical, synthetical, and syntaxonomical. In the analytical phase samples or relevés representing kinds of phytocoena to be studied are taken in the field. The relevés should be chosen to represent the phytocoena to be studied, and each should sample a uniform area of sufficient size to represent that phytocoenose adequately. A relevé includes information on environment and location and a list of all plant species in the sample area, with species recorded by strata and rating scales (for combined cover and abundance, and if possible sociability, vitality, and periodicity).

(20.5) In the synthetical phase the relevés are compared with one another to derive a preliminary classification. A set of relevés are first listed in their full species composition in a primary table. Species of intermediate presence values are emphasized in seeking groups of differentiating species that characterize some of the relevés but not others. Partial tables are prepared using these species to rearrange the relevés into groups characterized by the groups of differentiating species. When the rearrangement is satisfactory, the data from all species are recopied into a phytocoenon table in which boundaries may be drawn to mark off differential-species groups and phytocoena. A synoptic table may now be prepared in which each column summarizes data for species occurrence in the relevés of a phytocoenon. By use of synoptic tables the phytocoena under study may be compared with others from the same area in a search for character-species. Problems in the recognition of character-species, including geographic differences in fidelity of species to phytocoena, are discussed.

(20.6) In the syntaxonomical phase the studied phytocoena may enter the formal hierarchy. The phytocoenon table is revised into a formal syntaxonomic table indicating the character-species and differential-species groups represented. If the phytocoenon represents a new association, this is named and placed in the hierarchy in which associations are grouped into alliances, alliances into orders, orders into classes. Syntaxa on all these levels may be defined by character-species; in some cases character-genera are used for higher units. Apart from the hierarchy, associations (and other syntaxa) from different geographic areas may be grouped by floristic affinities, especially as marked by vicariant species. Associations are divided into subordinate units: subassociations and variants characterized by differential-species, and facies as the lowest-level units, characterized by quantitative differences in species representation. Associations are named on the basis of one or two characteristic species and the suffix -etum. Further rules of nomenclature for associations and other syntaxa are discussed, and an outline of the syntaxa is given.

(20.7) Various extensions of the approach are described. Applications to vegetation outside the European homeland are summarized. The approach is applied also to biotic and animal communities, and to stratal and life-form subcommunities to produce a separate classification of these into synusiae. Complexes (mosaics or patterns of contiguous or continuous communities) can be studied on various levels from the microrelief of bogs to whole landscapes. Mapping of phytocoena is valuable in applied phytosociology and land management. For applied work, diagnostic species groups can be used to indicate environment. Species can also be classed into ecological groups by their distributional responses to environmental gradients, and representation of ecological groups can be used to indicate position of a phytocoenose along an environmental gradient.

(20.8) Numerical techniques are being developed to store relevés, to measure homogeneity of phytocoenon tables, and to aid in sample rearrangement in tables and the earlier stages of classification. (20.9) Ordinations are possible on either a less formal basis, using relevés or phytocoena along known environmental gradients, or on a more formal basis using quantitative comparisons amongst samples to arrange them along abstract axes. Ordinations can be applied to phytocoena (each summarizing data from several relevés) as well as to individual relevés. Results to date indicate that this is a promising means of clarifying relationships amongst species and syntaxa.

(20.10) Numerical techniques and ordination are aids to the essential floristic-sociological procedures of the Braun-Blanquet approach. Despite the complexity of the procedures, the approach is felt to be flexible to different kinds of vegetation and research purposes, productive of understanding of the relations of species and samples to one another and environment, useful for applied purposes, and valuable for its coordination of the results from different areas into a single classification. It is judged the most fully developed and most widely useful approach to the classification and interpretation of vegetation.

20.11.2 Zusammenfassung

(20.1) Der Kern der Lehre von Braun-Blanquet umfaßt:

(i) Die Pflanzengesellschaften werden auf Grund ihrer vollständigen floristischen Zusammensetzung (Arten-Verbindung) klassifiziert und interpretiert.

(ii) Dabei werden die diagnostische Arten betont deren Auftreten die Gesellschaften kennzeichnet und zugleich Eigenschaften ihrer Standorte andeutet.

(iii) Mit Hilfe der diagnostischen Arten werden die Gesellschaften zu einem hierarchischen System geordnet.

(20.2) Diese floristisch-soziologische Betrachtungsweise der Vegetation fand ihren Ursprung in den Zentren der Pflanzensoziologie, Zürich und Montpellier, jedoch vor allem in de nIdeen von Braun-Blanquet.

(20.3) Wichtig ist der Unterschied zwischen bestimmten Gesellschafts- (Einzel-)beständen (sog. Phytozönosen) und abstrakten Gesellschaftstypen, für die der Begriff 'Phytozönon' empfohlen wird. Die Phytozöna des hierarchischen Systems von Braun-Blanquet werden Syntaxa genannt in Analogie zu den Taxa, mit denen Organismen klassifiziert werden. Syntaxa können durch drei Typen von diagnostischen Arten vertreten sein: Durch Charakter-(Kenn-)arten, die auf ein Syntaxon konzentriert oder beschränkt sind und darum das Syntaxon und dessen Standort kennzeichnen; durch Differential(-Trenn)arten, die zwei nahverwandte Syntaxa unterscheiden durch eine hohe Präsenz in den meisten Aufnahmen des einen und Abwesenheit in den Aufnahmen der anderen Syntaxons; ferner durch konstante Begleiter, die nicht auf ein bestimmtes Syntaxon beschränkt sind, jedoch bei der Charakterisierung des Syntaxons und beim Nachweis seiner Beziehungen zu höheren Einheiten mithelfen.

Syntaxa können auch von einfach organisierten bis hochdifferenzierten Gesellschaften nach einer soziologischen Progression angeordnet werden. Die Klassifikation von Braun-Blanquet wird in dem Sinne als natürlich angesehen, als ihre Syntaxa Beziehungen zwischen Standort, Artpopulationen und Biozönosen wiederspiegeln.

(20.4) Das Untersuchungsprogram umfaßt drei Phasen, eine analytische, eine synthetische und eine syntaxonomische. In der analytischen Phase werden im Gelände von Einzel-Beständen der zu untersuchenden Phytozönosen Aufnahmen gemacht. Für eine richtige Fassung der zu untersuchenden Phytozönose ist die Wahl der Aufnahmeflächen wichtig, sie sollen außerdem homogen und groß genug sein. Eine Aufnahme enthält Angaben über die topographische Lage und den Standort, sowie eine Liste aller Pflanzenarten der Aufnahmefläche. Die Pflanzenarten werden, entsprechend der Höhenstruktur der Vegetation nach Schichten notiert und nach ihrer Deckung, Abundanz und gegebenenfalls nach ihrer Soziabilität, Vitalität und Periodizität mit Hilfe von verschiedenen Skalen geschätzt.

(20.5) In der synthetischen Phase werden für die Aufstellung einer vorläufigen Klassifikation die Aufnahmen mit einander verglichen. Eine Reihe von Aufnahmen wird zuerst in einer Primärtabelle vereinigt. Arten mit mittlerer Präsenz werden verwendet für die Schaffung von Differentialarten-Gruppen, die jeweils nur einen Teil der Aufnahmen charakterisieren. Mit Hilfe dieser Gruppen werden die Aufnahmen in Teiltabellen neu geordnet. Nachdem die Neuordnung zufriedenstellend ist, werden die Angaben für alle Arten eingetragen; es entsteht eine Gesellschafts-Tabelle, in der die Gesellschaften bzw. die Differentialarten-Gruppen markiert werden können. Danach wird eine synoptische Tabelle erarbeitet, worin in Kolonnen zusammengefaßte Angaben über das Vorkommen der Arten in den Aufnahmen jeder Gesellschaft stehen. Mit Hilfe der synoptischen Tabellen können die Gesellschaften mit anderen des selben Gebiets verglichen und Charakterarten gefunden werden. Probleme über die Unterscheidung von Charakterarten und über geographische Abwandlungen der Gesellschaftstreue einzelner Arten betreffend werden besprochen.

(20.6) In der syntaxonomischen Phase werden die untersuchten Phytozöna in das hierarchische System eingeordnet. Die Gesellschafts-Tabelle ergibt eine syntaxonomische Tabelle welche die Gruppen der Charakter- und Differentialarten anzeigt. Falls die Gesellschaft eine neue Assoziation repräsentiert, wird ihr ein Name gegeben und ein Platz in der Hierarchie nachgewiesen, in der Assoziationen den Verbänden, Verbänden den Ordnungen und Ordnungen den Klassen untergeordnet werden. Syntaxa können durch Charakterarten definiert werden. Daneben können Assoziationen (und andere Syntaxa) aus verschiedenen geographischen Gebieten auf Grund floristischer Affinitäten zusammengefaßt werden. Sie werden durch vikariierende Arten unterschieden. Assoziationen werden in Unter-Einheiten gegliedert: in Subassoziationen und Varianten, die durch Differentialarten gekennzeichnet werden, oder in Fazies, die als niedrigste Einheiten nur noch durch quantitative Unterschiede im Vorkommen der Arten ausgewiesen werden. Assoziationen erhalten ihren Namen nach einer oder zwei ihrer kennzeichnenden und möglichst herrschenden Arten. Sie werden mit der Endung -etum bezeichnet. Weitere Regeln für die Nomen-klatur der Assoziationen und anderer Syntaxa werden besprochen.

(20.7) Verschiedene Erweiterungen dieser Betrachtungsweise werden beschrieben. Ihre Anwendung für Vegetationen außerhalb des ursprünglichen europäischen Bereichs wird dargestellt. Die gleiche Betrachtungsweise wird auch bei Biozönosen angewendet und bei Schichten- und Lebensformengemeinschaften, die gesondert in einer Synusien-Klassifikation geordnet werden. Komplexe von räumlich oder zeitlich aufeinander folgenden Gesellschaften können auf verschiedenen Niveaus untersucht werden.

Vegetationskartierungen sind wertvoll für die angewandte Pflanzensoziologie und die Natur- und Landschaftspflege. Bei praktischen Untersuchungen können diagnostische Artengruppen als Standortsindikatoren dienen. Arten können auch, auf Grund ihrer Beziehungen Milieu-Gradienten in ökologischen Gruppen zusammengefaßt werden, und die Anwesenheit bestimmter ökologischer Gruppen kann die Lage einer Gesellschaft entlang eines Milieugradienten verdeutlichen.

(20.8) Es werden mathematische Methoden entwickelt zur Speicherung von Aufnahmen, zur Bestimmung der Homogenität von Gesellschafts-Tabellen und zur Erleichterung der Neuordnung von Tabellen und der Beginnphasen der Klassifikation.

(20.9) Ordinationen sind möglich auf einfacher Basis, dann werden Aufnahmen oder Gesellschaften entlang bekannter Milieugradienten angeordnet, oder auf der Basis eines quantitativen Vergleichs der Aufnahmen für eine Platzierung der Aufnahmen entlang abstrakter Achsen. Ordinationen können mit Phytozöna (in denen Daten von verschiedenen Aufnahmen zusammengefaßt werden) sowie mit Einzelaufnahmen durchgeführt werden. Bisherige Ergebnisse zeigen, daß sich das Verfahren zur Erläuterung der Beziehungen zwischen Arten bzw. Syntaxa eignet.

(20.10) Mathematische Methoden der Klassifikation und Ordination stellen Hilfsmittel dar für die im wesentlichen floristisch-soziologische Betrachtungsweise der Lehre von Braun-Blanquet. Trotz der Komplexität des Verfahren ist die Betrachtungsweise flexibel im Bezug auf die verschiedenen Vegetationsbereiche und Forschungsziele, produktiv im Hinblick auf die Klärung der Beziehungen von Arten und Aufnahmen zueinander und zu ihren Standorten, nützlich für praktische Zielsetzungen und wertvoll im Hinblick auf die Möglichkeit einer Koordination der Ergebnisse verschiedener Gebiete in einer einzigen Klassifikation.

Diese Lehre darf mit Recht die am vollständigsten entwickelte und für die Klassifikation und Interpretation der Vegetation meist angewandte genannt werden.

REFERENCES

This bibliography does not intend to be complete. In particular, those papers have been omitted dealing with a similar topic as the first and (or) main paper of the same author referred to.

Adjanohoun, E. J., – 1962 – Étude phytosociologique des savanes de Basse Côte d'Ivoire (Savanes lagunaires). *Vegetatio* 11: 1—38.

Aichinger, E., – 1951 – Soziationen, Assoziationen und Waldentwicklungstypen. *Angew. Pfl. Soziol.*, Wien 1: 21—68.

Aichinger, E., – 1954 – Statische und dynamische Betrachtung in der pflanzensoziologischen Forschung. *Veröff. geobot. Inst. Rübel* 29: 9—28.

Alechin, W. W., – 1926 – Was ist eine Pflanzengesellschaft? *Repert. spec. nov. Regni veg., Beih.* 37: 1—50.

Bach, R., R. Kuoch & M. Moor, – 1962 – Die Nomenklatur der Pflanzengesellschaften. *Mitt. flor.-soz. Arbeitsgem. N. F.* 9: 301—308.

Balogh, J., – 1958 – Lebensgemeinschaften der Landtiere. Ihre Erforschung unter besonderer Berücksichtigung der zoozönologischen Arbeitsmethoden. Akademie Verlag GmbH, Berlin-Budapest. 560 pp.

Barkman, J. J., – 1953b – Some proposals for a phytosociological nomenclature. *Proc. 7th int. bot. Congr. Stockholm* 1950: 662—663; discussion: 663—666.

Barkman, J. J., – 1958a – Phytosociology and Ecology of Cryptogamic Epiphytes. Van Gorcum, Assen. 628 pp.

Barkman, J. J., – 1958b – La structure du Rosmarineto-Lithospermetum helianthemetosum en Bas-Languedoc. *Blumea suppl.* 6: 113—136.

Barkman, J. J., H. Doing, C. G. van Leeuwen & V. Westhoff, – 1958 – Enige opmerkingen over de terminologie in de vegetatiekunde. *Corr. Bl. Rijksherbarium* 8: 87—93.

Barkman, J. J., H. Doing & S. Segal, – 1964 – Kritische Bemerkungen und Vorschläge zur quantitativen Vegetationsanalyse. *Acta bot. neerl.* 13: 394—419.

Becking, R. W., – 1957 – The Zürich-Montpellier school of phytosociology. *Bot. Rev.* 23: 411—488.

Becking, R. W., – 1961 – Mathematical analysis of plant communities. *Rec. Adv. Botany*: 1346—1350.

Beeftink, W. G., – 1962 – Conspectus of the phanerogamic salt plant communities in the Netherlands. *Biol. Jb. Dodonaea* 30: 325—362.

Beeftink, W. G., – 1965 – De zoutvegetaties van ZW-Nederland beschouwd in Europees verband. *Meded. Landbouwhogeschool Wageningen* 65(1): 1—167.

Beeftink, W. G., – 1968 – Die Systematik der europäischen Salzpflanzengesellschaften. In, 'Pflanzensoziologische Systematik', ed. R. Tüxen, *Ber. Symp. int. Ver. Vegetationskunde*, Stolzenau 1964: 239—263. Junk, The Hague.

Benninghoff, W. S. & W. C. Southworth, – 1964 – Ordering of tabular arrays of phytosociological data by digital computer. *Abstr. 10. Int. bot. Congr. Edinburgh*: 331—332.

Bharucha, F. R., – 1956 – Methods for the study of tropical vegetation. *Proc. Kandy Symp. Study Tropical Vegetation*, UNESCO, Paris. p. 89—90.

Bolos, O., – 1968 – Tabula vegetationis Europae occidentalis. *Acta geobot. Barcinonensa* 3: 5—8.

Boudouresque, C. F., – 1971 – Contribution à l'étude phytosociologique des peuplements algaux des côtes varoises. *Vegetatio* 22: 83—184.

Braun-Blanquet, J., – 1913 – Die Vegetationsverhältnisse der Schneestufe in den Rätisch-Lepontischen Alpen. *Schweiz. Naturf. Gesellsch. Neue Denkschr.* 48: 1—347.

Braun-Blanquet, J., – 1915 – Les Cévennes méridionales. Étude phytogéographique. *Biblioth. Univ. Arch. Sci. Phys. Nat. Ser.* 4: 39: 72—81 a.f.

Braun-Blanquet, J., – 1918 – Eine pflanzengeographische Excursion durchs Unterengadin und in den Schweizerischen Nationalpark. *Beitr. geobot. Landesaufn. Schweiz* 4: 1–80.

Braun-Blanquet, J., – 1921 – Prinzipien einer Systematik der Pflanzengesellschaften auf floristischer Grundlage. *Jahrb. St. Gallen Naturw. Ges.* 57: 305—351.

Braun-Blanquet, J., – 1923 – Origine et Développement des Flores dans le Massif Central de France. Zürich-Paris.

Braun-Blanquet, J., – 1925 – Zur Wertung der Gesellschaftstreue in der Pflanzensoziologie. *Vierteljahrsschr. Naturf. Ges. Zürich* 70: 122—149.

Braun-Blanquet, J., – 1928 – Pflanzensoziologie. Grundzüge der Vegetationskunde. Biologische Studienbücher 7. 1. Ed. Berlin. x + 330 pp.

Braun-Blanquet, J. – 1930 – L'importance pratique de la sociologie végétale. *Comm. S.I.G.M.A.* 4: 1—8.

Braun-Blanquet, J., – 1932 – Plant Sociology. (Transl. by G. D. Fuller and H. S. Conard). New York, xviii + 439 pp. Reprint 1966.

Braun-Blanquet, J., – 1933a – Phytosociological nomenclature. *Ecology* 14: 315—317.

Braun-Blanquet, J., – 1933b – Prodrome des Groupements végétaux, Fasc. 1. Ammophiletalia et Salicornietalia medit. Com. Int. Prodrome Phytosociol. Montpellier. 23 pp.

Braun-Blanquet, J., – 1939 – Lineares oder vieldimensionales System in der Pflanzensoziologie. *Chronica bot.* 5: 391—395.

Braun-Blanquet, J., – 1946 – Über den Deckungswert der Arten in den Pflanzengesellschaften der Ordnung Vaccinio-Piceetalia. *Jahresber. Naturf. Gesell. Graubündens* 130: 115—119.

Braun-Blanquet, J., – 1950 – Sociologica vegetal. Versión espan. A. P. L. Gigilio & M. M. Grassi. Buenos Aires.

Braun-Blanquet, J., – 1951a – Pflanzensoziologie. Grundzüge der Vegetationskunde, 2nd ed. Springer, Wien. 631 pp.

Braun-Blanquet, J., – 1951b – Pflanzensoziologische Einheiten und ihre Klassifizierung. *Vegetatio* 3: 126—133.

Braun-Blanquet, J., – 1953 – Essai sur le classement des Biocénoses. *Comm. S.I.G.M.A.* 118: 8—12.

Braun-Blanquet, J., – 1955 – Zur Systematik der Pflanzengesellschaften. *Mitt. flor.-soz. Arbeitsgem. NF* 5: 151—154.

Braun-Blanquet, J., – 1959 – Grundfragen und Aufgaben der Pflanzensoziologie. *Vistas in Botany* 1: 145—171.

Braun-Blanquet, J., – 1962 – Zur pflanzensoziologischen Systematik, Erinnerungen und Ausblick. *Comm. S.I.G.M.A.* 159: 1—11.

Braun-Blanquet, J., – 1964 – Pflanzensoziologie, Grundzüge der Vegetationskunde, 3rd ed. Springer, Wien-New York. 865 pp.

Braun-Blanquet, J., – 1968 – L'école phytosociologique Zuricho-Montpelliéraine et la S.I.G.M.A. *Vegetatio* 16: 1—78.

Braun-Blanquet, J., – 1969 – Reinhold Tüxen, Meister-Pflanzensoziologe. *Vegetatio* 17: 1—25.

Braun-Blanquet, J. & E. Furrer, – 1913 – Remarques sur l'étude des groupements de plantes. *Bull. Soc. languedoc. Géogr.* 36: 20—41.

Braun-Blanquet, J. & H. Jenny, – 1926 – Vegetations-Entwicklung und Bodenbildung in der alpinen Stufe der Zentralalpen. *Schweiz. Naturf. Ges., Denkschr.* 63(2): 181—349.

Braun-Blanquet, J. & W. C. de Leeuw, – 1936 – Vegetationsskizze von Ameland. *Ned. kruidk. Arch.* 46: 359—393.

Braun-Blanquet, J. & M. Moor, – 1938 – Verband des Bromion erecti. Prodromus der Pflanzengesellschaften. *Comm. S.I.G.M.A.* 5: 1—64.

Braun-Blanquet, J. & J. Pavillard, – 1922/28 – Vocabulaire de sociologie végétale. 1. Ed. Montpellier, 16 pp. 3. Ed. Montpellier, 23 pp.

Braun-Blanquet, J. & J. Pavillard, – 1930 – Vocabulary of Plant Sociology. (Transl. by Bharucha, F. R.). Cambridge. 23 pp.

Braun-Blanquet, J., G. Sissingh & J. Vlieger, – 1939 – Prodrome des Groupements Végétaux. Fasc. 6 Klasse der Vaccinio-Piceetea. Montpellier 123 pp.

Braun-Blanquet, J. & R. Tüxen, – 1932 – Die Pflanzensoziologie in Forschung und Lehre. *Biologe* 1931/32: 175—186.

Bray, J. R. & J. T. Curtis, – 1957 – An ordination of the upland forest communities of southern Wisconsin. *Ecol. Monogr.* 27: 325—349.

Brockmann-Jerosch, H., – 1907 – Die Pflanzengesellschaften der Schweizeralpen. I. Die Flora des Puschlav und ihre Pflanzengesellschaften. Engelmann, Leipzig. 438 pp.

Cain, S. A. & G. M. de Oliveira Castro, – 1959 – Manual of Vegetation Analysis. Harper and Brothers, New York. xvi+325 pp.

Cajander, A. K., – 1903 – Beiträge zur Kenntniss der Vegetation der Alluvionen des nördlichen Eurasiens. I. Die Alluvionen des unteren Lena-Thales. *Acta soc. sci. Fenn.* 32 (1): 1—182.

Češka, A., – 1966 – Estimation of the mean floristic similarity between and within sets of vegetational relevés. *Folia geobot. phytotax.* 1: 92—100.

Češka, A., – 1968 – Application of association coefficients for estimating the mean similarity between sets of vegetational relevés. *Folia geobot. phytotax.*, 3: 57—64.

Češka, A. & H. Roemer, – 1971 – A computer program for identifying species-relevé groups in vegetation studies. *Vegetatio* 23: 255—276.

Chapman, V. J., – 1952 – Problems in ecological terminology. *Rep. Austr.-Nw Zealand Ass. Adv. Science* 29th Meeting 259—279.

Chapman, V. J,, – 1959 – Salt marshes and ecological terminology. *Vegetatio* 8: 215—234.

Clements, F. E., – 1936 – Nature and structure of the climax. *J. Ecol.* 24: 252—284.

Conard, H., – 1935 – The plant associations of Central Long Island. *Amer. Midl. Nat.* 16: 433—516.

Conard, H., – 1952 – The vegetation of Iowa. An approach toward a phytosociologic account. *Stud. nat. Hist. Iowa Univ.* 19 (4): 1—166.

Cottam, G. & J. T. Curtis, – 1948 – The use of the punched card method in phytosociological research. *Ecology* 29: 516—519.

Crawford, R. M. M. & D. Wishart, – 1966 – A multivariate analysis of the development of dune slack vegetation in relation to coastal accretion at Tentsmuir, Fife. *J. Ecol.* 54: 729—743.

Crawford, R. M. M. & D. Wishart, – 1967 – A rapid multivariate method for the detection and classification of groups of ecologically related species. *J. Ecol.* 55: 505—524.

Cristofolini, G. and S. M., – 1967 – Phytosociological study on some Liriodendron tulipifera L. forests of Tennessee, U.S.A. *N. Gior. bot. ital.* 101—6: 317—346.

Cristofolini, G., D. Lausi & S. Pignatti, – 1969 – Survey of the system for coding of plantsociological records used by the 'Trieste Group'. Report Working Group Data-Processing Phytosoc. (mimeographed) 13 pp.

Curtis, J. T., – 1959 – The vegetation of Wisconsin. An Ordination of Plant Communities. Univ. of Wisconsin Press. Madison, Wisconsin. xi + 657 pp.

Curtis, J. T. & R. P. McIntosh, – 1951 – An upland forest continuum in the prairie-forest border region of Wisconsin. *Ecology* 32: 476—496.

Daget, Ph. et al., – 1972 – Profils écologiques et information mutuelle entre espèces et facteurs écologiques. In, 'Grundfragen und Methoden in der Pflanzensoziologie', ed. E. van der Maarel & R. Tüxen, *Ber. Symp. int. Ver. Vegetationskunde*, Rinteln 1970: 121—148. Junk, The Hague.

Dagnelie, P., – 1960 – Contribution à l'étude des communautés végétales par l'analyse factorielle. *Bull. Serv. Carte Phytogéogr. CNRS* 5: 7—71, 93—195.

Dahl, E., – 1957 – Rondane; Mountain vegetation in South Norway and its relation to the environment. *Skr. norske Vidensk-Akad.*; *mat.-naturv. Kl.* 1956(3): 1—374.

Dahl, E., – 1960 – Some measures of uniformity in vegetation analysis. *Ecology* 41: 785—790.

Dahl, E. & E. Hadač, – 1941 – Strandgesellschaften der Insel Ostoy im Oslofjord. *Nytt Mag. Naturv.* 82: 251—312.

Dahl, E. & E. Hadač, – 1949 – Homogeneity of plant communities. *Studia Bot. Cechosl.* 10: 159—176.

Damman, A. W. H., – 1964 – Some forest types of central Newfoundland and their relation to environmental factors. *Forest Sci. Monogr.* 8: 1—62.

Damman, A. W. H. & D. M. de Vries, – 1954 – Testing of grassland associations by combinations of species. *Biol. Jb. Dodonaea* 21: 35—46.

Dansereau, R., – 1957 – Biogeography; an Ecological Perspective. Ronald Press Comp., New York. xiii + 394 pp.

Dansereau, P. & J. Arros, – 1959 – Essais d'application de la dimension structurale en phytosociologie. I. Quelques exemplaires européens. *Vegetatio* 9: 48—99.

Davis, P. H. & V. H. Heywood, – 1963 – Principles of Angiosperm Taxonomy. Oliver & Boyd, Edinburgh. xx + 556 pp.

Devred, R., – 1956 – Les savanes herbeuses de la région de Mvuazi (Bas Congo). *Publ. INEAC Bruxelles Sér. Scient.* 65: 1—115.

Dierschke, H., – 1971 – Stand und Aufgaben der pflanzensoziologischen Systematik in Europa. *Vegetatio* 22: 255—264.

Doing Kraft, H., – 1954 – L'analyse des carrés permanents. *Acta bot. neerl.* 3: 421—425.

Doing, H., – 1962 – Systematische Ordnung und floristische Zusammensetzung niederländischer Wald- und Gebüschgesellschaften. *Wentia* 8: 1—85.

Doing, H., – 1969a – Sociological species groups. *Acta bot. neerl.* 18: 398—400.

Doing, H., – 1969b – Assoziationstabellen von niederländischen Wäldern und Gebüschen. Rep. Lab. Plantensyst. Geogr. Wageningen 29 pp. + tab.

Donselaar, J. van, – 1965 – An ecological and phytogeographic study of northern Surinam savannas. *Wentia* 14: 1—163.

Donselaar, J. van, – 1969 – Observations on savanna vegetation types in the Guianas. *Vegetatio* 17: 271—312.

Drude, O., – 1890 – Handbuch der Pflanzengeographie, Stuttgart, 582 pp.

Du Rietz, G. E., – 1921 – Zur methodologischen Grundlage der modernen Pflanzensoziologie. Akad. Abhandl. Uppsala. 272 pp.

Du Rietz, G. E., – 1930a – Vegetationsforschung auf soziationsanalytischer Grundlage. *Handb. Biol. Arbmeth.* 11, 5: 293—480.

Du Rietz, G. E., – 1930b – Classification and nomenclature of vegetation. *Svensk Bot. Tidskr.* 24: 489—503.

Du Rietz, G. E., – 1936 – Classification and nomenclature of vegetation units 1930—1935. *Svensk Bot. Tidskr.* 30: 580—589.

Du Rietz, G. E., T. C. E. Fries, H. Oswald & T. A. Tengwall, – 1920 – Gesetze der Konstitution natürlicher Pflanzengesellschaften. *Vetensk. prakt. Unders. Lappl. Flora och Fauna* 7: 1—47.

Duvigneaud, P., – 1946 – La variabilité des associations végétales. *Bull. Soc. Roy. bot. belg.* 78: 107—134.

Duvigneaud, P., – 1949 – Les savannes du Bas-Congo. Essai de phytosociologie topographique. *Lejeunia Mem.* 10, 192 pp.

Egler, F. E., – 1954 – Philosophical and practical considerations of the Braun-Blanquet system of plant sociology. *Castanea* 19: 45—60.

Ellenberg, H., – 1939 – Über Zusammensetzung, Standort und Stoffproduktion bodenfeuchter Eichen- und Buchen-Mischwaldgesellschaften Nordwestdeutschlands. *Mitt. flor.-soz. Arbeitsgem. Niedersachsen* 5: 1—135.

Ellenberg, H., – 1950 – Landwirtschaftliche Pflanzensoziologie. I. Unkrautgemeinschaften als Zeiger für Klima und Boden. Stuttgart-Ludwigsburg. 141 pp.

Ellenberg, H., – 1952 – Landwirtschaftliche Pflanzensoziologie. II. Wiesen und Weiden und ihre standörtliche Bewertung. Stuttgart. 143 pp.

Ellenberg, H., – 1954a – Landwirtschaftliche Pflanzensoziologie. III. Naturgemässe Anbauplanung, Melioration und Landespflege. Stuttgart. 109 pp.

Ellenberg, H., – 1954b – Zur Entwicklung der Vegetationssystematik in Mitteleuropa. *Angew. Pfl. Soziol.*, Wien, Festschr. Aichinger 1: 134—143.

Ellenberg, H., – 1956 – Grundlagen der Vegetationsgliederung. 1. Teil: Aufgaben und Methoden der Vegetationskunde. In, 'Einführung in die Phytologie' IV-1, ed. H. Walter, Stuttgart, 136 pp.

Ellenberg, H., – 1963 – Vegetation Mitteleuropas mit den Alpen. In, 'Einführung in die Phytologie' IV-2, ed. H. Walter, Stuttgart, 943 pp.

Ellenberg, H., – 1968 – Sichtlochkarten zur Ordnung, Klassifikation und Analyse pflanzensoziologischer Waldaufnahmen. In, 'Pflanzensoziologische Systematik', ed. R. Tüxen, *Ber. Symp. int. Ver. Vegetationskunde*, Stolzenau 1964: 163—175. Junk, The Hague.

Ellenberg, H. & G. Cristofolini, – 1964 – Sichtlochkarten als Hilfsmittel zur Ordnung und Auswertung von Vegetationsaufnahmen. *Ber. geobot. Inst. ETH Rübel* 35: 124—134.

Emberger, L. (ed.), – 1968 – Code pour le relevé méthodique de la végétation et du milieu. Edit. CNRS, Paris. 292 pp.

Emberger, L. et al., – 1957 – Description et mode d'emploi d'une fiche de relevé pour l'inventaire de la végétation. *Bull. Serv. Carte Phytogéogr.* B-2: 7—23.

Eskuche, U., – 1968 – Fisionomía y sociologia de los bosques de Nothofagus dombeyi en la région de Nahuel Huapi. *Vegetatio* 16: 192—204.

Eskuche, U., – 1969 – Berberitzengebüsche und Nothofagus antarctica-Wälder in Nordwest-Patagonien. *Vegetatio* 19: 264—285.

Etter, H., – 1949 – De l'analyse statistique des tableaux de végétation. *Vegetatio* 1: 147—154.

Evans, F. C. & E. Dahl, – 1955 – The vegetational structure of an abandoned field in Southeastern Michigan and its relation to environmental factors. *Ecology* 36: 685—706.

Falinski, J., – 1960 – Zastosowanie taksonomii wroclawskiej do fitosocjologii. (Anwendung der sogenannten 'Breslauer Taxonomie' in der Pflanzensoziologie.). *Acta Soc. bot. pol.* 29: 333—361.

Falinski, J., – 1964 – O róznych sposobach rozumiena pojęcia typu w fitosocjologii. Dyskusje fitosocjologiczne (1). (Sur les différentes façons d'envisager la notion du type en phytosociology. Discussion phytosociologique (1). *Ekol. Polska Ser.* B 10: 297—306.

Flahault, Ch., – 1893 – Les zones botaniques dans le Bas-Languedoc et les pays voisins. *Soc. bot. France, Bull.* 40 (Sess. extraord.): 35—62.

FLAHAULT, CH., – 1898 – Introduction, pp. 11—48, in: GAUTIER, G., Catalogue raisonné de la flore des Pyrénées orientales. Perpignan, 550 pp.

FLAHAULT, CH., – 1901 – A project for phytogeographic nomenclature. *Bull. Torrey bot. Club* 28: 391—409.

FLAHAULT, CH. & C. SCHRÖTER, – 1910 – Rapport sur la nomenclature phytogéographique. (Phytogeographische Nomenklatur). *Actes III Congr. int. bot. Bruxelles* 1: 131—164.

FRESCO, L. F. M., – 1969 – Q-type factor analysis as a method in synecological research. *Acta bot. neerl.* 18: 477—482.

FRESCO, L. F. M., – 1971 – Compound analysis: a preliminary report on a new numerical approach in phytosociology. *Acta bot. neerl.* 20: 589—599.

FRESCO, L. F. M., – 1972 – A direct quantitative analysis of vegetational boundaries and gradients. In, 'Grundfragen und Methoden in der Pflanzensoziologie', ed. E. VAN DER MAAREL & R. TÜXEN, *Ber. Symp. int. Ver. Vegetationskunde*, Rinteln 1970: 99—111. Junk, The Hague.

FREY, A., – 1928 – Anwendung graphischer Methoden in der Pflanzensoziologie. *Handb. biol. ArbMeth.* 11, 5: 203—232.

FRIEDERICHS, K., – 1927 – Grundsätzliches über die Lebenseinheiten höherer Ordnung und den ökologischen Einheitsfaktor. *Naturwissenschaften* 15: 153—157, 182—186.

FRIEDERICHS, K., – 1958 – A definition of ecology and some thoughts about basic concepts. *Ecology* 39: 154—159.

FRYDMAN, J. & R. H. WHITTAKER, – 1968 – Forest associations of southeast Lublin Province, Poland. *Ecology* 49: 896—908.

FUKAREK, F., – 1964 – Pflanzensoziologie. Berlin. 160 pp.

FUKAREK, F., M. JASNOWSKI & R. NEUHÄUSL, – 1964 – Termini phytosociologici. Jena. 74 pp.

GAMS, H., – 1918 – Prinzipienfragen der Vegetationsforschung. *Vierteljahrsch. Naturf. Gesellsch. Zürich* 63: 293—493.

GAMS, H., – 1941 – Über neuen Beiträge zur Vegetationssystematik unter besonderer Berücksichtigung des floristischen Systems von Braun-Blanquet. *Bot. Arch.* 42: 201—238.

GAMS, H., – 1954 – Vegetationssystematik als Endziel oder Verständigungsmittel? *Veröff. geobot. Inst. Rübel* 29: 39-40,

GAMS, H., – 1961 – Erfassung und Darstellung mehr-dimensionaler Verwandtschaftsbeziehungen von Sippen und Lebensgemeinschaften. *Ber. geobot. Inst. Rübel* 32: 96—115.

GAMS, H., – 1967 – Anordnung (Ordination), Aufschlüsselung (Klavikation) und Systematik (Klassifikation) von Biozönosen und anderen Naturerscheinungen. *Aquilo, Ser. Bot.* 6: 9—17.

GAMS, H. – 1972 – Die floren- and vegetationsgeschichtliche Erforschung der Alpen. *Ber. dt. bot. Ges.* 85: 7—10.

GÉHU, J. M., – 1969 – Les associations végétales des dunes mobiles et des bordures de plages de la côte atlantique française. *Vegetatio* 23: 122—166.

GILLI, A., – 1969 – Afghanische Pflanzengesellschaften. *Vegetatio* 16: 307—375.

GILLI, A., – 1971 – Afghanische Pflanzengesellschaften. II. Die mesophilen und hygrophilen Pflanzengesellschaften im sommertrockenen Gebiet. *Vegetatio* 23: 199—234.

GITTINS, R., – 1965 – Multivariate approaches to a limestone grassland community. III. A comparative study of ordination and association analysis. *J. Ecol.* 53: 411—425.

GLAHN, H. VON, – 1968 – Der Begriff des Vegetationstyps im Rahmen eines allgemeinen naturwissenschaftlichen Typenbegriffes. In, 'Pflanzensoziologische

Systematik', ed. R. Tüxen, *Ber. Symp. int. Ver. Vegetationskunde*, Stolzenau 1964: 1—13. Junk, The Hague.

Gleason, H., – 1926 – The individualistic concept of the plant association. *Bull. Torrey bot. Club* 53: 7—26.

Godron, M., – 1966 – Application de la théorie de l'information à l'étude de l'homogénéité et de la structure de la végétation. *Oecol. Planta* 1: 187—197.

Goff, F. G. & G. Cottam, – 1967 – Gradient analysis: the use of species and synthetic indices. *Ecology* 48: 793—806.

Goodall, D. W., – 1952 – Quantitative aspects of plant distribution. *Biol. Rev.* 27: 194—245.

Goodall, D. W., – 1953a – Objective methods for the classification of vegetation. I. The use of positive interspecific correlation. *Austr. J. Bot.* 1: 39—63.

Goodall, D. W., – 1953b – Objective methods for the classification of vegetation. II. Fidelity and indicator value. *Austr. J. Bot.* 1: 434—456.

Goodall, D. W., – 1954 – Objective methods for the classification of vegetation. III. An essay int he use of factor analysis. *Austr. J. Bot.* 2: 304—324.

Goodall, D. W., – 1961 – Objective methods for the classification of vegetation. IV. Pattern and Minimal Area. *Austr. J. Bot.* 9: 162—196.

Goodall, D. W., – 1963 – The continuum and the individualistic association. *Vegetatio* 11: 297—316.

Goodall, D. W., – 1969 – A procedure for recognition of uncommon species combinations in sets of vegetation samples. *Vegetatio* 18: 19—35.

Gounot, M., – 1957 – Utilisation des fiches perforés pour la comparaison des relevés. *Bull. Serv. Carte Phytogéogr.* B,2: 37—50.

Gounot, M., – 1969 – Méthodes d'étude quantitative de la végétation. Masson, Paris. 314 pp.

Gradmann, R., – 1909/40 - Über Begriffsbildung in der Lehre von den Pflanzenformationen. *Bot. Jb.* 43, Beibl. 99: 91—103. Reprinted 1940.

Grandtner, M. M., – 1966 – La végétation forestière du Québec méridional. Presses Univ. Laval, Quebec. XXV + 216 pp.

Greig-Smith, P., – 1964 - Quantitative Plant Ecology, 2nd ed. Butterworths, London. xii + 256 pp.

Grisebach, A., – 1838 – Über den Einfluss des Climas auf die Begränzung der natürlichen Floren. *Linnaea* 12: 159—200.

Grisebach, A., – 1872 – Die Vegetation der Erde nach ihrer klimatischen Anordnung. Leipzig, 2 vols., 603 + 709 p. 2nd ed. 1884, 567 + 693 p.

Groenewoud, H. van, – 1965 – Ordination and classification of Swiss and Canadian forests by various biometric and other methods. *Ber. geobot. Inst. Rübel* Zürich 35: 28—102.

Guillerm, J. L., – 1971 – Calcul de l'information fournie par un profil écologique et valeur indicatrice des espèces. *Oecol. Plant.* 6: 209—225.

Guinochet, M., – 1955 – Logique et dynamique du peuplement végétal. Coll. Ecol. Sci., Paris. 143 pp.

Guinochet, M. & P. Casal, – 1957 – Sur l'analyse différentielle de Czekanowski et son application à la phytosociologie. *Bull. Serv. Carte phytogéogr.* B,2: 25—33.

Hadač, E., – 1962 – Übersicht der höheren Vegetationseinheiten des Tatragebirges. *Vegetatio* 11: 46—54.

Hadač, E., – 1967 – On the highest units in the system of plant communities. *Folia geobot. phytotax.* 4: 429—432.

Hartmann, H., – 1968 – Über die Vegetation des Karakorum. I. Gesteinsfluren, subalpine Strauchbestände und Steppengesellschaften im Zentral-Karakorum. *Vegetatio* 15: 297—387.

Hartmann, H., – 1972 – Über die Vegetation des Karakorum. II. Rasen- und Strauchgesellschaften im Bereich der alpinen und der höheren subalpinen Stufe des Zentral-Karakorum. *Vegetatio* 24: 91—157.

Hartog, C. den, – 1959 – The epilithic algal communities occurring along the coast of the Netherlands. *Wentia* I: 1—241.

Hartog, C. den & S. Segal, – 1964 – A new classification of the water plant communities. *Acta bot. neerl.* 13: 367—393.

Heer, O., – 1835 – Die Vegetationsverhältnisse des südöstlichen Teiles des Kantons Glarus; ein Versuch die pflanzengeographischen Erscheinungen der Alpen aus klimatischen und Bodenverhältnissen abzuleiten. Mitt. theor. Erdkunde Zürich.

Hegg, O., – 1965 – Untersuchungen zur Pflanzensoziologie und Ökologie im Naturschutzgebiet Hohgant (Berner Voralpen), mit einem Beitrag zur Methodik der floristisch-statistischen Erfassung pflanzensoziologischer Zusammenhänge. *Beitr. geobot. Landesaufn. Schweiz* 46: 1—188.

Heimans, J., – 1939 – Plantengeographische elementen in de Nederlandse flora. *Ned. kruidk. Arch.* 49: 416—436.

Hofmann, G. & H. Passarge, – 1964 – Über Homogenität und Affinität in der Vegetationskunde. *Arch. Forstwes.* 13: 1119—1138.

Holub, J., S. Hejný, J. Moravec & R. Neuhäusl, – 1967 – Übersicht der höheren Vegetationseinheiten der Tschechoslowakei. *Rozpr. čs. Akad. Věd, Ser. math. nat. Praha* 77(3): 1—75.

Hopkins, B., – 1955 – The species-area relations of plant communities. *J. Ecol.* 43: 409—426.

Hopkins B., – 1957 – The concept of minimal area. *J. Ecol.* 45: 441—449.

Hult, R., – 1881 – Försök till analytik behandling af växtformationerna. (Attempt towards an analytical treatment of plant formations.) *Meddn. Soc. Faun. Flor. fenn.* 8. Helsinki 155 pp.

Humboldt, A. von, – 1805 – Essai sur la géographie des plantes. Paris, 155 pp.

Ivimey-Cook, R. B. & M. C. F. Proctor, – 1966 – The application of association-analysis to phytosociology. *J. Ecol.* 54: 179—192.

Jakucs, P., – 1961 – Die phytozönologischen Verhältnisse der Flaumeichen-Buschwälder Südostmitteleuropas. Budapest, 313 pp.

Jakucs, P., – 1970 – Bemerkungen zur Saum-Mantel-Frage. *Vegetatio* 21: 29—47.

Jakucs, P., – 1972 – Dynamische Verbindung der Wälder und Rasen. Akademiai Kiadó, Budapest. 228 pp.

Janssen, C. R., – 1967 – A floristic study of forests and bog vegetation, northwestern Minnesota. *Ecology* 48: 751—755.

Janssen, J. G. M., – 1972 – Detection of some micropatterns of winter annuals in pioneer communities of dry sandy soils. *Acta bot. neerl.* 21: 603—610.

Janssen, J. G. M. & E. van der Maarel, – 1972 – A computer program for structuring a phytosociological table on the basis of relevé similarity (in prep.).

Kerner von Marilaun, A, – 1863 – Das Pflanzenleben der Donauländer. Innsbruck, 348 pp.

Klement, O., – 1955 – Prodromus der mitteleuropäischen Flechtengesellschaften. *Repert. Spec. nov. Regni veg.*, Beih. 135: 5—194.

Knapp, R., – 1942 – Zur Systematik der Wälder, Zwergstrauchheiden und Trockenrasen des eurosiberischen Vegetationkreises. Beil. 12, Rundbrief Zentralstelle Veget. kartierung. 183 pp.

Knapp, R., – 1948 – Einführung in die Pflanzensoziologie. 1: Arbeitsmethoden der Pflanzensoziologie und Eigenschaften der Pflanzengesellschaften. Stuttgart - Ludwigsburg. 100 pp.

KNAPP, R., – 1957 – Über die Gliederung der Vegetation von Nordamerika. Höhere Vegetationseinheiten. *Geobot. Mitt. Köln* 4: 1—63.

KNAPP, R., – 1958 – Einführung in die Pflanzensoziologie. 1: Arbeitsmethoden der Pflanzensoziologie und Eigenschaften der Pflanzengesellschaften. 2.Ed. Stuttgart. 112 pp.

KNAPP, R., – 1965 – Die Vegetation von Nord- und Mittelamerika und der Hawaii-Inseln. Fischer, Stuttgart xl + 373 pp.

KNAPP, R., – 1966a – Höhere Vegetationseinheiten von West-Afrika unter besonderer Berücksichtigung von Nigeria und Kamerun. *Geobot. Mitt.* 34: 1—16.

KNAPP, R., – 1966b – Höhere Vegetationseinheiten von S.-Patagonien und Feuerland. *Geobot. Mitt.* 35: 1—4.

KNAPP, R., – 1971 – Einführung in die Pflanzensoziologie. Eugen Ulmer, Stuttgart. 388 pp.

KNIGHT, D. H., – 1965 – A gradient analysis of Wisconsin prairie vegetation on the basis of plant structure and plant function. *Ecology* 46: 744—747.

KOCH, W., – 1925 – Die Vegetationseinheiten der Linthebene unter Berücksichtigung der Verhältnisse in der N.O. Schweiz. *Jb. St. Gall. Naturw. Ges.* 61(2): 1—146.

KOHLER, A., – 1968 – Beiträge zur Kenntnis der ephemeren Vegetation am Südrand der Atacama-Wüste. *Ber. dt. bot. Ges.* 80: 563—572.

KOHLER, A., – 1970 – Geobotanische Untersuchungen an Küstendünen Chiles zwischen 27 und 42 Grad südl. Breite. *Bot. Jb.* 90: 55—200.

KORNAS, J., – 1965 – Phytosociological observations on plant communities of the Duke Forest near Durham, N. Carolina, U.S.A. *Fragm. flor. geobot.* 11: 307—338.

KORTEKAAS, W. & E. VAN DER MAAREL, – 1972 – A numerical classification of Spartinetum vegetations. Preliminary paper int. colloquium Rinteln 1972, 9 pp. + tables.

KRAJINA, V. J., – 1969 – Ecology of forest trees in British Columbia. *Ecol. W.N. America* 2: 1—146.

KRAUSE, W., – 1952 – Das Mosaik der Pflanzengesellschaften und seine Bedeutung für die Vegetationskunde. *Planta* 41: 240—289.

KÜCHLER, A. W., – 1967 – Vegetation Mapping. Ronald Press, New York vi + 472 pp. 30 tab. 21 fig.

LACOSTE, A. & M. ROUX, – 1971 – L'analyse multidimensionnelle en phytosociologie et en écologie. Application à des données de l'étage subalpin des Alpes maritimes. 1. L'analyse des données floristiques. *Oecol. Plant.* 6: 353—369.

LAMBERT, J. M., M. B. DALE, – 1964 – The use of statistics in phytosociology. *Adv. ecol. Res.* 2: 59—99.

LEBRUN, J., – 1960 – Études sur la flore et la végétation des champs de Lave au Nord du Lac Kivu (Congo belge). Inst. Parcs nat. Congo Belge Bruxelles pp. 1—352.

LEBRUN, J. & G. GILBERT, – 1954 – Une classification écologique des forêts du Congo. *Publ. Inst. Nat. Étude Agron. Congo Belge, Ser. Sci.* 63: 1—89.

LECOQ, H., – 1844 – Traité des plantes fourragères, ou flore des prairies naturelles et artificielles de la France. Paris, 620 pp.

LECOQ, H., – 1855 – Études sur la géographie botanique de l'Europe et en particulier sur la végétation du plateau central de la France. Vol. 4. Paris, 563 pp.

LEEUWEN, C. G. VAN, – 1965 – Het verband tussen natuurlijke en anthropogene landschapsvormen, bezien vanuit de betrekkingen in grensmilieu's. (Engl. summary) *Gorteria* 2: 93—105.

LEEUWEN, C. G. VAN, – 1966 – A relation theoretical approach to pattern and process in vegetation. *Wentia* 15: 25—46.

LEEUWEN, C. G. VAN, – 1970 – Raum-zeitliche Beziehungen in der Vegetation. In, 'Gesellschaftsmorphologie', ed. R. TÜXEN, *Ber. Symp. int. Ver. Vegetationskunde*, Rinteln 1966: 63—68. Junk, The Hague.

LIETH, H. & G. W. MOORE, – 1971 – Computerized clustering of species in phytosociological tables and its utilization for field work. In: G. P. PATIL, E. C. PIELOU & W. E. WATERS (ed): Spatial patterns and statistical distributions. Statistical Ecology Vol. 1 Penn State Univ. Press p. 403—422.

LIPPMAA, T., – 1935 – La méthode des associations unistrates et le système écologique des associations. *Acta Inst. Horti bot. tartu.* 4 (1—2, art. 3): 1—7.

LOHMEYER, W. & W. TRAUTMANN, – 1970 – Zur Kenntnis der Vegetation der Kanarischen Insel La Palma. *Schr. Reihe Vegetationskde* 5: 209—236.

LONDO, G., – 1971 – Patroon en proces in duinvalleivegetaties langs een gegraven meer in de Kennemerduinen. *Thesis Nijmegen*, 279 pp.

LOOMAN, J., – 1969 – The fescue grasslands of western Canada. *Vegetatio* 19: 128—145.

LOOMAN, J. & J. B. CAMPBELL, – 1960 – Adaptation of Sorensen's K (1948) for estimating unit affinities in prairie vegetation. *Ecology* 41: 409—416.

LORENZ, J. R., – 1858 – Allgemeine Resultate aus der pflanzengeographischen und genetischen Untersuchung der Moore im präalpinen Hügellande Salzburgs. *Flora* 41: 209—221, 225—237, 241—253, 273—286, 289—302, 344—355, 360—376.

LOUCKS, O. L., – 1962 – Ordinating forest communities by means of environmental scalars and phytosociological indices. *Ecol. Monogr.* 32: 137—166.

MAAREL, E. VAN DER, – 1960 – Rapport inzake de vegetatie van het duingebied van de Stichting 'Het Zuid-Hollands Landschap' bij Oostvoorne. Delft, pp. 1-57.

MAAREL, E. VAN DER, – 1965 – Beziehungen zwischen Pflanzengesellschaften und Molluskenfaunen. In, 'Biosoziologie.', ed. R. TÜXEN, *Ber. Symp. int. Ver. Vegetationskunde*, Stolzenau 1960: 184—197. Junk, The Hague.

MAAREL, E. VAN DER, – 1966a – Dutch studies on coastal sand dune vegetation, especially in the Delta region. *Wentia* 15: 47—82.

MAAREL, E. VAN DER, – 1966b – On vegetational structures, relations and systems, with special reference to the dune grasslands of Voorne, The Netherlands. *Thesis Utrecht* 170 pp.

MAAREL, E. VAN DER, – 1969 – On the use of ordination models in phytosociology. *Vegetatio* 19: 21—46.

MAAREL, E. VAN DER, – 1971 – Basic problems and methods in Phytosociology. *Vegetatio* 22: 275—283.

MAAREL, E. VAN DER, – 1972a – Ordination of plant communities on the basis of their plant genus, family and order relationships. In, 'Basic problems and methods in phytosociology', ed. E. VAN DER MAAREL & R. TÜXEN, *Ber. Symp. int. Ver. Vegetationskunde*, Rinteln 1970: 183—190. Junk, The Hague.

MAAREL, E. VAN DER, – 1972b – On the transformation of cover-abundance values in phytosociology. Report Bot. Lab. Nijmegen.

MAAREL, E. VAN DER & J. LEERTOUWER, – 1967 – Variation in vegetation and species diversity along a local environmental gradient. *Acta bot. neerl.* 16: 211—221.

MAAREL, E. VAN DER & R. TÜXEN (eds), – 1972 – Grundfragen und Methoden in der Pflanzensoziologie (Basic problems and methods in phytosociology). *Ber. Symp. int. Ver. Vegetationskunde*, Rinteln 1970: xix + 523 pp. Junk, Den Haag.

MAAREL, E. VAN DER & V. WESTHOFF, – 1964 – The vegetation of the dunes near Oostvoorne, Netherlands (with one vegetation map). *Wentia* 12: 1—61.

MAAREL, E. VAN DER, & V. WESTHOFF & C. G. VAN LEEUWEN, – 1964 – European approaches to the variation in vegetation. Paper 10th. Int. bot. Congress Edinburgh.

MAC INTOSH, R. P., – 1967 – The continuum concept of vegetation. *Bot. Rev.* 33: 131—187.

MARGL, H., – 1967 – Ein Gerät zum raschen Ordnen einer Tabelle. Forstl. BVA Wien Informationsdienst 109, 2 pp.

MATUSZKIEWICZ, W., – 1948 – Roślinność lasów okolik Lwowa (Engl. summ.). *Ann. Univ. Mariae Curie-Skołdowska Sect.* C 3: 119—193.

MATUSZKIEWICZ, W., – 1962 – Zur Systematik der natürlichen Kiefernwälder des mittel-und osteuropäischen Flachlandes. *Mitt. flor.-soz. Arbeitsgem.* Stolzenau *NF* 9: 145—186.

MATUSZKIEWICZ, W. & M. BOROWIK, – 1957 – Materialy do fitosociologicznej systematyki losow legowych w Polsce. Zur Systematik der Auenwälder in Polen. *Acta Soc. bot. polon.* 26: 719—756.

MATUSZKIEWICZ, W. & T. TRACZYK, – 1958 – Zur Systematik der Bruchwaldgesellschaften (Alnetalia glutinosa) in Polen. *Acta Soc. bot. polon.* 27: 21—44.

MEIER, H. & J. BRAUN-BLANQUET, - 1934 - Prodrome des Groupements Végétaux. Fasc. 2 Classe des Asplenietales rupestres, groupements rupicoles. Montpellier, 47 pp.

MELTZER, J. & V. WESTHOFF, – 1942 – Inleiding tot de Plantensociologie. 's Graveland. 326 pp.

MEIJER DREES, E., – 1949 – Combined taxation and presence in analysing and comparing association tables. *Vegetatio* 2: 43—46.

MEIJER DREES, E., – 1951 – Capita selecta from modern plant sociology and a design for rules of phytosociological nomenclature. *Rapp. Bosb. proefst. Bogor* 52: 1-68.

MIYAWAKI, A., – 1960 – Pflanzensoziologische Untersuchungen über die Reisfeld-Vegetation auf den Japanischen Inseln mit vergleichender Betrachtung Mitteleuropas. *Vegetatio* 9: 345—402.

MOLINIER, R. fils, – 1960 – Études des biocénoses marines du Cap Corse. *Vegetatio* 9: 121—192, 217—312.

MOORE, G. W., W. S. BENNINGHOFF & P. S. DWYER, – 1967 – A computer method for the arrangement of phytosociological tables. Proc. Ass. Computing Mach. 1967, Washington: 297—299.

MOORE, J. J., – 1962 – The Braun-Blanquet system: a reassessment. *J. Ecol.* 50: 761—769.

MOORE, J. J., – 1971 – Phyto—A suite of programs in Fortran IV for the manipulation of phytosociological tables according to the principles of Braun-Blanquet. Mscr. Dept. Botany Univ. Coll. Dublin, 11 pp.

MOORE, J. J., – 1972 – An outline of computer-based methods for the analysis of phytosociological data. In, 'Grundfragen und Methoden in der Pflanzensoziologie', ed. E. VAN DER MAAREL & R. TÜXEN, *Ber. Symp. int. Ver. Vegetationskunde,* Rinteln 1970: 29—38. Junk, The Hague.

MOORE, J. J., P. FITZSIMONS, E. LAMBE & J. WHITE, – 1970 – A comparison and evaluation of some phytosociological techniques. *Vegetatio* 20: 1—20.

MOORE, J. J. & A. O'SULLIVAN, – 1970 – A comparison between the results of the Braun-Blanquet method and those of 'cluster-analysis'. In, 'Gesellschaftsmorphologie', ed. R. TÜXEN, *Ber. Symp. int. Ver. Vegetationskunde,* Rinteln 1966: 26—29. Junk, The Hague.

MORAVEC, J. – 1968 – Zu den Problemen der pflanzensoziologischen Nomenklatur. In, 'Pflanzensoziologische Systematik', ed. R. TÜXEN, *Ber. Symp. int. Ver. Vegetationskunde,* Stolzenau 1964. 142—151. Junk, The Hague.

MORAVEC, J., – 1969 – Die Anwendung der Typenmethode in der Phytosoziologischen Nomenklatur. *Folia geobot. phytotax.* 4: 23—31.

MORAVEC, J. – 1971 – A simple method for estimating homotoneity of sets of phytosociological relevés. *Folia geobot. phytotax.* 6: 147—170.

MÖRZER BRUYNS, M. F., – 1950 – On biotic communities. *Comm. S.I.G.M.A.* 96: 1—59, Montpellier.

MOSS, C. E., – 1971 – The fundamental units of vegetation. *New Phytol.* 9: 18—53.

MÜLLER, P. J., M. J. A. WERGER, B. J. COETZEE, D. EDWARDS, N. G. JARMAN, – 1972 – An apparatus for facilitating the manual tabulation of phytosociological data. *Bothalia* 10, (in the press).

MÜLLER, TH., – 1968 – Die Gliederung von Pflanzengesellschaften in Rassen und Formen als Beitrag zur Landschaftsökologie, dargestellt am Beispiel von wärmeliebenden Eichen-Hainbuchen-wäldern in Südwestdeutschland. In, 'Pflanzensoziologie und Landschaftsökologie', ed. R. TÜXEN, *Ber. Symp. int. Ver. Vegetationskunde*, Stolzenau 1963: 60—64. Junk, The Hague.

MÜLLER, TH., – 1970 – Mosaikkomplexe und Fragmentkomplexe. In R. TÜXEN (ed): Gesellschaftsmorphologie. *Ber. Symp. int. Ver. Vegetationskunde*, Rinteln 1966: 69—75, Junk, The Hague.

NEUHÄUSL R., – 1968 – Draft proposals for nomenclature principles in floristic phytosociology. *Folia geobot. phytotax.* 3: 47—55.

NEUHÄUSL, R. & Z. NEUHÄUSLOVÁ-NOVOTNÁ, – 1972 – Eine einfache Orientierungsmethode zur Beurteilung des Assoziations-ranges. In, 'Basic problems and methods in phytosociology', ed. E. VAN DER MAAREL & R. TÜXEN, *Ber. Symp. int. Ver. Vegetationskunde*, Rinteln 1970: 211—223. Junk, The Hague.

NORDHAGEN, R., – 1937 – Versuch einer neuen Einteilung der subalpinen - alpinen Vegetation Norwegens. *Bergens Mus. Årbok* Naturv.rekke 1936: 1—88.

NORDHAGEN, R., – 1943 – Sikilsdalen og Norges fjellbeiter. En plantesosiologisk monografi. *Bergens Mus. Skr.* 22: 1—607.

OBERDORFER, E., – 1957 – Süddeutsche Pflanzengesellschaften. Jena, XXXVIII + 564 pp.

OBERDORFER, E., – 1960 – Pflanzensoziologische Studien in Chile, ein Vergleich mit Europa. Flor. Veg. Mundi 2. 208 pp. Weinheim/Bergst.

OBERDORFER, E., – 1968 – Assoziation, Gebietsassoziation, Geographische Rasse. In, 'Pflanzensoziologisch Systematik', ed. R. TÜXEN, *Ber. Symp. int. Ver. Vegetationskunde*, Stolzenau 1964: 124—131. Junk, The Hague.

OBERDORFER, E., – 1970 – Pflanzensoziologische Strukturprobleme am Beispiel Kanarischer Pflanzengesellschaften. In, 'Gesellschaftsmorphologie.', ed. R. TÜXEN, *Ber. Symp. int. Ver. Vegetationskunde*, Rinteln 1966: 273—278. Junk, The Hague.

OBERDORFER, E. et al., – 1967 – Systematische Übersicht der westdeutschen Phanerogamen- und Gefässkryptogamen-Gesellschaften. Ein Diskussionsentwurf. *Schr. Reihe Vegetationskde* 2: 7—62.

ODUM, E. P., – 1969 – The strategy of ecosystem development. *Science* 164: 262—270.

OHBA, T., – 1972 – Übersicht über die Salzwiesengesellschaften Japans. In, 'Grundfragen und Methoden in der Pflanzensoziologie', ed. E. VAN DER MAAREL & R. TÜXEN, *Ber. Symp. int. Ver. Vegetationskunde*, Rinteln 1970: 413—418. Junk, The Hague.

ORLOCI, L., – 1966 – Geometric models in ecology. I The theory and application of some ordination methods. *J. Ecol.* 54: 193—215.

PACZOSKI, J. K., – 1930 – Życie gromadne róslin. Social life of plants. *Bibljot. Bot.* (*Krakow*) 2: 1—40.

PASSARGE, H., – 1966 – Die Formationen als höchste Einheiten der soziologischen Vegetationssystematik. *Repert. spec. nov. Regni veg.* 73: 226—235.

PASSARGE, H., – 1968 – Neue Vorschläge zur Systematik nord-mitteleuropäischer Waldgesellschaften. *Reprium nov. spec. Regni veg.* 77: 75—103,

PASSARGE, H. & G. HOFMANN, – 1968 – Pflanzengesellschaften des nordostdeutschen Flachlandes II. Pflanzensoziologie Bd XVI. Jena Fischer 288 pp.

PAVILLARD, J., – 1901 – Eléments de Biologie végétale. Paris-Montpellier 589 pp.

PAVILLARD, J., – 1912 – Essai sur la nomenclature phytogéographique. *Bull. soc. langued. Géogr.* 35: 165—176.

PAVILLARD, J., – 1919 – Remarques sur la nomenclature phytogéographique. Montpellier 27 pp.

PAVILLARD, J., – 1920 – Espèces et associations. Essai phytosociologique. Montpellier, 34 pp.

PAVILLARD, J., – 1921 – L'Association végétale, unité phytosociologique. Montpellier, 11 pp.

PAVILLARD, J., – 1935a – Eléments de sociologie végétale (Phytosociologie). *Actual. scient. ind.* 251: 3—102, Paris.

PAVILLARD, J., – 1935b – The present status of the plant-association. *Bot. Rev.* 1: 210—232.

PFEIFFER, H. H., – 1957 – Betrachtungen zum Homogenitätsproblem in der Pflanzensoziologie. *Mitt. flor. soz. Arbeitsgem.* Stolzenau, *NF* 6/7: 103—117.

PIGNATTI, S., – 1964 – Ein neues Verfahren zur Bearbeitung von Assoziationstabellen. *Acta bot. croat. Vol. extraord.* 89—93.

PIGNATTI, S., – 1968a – Die Verwertung der sogenannten Gesamtarten für die floristische Systematik. In, 'Pflanzensoziologische Systematik', ed. R. TÜXEN, *Ber. Symp. int. Ver. Vegetationskunde*, Stolzenau 1964: 71—73. Junk, The Hague.

PIGNATTI, S., – 1968b – Die Inflation der höheren pflanzensoziologischen Einheiten. In, 'Pflanzensoziologische Systematik', ed. R. TÜXEN, *Ber. Symp. int. Ver. Vegetationskunde*, Stolzenau 1964: 85—97. Junk, The Hague.

POORE, M. E. D., – 1955a – The use of phytosociological methods in ecological investigations. I. The Braun-Blanquet system. *J. Ecol.* 43: 226—244.

POORE, M. E. D., – 1955b – The use of phytosociological methods in ecological investigations. II. Practical issues involved in an attempt to apply the Braun-Blanquet system. *J. Ecol.* 43: 245—269.

POORE, M. E. D., – 1955c – The use of phytosociological methods in ecological investigations. III. Practical applications. *J. Ecol.* 43: 606—651.

POORE, M. E. D., – 1956 – The use of phytosociological methods in ecological investigations. IV. General discussion of phytosociological problems. *J. Ecol.* 44: 28—50.

POORE, M. E. D., – 1962 – The method of successive approximation in descriptive ecology. *Adv. ecol. Res.* 1: 35—68.

POST, H. VON, – 1862 – Försök till en systematisk uppställning af vextställena i mellersta Sverige. Bonnier, Stockholm. 42 p.

PRAEGER, R. LLOYD, – 1934 – The botanist in Ireland. Dublin.

PRESTON, F. W., – 1962 – The canonical distribution of commonness and rarity. *Ecology* 43: 185—215, 410—432.

QUÉZEL, P., – 1965 – La végétation su Sahara, du Tchad à la Mauretanie. Geobotanica Selecta II, xii + 333 pp., Fischer, Stuttgart.

QUÉZEL, P. & P. VERDIER, – 1953 – Les méthodes de la phytosociologie sont-elles applicables à l'étude des groupements animaux? *Vegetatio* 4: 165—181.

RAABE, E. W., – 1952 – Über den 'Affinitätswert' in der Pflanzensoziologie. *Vegetatio* 4: 53—68.

RABELER, W., – 1937 – Die planmässige Untersuchung der Soziologie, Ökologie und Geographie der heimischen Tiere, besonders der land- und forstwirtschaftlich wichtigen Arten. :*JBer. naturhist. Ges. Hannover* 81—87: 236—247.

Rabeler, W., – 1960 – Biozönotik auf Grundlage der Pflanzengesellschaften. *Mitt. flor.-soz. Arbeitsgem.*, Stolzenau, *NF* 8: 311—332.

Rabeler, W., – 1962 – Die Tiergesellschaften von Laubwäldern (Querco-Fagetea) im oberen und mittleren Wesergebiet. *Mitt. flor.-soz. Arbeitsgem.*, Stolzenau, *NF* 9: 200—229.

Rabeler, W., – 1965 – Die Pflanzengesellschaften als Grundlagen für die landbiozönotischen Forschung. In R. Tüxen (ed): Biosoziologie. *Ber. Symp. int, Ver. Vegetationskunde,* Stolzenau 1960: 43—49. Junk, The Hague.

Ramensky, L. G., – 1930 – Zur Methodik der vergleichenden Bearbeitung und Ordnung von Pflanzenlisten und anderen Objekten, die durch mehrere, verschiedenartig wirkende Faktoren bestimmt werden. (transl. from the -Russian paper in Trudy soc. geobot.-lugov, 15—20 Jan. 1929 p. 11—36) *Beitr. Biol. Pfl.* 18: 269—304. Breslau.

Raunkiaer, C., – 1934 – The life forms of plants and statistical plant geography. Oxford. xvi + 632 pp.

Rauschert, S., – 1963 – Beitrag zur Vereinheitlichung der soziologischen Nomenklatur. *Mitt. flor.-soz. Arbeitsgem.*, Stolzenau, *NF* 10: 232—249.

Rauschert, S., – 1969 – Über einige Probleme der Vegetationsanalyse und Vegetationssystematik. *Arch. Naturschutz Landschaftsforsch.* 9: 153—174.

Rogers, D. J., – 1970 – A preliminary ordination study of forest vegetation in the Kirchleerau area of the Swiss midlands. *Ber. geobot. Inst. Rübel* 40: 28—78.

Romane, F., – 1972 – Un exemple d'analyse factorielle des correspondances en écologie végétale. In, 'Grundfragen und Methoden in der Pflanzensoziologie', ed. E. van der Maarel & R. Tüxen, *Ber. Symp. int. Ver. Vegetationskunde,* Rinteln 1970: 151—162. Junk, The Hague.

Rübel, E., – 1912 – Pflanzengeographische Monographie des Berninagebietes. *Bot. Jb.* 47: 1—616.

Scamoni, A. & H. Passarge, – 1959 – Gedanken zu einer natürlichen Ordnung der Waldgesellschaften. *Arch. Forstwes.* 8: 386—426.

Scamoni, A. & H. Passarge, – 1963 – Einführung in die praktische Vegetationskunde. 2. Ed. Jena xi + 236 pp.

Scamoni, A., H. Passarge & G. Hofmann, – 1965 – Grundlagen zu einer objektiven Systematik der Pflanzengesellschaften. *Repert. Spec. nov. Regni veg.*, Beih. 142: 117—132.

Schmid, E., – 1923 – Vegetationsstudiën in den Urner Reusstälern. Thesis, Zürich.

Schmid, P. & N. Kühn, – 1970 – Automatische Ordination von Vegetationsaufnahmen in pflanzensoziologischen Tabellen. *Naturwissenschaften* 57: 462.

Schmithüsen, J., – 1960 – Die Nadelhölzer in den Waldgesellschaften der südlichen Anden. *Vegetatio* 9: 313—327.

Schmithüsen, J., – 1961 – Allgemeine Vegetationsgeographie. In: Obst, ed.: Lehrbuch der Allgemeinen Geographie 4, Berlin. (3 ed. xxiii + 463 pp. 1968).

Schnell, R., – 1952 – Contribution à une étude phytosociologique et phytogéographique de l'Afrique occidentale: les groupements et les unités géobotaniques de la Région Guinéenne. *Mem. Inst. Franc. Afrique Noir Mélanges bot.* 18.

Schouw, J. F., – 1823 – Grundzüge einer allgemeinen Pflanzengeographie. Reimer, Berlin. 524 pp.

Schröter, C., – 1894 – Notes sur quelques associations de plantes rencontrées pendant les excursions dans la Valais. *Soc. bot. France, Bull.* 41 (sess. extraord.): 322—325.

Schröter, C. & O. Kirchner, – 1902 – Die Vegetation des Bodensees. *Schr. Ver. Gesch. Bodensees Lindau* 9 (2): 1—86.

SCHUBERT, R., – 1960 – Die Zwergstrauchreichen azidiphilen Pflanzengesellschaften Mitteldeutschlands. Pflanzensoziologie Band 11, Fischer, Jena. 235 pp.

SCHWICKERATH, M., – 1931 – Die Gruppenabundanz (Gruppenmächtigkeit); ein Beitrag zur Begriffsbildung der Pflanzensoziologie. *Bot. Jb.* 64: 1—16.

SCHWICKERATH, M., – 1940 – Ausgleich – und Richtungsprinzip als Grundlage der Pflanzengesellschaftslehre. *Repert. Spec. nov. Regni veg.*, Beih. 121: 53–67.

SCHWICKERATH, M., – 1942 – Bedeutung und Gliederung des Differentialartenbegriffs in der Pflanzengesellschaftslehre. *Beih. bot. Zentralbl.* 61B: 351—383.

SCHWICKERATH, M., – 1954a – Lokale Charakterarten - geographische Differentialarten. *Veröff. geobot. Inst. Rübel* 29: 96—104.

SCHWICKERATH, M., – 1954b – Die Landschaft und ihre Wandlung auf geobotanischer und geographischer Grundlage entwickelt und erläutert im Bereich der Messtischblattes Stolberg. Aachen.

SCHWICKERATH, M., – 1963 – Assoziationsdiagramme und ihre Bedeutung für die Vegetationskartierung. In R. TÜXEN (ed): Ber. Int. Symposium für Vegetationskartierung 1959 p. 11—35. Cramer, Weinheim.

SEGAL, S. & V. WESTHOFF, – 1959 – Die Vegetationskundliche Stellung von Carex buxbaumii Wahlenb. in Europa, besonders in den Niederlanden. *Acta bot. neerl.* 8: 304—329.

SEIBERT, P., – 1968 – Gesellschaftsring und Gesellschaftskomplex in der Randschaftsgliederung. In, 'Pflanzensoziologie und Landschaftsökologie', ed. R. TÜXEN, *Ber. Symp. int. Ver. Vegetationskunde*, Stolzenau 1963: 48—60. Junk, The Hague.

SHIMWELL, D. W., – 1972 – Description and Classification of Vegetation. Sidgwick & Jackson, London xiv + 322 pp.

SISSINGH, G., – 1950 – Onkruid-associaties in Nederland. Thesis Wageningen Versl. Landbouwk. Onderz. 56, 14: 1—24.

SOKAL, R. R. & P. H. A. SNEATH, – 1963 – Principles of Numerical Taxonomy. San Francisco xvi + 359 pp.

SØRENSEN, TH. A., – 1948 – A method of establishing groups of equal amplitude in plant sociology based on similarity of species content. *Biol. Skr. K. danske Vidensk. Selsk.* 5 (4): 1—34.

SPATZ, G., – 1969 – Elektronische Datenverarbeitung bei pflanzensoziologischer Tabellenarbeit. *Naturwissenschaften* 56: 470—471.

SPATZ, G., – 1972 – Eine Möglichkeit zum Einsatz der elektronischen Datenverarbeitung bei der pflanzensoziologischen Tabellenarbeit. In, 'Grundfragen und Methoden in der Pflanzensoziologie', ed. E. VAN DER MAAREL & R. TÜXEN, *Ber. Symp. int. Ver. Vegetationskunde*, Rinteln 1970: 251—258. Junk, The Hague.

STEBLER, F. G. & C. SCHRÖTER, – 1893 – Beiträge zur Kenntnis der Matten und Weiden der Schweiz. 10. Versuch einer Übersicht über die Wiesentypen der Schweiz. *Landw. Jahrb. Schweiz* 6: 95—212, Bern.

STOCKINGER, J. J. & W. F. HOLZNER, – 1972 – Rationelle Methode zur Auswertung pflanzensoziologischer Aufnahmen mittels Elektronenrechner. In, 'Grundfragen und Methoden in der Pflanzensoziologie', ed. E. VAN DER MAAREL & R. TÜXEN, *Ber. Symp. int. Ver. Vegetationskunde*, Rinteln 1970: 239—248. Junk, The Hague.

SUKATSCHEW, W. N. (SUKACHEV, V. N.) – 1929 – Über einige Grundbegriffe in der Phytosoziologie. *Ber. dt. bot. Ges.* 47: 296—312.

SUKATSCHEW, W. N. (SUKACHEV, V. N.), – 1954 – Die Grundlagen der Waldtypen. *Angew. Pfl. Soziol.* Wien. Festschr. Aichinger 2: 956—964.

SUKACHEV, V. N., – 1960 – The correlation between the concept 'forest ecosystem' and 'forest biogeocenose' and their importance for the classification of forests. *Silva fenn.* 105: 94—97.

SUZUKIO-TOKIO, – 1954 – L'Alliance du Shiion sieboldi. *Vegetatio* 5/6: 361—372.

SZAFER, W. & B. PAWŁOWSKI, – 1927 – Die Pflanzenassoziationen der Tatra-Gebirges. A. Bemerkungen über die angewandte Arbeitsmethodik (zu den Teilen III, IV, und V). *Bull. Int. Acad. pol. Sci. Lett.*, Cl. Sci. Nat. Math., sér B, 1926 (12) Suppl: 1—12.

TAKHTAJAN, A., – 1969 – Flowering Plants, Origin and Dispersal. Transl. by C. JEFFREY. Edinburgh Oliver & Boyd. x + 310 pp.

TANSLEY, A. G., – 1920 – The classification of vegetation and the concept of development. *J. Ecol.* 8: 118—149.

TANSLEY, A. G., – 1922 – The new Zürich-Montpellier school. *J. Ecol.* 10: 241—243.

TANSLEY, A. G., – 1935 – The use and abuse of vegetational concepts and terms. *Ecology* 16: 284—307.

THIENEMANN, A., – 1956 – Leben und Umwelt. Vom Gesamthaushalt der Natur. Hamburg, 153 pp.

THURMANN, J., – 1849 – Essai de phytostatique appliqué à la chaîne du Jura et aux contrées voisines. Bern.

TRENTEPOHL, M. W., – 1968 – Ein mechanisch-elektromagnetisches Gerät zur Schnellbearbeitung pflanzensoziologischer Tabellen. Ref. Symposium Tatsachen und Probleme der Grenzen in der Vegetation Rinteln 1968.

TUOMIKOSKI, R., – 1942 – Untersuchungen über die Untervegetation der Bruchmoore in Ostfinnland. I. Zur Methodik der pflanzensoziologischen Systematik. *Ann. bot. Soc. zool. bot. fenn. Vanamo* 17 (1) vi + 203 pp.

TÜXEN, R., – 1933 – Klimaxprobleme des nw. europäischen Festlandes. *Ned. kruidk. Arch.* 43: 293—309.

TÜXEN, R., – 1937 – Die Pflanzengesellschaften Nordwestdeutschlands. *Mitt. Flor.-soz. Arbeitsgem. Niedersachsen* 3: 1—170.

TÜXEN, R., – 1950 – Grundriss einer Systematik der nitrophilen Unkrautgesellschaften. *Mitt. flor.-soz. Arbeitsgem.*, Stolzenau, *N.F.* 2: 94—175.

TÜXEN, R., – 1952 – Hecken und Gebüsche. *Mitt. geogr. Ges. Hamburg* 50: 85—117.

TÜXEN, R., – 1955 – Das System der nordwestdeutschen Pflanzengesellschaften. *Mitt. flor.-soz. Arbeitsgem.*, Stolzenau, *N.F.* 5: 155—176.

TÜXEN, R., – 1956 – Die heutige potentielle natürliche Vegetation als Gegenstand der Vegetationskartierung. *Angew. PflSoziol.*, Stolzenau 13: 5—42.

TÜXEN, R., – 1957 – Entwurf einer Definition der Pflanzengesellschaft (Lebensgemeinschaft). *Mitt. flor.-soz. Arbeitsgem.*, Stolzenau, *N.F.* 6/7: 112—113.

TÜXEN, R., – 1962 – Zur systematischen Stellung von Spezialisten-Gesellschaften. *Mitt. flor.-soz Arbeitsgem.*, Stolzenau, *N.F.* 9: 57—59.

TÜXEN, R. (ed.), – 1963 – Bericht über das Internationale Symposion für Vegetationskartierung Stolzenau 1959. Cramer, Weinheim. vii + 500 pp.

TÜXEN, R. (ed), – 1965a – Biosoziologie. *Ber. Symp. int. Ver. Vegetationskunde,* Stolzenau 1960. xvi + 350 pp. Junk, The Hague.

TÜXEN, R., – 1965b – Wesenszüge der Biozönose: Gesetze für das Zusammenleben von Pflanzen und Tieren. In, 'Biosoziologie', ed. R. TÜXEN, *Ber. Symp. int. Ver. Vegetationskunde,* Stolzenau 1960: 10—13. Junk, The Hague.

TÜXEN, R., – 1966a – Über nitrophile Elymus-Gesellschaften an nordeuropäischen, nordjapanischen und nordamerikanischen Küsten. *Ann. bot. fenn.* 3: 358—367.

TÜXEN, R., (ed.), – 1966b – Anthropogene Vegetation. *Ber. Symp. int. Ver. Vegetationskunde,* Stolzenau 1961. xvi + 398 pp. Junk, The Hague.

TÜXEN, R., – 1967a – Ausdauernde nitrophile Saumgesellschaften Mitteleuropas. Contributii Bot. Cluj p. 431—453.

TÜXEN, R., – 1967b – Pflanzensoziologischen Beobachtungen an südwestnorwegischen Küsten-Dünengebieten. *Aquilo Ser. Bot.* 6: 241—272.

TÜXEN, R., (ed), – 1967c – Pflanzensoziologie und Palynologie. *Ber. Symp. int. Ver. Vegetationskunde,* Stolzenau 1962, xvii + 275 pp.

TÜXEN, R. (ed), – 1968a — Pflanzensoziologie und Landschaftsökologie. *Ber. Symp. int. Ver. Vegetationskunde,* Stolzenau 1963. xvii+426 pp. Junk, The Hague.

TÜXEN, R. (ed), – 1968b – Pflanzensoziologische Systematik. *Ber. Symp. Int. Ver. Vegetationskunde,* Stolzenau 1964. xii+348 pp. Junk, The Hague.

TÜXEN, R., – 1969a – Stand und Ziele geobotanischer Forschung in Europa. *Ber. geobot. Inst. Rübel* 39: 13—26.

TÜXEN, R. (ed), – 1969b – Experimentelle Pflanzensoziologie. *Ber. Symp. Int. Ver. Vegetationskunde,* Rinteln 1965, xvii + 256 pp. Junk, The Hague.

TÜXEN, R., – 1970a – Entwicklung, Stand und Ziele der pflanzensoziologischen Systematik (Syntaxonomie). *Ber. dt. bot. Ges.* 83: 633—639.

TÜXEN, R., – 1970b – Pflanzensoziologie als synthetische Wissenschaft. *Misc. Papers Landbouwhogeschool Wageningen* 5: 141—159.

TÜXEN, R., – 1970c – Einige Bestandes- und Typenmerkmale in der Struktur der Pflanzengesellschaften. In, 'Gesellschaftsmorphologie', ed. R. TÜXEN, *Ber. Symp. int. Ver. Vegetationskunde,* Rinteln 1966. p. 76—98. Junk, The Hague.

TÜXEN, R., – 1971 – Besprechung von H. Passarge & G. Hofmann 1968 - Pflanzengesellschaften des nordostdeutschen Flachlandes II. *Vegetatio* 23: 382—383.

TÜXEN, R., – 1972 – Kritische Bemerkungen zur Interpretation pflanzensoziologischer Tabellen. In, 'Grundfragen und Methoden in der Pflanzensoziologie', ed. E VAN DER MAAREL & R. TÜXEN, *Ber. Symp. int. Ver. Vegetationskunde,* Rinteln 1970: 168—173. Junk, The Hague.

TÜXEN, R. & W. H. DIEMONT, – 1937 – Klimaxgruppe und Klimaxschwarm. Ein Beitrag zur Klimaxtheorie. *JBer. naturhist. Ges. Hannover* 88,89: 73—87.

TÜXEN, R. & H. ELLENBERG, – 1937 – Der Systematische und der ökologische Gruppenwert. Ein Beitrag zur Begriffsbildung und Methodik der Pflanzensoziologie. *Mitt. flor.-soz. Arbeitsgem. Niedersachsen* 3: 171—184.

TÜXEN, R., A. MIYAWAKI & K. FUJIWARA, – 1972 – Eine erweiterte Gliederung der Oxycocco-Sphagnetea. In, 'Grundfragen und Methoden der Pflanzensoziologie', ed. E. VAN DER MAAREL & R. TÜXEN, *Ber. Symp. int. Ver. Vegetationskunde,* Rinteln 1970: 500—509. Junk, The Hague.

TÜXEN, R. & E. PREISING, – 1951 – Erfahrungsgrundlagen für die pflanzensoziologische Kartierung des westdeutschen Grünlandes. *Angew. Pfl.Soziol.* Stolzenau 4: 1—28.

VARESCHI, V. – 1931 – Die Gehölztypen des obersten Isartales. *Ber. naturw. -med. Ver. Innsbruck* 42: 79—184.

VESTAL, A. G., – 1949 – Minimum areas for different vegetations. *Univ. Illinois biol. Monogr.* 20 (3):1—129.

VRIES, D. M. DE, J. P. BARETTA & G. HAMMING, – 1954 – Constellation of frequent herbage plants, based on their correlation in occurrence. *Vegetatio* 5/6: 106—111.

WAGNER, H., – 1954 – Gedanken zur Berücksichtigung der mehr-dimensionalen Beziehungen der Pflanzengesellschaften in der Vegetationssystematik. *Rapp. Comm. 8th Congr. Int. Bot., Paris,* Sect. 7—8: 9—11.

WAGNER, H., – 1968 – Prinzipienfragen der Vegetationssystematik. In. 'Pflanzensoziologische Systematik', ed. R. TÜXEN, *Ber. Symp. int. Ver. Vegetationskunde* Stolzenau 1964: 15—20. Junk, The Hague.

WAGNER, H., – 1972 – Zur Methodik der Erstellung und Auswertung von Vegetaitonstabellen. In, 'Grundfragen und Methoden in der Pflanzensoziologie', ed. E. VAN DER MAAREL & R. TÜXEN, *Ber. Symp. int. Ver. Vegetationskunde* Rinteln 1970: 225—233. Junk, The Hague.

WALTER, H. & H. STRAKA, – 1970 – Arealkunde. Floristisch-historische Geobotanik. 2. Aufl. Ulmer, Stuttgart 478 pp.

WARMING, E., – 1895 – Plantesamfund. Kjøbenhavn. 335 pp.

WEAVER, J. E., & F. E. CLEMENTS, – 1938 – Plant Ecology. 2.ed. New York-London. xxii + 601 pp.

WENDELBERGER, G., – 1951 – Das vegetationskundliche System Erwin Aichingers und seine Stellung im pflanzensoziologischen Lehrgebäude Braun-Blanquets. *Angew. Pfl. Soziol.* Wien. 1: 69—92.

WERGER, M. J. A., – 1973 – Phytosociology of the upper Orange River Valley, South Africa. A syntaxonomical and synecological study. Thesis Nijmegen. Pretoria, 222 pp.

WERGER, M. J. A., F. J. KRUGER & H. C. TAYLOR, – 1972 – A phytosociological study of the Cape Fijnbos and other vegetation at Jonkershoek, Stellenbosch. *Bothalia* 10 (4): 1—19, German version *Vegetatio* 24: 71—89 (1972).

WESTHOFF, V., – 1947 – The vegetation of dunes and salt marshes on the Dutch islands of Terschelling, Vlieland and Texel. Thesis Utrecht, 131 pp.

WESTHOFF, V., – 1949 – De plantengezelschappen van Botshol. In: Landschap, flora en vegetatie van de Botshol nabij Abcoude. Baambrugge p. 43—102.

WESTHOFF, V., – 1950 – Het associatiebegrip in geografisch verband. *Ned. kruidk. Arch.* 57: 98—100.

WESTHOFF, V., – 1951 – An analysis of some concepts and terms in vegetation study or phytocenology. *Synthese* 8: 194—206.

WESTHOFF, V., – 1954 – Some remarks on synecology. *Vegetatio* 5—6: 120—128.

WESTHOFF, V., – 1959 – The vegetation of Scottish pine woodlands and Dutch artificial coastal pine forests; with some remarks on the ecology of Listera cordata. *Acta bot. neerl.* 8: 422—448.

WESTHOFF, V., – 1965 – Plantengemeenschappen. In: 'Het leven der planten'. 2e druk Zeist-Arnhem p. 288—349.

WESTHOFF, V., – 1967 – Problems and use of structure in the classification of vegetation. The diagnostic evaluation of structure in the Braun-Blanquet system. *Acta bot. neerl.* 15: 495—511.

WESTHOFF, V., – 1970 – Vegetation study as a branch of biological science. *Misc. Papers Landbouwhogeschool Wageningen* 5: 11—30.

WESTHOFF, V., – 1971a – The dynamic structure of plant communities in relation to the objectives of conservation. In E. DUFFEY & A. S. WATT (ed) The scientific management of animal and plant communities for conservation. Oxford. p. 3—14.

WESTHOFF, V., – 1971b – Choice and management of nature reserves in the Netherlands. *Bull. Jard. Bot. Nat. Belgique* 41 (1): 231—245.

WESTHOFF, V., J. J. BARKMAN, H. DOING & C. G. VAN LEEUWEN, – 1959 – Enige opmerkingen over de terminologie in de vegetatiekunde. *Jaarb. Kon. ned. bot. Ver.* 1958: 44—46.

WESTHOFF, V. & A. J. DEN HELD, – 1969 – Plantengemeenschappen in Nederland. Zutphen Thieme, 324 pp.

WESTHOFF, V. & C. G. VAN LEEUWEN, – 1966 – Ökologische und systematische Beziehungen zwischen natürlicher und anthropogener Vegetation. In, 'Anthropogene Vegetation', ed. R. TÜXEN, *Ber. Symp. int. Ver. Vegetationskunde*, Stolzenau 1961: 156—172. Junk, The Hague.

WESTHOFF, V., H. PASSCHIER & G. SISSINGH, – 1946 – Overzicht der plantengemeenschappen in Nederland. 2e dr. Breughel, Amsterdam 118 pp.

WESTHOFF, V. & J. N. WESTHOFF-DE JONCHEERE, – 1942 – Verspreiding en nestoecologie van de mieren in de Nederlandse bossen. *Tijdschr. Plantenziekten* 48: 138—212.
WHITTAKER, R. H., – 1951 – A criticism of the plant association and climatic climax concepts. *NW. Sci.* 25: 17—31.
WHITTAKER, R. H., – 1953 – A consideration of climax theory; the climax as a population and pattern. *Ecol. Monogr.* 23: 41—78.
WHITTAKER, R. H., – 1954a – Plant populations and the basis of plant indication. *Angew. Pfl. Soziol.* Wien, Festschr. Aichinger 1. 183—206.
WHITTAKER, R. H., – 1954b – The ecology of serpentine soils. IV. The vegetational response to serpentine soils. *Ecology* 35: 275—288.
WHITTAKER, R. H., – 1956 – Vegetation of the Great Smoky Mountains. *Ecol. Monogr.* 26: 1—80.
WHITTAKER, R. H., – 1960 – Vegetation of the Siskiyou Mountains, Oregon and California. *Ecol. Monog.* 30: 279—338.
WHITTAKER, R. H., – 1962 – Classification of natural communities. *Bot. Rev.* 28: 1—239.
WHITTAKER, R. H., – 1967 – Gradient analyses of vegetation. *Biol. Rev. London* 42: 207—264.
WHITTAKER, R. H., – 1970 – The population structure of vegetation. In R. Tüxen (ed) Gesellschaftsmorphologie. *Ber. Symp. int. Ver. Vegetationskunde*, Rinteln 1966: 39—59. Junk, The Hague.
WHITTAKER, R. H., – 1972 – Convergences of ordination and classification. In, 'Basic problems and methods in phytosociology', ed. E. VAN DER MAAREL & R. TÜXEN, *Ber. Symp. int. Ver. Vegetationskunde* Rinteln 1970: 39—57. Junk, The Hague.
WILLIAMS, C. B., – 1964 – Patterns in the Balance of Nature and Related Problems in Quantitative Ecology. London-New York Academic Press. vii + 324 pp.
WILLIAMS, W. T. & J. M. LAMBERT, – 1959 – Multivariate methods in plant ecology. I Association analysis in plant communities. *J. Ecol.* 47: 83—101.
WILLIAMS, W. T. & J. M. LAMBERT, – 1961 – Multivariate analysis in plant ecology. III Inverse association analysis. *J. Ecol.* 49: 717—729.
WILLIAMS, W. T., J. M. LAMBERT & G. N. LANCE, – 1966 – Multivariate Methods in Plant Ecology. V Similarity Analyses and information-analysis. *J. Ecol.* 54: 427—445.
WILMANNS, O., – 1959 – Ein Gerät zur Mechanisierung von Tabellenarbeit. *Ber. dt. bot. Ges.* 72: 419—420.
WILMANNS, O., – 1970 – Kryptogamen - Gesellschaften oder Kryptogamen - Synusien. In 'Gesellschaftsmorphologie', ed. R. TÜXEN, *Ber. Symp. int. Ver. Vegetationskunde* Rinteln 1966: 1—6. Junk, The Hague.
WISHART, D., – 1969 – Clustan Ia. User manual. St. Andrews Computing Centre.
ZINDEREN BAKKER JR., E. M. VAN, – 1971 – Ecological investigations on ravine forests of the Eastern Orange Free State. Thesis Bloemfontein, 123 pp.
ZOLLER, H., – 1954 – Die Typen der Bromus erectus-Wiesen des Schweizer Juras. Ihre Abhängigkeit von den Standortsbedingungen und wirtschaftlichen Einflüssen und ihre Beziehungen zur ursprünglichen Vegetation. *Beitr. geobot. Landesauf. Schweiz* 33: 1—309.
ZONNEVELD, I. S., – 1960 – De Brabantse Biesbosch. Een studie van bodem en vegetatie van een zoetwatergetijdendelta. With English summary, Thesis, Wageningen, part A, p. 1—210 - part B, p. 1—396.

INDEX

T

ERRATUM

Handbook V, article 8, page 202

Formula should read:

$x = (L^2 + D_1^2 - D_2^2)\,/2L$, and $e = \sqrt{(D_1^2 - x^2)}$,

instead of:

$x = (L^2 + D_1^2 - D_2^2)\,/2L$, and $e = \sqrt{(D_1^2 - x_2)}$,

MIX
Papier aus verantwortungsvollen Quellen
Paper from responsible sources
FSC® C105338

If you have any concerns about our products,
you can contact us on
ProductSafety@springernature.com

In case Publisher is established outside the EU,
the EU authorized representative is:
Springer Nature Customer Service Center GmbH
Europaplatz 3, 69115 Heidelberg, Germany

Printed by Libri Plureos GmbH
in Hamburg, Germany